KITAABA WAL- HIKMATA

'MANIFESTED NATURE AND THE UTILITY OF ONE'S UPRIGHT LOGIC'
VOL 1

Jamir Ahmed Choudhury

PARTRIDGE

To order additional copies of this book, contact
Partridge India
000 800 10062 62
orders.india@partridgepublishing.com

www.partridgepublishing.com/india

KITAABA WAL-HIKMATA
[MANIFESTED NATURE AND THE UTILITY OF ONE'S UPRIGHT LOGIC]

VOLUME - I

AN INSPIRED SHARING
CONCERNING MANIFESTED NATURE
&
NATURAL MAGNETISM

SHARER

Jamir Ahmed Choudhury
Asst. Prof. – Dept. of Philosophy
S. S. College, Hailakandi
Assam, Hindustan

A Research Scholar
On
Solidarity Rights in Islam
Under Assam University, Silchar [AUS]
Registration No. – Ph. D/2491/14
Dated - 21 – 03 – 14

An inhabitant of
Abdul Mazid Road [Bonomali Road]
W/No. – 13
P. O. & District – Karimganj
Assam, Hindustan

A son of the soil of Hindustan
Bharat, Asia
Middle Stair out of three ascending stairs of the Pentagonal Earth
Western [North-West] Continent of the Appearing Pentagonal Earth
North-West Region of the Manifested Hexagonal World within the
Black Square [Non-luminous Moon or Dark Moon] of East Horizon
[Sirius B of Sirius Binary System / Star Saturn]

UNIQUE UTILITARIAN PRAYER

FASTA-IZ BILLAAHI MINASH-SHAYTANIR-RAJIM
When you recite Quran seek refuge in Allah from Shaytan the outcast
Fa-'izaa qara'-tal-Qur-'aa-na **fasta-'iz billaahi minash-Shaytaanir-Rajim**
[98]. Innahuu laysa lahuu sul-taanan 'alallaziina 'aamanuu wa 'alaa Rabbihim
yatawak-kaluun [99]. 'Innamaa sultaanuhuu'alallaziina yata-wallaw-nahuu
wallaziina hum-bihii mushri-kuun [100]. [Trans.] **When you recite Quran
seek refuge in Allah from Shaytan the outcast** [98]. No authority has he over
those who believe and put their trust in their Rab [99]. His authority is over
those only, who take him as patron and **who join partners with Allah.** [100]
[Sura No. – 15\16 – Nahl, Ayat\Verses – 98 to 100]

BISMILLAAHIR-RAHMAANIR-RAHIM
In the name of Allah, The Beneficent, The Merciful
Qaala sananguru asadaqta am kunta minal-kaazibim. Izhab-bi-Kittabii haazaa
fa-alqih ilayhim summa tawalla anhum fanzur maa zaayarji-uun. Qaalat yaa-
ayyuhal-mala-u inniii ulqiya ilayya Kitabun-Karim. Innahu min-Sulaymaana
wa innahu **Bismillaahir-Rahmaanir- Rahim.** Allaa ta-luu alayya wa-tuunii
Muslimin. [Trans.] [Sulaymaan] said: Soon we will see whether you have told
the **truth or lied! Go you, with this letter of mine,** and **deliver** it to them.
Then **draw back** from them then **turn away** and [wait to] see what **answer
they return.** [The queen of Saba] said (when she received the latter): O chiefs!
Here is delivered to me - a letter **worthy of respect.** Lo! It is from Sulaymaan,
and lo! It is **Bismillaahir-Rahmaanir-Rahim / In the name of Allah as** The
Beneficent, The Merciful. Be you **not arrogant against me,** but come to me as
those who **surrender.** [Sura No. – 26\27 – Namal, Ayat No. – 27 to 31]

FATIHA
[Unique Utilitarian Prayer]
Fasta-'iz billaahi minash-Shaytaanir-Rajim
Bismillaahir-Rahmaanir-Rahim
Praise be to Allah, Rab of the Universe [Alamin]
The Beneficent, the Merciful
Owner of the Day of Judgment
You (alone) we devote; we ask for [seek] Your help (alone).
Show us Right Path [towards Upright West / Arsh]
The Path of those on whom You have bestowed Your Grace.
Not (the path) of those who earn Your anger nor of those who go astray.
AAMIN

Prologue
Assa-la-mualaikum,

Respected Sir / Madam
    &
Dear Friends in Deed [Students],

With due respect and a package of anticipation, the sharer of 'Kitaaba Wal-Hikmata' or 'Manifested Nature and the Utility of One's Upright Logic' would like to share with you several manifest truths which are being concealed from mankind in general and particularly from followers of Upright West facing each other. The searcher & sharer of manifest truth had firm faith on One Dot Leader [from scientific point of view] and pragmatic belief on Diamond Operator [from scientific point of view], yet he was far from the equal & opposite right direction towards Upright West [from the point of view of intrinsic values] due to binary nature and functions of the Epistemic Uniqueness of Scientific Certainty and Researchers of Quran in the name of 'The Holy Quran' [particularly researchers of IFTA]. The Epistemic Uniqueness of Scientific Certainty has **introduced Manmade Nature** within the domain of Formal Education with a view to conceal several manifest truths. The researchers of IFTA have been gifting mankind the package of 'translation of the meanings and commentary on Quran [Revelation & Manifestations or Truth-in-itself & Tautologies] in the name of 'The Holy Quran' in resemblance with Introduced Manmade Nature of the Epistemic Uniqueness of Scientific Certainty. From the free gift package of 'translation of the meanings and critical remarks [6310 commentaries] on Quran in the name of 'The Holy Quran' by the researchers of IFTA in resemblance with Introduced/Projected Manmade Nature of Epistemic Uniqueness of Scientific Certainty, it can validly be inferred that there is a Golden Mean or a Permanent Bridge or an Iron-Bond or a Two-in-One partnership or a ~sth between two extremes i.e. Upright West Region of Arabian Peninsula and Straight Middle East Region of Eartha 3D [Black Hole] or the Region of White and the Region of Black. In other words, the two extremes are the two faces of the same thread in resemblance with Pythagorean Number – 'Line – Two'.

Introduced/Projected Manmade Nature comprises the Trinity of (i) Mechanical Globalisation, (ii) Circular Rotation & Revolution System, and (ii) Manmade

Magnetism. This Trinity is neither manifest truth nor manifested nature in resemblance with revelation. On the contrary, this Trinity is the Vera Causa of Activism, Terrorism, Persecution, Evidence Sorcery, Hypocrisy, Tyranny, Conspiracy, and the like. This Trinity is the wise policy of Firawn with reference to Kitaab with Truth. This Trinity is leading believers towards wrong direction without Broken Bar in resemblance with the wrong dialling shown in the film PK. Consequently, believers of Middle Stair [Europe, Asia, Africa, and Australia] of the appearing Pentagonal Earth are performing Salat [compulsory prayer] towards Haiyalas-Swalaah; while believers of the Ground Stair [North America and South America] of the appearing Pentagonal Earth are performing Salat [compulsory prayer] towards Haiyalal-Falaah. Moreover, global trinity as well as quaternary i.e. global universe, global world, global earth projected through global map [of the projected manmade nature] are leading believers towards Straight Middle East Region of Eartha 3D [Recycling Region, Black Hole, Bermuda Triangle of Titanic, penetrated hole in the journey boat of Muusa (ass), hollow in the hand (pentagon), projected planet Mars as the peg of the immovable hexagonal world and appearing pentagonal earth].

Now, Rab of the universe has shown the searcher and sharer of manifest truth one of the Great Signs with a view to confirm equal & opposite right direction towards Upright West [Arsh] in resemblance with Revelations. In this connection, the sharer of 'Kitaaba Wal-Hikmata' or 'Manifested Nature and the Utility of One's Upright Logic' has searched out several manifest truths in resemblance with revelations, equal & opposite manifested signs of natural magnetism, manifested nature, verifiable truth of facts, justifiable necessary truth, searched out real scientific symbols, acquired formal knowledge, and concealed terminologies [Ref. Insert - Symbols]. These searched out findings are free from subjective selfcontradictions and objective paradoxes, fallacy of arguing in a circle and fallacy of infinite regress, existential fallacy and post hoc ergo propter hoc, characteristic imperfection and practical imperfection, fallacy of non-observation and fallacy of wrong observation, fallacy of plurality of causes, and the like. Moreover, these searched out findings are verifiable in resemblance with manifested signs & real scientific symbols [Injiil or Minor Premises], as well as justifiable in resemblance with Truths-in-themselves [Revelations or Fundamental Ayat] & Coded Shared Tautologies [Detail Explanations of Fundamental Ayat or Laws or Tawraat or Major Premises]. So, the shared findings are selfevident truths.

An attempt has been made to compile searched out findings of the re-search project on Solidarity Rights grounded on Equal & Opposite Manifested Signs of Natural Magnetism and Manifested Nature in the form of Dictum De Omni

ET Nullo [Deductive Method or Cartesian Method] in the name of 'Kitaaba Wal-Hikmata' or 'Manifested Nature and the Utility of One's Upright Logic' in resemblance with contents and contexts of this inspired sharing. The sharer of this inspired sharing is also a re-search scholar on Solidarity Rights in Islam. So, to fulfil two-in-one teleology as well as ontology, an attempt has been made to search out **Solidified Solid Human Rights** under the supervision of Dot Operator [from Scientific point of view] so that mortal beings like Researchers of IFTA and Epistemic Uniqueness of NASA can neither reject nor falsify searched out and shared manifest truths of this inspired sharing. From the point of view of the sharer of 'Kitaaba Wal-Hikmata' or 'Manifested Nature and the Utility of One's Upright Logic', the searched out findings are the selfevident truths. Consequently, to get confirmation as well as recognition of the searched out selfevident truths are the Solidified Solid Human Rights not only of the searcher & sharer of 'Kitaaba Wal-Hikmata' or 'Manifested Nature and the Utility of One's Upright Logic', but also the Solidified Solid Human Rights of each and every human being of the manifested hexagonal world and appearing pentagonal earth within the Black Square [Non-luminous Moon or dark Moon] of East Horizon [Sirius B].

Moreover, shared findings are grounded on Manifested Nature and Equal & Opposite Manifested Signs of Natural Magnetism in resemblance with Revelation. So, all pen-paper-pencil works & activities, established laws, formalities, guidelines, rules, theories, suppositions, imaginations, predictions, opinions, hypotheses, discoveries, and the like which are either contrary or contradictory to Manifested Signs of Equal & Opposite Natural Magnetism and Manifested Nature can be categorised as Teleological Evidence Sorceries and Unique Epistemic Persecutions i.e. Historical Black Magic. In other words, shared searched out findings are both contrary as well as contradictory to introduced/projected Manmade Nature i.e. the Trinity of Mechanical Globalisation, Circular Rotation & Revolution System, and Manmade Magnetism. The sole purpose of this inspired sharing is to get rid of Projected Manmade Nature and Historical Black Magic i.e. to get rid of the fallacy of subjective selfcontradiction, fallacy of objective paradox, existential fallacy, fallacy of arguing in a circle, fallacy of infinite regress, fallacy of non-observation of instances, fallacy of non-observation of essential circumstances, fallacy of universal mal-observation, fallacy of individual mal-observation, fallacy of plurality of causes, characteristic imperfection, practical imperfection, fallacy of post hoc ergo propter hoc, and the like.

Ontology – To get confirmation of the verifiable truth of facts & justifiable necessary truth i.e. searched out findings / manifest truths shared as **selfevident truths** as well as recognition of shared selfevident truths as **Solidified Solid**

Human Rights from the Respective Authorities / Honourable Chairs for the Greatest Happiness of the Greatest Number, confirmation of the Equal & Opposite Right Direction towards Upright West, confirmation of the appointed day of observing Idd uniformly, and the like.

Teleology – To do the final step of the re-search work for name & fame as well as carrier advancement on those searched out findings which are neither verifiable truth of facts nor justifiable necessary truth in resemblance with one or the other established criterion of truth, revelations, manifested signs, equal & opposite manifested signs of natural magnetism, manifested nature, verifiable scientific laws, and justifiable philosophical theories. For this reason, an attempt has been made to share 'Kitaaba Wal-Hikmata' or 'Manifested Nature and the Utility of One's Upright Logic' through e-mail accounts [available in the Internet] with Epistemic Persons, Religious Scholars, Faculty Members of the Universities, Scientists of the Recognised Scientific Research Institutions, Member State/Country of UNO, Media Offices & Journalists, National as well as International Human Rights Organizations, World Peace Mission Organization, Govt. Officials, Public Representatives with the hope that possessors of upright logic will take the initiative/responsibility to verify as well as to justify searched out manifest truths i.e. shared selfevident truths. If someone or some institution is able to search out lacunae in the searched out manifest truth shared as Solidified Solid Human Rights with reference to four witnesses i.e. four revealed and established criterions of truth, namely, Coherence Truth [Tawraat], Correspondence Truth [Injiil], Pragmatic Truth [Zabuur], and Selfevident Truth [Furqan], then the searcher & sharer of this inspired sharing will try to do the final stage of his re-search work on those lacunae with a view to become a practical follower of the imposed UGC Guidelines.

Ayat\Verses - Say: O **People of the Kitaab [Ahlal Kitaab]! You have no ground to stand upon [upright] unless you stand fast by the Tawraat [Revelation or Fundamental Ayat or Formal Ground or Coherence Truth], the Injiil [Correspondence Truth or Manifested Signs or Detail Explanation of Fundamental Ayat], and all the revelations [Manifest Truths] that have come to you from your Rab.** It is the revelation that comes to you from your Rab, that increases in most of them their obstinate rebellion and blasphemy. **But sorrow you not over (those) people without Faith.** Those who believe, those who are **Haaduu, Saabiuu-nas** and **Nasaaras**, and any who **believe in Allah** and the **Last Day** [Yawmil-Aakhiri], and **work righteousness [do tabligh]**, -- on them shall be no fear, nor shall they grieve. [Sura No. – 4\5 – Maaaidah, Ayat\Verses – 68 & 69]

So, kindly provide a scope to the searcher & sharer of manifest truth to do the final step of his **re-search work** on 'Solidarity Rights'. This will be the Survival of the Fittest [Extrinsic Value] of the searcher & sharer of 'Kitaaba Wal-Hikmata' or 'Manifested Nature and the Utility of One's Upright Logic' as an individual being. To fulfil the Ontology of life as a human person i.e. 'Survival of the Truest' or 'Intrinsic Value' or 'Duty for Duty's Shake' or 'Nish-Kama Karma' or to do righteous work and to give good tidings to those who believe [Tabligh] or Utilitarianism, or Altruistic Hedonism i.e. Greatest Happiness of the Greatest Number, the searcher & sharer of this inspired sharing is trying to get his searched out findings self-published. In this regard, the searcher & sharer of this inspired sharing is not only the so-called Fundamentalist or Jihadist or Egoist or Essentialist for the 'Survival of the Fittest' in resemblance with the concept of non-human person of Peter Singer, but a Solidified Solid Fundamentalist or an Akbari-Jihadist or an Altruist or a Karma-Yogi or a Satyagrahi for 'Survival of the Truest' as the possessor of upright logic in resemblance with the concept of human person of Peter Singer.

[Note: This inspired work / re-search project has been done individually as well as independently without taking financial assistance from an NGO / approval or sanctioning or recognition from an authority / guidance of a mortal being like David Hume, and the like. In this regard, this sharing is free from illicit formalities & guidelines, imposed laws in resemblance with Manmade Nature, any concealed agreement or wise policy of Firawn. In other words, this sharing is free from the Trinity of Conspiracy – Hypocrisy – Tyranny as well as Duality of Activism & Terrorism. So, anyone can freely verify, justify, share, review, and comment on the searched out manifest truths shared as Solidified Solid Human Rights in the name of 'Kitaaba Wal-Hikmata' or 'Manifested Nature and the Utility of One's Upright Logic'.]

It is tautologous that Laws are made and established not to regulate manifested nature and equal & opposite manifested signs of natural magnetism as manifested nature is beyond the control of finite mortal beings. On the contrary, laws are made and established to protect the sanctity of manifest truth [Manifested Nature], Solidified Solid Human Rights, Justified Animal Rights, and the like from the Trinity of Conspiracy – Hypocrisy – Tyranny, Duality of Activism & Terrorism, Teleological Evidence Sorceries, Unique Epistemic Persecutions, Historical Black Magic, Concealment of Truth and Projection of Falsehood, and the like i.e. from the wicked hands of Firawn and Devil as well as their wise policies. So, if you find any concept of this inspired sharing in resemblance with conspiracy or evidence sorcery or persecution or manmade nature or the like, then the respective authorities are requested to provide the searcher &

sharer of 'Kitaaba Wal-Hikmata' or 'Manifested Nature and the Utility of One's Upright Logic' verifiable as well as justifiable punishment. Simultaneously, it is logically justified to place appeals/demands before the respective authorities to provide justifiable preventive and educative punishments, as well as verifiable retributive punishment [in its both forms] to those who are consciously – knowingly – purposefully and without Broken Bar concealing right directions of equal & opposite manifested signs of Natural Magnetism, reality of the manifested Upright Rectangular Universe, reality of the Manifested Hexagonal World within the Black Square of East Horizon, reality of the appearing Pentagonal Earth with Three Ascending Stairs, reality of the Upright West Region of the manifested hexagonal world as well as Uppermost Land of the Appearing Pentagonal Earth where there is the appointed Black & White Imam of the City called Kaba on the right side of the Mount Tuur, equal & opposite right direction of performing Salat towards Upright West, appointed day of observing Idd Uniformly, and the like through the projections of mechanical globalisation, circular rotation & revolution system, manmade magnetism i.e. constituents of Introduced Manmade Nature as well as Copernican Revolution, Kepler's Axis, and Brahe's Data i.e. components of Historical Black Magic.

Further, it is an appeal from the core of the psyche of the searcher & sharer of manifest truths before each and every matured member of the Manifested Hexagonal World [Asterisk] and appearing Pentagonal Earth [House or Spiders' Net] with three ascending stairs within the Black Square of East Horizon to verify and to justify shared findings of 'Kitaaba Wal-Hikmata' or 'Manifested Nature and the Utility of One's Upright Logic' with a view **to recognise shared selfevident truths as Solidified Solid Human Rights as well as to judge the standard of this re-search project i.e.** the quality of 'Kitaaba Wal-Hikmata' or 'Manifested Nature and the Utility of One's Upright Logic'. Your verified as well as justified recognition will help the searcher & sharer of this re-search project to become an altruist [an akbari-jihadist or a karma-yogi] and to attain Summum Bonum of Life or Liberation or Mukti or Muksa or Nirvana or to become the follower of equal & opposite right direction towards Upright West [Arsh] as well as to observe Idd on the appointed day uniformly. This will be the Survival of the Truest [Intrinsic Value or Niskama-Karma or Jihade-Akbar] of the searcher & sharer of 'Kitaaba Wal-Hikmata' or 'Manifested Nature and the Utility of One's Upright Logic'.

Ummat: - The sole duty of the possessor of Upright Logic or an Ummat or a human person of Peter Singer is to do righteous work and to give good tidings to those who believe. This is what is called Tabligh or Niskama Karma of Gita or Renunciation of the fruits of action i.e. Duty for Duty's Sake of I. Kant or

Altruistic Hedonism of J. S. Mill. So, this sharing is nothing but just shouting as well as announcement of the Greatest War [Jihade-Akbar] for the Greatest Happiness of the Greatest Number [Altruistic Hedonism / Duty for Duty's Sake / Nish-Kama Karma / Survival of the Truest] against Teleological Evidence Sorceries [Introduced Manmade Nature] and Unique Epistemic Persecutions [Historical Black Magic].

But the searcher & sharer of this inspired sharing is an ordinary man or a common man lack of certified pieces of paper / recognised identity [shane-nuzuul] save a univocal sharer [free from the fallacies of equivocations] of selfevident truth or so-called philosophical knowledge with his friends in deed [students]. For this reason, he is not well acquainted with Religious Scholars & Group Leaders, so-called pen-paper-pencil workers, mechanical artists, possessors of true knowledge of the Kitaab with Truth, Certified Activists of the Global Trinity, Recognised Persecutors [Man-made Magnetists like AK47 Laden] as the possessors of Common Run, Epistemic Uniqueness of Mystic Religious Experience, Real Scientists, Human Rights Authorities, Honourable Chairs of Human Rights Organisations, Honourable Chairs of Law & Justice, Honourable Govt. Officials, Honourable Public Representatives, impartial Journalists & Media Persons, and the like. In brief, though it is unknown who will take the responsibility for verification, justification, confirmation, and broadcasting of the searched out manifest truths and to recognise those manifest truths as **Solidified Solid Human Rights, yet it is certain that there are possessors of upright nature** who will reflect on searched out manifest truths shared in the 'Kitaaba Wal-Hikmata' or 'Manifested Nature and the Utility of One's Upright Logic'. Through this sharing an attempt has been made to become the follower of three core teachings of Gita – (i) Do what is right, (ii) Choose what is good, and (iii) Sticking to what is truth. So, with a view to share the searched out manifest truths with each and every matured member of the manifested hexagonal world and appearing pentagonal earth with three ascending stairs within the Black Square of East Horizon [Sirius B or Star Saturn], the searcher & sharer of this inspired sharing is trying to get the compiled preliminary findings of the re-search project on 'Solidarity Rights' grounded on Natural Magnetism self-published. Kindly try to verify & justify each and every searched out finding shared in the 'Kitaaba Wal-Hikmata' or 'Manifested Nature and the Utility of One's Upright Logic'. If you find any lacuna in the shared searched out finding, please share the same with four witnesses i.e. with reference to four revealed and established criterions of truth. The re-search scholar as well as sharer of 'Kitaaba Wal-Hikmata' or 'Manifested Nature and the Utility of One's Upright Logic' will try to do his final re-search work on your pointed lacunae and shared searched out finding.

Prologue

Simultaneously, try to recognise and share those findings which are verifiable truth of facts as well as justifiable necessary truth i.e. free from subjective selfcontradictions and objective paradoxes with those whom you suppose / think right persons [or places] to share / forward / refer the same on behalf of the searcher of manifest truth, re-search scholar of Solidarity Rights, and sharer of 'Kitaaba Wal-Hikmata' or 'Manifested Nature and Utility of One's Upright Logic [Solidified Solid Human Rights]. In this regard, the searcher of manifest truth & sharer of 'Kitaaba Wal-Hikmata' or 'Manifested Nature and the Utility of One's Upright Logic' will remain grateful to you forever.

With Regards,
Searcher of Manifest Truth & Sharer of 'Kitaaba Wal-Hikmata'
[Manifested Nature and the Utility of One's Upright Logic]
&
Re-search Scholar on 'Solidarity Rights in Islam'

Contents

Contents

SECTION – I
HIGHLIGHTS – 1.1
SIGNIFICANT FINDINGS

Fasta-'iz billaahi minash-Shaytaanir-Rajim
Bismillaahir-Rahmaanir- Rahim

CONCLUSIONS / SEARCHED OUT MANIFEST TRUTHS

1. There are three periods of time, namely, conceptual period, psychic period, and reproductive period.

2. Conceptual period is monodimensional or windowless i.e. Unique. Pythagorean Number – 'Point – One' represents conceptual period or Uniqueness.

3. Psychic period is bi-dimensional or binary ~sth or Formal Relation or Formal Ground only. Pythagorean Number – 'Line – Two' represents psychic period. Moreover. 'Line – Two' also represents converse and converted obverse of the Staff of Muusa (ass) into Snake and White.

4. Reproductive period is the manifestation of the equal & opposite ongoing 3D command or three dimensional arrow of science or trinity of philosophy or three minutes after Big Bang in resemblance with Pythagorean Number – 'Plain – Three' and the concept of past – present – future. With respect to Kitaab with Truth, Reproductive Period represents 'Two-fold Mercy and a Light' i.e. Two & Half or Odd integer. Equal & Opposite 3D represents Animation and the ~sth between equal & opposite 3D represents manifestation in resemblance with Pythagorean Numbers – 'Animation – Six' and 'Manifestation – Seven'.

5. Established Philosophical concept/theory that 'Time has only one dimension called succession' represents the 'Line' of Pythagorean Number 'Line – Two' or Staff of Muusa (ass). The psychic period represents 'Equal & Opposite' faces. The reproductive period represents the manifested sign of equal & opposite relation. So, the philosophical concept – 'Time has only one dimension' with respect to reproductive period is a misconception. With respect to 'span of life', time has only one dimension called succession.

6. There is no monism in Islam. Allah is the Manifest Truth. The development of Islamic Monism is a game of Evidence Sorcery i.e. concealment of equal & opposite Trinity [Manifestation] and projection of Mysticism, Spiritualism, Sufism, Awaliaism, Imamgiris, and the like. So, all pen-paper-pencil works & activities in resemblance with Monism, Spiritualism, Mysticism, Sufiism etc. are either contrary or contradictory to Truths-in-themselves as well as coded shared

1

tautologies. Search for the Revealed Truth – Which Ayat\Verses share with us the concepts of Spiritualism, Monism, Mysticism, Sufism, and the like in Islam?

7. Universe is an intrinsically luminous star called Shi-rra or Sirius or Diamond Operator in resemblance with Pythagorean Number – 'Solid – Four'.

8. Universe has been revealed as an equal & opposite Trinity. This equal & opposite revelation is called Tawraat or Law or Coherence Truth. This equal & opposite revelation has been developed as Newton's Third Law – 'Equal & opposite'. So, Newton's Third Law is a Revealed Truth [Tawraat or Law]. Each Fundamental Ayat or abbreviated Ayat is a Law or Tawraat or Formal Ground of Manifestation; while detail explanation of a fundamental Ayat is called manifestation or manifest truth or Injiil or correspondence truth.

9. Four Revealed and Established criterions of truth or four Kitaabs are not independent Revelations or excluded from Kitaab with Truth. On the contrary, four Kitaabs or criterions of truth are the four witnesses for verification as well as justification of manifest truth in resemblance with revelation with a view to confirm one's faith & belief i.e. faith on One Dot Leader and belief on Diamond Operator. It has been inspired repeatedly to confirm manifest truth by searching resemblance between manifested sign and revelation or between detail explanation of Fundamental Ayat [Injiil or Manifested Signs or Correspondence Truth] and Fundamental Ayat [Tawraat / Laws / Formulae / Formal Grounds]. Search for the Revealed Truth – Which Ayat\Verses inspire us to do re-search work on Revelations? Which Ayat\Verses share with us that Fundamental Ayat [Tawraat / Laws / Formal Grounds] are inconceivable? Which Ayat\Verses inspire us to believe either on detail explanations of Fundamental Ayat only [Injiil] or on further explanations of what have been shared explaining in detail by Allah and His Messenger only [Tafsir or Opinions] or on both Injil and Tafsir excluding Fundamental Ayat?

10. Four witnesses without reference to male and female represent four criterions of truth. These are - Tawraat [Coherence Truth], Injiil [Correspondence Truth], Zabur [Pragmatic Truth], and Furqan [Selfevident Truth].

11. Revealed Universe is a Trinity of West Horizon – cloud & sky – East Horizon or Sirius Binary System or Star System. Epistemic Uniqueness of Scientific Certainty has converted star system into Lunar System [Solar System or Planetary System or Astrophysical System or Astral System or Stellar System or Terrestrial System, and the like].

Consequently, each guiding star of the star system becomes planet or moon or moonlet. West Horizon is the main sequence Sirius A [Star Jupiter] and East Horizon is the white dwarf companion Sirius B [Star Saturn]. Cloud & Sky represent equal & opposite ~sth between West and East as well as Veto [Screen / Veil] between Wash-Shams Wal Qamar [so-called sun and so-called moon].

12. Universe has been manifested as an upright rectangle in coherence with the 'End of Proof' of Real Science [Tawraat] and in correspondence with the appointed Kaba [Injiil].

13. West Horizon is called Heaven or Galaxy of Stars or Ashabikan-Nujjum or Star Jupiter or Main Sequence Sirius A of the Sirius Binary System. Rain comes from the West.

14. East Horizon has been revealed further within the four basic forces. East Horizon within four basic forces is called Black Square or Non-luminous Moon or Dark Moon. Earthquake emerges from the East.

15. East Horizon has been revealed into two Zones and an octagon. So, matter particles are octagonal black spots or so-called sun-spots.

16. There are twelve matter particles in resemblance with equal & opposite months of the two Mercurian Years i.e. Lunar Year and Calendar Year.

17. Two Zones within the East Horizon are West Zone [Upper Seashore] and East Zone [Lower Seafore]. Upright West Region of the appointed Kaba represents West Zone and Straight Middle-East Region of Eartha 3D represents East Zone.

18. There are two West, namely, West Horizon and West Zone.

19. There are two East, namely, East Horizon and East Zone.

20. Revealed Octagon has been manifested as a hexagon after equal & opposite psychic sacrifices in resemblance with Ibrahim (ass) and Ismail (ass), and in correspondence with the sacrifices of the converted green and blue shades of Windows'7 Ultimate. This manifested hexagon within the Black Square [Non-luminous Moon or Dark Moon] of East Horizon is called World or Asterisk or Star Pluto. Hexagonal Symbols in the converted green shade of Windows'7 Ultimate, hexagonal units, creation of the world in six days etc. are the clear proofs as well as valid references of the Manifested Hexagonal World. So, manifested hexagonal world [Asterisk] has six regions [triangles].

21. There are four revealed or cardinal directions, namely, Down [east], Up [West], Revealed Left [Northern Hemisphere], and Revealed Right [Southern Hemisphere]. Crucified Sign is the clear proof as well as correspondence truth [Injiil] of the four revealed [cardinal] directions. Isa (ass) brought Clear Proofs of the four cardinal directions. Revelation resembles with Plus Sign of Mathematics; while

Manifestation resembles with Multiplication Sign of Mathematics [i.e. declining towards Revealed Right in correspondence with the appointed Kaba on the right side of the Mount Tuur]. So, there are four revealed [cardinal] directions in correspondence with Crucified Sign and Diamond Operator [Helium-4 or 2 Horizons & 2 Hemispheres], and six manifested directions i.e. Helium-4 atom consisting of 2 proton [West] – 2 Neutron [Hemispheres] – 2 Electron [East] [Hexagon or Asterisk]. Four cardinal [revealed] directions are Up [West Horizon], Down [East Horizon], Revealed Left [Northern Hemisphere], and Revealed Right [Southern Hemisphere]. Six manifested directions are North-East, Middle-East, South-East, South-West, Upright-West, and North-West.

22. There are four basic forces in resemblance with four cardinal [revealed] directions. East or Downward Direction represents Gravitational Force. West or Upward Direction represents Strong Force. North or Reveal Left or E-Point [Manifested North-East & North-West] represents Magnetic Force. South or Revealed Right or T-Point [Manifested South-East & South-West] represents Weak Force.

23. Revelation resembles with Plus Sign of Mathematics. In other words, revelation represents arithmetic progression and Axis [Major Axis, Semi Major Axis, and Minor Axis]. Moreover, revelation represents Night Sky.

24. World has been manifested declining towards revealed right. In other words, manifestation resembles with Multiplication Sign, geometric progression, and Orbit. Moreover, Manifestation represents Earth's Sky.

25. There are six manifested directions, namely, North-East, Middle-East, South-East, South-West, Upright-West, and North-West. Each manifested direction represents a Region or a Triangle of the manifested hexagonal world [Asterisk] within the Black Square of East Horizon.

26. Each Zone of the manifested hexagonal world comprises three Regions [Triangles]. East Zone comprises North-East Region of North America, Straight Middle-East Region of Eartha 3D, and South-East Region of South America.

27. West Zone [Upper Seashore] comprises South-West Region of Europe, Upright-West Region of Arabian Peninsula, and North-West Region of Asia, Africa, and Australia.

28. Due to equal & opposite ~sth between Upright West Region of Arabian Peninsula and Straight Middle-East Region of Eartha 3D [Black Hole or Recycling Region or penetrated hole in the journey boat of Muusa (ass)], the manifested hexagonal world [Asterisk] within the Black Square of East Horizon is appearing as a Pentagon. This appearing

pentagon is called the Earth or Star Operator. So, the appearing earth is a Pentagon in resemblance with House or Spiders' Net and in correspondence with 9/11 pentagon.

29. Due to Star Saturn as the Revealed Peg [1st peg] and Star Mars as the manifested peg [2nd peg], neither the manifested Hexagonal World nor the Appearing Pentagonal Earth is moving. Moreover, the intrinsically luminous star is a Fixed Star with reference to real science. The revealed concepts '**Each one is moving**' with reference to Ayat No. – 40 of Sura Yaa-siiin & Ayat No. 1 to 4 of Sura – Rad, and '**They float, each in an orbit/axis**' with reference to Ayat No. – 30 to 33 of Sura – Ambiyaaa, and the like represent Wash-Shams and Qamar or so-called sun and so-called moon. Each one is moving following equal & opposite Orbit and Axis. On the contrary, shared concepts like '**We have set in the world mountains standing firm, lest it should shake with them**' with reference to Ayat No. - 30 to 33 of Sura – Ambiyaaa and '**When the world will be shaken [future event] to her [utmost] convulsion [earthquake]**' with reference to Ayat No. 1 to 8 of Sura - **Izaa zulzilatil-arzu zilzalahaa**, are the revealed truths [clear proofs] that neither the manifested hexagonal world nor the appearing pentagonal earth is moving. Searching question – Which Ayat\Verses share with us that the universe, the world, and the earth are moving in resemblance with Wash-Shams Wal Qamar [so-called sun and so-called moon]?

30. Appearing Pentagonal Earth has three ascending stairs, namely, Ground Stair [Tin or Township or Super Script 1], Middle Stair [Zaytuun or Median or Super Script 2], and Topmost Stair [Tuur or City or Super Script 3].

31. North America and South America of the East Zone [Lower Seashore] represent Ground Stair [Tin or Township or Super Script 1] of the appearing Pentagonal Earth.

32. Europe, Asia, Africa, and Australia of the West Zone [Upper Seashore] represent Middle Stair [Zaytuun or Median or Nation or Super Script 2] of the appearing Pentagonal Earth.

33. Arabian Peninsula represents Topmost Stair or Uppermost Land [Tuur or City or Super Script 3] of the appearing Pentagonal Earth.

34. Appearing Pentagonal Earth has seven continents or windows. North America and South America of the Ground Stair are called Eastern Continents or Eastern Windows. North America represents North-Eastern Continent [Window]. South America represents South-Eastern Continent [Window].

35. Europe, Asia, Africa, and Australia of the Middle Stair represent Western Continents or Western Windows. Europe represents

5

South-Western Continent [Window]. Asia is not an Eastern Continent [Window]. On the contrary, Asia is a North-Western Continent [Window]. Africa is neither in the Southern Hemisphere nor in the East Zone [Lower Seashore]. On the contrary, Africa is a North-Western Continent [Window]. Australia is also a North-Western Continent [Window].

36. Arabian Peninsula is neither in the Middle-East Region of Eartha 3D nor in the East Zone within the Black Square of East Horizon. On the contrary, Arabian Peninsula is Upright West Region, Topmost Stair out of three ascending stairs, and Uppermost Land of the appearing Pentagonal Earth within the Black Square of East Horizon. Projection of Arabian Peninsula in the Middle-East Region of Eartha 3D is a conspiracy against believers & followers of Upright West.

37. The Black & White Imam of the City called Kaba as the manifested leader of right direction [standard] for mankind has been appointed neither in the Middle-East Region of Eartha 3D, nor at the centre of Revelation & Manifestation, nor in the East Zone of the manifested East Horizon. On the contrary, Kaba has been appointed on the right side of the Mount Tuur in the Upright West Region of the manifested hexagonal world, topmost stair out of three ascending stairs, and uppermost land of the appearing pentagonal earth within the Black Square of East Horizon.

38. Newton's Three Laws of Motion and Law of Gravitation are the Revealed Truths. These laws represent Axis i.e. Major Axis, Minor Axis, and Semi Major Axis. Einstein's Gravitational Lensing and Binary Pulsar represent Manifested Truths and Orbits. Newton has the proper place in revelation i.e. accelerating masses (m) and accelerating charges (a). Einstein has the proper place in manifested nature i.e. equal & opposite manifested signs of natural magnetism [what accelerating charges emit and what accelerating masses emit]. So, Newton and Einstein are the two faces of Muusa's Staff in resemblance with Pythagorean Number – 'Line – Two'.

39. Copernican Hypothesis, Kepler's Axis, and Brahe's Data are either contraries or contradictories with respect to both revelation and manifestation. On the contrary, Copernican Hypothesis, Kepler's Axis, and Brahe's Data are the components of Historical Black Magic as well as Vera Causa of Activism and Persecution. Copernican Hypothesis, Kepler's Axis, and Brahe's Data have the significant places in Projected Mechanical Globalisation through Napier Bones, Projected Petitio Principii or Circular Rotation & Revolution System, and Projected Post Hoc Ergo Propter Hoc or Manmade Magnetism. In brief, Copernican

Hypothesis, Kepler's Axis, and Brahe's Data are the Magical Keys of Projected Manmade Nature.

40. There is an equal & opposite resemblance and difference between Revelation and Manifestation, between Plus Sign and multiplication Sign, between Newton and Einstein, between Axis and Orbit, between accelerating charges and what accelerating charges emit [release], between accelerating masses and what accelerating masses emit [release].

41. There is **only one kind of Magnetism**. This sole Magnetism is called **Natural Magnetism**.

42. Introduction of **Man-made Magnetism** in the domain of **Formal Education** [General Education] is a **Black Magic** i.e. **Teleological Evidence Sorcery and Unique Epistemic Persecution** of faith and belief of mankind in general and particularly faith & belief of the followers of equal & opposite Middle Course towards Upright West [Throne of Authority or Arsh].

43. Globalisation or Common Run is neither revealed truth, nor Manifested Nature [Manifest Truth]. On the contrary, Globalisation or Common Run is the Projected Mechanism through Napier Bones with a view to play the game of Black Magic. The revealed truth is that Globalisation or Common Run is an impossible task. Search for revealed truth – Which Ayat\Verses share with us crookedness or global concept or common run?

44. The so-called sun [Bullet / Electromagnetic Wave / converted snake of Muusa's Staff] as the manifested sign of Natural Magnetism rises [once] from the East [Gravitational Force] and sets [once] in the West [Strong Force] to cause the alteration of day and night for the equal & opposite Upper Seashore [West Zone] and Lower Seashore [East Zone] within the Black Square of East Horizon is a **Post Hoc Ergo Propter Hoc Statistics**.

45. The so-called sun [Bullet / Electromagnetic Wave / Sister Planet Venus / Maryam supplied with sustenance] neither rises from the East nor sets in the West. These are the Manifest Truths in resemblance with Revelation [Ref. - Ayat\Verse No. – 35 of Sura -Nuur]. The Electromagnetic Wave [so-called sun as the manifested sign of natural magnetism / Bullet] enters from North and ends the equal & opposite stages of journey in North. These are the Manifest Truths in resemblance with Revelation [Ref. - Ayat\Verse No. – 17 of Sura – Khaf].

46. The so-called sun [Bullet / Electromagnetic Wave] as the manifested sign of natural magnetism rises twice and sets twice to cause the alteration of day and night for the equal & opposite Upper Seashore [West Zone] and Lower Seashore [East Zone].

47. For the Lower Seashore [Ground Stair of North America, and South America], the so-called sun [Bullet / Electromagnetic Wave] as the manifested sign of natural magnetism enters from North-East in correspondence with North America and sets in South-East in correspondence with South America. This is the Semi-anti-clockwise journey of the so-called sun [Bullet] and the day light for the Ground Stair of the appearing pentagonal earth and the East Zone within the Black Square [Non-luminous Moon or Dark Moon] of East Horizon. As the so-called sun [Bullet / Electromagnetic Wave] enters from North-East, so, the concept of day [am] manifests [origins] from the East Zone and begins from the West Zone.

48. For the Upper Seashore [Middle Stair of Europe, Asia, Africa, and Australia as well as Topmost Stair of Arabian Peninsula], the so-called sun [Bullet / Electromagnetic Wave] as the manifested sign of natural magnetism rises from South-West in correspondence with Europe and sets in North-West in correspondence with Australia. This is the Semi-clockwise journey of the so-called sun [Bullet] and the day light for the Middle Stair and Topmost Stair of the appearing pentagonal earth and the West Zone within the Black Square [Non-luminous Moon or Dark Moon] of East Horizon.

49. So-called new moon [Triangular Bullets / Planets Uranus & Neptune i.e. twin / Muzzammil & Muddassir] arises from North-West for the West Zone. So, the concept of Night [pm] manifests [origins] from the West Zone [Upper Seashore] and begins from the East Zone [Lower Seashore]. The arising moon [Triangular Bullet] of the West Zone follows Axis & Right Double Quotation Mark i.e. North-West to South-West for the 1st & 3rd periods/quarters, and Left Double Quotation Mark i.e. South-West to North-West for the 2nd & 4th periods/quarters. Neither the so-called sun nor the so-called new moon enters/arises at mid-night or at noon. So, the projected concepts of am and pm are nothing save post hoc ergo propter hoc.

50. Appointed day of observing Idd uniformly depends on the arising of new moon [Star Uranus or Muzzammil] for the West Zone. So, people of Australia, Africa, Asia, Arabian Peninsula, and Europe will observe Idd on the equal & Opposite Break of the Day & Night Fall i.e. Moring from the point of view of West Zone [Upper Seashore] and prior to people of the East Zone. People of North America and South America will observe Idd following West Zone i.e. equal & Opposite Early Hours of Night & Morning [Evening from the point of view of West Zone but Morning from the point of view of East Zone]. This is called Unity-in-Diversity. Either contrary or contradictory pen-paper-pencil

works and mechanical activities concerning the appointed day of observing Idd uniformly is a conspiracy against Muslims. Due to this conspiracy, I had failed to observe a single Idd on the appointed day. In this regard, a list of searched out pen-paper-pencil works and mechanical activities will be shared as experimentum crusis as well as crucial instance for verification, justification, confirmation, general recognition, necessary action, and the like by the respective authorities of Human Rights.

51. So-called new moon [Triangular Bullet/ Planets Uranus & Neptune i.e. twin / Muzzammil & Muddassir] appears from South-East for the ground stair of East Zone [South America and North America]. The appearing moon [Triangular Bullet] of the East Zone follows Axis & Left Double Quotation Mark i.e. South-East to North-East for the 1st & 3rd periods/quarters, and Right Double Quotation Mark i.e. North-East to South-East for the 2nd & 4th periods/quarters.

52. Star [Planet/Moon] Neptune or Muddassir or Triangular Black Bullet is not visible to the unaided eye and its Axis is subject to Gravitational Perturbation or Enveloped. The Axis of Neptune is called Minor Axis. Star [Planet/Moon] Uranus or Muzzammil or Triangular White Bullet [ice giants / white] carries the Blessed Message.

53. Four Galilean Moons represent Twin Triangular Bullets of the West Zone [Upper Seashore] and Twin Triangular Bullets of the East Zone. Searching questions – Which one among four Galilean Moons is the Nil Arm Strong's visited Moon? In which Moon the so-called Scientists often Visit?

54. The so-called sun or Bullet [Electromagnetic Wave or converted Snake of Muusa's Staff] is the Manifested Sign as well as Clear Proof of the sole magnetism called Natural Magnetism.

55. The direction from which the so-called sun [Bullet / Electromagnetic wave] as the manifested sign of natural magnetism enters as the Morning Show in the name of sister planet Venus or Maryam supplied with sustenance for the East Zone [Lower Seashore] within the Black Square of East Horizon is the Revealed Left and Manifested North Direction or Haiyalas-Swalaah for the Ground Stair of North America and South America. The direction towards which the so-called sun [Bullet] as the manifested sign of natural magnetism sets or the direction from which the so-called new moon [Triangular White Bullet] appears for the East Zone is the Revealed Right and Manifested South Direction or Haiyalal-Falaah for the Ground Stair of North America and South America. In other words, **the so-called sun or Bullet rises [enters] from North and sets in South for the Ground**

Stair of the Pentagonal Earth and the East Zone of the Hexagonal World within the Black Square of East Horizon. These are the unalterable manifest truths in resemblance with revelations.

56. The direction from which the so-called sun [Bullet / Electromagnetic wave] as the manifested sign of natural magnetism rises as the Evening Show in the name of sister planet Venus or Maryam supplied with sustenance but Day Break [Morning Show] for the West Zone [Upper Seashore] within the Black Square of East Horizon is the Revealed Right and Manifested South Direction or Haiyalal-Falaah for the Middle Stair of Europe, Asia, Africa, Australia as well as Topmost Stair of Arabian Peninsula. The direction towards which the so-called sun [Bullet] as the manifested sign of natural magnetism sets or the direction from which the so-called new moon [Triangular White Bullet] arises for the West Zone is the Revealed Left and Manifested North Direction or Haiyalas-Swalaah for the Middle Stair of Europe, Asia, Africa, Australia as well as Topmost Stair of Arabian Peninsula. In other words, **the so-called sun or Bullet rises from South and sets [ends] in North for the Middle Stair as well as Topmost Stair** of the Pentagonal Earth and the West Zone of the Hexagonal World within the Black Square of East Horizon. These are the unalterable manifest truths in resemblance with revelations.

57. Ring Operator Zakariya, the Guardian of Maryam [Gravitational Wave or Ether or Electron Cloud or Bish-Shamsi or so-called sun-spots] has been appointed to rise from the East as a Screen [veil / veto] between equal & opposite Early Hours of Night & Morning as well as between equal & opposite Break of the Day & Fall of Night.

58. Introduction of Manmade Magnetism in the domain of formal education by the Epistemic Uniqueness of Scientific Certainty is the Vera Causa of Activism and Persecution. As there is only one Revealed Left or Magnetic Force, so, there is only one Northern Hemisphere. Similarly, there is only one Revealed Right or Weak Force or Southern Hemisphere. In brief, there is only one kind of Magnetism called Natural Magnetism. The so-called sun [Bullet] is the manifested sign of equal & opposite Natural Magnetism. With a view to conceal the reality of Natural Magnetism, the Don of Historical Conspiracy Science or Epistemic Uniqueness of Scientific Certainty has introduced Manmade Magnetism. North and South directions determined on the basis of Manmade Magnetism neither represent Manifested North and South Directions, nor represent Revealed Left and Revealed Right, nor represent Haiyalas-Swalaah and Haiyalal-Falaah, nor represent Northern Hemisphere and Southern Hemisphere. On the contrary, North

and South Directions determined on the basis of Manmade Magnetism represent Manmade North and Manmade South or Conspiracy North & South Directions. Moreover, the projection of Revealed Left [Northern Hemisphere] as top of the globe in resemblance with the head of an upright individual as well as the projection of Revealed Right [Southern Hemisphere] as bottom of the globe in resemblance with the leg of an upright individual with a view to conceal existential imports of West Direction and East Direction are not only teleological evidence sorceries but also post hoc ergo propter hoc projections. In projected Global Trinity as well as Quaternary called Manmade Nature, there are several fictitious North and South Directions. That means there are neither East, nor West, nor Revealed Left [Northern Hemisphere / True North], nor Revealed Right [Southern Hemisphere / True South]. Consequently, those who have rejected faith on equal & opposite revelation and belief on manifested signs of equal & opposite natural magnetism become the followers of Manmade Magnetism, Manmade Nature, Manmade Directions, Haiyalas-Swalaah & Haiyalal-Falaah. Believers of the Ground Stair of North America and South America are performing Salat towards setting direction of the so-called sun [Bullet] as the manifested sign of natural magnetism i.e. Revealed Right or Haiyalal-Falaah or Manifested North Direction from the point of view of East Zone. Believers of the Middle Stair of Europe, Asia, Africa, and Australia are performing Salat towards ending direction of the equal & opposite stages of journey of the so-called sun [Bullet] i.e. Revealed Left or Haiyalas-Swalaah or Manifested North Direction from the point of view of West Zone.

59. Mechanical Globalisation, Circular Rotation & Revolution System, and Manmade Magnetism are the constituents of Projected Manmade Nature, Teleological Evidence Sorceries, Unique Epistemic Persecutions, Real Activism, Historical Black Magic, Veil of Ignorance, and the like in the name of Scientific Certainty & Formal Education. Due to overspreading of projected falsehood i.e. Manmade nature like the advertisement of a product, believers of Upright West become followers of Straight Middle-East Region of Earha 3D [Black Hole, Recycling Region, penetrated hole in the journey boat of Muusa (ass), Bermuda Triangle of Titanic, Hollow in the Hand, Projected Planet Mars].

60. Real Scientific works like Diamond Operator, End of Proof, Black Square, Star Operator etc. are available in Insert – 'Symbol' of your PC. Projected Nine Planets are the nine clear proofs given to Muusa (ass) as eye opening evidences.

61. Pen-Paper-Pencil Works and Mechanical Activities which are either contrary or contradictory to Manifested Nature, and neither verifiable nor justifiable by one or the other revealed & established criterion of truth cannot be called Knowledge i.e. neither right knowledge nor wrong knowledge with reference to Plato's Theory of Knowledge. On the contrary, projections of such works and activities in the domain of Formal Education [General Education] can be categorised as Teleological Evidence Sorceries and Unique Epistemic Persecutions or Historical Black Magic.

62. Massive Publications and Impressive Broadcasting of the Teleological Evidence Sorceries and Unique Epistemic Persecutions or Historical Black Magic are the means of the Concealment of Manifested Nature and Projection of Man-made Nature, Concealment of Intrinsically Luminous Star and Projection of intrinsically luminous Moon, Concealment of Manifested Hexagon within the Black Square and Projection of Circular Rotation and Revolution System, Concealment of Appearing Pentagonal Orange and Projection of global orange, Concealment of Three Ascending Stairs [Stages of Journey] and Projection of common run [binary international harmony], Concealment of Equal & Opposite Middle Course towards Upright West [Arsh] and Projection of Straight way towards Middle East Region of Eartha 3D [Mars or Black Hole], Concealment of the appointed day of observing IDD uniformly i.e. Unity-in-diversity and Projection of fictitious days of observing Idd i.e. Diversities-in-Unity, and the like.

The above sixty two shared significant findings are neither suppositions, nor imaginations, nor hypotheses, nor predictions, nor opinions, nor discoveries. On the contrary, most of the above shared findings are the manifest truths in resemblance with revelations. The rest searched out findings are the projected falsehoods. Shared manifest truths are unalterable truths. So, shared manifest truths are certain like 'Cogito Ergo Sum' of Rene Descartes and Conclusion of Dictum De Omni ET Nullo of Aristotle i.e. unerring or infallible truths or selfevident truths. Projected falsehoods are verifiable by the uncountable pen-paper-pencil works and activities. All pen-paper-pencil works and activities which are either contrary or contradictory to searched out manifest truths must be categorised as Evidence Sorceries. So, those who have the capability to argue either contrary to manifest truth or contradictory to manifest truth or in favour of projected falsehoods [Historical Black Magic, Teleological Evidence Sorceries, Unique Epistemic Persecutions, Real Activism, Hypocrisy, and the like], as well as Owner & possessors of Manmade Nature are sincerely invited to play the game of 'Magic with Natural Magnetism and Upright Logic' openly and publicly.

SECTION – I
HIGHLIGHTS – 1.2
AHAD OR UNIQUENESS

Fasta-'iz billaahi minash-Shaytaanir-Rajim
Bismillaahir-Rahmaanir- Rahim

THREE PERIODS OF TIME

There are **three periods of time**, namely, **Conceptual Period, Psychic Period**, and **Reproductive Period**. Conceptual period is **Monodimensional and Windowless**. It is one, unique, eternal, beyond space-time relation, beyond categories of knowledge, and consequently beyond the octagonal rings and quadrilateral limitations. Neither **prefix** nor **suffix** nor shane-nuzzul can be added with the conceptual period or one or uniqueness save the concept of **'Neti-Neti'** or **'not this – not this'** of Sankara, the founder of Advaita Vedanta. This very concept of neti-neti resembles with the concept of **laa ilaaha**. This means that there is **no unbreakable reality save Monodimensional or Windowless One [Ref. Monadology of Leibnitz]. This Unbreakable Reality** is the Uniqueness or Ahad who has no attribute save that He is the **owner** of all attributes. The dimension of the Uniqueness is unknown and unknowable since it is monodimensional or windowless i.e. unique. The concept of Uniqueness can also be understood in resemblance with the concepts like **Absolute Reality, Unbreakable Reality**, Unity, all inclusive etc. Ahad or Uniqueness is called One Dot Leader in Real Science and Rabbus-Shi-raa in the Kitaab with Truth.

Psychic period is **bi-dimensional**. The Pythagoreans drew up a list of ten opposites of which the universe is composed. These are – (i) **Limited** and **Unlimited**, (ii) **Odd** and **Even**, (iii) **One** and **Many**, (iv) **Right** and **Left**, (v) Masculine and Feminine, (vi) **Rest** and **Motion**, (vii) **Straight** and **Crooked**, (viii) **Light** and **Darkness**, (ix) **Good** and **Evil**, (x) **Square** and **Oblong**. In reality, Pythagorean Number **'Line – Two'** represents two opposite faces of the same line as the formal truths or formal grounds only. Neutrino and Neutron of an atom represent psychic period. In other words, psychic period is the conflict of desires between two opposite faces of the same coin or binary threads of a niche. The root cause of this conflict is nothing but the sense of recognition. I have conceptualised in resemblance with the ten opposites of Pythagoreans that the universe is composed of hundred minus one i.e. ninety and nine beautiful attributive names. Science has also discovered ninety-two [92] attributes and seven [7] Windows [force carrier particles or eye opening evidences out of nine clear proofs given to Muusa (ass)]. The remaining one is the Substratum or Uniqueness in resemblance with the concept of a Pronoun. For instance, the pronoun 'I' is a substratum of the trinity of my body, my mind, and my soul. As the pronoun 'I' has no gender, so also the Uniqueness. So, the Uniqueness is beyond the question of gender biased. The reference of twenty five more represents twenty five names of the messengers and warners as signs of formal relation.

Reference from Science - All **observed elementary particles are either fermions or bosons,** elementary bosons are all <u>gauge bosons</u>: photons, <u>W and Z bosons</u>, gluons, and the <u>Higgs boson</u>, **Ninety-two elements, and** around **25 more** have been made **artificially.**

Reproductive period is **three dimensional**. This three dimensional concept resembles with the concept of past, present, and future. The concept of present is the ~sth or the line of Pythagorean Number 'Line – Two' between two angular faces - past and future. In reproductive period, the ~sth establishes a relation between opposite particles or constituents. This established relation is called Trinity or Equal and Opposite Relation or Pythagorean Number – 'Plane – Three'. So the concept of trinity is the concept of an established ~sth as a binary sign of relation between opposite faces [snake & white] of Muusa's Staff. The result of the establishment of a binary ~sth between two angular faces is a triangle or three dimensions or 3D or triangular arrow or trinity or three minutes after Big Bang. The philosophical concept that time has only one dimension is a misconception with respect to psychic period and three dimensional space-time relations or Reproductive Period. Time has only one dimension called succession represents the span of life.

The detail explanations of Fundamental Ayat are called Coded Shared Tautologies or Kitaab with Truth. These Tautologies are the Universal Major Premises of the Dictum de Omni ET Nullo of Aristotle. To avoid the **fallacy of arguing in a circle** or the fallacy of **infinite regress**, we must have to rely on Coded Shared Tautologies. This reliance is the reliance on Uniform Principle or Principle of the Uniformity of Nature. The Kitaab with Truth includes both Truth [Revelation or Fundamental Ayat] as well as Tautologies [Manifestations or Criterions of Truth]. The purpose of this sharing is to get recognised confirmation from the respective authorities of the searched out manifest truths [resemblance between Revelation and Manifestation] which are being shared in the 'Kitaaba Wal-Hikmata' or 'Manifested Nature and the Utility of One's Upright Logic'.

It is selfevident that further interpretation of the resemblance between Revelation and Manifestation with categories of knowledge will involve the existential fallacy and the dialectic of **thesis – anti-thesis - synthesis** will commit either the **fallacy of infinite regress or the fallacy of arguing in a circle**. With

14

a view to share selfevident truth free from **subjective selfcontradictions and objective paradoxes**, I must have to avoid critical remarks of IFTA on detail explanations in the name of Tafsir or further explanations of what have been shared explaining in detail by Allah and His Messenger. Moreover, the critical remarks of IFTA on Revelation are neither truth nor criterions of truth nor established laws. On the contrary, translation of the meanings and commentary on Quran are black & white works and activities in resemblance with Copernican Hypothesis, Kepler's Imagination, Newton's Supposition, Aristotle's perception, Galileo's Observation, and Einstein's Prediction.

Here I am not denying Newton, Einstein, Aristotle, and Galileo. There are two kinds of Sciences with respect to the searched out manifest truth of this sharing. One is the **progressive science** which includes Biology, Chemistry, Physics, Mathematics, Statistics etc. Progressive Science is not the concern of this research project as well as inspired sharing. There is another kind of science which has projected rectangular Universe, non-luminous East Horizon [Black Square], hexagonal world, and pentagonal earth with three ascending stairs as global universe – global world -global earth. This projection is contrary as well as contradictory to both revelations and manifestations. So, this sharing is concerned with pen-paper-pencil works and activities which are concealing manifest truth and manifested nature with a view to project manmade nature and falsehood as truths in the name of theorisation with reference to X, Y, and Z. Such pen-paper-pencil works and activities have been termed as **Conspiracy Science or Historical Black Magic or Projected Manmade Nature**. This sharing is concerned with Historical Black Magic or Teleological Evidence Sorcery and Unique Epistemic Persecution.

Through the projection of **Mechanical Globalisation**, the **Conspiracy Science** has been concealing **Revealed & Manifest Truths** from mankind without **Broken Bar**. Concealment of manifest truth and projection of man-made falsehood are the binary functions of **Historical Black Magic in resemblance with the functions of Maya**. Through the projection of Global Trinity, **Black Magicians** have been concealing the realities of the manifested Universe or 'End of Proof' of the **Real Science** or the appointed Kaba, Non-luminous **East Horizon** or **Black Square** of the **Real Science, Manifested Hexagonal World** or **Asterisk** within the Black Square of East Horizon, appearing **Pentagonal Earth** with three ascending stairs, and the like. Through the projections of top and bottom of the Global Trinity as Northern Hemisphere and Southern Hemisphere, left and right as North and South respectively, Black Magicians have been successfully concealing true concepts of **two West and two East as well as Revealed Left & Revealed Right**. Through the projection of man-made magnetism, **Black Magicians** have been successfully snatching away **Manifested North Direction** and **Manifested South Direction** from both literate & illiterate human persons as well as from the domain of Formal Education. Further, through the projection of Circular Rotation & Revolution System, Historical Black Magicians have snatched away equal & opposite manifested sign of natural magnetism, equal & opposite right direction of Qibla of the believers as well as appointed day of observing Idd uniformly.

The sole purpose of education is search for truth. I have searched out manifest truths. But I have failed to share searched out manifest truths with the so-called epistemic persons as some are the representatives of one or the other Historical Black Magician, some are the blind followers of several Black Magicians, and some are self-sensed epistemic persons as they are the possessors of recognised pieces of paper called Degrees & Certificates, and the like. So, I have placed my searched out findings before the tables of some respective authorities. I hope those authorities will verify and justify searched out manifest truths with a view to rescue Manifested Nature from the wicked hands as well as Books of Black Magic. For open & public sharing, I am trying to get my searched out findings self-published.

ONE WAHID AND BEAUTIFUL NAMES

Ayat\Verses - "Huwallaa-hullazii laaa-ilaaha illaa huu –Aalimul-gaybi wash-ahaadah, Huwar-Rahmaanur-Rahim." [Trans.] He is Allah, than Whom there is no other reality (ilaaha) the Knower of the invisible and the visible. He is The Beneficent, The Merciful. [Sura No. – 58\59 – Awwalil-Hashr, Ayat No. – 22]

Ayat\Verses - "Huwallaa-hullazii laaa ilaaha illaa huu, – **Al-Malikul Qudduusus**–Salaamul-Mu-minul-Muhay-minul Azizul–Jabbaarul-Mutakabbir. Subhaanallaahi ammaa yushrikuun." [Trans.] He is Allah, than Whom there is no other reality (ilaaha) **The Sovereign Rab, the Holy One, Peace, the keeper of Faith, the Guardian, the Compeller (compel – arouse irresistibly), the Superb. Glorified be Allah from all that they ascribe as partner (to Him) [manmade magnetism with natural magnetism].** [Sura No. – 58\59 – Awwalil-Hashr, Ayat No. – 23]

Ayat\Verses - "Huwal-laahul-Khaaliqul-Baari-ul-Musawwiru l a h u l Asmaaa-ul-Husnaa; yusabbihu lahuu maa fis-samaawaati wal-arz; wa Huwal-Azizul-Hakim." [Trans.] He is Allah, **the Creator, the Evolver out of nothing, the Fashioner of Forms.** To Him belong the **Most Beautiful Names.** All that is in the heavens [West Horizon] and in the world [East Horizon] glorify Him, and He is The Mighty, The Wise. [Sura No. – 58\59 – Awwalil-Hashr, Ayat No. – 24]

16

NINETY AND NINE, AND ONE

Ayat\Verses – How they burst in upon **Daawuud**, and he was afraid of them. They said: Be not afraid! (We are) **two litigants** (petitioners), one of whom has wronged the other. Therefore, judge aright between us, be not unjust; and show us the fair way. Lo! This man is my brother. He has **ninety and nine** ewes (female ships / attributes / shane-nuzuul), and I have [but] **one**. Yet he says: Entrust it to me, and he is harsh to me in speech. [Sura No. – 37\38 – Swad, Ayat No. – 22 & 23]

TO HIM BELONG THE MOST BEAUTIFUL NAMES

Ayat\Verses - Say: **Call unto Allah**, or **cry unto Rahman by whatever name** you call upon Him, [it is well], for to **Him belong the Most Beautiful Names**. **Neither** you speak **aloud** in Prayer, **nor** speak it in a **low tone**, but **seek a middle course between**. And say: Praise be to Allah, **who begets no son**, and has **no partner in [His] dominion**. Nor [needs] He any to protect Him from humiliation: yea, magnify Him for His greatness and glory! [Sura No. – 16\17 - Banii-Israai-iil, Ayat No. – 110 & 111]

REJECT SUCH MEN AS USE BAD LANGUAGE

Ayat\erses - **The most beautiful names belong to Allah: so call on Him by them; but reject such men as use bad language in His names: for what they do, they will soon be requited?** And of those whom We have created there is a nation (ummat or median people) who guide with **truth** and establish **justice** therewith. [Sura No. – 6\7 - As-haabul- a' raaf, Ayat No. – 180 & 181]

AHAD OR UNIQUENESS

Ayat\Vreses – "Qul-Hu-wallaahu Ahad. Allaahus-Samad. Lam-yalid wa lam-yuulad. Walam yakul-la-Huu kufu-wan Ahad." [**Selfevident Concept**] Say: He is Allah, the UNIQUENESS [AHAD]! Allah, the eternally besought of all; He begets not, nor is He begotten. And there is none comparable to UNIQUENESS [AHAD]. [Sura No. – 111 \112 – Ahad (Prev. Iklas), Ayat No. – 1 to 4].

Note: There is no reference of the term 'Iklas' in the Sura. The term 'Ahad' has been referred twice in the Sura. Moreover, the term 'Wahid' is a connotative term which denotes 'One' and connotes relativity, but the term 'Ahad' is a non-connotative term which only connotes 'Uniqueness'. So, suggest us which name is consistent with the Sura. Options are – (i) AHAD [UNIQUENESS] or (ii) IKLAS [The Unity – provided meaning in translations & transliterations of Quran]. Further, whether the term 'Iklas' means 'unity' or inner state of mind of an individual [i.e. one's Teleology as well as Ontology]?

17

A human being is a bond of the Trinity of Soul-Psyche-Body limited in space & time. So, the philosophical question is – 'Can a physically existent being be called a metaphysical entity?' A Metaphysical Entity means a non-physical entity, a neoumena, a so-called spiritualist, a so-called mystic, pure form, a so-called awliyaaa, a so-called sufi, an atman, a soul without body, Paramartika Sattve, Idea of Good, Truth-in-itself, and the like.

Searching Questions: Is there any existential import of Mysticism or Sufism or Spiritualism or Awliyaaism in Islam? Which Ayat\Verses of Quran [Kitaab with Tuth] justify Mysticism or Spiritualism or Sufism in Islam going contradictory to the following Ayat\Veses? Whether Allah is the Spiritual Truth or Allah is the Manifest Truth? If there is an Ayat\Verse contradictory to the following Ayat\Verses, then it will imply that Quran is a Kitaab of Selfcontradictions and Paradoxes. Do you accept that Quran is a Kitaab of Selfcontradictions and Paradoxes in resemblance with Copernican Revolution i.e. Post Hoc Ergo Propter Hoc? It is a supposition that you will try to prove that Quan is a Kitaab with Truth free from Subjective Selfcontradictions and Objective Paradoxes.

It is absolutely certain that a physically existent being cannot be called a mystic. Moreover, Allah is not the spiritual truth. On the contrary, **Allah is the manifest truth**. The establishment of mysticism in Islam is a conspiracy against manifest truth. If so-called mystics have no concept of the source of light as well as equal & opposite right direction of Qibla in this life, then on the basis of which ground I will believe that they will lead me towards Light as well as Arsh in the life hereafter? So, the question is – Who are they?

FUNDAMENTAL AYAT [QURAN] HAVE BEEN SHARED EXPLAINING IN DEATAIL [TAFSIR]
Who is explaining further detail revelations?
Who is the ultimate knower? What is meant by Existential Fallacy?
It has been inspired to confirm revelations in
resemblance with manifested signs.

ILLUSTRIOUS MESSENGER
"It [Kitaab with Truth] is not the speech of a poet - little is it that you believe! Nor is it the **diviner's speech** - little is it that you **remember**. [This is] a **revelation** from the Rab of the Universe [Alamiin]. --- It is **Absolute Truth**." Ayat\Verses - But, nay! I swear by Allah that you see and all that you see not, that this is verily **the speech of an illustrious messenger**. **It is not the speech of a poet** - little is it that you believe! Nor is it the **diviner's speech** - little is it that you **remember**. [This is] a revelation from the Rab of the Universe

[Aalamiin]. And if he had invented any sayings in Our name, We should certainly take him by the right hand, and then severed his life-artery, nor could any of you withhold him. But verily this is a **warrant** for those who ward off evil. And We certainly know that some amongst you will deny. And lo! It is a **cause of sorrow for the disbelievers**. And lo! It is **Absolute Truth**. So glorify the name of your Rab **Most High**. [Sura No. – 68\69 – Al-Haaaqqatu, Ayat No. – 38 to 52]

ILLUSTRIOUS BUT ILLITERATE PROPHET

Ayat\Verses - Those who follow the messenger, the **illiterate Prophet whom they will find mentioned in their own [Kitab], Tawraat and Injiil** (which are) **with them. He will enjoin on them that what is right and forbid them what is wrong.** He allows them as lawful what is good [Manifest Truth in resemblance with Revelation] and prohibits them from what is foul [Manmade Nature or projected falsehood]. He **releases** them from their **heavy burdens** and from the **chains that they used to wear [Teleological Evidence Sorceries and Unique Epistemic Persecutions].** So it is those who **believe** in him, **honour** him, **help** him, and **follow the light [manifested sign of equal & opposite natural magnetism] which is sent down with him. It is they who will prosper.** Say: O mankind! I am sent unto you all, as the Messenger of Allah, to Whom belongs the dominion of the heavens [West Horizon] and the world [East Horizon]. There is no reality but He. **It is He Who gives both life and death.** So **believe in Allah and His Messenger**, the **Unlettered [illiterate] Prophet**, who **believes in Allah** and **His words: follow him [so] that you may be guided**. [Sura No. – 6\7 - As-haabul- a' raaf, Ayat No. – 157 & 158]

Ayat\Verses - **"Ha-Miim. Tanziilum-minar-Rahmaanir-Rahiim. Kitaabun-fussilat Aayaatuhuu Quraanan Arabiyyal-liqawminy-ya-lammuun. Bashiranwwa Naziiraa; fa-a-raza aksaruhuum fahum laa yasma-uun."**

Ha Mim - [Sura No. – 40\41 – Haa-Mim-Sajdah, Fundamental Ayat No. -1] (Revelation) from the **Beneficent, Most Merciful, A Kitaabun-fussilat whereof the Ayats are explained in detail; a Qur'an in Arabic, for people who understand,** giving good tidings and admonition; **yet most of them turn away, and so they hear not.** [Sura No. – 40\41 – Haa-Mim-Sajdah, Ayat No. – 2 to 4]

Ayat\Verses - And verily **We have explained in detail in this Qur'an**, for the benefit of mankind, **every kind of similitude [resemblance]**. But man is, in most things, contentious. And what is there to keep back mankind from believing, when the Guidance has come to them, nor from praying for forgiveness from their Rab, but that [they ask that] **the ways of the ancients be repeated with them,** or the wrath be brought to them **face to face**? We only send the messengers to give Good Tidings and to give warnings. But the **disbelievers dispute with vain argument, in order therewith to weaken the truth,** and they treat **My Signs as a jest,** as also the fact that they are warned! And who does greater wrong than one who is reminded of the **Signs of his Rab**; but **turns away** from them, forgetting the [deeds] which his hands have sent forward (to the Judgment)? Verily We have **set veils** over their hearts lest they should understand this, and over their ears, deafness, if you call them to guidance, even then they will never accept guidance. [Sura No. – 17\18 – Khaf – Ayat No. – 54 to 57]

"Thus do **We expound the Signs in detail for those who reflect?**" **Ayat\Verses -** The likeness (similitude) of the life of the present is as **the rain which We send down from the sky [West],** then the world's growth [East] of that which men and cattle eat mingle with it till, when the earth has taken on her ornaments and is embellished, and her people deem that they are masters of her. There reaches Our commandment by night or by day [equal & opposite revealed magnetism], and We make it like a reaped harvest, as **if it had not flourished yesterday.** Thus do **We expound the Signs (Manifest Truth) in detail for those who reflect**? But Allah summons to the abode of Peace. He guides whom He will to **a way that is right.** [Sura No. – 9\10 – Yuunus, Ayat No. – 24 & 25]

Ayat\Verses - When those **who believe in Our signs** come to **you,** Say: **Peace be upon you**! Your Rab has prescribed for Himself **mercy.** Verily, if any of you **did evil** in **ignorance,** and thereafter **repented,** and **amend** [your conduct], lo! He is Ever Forgiving, Most Merciful. Thus **We expound the Signs (Manifest Truth) in detail** that the way of **unrighteousness may be manifested.** [Sura No. – 5\6 – An-aam, Ayat No. – 54 to 55]

Ayat - Alif-Lam-Mim [Sura No. – 2\3 – Imran, Fundamental Ayat – 1] **Ayat\Verses -** Allah! There is no unbreakable reality save Him, the Alive, the Eternal. He has revealed unto you **a Kitaab with truth, confirming** that **which were revealed before it,** even as He revealed the **Tawraat** and the **Injiil** before this, as a **guide** to mankind, and **He has revealed the Criterions** (scales of justifications as well as verifications). Lo! **Those who reject Faith on (or disbelieve) the Signs of Allah (Manifested Signs of Natural Magnetism) theirs will be a severest chastisement** and Allah is Mighty and Able to Requite (the wrong). [Sura No. – 2\3 – Imran, Ayat\Verses – 2 to 4]

Ayat\Verses - From Allah, verily **nothing is hidden** in the world [East Horizon] or in the heavens [West Horizon]. He it is Who **shapes** you in the wombs as He pleases. There is no unbreakable reality save Him, the Almighty, the Wise. It is He Who has revealed to you the **Alaykal-Kitaab** [Fundamental Ayat or Revelation or Abbreviated Ayat] wherein there are **Clear Signs** [Clear Proofs]. These are the **Substances** [foundations or substratum] of the **Kitaab**; and others [which are] **allegorical** [Symbolic or Phonetic presentations or manifested signs or detail explanations]. But those in whose hearts are **doubts, pursue** forsooth (related to), that which is symbolic (**allegorical or manifested signs or detail explanations**) seeking (to cause) **dissension by seeking to explain it. None knows its explanation [further explanations of what have been shared explaining in detail or Tafsir]** except Allah. And those who are of **sound instruction** say: We believe therein, the **whole** [both fundamental ayat and detail explanations of fundamental ayat] is from our Rab; but **only men of understanding really notice**. [Sura No. - 2\3 – Imran, Ayat No. – 5 to 7]

Ayat\Verses - Verily We have sent you in truth as **a bringer of good tidings and a warner.** And you will not be asked about the **owners of hell-fire.** Never will the Yahuudis or the Nasaaras be satisfied with you unless you follow their creed. Say: Lo! The **Guidance of Allah**, which is the [only] **Guidance.** And if **you** follow their desires after the knowledge which has come to you, then you would have from Allah neither protecting friend nor helper. **Those, to whom We have given the Kitaab, study it as it should be studied.** They are the ones that **believe** therein. Those who **reject faith** therein, the **loss** are their own. [Sura No. 1\2 – Baqarah, Ayat No. – 119 to 121]

What a Kitaab is this!

"What a Kitaab is this! It leaves out nothing small or great, but takes account thereof!"

Ayat\Verses - In this case protection comes only from Allah, the True One. He is the Best to reward, and the Best to give success. Set forth to them **the similitude of the life of this world as water** which We send down from the skies [West} and the world's vegetation absorbs it [East], but soon it becomes **dry stubble**, which the winds do scatter. It is [only] Allah who prevails over all things. **Wealth and sons are allurements of the life of this world**. But the things that endure, good deeds, are best in the sight of your Rab, as rewards, and best as [the foundation for] hopes. And (bethink you of) the Day when **We will remove the mountains** and you will see the **world as a level stretch** and We will gather them all together, We will not leave out any one of them. And they will be marshalled before your **Rab in ranks**, [with the announcement]: Now have you come to Us [bare] as We created you first. But you thought We shall not fulfil the appointment made to you to meet [Us]! And the Kitaab [of Deeds] will be placed [before you]; and **you will see the sinful in great terror because of what is [recorded] therein.** They will say: Ah! Woe to us! **What a Kitaab is this! It leaves out nothing small or great, but takes account thereof!** They will find all that they did confronting them [in conformity]. And not one will your Rab treat with injustice. [Sura No. – 17\18 – Khaf – Ayat No. – 44 to 49]

This is a wonderful thing!
They are in a confused state.
Q A A A F: Wal-Quraanil-Majiid

But they wonder that there has come to them a warner from among themselves. So **the disbelievers say: This is a wonderful thing!** What! When we die and become dust, [shall we live again?] That is a [sort of] return far [from our understanding]. We already know how much of them the world takes away. With Us is a record guarding [the full account]. But **they deny the Truth when it comes to them.** So **they are in a confused state. [Sura No. 49\50 – Qaaaf Wal-Quraanil-Majiid, Ayat No. – 1 to 5]**

Do they not look at the sky above them? How We have made it and adorned it, and there are no flaws (rifts) in it? And **the world We have spread it out**, and **set thereon mountains standing firm**, and have caused of every lovely kind to grow thereon a vision and a reminder for every devotee turning [to Allah]. And We send down from the sky rain charted with blessing, and We produce therewith gardens and grain for harvests; **and tall [and stately] palm-trees,** with **shoots of fruit-stalks, piled one over another; Sustenance** for [Allah's] servants; and We give life therewith to land [earth] that is dead. Thus will be the Resurrection **[Sura No. 49\50 – Qaaaf Wal-Quraanil-Majiid, Ayat No. – 6 to 11]**

The folk of Nuuh denied (the truth) before them, and so did (the dwellers at) As-haabur-Rassi and (the tribe of) Samuud. And (the tribe of) Aad, Firawn, the brethren of Luut, and the Companions (dwellers) of the Wood, and the People of Tubba'; each one [of them] rejected the messengers, and My warning took effected. Were We then worn out by the first creation? Yet they are in doubt about a new creation. **[Sura No. 49\50 – Qaaaf Wal-Quraanil-Majiid, Ayat No. – 12 to 15]**

We verily created a man, and We know **what his soul whispers to him, and We are nearer to him than [his] jugular vein.** When the **two [guardian angels or binary pulsar] appointed to read** [his doings] **and to write** [noted them], **seated on the right and on the left. And not a word does he utter but there** is with him an **observer** ready. And the stupor of **death comes in Truth**. (And it is said to him): **This was that which you were trying to escape. [Sura No. 49\50 – Qaaaf Wal-Quraanil-Majiid, Ayat No. – 16 to 19]**

And the Trumpet shall be blown: That will be the Day whereof Warning [had been given]. And **there will come forth** every soul with each will be an [angel] to drive, and an [angel] to bear witness. [**And to the evil-doers** it will be said]: You were **heedless** of this. Now **We have removed from you your veil (covering) of ignorance** and **sharp is you sight this Day. And (unto the devil-doer) his comrade (friend) will say: Here is [his Record] ready with me. [Sura No. 49\50 – Qaaaf Wal-Quraanil-Majiid, Ayat No. – 20 to 23]**

[And it is said): Do **you twain [Two-in-one partnership] throw to hell each revolutionary ingratiate, hinder of good, transgressor of all bounds, spreader of doubts and suspicions; who set up [manmade magnetism] another reality along with Allah. Do you twain throw him to the dreadful doom?**

His comrade will say: Our Rab! **I did not make him transgress,** but **he was [himself] far astray.** He will say: **Dispute (contend) not with each other** in My presence **when I had already in advance sent you the warning. The sentence that comes from Me cannot be changed,** and I do not the least injustice to slaves. **[Sura No. 49\50 – Qaaaf Wal-Quraanil-Majiid, Ayat No. – 24 to 29]**

On the Day when We say to hell: **Are you filled? It will say: Are there any more [to come]?** And the **Garden will be brought near** to those who kept from evil (the Righteous), no longer distant. [And it is said]: **This is what We were promised. (It is) for every repentant and thoughtful one who fears The Beneficent in secret,** and comes with a **regretful heart.** Enter you therein in Peace and Security; this is a Day of Immortality. There will be for them therein all that they wish, and there is more with Us. **[Sura No. 49\50 – Qaaaf Wal-Quraanil-Majiid, Ayat No. – 30 to 35]**

And how many generations We destroyed before them who were mightier than these in skill so that they wander through the land! Had there any place of escape (when the Judgment came)? Lo! Verily **therein a reminder for him who has a heart, or gives ear with full intelligence.** And verily **We created the heavens [West Horizon] and the world [East Horizon] and all between them** in **Six Days [Hexagon]**, nor did any sense of weariness touch Us. So bear with what they say, and hymn the praise of your Rab **qabla**

tuluu-ish-shamsi wa qabla-lguruub (before the rising and before the setting of manifested sign of natural magnetism). And during part of the night, [also,] hymn His praises, and [so likewise] after the **postures** of adoration. And **listen on the Day when the caller will call out from a quiet near place**, The Day when they will hear the [awful] Cry in [very] truth that is **the Day of Renaissance coming forth.** Verily it is We Who give Life and Death; and to Us is the Final Goal. On the Day when the world will be rent asunder (apart) from them, **hastening forth (they come). That is a gathering easy for Us (to make).** We know best what they say; and you are not one to subdue them by force. So **warn by (admonish with) the Qur'an him who fears My Warning.** [Sura No. 49\50 – Qaaaf Wal-Quraanil-Majiid, Ayat No. – 36 to 45]

EXISTENTIAL IMPORT OF AWLIYAAA OR MYSTIC OR SPIRITUALIST
Allah is the Manifest Truth

Ayat\Verses - As for those who insult virtuous women, indiscreet (careless) but believing, cursed are they in this life and in the Hereafter. For them is a grievous Penalty. On the Day when their tongues, their hands, and their feet will bear witness against them as to what they used to do. On that Day Allah will pay them their just dues, and they will realise that **Allah is the Manifest Truth**. Vile women are for vile men, and vile men for vile women and good women are for good men, and good men are for good women. Such are innocent (not affected) of that which people say. For them there is forgiveness and a beautiful provision. [Sura No. – 23\24 – Nuur, Ayat No. – 23 to 26]

UNIQUENESS HAS NO COMPANION

Ayat\Verses - To Him is due the primal origin of the heavens [West Horizon] and the world [East Horizon]: **How can He have a son when He has no consort (companion)?** He created all things, and He has full knowledge of all things. That is Allah, your Rab! There is no unbreakable reality but He, the Creator of all things: then serve you Him: and He has power to dispose of all affairs. **No vision can grasp** Him, but His grasp is over all vision: **He is above all comprehension, yet is acquainted with all things. Now have come to you, from your Rab, proofs [to open your eyes]: if any will see, it will be for [the good of] his own soul; if any will be blind, it will be to his own [harm]: I am not [here] to watch over your doings. Thus do we explain the signs by various [Manifest Truths]: that they may say, You have taught [us] diligently, and that We may make the matter clear to those who have knowledge?** Follow what you are taught by inspiration from your Rab: there is no unbreakable reality but He: **and turn aside from those who join breakable realities [manmade magnetism] with Allah.** [Sura No. – 5\6 – An-aam, Ayat No. –102 to 107]

24

WHO IS AN AWLIYAA SAVE MY RAB AND YOUR RAB?

Ayat\Verses - "Tanziilul-Kitaabi minallaahil – 'Azizul-Hakim" – [Trans.} The revelation of this Kitaab is from Allah, the Exalted in Power, full of Wisdom. Verily it is We Who have revealed the Kitaab to you in Truth. So serve (worship) Allah, making **Mukhlisal-lahud-din** (confirming right direction free from projected conspiracies). "Alaa lillaahid-dinul-khaalis; Wallaziinattakhazuu min duunihiii **awliyaaa.** Maa na-buduhum illaa liyuqarribuunaaaa ilallaahi zulfa. Innallaaha yahkumu baynahum fii maa hum fiihi yakhtalifuun. Innallaaha laa yahdii man huwa kaazibun-kaffaar." [Trans.] **Surely pure religion is for Allah only.** And **those who choose protecting friends (awliyaaa) besides Him** (say): **We worship them only that they may bring us near to Allah.** Lo! Allah will **judge** between them **concerning that wherein they differ.** But Allah guides not such as are **liars (false) and ungrateful.** If Allah had willed to choose a **son,** He could have chosen whom He pleased out of those whom He created. Be He Glorified! [He is above such things.] He is **Allah, the One, the Irresistible** (Absolute) ["Huwallaahul-Waahidul-Qahaar"] [Sura No. – 38\39 – Zumar, Ayat No. – 1 to 5]

Ayat\Verses - Yaaa-'ayyu-hallaziina 'aa-manuu laa tataa-khizul-**Yahuu-da** wan-**Nasaaraa** 'awli-yaaa'. Ba-zuhum **awli-yaaa-'u** ba'-z. Wa many-yata-wallahum-min-kum fa-'innahuu minhuum. 'In- nallaaha laa yahdil-qaw-maz-zaalimiin. [Trans.] O you who believe! Take not the Yahuud and the Nasaara [followers of manmade nature] for your **'awli-yaa** [friends and protectors]. They are but friends and protectors to each other [Ba-zuhum **awli-yaaa-'u** ba'-z]. And **he** amongst you takes them for friends is [one] of them. Verily Allah guides not wrong-doing folk. [Sura No. – 4\5 – Maaaidah, Ayat\Verses – 51]

Ayat\Verses - O you who believe! If any from among you **turn back from his Faith,** soon will Allah produce a people whom He will love as they will love Him, --**Lowly with the Believers, Mighty against the Rejecters, Fighting in the Way of Allah,** and **never afraid of the reproaches of such as find fault.** That is the **Grace of Allah,** which He will **bestow** on whom He pleases. And Allah **encompasses** all, and He knows all things. Your (real) friends are (no less than) **Allah,** His **Prophet,** and the (Fellowship of) **Believers,** --those who establish regular prayers and regular charity, and they bow down humbly (in prayer). As to those who turn (for friendship) to Allah, His Prophet, and the (Fellowship of) Believers, --it is the Fellowship of Allah that must certainly triumph. [Sura No. – 4\5 – Maaaidah, Ayat\Verses – 54 to 56]

Ayat\Verses - **O you who believe! Choose not for friends and protectors such of those who take your religion for a mockery or sport, whether** among those who received the uutul-Kitaab before you, or among those who reject Faith; but fear you Allah, if you have Faith (indeed). And when you proclaim your call to prayer, they take it (but) as mockery and sport [as you are following wrong direction]; that is because they are a people without understanding. [Sura No. – 4\5 – Maaaidah, Ayat\Verses – 57 & 58]

Ayat\Verses - Why do not the Rabbis and the Priests [humur-Rabbaa-niyyuuna wal-abbaaru] forbid them from their (habit of) **uttering sinful words and their devouring of illicit gain? Evil indeed are their handworks.** [Sura No. – 4\5 – Maaaidah, Ayat\Verses – 63]

INVENTION / MECHANISM
"You do nothing but invent."

Ayat\Verses - To the Aad People [We sent] **Huud**, one of their own **brethren**. He said: O my people! **Worship Allah!** You have no other reality except Him. **You do nothing but invent.** O my people! I ask of you **no** reward for this [Message]. My reward is from none but Him who created me: **Will you not then understand?** And O my people! Ask forgiveness of your Rab, and **turn to Him** [in repentance]: He will cause the sky to rain abundance on you, and add strength to your strength: so **turn you not back in sin.** [Sura No. – 10\11 – Kitaabun-uh-kimat (Prev. – Huud), Ayat No. – 50 to 52]

Wasilah
[Avoid invalid references]

Ayat\Verses - "Ma ja-'alal-laahu mim-bahii-ratinw-wallaasaaa-'iba-tinw-wa **laa wasiilatinw**-wa laa haaminw wa laa kinnal-laziina kafaruu yaftaruuna 'alal-laahi-kazib; wa 'aksaruhum laa ya'- qiluun." [Sura No. – 4\5 – Maaaidah, Ayat\Verses – 103]

It was not Allah who instituted (superstitions like those of) a slit-ear **she-camel,** or a **she-camel let loose for free pasture,** or **idol sacrifices for twin-births in animals,** or **stallion-camels** freed from work: It is blasphemers who invent a lie against Allah; but most of them lack wisdom. (IFTA) [Sura No. – 4\5 – Maaaidah, Ayat\Verses – 103]

Or

Allah has not appointed anything in the nature of a Bahirah or a Saibah or **Wasilah** or a Hami, but those who disbelieve invent a lie against Allah. Most people have no sense. (IBPLtd) [Sura No. – 4\5 – Maaaidah, Ayat\Verses – 103]

Ayat\Verses - **When it is said to them: Come to what Allah has revealed [equal & opposite law]; come to the Prophet: They say: Enough for us are the ways we found our fathers following [manmade nature]. What!**

Even though their fathers were void of knowledge and guidance? O you who believe! Guard your own souls: If you follow (right) guidance, no err can come to you from those who go astray. The goal of you all is to Allah: It is He Who will show you the truth of all that you do. [Sura No. – 4\5 – Maaaidah, Ayat\Verses – 104 & 105]

WHAT MOST OF THEM BELIEVE AS WELL AS WORSHIP?
[Projected Falsehood and Man-made Magnetism]
"Nay, but they worshipped the Jinns; most of them were believers in them."
Ayat\Verses - One Day He will gather them all together, and say to the angels: Did these worship you? They will say: Glory to You! You are our Protector from them. Nay, but they worshipped the Jinns [Middle-East Region of Eartha 3D]; most of them were believers in them. So on that Day no power [projected spiritualism] shall they have over each other, for profit or harm; and We shall say to the wrong-doers: Taste you the Penalty of the Fire which you used to deny! [Sura No. – 33\34 – Saba, Ayat No. – 40 to 42]

EVIL AND GOOD
Ayat\Verses - It is not (the purpose) of Allah to leave you in your present state till He shall separate what is evil [projected falsehood] from what is good [manifest truth]. And it is not the purpose of Allah to let you know the unseen. But Allah chooses of his messengers whom He will (to receive knowledge thereof). So believe in Allah and His messengers. And if you believe and do right, you have a reward without measure. [Sura No. 2\3 – Imran, Ayat No. – 179]

"Lo! I know full well all that you conceal and all that you reveal."
Ayat\Verses - "Yaaa-ayyu-hallaziina aa-manuu laa tattakhizuu aduw-wii wa aduw-wakum awliyaaa –a tulquuna ilay-him-bil-mawaddati wa qad kafaruu bi-maa-jaa-akum-minal-haqqi, yukhrijuunar- Rasuula wa iyyakum an-tu-minuu billaahi Rabbikum. In kuntum kha-rajtum jihaadan-fii sabiilii wab-tigaaa-a marzaatii tusirruuna ilay-him bil-mawaddati wa ana a-lamu bimaaa akh-fay-tum wa maaa a-lantum. Wa many-yaf-alhu minkum faqad galla sawaaa-as-sabil." [Trans.] O you who believe! Choose not My enemies and yours as friends [or protector as an awliyaa]. Do you give them friendship when they disbelieve in that truth which has come to you, driving out the Prophet and you (from your houses i.e. right direction) because you believe in Allah, your Rab? If you have come out to strive in My way and to seek My good pleasure, [take them not as friends], holding secret converse of love [and friendship] with them. Lo! I know full well all that you conceal and all that you reveal. And whosoever does this among you, verily he has strayed from

the Right Path. [Sura No. 59\60 – Yuhibbul-Muqsitin (prev. Mumtahana), Ayat No. – 1]

PARTNERS

Ayat\Verses - [Inevitable] **comes** [to pass] the **Command of Allah:** seek you **not** then to hasten it. Glory to Him, and far is He above having the **partners** they ascribe unto Him! [Sura No. – 15\16 – Nahl, Ayat\Verses – 1]

Ayat\Verses - He has created the **heavens and the world** for **just ends (with truth).** Far is He above having the **partners** they ascribe to Him! [Sura No. – 15\16 – Nahl, Ayat\Verses – 3]

PARTNERS' SHARE

Ayat\Verses - Say: O my people! Do whatever you can: I will do [my part]: soon will you know who it is whose end will be [best] in the Hereafter. It is certain that the wrong-doers will not prosper. Out of what Allah has produced in abundance in crops and in cattle, they assigned Him a share. They say, according to their fancies: **This is for Allah, and this is for our partners!** But the share of their partners reaches not unto Allah, while the share of Allah reaches their partners! Evil [and unjust] is their assignment! [Sura No. – 5\6 – An-aam, Ayat No. –136 & 137]

[If you have any certain knowledge either contrary or contradictory to the shared selfevident truth, then produce the same.]

Ayat\Verses - Those who give **partners [to Allah] will say: If Allah had wished, we should not have given partners to Him nor would our fathers; nor should we have had any taboos (forbidden).** So did their ancestors **argue falsely,** until they tasted Our punishment? Say: **Have you any [certain] knowledge?** If so, produce it before us. You **follow nothing but conjecture. You do nothing but lie. Say: With Allah is the argument that reaches home: if it had been His will, He could indeed have guided you all.** Say: Come, bring your witnesses who can bear witness that Allah forbade (all) this. And if they bear witness, do not you bear witness with them. Follow you not the whims of **those who deny Our Signs (Manifest Truth),** those who believe not in the Hereafter and deem (others) equal with their Rab.. [Sura No. – 5\6 – An-aam, Ayat No. –149 to 151]

And to Allah do bow all that is in the **heavens [West Horizon]** and in the **world [East Horizon],** whether **moving** [living] **creatures** or the **angels;** for **none are proud** [before their Rab]. They all **admire** their Rab, **high above them,** and they do all that they are commanded. Allah has said: **Choose not** two realities [both manmade magnetism and natural magnetism]; for He is **just One Allah**. So of Me, Me only, be in awe. **To Him belongs whatever is in the heavens and in the world, and Diin (Religion) is His forever. Will you then fear any other than Allah?** And whatever of comfort you enjoy, it is from Allah. Then when you are touched by distress, unto Him you cry for help! And when He removes the distress from you, behold! **Some of them [you] attribute partners to their Rab.** [As if] to show their ingratitude for the favours we have bestowed on them. Then enjoy [your brief day]: but **soon you will come to know** [your folly]! And they [even] assign, to things **a portion out of that** what they do not know **which We have bestowed for their sustenance!** By Allah, **you will certainly be called to account for your false inventions [Manmade Nature].** [Sura No. – 15\16 – Nahl, Ayat\Verses – 49 to 55]

Call your 'God-partners'
Owner and Possessors of Manmade Nature are invited to project manifested signs of Electromagnetic Wave in resemblance with their Manmade Magnetism.
[Try to confirm right direction in resemblance with equal & opposite manifested signs of natural magnetism]

"Hold to forgiveness; command what is right; But turn away from the ignorant."

Ayat\Verses - Do they indeed **ascribe to Him as partners** [manmade magnetism] things that can **create nothing**, but are themselves created? **No aid** can they give them, nor can they aid themselves! If you call them to guidance, they will **not follow you.** Whether you call them or are silent is all one to them. Verily those whom you call upon besides Allah are servants like unto you. Call upon them, and let them listen to your prayer, if you are [indeed] truthful. **Have they feet to walk with? Or hands to lay hold with? Or eyes to see with? Or ears to hear with? Say: Call your 'god-partners', scheme [your worst] against me, and give me no relief.** [Sura No. – 6\7 - As-haabul- a' raaf, Ayat No. – 191 to 195]

Ayat\Verses - For my Protector is Allah, Who revealed the Kitab [from time to time], and He will choose and befriend the righteous. But those you call upon besides Him, are unable to help you, and indeed to help themselves. **If you call them to guidance, they will hear not. You will see them looking at you, but they see not. Hold to forgiveness; command what is right; but turn away from the ignorant. If a slander from Shaytan wound your [mind], seek refuge with Allah; for He hears and knows [all things].** [Sura No. – 6\7 - As-haabul- a' raaf, Ayat No. – 196 to 200]

WHO ATTRIBUTE PARTNERS WITH ALLAH?

"See **how We display the Signs (manifested signs of equal & opposite magnetism) so that they may understand.**"

Ayat\Verses - Say: **Allah delivers you from this and from all [other] distresses: and yet you attribute partners unto Him.** Say: **He has power to send calamities on you, from above [West] and below [East], or to puzzle you with dissension giving you a taste of mutual revenge each from the other [equal & opposite].** See **how We display the Signs (Manifest Truth) so that they may understand.** [Sura No. – 5\6 – An-aam, Ayat No. –64 & 65]

TREMENDOUS SIN AND MANIFEST SIN

Evidence Sorcery [Invention] and Persecution [Projection of Falsehood] Manmade Magnetism [sin] and Manmade Nature [lie]

Ayat\Verses - Lo! Allah forgives **not** that **partner** should be **ascribed** unto Him. He forgives except (anyone else) to whom He will. Whoso ever **ascribes partner** to Allah, he has indeed **invented** a **tremendous sin.** Have you **not seen** those who claim sanctity for themselves (praise themselves for purity)? Nay-but Allah **sanctifies** (purifies) whom He will, and they **will not** be wronged **even the hair upon a date-stone.** See, how they **invent a lie against** Allah! That by itself is a **manifest sin.** [Sura No. – 3\4 – Nisaa, Ayat No. – 48 to 50]

Ayat\Verses - And verily **We gave Luqmaan wisdom** saying: **Give thanks unto Allah and whosoever give thanks, he gives thanks (for the good of) his soul.** And whosoever refuses, lo! Allah is Absolute Owner of Praise. And (remember) **Luqmaan** said to his son by way of instruction: O my son! **Ascribe no partners unto Allah.** Lo! **To ascribe partners** (unto Him) is indeed the **highest wrong-doing.** [Sura No. – 30\31 – Luqmaanal-Hikmata, Ayat\Verses – 12 & 13]

"Or **have We sent down authority to them, which points out to them that which they associate with Him.**"

Ayat\Verses - And when trouble touches men, they cry to their Rab, turning back to Him in repentance. But when He gives them a taste of Mercy as from Himself, behold! Some of them **attribute partners to their Rab.** [As if] to show their ingratitude for the [favours] We have bestowed on them! Then enjoy [your brief day]; **but soon you will know [your folly]. Or have We sent down authority to them, which points out to them that which they associate with Him.** [Sura No. – 29\30 – Ruum, Ayat\Verses No. – 33 to 35]

Ayat\Verses - Those before them **plotted [against Allah's Way through the projections of mechanical globalisation, circular rotation & revolution system, and man-made magnetism],** so Allah **struck** at the foundations of their **structures,** and then the roof **fell down** on them from above [West]; and the **wrath seized** them from directions **they did not perceive.** Then, on the **Day of Resurrection,** He will **cover them with shame,** and will say: **Where are My 'partners' for whose sake you opposed (My Guidance)?** Those who have been given **knowledge** will say: **This Day, indeed, are the unbelievers covered with shame and misery.** [Namely] those whose lives the angels take in a **state of wrong-doing** to their **own souls.** Then would they **offer submission** [with the pretence], We did no evil [knowingly]. [The angels will reply]: Nay, but verily Allah knows all **that you did. So enter the gates of Hell, to dwell therein. Thus evil indeed is the abode of the arrogant.** [Sura No. – 15\16 – Nahl, Ayat\Verses – 26 to 29]

Ayat\Verses - Await they aught save that the angels should come unto them or the command of their Rab should pass. So did those who went before them. But Allah wronged them not; but, they did wrong themselves. But the evil results of their deeds overtook them, and that very [wrath] at which they used to mock surround them. [Sura No. – 15\16 – Nahl, Ayat\Verses – 33 & 34]

Ayat\Verses - **When those who gave partners to Allah will see their "partners",** they will say: Our Rab! These are our **'partners, those whom we used to invoke besides You.** But they will throw back their word at them [and say]: **Indeed you are liars!** That Day shall they [openly] show [their] submission to Allah; and **all their inventions shall leave them in the lurch. Those who reject Allah** and **hinder [men] from the Path of Allah,** for them We will add **penalty to penalty;** for that **they used to spread mischief.** One day We shall raise from every nation [Middle West or Upper Seashore of Europe, Asia, Africa, and Australia] a witness against them, from amongst themselves; and We shall bring you as a witness against these [your people]; and We have sent down to you the Kitaab explaining all things, a Guide, a Mercy, and Glad Tidings to Muslims. [Sura No. – 15\16 – Nahl, Ayat\Verses – 86 to 89]

Lo! Allah pardons not that partners should be ascribed unto Him. But He pardons whom He wills other sins than this. One, who ascribes partners with Allah, has wandered far astray.
[The Pagans], leaving Him, call but upon female deities: They call but upon Satan the persistent rebel! (IFTA)
Or
They invoke in His stead only females, they pray to none else than shaytan, a rebel. (IBPLtd.)

Ayat\Verses - Whom Allah cursed, and he (shaytan) said: Surely I will take of Your bondmen an appointed portion. And surely I will mislead them, and I will arouse in them false desires and surely I will order them and they will cut the ears of cattle, and surely I will command them and they will deface the [fair] nature created by Allah. **Whoso chooses shaytan for a patron instead of Allah is verily a loser and his loss is manifest. Shaytan makes them promises, and creates in them false desires [project falsehood]; but Shaytan's promises are nothing but deception.** They [his dupes] will have their dwelling in Hell, and from it they will find no way of escape. [Sura No. – 3\4 – Nisaa, Ayat No. – 116 to 121]

ASCRIBED PARTNERS

Ayat\Verses - And behold! You come to us bare and alone as We **created you for the first time**. You have left behind you all [the favours] whom We bestowed on you: We see not with you **your intercessors** whom **you thought to be partners in your affairs**: so now all relations between you have been cut off, and your [pet] fancies have left you in the lurch! [Sura No. – 5\6 – An-aam, Ayat No. –95]

WHERE?

"Where are the partners whom you [invented and] talked about (make-believe)?"

Ayat\Verses - One day We shall gather them all together: We shall say to those who ascribed partners [to Us]: **Where are the partners whom you [invented and] talked about (make-believe)?** [Sura No. – 5\6 – An-aam, Ayat No. – 22]

PENALTY OF SHAME

Ayat\Verses - **Who can be wicked than one who invents a lie against** Allah, or says: **I have received inspiration, when he has received none**, or [again] who says: **I can reveal the like of what Allah has revealed**? If you could but see how the wicked [do fare] in the flood of confusion at death! The angels stretch forth their hands, [saying], yield up your souls: this day shall you receive your reward, a **penalty of shame**, for that you used to tell lies against Allah, and **scornfully to reject of His signs**! [Sura No. – 5\6 – An-aam, Ayat No. –94]

Origination – Psychic Period and Repetition – Reproductive Period
Origination – Revelation and Repetition – Manifestation [Beginning]

Ayat\Verses - Say: Of your **'partners' [manmade magnetism]**, can any **originate** creation and **repeat** it? Say: It is Allah Who **originates creation and repeats it**: how then are you deluded away / **misled** [from the **truth**]? Say: Of your **'partners'** is there any that can give any **guidance towards truth**?

Say: It is Allah Who **gives** guidance towards truth, is then He Who gives guidance to truth more **worthy** to be followed, or he who **finds not guidance** [himself] **unless** he is guided? What then is the **matter** with you? **How judge you? But most of them follow nothing but fancy: truly fancy can be of no avail against truth**. Verily Allah is well **aware** of all that **they do**. [Sura No. – 9\10 – Yuunus, Ayat No. – 34 to 36]

"Fa-lillaahil-Hamdu Rab-bis-samaawaati wa Rabbil-arzi, Rabbil-Aalamin" {Trans.} **Ayat\Verses -** Then Praise be to **Allah**, **Rab** of the heavens and **Rab** of the world, **Rab** of the Universe [Alamin]! [Sura No. – 44\45 – Tanziilul Kitaab (Prev. Jaasiya), Ayat No. – 36].

O you who believe! If you fear Allah, **He will grant you a criterion [to justify truth],** remove from you [all] evil thoughts, and forgive you, for **Allah is the Rab of Infinite grace.** Remember how the disbelievers plotted against you, to keep you in bonds, or slay you or get you out [of your House]. They plot and plan, and Allah too plans; but the best of planners is Allah. When Our Signs are rehearsed to them, they say: We have heard this [before]. **If we wished, we could say [words] like these. These are nothing but tales of the ancients. [Sura No. – 7\8 – Anil-Anfaal, Ayat No. – 29 to 31]**

Alaa Kulli say-in-Qadiir - Omnipotent
Huwal-Hakiimul-Khabiir – Wise & Knower

If Allah touches you with affliction, none can remove it save Him. If He touches you with happiness, He has power over all things. He is the **Omnipotent** over His slaves; and He is the **Wise**, the **Knower**. [Sura No. – 5\6 – An-aam, Ayat No. – 17 & 18]

Say: The [Qur'an] was revealed by Him **who knows the mystery** [that is] in the heavens and the world. Verily He is Ever Forgiving, Most Merciful. [Sura No. – 24\25 - Nazzalal-Furqaan, Ayat No. – 6]

Say: You will **not be asked** of what we committed, **nor shall we be asked** of what you do. [Sura No. – 33\34 – Saba, Ayat No. - 25]

Say: Our Rab will **gather us together** and will in the end **decide the matter between us** [and you] in **truth and justice;** and He is the One to decide, the One Who knows all. [Sura No. – 33\34 – Saba, Ayat No. - 26]

SECTION – I
HIGHLIGHTS – 1.3
A WORD

Fasta-'iz billaahi minash-Shaytaanir-Rajim
Bismillaahir-Rahmaanir- Rahim

What is meant by Magnetism?
A WORD HAS BEEN CHANGED FOR ANOTHER SAYING
QUESTIONS – WHO? WHICH? WHY?
WHEN? WHAT? WHERE? HOW?
Do you know which word has been changed and all the wh-questions related with it?

O Messenger! Let not them **grieve you, who vie one with another in the race to disbelief among those who say "We believe" with their lips** but on **whose hearts have no faith**; men who will listen to any **lie**, will listen even to others who have never so much as come to you. **They change the words from their context and waying**. They say: If you are given this, take it, but if not, beware! If any one's trial is intended by Allah, you have no authority in the least for him against Allah. For such--it is not Allah's will to purify their hearts. For them there is disgrace in this world, and in the Hereafter a heavy punishment. (They are fond of) listening for the sake of **falsehood! Greedy for illicit** [unlawful] **gain**! If they do come to you, either **judge** between them, or **decline** to interfere. If you **decline**, they cannot hurt you in the least. If you judge, **judge in equity between them**; for **Allah loves those who judge in equity**. But how do they come to you for decision, when they have Tawraat [Equal & Opposite Law or Revelation] before them wherein Allah has delivered judgment [for them]? Yet even after that **they turn away**. Such [folk] are **not believers**. [Sura No. – 4\5 – Maaaidah, Ayat\Verses – 41 to 43]

And remember, We **divided the sea** [West Zone & East Zone within the Black Square of East Horizon] for you and saved you and drowned Firawn's people within your very sight. And remember We appointed **forty nights** for Muusa, and **in his absence you took the calf** [manmade magnetism for determining right direction] and you did grievous wrong. Even then We did forgive you; **there was a chance for you to be grateful**. And remember We gave **Muusa the Kitaab** and **the Criterions [scales of justification and verification]**. There was a **chance for you to be guided aright**. And when Muusa said to his people: O my people! You have indeed **wronged** yourselves by your **worship of the calf**. So **turn [in repentance] to your Creator**, and **kill (the calf) yourselves** that will be the best for with your Creator and He will relent toward you, for He is the Relenting, the Merciful. And remember you said: O Muusa! We shall never believe in you until **we see Allah manifestly**, but you were **confused with thunder and lightning seized you**. Then We received you after your extinction, that you might give thanks. And We gave you the **shade of clouds** and sent down to you **manna** and **quails**, saying: Eat of the good things We have provided for you. We wronged them not, but they did wrong themselves. And when We said: Enter this township and eat of the plenty therein as you wish; but **enter the gate prostrate**, and say: Repentance. We will forgive you your sins and will increase (reward) for the right-doers. **But those who did wrong changed the word which had been told them for another saying**, and We sent on the transgressors a plague from heaven, for evil doing. [Sura No. – 1\2 – Baqarah, Ayat No. – 50 to 59]

Have you any hope that **they** will be **true** to you when a **party** of **them** used to listen to **the word of Allah**, and **then used to change it, after they had understood it, knowingly**? And when they **fall** in with the men of **Faith**, they say: We **believe**. But when they **meet each other in private**, they say: **Shall you tell them what Allah has revealed to you**, that they may engage **you in argument** about it before your Rab? Have you then **no sense**? Know they not that **Allah knows what they conceal and what they reveal**? And there are among them **illiterates, who know not the Kitaab**, but [**see therein their own] desires**, and they **do nothing** but **conjecture the truth**. [Sura No. – 1\2 – Baqarah, Ayat No. – 75 to 78]

See you not those unto whom a portion of the Kitaab has been given, how they purchase error, and seek to make you (believers) err from the right way? Allah knows best (who are) your enemies. Allah is sufficient as a Friend and Allah is sufficient as a Helper. Some of those who are **haaduu change words** from their context and say: We hear and disobey; hear you as one who hears not and listen to us, **distorting** with their tongues and **slandering** religion. If they had said: We hear and we obey, hear you and look at us, it had been better for them and **more upright**. But Allah has **cursed** them for their disbelief, so they believe not save a few. [Sura No. – 3\4 – Nisaa, Ayat No. – 44 to 46]

And when it was said unto them: Dwell in this township and eat there from whence you will, and say: **'Repentance', and enter the gate prostrate**, We shall forgive you, your sins. We shall increase (reward) for the right doers. But **the transgressors among them changed the word from that which had been told them**. So we sent on them a plague from heaven. For that they repeatedly transgressed. [Sura No. – 6\7 - As-haabul- a' raaf, Ayat No. – 161 & 162]

We certainly gave the **Kitab** to **Muusa**, but **differences** arose therein. Had it not been that **a word** had gone forth before from your Rab, the matter would have been decided between them? But they are in **suspicious doubt concerning it**. And, of a surety, to all will your Rab pay back [in full the recompense] of their deeds, for He knows well all that they do. Therefore, **stand firm** [upright] **as you are commanded, you and those who with you turn [unto Allah];** and **transgress not** [from the equal & opposite Middle Course]; for He sees well all that you do. And **incline not to those who do wrong lest the Fire will seize you; and you have no protectors other than Allah, nor shall you be helped**. And establish regular prayers **at the two ends of the day** and **at the approaches of the night**; for **those things that are good remove those that are evil. This is a reminder for the mindful. And be steadfast in patience; for verily Allah loses not the wages of the good**. [Sura No. – 10\11 – Kitaabun-uh-kimat (Prev. – Huud), Ayat No. – 110 to 115]

LET US INVITE THE OWNER AND THE POSSESSORS OF MANMADE NATURE [Mechanical Globalisation, Circular Rotation & Revolution System, and Manmade Magnetism] TO CONVERT EQUAL & OPPOSITE MANIFESTED SIGNS OF ELECTROMAGNETIC WAVE IN RESEMBLANCE WITH THEIR PROJECTED NORTH AND SOUTH DIRECTIONS [Manmade Magnetism]

Above, Below, Mutual Revenge Each from the Other
Ayat\Verses - Say: **Allah delivers you from this and from all [other] distresses: and yet you attribute partners unto Him. Say: He has power to send calamities on you, from above [West] and below [East], or to puzzle you with dissension giving you a taste of mutual revenge each from the other [equal & opposite – North & South]. See how We display the Signs (Manifest Truth) so that they may understand.** [Sura No. – 5\6 – An-aam, Ayat No. –64 & 65]

NONE CAN CHANGE THE WORDS OF ALLAH
Say: Allah knows best how long they stayed. With Him is [the knowledge of] the secrets of the **heavens** and the **world**; how clearly He sees, how finely He hears [everything]! They have **no protector** other than Him; **nor does** He share His Command with any person whatsoever. And recite what has been revealed to you of the Kitaab of your Rab. **None can change His Words, and none of you will find as a shelter other than Him.** [Sura No. – 17\18 – Khaf – Ayat No. – 26 to 27]

"There can be no change in the words of Allah"
[Good tidings for those who believe on Manifest Truth /
Manifested Sign of Equal & Opposite Natural Magnetism]

Behold! Verily on the friends of Allah there is no fear, nor shall they grieve. Those who believe and [constantly] guard against evil; for them are good tidings, in the life of the present and in the Hereafter. **There can be no change in the words of Allah**. This is indeed the supreme felicity. Let not their speech grieves you, for all power and honour belong to Allah. It is He Who hears and knows [all things]. [Sura No. – 9\10 – Yuunus, Ayat No. – 62 to 65]

Every man's fate We have fastened on his own neck. On the Day of Resurrection [Yawmal-Qiyaamati] We shall bring out for him a scroll, which he will **see spread open**. [It will be said to him:] Read your [own] record. **Sufficient is your soul that day to make out an account against you.** Who receives guidance, receives it for his own benefit; he who will go astray, so to his own loss. **No bearer of burdens** can bear the burden of another, **nor would We visit with Our Wrath** until We had sent a messenger [**to give warning**]. And when We decide to **destroy** a **township**, We [first] **send a definite order** to those among them who are given the good things of this life and yet transgress; so that the **word is proved true** against them, then [it is] We destroy them **utterly.** [Sura No. – 16\17 - Banii-Israai-iil, Ayat No. – 13 to 16]

Mankind were but one community, and they differed later. **Had it not been for a word that had already gone forth from your Rab**, it had been judged between them in respect of that wherein they differ? And they will say: if only a **Clear Sign** were sent down upon him from his Rab! Say: The Unseen belongs to Allah. So, wait! I am also waiting with you. [Sura No. – 9\10 – Yuunus, Ayat No. – 19 to 20]

But [now that the Qur'an has come], they reject it. But soon will they know! And verily **Our word went forth** before [this] to our bondmen sent that they would certainly be assisted, and that Our hosts, they surely would be the victors. So **turn you away** (withdraw) from them for a little while, and watch them, for they will soon see. Do they wish [indeed] to hurry on our Punishment? But when it comes home to them, then it will be a hapless morn for those who have been warned. So **turn you away** (withdraw) from them for a little while, and watch them, for they will soon see. [Sura No. – 36\37 – Was-saaaffaati, Ayat No. – 170 to 179]

He has ordained for **you [Prophet]** that religion which He ordained unto **Nuuh**, and that which We inspire in **you**, and that We commanded to **Ibrahim**, and **Muusa**, and **Iisa** saying: **Establish the religion**, and **be not divided therein.** Hard for the **idolaters** (those who follow projected mechanical magnetism) is the (way) to which you call them. Allah chooses for Himself whom He pleases, and guides to Himself those **who turn** [towards equal & opposite way towards Arsh or Upright West]. And **they became divided only after Knowledge reached them through selfish envy [projected falsehood]** as between themselves. Had it not been for **a Word that went forth** before from your Rab, [tending] to a **Term appointed** that the matter would have been settled between them. But truly those who have inherited the Kitaab after them are in suspicious [disquieting] doubt concerning it. [Sura No. – 41\42 – Shuuraa, Ayat No. – 13 & 14]

(And the warner) said: What! Even if I brought you better guidance than that which you found your fathers following? They answered: Lo! In what you bring we are disbelievers. So We requited them. Then see the nature of the consequence for the rejecters! / (Then see the end of those who rejected [Truth]. [Sura No. 42\43 – Ummil-Kitaab \ (Zukhruf), Ayat No. – 24 & 25]

And when **Ibrahim** said unto his father and his folk: Lo! I am innocent of what you worship, except Him Who created me, for He will surely guide me. And he made it as **a Word enduring** among his seed, that they may return [turn back to Allah]. [Sura No. 42\43 – Ummil-Kitaab \ (Zukhruf), Ayat No. – 26 to 28]

Or have they partners (of Allah) who have made lawful for them in religion that [mechanical magnetism] which Allah allowed not? And but for **a crucial Word** (gone forth already), it would have been judged between them. Lo! For wrong doers is a painful doom. [Sura No. – 41\42 – Shuuraa, Ayat No. - 21]

So the **Ark floated with them on the waves like mountains**, and Nuhh cried out unto his son – and he was standing aloof - **O my son! Come with us, and be not with the unbelievers.** The son replied: I will betake myself to some mountain [for instance a fictitious planet called Mars], it will save me from the water. Noah said: this day nothing can save, from the command of Allah, save him whom He has had mercy! And the waves came between them, and the son was among drowned in the Flood. **Then the word went forth: O world! Swallow up your water** and **O sky! Be cleared of clouds** and **the water was made to subside**. And the Commandment was fulfilled. The **Ark rested on Mount Alal-Ju-diyyi**, and it was said: **Away with those who do wrong.** [Sura No. – 10\11 – Kitaabun-uh-kimat (Prev. – Huud), Ayat No. – 42 to 44]

Say: Who it is that sustains you [in life] from the sky and from the world? Or who it is that has power over hearing and sight? And who it is that brings out the living from the dead and the dead from the living? And who it is that rules and regulates all affairs? They will soon say, "Allah". Say, will you then not keep your duty [unto Him]? Such is Allah, your real Cherisher and Sustainer. After the truth, what is there except error? **How then are you turned away** [from revealed magnetism]? Thus **the word of your Rab is proved true against those who do wrong** [projected mechanical magnetism] **that they will not believe**? [Sura No. – 9\10 – Yuunus, Ayat No. - 31 to 33]

Ya Sin. By the Qur'an, full of Wisdom, you are of those sent [la-minal-mursaliin], in an equal & opposite **right direction towards Upright West [Siraatim-Mustaqiim]**. It is a Revelation sent down by [Him], the Exalted in Might, Most Merciful in order that you may warn a folk, whose fathers were not warned, so they are heedless. The **word is proved true** against the greater part of them, for they do not believe [Sura No. – 35\36 – Yaa-siiin, Ayat No. – 1 to 7]

And when We said: **Enter** this **township** and **eat** of the plenty **therein** as you wish; but **enter** the **gate prostrate**, and say: **Repentance**. We will **forgive you your sins** and will increase (reward) for the **right-doers**. But those **who did wrong changed the word** which had been told them **for another saying**, and We **sent** on the transgressors a **plague** from heaven, **for evil doing**. [Sura(1)-Baqara - A-No.- 58 & 59]

Lo! They were indeed in suspicious doubt.
Could you but see when they are terrified with no escape; but then there will be no escape [for them], and are seized from near of hand. And they will say: We do believe [now] in the [Truth]; but how could they receive [Faith] from a position [so far off. Seeing that they rejected Faith [entirely] before, and that they [continually] cast [slanders] on the unseen from a position far off. And a gulf is set between them and their desires, as was done for people of their kind of old. **Lo! They were indeed in suspicious doubt.** [Sura No. – 33\34 – Saba, Ayat No. – 51 to 54]

SPOILS OF WAR
"That He might justify Truth and prove Falsehood false, distasteful though it is to those in guilt."
Ayat\Verses - They ask you concerning [things taken as] **spoils of war**. Say: The spoils of war belong to Allah and the Prophet. So keep our duty to Allah, and adjust the matter of your difference, and obey Allah and His Prophet, if you are believers. For, believers are those who, when Allah is mentioned, feel a vibration in their hearts, and when they hear His signs rehearsed, find their faith strengthened, and put [all] their trust in their Rab. Who establish regular prayers and spend [freely] of that We have bestowed on them for sustenance. Such in truth are the believers. They have **grades of dignity** [stages of journey] with

their Rab, and forgiveness, and **generous sustenance just as your Rab ordered you to go forth from your house with the truth, even though a party among the believers disliked it.** Disputing with you concerning the truth after it was made manifest, as if they were being driven to death and they [actually] saw it; and when Allah promised you one of the **two bands** that it should be yours. You wished that the **one unarmed should be yours**. And **Allah willed to justify the Truth according to His words and to cut off the roots of the disbelievers**. That **He might justify Truth and prove Falsehood false, distasteful though it is to those in guilt.** [Sura No. – 7\8 – Anil-Anfaal, Ayat No. 1 to 8]

"This is because they challenged against Allah and His Prophet. If there is any challenge against Allah and His Prophet, Allah is strict in punishment."
   (i)   Allah will assist us with a thousand of the angels, ranks on ranks.
   (ii) Give firmness to the believers: I will implant terror into the hearts of the disbelievers / unbelievers.

Ayat\Verses - Remember you implored **the assistance of your Rab**, and He answered you: I will assist you with a **thousand of the angels, ranks on ranks**. Allah appointed it only a message of hope (as good tidings), and an assurance to your hearts, [in any case] there is no help except from Allah; and Allah is Exalted in Power, Wise. Remember He covered you with a sort of drowsiness, to give you calm as from Him, and he caused rain to descend on you from heaven, to clean you therewith, to remove from you the stain of Shaytan, to strengthen your hearts, and to plant your feet firmly therewith. Remember your Rab inspired the angels [with the message]: I am with you. Give firmness to the believers: **I will implant terror into the hearts of the disbelievers / unbelievers**. Then you smite their necks and smite all their finger-tips off them. This is because they challenged against Allah and His Prophet. If there is any challenge against Allah and His Prophet, Allah is strict in punishment. Thus (is the award), so taste it, and (know) that for disbelievers is the suffering of the fire. [Sura No. – 7\8 – Anil-Anfaal, Ayat No. – 9 to 14]

STRATAGEM OF WAR
"Allah is He Who makes weak the plans and stratagem of the unbelievers / disbelievers"

Ayat\Verses - O you who believe! When you meet the unbelievers / disbelievers in hostile array, **never turn your backs to them**. If any do turn his back to them on such a day, unless it be in a stratagem of war, or to retreat to a troop [of his own], he draws on himself the wrath of Allah, and his abode is hell, an evil refuge [indeed]! It is not you who slew them; it was Allah: when you threw [a handful of dust], it was not your act but Allah's in order that He might test the believers by a **gracious trial** from Himself, for Allah is He Who hears and knows [all things]. That (is the case) and (know) that Allah is He Who makes weak the plans and stratagem of the unbelievers / disbelievers. [Sura No. – 7\8 – Anil-Anfaal, Ayat No. – 15 to 18]

"Sufficient is your soul that day to make out an account against you." Ayat\Verses - Every man's fate We have fastened on his own neck. On the Day of Resurrection [Yawmal-Qiyaamati] We shall bring out for him a scroll, which he will **see spread open**. [It will be said to him:] Read your [own] record. **Sufficient is your soul that day to make out an account against you.** Who receives guidance, receives it for his own benefit. He who will go astray that is to his own loss. **No bearer of burdens** can bear the burden of another, **nor would We visit with Our Wrath** until We had sent a messenger [**to give warning**]. And when We decide to **destroy** a **township**, We [first] **send a definite order** to those among them who are given the good things of this life and yet transgress; so that the **word** is proved true against them, then [it is] We destroy them **utterly.** [**Sura No. – 16\17 - Banii-Israai-iil, Ayat No. – 13 to 16**]

Your **Ultimate-Rab is Allah,** Who created the heavens and the world in **six** days, and then He **firmly mounted the Arsh** [Throne]. **He draws the night as a veil over the day, each seeking the other in rapid succession.** He made **Wash-shamsa, Wal-qamara** and **Wan-nujuuma** (galaxy stars) **subservient by His command**; His verily all **creations and commandment.** Blessed be Allah, the Rab of the Universe! [Sura No. – 6\7 - As-haabul- a' raaf, Ayat No. – 54]

ALLAH WILL BRING TO LIGHT ALL THAT YOU FEAR

Among them are men **who assault the Prophet** and say: He is only a **hearer.** Say: He listens to what is best for you, he believes in Allah, has faith in the believers, and is a Mercy to those of you who believe. But those who assault the Messenger will have a grievous penalty. To you they swear by Allah in order to please you. But it is more fitting that they should please Allah and His Messenger, if they are believers. Know they not that for those who oppose Allah and His Messenger, is the Fire of Hell wherein they shall dwell? That is the supreme disgrace. **The Hypocrites are afraid lest a Sura should be sent down about them, showing them what is [really passing] in their hearts**. Say: Mock you! But verily **Allah will bring to light** all that you fear [should be revealed]. If you question them, they declare [with emphasis]: We were only talking idly and in play. Say: **Was it at Allah, and His Signs, and His Messenger that you were mocking?** Make you no excuses: You have rejected Faith after you had accepted it. If We pardon some of you, We will punish others amongst you, for that they are in sin. [Sura No. – 8\9 – Tauba, Ayat No. – 61 to 66]

SECTION – I
HIGHLIGHTS – 1.4
GOOD TIDINGS AND ADMONITIONS

Fasta-'iz billaahi minash-Shaytaanir-Rajim
Bismillaahir-Rahmaanir- Rahim

[REFORMATIVE OR EDUCATIVE THEORY]
DARKNESS TO LIGHT
Ha Mim - [Sura No. – 40\41 – Haa-Mim-
Sajdah, Fundamental Ayat No. -1]

(Revelation) from the Beneficent, the Merciful, a Kitaab whereof the Ayats are explained in detail; a **Qur'an in Arabic** for people who understand. Giving **good tidings and admonition**; yet most of them **turn away**, and so they hear not. [Sura No. – 40\41 – Haa-Mim-Sajdah, Ayat No. – 2 to 4]

Alif-Lam-Ra. A Kitaab which We have revealed unto you, in order that you might lead mankind out of the **depths of darkness into light** by the permission of their Rab to the Way of [Him] the Exalted in power, worthy of all praise! [Sura No. – 13\14 – Ibrahim, Fundamental Ayat No. - 1]

And they say: **Our hearts are protected from that unto which you call us, and in our ears there is deafness, and between us and you is a screen (veil).** So **act you** [what you will]; for us, **we shall do [what we will!]** [Sura No. – 40\41 – Haa-Mim-Sajdah, Ayat No. - 5]

Say: **I am but a man (mortal) like you**. It is **inspired** in me that **your Allah is one Allah.** So **stand true to Him,** and ask for His Forgiveness; **and woe to those who join breakable reality [manmade magnetism] with Allah.** Those who give not the poor-due, and **who even deny (disbelieve)** the **Hereafter.** [Sura No. – 40\41 – Haa-Mim-Sajdah, Ayat No. – 6 & 7]

For those **who believe** and **work deeds of righteousness** are a **reward** that will **never fail (enduring).** [Sura No. – 40\41 – Haa-Mim-Sajdah, Ayat No. - 8]

Say: **Disbelieve** you verily in Him **Who created the world in two days [represent two Zones]**, and **ascribe** you unto Him **rivals [equal & opposite]**? He is the Rab of the Universe. He set in the [world], mountains **standing firm,** high above it, and bestowed blessings in the world, and **measure therein all things to give them nourishment in due proportion, in four Days [represent** four manifested directions], in accordance with [the needs of] those who seek [right way]. Then **He turned to the heavens when it was smoke**, and said unto it and the world: **Come both of you, willingly or unwillingly [positive & negative ~sth]**. They said: We come, obedient. So He **ordained** them as **seven firmaments (windows) in two Days**, and **He inspired in each heaven its mandate**. And **We adorned the lower heaven with lights**, and **[provided it] with guard [lamp or planetary barrier]**. Such is the Measuring of the Mighty, The Knower. [Sura No. – 40\41 – Haa-Mim-Sajdah, Ayat No. – 9 to 12]

But **if they turn away**, then say: **I have warned you of a stunning Punishment [as of thunder and lightning] like that which [overtook] the Aad and the Samuud!** When the messengers came to them, from **before** them and **behind** them, [preaching]: **Serve none but Allah**; they said: If our Rab had so pleased, He would certainly have sent down angels [to preach]. So lo! **We are disbelievers in that wherewith you have been sent. [Sura No. – 40\41 – Haa-Mim-Sajdah, Ayat No. – 13 & 14]**

As for the **Aad**, they behaved arrogantly through the land **without right, against truth and reason**, and said: **Who is mightier than us in power? What! Did they not see that Allah, Who created them, was Mightier than them in Power?** But **they continued to reject Our Signs!** So We sent against them **a furious Wind** through **days of disaster**, that We might give them a taste of **a Penalty of humiliation** in this life; but the Penalty of a Hereafter will be **more humiliating still**; and they will find **no help. [Sura No. – 40\41 – Haa-Mim-Sajdah, Ayat No. – 15 & 16]**

And as to the **Samuud**, We gave them **Guidance**, but **they preferred blindness** [of heart] **to Guidance**. So the **stunning Punishment of humiliation** seized them, because of **what they had earned**. And **We delivered those who believed and practised righteousness**. [Sura No. – 40\41 – Haa-Mim-Sajdah, Ayat No. – 17 & 18]

And on the **Day** that the **enemies** of Allah will be gathered together to the **Fire, they are driven on**. Till when they reach it [the Fire], their **hearing**, their **sight**, and their **skins** will bear **witness** against them, as to **what they used to do**. And they will say to their **skins: Why bear you witness against us?** They will say: Allah has given us **speech**, [He] Who gives speech to everything, and Who created you at the **first**, and unto Whom you are **returned**. And you did not hide yourselves, lest your hearing, your sight, and your skins should bear witness against you! But **you deemed that Allah knew not many of the things that you used to do!** And this thought of yours which you did entertain concerning your Rab, has brought you to destruction, and **you find yourselves among the lost. [Sura No. – 40\41 – Haa-Mim-Sajdah, Ayat No. – 19 to 23]**

If, then, they have patience, the **Fire will be a home for them**! And if they beg to be received into favour, yet **they are not of those** unto whom favour will be shown. And We have assigned for them comrades (of the like nature) **who made alluring to them what was before them and behind them**; and **the Word** concerning among the previous generations of jinns and humankind, who have passed away, is proved against them; for **they are utterly lost. [Sura No. – 40\41 – Haa-Mim-Sajdah, Ayat No. – 24 & 25]**

And the disbelievers say: Listen not to this Qur'an, but talk at random in the midst of its [reading], that you may gain the upper hand! **[Sura No. – 40\41 – Haa-Mim-Sajdah, Ayat No. - 26]**

But We will certainly **cause** the **disbelievers** to **taste** a severe Penalty, and We will avenge them **for the worst of their deeds**. Such is the requital of the enemies of Allah, the Fire. Therein will be for them the eternal home, **requital for as much as they denied Our Signs (Revelations). [Sura No. – 40\41 – Haa-Mim-Sajdah, Ayat No. – 27 & 28]**

And the **disbelievers will say**: Our Rab! **Show us those**, among jinns and humankind, **who beguiled (misled) us. We shall crush them beneath our feet that they become among the disgusting. [Sura No. – 40\41 – Haa-Mim-Sajdah, Ayat No. - 29]**

In the case of those **who say: Our Rab is Allah**, and afterward **Upright**, the angels descend on them [from time to time] saying: Fear you not! [They suggest]: Nor grieve! But receive the Good Tidings of the Garden [of Bliss], which you are promised. **We (angels) are your protecting friends in this life of the world and in the Hereafter.** Therein you will have all that your souls shall desire. Therein you will have all that you ask for! **A gift of welcome** from one Ever Forgiving, Most Merciful! **[Sura No. – 40\41 – Haa-Mim-Sajdah, Ayat No. – 30 to 32]**

Who is better in speech than one who calls [men] to Allah, works righteousness, and says: Lo! **I am of those (Muslimiin) who surrender (unto Him).** The good deed and the evil deed are not alike. Repel the evil deed with what is better. Then lo! **He, between whom and you was hatred become as it were your friend and intimate! [Sura No. – 40\41 – Haa-Mim-Sajdah, Ayat No. – 33 & 34]**

And no one will be granted such goodness except those who exercise patience and self-restraint, and none but owners of great happiness. And if [at any time] an incitement to discord is made to you by the evil one (shaytan / devil), seek refuge (safe heaven) in Allah. He is the One Who hears and knows all things. [Sura No. – 40\41 – Haa-Mim-Sajdah, Ayat No. – 35 & 36]

"Wa min Aayaatihil-laylu wan-nahaaru **wash-shamsu** wal qamar. Laa tasjuduu **L i s h-shamsi** wa laa lil-qamari wasjuduu lilaahil-lazii khalaqahunna in-kuntum iyyaahu ta-budun." [Sura No. – 40\41 – Haa-Mim-Sajdah, Ayat No. - 37]

And among **His Signs (Manifest Truths)** are the **Night and the Day, wash-shamsu wal qamar. Adore not Lish-shamsi wa laa lil-qamari**, but **adore Allah Who created them, if it is in truth Him Whom you worship.** [Sura No. – 40\41 – Haa-Mim-Sajdah, Ayat No. - 37]

But If the [disbelievers] are too proud, [**no matter**], still those who are with your Rab glorify Him night and day and tire not. And **of His Signs (portents)** (is this): That you see the **world lowly**, but when We **send down water to it, it is stirred to life** and yields increase. **Truly, He Who gives life to the [dead] earth can surely give life to [men] who are dead**; for He has power over all things. [Sura No. – 40\41 – Haa-Mim-Sajdah, Ayat No. – 38 & 39]

Lo! **Those who distort the Truth in Our Signs (Manifested Sign of Natural Magnetism) are not hidden from Us. Is he who is hurled (cast) into the fire better or he who comes secure on the Day of Renaissance (Resurrection)? Do what you will. Verily He sees [clearly] all that you do.** [Sura No. – 40\41 – Haa-Mim-Sajdah, Ayat No. - 40]

Lo! Those **who reject (disbelieve) in the reminder when it comes to them** [are not hidden from Us]; and indeed it is an **unassailable (unquestionable) Kitaab. No falsehood can approach it from before or behind it. It is sent down by One Full of Wisdom, Worthy of all Praise Nothing is said to you that were not said to the messengers before you.** Lo! Your Rab is **Owner** of forgiveness and **Owner** of terrible punishment. [Sura No. – 40\41 – Haa-Mim-Sajdah, Ayat No. – 41 to 43]

Had We sent this as a **Qur'an [in the language] other than Arabic**, they would assuredly have said: **Why are not its verses explained in detail?** What! [A Kitaab] **not in Arabic** and [**a messenger] an Arab**! Say unto them: **It is Guidance and a Healing to those who believe; and for those who believe not** there is deafness in their ears, and it is blindness in their [eyes]. They are [as it were] being called from a place far distant! [Sura No. – 40\41 – Haa-Mim-Sajdah, Ayat No. - 44]

And We verily gave Musa the Kitaab, but disputes arose therein. But there has been dispute concerning it, and but for **a Word** that had already gone from your Rab, it would ere, now have been judged between them. But lo! **They are in hopeless doubt concerning it.** [Sura No. – 40\41 – Haa-Mim-Sajdah, Ayat No. - 45]

Whoever works righteousness benefits his own soul; whoever works evil, it is against his own soul. And your Rab is not at all a tyrant to His slaves. To Him is referred (all) Knowledge of the Hour. No fruit comes out of its sheath, nor does a female conceive [within her womb] nor bring forth, but with His Knowledge. And on the Day when He will call to them: **Where are now My partners**? They will say: We confess to You, **not one of us can bear witness. And those to whom they used to cry (invoke) of old have failed them, and they will perceive that they have no way of escape.** [Sura No. – 40\41 – Haa-Mim-Sajdah, Ayat No. – 46 & 48]

Man tries **not of praying for good** [things], but if **ill touches him**, he gives up all hope [and] is lost in despair. And verily, **if We cause him to taste mercy after some hurt that has touched him**, he will say: This is **my own**; and I deem not that the Hour will ever rise, and if I brought back to my Rab, I surely shall be better off with Him. But We verily shall tell those who disbelieve (all) they did, and We verily shall make them taste hard punishment. [Sura No. – 40\41 – Haa-Mim-Sajdah, Ayat No. – 49 & 50]

And when We bestow favours on man, he turns away, and gets himself remote on his side [instead of coming to Us]; and when evil seizes him, [he comes] full of prolonged prayer! Bethink you. If it is from Allah and you reject it – who is further astray than one who is at open feud (argument with Allah)? [Sura No. – 40\41 – Haa-Mim-Sajdah, Ayat No. – 51 & 52]

Soon We will show them Our Signs (Clear Proofs) in the [furthest] regions [of the world], and within themselves, until it will manifest to them that this is the Truth. Does not your Rab enough **since He is Witness** over all things? Ah indeed! **Are they still in doubt concerning the Meeting with their Rab?** Lo! Is not He encompasses all things? [Sura No. – 40\41 – Haa-Mim-Sajdah, Ayat No. – 53 & 54]

Alif-Lam-Mim [Sura No. – 2\3 – Imran, Fundamental Ayat – 1]

Allah! There is no unbreakable reality save Him, the Alive, the Eternal. He has revealed unto you **a Kitaab with truth, confirming** that **which were revealed before it**, even as He revealed the **Tawraat** and the **Injiil** before this, as a **guide to mankind**, and **He has revealed the Criterions** (scales of justification as well as verification). Lo! **Those who reject Faith in (or disbelieve) the Signs of Allah (Manifested Nature) theirs will be a severest chastisement** and Allah is Mighty and Able to Requite (the wrong). [Sura No. – 2\3 – Imran, Ayat\Verses – 2 to 4]

O Prophet! Keep your **duty** to Allah and **obey not the disbelievers** and the **hypocrites**. Verily Allah is full of Knowledge and Wisdom. But **follow** that which **comes** to you by **inspiration** from your Rab. Lo! Allah is well **acquainted** with [all] that you do. And **put your trust** in Allah, and **enough is Allah as a disposer of affairs**. [Sura No. – 32\33 – Yahsabuunal-'Ahzaaab, Ayat\Verses – 1 to 3]

Alif Lam Mim. [Sura No. 1\2 – Baqarah, Fundamental Ayat – 1]

This is the **Kitab**. In it is **guidance sure, without doubt,** to those who fear Allah (who ward off evil). Who believe in the Unseen, and steadfast in prayer, and spend out of what We have bestowed upon them. And who believe in that which is **revealed unto you**, and that which was **revealed before your** time and [in their hearts] has the **assurance of the Hereafter**. They depend on guidance from their Rab, and it is they who will **succeed**. [Sura No. 1\2 – Baqarah, Ayat\Verses – 2 to 5]

Ayat\Verses – Believe in Allah and His messenger, and spend of that whereof He has made you **trustees**. And those of you who believe and **spend** [a right]; for them is a great reward. [Sura No. – 56\57 - Anzalnal-Hadiid, Ayat No. – 7]

"If I receive guidance, it is because of the inspiration of my Rab to me." **Ayat\Verses –** Say: If I am astray, I only stray to the loss of my own soul. But if I receive guidance, it is because of the **inspiration** of my Rab to me. It is He Who hears all things, and is [ever] near. [Sura No. – 33\34 – Saba, Ayat No. - 50]

Ha-Mim [Revelation – Sura No. – 41\42 – Shuuraa, Fundamental Ayat No. - 1] **Ain-Sin-Qaf.** [Revelation – Sura No. – 41\42 – Shuuraa, Fundamental Ayat No. - 2] Thus [He] has sent **inspiration to you** as [He did] **to those before you**. Allah is Exalted in Power, Full of Wisdom. To Him belongs all that is in the heavens and in the world: and He is The Sublime, The Tremendous. [Sura No. – 41\42 – Shuuraa, Ayat No. – 3 & 4]

SECTION – I
HIGHLIGHTS – 1.5
SINGLE FACT [MAGNETISM]

Fasta-'iz billaahi minash-Shaytaanir-Rajim
Bismillaahir-Rahmaanir- Rahim

- There is only one kind of Magnetism called Natural Magnetism.
- Revealed Left [North Pole or Magnetic Force] and Revealed Right [South Pole or Weak Force] represent Natural Magnetism.
- There are **four** cardinal [revealed] directions in correspondence with **Crucified Sign**, namely, **Down** [East], **Up** [West], **Revealed Left** [Manifested Northern Hemisphere], **Revealed Right** [Manifested Southern Hemisphere].
- There are **six** manifested directions, namely, **North-East** [Region of North America], **Straight Middle-East** [Region of Eartha 3D or penetrated hole in the journey boat of Muusa (ass)], **South-East** [Region of South America], **South-West** [Region of Europe], **Upright West** [Region of Arabian Peninsula], and **North-West** [Region of Asia – Africa – Australia].
- There are four basic forces, namely, **Gravitational Force**, **Strong Force**, **Magnetic Force**, and **Weak Force**. Gravitational Force represents downward direction towards **East**. Strong Force represents upward direction towards **West**. Magnetic Force represents entering and ending of the equal & opposite stages of journey i.e. Revealed Left or **North Pole**. Weak Force represents turning of the stages of journey of the manifested sign of natural magnetism from Revealed Right **South Pole**.
- There are two East and two West.
- Though there are manifested North-East and Manifested North-West, yet there is only one Northern Hemisphere or Haiyalas-Swalaah or Revealed Left.
- Though there are manifested South-East and Manifested South-West, yet there is only one Southern Hemisphere or Haiyalal-Falaah or Revealed Right.
- Projected Northern Hemisphere and Southern Hemisphere as the Top and Bottom of the Globe are not manifest truths in resemblance with revelations. On the contrary, these are the teleological evidence sorceries i.e. concealment of the true concepts of West and East and projection of falsehoods.
- Projected North and South directions on the basis of manmade magnetism are neither manifest truths, nor manifested nature, nor

manifested Haiyalas-Swalaah and Haiyalal-Falaah. On the contrary, these are the Unique Epistemic Persecutions of faith and belief of mankind in general and particularly believers & followers of Upright West.

- The so-called Sun [Bullet or Electromagnetic Wave] is the Manifested Sign of Natural Magnetism i.e. Manifested Sign of Revealed Left [Haiyalas-Swalaah or E-Point] and Revealed Right [Haiyalal-Falaah or T-Point]
- Projected Manmade Nature [Mechanical Globalisation, Circular Rotation & Revolution System, and Manmade Magnetism] is a conspiracy against followers of equal & opposite right direction towards Upright West [Arsh]
- The Revealed Truth is that the so-called Sun [Bullet or Electromagnetic Wave] neither rises from the East [Gravitational Force] nor sets in the West [Strong Force].
- The so-called Sun [Bullet] as the manifested sign of natural magnetism rises once from the East and sets once in the West to cause the alteration of day and night for the equal & opposite Seashore [East Zone and West Zone] is a **Projected Conspiracy** as well as **Post Hoc Ergo Propter Hoc Statistics.**
- The Revealed Truth is that the so-called Sun [Bullet or Electromagnetic Wave] **rises twice** and **sets twice** to cause the alteration of day and night for the equal & opposite Seashore [East Zone and West Zone].
- Manifested North-East in correspondence with North America and Manifested South-East in correspondence with South America represent Semi-anticlockwise journey of the so-called sun [Bullet] for the East Zone or Lower Seashore. In other words, the so-called sun [Bullet or Electromagnetic Wave or converted snake of Muusa's Staff] enters [rises] from Manifested North-East and sets in Manifested South-East for the Ground Stair of North America and South America of the appearing Pentagonal Earth. That means the direction from which the so-called sun [Bullet] enters for the East Zone is the Manifested North [Haiyalas-Swalaah] and the direction towards which the so-called sun [Bullet] sets for the East Zone is the Manifested South [Haiyalal-Falaah] of the Ground Stair of North America and South America of the appearing Pentagonal Earth within the Black Square of East Horizon.
- Manifested South-West in correspondence with Europe and Manifested North-West in correspondence with Asia - Africa - Australia represent Semi-clockwise stages of journey of the so-called sun [Bullet] for the West Zone or Upper Seashore. In other words, the so-called (setting)

sun [Bullet or Electromagnetic Wave or converted snake of Muusa's Staff] due to Revealed Magnetic Force runs away from Manifested South-East and rises from Manifested South-West and ends [sets] in Manifested North-West. That means the direction from which the so-called sun [Bullet] rises for West Zone is the Manifested South [Haiyalal-Falaah] and the direction towards which the so-called sun [Bullet] ends (sets) for the West Zone is the Manifested North [Haiyalas-Swalaah] of the Middle Stair of Europe, Asia, Africa, and Australia as well as Topmost Stair of Arabian Peninsula of the appearing Pentagonal Earth within the Black Square of East Horizon.

• Epistemic Uniqueness of Scientific Certainty, Owner of Manmade Magnetism, and Possessors of Projected Manmade Nature have introduced two kinds of magnetism i.e. **two Realities**, namely, manmade magnetism and natural magnetism with a view to conceal manifest truth of Magnetism [so-called Natural Magnetism] and manifested cardinal directions [East, West, North, and South]. Due to introduced manmade magnetism within the domain of formal education, we [so-called epistemic persons] have conceptualised manmade directions as natural directions [manifested directions in resemblance with revelations]. Consequently, we are using our pen-paper-pencil in resemblance with manmade magnetism **forgetting sole magnetism [fish]**. So, for us the Electromagnetic Wave [so-called sun or Bullet or Magnetic Force] rises from East or Gravitational Force and sets in the West or Strong Force going both contrary as well as contradictory to Newton's Three Laws of Motion, Newton's Law of Gravitation, Einstein's Gravitational Lensing, Einstein's Binary Pulsar, Manifested Signs of Natural Magnetism in resemblance with Revelations, and Upright Logic. **If Bullet or Electromagnetic Wave or so-called sun as the equal & opposite manifested sign of magnetism rises from the Gravitational Force [East] and sets in the Strong Force [West], then where are the existential imports of Newton and Einstein in Scientific Certainty?** Why are you teaching their Laws and Theories through formal education? What is the existential import of Newton's Third Law – 'Equal & Opposite' with respect to the alteration of day and night? Whether the Epistemic Uniqueness of Scientific Certainty teaching us falsehood in resemblance with manmade nature or manifest truth in resemblance with manifested nature? Whether the Epistemic Uniqueness of Scientific Certainty concealing from us manifest truths or projecting manifest truths through pen-paper-pencil works and black & white activities? ? ? ? ? ? ? ? ? ? ? ? ? ? ? ? ? ?

Nothing but let us invite the Epistemic Uniqueness of Scientific Certainty, Owners of Manmade Magnetism, and Possessors of

Manmade Nature to share with us manifest truth of a single fact i.e. equal & opposite manifested signs of Magnetic Force or Electromagnetic Wave or Natural Magnetism.

FISH

Ayat\Verses - But your Rab is Most forgiving, full of Mercy. If He were to call them [at once] to account for what they have earned, then surely He would have hastened their punishment; but they have **their appointed time,** beyond which they will find no refuge. Such were the **townships** [Ground Stair] we destroyed when they **committed iniquities;** but we fixed an **appointed time** for their **destruction.** And when, Muusa said to his attendant: I will not give up until I reach **the junction of the two seas,** though I march on for ages. But **when they reached the Junction, they forgot [about] their Fish, which took its course through the sea [wave] as in a tunnel.** [Sura No. – 17\18 – Khaf – Ayat No. – 58 to 61]

WE ARE TO PONDER OVER THE FOLLOWING AYAT\VERSES

We created you. **Will you then admit the truth? [57]**

Have you seen that which you **emit**? [58]

Do you create it, or are We the Creator? [59]

We have decreed Death to be your common lot, and We are not to be frustrated that We may transfigure you and make you what you know not. [60 & 61]

And certainly you know the first form of creation, why then do you not reflect? [62]

Have you seen that which you cultivate? [63]

Is it you who faster it, or are We the Fasterer? [64]

If We willed, We may verily could make it chaff, then would you cease not to exclaim. Lo! We are laden with debt. Nay, but we are deprived. [65 to 67]

Have you observed the water which you drink? [68]

Is it you **who shed it from the raincloud [West]**, or are We the shedder? [69]

If We willed We could verily make it bitter. Why then give you not thanks? [70]

Have you seen the Fire which you kindle? [71]

Is it you who grow the tree which feeds the fire, or do We grow it? [72]

We, even, We appointed it a memorial and a comfort for the dwellers in the wilderness. Therefore praise the name of your Rab, the Supreme! [73 & 74] [Sura No. – 55\56 - Waqa-atil-waaqiah, Ayat No. – 57 to 74]

TO CONFIRM WITH THE REVELATIONS

There was indeed a Sign for **Saba,** in their dwelling place. Two gardens to the right and to the left. Eat of the Sustenance [provided] by your Rab, and be grateful to Him: A territory fair and happy, and a Rabbun Gafuur. [Sura No. – 33\34 – Saba, Ayat No. - 15]

But they **turned away** [from manifested signs of equal & opposite magnetism], and We sent against them the **Flood** [released] from the dams, and We **converted their two garden into "gardens"** producing bitter fruit, and tamarisks, and some few [stunted] lote- trees. That was the requital We gave them because they ungratefully rejected Faith, and never do We give [such] requital except to such as be ungrateful rejecters. **Between them and the Cities** on which We had **poured our blessings**, We had placed **Cities in prominent position**, and between them We had **appointed stages of journey** in due **proportion**. Travel therein, secure, by night and by day. [Sura No. – 33\34 – Saba, Ayat No. – 16 to 18]

But they said: Our Rab! Place longer distances between our journey- stages. But they wronged themselves [therein]. At length We made them as a tale, and We dispersed them all in scattered fragments. Verily in this are **Signs** for every [soul that is] patiently constant and grateful. And Shaatyan indeed found his calculation true concerning them, for they follow him all except a group of **true believers.** But he had no authority over them except that We might test the man who believes in the Hereafter from him who is in doubt concerning it, and takes note of all things. [Sura No. – 33\34 – Saba, Ayat No. – 19 to 21]

Say: It is Allah Who **originates creation and repeats it**
Say: Of your 'partners', can any originate creation and repeat it? Say: It is Allah Who **originates creation and repeats it**: how then are you deluded away / misled [from the manifest truth]? Say: Of your 'partners' is there any that can give any guidance towards truth? Say: It is Allah Who gives guidance towards truth, is then He Who gives guidance to truth more worthy to be followed, or he who finds not guidance [himself] unless he is guided? What then is the matter with you? How judge you? But most of them follow nothing but fancy: truly fancy can be of no avail against truth. Verily Allah is well aware of all that they do. [Sura No. – 9\10 – Yuunus, Ayat No. – 34 to 36]

"Say: Has Allah indeed permitted you, or do you invent [manmade magnetism] to attribute to Allah?"
Say: See you what things Allah has sent down to you for sustenance? Yet you hold forbidden some things thereof and [some things] lawful. **Say: Has Allah indeed permitted you, or do you invent [manmade magnetism] to attribute to Allah?** And what think those who invent lies against Allah, of the Day of Resurrection [Yawmal-Qi-yaamah]? Verily Allah is full of bounty to mankind, but most of them are ungrateful. In whatever business you may be, and whatever portion you may be reciting from the Qur'an, and whatever deed you [mankind] may be doing, We are witnesses thereof when you are deeply

engrossed therein. Nothing is hidden from your Rab [so much as] the weight of an **atom** in the world or in heaven. And not the least and not the greatest of these things but are recorded in a clear record. [Sura No. – 9\10 – Yuunus, Ayat No. – 59 to 61]

A SINGLE FACT – NATURAL MAGNETISM

Have you not seen those unto whom it was said: **Hold back your hands**, establish worship and pay the poor-due? But when fighting was prescribed for them, behold! A section of them feared men as - or even more than - they should have feared Allah. They said: Our Rab! Why have you ordained fighting for us? If only You give us respite yet a while! Say: Short is the enjoyment of this world [present life]; the Hereafter is the better for those who ward off evil. Never will you be dealt with unjustly in the very least (down upon a date-stone). **Wherever you are, death will find you out, even if you are in towers built up strong and high!** If some good befalls them, they say: This is from Allah; but if evil, they say, this is from you. Say: All things are from Allah. But **what has come to these people that they fail to understand a single fact?** Whatever good, [O man!] happens to you, is from Allah; but whatever evil happens to you, is from yourself. We have sent you as a messenger unto mankind. And **enough is Allah for a witness**. [Sura No. – 3\4 – Nisaa, Ayat No. – 77 to 79]

PAIRS [EQUAL & OPPOSITE REVELATION] IS AN INSTRUCTION
"And **all things We have created by pairs**
that you may **receive instruction**."
SET NOT ANOTHER OBJECT
[MANMADE MAGNETISM] OF WORSHIP

With power and skill We have constructed the Firmament; for it is We Who create the vastness of rapidity. And We have spread out the world; How excellently We do spread out! And **all things We have created by pairs** that you may **receive instruction**. Hasten you then [at once] to Allah. **I am a plain warner to you from Him.** And **set not another object of worship** (i.e. the introduced calf or manmade magnetism) along with Allah. **I am a plain warner to you from Him.** Similarly, **no messenger** came to the peoples before them, but they [the guilty folk] said [of him]: **A sorcerer or a madman.** [Sura No. – 50\51 – Waz-Zaariyaat, Ayat No. – 47 to 52]

Revealed Truth & Manifested Tautologies
The so-called sun [Bullet or Niche] as the manifested sign of natural magnetism is kindled each day neither from the East nor from the West
Or
So-called Sun [Bullet] neither rises from the East nor sets in the West

"**Allaahu Nuurus-samaa-waati wal-arz** – Masalu **Nuu-rihi** kaMishkaatin-fiihaa **Mis-baah** – Al-**Misbaahu** fii **Zujaa-jah** – **azzuujaajatu** ka-annahaa – kawkabun durriyyuny-yuuqa-du – min Shajaratimmubaara-katin-**Zaytuunatil-**laa **Sharqiy**-yatinw wa laa **Garbiyyatiny**-yakaadu **Zaytuhaa** yuziii- u wa law lam tamsas-hu naar. **Nuu-run alaa Nuur! Yahdillaahu** li-**Nuurihii** many-yashaaa – wa-yazribullaahul-amsaala linnaas – wallahu bi-kulli shay-in Alim." [Sura No. – 23\24 – Nuur, Ayat No. – 35] [Trans.] **Allah** is the **Light** of the **heavens** and the **world**. The **similitude** (likeness) of **His Light** is as a **niche** (binary thread of a candle or binary pulsar of Einstein) **in a lamp [Black Square]**. The **lamp** is in a **glass [Diamond Operator]**. The **glass** is as it is a **shining star [Shi-rra]**. (The niche is) kindled from a **blessed tree**, an **olive** (Zaytuun tree), **neither from the East nor from the West**, whose (niche's) oil will almost glow forth (of itself) though **no fire touches** it. **Light upon Light!** Allah guides unto His light whom He wills. And Allah speaks to mankind in **parables**, for Allah is Knower of all things. [Sura No. – 23\24 – Nuur, Ayat No. – 35]

MANIFESTED SIGNS & NATURAL MAGNETISM
NORTH to SOUTH and SOUTH to NORTH
Revealed Truth & Manifested Tautologyies

The so-called sun [Bullet / Niche] as the manifested sign of natural magnetism **enters** from Manifested **North-East** [Haiyalas-Swalaah] in correspondence with **North America** and **sets** in Manifested **South-East** [Haiyalal-Falaah] in correspondence with **South America** for the **Ground Stair** of the appearing **pentagonal earth** and **East Zone** of the manifested **hexagonal world** within the **Black Square** of **East Horizon**. The so-called sun [setting sun or Bullet or Niche] **rises** from Manifested **South-West** [Haiyalal-Falaah] in correspondence with **Europe** and **ends** in Manifested **North-West** [Haiyalas-Swalaah] in correspondence with **Australia** for the **Middle Stair** of Europe, Asia, Africa, and Australia as well as **Topmost Stair [Upright West Stair]** of Arabian Peninsula of the appearing **pentagonal earth** and **West Zone** of the manifested **hexagonal world** within the **Black Square** of **East Horizon**. The so-called sun [Bullet] as the manifested sign of natural magnetism **rises twice** and **sets twice** to cause the alteration of day and night for the equal & opposite Seashore [East Zone and West Zone].

The so-called new moon [White] arises from Manifested North-West [Revealed Left] in correspondence with Australia for the West Zone and follows Semi Major Axis and Right Double Quotation Mark. The so-called new moon appears from Manifested South-East [Revealed Right] for the East Zone and follows Semi Major Axis and Left Double Quotation Mark. Both the so-called sun and the so-called moon have the single root.

"Wa **tarash-shamsa** izaa tala-at-lazaawaru an-kahfihim zaatal-**yamimi** wa izaa gara-at-taqri-zuhum **zaatash-shimaali** wa hum fi fajwatim-minh. Zaalikka min aayaatillaah; many-yahdillaaahu fahuwal-muhtad; wa many-yuzlil falan-tajida lahuu waliyyam-murshidaa". {Trans.} And might have seen **tarash-shamsa** when it **rises, declining to the right** [semi-anticlockwise] from their Cave, **when it sets, turning away from them to the left** [semi-clockwise], while they lay in the **open space** in the midst of the **Cave** [within the Black Square of Manifested East Horizon]. **Such** [equal and opposite i.e. half anti-clockwise and half clockwise stages of journey as the glittering show for the earth] are among the **Signs of Allah**. He whom Allah, guides is **rightly guided**; but he whom Allah **leaves to stray**, for him **you will not find protector to lead him to the Right Way**. [Sura No. – 17\18 – Khaf – Ayat No. – 17]

SINGLE ROOT

Alif-Lam-Mim-Ra. These are the **Clear Signs** of the Kitaab. That which has been revealed to you from your Rab is the **Truth**. But **most men believe not**. Allah is He Who raised the heavens without any pillars that you can see; is firmly established on the **throne** [of authority]. He has compelled **sakh-kharash-shamsa wal-qamar to be of service. Each one runs [its course] for a term appointed.** He regulates all affairs, explaining the **signs in detail** that you may **believe with certainty** in the meeting with your Rab. And it is He who spread out the world, and set thereon mountains standing firm and [flowing] rivers: and fruit of every kind He made in **pairs, two and two**: He draws the **night as a veil over the day**. Behold, verily in these things there are **signs** for those **who give thought**! And in the manifested world are tracts [ascending stairs] neighbouring, and gardens of vines and fields sown with corn, and palm trees - growing out of **single root** or otherwise (like and unlike): watered with the same water, yet some of them We make more excellent than others to eat. Behold, verily in these things there are **signs** for those **who understand**! [Sura No. – 12\13 – Rad, Ayat No. – 1 to 4]

A BARRIER OR A DANGER LINE OR A VETO

It is He Who has **let free (independent) the two bodies of flowing water.** One palatable and sweet and the other **salt** and **bitter; yet He has appointed a barrier between them, a veto that is forbidden to be passed.** It is He Who has created man from **water**. Then He has established **relationships** of lineage and marriage. Lo! Your Rab has power [over all things]. Yet **they worship instead of (besides) Allah that which can neither benefit them nor hurt them**, and **the disbeliever is a helper [of devil / Shaytan] against his own Rab!** [Sura No. – 24\25 - Nazzalal-Furqaan, Ayat No. – 53 to 55]

CLEAR PROOFS

Twa-Sin-Mim.

These are the **shared ideograms** revealed in the alaykal-**Kitaabim-Mubin** as **Clear Proofs [Revelations]**. We narrate to you some of the tales of Muusa and Firawn in Truth, for **people who believe**. [Sura No. – 27\28 – Qasas, Ayat\ Verses – 1 to 3]

Twa-Sin-Mim.

These are the shared ideograms revealed in the Kitaab as Clear Proofs which makes things clear. [Sura No. 25\26 – Shuraa, Ayat No. – 1 & 2]

TWO PROOFS
Snake and White [Natural Magnetism / Niche / Binary Pulsar]

Now when Muusa had **fulfilled the term**, and was **travelling** with his family, he perceived **a fire [light] in the direction of Mount Tuur**. He said to his family: **Tarry** you; I **perceive** a fire [light]; I hope to **bring you** from there some **information** (tidings), or a brand from the fire that you may **warm yourselves**. And when he reached it, he was called **from the right side of the valley in the blessed field, from the tree**: O Muusa! Lo! I, even I, am Allah, the Rab of the Universe. **Throw down your staff**. And **when he saw it moving as it had been a demon (Electromagnetic Wave / snake), he turned to flee headlong** (and it was said unto him): **Draw near, and fear not**. Lo! **You are of those who are secure. Thrust your hand into the bosom of your robe, it will come forth white without hurt. And guard your heart from fear. These are the two proofs from your Rab to Firawn and his chiefs**, for **truly they are a people rebellious and wicked**. [Sura No. – 27\28 – Qasas, Ayat\Verses – 29 to 32]

SNAKE AND WHITE - CLEAR SIGNS

Then after them **We sent Muusa with Our signs** to **Firawn** and **his chiefs**. But they **wrongfully rejected** them. So see what was the **end** of those who made **mischiefs**? Muusa said: O Firawn! I am a messenger from the Rab of the universe [Aalamin]. One for whom **it is right to say nothing but truth** about Allah. Now I have come unto you [people], from your Rab, with a **Clear Sign**. So let the **Baniii-Israaa-il [believers] depart along with me**. [Firawn] said: If indeed you have **come with a Sign**, then **produce** it, if you are telling the truth. Then [Muusa] threw his staff, and behold! It was a **large snake**, plain [for all to see]! And he drew out his hand, and behold! It was **white** to all beholders! [Sura No. – 6\7 - As-haabul- a' raaf, Ayat No. – 103 to 108]

[Muusa] said: **Rab of the East and the West**, and all between! **If you only had sense!** [Firawn] said: If you put forward any god other than me, I will certainly put you in prison! [Muusa] said: **Even if I showed you something clear [and] convincing**? [Firawn] said: Show it then, **if you tell the truth**! So [Muusa] flung down his staff, and behold, it was a **serpent**, manifest [for all to see]! And he drew out his hand, and behold, it was **white** to all beholders! [Sura No. 25\26 – Shuraa, Ayat No. – 28 to 33]

UNTIL THE WHITE THREAD BECOMES DISTINCT FROM THE BLACK THREAD OF DAWN

It is made lawful for you to go unto your wives on the night of fasting. They are your garments and you are their garments. Allah is aware that you were deceiving yourselves in this respect and He has turned in mercy toward you and relieved you. So associate with them, and seek that which Allah has ordained for you, and eat and drink, until **the white thread becomes distinct to you from the black thread of the dawn then strictly observe the fast till the nightfall; but do not associate your wives at your devotions in the mosques. These are limits imposed by Allah. So, approach them not. Thus, Allah expounded His Clear Signs (Manifested Signs of Natural Magnetism)** to mankind that they may learn **self- restraint. And do not eat up your property among yourselves for vanities (falsehood), nor seek by it to gain hearing of the judges, with the plan that you may knowingly devour a portion of the property of others' wrongfully.** [Sura No. 1\2 – Baqarah, Ayat No. – 187 & 188]

WHITE AND RED

Have you not seen that Allah causes water to fall from the sky [West]? With it We then bring out produce of various **colours**. And in the mountains are tracts **white and red**, of various shades of **colour**, and **raven-black** hues. [Sura No. – 34\35 - Faatiris-Samaawaati wal-'arz, Ayat\Verses – 27]

By the **Star when it sets, your comrade errs not, nor is misled, nor does he speak** of (his own) **desire. It is nothing save an inspiration that is inspired.** He was **taught** by **one Mighty in Power, endowed with Wisdom and clear to view. While he was in the upper part of the horizon [West]** then **he drew neigh (near) and came down**, and was at a distance of but **two bows length** or even **nearer**. So did [Allah] **convey** the **inspiration to His slave** [conveyed] what He [meant] to convey. [Sura No. – 52\53 – Wan-Najm, Ayat No. – 1 to 10]

To **Allah** belong the **Mystery** of the **heavens** and the **world**. And the Decision of the Hour [of Judgment] is as the twinkling of an eye, or even quicker: for Allah has power over all things. **It is He Who brought you forth from the wombs of your mothers when you knew nothing; and He gave you hearing** and sight and intelligence and affections that you may give thanks [to Allah]. Have they not seen the birds obedient in mid-air? None holds them save Allah. Lo! Herein, verily are signs for those who believe. [Sura No. – 15\16 – Nahl, Ayat\Verses – 77 to 79]

"So the disbelievers will say: This is a wonderful thing!"
"But they deny the Truth when it comes to them. So they are in a confused state."
Fundamental Ayat [Revelation] – Q A A F: Wal-Quraanil-Majiid
But they **wonder** that there has come to them a **warner** [reminder] from among themselves. **So the disbelievers say: This is a wonderful thing!** What! When we die and become dust, [shall we live again?] That is a [sort of] return far [from our understanding]. We already know how much of them the earth takes away. With Us is a record guarding [the full account]. **But they deny the Truth when it comes to them. So they are in a confused state.** [Sura No. 49\50 – Kaaaf, Ayat No. – 1 to 5]

Rabbukum wa Rabbu
"Laaa ilaaha illa Huwa yuhyu wa yumiit – Rabbukum wa Rabbu aabaaa-ikumul-aawalin." [Self-evident Concept] There is no unbreakable reality except Him. It is He Who gives life and gives death. The **Rabbukum wa Rabbu** of you and your forefathers. **Nay but they play in doubt.** Then **watch you for the Day when the sky** [psychic or planetary barrier] **will bring forth a kind of smoke [or mist] plainly visible, that will envelop the people. This will be a painful torment.** [Then they will say]: Our Rab! **Relieve us from the torment! Lo! We are believers.** [Sura No. – 43\44 - Bi-Dukhaanim-Mubiin, Ayat No. – 8 to 12]

"Will they not ponder on the Qur'an [with care]?"
Whoso obeys the Messenger, obeys Allah, and those [whoso] turn away, We have not sent you to watch over their [evil deeds]. They have obedience on their lips; but when they leave you, a section of them **meditate all night on things very different from what you tell them.** But Allah records their nightly [plots]. So **keep clear of (oppose) them, and put your trust in Allah, and enough is Allah as a disposer of affairs (trustee). Will they not ponder on the Qur'an [with care]?** If it had been from other than Allah, **they would surely have found therein much discrepancy (incongruity).** [Sura No. – 3\4 – Nisaa, Ayat No. – 80 to 82]

"Will you not then understand?"
"It is **not you** they reject. It is the **signs of Allah**, which the **wicked flouted.**"

What is the **life of this world [present life]** but play and amusement? But **best** is the home in the **hereafter**, for those who are **righteous. Will you not then understand? We know** indeed the **grief** which **their words do cause you.** It is **not you** they reject. It is the signs of Allah, which the wicked flouted [disobeyed]. [Sura No. – 5\6 – An-aam, Ayat No. – 32 &33]

Think you to yourselves - if you are truthful.

Say: **Think you to yourselves,** if there come upon you the punishment of Allah, or the Hour [that you fear], would you then call upon other than Allah, **if you are truthful**? Nay, On Him would you call, and if it be His will, He would remove [the distress] which occasioned your call upon Him, and you would forget [projected falsehood & manmade magnetism] **which you join with Him.** [Sura No. – 5\6 – An-aam, Ayat No. – 40 to 41]

But [now that the Qur'an has come], they reject it. But soon will they know! And verily **Our word went forth before [this] to our bondmen** sent that they would certainly be assisted, and that Our hosts, they surely would be the victors. So **turn you away (withdraw) from them for a little while,** and watch them, for they will soon see. [Sura No. – 36\37 – Was-saaaffaati, Ayat No. – 170 to 175]

Do they wish [indeed] to hurry on our Punishment? **But when it comes home to them, then it will be a hapless morn for those who have been warned. So turn you away (withdraw) from them for a little while,** and watch them, for they will soon see. [Sura No. – 36\37 – Was-saaaffaati, Ayat No. – 176 to 179]

"Say: Wait you then, for I, too, will wait with you."

Do they then expect [anything] but [what happened in] the days of the men who passed away before them? Say: Wait you then, for I, too, will wait with you. In the end We save Our messengers and those who believe: Thus, it is incumbent upon Us to save believers. [Sura No. – 9\10 – Yuunus, Ayat No. – 102 & 103]

"One day the world will be changed to a different world"

Yawma tubad-dalul- Arzu gayral – Arzi was-Samaa-waatu wa barazuu lillaahil- Waa-hidil-Qahhaar. {trans.] One day the world will be changed to a different world [East Horizon] and so will be the heavens [West Horizon] and they will come forth unto Allah, the One-Almighty. [Sura No. – 13\14 – Ibrahim, Ayat No. – 48]

A logician or a madman
Vs
A two-in-one Rascal or the Don of Historical Black Magic

And in **Muusa [too there is a Sign]**: When, We sent him to Firawn, with **authority manifest**. But [Firawn] withdrew (confiding) in his might, and said: **A sorcerer, or a madman.** So We **seized** him and his **hosts**, and **threw** them into the **sea**; for he was a **rascal** (mischief). [Sura No. – 50\51 – Waz-Zaariyaat, Ayat No. – 38 to 40]

A WARNER AGAINST GREAT TYRANTS

Glorified be He Who created His servant by night **from the Inviolable Place of Worship to the far distant place of worship the neighbourhood whereof We have blessed that We might show him of (Our Signs)**! Lo! He, only He, the Hearer, the Seer. We gave unto Muusa the Kitaab, and We appointed it guidance to the Banii-Israai-iil [commanding]: **Take not other than Me as disposer of [your] affairs.** (They were) the seed of those whom We carried [in the Ark] along with Nuuh! Verily he was a most grateful devotee. And We gave [Clear] warning to the Banii-Israa-iil in the Kitaab, that **twice** you would do mischief in the world and you will become **great tyrants.** [**Sura No. – 16\17 -** Banii-Israai-iil, Ayat No. – 1 to 4]

EVIDENCE SORCERIES AND WISE POLICIES OF FIRAWN

O my people! Yours is the **Kingdom** (dominion) today. **You are the uppermost in the land.** But who will help us from the Punishment of Allah, should it befall us? **Firawn** said: I do but **show you what I think**, and I do but **guide you to wise policy [Mechanical Globalisation, Manmade Magnetism, and Circular Rotation & Revolution System i.e. Manmade Nature].** [Sura No. – 39\40 – Mu-Min, Ayat No. –29]

PLOT OF SCHEMING NIGHT & DAY
[Reference of Post Hoc Ergo
Propter Hoc Statistics and Historical Black Magic]

Those who had been despised will say to the proud ones: Nay! **It was a plot [of yours] scheming night and day when you [constantly] ordered us to be ungrateful to Allah and to attribute equals (manmade magnetism and natural magnetism) to Him! They will declare [their] repentance when they see the Penalty. We shall put yokes on the necks of the disbelievers. It would only be requital for their [ill] deeds.** [Sura No. – 33\34 – Saba, Ayat No. - 33]

SEPARATE SHEET OF REAL SCIENTIFIC SYMBOLS

No just estimate of Allah do they make when they say: Nothing Allah sends down to man [by way of revelation]. Say: Who then sent down the Kitab which Muusa brought - **a light and guidance to man**? **But you make it into [separate] sheets for show, while you conceal much [of its contents]**: therein were you taught that which you knew not- neither you nor your fathers. Say: Allah [sent it down]: Then leave them to plunge in vain discourse and trifling. And this is a Kitab which We have **sent down** [towards East], bringing blessings, and confirming which came before it: that you may warn **the mother of cities [Arabian Peninsula] and all around her**. Those who believe in the Hereafter believe in this and **they are careful in guarding their prayers**. [Sura No. – 5\6 – An-aam, Ayat No. –92 & 93]

Follow not Shaytan's footsteps
[Mechanical Globalisation, Manmade Magnetism, and Circular
Rotation & Revolution System i.e. Manmade Nature]

O you who believe! **Follow not Shaytan's footsteps. If any will follow the footsteps of Shaytan, he will command what is filthiness and wrong;** and had it not been for the grace and mercy of Allah unto you, not one of you would ever have **grown pure**. But Allah purifies whom He pleases, and Allah is One Who hears and knows [all things]. [Sura No. – 23\24 – Nuur, Ayat No. – 21]

PERSECUTIONS ARE WORSE THAN KILLING

Fight in the way of Allah against those who fight against you, but begin not hostilities. Lo! Allah loves not the aggressors. And slay them wherever you find them, and drive them out from the (sacred / inviolable) places whence **they drove you out**, for **persecutions / maltreatments are worse than killing**. And fight not with them at the Sacred Mosque, until they [first] attack you there. But if they attack you (there), then slay them. Such is the reward of those who suppress [conceal] manifest truth. But if they cease [confess manifest truth], Allah is Ever Forgiving, Most Merciful. And **fight them on until persecutions / maltreatments / oppressions are no more**, and there **prevail Justice and faith in Allah**. But if they cease, let there be no hostility except to those who practise oppression. [Sura No. – 1\2, Baqarah, Ayat\Verse No. – 190 to 193]

PERSECUTION IS WORSE THAN KILLING

Or do you think that you shall enter the Garden [of bliss] without such [trials] as came to those who passed away before you? They encountered suffering and adversity, and were so shaken in spirit that even the Messenger and those of faith who were with him cried. When [will come] the help of Allah? Ah! Verily, the help of Allah is [always] near! They ask you what they should spend [in charity]. Say: Whatever you spend for good to parents and kindred and orphans and those in want and for wayfarers; and whatever you do that is good, lo! Allah knows it well. **Warfare (Tabligh or selfpurification) is prescribed for you** and **you dislike it**. But **it is possible that you dislike a thing which is good for you, and that you love a thing which is bad for you**. But **Allah knows**, and **you know not**. They ask you concerning warfare in the Sacred Month (in the sense of both time and place). Say: **Warfare therein** is a **great (transgression)** but to **track off [believers] from the way of Allah, and from the Inviolable Place of Worship**, and to expel his people thence, is **greater in the sight of Allah, for persecution is worse than killing**. Nor will they cease fighting you until they **turn you back from your faith if they can**. And whosoever Turn back from their faith and die in unbelief, their works will bear no fruit in this life and in the Hereafter. They will be companions of the Fire and will abide therein. Lo! Those who **believe** and those who **emigrate** and **strive in the way of Allah, they have hope of Allah's mercy**. Allah is Ever Forgiving, Most Merciful. [Sura No. – 1\2, Baqarah, Ayat\Verse No. – 214 to 218]

THE POWER OF THE
HISTORICAL BLACK MAGICIAN WILL PERISH

The power [hands] of the **father of flame** will perish and **he** will perish. No profit to him from all his wealth, and all his gains! **He will soon be burnt in flaming fire**! And his wife [activity], the wood carrier as fuel, a twisted rope of palm-leaf fibre round her [own] neck! [Sura No. – 110\111 - Abii lahabinw-wa tabb, Ayat – 1 to 5]

WHEN

When the sky [planetary barrier] is rent asunder [apart] [1], and attentive to **his** Rab in fear [2], and when **the world is flattened out**, [3], and casts forth what is within it [buried treasure] and becomes [clean] empty [4], and attentive to **her** Rab in fear [5]. O you man! Verily you are ever working on towards your Rab painfully toiling, but you shall meet Him [Sura No. – 83 \ 84 - Izas-Samaaa-unshaqqat, Ayat No. – 1 to 6]

"But verily, many among mankind are **heedless of Our Signs.**"
We took the Bani-Israa-iil **across the sea**: Firawn and his hosts followed them in insolence and spite. At length, when overwhelmed with the flood, he said: I believe that there is no reality except Him Whom the **Children of Israa-iil believe in**. I am of those who submit unto Him. [It was said to him]: Ah now! But a little while before, was you in uprising, and you did **mischief** [and violence]. But this day We save you in your body, that you may be **a sign** to those who come after you! But verily, many among mankind are **heedless of Our Signs [Manifested Signs of Natural Magnetism].** [Sura No. – 9\10 – Yuunus, Ayat No. 90 to 92]

62

"Say: Behold all that is in the heavens and in the world; but neither Signs nor Messengers profit those who believe not."

Why was there not a single township [among those We warned], which believed, so its faith should have profited it, except the people of Yuunus? When they believed, We removed from them the penalty of ignominy in the life of the present, and permitted them to enjoy [their life] for a while. If it had been the will of your Rab, they would all have believed, all who are on earth! With you then come mankind, against their will to believe? No soul can believe, except by the will of Allah, and He will place doubt [or obscurity] on those who will not understand. Say: Behold all that is in the heavens and in the world; but **neither Signs nor Messengers profit those who believe not.** [Sura No. – 9\10 – Yuunus, Ayat No. – 98 to 101]

Then he who is given his **Record in his right hand [7],** soon his account will be taken by an easy reckoning [8], and **he will turn to his people, rejoicing!** [Sura No. – 83 \ 84 - Izas-Samaaa-unshaqqat, Ayat No. – 7 to 9]

But he who is given his **Record behind his back** [10], he surely will invoke destruction [11], and be thrown to scorching fire [12]. He verily lived with his folk joyous. He verily deemed that he would never return [to Us]! Nay, but lo! His Rab is [ever] watchful on him! [Sura No. – 83 \ 84 - Izas-Samaaa-unshaqqat, Ayat No. – 10 to 15]

"And they wait for but **one Shout**. There will
be **no second Day of Reckoning**."

And **the chiefs (leaders) among them** go away [**impatiently**], [saying]: Walk you away, and **remain constant to your reality [manmade nature]! For this is truly a thing designed [against you]! We never heard** [the like] of this among the people of later religion. This is **nothing** but a **made-up tale** (invention)! What! **Has the reminder been revealed unto him [of all persons] among us?** But **they are in doubt concerning My Message!** Nay, **they have not yet tasted My Punishment! Or have they the treasures of the mercy of your Rab, the Exalted in Power, The Grantor of Bounties without measure?** Or have they the **Sovereignty of the heavens and the world and all between?** If so, **let them ascend by ropes. A defeated host (confederates) are (all) the factions that are there.** The folk of Nuuh before them denied (their messenger) and (so did the tribe of) Aad, and **Firawn firmly planted**. And Samuud, and the people of Luut, and the Companions of the Wood; **such were the Confederates** Not one [of them] but **rejected the messengers**. Therefore My **punishment was inevitably justified [on them]**. And they wait for but **one Shout**. There will be **no second Day of Reckoning**. [Sura No. – 37\38 – Swad, Ayat No. – 6 to 15]

And **this is a blessed Message which We have
sent down: Will you then reject it?**
We shall set up **scales of justification** for **the Day of Resurrection [Yawmil-
Qiyaamati]**, so that not a soul will be dealt with unjustly in the least, and if
there be [no more than] **the weight of a mustard seed**, We will bring it [to
account]: and enough are We to take account. In the past We granted to Muusa
and Haarun **the criterion [for justification]**, and a **Light** and a **Message for
those who would do right.** Those who fear their Rab in their most secret
thoughts, and who hold the Hour in awe; and **this is a blessed Message which
We have sent down: Will you then reject it?** [Sura No. – 20\21 – Ambiyaaa,
Ayat No. – 47 to 50]

"There can be no change in the words of Allah."
Behold! Verily on the friends of Allah there is no fear, nor shall they grieve.
Those who believe and [constantly] guard against evil; for them are good
tidings, in the life of the present and in the Hereafter. **There can be no change
in the words of Allah.** This is indeed the supreme felicity. Let not their speech
grieves you, for all power and honour belong to Allah. It is He Who hears and
knows [all things]. [Sura No. – 9\10 – Yuunus, Ayat No. – 62 to 65]

"What do they follow who worship as His "partners" other than Allah?"
Behold! Verily to Allah belong all creatures, in the heavens and in the world.
What do they follow who worship as His "partners" other than Allah?
They follow nothing but fancy, and they do nothing but lie. He it is Who has
appointed for you the night that you may rest there in and the day to make things
visible [to you]. Verily in this are signs for those who listen [to His Message].
They say: Allah has begotten a son. Glory is to Him! He is self- sufficient. His
are all things in the heavens and in the world. You have no warrant for this.
You tell about Allah what you know not? Say: Those who invent a lie against
Allah will never prosper. A little enjoyment in this world and then, to Us will
be their return, then shall We make them taste the severest penalty for their
wickedness. [Sura No. – 9\10 – Yuunus, Ayat No. – 66 to 70]

"They brought them **Clear Signs, but they would not believe what they had
already rejected beforehand.**"
Recite unto them the parable of **Nuuh.** He said to his people: O my people, if it
be hard on your [mind] that I should stay [with you] and commemorate the signs
of Allah, yet I put my trust in Allah. So, decide upon your course of action,
you and your partners. Let not your course of action be in doubt for you. Then
have at me, give me no relief. But if you turn back, [consider]; no reward have

I asked of you. My reward is only due from Allah, and I have been commanded to be of those who submit to Allah's will. They rejected Him, but We delivered him, and those with him, in the Ark, and We made them inherit [the earth], while We overwhelmed in the flood those **who rejected Our Signs**. See then the nature of the consequence of those who were **warned** [but heeded not]. Then after him We sent [many] **Messengers to their peoples**. Thus We seal the hearts of the transgressors. [Sura No. – 9\10 – Yuunus, Ayat No. 71 to 74]

"Lo! Firawn was verily a tyrant in the land [earth], and lo! He verily was of the wanton [reckless]."

But none trusted Muusa save some scions [children] of his people in fear of Firawn and their chiefs, that they would persecute them. **Lo! Firawn was verily a tyrant in the land [earth], and lo! He verily was of the wanton [reckless].** Muusa said: O my people! If you do [really] believe in Allah, then in Him put your trust if you submit [your will to His]. They said: In Allah do we put out trust. Our Rab! Oh, make us not a trial for those who practise tyranny and deliver us by Your Mercy from those who reject [You]. [Sura No. – 9\10 – Yuunus, Ayat No.-83 to 86]

"Those who invent a lie against Allah will never prosper."
Behold! Verily to Allah belong all creatures, in the heavens and in the world. What do they follow who worship as His "partners" other than Allah? **They follow nothing but fancy, and they do nothing but lie**. He it is That has appointed for you the night that you may rest there in and the day to make things visible [to you]. Verily in this are signs for those who listen [to His Message]. They say: Allah has begotten a son. Glory is to Him! **He is self-sufficient**. His are all things in the heavens and in the world. You have no warrant for this. You tell about Allah what you know not? Say: Those who invent a lie against Allah will never prosper. A little enjoyment in this world and then, to Us will be their return, then shall We make them taste the severest penalty for their wickedness. [Sura No. – 9\10 – Yuunus, Ayat No. – 66 to 70]

Has he invented a lie concerning Allah, or is there in him madness?

Nuuun. By the pen and that which they write (therewith), you are not, **by the Grace of your Rab unto you**, a **madman**. [Sura No. – 67\68 – Qalam, Ayat No. 1 & 2]

Has he invented a lie concerning Allah, or is there in him madness? Nay, it is those who believe not in the Hereafter that is in [real] Penalty, and in farthest error. Have they not observed what is **before** them and **behind** them, of the sky and in the world? If We wished, We could cause the earth to swallow them up, or cause a piece of the sky to fall upon them. Verily in this is a Sign for every devotee that turns to Allah [in repentance]. [8 & 9] [Sura No. – 33\34 – Saba, Ayat No. – 8 & 9]

"We do indeed know how your heart is distressed at what they say."
"And serve your Rab **until there come unto You the Hour that is Certain.**" For **sufficient** are We unto you against those who scoff [mock]. Who set some other ilaaha (breakable reality); but **soon they will come to know.** We do indeed know how your heart is distressed at what they say. But **remember** your Rab, and be of **those** who **prostrate** themselves in adoration. And serve your Rab **until there come unto You the Hour that is Certain.** [Sura No. – 14\15 = Hur, Ayat No. – 95 to 99]

"Say: Each one [of us] is waiting, wait you, therefore, and soon shall you know who it is that is on the straight and even way, and who it is that has received Guidance."
And they say: Why does he not bring us **Clear Sign** from his Rab? Has not a Clear Sign come to them of all that was in the **former Kitaabs of revelation**? And if We had **inflicted** on them a penalty **before this**, they would have said: Our Rab! If only You had sent us a **messenger**, we should certainly have followed **Your Signs** before we were humbled and put to shame. **Say: Each one [of us] is waiting, wait you, therefore, and soon shall you know who it is that is on the straight and even way, and who it is that has received Guidance.** [Sura No. – 19\20 – Twa-Haa, Ayat No. – 133 & 135]

PENALTY OF HELL
OR
GREAT SUCCESS

By the sky holding mansions [halls] of the stars [1], and by the **Promised Day [2]**, and by **one** who witnesses, and the **subject** of the witness [testimony] [3]; woe to the makers of the **pit** [of fire] [4], of the **fuel-fed fire [5]**, when they sat by it [6], and they witnessed [all] that **they were doing against the believers [7],** and **they ill-treated them for no other reason than that they believed in Allah**, Exalted in Power, Worthy of all Praise [8]! Him to Whom belongs the Sovereignty of the heavens and the world! And Allah is Witness to all things [9]. Those who persecute [or draw into temptation] the believing men and believing women, and **do not turn in repentance**, will have the penalty of hell. They will have the penalty of the **burning fire [10]**. Those who believe and do righteous deeds [tabligh], will be gardens for them; beneath which rivers flow.

That is the great success. [Sura No. – 84\85 – Was-Samaaa-i Zaatil-Buruuj, Ayat No. – 1 to 11]

INSPIRATION

"We granted **inspiration**: If you realise this not, **ask of those who possess the message**."

[Sign, Mixture of dream, Forgery, Township, Belief, Inspiration, Message, Possessor of Message, Promise of Allah, Punishment of the Transgressors]

Ayat\Verses – Nay, they say: [These are] **mixture** of dream! Nay, He **forged** it! Nay, He is [but] a **poet**! Let him then bring us **a Sign like the ones** [similitude] that was sent to [Prophets] of old! [As to those] before them, not a **township** which We destroyed believed: **Would they then believe**? Before you, also, the messengers We sent were but men, to whom We granted **inspiration**: If you realise this not, **ask of those who possess the message**. **Nor** did We give them bodies that ate no food, **nor** were they exempt from death. In the end We fulfilled to them **Our Promise**, and We **saved them** and those whom We **pleased**, but We **destroyed those who transgressed beyond bounds**. [Sura No. – 20\21 – Ambiyaaa, Ayat No. – 5 to 9]

"Follow what you are taught by inspiration from your Rab: there is no unbreakable reality but He: and turn aside from those who join breakable realities with Allah."

Ayat\Verses – To Him is due the primal origin of the heavens and the world: **How can He have a son when He has no consort (companion)?** He created all things, and He has full knowledge of all things. That is Allah, your Rab! There is no unbreakable reality but He, the Creator of all things: then serve you Him: and He has power to dispose of all affairs. **No vision can grasp Him**, but His grasp is over all vision: He is above all comprehension, yet is acquainted with all things. Now have come to you, from your Rab, proofs [to open your eyes]: if any will see, it will be for [the good of] his own soul; if any will be blind, it will be to his own [harm]: I am not [here] to watch over your doings. Thus do we explain the signs by various [symbols]: that they may say, You have taught [us] diligently, and that We may make the matter clear to those who have knowledge? **Follow what you are taught by inspiration from your Rab**: there is no unbreakable reality but He: **and turn aside from those who join breakable realities with Allah**. [Sura No. – 5\6 – An-aam, Ayat No. –102 to 107]

SECTION – I
HIGHLIGHTS – 1.6
AN ANT MAY BE A WARNER

Fasta-'iz billaahi minash-Shaytaanir-Rajim
Bismillaahir-Rahmaanir- Rahim

Ayat\Verses - O mankind! There has come to you a **direction** from your Rab and a **slave** for the [diseases] in your hearts, and for those who believe guidance and a Mercy. Say: In the **bounty of Allah** and in **His Mercy**, in that let them **rejoice**, that is **better than what they stored**. [Sura No. – 9\10 – Yuunus, Ayat No. – 57 to 58]

Who is Ja-alnabna Maryam?

Then We sent Muusa and his brother Haarun, with **Our Signs and a clear warrant to Firawn and his Chiefs**. But they scorned (them) and they were **dictatorial folk**. They said: Shall we believe in two men like ourselves? And their people are subject to us! So they accused them of falsehood, and they became of those who were destroyed. And We gave Muusa the Kitaab, in order that they might receive guidance. And We made **Ja-alnabna Maryam** and his mother as a Sign. We gave them refuge on a height, a place of thanks and water spring. [Sura No. – 22\23 – Mu-Minuun, Ayat\Verses – 45 to 50]

UNIVOCAL WARNING
Who is breaking Quran into parts save sceptics and hypocrites?
And say: **I am indeed he who warns openly and without ambiguity,** such as We sent down for those **who make division, those who break the Quran into parts.** Therefore, by the Rab, We will, of a surety, call them to account, for all their deeds. **Therefore expound openly what you are commanded**, and **turn away from those who join false reality with Allah [Mechanism with Manifest Truth].** [Sura No. – 14\15 = Hur, Ayat No. – 90 to 94]

RANDOM TALKING IS THE INSTRUMENT OF HYPOCRISY OR CONCEALMENT OF TRUTH AND PROJECTION OF FALSEHOOD
[Reference or Concrete Example – Preface of 'The Holy Quran' of IFTA]
And the disbelievers say: **Listen not to this Qur'an, but talk at random in the midst of its [reading], that you may gain the upper hand!** But We will certainly **cause** the **disbelievers** to **taste a severe Penalty**, and We will punish them **for the worst of their deeds**. Such is the **requital of the enemies of Allah**, the **Fire**. Therein will be for them the eternal home, **requital for as much as they denied Our Signs (Manifested Signs of Natural Magnetism).**

And the **disbelievers will say**: Our Rab! **Show us those**, among jinns and humankind, **who beguiled (misled) us. We shall crush them beneath our feet that they become among the disgusting. [Sura No. – 40\42 – Haa-Mim-Sajdah, Ayat No. 26 to 29]**

Good Tidings and Admonitions
Challenges before Researchers of IFTA and Two-in-one Partnership
To convert equal & opposite manifested signs of Natural Magnetism
in resemblance with their projected Manmade Magnetism

Say: If the whole of mankind and jinns were to gather together to produce the like of this **Qur'an [Revelation]**, they could not produce the like thereof, even if **they backed up each other with help and support**. And We have explained to man, in this **Qur'an,** every kind of **similitude (resemblance)**; yet the greater part of men refuse [to receive it] except with **ingratitude!** [Sura No. – 16\17 – Banii-Israai-iil, Ayat No. – 88 to 89]

FAITH AND BELIEF [SOLIDARITY RIGHTS]

This **Qur'an** is not such as **can be produced by other than Allah**; on the **contrary** it is a **confirmation** of [Manifested Nature] that went before it, and a **fuller explanation** of that which is **decreed** for mankind wherein there is **no doubt** from the **Owner** of the universe. **Or do they say He has forged it? Say: Bring then a Sura like unto it, and call [to your aid] anyone you can besides Allah, if you are truthful.** Nay, but they **denied** that whose knowledge they cannot compass, and where of the **interpretation** (in events) has **not yet come unto them. Thus did those before them make charges of falsehood:** but see **what the end** of those who did wrong was? Of them there are some who believe therein, and some who do not: and your Rab knows best those who are out for mischief. **If they charge you with falsehood, say: My work to me, and yours to you. You are free from responsibility for what I do, and I for what you do.** [Sura No. – 9\10 – Yuunus, Ayat No. – 37 to 41]

Never be then of those who doubt.

To such [deceit] let the hearts of those incline, **who have no faith in the hereafter**: let them **delight** in it, and let them **learn** from it what they may. **Say: Shall I seek for a judge other than Allah?** When He it is Who has sent unto you **the Kitab, explained in detail.** They know full well, to whom We have given the Kitab that it has been sent down from your Rab in truth. **Never be then of those who doubt. The word** of your Rab **finds its fulfilment in truth** and in **justice: None can change His words:** for He is the one who hears and knows all. [Sura No. – 5\6 – An-aam, Ayat No. –114 to 116]

WHAT DOES A SCEPTIC EARN?
Would any of you like to eat the flesh of his dead brother?

O you who believe! **Avoid suspicion** as much [as possible]; for suspicion in some cases is a **sin**. And **spy not** on each other behind their backs. Would any of you like **to eat the flesh of his dead brother**? Nay, you would abhor it (so abhor the other) and **keep your duty** to Allah. For Allah is Ever Returning, Most Merciful. [Sura No. – 48\49 - Minw-waraaa-il-Hujuraat, Ayat No. – 12]

"Let him stretch out a rope up to the roof (of his dwelling) and let him hang himself."

Verily Allah will admit those who believe and do tabligh (work of righteous deeds) to Gardens, beneath which rivers flow. For Allah carries out all that He plans. **If any think that Allah will not help him in this world and the Hereafter, let him stretch out a rope up to the roof (of his dwelling) and let him hang himself. Then let him see whether his plan will remove that which enrages [him]!** Thus We have **revealed it** as **Clear Signs (manifested truth)**; and verily Allah guides whom He **will**! [Sura No. – 21\22 – [Hajj], Ayat No. – 14 to 16]

FLEE NOT BUT RETURN TO THE GOOD THINGS
We have revealed for you [O men!] a Kitaab in which is a Message for you: **Have you then no sense?** How many were the communities We utterly destroyed because of their **iniquities**, setting up in their places other folks? Yet, when they felt **Our Punishment** [coming], behold, they [tried to] flee from it. **Flee not, but return to the good things of this life** which emasculated you, and to your **dwellings** in order that you may be called to account. They cried: Ah! Woe to us! We were indeed **wrong-doers**! And that cry of theirs ceased **not**, till We made them as a field that is **mown**, as **ashes** silent and **quenched**. [Sura No. – 20\21 – Ambiyaaa, Ayat No. – 10 to 15]

To those who **do good** is the best (reward) and more will be there. **Neither darkness nor shame shall cover their faces.** They are **rightful owners of the Garden**; they will abide therein. But those who have **earned evil** will have a reward of like **evil: ignominy** will cover their [faces]: **No defender will they have from [the wrath of] Allah**: Their faces will be **covered**, as it were, with pieces from the **depth of the darkness of night**: they are **companions of the Fire**: they will abide therein. One day when We shall gather them all together, then We say to those who joined **breakable reality or partners** [with Us], stand back to your place! You and those you joined as **'partners'**. We shall separate them, and their **'Partners'** shall say: It was not us that you worshipped.

70

Enough is Allah for a witness between us and you, that we were unaware of your worship. There will **every soul experience the fruits of the deeds it sent before**: they will be brought back to Allah their rightful Rab, and their **invented falsehoods** will leave them in the rock. [Sura No. – 9\10 – Yuunus, Ayat No.-26 to 30]

AN ANT MAY BE A WARNER

And there were gathered together unto **Sulaymaan** his armies of the **jinns** and **men** and **birds**, and they were all set in a battle order. Till when they reached the **valley of the ants**, **an ant exclaimed**: O ants! **Enter your dwellings** lest Sulaymaan and his armies crush you, unperceiving. And (Sulaymaan) smiled, amused at her speech; and he said: O my Rab! **Arouse me** that **I may be grateful** for Your favours, which You had bestowed on me and on my parents, and that I may work the righteousness (Tabligh) that will please You. And **admit me**, by Your Grace, to the ranks of Your **righteous Servants**. [Sura No. – 26\27 – Namal, Ayat No. – 17 to 19]

Glorified be He Who created His servant by night **from the Inviolable Place of Worship to the far distant place of worship the neighbourhood whereof We have blessed that We might show him of (Our Signs)**! Lo! He, only He, is the Hearer, the Seer. **[Sura No. – 16\17 - Banii-Israai-iil, Ayat No. – 1]**

"I shall show you My Signs, but ask Me not to hasten."
Man is a creature of haste. **I shall show you My Signs**, but **ask Me not to hasten.** They say: **When will this promise come to pass, if you are telling the truth?** If only the disbelievers knew [the time] when, they will **not** be able to **ward off the fire from their faces, nor yet from their backs,** and [when] **no help can reach them!** Nay, **it may come to them all of a sudden** and **confound** them: **no power will they have then to avert it, nor will they [then] get respite.** Mocked were [many] **messengers** before you. But their scoffers were hemmed in by the thing that they mocked. [Sura No. – 20\21 – Ambiyaaa, Ayat No. – 37 to 41]

He is the One Who sends to His slave **Manifest Signs,** that **he may lead you [guide] from the depths of darkness into the light** and verily for you Allah is full of Pity, Merciful. [Sura No. – 56\57 - Anzalnal-Hadiid, Ayat No. – 9]

One House and a Sign

They said: **Even so has said your Rab; and He is full of Wisdom and Knowledge.** [Ibrahim] said: And (afterward) what is your duty, O you sent (from Allah)? They said: Lo! We are sent to a **guilty folk** that we may send upon them **stones of clay**, marked as from your Rab for those **who trespass beyond bounds**. Then We **brought forth** such believers who were there. But **We found not there any just person** except **in one house** and **We left there a Sign** for such as fear the **Grievous** Penalty. [Sura No. – 50\51 – Waz-Zaariyaat, Ayat No. – 30 to 37]

"Qul innii nuuhiitu an a'-budallaziina tad-uuna **min-duunillaahi** lammaa **jaa-anil-yal-bayyinaatu mir-Rabbii**, wa umirtu an uslima li-Rabbil-Aalamiin." [Trans.] I am **forbidden to invoke [mechanical globalisation, manmade magnetism, circular rotation and revolution system i.e. Manmade Nature] those whom you cry besides Allah** since there have come unto me **clear signs [clear proofs of equal & opposite right direction] from my Rab**; and I am commanded to surrender [completely] to the Rab of the Universe [Alamiin]. [Sura No. – 39\40 – Mu-Min, Ayat No. – 66]

ONE OF THE GREATER SIGNS OF MY RAB
The heart **in no way falsified that which he saw. Will you then dispute with him concerning what he saw**? And indeed he **saw** Him yet **another time**, by the **lot-tree** (Sidratil) of the outmost boundary near to which is the Garden of Abode. Behold, the **Lot-tree** was covered **[in mystery unspeakable!]** The **eye turned not aside, nor yet** was overbold (did it go wrong); for **truly he saw one of the greater Signs of his Rab.** [Sura No. – 52\53 – Wan-Najm, Ayat No. – 11 to 18]

A DISCIPLE OF ALLAH, A LIGHT AND A PLAIN KITAAB
Praise be to Allah, Who has revealed the Kitaab unto His slave (servant), and has placed therein **no crookedness or global concept. [He has made it] clearly and distinctly** in order that He may **warn** of a terrible **punishment** from Him, and that He may give **good tidings** to the **believers** who work righteous deeds [tabligh], that they will have a fair **reward;** wherein they will abide forever. Further, that He may **warn those** [also] who say: Allah **has chosen a son. No knowledge they have of such a thing, nor had their fathers. Dreadful** is the **word** that comes out of their mouth. **What they say is nothing but falsehood!** Yet it may be trouble yourself to death, following after them, in grief, if they believe not in this Message. That which is on earth **we have made but as a glittering show for the world [earth]**, in order that We may test them - as to which of them are best in conduct. [Sura No. – 17\18 – Khaf – Ayat No. – 1 to 7]

O People of the Kitaab [Yaaa-Ahlal-Kitaabh]! Now has come to you our **disciple**, revealing to you much that you used to hide in the Kitaab, and **passing over** much. There has come to you from Allah a **light and a plain**

Kitaab [Kitabbum-Mubin]. **Wherewith Allah guides all who seek His good pleasure to ways of peace and safety, and leads them out of darkness, by His Will, unto the light, -- guides them to a Path that is Right.** [Sura No. – 4\5 – Maaaidah, Ayat\Verses – 15 & 16]

O People of the Kitaab [Yaaa-Ahlal-Kitaabh]! Now has come unto you, making (things) clear unto you, our **disciple, after an interval of the Messengers,** lest you should say: There came unto us **no bringer of good tidings** and **no warner** (from evil): But now has come unto you **a bringer of good tidings and a warner** (from evil). And Allah has power over all things. [Sura No. – 4\5 – Maaaidah, Ayat\Verses – 19]

And when I **inspired** the **disciple** (saying): Believe in Me and in My Messenger: They said: We **believe. Bear witness** then we have **surrendered** (unto You). When the **disciple** said: Is your Rab able to **send down** for us **a table spread with food from heaven [West]**? He said: **Observe your duty to Allah if you are true believers.** [They said]: We wish to eat thereof and satisfy our hearts, and to know that **you have indeed told us the truth**; and that we ourselves may be **witnesses.** [Sura No. – 4\5 – Maaaidah, Ayat\Verses – 111 to 113]

NO NEW REMINDER
**"Say: My Rab knows [every] word [spoken]
in the heavens and in the world."**
[Turn away, wrong-doers, private counsels]

Closer and closer to mankind comes their **Reckoning**; yet they heed **not** and they **turn away.** Never comes [aught] to them of **a new reminder** from their Rab, but they listen to it as in **jest.** Their hearts are toying as with trifles. The **wrong-doers conceal** their **private counsels,** [saying]: **Is this other than mortal like you? Will you go to witchcraft with your eyes open?** Say: My Rab knows [every] **word** [spoken] in the heavens [West Horizon] and in the world [East Horizon]. He is the Hearer, the Knower. [Sura No. – 20\21 – Ambiyaaa, Ayat No. – 1 to 4]

"Say: It is not for me, to change it of my own accord. I only follow that which is inspired in me."

Then We appointed you **heirs** on the land [earth] after them, **to see** how you would behave? But when **Our Clear Signs** are **rehearsed** unto them, those who expect **not** their hope on their meeting with Us. **Say: Bring us a reading other than this, or change this.** Say: It is not for me, to change it of my own accord. I only follow that which is inspired in me. If I were to disobey my Rab,

I should myself fear the penalty of a Great Day [Azaaba Yaw-min 'Aziim] [to come]. [Sura No. – 9\10 – Yuunus, Ayat No. – 14 to 15]

"We have given the same name to none before (him)."

[His prayer was answered]: O Zakariya! We give you good news of a **son**: His name shall be **Yahya. We have given the same name to none before (him).** He said: O my Rab! **How shall I have a son, when my wife is barren and I have grown quite feeble from old age**? He said: So [it will be] the sayings of your Rab. It is easy for Me, even as I have created you before, **when you had been nothing**! [Zakariya] said: O my Rab! Give me a Sign (Token). Your **Sign or Token** is that you with no bodily defect shall **not speak unto mankind three nights. So Zakariya came out to his people from the sanctuary and signified to them Glorify your Rab at break of the day and fall of night.** [Sura No. – 18\19 – Maryam, Ayat\Verses – 7 to 11]

"Who does more wrong than he who invents a lie against Allah, or denies His Signs (Manifested Sign of Natural Magnetism)?"

Say: If Allah had so willed, I should **not** have **rehearsed** it to you, **nor** would He have made it known to you. I **dwell** among you a whole life-time before it. Have you then **no sense**? Who does more wrong than he who invents a lie against Allah, or denies His Signs (Manifested Nature)? But never will prosper those who sin. **They serve, besides Allah, things that neither hurt them nor profit them, and they say:** These are our appeals with Allah. Say: **Do you indeed inform Allah of something He knows not, in the heavens or in the world?** Glory to Him! And far is He above the partners that **you associate** [with Him]. [Sura No. – 9\10 – Yuunus, Ayat No.16 to 18]

"After truth, what is there except error?"
"How then are you turned away?"

Say: Who is it that **sustains** you [in life] from the **sky** and from the **earth**? Or who is it that has power over **hearing and sight**? And who is it that **brings** out the living from the dead and the dead from the living? And who is it that rules and regulates all affairs? They will soon say, "Allah". Say, will you then not keep your duty [unto Him]? Such is Allah, your real Cherisher and Sustainer. **After truth, what is there except error? How then are you turned away?** Thus the word of your Rab is proved true against those who do wrong that they will not believe. [Sura No. – 9\10 – Yuunus, Ayat No. – 31 to 33]

Say: No reward do I ask of you. It is [all] in your interest. **My reward is only due from Allah**: And He is **witness** to all things. Say: Verily my Rab will cast the **Truth**. (He is) the Knower of the things **that is hidden**. Say: The **Truth** has arrived, and **Falsehood neither creates anything new, nor restores anything**. Say: If I am **astray**, I only **stray to the loss of my own soul**. But if I **receive guidance**, it is because of the **inspiration of my Rab to me**. It is He Who **hears** all things, and is [ever] **near**. [Sura No. – 33\34 – Saba, Ayat No. – 47 to 50]

O you who believe! **Stand out firmly for justice, as witnesses to Allah, even as against yourselves, or your parents, or your kin, and whether it (the case be of) a rich man or a poor man, for Allah is nearer to both (than you are). So follow not the lusts [of your hearts], lest you swerve, and if you distort [justice] or decline to do justice,** verily Allah is well-informed of all that you do. [Sura No. – 3\4 – Nisaa, Ayat No. – 135]

OWNER OF TAUTOLOGIES
(i) The decision rests with none but Allah.
(ii) He [Allah] declares the truth.
(iii) He [Allah] is the best of judges.
(iv) He is the sole Mystic.

Say: For me, I am [**relying**] on a **Clear Sign** (Proof) **from my Rab, while you reject (deny) Him. I have not that for which you are impatient.** The **decision rests with none but Allah. He declares the truth** and **He is the best of judges.** [Sura No. – 5\6 – An-aam, Ayat No. –57]

Say: The [Qur'an] was revealed by Him **who knows the mystery** of the heavens [West Horizon] and the world [East Horizon]. Verily He is Ever Forgiving, Most Merciful. [Sura No. – 24\25 - Nazzalal-Furqaan, Ayat No. – 6]

To **Allah** belong the **Mystery** of the heavens [**West Horizon**] and the world [**East Horizon**]. And the Decision of the Hour [of Judgment] is as the twinkling of an eye, or even quicker: for Allah has power over all things. It is He Who brought you forth from the wombs of your mothers when you knew nothing; and He gave you hearing and sight and intelligence and affections that you may give thanks [to Allah]. **Have they not seen the birds obedient in mid-air?** None holds them save Allah. Lo! Herein, verily are **signs** for those who believe. [Sura No. – 15\16 – Nahl, Ayat\Verses – 77 to 79]

"I am no bringer of new-thing / doctrine among the messengers,
nor do I know what will be done with me or with you."
"I am but a plain & clear warner."
"Believe you if it is from Allah."

Or they will say: He has forged (invented) it. Say: If I have forged (invented) it, still you have no power to support me against Allah. He is best aware of what you say among yourselves concerning it. Enough is He for a witness between me and you! And he is Ever Forgiving, Most Merciful. Say: **I am no bringer of new-thing / doctrine among the messengers, nor do I know what will be done with me or with you.** I do follow that which is inspired in me; and I am but a **plain & clear** warner. **Believe you if it is from Allah.** And you disbelieve therein and a witness of the Bani-Israa-iil has already testified to its similarity (to the like thereof) and has believed, and you are too proud (how unjust you are!). Lo! Allah guides not the wrongdoing folk. [Sura No. – 45\46 – Bil-ahqaafi, Ayat No. – 8 to 10]

With power and skill **We** have constructed the firmament [windows]; for it is **We** Who create the vastness of **rapidity [succession]**. And We have spread out the world; How excellently We do spread out! And **all things We have created by pairs** [equal & opposite] that you may **receive instruction**. Hasten you then [at once] to Allah. **I am a plain warner to you from Him.** [Sura No. – 50\51 – Waz-Zaariyaat, Ayat No. – 47 to 50]

And **set not another object of worship** along with Allah [Manmade Magnetism instead of Natural Magnetism]. **I am a plain warner to you from Him.** Similarly, **no messenger** came to the Peoples before them, but they [the guilty folk] said [of him]: **A sorcerer or a madman.** [Sura No. – 50\51 – Waz-Zaariyaat, Ayat No. – 51 & 52]

"And say: I am indeed he who warns openly and without ambiguity."
We are to search out - Who are breaking Quran [Truth] into Parts?
And say: **I am indeed he who warns openly and without ambiguity.** Such as We sent down for those **who make division, those who break the Quran into parts.** Therefore, by the Rab, We will, of a surety, **call them to account**, for all their deeds. **Therefore expound openly what you are commanded**, and **turn away from those who join breakable reality [falsehood] with Allah.** [Sura No. – 14\15 = Hur, Ayat No. – 90 to 94]

"My reward is only due from Allah."
Say: **No reward do I ask of you**. It is [all] in your interest. My reward is only due from Allah: And He is witness to all things. [Sura No. – 33\34 – Saba, Ayat No. - 47]

"Loyal [sincere lovers of Truth], and the martyrs are with their Rab."
And those who believe in Allah and His messengers, they are the **loyal [sincere lovers of Truth]**, and **the martyrs are with their Rab**. They have their reward

and their **light. But those who reject Allah and deny Our Signs, they are the Companions of Hell-Fire.** [Sura No. – 56\57 - Anzalnal-Hadiid, Ayat No. – 19]

"Shall I inform you good tidings of things far better than those?"

Say: Shall I inform you good tidings of things far better than those? For the **righteous** are **Gardens [West Horizon]** in **nearness** to their Lord, with rivers flowing beneath; therein is their **eternal home**; with **companions pure** [and holy]; and the **good pleasure of Allah,** for in Allah's sight are [all] His bondmen. [Namely], those who say: Our Rab! We indeed believe, so, forgive us our sins, and guard us from the punishment of fire. Those who show patience, firmness and self-control; who are true [in word and deed]; who serve devoutly; who spend [in the way of Allah]; and who pray for forgiveness in the **early hours of the morning.** Allah (Himself) is witness that **there is no ilaaha** (unbreakable reality) save Him. And His angels and those endowed with knowledge, standing **firm on justice. There is no ilaaha** (unbreakable reality) save Him, the Almighty, the Wise. **[Sura No, 2\3 – Imran, Ayat No. –15 to 18]**

"Is it then they who will win?"

Say: Who can keep you safe by night and by day from [the Wrath of] [Allah] Most Gracious? Yet they **turn away** from the mention of their Rab. Or have they absolutes that can guard them from Us? They have no power **to aid themselves, nor** can they be defended from Us. Nay, We gave the good things of this life to these men and their fathers **until the period grew long for them. They do not see how We visit the land reducing it of its outlying borders.** Is it then they who will win? [Sura No. – 20\21 – Ambiyaaa, Ayat No. – 42 to 44]

"Say: I do but warn you in resemblance with revelations."

Say: I do but **warn** you **in resemblance with revelations.** But the deaf will not hear the call, [even] when they are warned! If but a breath of the wrath of your Rab do touch them, they will then say: Woe to us! We did **wrong indeed!** [Sura No. -20\21- Ambiyaaa, Ayat No. – 45 & 46]

"Say: I have no power over any harm or profit to me except as Allah wills."

To every people / nation [was sent] a messenger. When their messenger comes [before them] the matter will be judged between them with **justice**, and they will not be **wronged.** They say: When will this **promise** come to pass, **if you speak the truth? Say: I have no power over any harm or profit to me except as Allah wills.** To every people is a **term appointed.** When their term comes, then **they cannot put it off an hour, nor haste** (it). Say: Do you see, if His punishment should come to you by night or by day, what portion of it would the sinners wish to hasten? Would you then believe in it at last, when it actually comes to pass? [It will then be said]: 'Ah! Now? And you wanted [afore time] to hasten it on. At length will be wrong-doers; taste you the enduring punishment! You get but the recompense of what you used to earn! [Sura No. – 9\10 – Yuunus, Ayat No. – 47 to 52]

WE ARE INSPIRED TO SAY 'ANY-YASHAAA-ALLAH' [IF ALLAH WILLS]

And **say not of anything**: I shall be sure to do so and so tomorrow, except **'If Allah wills [Any-yashaaa-Allah]**!' And **call your Rab to mind** when you forget and say: I hope that **my Rab will guide me ever closer** [even] than this to the **right path** (a near way of truth). So, they stayed in their cave **three hundred years**, and [some] add **nine** [more] [Sura No. – 17\18 – Khaf – Ayat No. – 25]

SINGLE-MINDED SLAVES

And verily We sent, among them warners. Then **see** the nature of the consequence of those **who were warned** (admonished) except the **single-minded slaves** of Allah. And **Nuuh** prayed to Us, and gracious was the Hearer of his prayer; and We saved him and his people from the Great Calamity, and made his progeny the survivors, and left for him among the later folk. Peace be unto Nuuh among the nations (peoples)! Thus indeed we reward those who do right, for he was one of our **believing slaves**. Then the rest we drowned in the Flood. [Sura No. – 36\37 – Was-saaaffaati, Ayat No. – 72 to 82]

Glorified be Allah from that which they ascribe [to Him]! Except the **single-minded slaves of Allah** / (Not so do the Servants of Allah, sincere and devoted). For, verily, neither you nor those you worship can lead [any] into temptation concerning Allah, except such as are [themselves] going to the **blazing Fire!** And there is not one of us but has a place appointed, and we are verily ranged in ranks [for service]; and we are verily those who hymn His praise. And there were those who said: If we had but a reminder from the men of old, **we would be single minded slaves of Allah**. [Sura No. – 36\37 – Was-saaaffaati, Ayat No. – 159 to 169]

They seek to be informed by you: **Is that true**? Say: Yes! By my Rab! **It is the very truth!** And you cannot **frustrate** it! Every soul that has sinned, if it possessed all that is on earth, would fain give it in ransom. They would declare [their] repentance when they see the penalty. But the judgment between them will be with justice, and no wrong will be done unto them. Is it not [the case] that to Allah belongs whatever is in the heavens and in the world? Is it not [the case] that **Allah's promise is assuredly true**? Yet most of them **understand**

not. It is He Who gives life and who takes it, and to Him shall you all be brought back. [Sura No. – 9\10 – Yuunus, Ayat No. – 53 to 56]

NINETEEN – 19 [ONE NINE]

Soon I will cast him into **Hell-Fire**! Ah, what will explain to you what Hell-Fire is! It leaves none, it spears none! It dried-up man! Above it are **Nineteen**. [Sura No. – 73\74 – Yaaa-ayyuhal – Muddassir, Ayat No. – 26 to 30]

ONE POINT ONLY

Say: I do admonish you on **one point only**: **That you awake for the sake of Allah, by twos [top stair & middle stair] or singly [ground stair], and you reflect** (within yourselves / give thought). **There is no madness in your friend**. He is no less than a **warner** to you, in face of a **terrible Penalty.** [Sura No. – 33\34 – Saba, Ayat No. – 46]

A PRAYER

Our Rab! **Cause not our hearts to go astray after** You have **guided us**, and **bestow upon us mercy** from Your Presence. Lo! Only You are the Bestower. Our Rab! It is You Who will gather mankind together to a **Day** of which there is **no doubt**. Lo! Allah fails not to keep the assignation (promise). **[Sura No. - 2\3 – Imran, Ayat No. – 8 & 9]**

ALLAH IS THE BEST TO DECIDE

Say: O you men! **Now Truth has reached you from your Rab**. Those **who receive guidance do so for the good of their own souls; those who stray, do so to their own loss, and I am not [set] over you to arrange your affairs.** Follow you the **inspiration** sent unto you, and be patient and constant, till **Allah decides,** for He is the best to decide. [Sura No. – 9\10 – Yuunus, Ayat No. – 108 & 109]

SECTION – I
HIGHLIGHTS – 1.7
SEARCH FOR MANIFEST TRUTHS

Fabi-ayyi aalaaa – i Rabbikumaa tukazzibaan?
Which is it of the **favours [manifest truths /
manifested signs]** of your Rab that you **deny?**
**THIRTY ONE QUESTIONS ASKED CONCERNING MANIFEST
TRUTHS [MANIFESTED SGNS] IN SURA – AR-RAHMAN**

**Fasta-'iz billaahi minash-Shaytaanir-Rajim
Bismillaahir-Rahmaanir- Rahim**

Fabi-ayyi aalaaa – i Rabbikumaa tukazzibaan?
Which is it of the favours [manifested signs] of your Rab that you deny?
Let us search answers of the **Thirty One Questions** asked in Sura 'Ar-Rahman'
concerning Manifest Truths [Manifested Signs of Equal & Opposite Magnetism]
from those who have clear concepts of the language of the Kitaab with Truth as
well as Manifested Nature in resemblance with Revelations, possessors of true
knowledge, interpreters of manifest truth, and those who have the capability to
compare manifest truth on the basis of four Kitaabs [criterions of truth]. If you want
to argue with **the searched out Manifest Truths shared in the 'Kitaaba Wal-
Hikmata' or 'Manifested Nature and the Utility of One's Upright Logic',** then
try to make clear your concept concerning each question asked in Sura 'Ar-Rahman'
on the basis of equal & opposite revelation, clear signs, clear proofs, revelations,
manifested signs, signs of indication, ideographic presentation, symbolic as well
as phonetic representations, justifiable as well as verifiable by one or the other
criterion of truth in resemblance with coded shared tautologies. If you are able to
conceptualise the Manifest Truth [Right Direction of the Equal & Opposite stages
of journey of the Manifested Signs of Natural Magnetism] on the basis of Tawraat,
Injiil, and all the criterions of truth in resemblance with the Kitaab with Truth,
then there will remain nothing to argue with the selfevident truth or searched out
Manifest Truth, and your selfevident concept will become the existential import
for generalisation of **the shared findings as Solidified Solid Human Rights.**

References
**Allah has appointed (manifested) the Kaba, the Sacred House, as a
standard [criterion] for mankind,** and the **Sacred Months**, the **offerings**, and
the **garland** (Festoon): That you may know that Allah has knowledge of what is
in the heavens [West Horizon] and in the world [East Horizon] and that Allah is
well acquainted with all things. [Sura No. – 4\5 – Maaaidah, Ayat\Verses – 97]

Say: O **People of the Kitaab [Ahlal Kitaab]! You have no ground to stand upon unless you stand fast by the Tawraat, the Injiil, and all the revelations that have come to you from your Rab.** It is the revelation that comes to you from your Rab, that increases in most of them their obstinate **rebellion** and **blasphemy. But sorrow you not over (these) people without Faith.** Those who believe, those who are **Haaduu, Saabiuu-nas** and **Nasaaras,** and any who **believe in Allah** and the **Last Day** [Yawmil-Aakhiri], and **work righteousness [do tabligh],** --on them shall be **no fear, nor** shall they **grieve.** [Sura No. – 4\5 – Maaaidah, Ayat\Verses – 68 & 69]

We shall set up **scales of justification** for **the Day of Resurrection [Yawmil-Qiyaamati],** so that not a soul will be dealt with unjustly in the least, and if there be [no more than] **the weight of a mustard seed,** We will bring it [to account]: and enough We are to take account. In the past We granted to Muusa and Haarun **the criterion [for justification],** and **a Light** and **a Message for those who would do right.** Those who fear their Rab in their most secret thoughts and who hold the Hour in awe; and **this is a blessed Message which We have sent down: Will you then reject it?** [Sura No. – 20\21 – Ambiyaaa, Ayat No. – 47 to 50]

SURA - AR-RAHMAN – Question: - No. 1
Ayat\Verses - "Fabi-ayyi aalaaa – i Rabbikumaa tukazzibaan?" [Trans.] Which is it of the **favours [manifested signs]** of your Rab that you **deny**? [Sura No. 54\55, Ar-Rahman, Ayat\Verses 13]
Ayat\Verses - The Most Gracious! (The Beneficent!), it is He Who has **Revealed** (taught) the **Quran.** He has **created man.** He has **taught** him an **intelligent speech** (utterance). "**Ash-shamsu wal-qamaru** bi-husbaan" – **The Sign of day-break** and **the sign of night-fall** are made **punctual** i.e. they **follow** the **courses exactly computed.** The **Stars** and the **trees** adore. And the **skies or canopies** He has raised high (uplifted), and He has set the **balance (measure)** In order that you **may not exceed the measure** (limit) or **transgress the balance** (justice). But **observe the measure (balance) strictly,** and fall not short thereof. And **He has appointed** the **earth for (His) creatures** wherein are fruits and sheathed **palm-trees, husked grain** and **scented herb.** [Sura No. – 54\55, Ar-Rahman, Ayat\Verses – 1to 12]

Say: **Disbelieve** you verily in Him **Who created the world in two days,** and **ascribe** you unto Him **rivals**? He is the Rab of the universe. He set in the [world], mountains **standing firm,** high above it, and bestowed blessings in the world, and **measure therein all things to give them nourishment in due proportion, in four Days,** in accordance with [the needs of] those who seek [Sustenance]. Then **He turned to the heavens [West Horizon] when it was** smoke [white dwarf], and said unto it and unto the earth: **Come both of you, willingly or unwillingly**. They said: We come, obedient. So He **ordained** them as **seven firmaments (canopies) in two Days**, and **He inspired in each heaven its mandate [Seven Windows]**. And **We adorned the lower heaven with lights** and [**provided it**] **with guard**. Such is the Measuring of the Mighty, The Knower. [**Sura No. – 40\42 – Haa-Mim-Sajdah, Ayat No. – 9 to 12**]

Alif-Lam-Mim-Ra. These are the Clear Signs of the Kitaab. That which has been revealed to you from your Rab is the Truth (Tautology). **But most men believe not. Allah is He Who raised the heavens without any pillars that you can see; is firmly established on the throne [Arsh]. He has compelled sakh-kharash-shamsa wal-qamar to be of service. Each one runs [its course] for a term appointed. He regulates all affairs, explaining the signs in detail that you may believe with certainty in the meeting with your Rab.** And it is He who **spread out the Earth**, and **set thereon mountains standing firm** and [**flowing**] **rivers**: and fruit of every kind He **made in pairs, two and two**: He draws the **night as a veil** over the day. Behold, verily in these things **there are signs for those who give thought!** And **in the earth** are **tracts [stages] neighbouring**, and **gardens of vines** and **fields sown with corn**, and **palm trees - growing out of single roots** or otherwise (**like and unlike**): **watered with the same water**, yet some of them We make **more excellent than others** to eat. Behold, verily in these things **there are signs for those who understand!** [Sura No. – 12\13 – Rad, Ayat No. – 1 to 4]

SURA - AR-RAHMAN – Question: - No. 2
CREATION OF MAN AND JINN
[Manifested signs of clay and fire]
Ayat\Verses - "Fabi-ayyi aalaaa – i Rabbikumaa tukazzibaan?" [Trans.] Which is it of the **favours [manifested signs]** of your Rab that you **deny**? [Sura No. –54\55, Ar-Rahman, Ayat\Verses 16]
Ayat\Verses - He has **created man from sounding clay** (potter's clay) and He has **created jinn from smokeless fire**. [Sura No. 54\55, Ar-Rahman, Ayat\ Verses.14 & 15]
Signs of Allah – Sounding clay, smokeless fire

SURA - AR-RAHMAN – Question: - No. 3
Ayat\Verses - "Fabi-ayyi aalaaa – i Rabbikumaa tukazzibaan?" [Trans.] Which is it of the **favours [manifested signs]** of your Rab that you **deny**? [Sura No. 54\55, Ar-Rahman, Ayat\Verses 18]
Ayat\Verses - "Rabbul-Mashri-qayni wa Rabbul-Magribayn" Rab of the **two Easts** and Rab of the **two Wests**. [Sura No. – 54\55, Ar-Rahman, Ayat\ Verses – 17]

TRY TO CONFIRM MANIFESTATIONS
IN RESEMBLANCE WITH THE REVELATIONS
Where are the two East and two West in projected Globalisation?
"You take Him alone for disposer of affairs."
"Rabbul-Mashriqi wal-Magribi Laaa ilaaha illa Huwa fattakiz-hu Wakiilaa"
Ayat\Verses – **Rab of the East and the West**. There is no unbreakable reality except Him. Therefore, you take Him alone for disposer of affairs. [Sura No. 72\73 – Yaaa-ayyuhal Muzzamil, Ayat No. - 9]

Ayat\Verses – And We made a people, considered weak [and of no account], inheritors of in **east zone** [lower part of the land] and **west zone** [upper part of the land], **lands** [middle parts] whereon We sent down Our blessings. The fair promise of your Rab was fulfilled for the Baniii - Israaa -il, because they had patience and constancy, and We levelled to the ground the great works and fine buildings which Firawn and his people erected [with such pride]. [Sura No. – 6\7 - As-haabul- a' raaf, Ayat No. – 137]

DECLINING TO THE RIGHT, TURNING
AWAY FROM THEM TO THE LEFT
FROM NORTH TO SOUTH AND FROM THAT TO THE END POINT
Ayat\Verses – "Wa **tarash-shamsa** izaa tala-at-lazaawaru an-kahfihim zaatal-yamimi wa izaa gara-at-taqri-zuhum zaatash-shimaali wa hum fi fajwatim-minh. Zaalikka min aayaatillaah; many-yahdillaaahu fahuwal-muhtad; wa many-yuzlil falan-tajida lahuu waliyyam-murshidaa". [Trans.] And might have seen wa tarash-shamsa, **when it ross, declining to the right** [semi-anticlockwise] from their Cave, and **when it sets, turning away from them to the left** [semi-clockwise], **while they lay in the open space in the midst of the Cave.** Such equal and opposite i.e. half-anti-clockwise and half-clockwise circulation of Allah's glittering show for the world **are among the Signs of Allah**. He whom Allah, guides is rightly guided; but he whom Allah **leaves to stray**, for him you will **not find** protector to lead him to **the Right Way**. [Sura No. – 17\18 – Khaf – Ayat No. – 17]

Niche [Binary Thread] is kindled neither
from the East nor from the West
"**Allaahu Nuurus-samaa-waati wal-arz** – Masalu **Nuu-rihi** ka **Mishkaatin-**fiihaa **Mis-baah** – Al-**Misbaahu** fii **Zujaa-jah** – azzuujaajatu ka-annahaa – kawkabun durriyyuny-yuuqa-du – min Shajaratimmubaara-katin-**Zaytuunatil-**laa **Sharqiy-**yatinw wa laa **Garbiyyatiny-**yakaadu **Zaytuhaa** yuziii- u wa law lam tamsas-hu naar. **Nuu-run alaa Nuur! Yahdillaahu** li-**Nuurihii** many-yashaaa – wa-yazribullaahul-amsaala linnaas – wallahu bi-kulli shay-in Alim."

83

[Sura No. – 23\24 – Nuur, Ayat No. – 35] **[Trans.] Allah** is the **Light** of the **heavens** and the **world**. The **similitude** (likeness) of **His Light** is as a **niche** (binary thread of a candle) **in a lamp [world]**. The **lamp [wax of the candle]** is in a **glass [Diamond Operator]**. The **glass** is as it is a **shining star**. (This niche is) kindled from a **blessed tree**, an **olive** (Zaytuun tree), **neither from the East nor from the West**, whose (niche's) oil [luminosity] will almost glow forth (of itself) though **no fire touches** it. **Light upon Light!** Allah guides unto His light whom He wills. And Allah speaks to mankind in **parables**, for Allah is Knower of all things. [Sura No. – 23\24 – Nuur, Ayat No. – 35]

SURA - AR-RAHMAN – Question: - No. 4

Ayat\Verses - "Fabi-ayyi aalaaa – i Rabbikumaa tukazzibaan?" [Trans.] Which is it of the **favours [manifested signs]** of your Rab that you **deny**? [Sura No. – 54\55, Ar-Rahman, Ayat\Verses – 21]

Ayat\Verses - "Marajal-bah-rayni yal-taqiyaan" He has let free the **two seas** meeting together. "Baynahumaa Barazakhul-laa yabgi-yaan" There is a **barrier between them**. They **encroach not** (one upon the other) [Sura No. – 54\55, Ar-Rahman, Ayat\Verses – 19 & 20]

Key Concepts
- i)   Let free the **two seas** meeting together
- ii) A **barrier between them**
- iii) **They encroach not**

MANIFESTATION – THE JUNCTIONS OF THE TWO SEASHORES

Ayat\erses - But your Rab is Most forgiving, full of Mercy. If He were to call them [at once] to account for what they have earned, then surely He would have hastened their punishment; but they have **their appointed time**, beyond which they will find no refuge. Such were the townships [east zones] we destroyed when they **committed iniquities**; but we fixed an **appointed time** for their **destruction**. And when, Muusa said to his attendant: I will not give up until I reach **the junction of the two seashores**, though I march on for ages. But **when they reached the Junction, they forgot [about] their Fish, which took its course through the sea [straight] as in a tunnel**. [Sura No. 17\18 – Khaf – Ayat No. – 58 to 61]

SURA - AR-RAHMAN – Question: - No. 5

Ayat\Verses - "Fabi-ayyi aalaaa – i Rabbikumaa tukazzibaan?" [Trans.] Which is it of the **favours [manifested signs]** of your Rab that you **deny**? [Sura No. 54\55, Ar-Rahman, Ayat\Verses 23]

Ayat\Verses - "Yakhruju **minhumal-Lu'lu-u** wal-**Marjaan." From both of them come Pearls (?) and Coral (?).** [Sura No. – 54\55, Ar-Rahman, Ayat\ Verses – 22]

SURA - AR-RAHMAN – Question: - No. 6

Ayat\Verses - "Fabi-ayyi aalaaa – i Rabbikumaa tukazzibaan?" [Trans.] Which is it of the **favours [manifested signs]** of your Rab that you **deny**? [Sura No. 54\55, Ar-Rahman, Ayat\Verses 25]

Ayat\Verses - And His are the **ships displayed upon the seas**, like **banners**. (Or) And His are the **ships sailing smoothly** through the seas, lofty as **mountains**. [Sura No. – 54\55, Ar-Rahman, Ayat\Verses – 24]

SURA - AR-RAHMAN – Question: - No. 7

Ayat\Verses - "Fabi-ayyi aalaaa – i Rabbikumaa tukazzibaan?" [Trans.] Which is it of the **favours [manifested signs]** of your Rab that you **deny**? [Sura No. 54\55, Ar-Rahman, Ayat\Verses 28]

Ayat\Verses - "Kullu man alay-haa faan." **All** that is in the world will **perish**. There **remains nothing** except the **countenance** of your Rab, full of **Majesty**, **Bounty** and **Honour**. [Sura No. – 54\55, Ar-Rahman, Ayat\Verses – 26 & 27]

SURA - AR-RAHMAN – Question: - No. 8

Ayat\Verses - "Fabi-ayyi aalaaa – i Rabbikumaa tukazzibaan?" [Trans.] Which is it of the **favours [manifested signs]** of your Rab that you **deny**? [Sura No. 54\55, Ar-Rahman, Ayat\Verses 30]

Ayat\Verses - **All** that are in the **heavens** and the **world entreat** Him. **Every day He exercises power.** (or) Of Him **seeks** every creature (its need) in the **heaven** and in the **world. Every day** in (**new**) **Splendour** He **Shines**. [Sura No. 54\55, Ar-Rahman, Ayat\Verses – 29]

SURA - AR-RAHMAN – Question: - No. 9

Ayat\Verses - "Fabi-ayyi aalaaa – i Rabbikumaa tukazzibaan?" [Trans.] Which is it of the **favours [manifested signs]** of your Rab that you **deny**? [Sura No. 54\55, Ar-Rahman, Ayat\Verses 32]

Ayat\Verses - "Sanafrugu lakum ayyuhas-saqalaan." We **dispose** of you **two dependents** [Sura No. – 54\55, Ar-Rahman, Ayat\Verses. 31]

SURA - AR-RAHMAN – Question: - No. 10

Ayat\Verses - "Fabi-ayyi aalaaa – i Rabbikumaa tukazzibaan?" [Trans.] Which is it of the **favours [manifested signs]** of your Rab that you **deny**? [Sura No. 54\55, Ar-Rahman, Ayat\Verses 34]

Ayat\Verses - O **company of jiin and men**, if you have **power** to penetrate the **regions** of the heavens and the world, then **penetrate. You will never penetrate them without authority.** [Sura No. – 54\55, Ar-Rahman, Ayat\ Verses – 33]

SURA - AR-RAHMAN – Question: - No. 11
Ayat\Verses - "Fabi-ayyi aalaaa – i Rabbikumaa tukazzibaan?" [Trans.] Which
is it of the **favours [manifested signs]** of your Rab that you **deny**? [Sura No.
54\55, Ar-Rahman, Ayat\Verses. 36]
Ayat\Verses - There **will be sent, against** you **both, heat of fire** and (flash of)
molten brass, and you will not escape. [Sura No. – 54\55, Ar-Rahman, Ayat\
Verses – 35]

SURA - AR-RAHMAN – Question: - No. 12
Ayat\Verses - "Fabi-ayyi aalaaa – i Rabbikumaa tukazzibaan?" [Trans.] Which
is it of the **favours [manifested signs]** of your Rab that you **deny**? [Sura No.
54\55, Ar-Rahman, Ayat\Verses 38]
Ayat\Verses - And when the **sky** is **split** asunder (apart), and becomes **rosy** like
red hide [Sura No. – 54\55, Ar-Rahman, Ayat\Verses – 37]

SURA - AR-RAHMAN – Question: - No. 13
Ayat\Verses - "Fabi-ayyi aalaaa – i Rabbikumaa tukazzibaan?" [Trans.] Which
is it of the **favours [manifested signs]** of your Rab that you **deny**? [Sura No.
54\55, Ar-Rahman, Ayat\Verses 40]
Ayat\Verses - **On that Day** neither **man** nor **jiin** will be **questioned** of his **sin.**
[Sura No. – 54\55, Ar-Rahman, Ayat\Verses – 39]

SURA - AR-RAHMAN – Question: - No. 14
Ayat\Verses - "Fabi-ayyi aalaaa – i Rabbikumaa tukazzibaan?" [Trans.] Which
is it of the **favours [manifested signs]** of your Rab that you **deny**? [Sura No.
54\55, Ar-Rahman, Ayat\Verses. 42]
Ayat\Verses - The **sinners** will be **known** by their **marks** and they will be **seized**
by their **forelocks** and their **feet.** [Sura No. – 54\55, Ar-Rahman, Ayat\Verses – 41]

SURA - AR-RAHMAN – Question: - No. 15
Ayat\Verses - "Fabi-ayyi aalaaa – i Rabbikumaa tukazzibaan?" [Trans.] Which
is it of the **favours [manifested signs]** of your Rab that you **deny**? [Sura No.
54\55, Ar-Rahman, Ayat\Verses.45]
Ayat\Verses - **This is the Hell which the sinners deny. They go circling round
between its fierce and boiling water.** [Sura No. – 54\55, Ar-Rahman, Ayat\
Verses – 43 & 44]

SURA - AR-RAHMAN – Question: - No. 16
Ayat\Verses - "Fabi-ayyi aalaaa – i Rabbikumaa tukazzibaan?" [Trans.] Which
is it of the **favours [manifested signs]** of your Rab that you **deny**? [Sura No.
54\55, Ar-Rahman, Ayat\Verses.47]

Ayat\Verses - But for **him** who **fears standing before his Rab**, there are **two gardens**. [Sura No. – 54\55, Ar-Rahman, Ayat\Verses – 46]

SURA - AR-RAHMAN – Question: - No. 17
Ayat\Verses - "Fabi-ayyi aalaaa – i Rabbikumaa tukazzibaan?" [Trans.] Which is it of the **favours [manifested signs]** of your Rab that you **deny**? [Sura No.54\55, Ar-Rahman, Ayat\Verses. 49]
Ayat\Verses - Of **spreading branches** -[Sura No.54\55, Ar-Rahman, Ayat\Verses – 48]

SURA - AR-RAHMAN – Question: - No. 18
Ayat\Verses - "Fabi-ayyi aalaaa – i Rabbikumaa tukazzibaan?" [Trans.] Which is it of the **favours [manifested signs]** of your Rab that you **deny**? [Sura No.54\55, Ar-Rahman, Ayat\Verses – 51]
Ayat\Verses - **Wherein two fountains** are **flowing** [Sura No. – 54\55, Ar-Rahman, Ayat\Verses – 50]

SURA - AR-RAHMAN – Question: - No. 19
Ayat\Verses - "Fabi-ayyi aalaaa – i Rabbikumaa tukazzibaan?" [Trans.] Which is it of the **favours [manifested signs]** of your Rab that you **deny**? [Sura No.54\55, Ar-Rahman, Ayat\Verses – 53]
Ayat\Verses - **Wherein** is **every kind of fruit in pairs** [Sura No. – 54\55, Ar-Rahman, Ayat\Verses – 52]

SURA - AR-RAHMAN – Question: - No. 20
Ayat\Verses - "Fabi-ayyi aalaaa – i Rabbikumaa tukazzibaan?" [Trans.] Which is it of the **favours [manifested signs]** of your Rab that you deny? [Sura No.54\55, Ar-Rahman, Ayat\Verses – 55]
Ayat\Verses - **Recycling** upon **couches** lined with **silk brocade**, the fruits of **both gardens near to hand**. [Sura No. – 54\55, Ar-Rahman, Ayat\Verses – 54]

SURA - AR-RAHMAN – Question: - No. 21
Ayat\Verses - "Fabi-ayyi aalaaa – i Rabbikumaa tukazzibaan?" [Trans.] Which is it of the **favours [manifested signs]** of your Rab that you **deny**? [Sura No.54\55, Ar-Rahman, Ayat\Verses – 57]
Ayat\Verses - **Therein** are those of **modest gaze**, whom **neither man nor jinn will have touched** before them. [Sura No. – 54\55, Ar-Rahman, Ayat\Verses – 56]

SURA - AR-RAHMAN – Question: - No. 22
Ayat\Verses - "Fabi-ayyi aalaaa – i Rabbikumaa tukazzibaan?" [Trans.] Which
is it of the **favours [manifested signs]** of your Rab that you **deny**? [Sura
No.54\55, Ar-Rahman, Ayat\Verses – 59]
Ayat\Verses - (In beauty) like the **jacinth (reddish orange)** and **coral (white
calcareous substance)** [Sura No. – 54\55, Ar-Rahman, Ayat\Verses – 58]

SURA - AR-RAHMAN – Question: - No. 23
Ayat\Verses - "Fabi-ayyi aalaaa – i Rabbikumaa tukazzibaan?" [Trans.] Which
is it of the **favours [manifested signs]** of your Rab that you **deny**? [Sura
No.54\55, Ar-Rahman, Ayat\Verses – 61]
Ayat\Verses - Is there any **reward** for **good other than good**? [Sura No. – 54\55,
Ar-Rahman, Ayat\Verses – 60]

SURA - AR-RAHMAN – Question: - No. 24
Ayat\Verses - "Fabi-ayyi aalaaa – i Rabbikumaa tukazzibaan?" [Trans.] Which
is it of the **favours [manifested signs]** of your Rab that you **deny**? [Sura
No.54\55, Ar-Rahman, Ayat\Verses – 63]
Ayat\Verses - And **besides them** (the above mentioned two gardens), there are
two other gardens. [Sura No. – 54\55, Ar-Rahman, Ayat\Verses – 62]

SURA - AR-RAHMAN – Question: - No. 25
Ayat\Verses - "Fabi-ayyi aalaaa – i Rabbikumaa tukazzibaan?" [Trans.] Which
is it of the **favours [manifested signs]** of your Rab that you **deny**? [Sura
No.54\55, Ar-Rahman, Ayat\Verses – 65]
Ayat\Verses - **Dark green foliage** (under growth) [Sura No. – 54\55, Ar-
Rahman, Ayat\Verses – 64]

SURA - AR-RAHMAN – Question: - No. 26
Ayat\Verses - "Fabi-ayyi aalaaa – i Rabbikumaa tukazzibaan?" [Trans.] Which
is it of the **favours [manifested signs]** of your Rab that you **deny**? [Sura
No.54\55, Ar-Rahman, Ayat\Verses – 67]
Ayat\Verses - **Wherein** is two **abundant springs,** [Sura No. – 54\55, Ar-
Rahman, Ayat\Verses – 66]

SURA - AR-RAHMAN – Question: - No. 27
Ayat\Verses - "Fabi-ayyi aalaaa – i Rabbikumaa tukazzibaan?" [Trans.] Which
is it of the **favours [manifested signs]** of your Rab that you **deny**? [Sura
No.54\55, Ar-Rahman, Ayat\Verses – 69]
Ayat\Verses - **Wherein** are **fruits**, the **palm** and **pomegranate** [Sura No. –
54\55, Ar-Rahman, Ayat\Verses – 68]

SURA - AR-RAHMAN – Question: - No. 28

Ayat\Verses - "Fabi-ayyi aalaaa – i Rabbikumaa tukazzibaan?" [Trans.] Which is it of the **favours [manifested signs]** of your Rab that you **deny**? [Sura No.54\55, Ar-Rahman, Ayat\Verses – 71]

Ayat\Verses - **Wherein** are **found** (maidens) **good** and **beautiful;** [Sura No. – 54\55, Ar-Rahman, Ayat\Verses – 70]

SURA - AR-RAHMAN – Question: - No. 29

Ayat\Verses - "Fabi-ayyi aalaaa – i Rabbikumaa tukazzibaan?" [Trans.] Which is it of the **favours [manifested signs]** of your Rab that you **deny**? [Sura No.54\55, Ar-Rahman, Ayat\Verses – 73]

Ayat\Verses - **Maidens restrained (close guarded),** in **pavilions,** [Sura No. – 54\55, Ar-Rahman, Ayat\Verses – 72]

SURA - AR-RAHMAN – Question: - No. 30

Ayat\Verses - "Fabi-ayyi aalaaa – i Rabbikumaa tukazzibaan?" [Trans.] Which is it of the **favours [manifested signs]** of your Rab that you **deny**? [Sura No.54\55, Ar-Rahman, Ayat\Verses – 75]

Ayat\Verses - **Whom neither man nor jinn will** have **touched before,** [Sura No. – 54\55, Ar-Rahman, Ayat\Verses – 74]

SURA - AR-RAHMAN – Question: - No. 31

Ayat\Verses - "Fabi-ayyi aalaaa – i Rabbikumaa tukazzibaan?" [Trans.] Which is it of the **favours [manifested signs]** of your Rab that you **deny**? [Sura No.54\55, Ar-Rahman, Ayat\Verses – 77]

Ayat\Verses - Reclining on **green** cushions and **red** carpets, [Sura No. – 54\55, Ar-Rahman, Ayat\Verses – 76]

Blessed be the name of your Rab, full of **Majesty, Bounty** and **Honour.** [Sura No. – 54\55, Ar-Rahman, Ayat\Verses.-78]

Ayat\Verses – And O my people! What ails me that I call you to **release** while **you call me to the Fire! You call me to disbelieve in Allah, and to join with Him partners of whom I have no knowledge; while I call you to the Exalted in Power, Who forgives again and again! Without doubt that you are calling me to one [sun / constructed leader] who has no claim in the world or in the hereafter; and the transgressors will be companions of the fire! Soon you will remember what I say to you. I confide my cause to Allah.** Lo! Allah [ever] watches over His slaves. [Sura No. – 39\40 – Mu-Min, Ayat No. –41to 44]

O our people! **Respond to Allah's summoner (one who invites you to Allah) and believe in him. He will forgive you your faults and deliver you from a Penalty Grievous.** And whoso responds not to Allah's summoner, **he cannot frustrate** (escape Allah's plan) in the earth, and **you (can find) no protectors (or friends)** instead of him. **Such are in manifest error.** [Sura No. – 45\46 – Bil-ahqaafi, Ayat No. – 31 to 32]

SECTION – II
METHODOLOGY – 2.1
SELF-CONTRADICTION & SELF-EVALUATION

Fasta-'iz billaahi minash-Shaytaanir-Rajim
Bismillaahir-Rahmaanir- Rahim

PUBLIC ROAD

Myself Jamir Ahmed Choudhury, a re-search scholar on 'Solidarity Rights' & sharer of searched out Solidified Solid Human Rights in the name of 'Kitaaba Wal-Hikmata' or 'Manifested Nature and the Utility of One's Upright Logic', with due respect and honest submission would like to state that I have searched out several Manifest Truths which are very much related with my re-search area i.e. Third Generation Human Rights [Solitary Rights or Solidarity Rights or Solid Human Rights]. I have compiled my searched out verifiable as well as justifiable Solitary Rights in **'KITAABA WAL-HIKMATA' OR 'MANIFESTED NATURE AND THE UTILITY OF ONE'S UPRIGHT LOGIC'** in a Logical Method called **Dictum De Omni ET Nullo** of Aristotle or **Cartesian Method** of Rene Descartes with a view to share the searched out manifest truths in a Deductive Method with each and every matured member of the Manifested Hexagonal World and Appearing Pentagonal Earth within the Black Square of East Horizon. But it is very surprising as well as shameful fact to express that I have failed to search out proper means to share searched out manifest truths with the so-called Epistemic Uniqueness of Scientific Certainty and Researchers on Truth-in-itself as well as coded shared Tautologies. There are innumerable writers & editors, publishers & publications, journalists & media persons, news papers & news channels, verifiers & justifiers, recognisers & certifiers, University Presses & Re-search Institutions, and the like. All are busy with **Mechanical Globalisation, Man-made Magnetism**, and **Circular Rotation & Revolution System [Manmade Nature]**. There is none to pay attention to the **Manifested Nature** and **Natural Magnetism** on the ground that **Manifested Nature does not fit with his/her pen – paper – pencil works and mechanical activities [Black Magic]**. But I must have to try all possible means with a view to get rid of **Subjective Selfcontradictions** as well as **Objective Paradoxes** as a human person [First Generation Human Rights], as a teacher [Second Generation Human Rights], as a follower of equal & opposite right direction towards Upright West [Third Generation Human Right] verifiable by the Crucified Sign and Equal & Opposite stages of journey of the Manifested Signs of Natural Magnetism and justifiable by Universal Major Premises and Equal & Opposite Laws of Revelation [Tawraat or Coherence Truth], and Upright Logic [Furqan or Selfevident Truth]. So, I had to take the help of **Public Roads** for the first open & public sharing of the searched out Manifest Truths. Now, I am trying to share several searched out verifiable as well as justifiable manifest truths through self-publication.

GO WITH ME – AGREE WITH ME
Subjective Selfcontradictions and Objective Paradoxes

Two contradictories cannot be true together or false together from the point of view of four fundamental categories of knowledge [space, time, substance, and causality], four cardinal directions [up, down, right, and left], and six manifested directions [North-East, Middle-East, South-East, South-West, Upright-West, North-West]. For example – the concepts of North and South directions grounded on man-made magnetism as well as the concepts of North and South directions grounded on Equal & Opposite stages of journey of the Manifested Signs of Natural Magnetism cannot be true together or false together. If the concept of man-made magnetism is true, then the concept of Natural Magnetism must be false. Conversely, if the equal & opposite [Semi-anticlockwise & Semi-clockwise] stages of journey of the so-called sun [Bullet] as the manifested sign of natural magnetism is true, then the concept of man-made magnetism must be false. To conceptualise both the contradictories either true together or false together is selefcontradiction from subjective point of view and paradox from objective point of view. Such kind of subjective selfcontradictions and objective paradoxes are not true knowledge. The purpose of teaching/learning is to share/acquire true knowledge. So, the morality of a teacher suggests to get confirmation from those who have true concepts as well as to get recognition of the true concepts from the respective authorities concerning subjective selfcontradictions and objective paradoxes. This is one of the purposes of the re-search project on 'Solidarity Rights in Islam' and self-publishing of the searched out Solidified Solid Human Rights in the name of 'Kitaaba Wal-Hikmata' or 'Manifested Nature and the Utility of One's Upright Logic'.

It has been projected in the domain of formal education that there are two kinds of magnetism, namely, Manmade Magnetism and Natural Magnetism. As there are only one Revealed Left [Northern Hemisphere / Haiyalas-Swalaah] and only one Revealed Right [Southern Hemisphere / Haiyalal-Falaah], so, there is only one kind of Magnetism. This sole Magnetism is called Natural Magnetism. The Don of Epistemic Uniqueness of Scientific Certainty or the Don of Historical Black Magic or the Don of Two-in-one Partnership has introduced manmade magnetism with a view to conceal equal & opposite manifested signs of natural magnetism, to track off believers from the equal & opposite right direction, to dominate mankind of Spiders' Net from Spiderman's Net Work. So, the **introduction of manmade magnetism within the domain of formal education is the Vera Causa of Activism and Persecution.**

Moreover, the projection of two contradictories i.e. manmade magnetism and natural magnetism as true together is nothing save Teleological Evidence Sorcery and Unique Epistemic Persecution or Black Magic. Those who are using their pen – paper – pencil in resemblance with Black Magic such as two kinds of Magnetism, they are Ignorant Evidence Sorcerers. Those who are consciously trying to project two kinds of Magnetism or subjective selfcontradictions and objective paradoxes as true knowledge with a view to conceal manifest truths they are Teleological Evidence Sorcerers and Unique Epistemic Persecutors of Faith and Belief of mankind in general on Revelations [Laws], Manifestations [Nature], Scientific Certainty [Verification], and Formal Education [Justification]. They are none but Black Magicians. The possessors of upright nature will never justify subjective selfcontradictions and objective paradoxes as true knowledge or manifest truths or manifested nature.

I am suffering from subjective selfcontradictions and objective paradoxes. With a view to get liberation, I have searched out several manifest truths free from subjective selfcontradictions and objective paradoxes. In other words, I have searched out several manifest truths which are scientifically verifiable in pragmatic correspondence with Manifested Nature [Injiil or Affirmative Minor Premise] and logically justifiable in coherence with Equal & Opposite Law [Tawraat or Revelation or Universal Major Premise]. So, the searched out manifest truths shared in the 'Kitaaba Wal-Hikmata' or 'Manifested Nature and the Utility of One's Upright Logic' are the selfevident truths [Furqan] from the point of view of re-searcher/sharer. To get confirmation through recognition of the searched out manifest truths from respective authorities are the Solidified Solid Human Rights of the re-searcher & sharer of manifest truths as a human person, as a teacher, and as a follower of equal & opposite right direction [Middle Course] towards Upright West [Arsh].

Here, I am trying to share 20 [twenty] subjective selfcontradictions and objective paradoxes in two sets. Both the sets cannot be true together or false together. The truth of one set implies the falsity of the other set, and conversely. If set – A is true, then set – B must be false. Conversely, if set – B is true, then set – A must be false. Set – A consists of what are being taught as scientific truths through formal education as well as which are available in the pen – paper – pencil works of the people of the book, people of the Kitaab and the like grounded on Suppositions, Imaginations, Predictions, Hypotheses, Fictitious Theorisations, Popular Revolutions, Mechanical Globalisation, Circular Rotation & Revolution System, Manmade Magnetism, illicit formalities & guidelines etc. In brief, Set – A is the Vera Causa of Activism and Persecution. Set – B consists of verifiable as well as justifiable Manifest Truths free from subjective selfcontradictions and objective paradoxes grounded on Manifested Signs of Natural Magnetism [Manifested Nature], Equal & Opposite Law, Separate Sheet of Real Scientific Symbols, and Upright Logic. In brief, Set – B not only represents manifested nature but also the experimentum crusis as well as crucial instances against Historical Black Magic or Teleological Evidence Sorceries and Unique Epistemic Persecutions. Now, with a view to get confirmation and recognition of true knowledge free from subjective selfcontradictions and objective paradoxes from the possessors of true knowledge and the respective authorities, two contradictory columns are being placed before you. You are the best verifiers, justifiers, certifiers, recognisers, sharers of Manifest Truths free from subjective selfcontradictions and objective paradoxes.

20 Major Findings shared in the 'Kitaaba Wal-Hikmata' or 'Manifested Nature and the Utility of One's Upright Logic'	
SET – A **TELEOLOGICAL EVIDENCE SORCERIES** & **UNIQUE EPISTEMIC PERSECUTIONS**	SET – B **MANIFESTED NATURE** & **NATURAL MAGNETISM**
Grounded on Mechanical Globalisation, Circular Rotation & Revolution, and Man-made Magnetism	Grounded on Revelation, Manifested Nature, and Four Revealed & Established Criterions of Truth
1 Universe is like the projected Globe and a Lunar System or a Solar System or a Planetary System or an Astrophysical System, and the like.	Universe is a Diamond Operator revealed as an equal & opposite Trinity and manifested as an Upright Rectangle in resemblance with the 'End of Proof' of Real Science and in correspondence with the appointed Kaba. Moreover, Universe is a Star System or Sirius Binary System.
2 The World is like the projected Globe.	East Horizon has been manifested as a Hexagon within the Black Square [Non-luminous Moon]. This manifested Hexagon is called World or Asterisk.
3 The Earth is also like the projected Globe. So, the projected Globe is a Trinity of the Universe, the World, and the Earth. This Trinity is being guided by the Mechanism of Circular Rotation & Revolution System i.e. the will of the Don of Projected Science.	Due to the ~sth between Upright [Top] West Region of Arabian Peninsula and Straight Middle East Region of Eartha 3D, the Manifested Hexagonal World is appearing as a Pentagonal Star. This Pentagonal Star or Star Operator is called the Earth. So, the appearing Earth is a Pentagon in correspondence with 9/11 Pentagon, Spiders' Net, House, Pentagonal Star, and Star Operator.

		The Universe is an Upright Rectangle, the World is a Black Square manifested as a Hexagon within the East Horizon, and the appearing Earth is a Pentagon.
4	The projected Global Trinity of the Universe – the World – the Earth has no ascending stairs.	There is no Global Concept or Common Run either in revelation or in Manifestation. The Revealed Truth is that Globalisation or Common Run is an Impossible Task. The appearing Pentagonal Earth has three ascending stairs – Superscripts 1, 2, 3 [Tin, Zaytuun, and Tuur]
5	The Upper Half of the projected Global Trinity is the Northern Hemisphere and the lower Half of the projected Global Trinity is the Southern Hemisphere. However, the left is North & Eastern Hemisphere and right is South & Western Hemisphere. Two absurd End points are North and South. There is neither West nor East in the projected Global Trinity.	The Universe has been revealed as an equal & opposite Trinity. So, the Universe is a Diamond Operator or a Four-figured Shining Star in resemblance with Crucified Sign. The West Horizon is called Heaven or Night Sky and the East Horizon is called world or Earth Sky. Moreover, the West Horizon is called Galaxy of Stars; while the East Horizon is called Black Square or Non-luminous Moon. Cloud/Rain is the clear proof of West Horizon. Volcano/Earthquake is the clear proof of East Horizon.
6	Do	The East Horizon has been revealed into two Zones and an Octagon in coherence with the sacrifice of a quadruped [Baqarah]. So, springs or matter particles are octagonal black spots or so-called sun-spots. The two Zones of the Manifested East Horizon are West Zone [Upper Seashore] and East Zone [Lower Seashore].
7	Do	The Black Square [Non-luminous Moon] of the East Horizon has been manifested as a Hexagon after two sacrifices/divorces in correspondence with the sacrifices in the converted green and blue shades of Windows'7 Ultimate and in coherence with the equal & opposite sacrifices of Ibrahim [ass] and Ismail [ass] from psychic point of view. So, the manifested East Horizon within the four cardinal directions or four basic forces or four springs for straight usage or Black Square is a Hexagon. This manifested Hexagon is called the World or Asterisk.

8	Do	The World has been Manifested within the four basic forces of East Horizon [Black Square] declining towards Revealed Right. So, there are four Revealed Directions and six Manifested Directions. Four Revealed Directions are the four walls of the manifested East Horizon as well as fundamental categories of knowledge. There are four basic forces. These are Strong Force, Gravitational Force, Magnetic Force, and Weak Force. Strong Force represents upward direction towards West [Upright West]. Gravitational Force represents downward direction towards East [Straight Middle East]. Magnetic Force represents entering and ending directions [E-Point] of the manifested sign of natural magnetism or Revealed Left [Manifested North-East & North-West]. Weak Force represents turning point [T-Point] of the manifested sign of natural magnetism or Revealed Right [Manifested South-East & South-West]. Six manifested directions are North-East, Middle-East, and South-East of the East Zone, and South-west, Upright West, and North-West of the West Zone.
9	Do	The East Horizon within four cardinal directions has been manifested as a Hexagon. West Zone [Upper Seashore] of the manifested hexagon comprises three Regions or Triangles. Manifested Right Side from the point of view of West Zone [Upper Seashore] is the South-West Region [Triangle] in correspondence with the existence of Europe. Manifested Left Side from the point of view of West Zone [Upper Seashore] is the North-West Region [Triangle] in correspondence with the existence of Asia, Africa, and Australia. So, Asia is not in the Eastern Region but in the North-West Region of the Manifested Hexagon. Africa is not in the Southern Region of the Manifested Hexagon. On the contrary, Africa is in the North-West Region of the Manifested Hexagon. Australia is also in the North-West Region of the Manifested Hexagon. The remaining region of the West Zone which is neither towards South [Revealed Right],

		nor towards North [Revealed Left], but which is in the Upright Middle as well as above South-West and North-West Regions [Nations or Middle Stair or Median] is the Arabian Peninsula. So, Arabian Peninsula is the Upright West Region, Topmost Triangle of the Manifested Hexagonal World, and the Uppermost Land of the appearing Pentagonal Earth.
10	Do	East Zone [Lower Seashore] of the manifested hexagon also comprises three Regions [Triangles]. Manifested Right Side from the point of view of East Zone [Lower Seashore] is the South-East Region [Triangle] in correspondence with the existence of South America. Manifested Left Side from the point of view of East Zone [Lower Seashore] is the North-East Region [Triangle] in correspondence with the existence of North America. The remaining Region [Triangle] of the East Zone [Lower Seashore] which is neither towards South [Revealed Right], nor towards North [Revealed Left], but which is in the Straight Middle of both South-East Region and North-East Region is the Middle East Region. So, Middle East is the Straight Downward Region, Eartha 3D, Recycling Region, Bermuda Triangle in resemblance with the story of Titanic, Hole in the journey boat of Muusa [ass], hollow in the hand, Black Hole, Peg [projected planet Mars] of the immovable Pentagonal Earth. Appointed Kaba or Arabian Peninsula is neither in the East Zone nor in the Middle East Region of Eartha 3D. Projection of Arabian Peninsula or appointed Kaba in the Middle East Region of the East Zone is a self-explained and self-proved conspiracy against followers of Upright West [Arsh].
11	Do	There is an Equal & Opposite ~sth between Upright West Triangle of Arabian Peninsula and Straight Downward Triangle of the Middle East [Black Hole] of the Manifested Hexagonal World. Due to this Equal & Opposite ~sth between Upright West Region and Straight Middle East Region, the manifested Hexagon is appearing as a Pentagon. This appearing

		Pentagon is like Spider's Net or House or Pentagonal Orange or Pentagonal Star in correspondence with the 9/11 Pentagon. This appearing Pentagon is called the Earth. So, the Universe has been revealed as a Trinity and manifested an Upright Rectangle, the World has been revealed as an Octagon and manifested as a Hexagon within the Black Square of East Horizon, and the Earth is appearing as a Pentagon out of manifested hexagon.
12	Do	There are three ascending stairs and seven continents of the appearing pentagonal earth. The West Zone [Upper Seashore] comprises Upright [Top] West Region of Arabian Peninsula, South-West Region of Europe, and North-West Region of Asia, Africa, and Australia. The East Zone [Lower Seashore] comprises North-East Region of North America, South-East Region of South America, and Eartha 3D of the Straight Middle East Region. The Ground Stair of the appearing Pentagonal Earth consists of North-Eastern Continent of North America and South-Eastern Continent of South America. The Middle Stair of the appearing Pentagonal Earth consists of South-Western Continent of Europe, and North-Western Continents of Asia, Africa, and Australia. The Uppermost Stair of the appearing Pentagonal Earth is the Arabian Peninsula or Upright Western Continent.
13	There are Contrary & contradictory North [two North] and Contrary & contradictory South [two South] in Projected Mechanical Globalisation. Projected two North represent North direction determined by man-made magnetism and projected Northern Hemisphere as the Top of the Globe. Projected two South represent South direction determined by man-made magnetism and projected Southern Hemisphere as the Bottom of the Globe. None of the above contraries or contradictories represents Revealed & Manifest Direction or Manifested Nature. North and South grounded on man-made magnetism are Man-made North and	Revealed & manifest Truths - There are two West and two East. "Rabbul-Mashri-qayni wa Rabbul-Magribayn" Rab of the two East and Rab of the two West. [Sura No. – 54\55, Ar-Rahman, Ayat\Verses. 17] With respect to Revelation West Horizon and West Zone represent two West. With respect to Manifestation Upright West Region of Arabian Peninsula and Middle West Regions represent two West. With respect to Revelation East Horizon and East Zone represent two East. With respect to Manifestation, Straight Middle East Region and Ground Stair of the appearing Pentagonal Earth represent two East.

	Manmade South. Projected Northern Hemisphere and Southern Hemisphere are conspiracies against two East and two West.	There is only one North and only one South. The equal & opposite stages of journey of the so-called sun as the manifested sign of natural magnetism represent Revealed Left [Manifested North-East & North-West] and Revealed Right [Manifested South-East & South-West].
14	There is no barrier in the projected Global Trinity [Mechanical Globalisation]. So, the concept of **am** origins at mid night and the concept of **pm** begins at noon. Circular Rotation and Revolution System of the immovable pentagonal earth causes the alteration of day and night for the projected Northern Hemisphere of North America, Europe, and Asia, and for the projected Southern Hemisphere of South America, South Africa, and Australia. Arabian Peninsula has been projected in the Middle-East of Eartha 3D. So, Arabian Peninsula has not been found in projected global projection.	There are barriers among / between Revelation & Manifestation as well as Manifestation & Appearance. In other words, there are Danger and Double Danger, skies & clouds. In brief, there is a Lakshman Rekha between West and East. This Lakshman Rekha is the barrier between Semi-anticlockwise and Semi-clockwise stages of journey of the so-called sun as the Manifested Sign of Natural Magnetism which causes the alteration of day and night for the equal & opposite Seashore.
15	The so-called sun as the manifested sign of natural magnetism [Bullet or Niche] rises once from the East and sets once in the West but that once rising and once setting cause day & night for the equal & opposite seashore. [**What a post hoc ergo propter hoc statistics of day & night calculation for the equal & opposite seashore!**] **Contradictory Findings** The so-called sun as the manifested sign of natural magnetism rises from the East implies that East [downward direction or Gravitational Force] represents Magnetic Force [Electromagnetic Wave] for the so-called epistemic persons. So, East represents both projected Magnetic Force as well as revealed Gravitational Force in resemblance with the binary nature of the Historical Don of Trinity & Duality [Black Magician]. This is called **Un-contradicted Selfcontradiction.** Simultaneously, West [upward direction or Strong Force] represents Weak Force for the so-called epistemic persons.	The so-called sun as the manifested sign of natural magnetism [Bullet or Electromagnetic Wave or Niche] is kindled each day neither from the East nor from the West [Ayat No. – 35 of Sura Nuur]. Is there an Ayat in Quran either contrary or contradictory to Ayat No. – 35 of Sura Nuur? If there is a contrary or a contradictory Ayat in Quran, then Quran cannot be believed / considered as a Kitaab with Truth. So, it is certain like 'Cogito Ergo Sum' of Descartes that the so-called sun as the Manifested Sign of Natural Magnetism [Bullet or Niche] neither rises from the East, nor sets in the West. In other words, the direction from which the so-called sun enters / rises cannot be called East. Simultaneously, the direction towards which the so-called sun sets / ends the Equal & opposite stages of journey cannot be called West. These are not the opinions of the sharer of 'Kitaaba Wal-Hikmata' or 'Manifested Nature and the Utility of One's Upright Logic'. On the contrary, these are the revealed truths verifiable by the equal & opposite manifested signs of natural magnetism.

So, West represents both projected Weak Force and revealed Strong Force in resemblance with the binary nature of the Epistemic Uniqueness of Scientific Certainty, Owner and Possessors of Manmade Nature. This is called **Un-contradicted Selfcontradiction.**

So, in this regard it will be the best justifiable solution if we make our concept clear of the Ayat No. – 35 of Sura Nuur in resemblance with four cardinal directions, four basic forces, true concept of Natural Magnetism, Equal & Opposite Manifested Signs of Natural Magnetism etc. from the people of the Kitaab with Truth, people of the book, Religious Scholars, Islamic Research Institutions, real scientists, owner as well as possessors of mechanical globalisation & man-made magnetism, Father of Circular Rotation & Revolution System, Conspiracy Scientists, Epistemic Uniqueness of Scientific Certainty, so-called Epistemic Persons, Truth Verifiers & Truth Justifiers, Truth Recogniser & Ultimate Decision Taker, Mystics & Spiritualists, Chiefs & Leaders of different isms, and the like. If you have true concept of natural magnetism and you are able to prove that the so-called sun as the manifested sign of natural magnetism [Bullet or Electromagnetic Wave or Niche or converted snake of Muusa's Staff] is kindled each day from the East and put out in the West going contradictory to the Revealed Ayat No. – 35 of Sura Nuur, Ayat No. – 17 of Sura Khaf, and Equal & Opposite Manifested Signs of Natural Magnetism, then you are invited to play the game of 'Magic with Manifested Sign of Natural Magnetism and Upright Logic'. Otherwise, let the **Respective Authorities play their Significant Roles Concerning searched out manifest truths shared as 'SOLIDIFIED SOLID HUMAN RIGHTS'.**

I have submitted, placed, and forwarded searched out manifest truths or 'Kitaaba Wal-Hikmata' / 'Manifested Nature and the Utility of One's Upright Logic' before the tables of Epistemic Persons, Respective Higher Authorities, Higher Educational Institutions, Scientific Research Institutions, Publishers, Media Press Offices, and the like so that the searched out manifest truths shared in detail with valid references in the 'Kitaaba Wal-Hikmata' \ 'Manifested Nature and the Utility of One's Upright Logic'

		I become Experimentum Crusis as well as Crucial Instances with a view to manifest Teleological Evidence Sorceries & Unique Epistemic Persecutions as well as the Historical Don of the Trinity of Hypocrisy – Tyranny – Conspiracy and the Duality of Activism & Terrorism i.e. Black Magician and his Mechanical Trinity.
16	North represents Revealed Magnetic Force. [What are being taught as North and South Directions on the basis of man-made magnetism are not manifested North and South directions in resemblance with Revealed Left and Revealed Right. On the contrary projected North & Northern Hemisphere and South & Southern Hemisphere represent Manmade Nature or Mechanical Nature or Historical Black Magic. Those who are using their pen – paper – pencil unconsciously / consciously in resemblance with Historical Black Magic, they are called Black & White Artists or Real Activists. Those who are trying to project Black Magic creating terror in the psyche of mankind by establishing illicit guidelines & formalities, they are called Epistemic Persecutors or Vera Causa of Original Terrorism.]	North represents Revealed Magnetic Force and Manifested Northern Hemisphere. Equal & Opposite stages of journey of the so-called sun as the manifested signs of natural magnetism [Bullet or Electromagnetic Wave or Niche] are the Clear Signs of North and South Directions in resemblance with Revelation. There is only one kind of revealed & manifested magnetism. This sole magnetism is called Natural Magnetism. The so-called sun [Bullet] is the Manifested Sign of Natural Magnetism. The so-called sun [Bullet or Electromagnetic Wave or Niche] enters from Manifested North-East [Revealed Left / Northern Hemisphere / Haiyalas-Swalaah] in correspondence with North America and converted red shade of Windows'7 Ultimate as the day-light for the East Zone and sets in Manifested South-East [Revealed right / Southern Hemisphere / Haiyalal-Falaah] in correspondence with South America and converted green shade of Windows'7 Ultimate. In other words, the direction from which the so-called sun as the Manifested Sign of Natural Magnetism [Niche] is kindled [enters] for the ground stair of North America and South America is the Manifested North Direction or Haiyalas-swalaah of the East Zone. The direction towards which the so-called sun as the Manifested Sign of Natural Magnetism [Niche] sets for the ground stair of North America and South America is the Manifested South Direction or Haiyalaal-Falaah of the East Zone. The so-called sun as the Manifested Sign of Natural Magnetism [Bullet / Electromagnetic Wave or Niche] runs away from setting point and rises from Manifested South-West [Revealed Right

/ Southern Hemisphere / Haiyalal-Falaah] in correspondence with Europe and converted yellow shade of Windows'7 Ultimate as the day-light for the Middle Stair of Europe, Asia, Africa, and Australia as well as Topmost Stair of Arabian Peninsula, and ends the stages of journey in Manifested North-West [Revealed Left / Northern Hemisphere / Haiyalas-Swalaah] in correspondence with Australia and converted blue shade of Windows'7 Ultimate for the West Zone within the Black Square of East Horizon. In other words, the direction from which the so-called sun as the Manifested Sign of Natural Magnetism [Niche] rises for the Middle Stair of Europe, Asia, Africa, and Australia as well as Topmost Stair of Arabian Peninsula is the Manifested South Direction or Haiyalaal-Falaah of the West Zone. The direction towards which the so-called sun as the Manifested Sign of Natural Magnetism sets for the Middle Stair of Europe, Asia, Africa, and Australia as well as Topmost Stair of Arabian Peninsula is the Manifested North Direction or Haiyalas-Swalaah of the West Zone. [Ayat No. – 17 of Sura Khaf]

The so-called sun as the Manifested Sign of Natural Magnetism [Bullet / Niche] enters from North-East for the East Zone [Lower Seashore] declining towards South-East and sets in South-East covering Semi-anti-clock-wise [4/8 or 1/2] stages of journey. The so-called sun as the Manifested Signs of Natural Magnetism [Bullet / Niche] runs away from there and rises from South-West and ends the Semi-clock-wise [1/3 & 3/4] stages of journey covering Arabian Peninsula for one whole day, and finally sets in North-West for the West Zone. This is called **Natural Magnetism.**

In brief, the so-called sun as the manifested sign of Natural Magnetism enters from North-East and sets in South-East for the Ground Stair of North America and South America of East Zone within the Black Square of East Horizon. The so-called sun [setting sun] or Bullet as the manifested

		sign of Natural Magnetism rises from South-West and ends in North-West for the Middle Stair of Europe, Asia, Africa, and Australia as well as Topmost Stair of Arabian Peninsula of the West Zone within the Black Square of East Horizon. These Equal & Opposite Semi-anticlockwise and Semi-clockwise stages of journey of the so-called sun [Bullet] from North-East to South-East and South-West to North-West are the Manifested Signs of Natural Magnetism
17	The so-called moon is **non-luminous but it has eclipse.** Contradictory Findings This implies the transition **from darkness to the depth of darkness.** This is the projected Veil of Ignorance or the game of selfcertified Evidence Sorcery as well as selfproved Persecution of faith and belief of mankind with a view to make them the Trinity of blind – deaf – dumb. With respect to 'Kitaaba Wal-Hikmata' or Manifested Nature and the Utility of One's Upright Logic', this game is called Historical Black Magic. Universe is an Intrinsically Luminous Quadrilateral Star. Universe has been revealed as an equal & opposite Trinity called Sirius Binary System or Star System. Conspiracy Science has converted Intrinsically Luminous Star into Intrinsically Luminous Moon. Through this conversion, the Universe has been projected as a Lunar System or a Solar System or a Planetary System, and the like. Consequently, each star is being called a planet or a moon. The non-luminous moon represents Black Square or Manifested East Horizon within four walls of the basic forces. So, what we perceive or commonly conceive as moon is not Intrinsically Luminous Moon, but the white of the obversion of the converted snake of Muusa's Staff.	The revealed East Horizon within the four basic forces is called Black Square. So, **Black Square [projected Dark Moon] is Non-luminous.** Manifested Hexagonal World within the Black Square of East Horizon is also **Non-luminous.** The appearing Pentagonal Earth within the Black Square of East Horizon is also **Non-luminous.** **The Moon is Earth's only permanent natural satellite.** [Source – Internet]. This projected Moon represents Intrinsically Luminous Star called Diamond Operator. The so-called moons are the signs to mark fixed period. The so-called new moon arises from Manifested North-West [Revealed Left / Northern Hemisphere] for the West Zone and moves towards Manifested South-West [Revealed Right / Southern Hemisphere] and appears from Manifested South-East for the East Zone. Both the so-called sun [Bullet] and the so-called new moon [Triangular Bullet] have the single root. The new moon arises from North-West [Northern Hemisphere] for the West Zone, so the concept of Night [pm] origins from the West Zone and begins from East Zone. Similarly, the so-called sun as the manifested sign of natural magnetism enters from North-East [Northern Hemisphere] for the East Zone, so the concept of day [am] origins from East Zone and begins from West Zone. [Sura No. – 12\13 – Rad, Ayat No. – 1 to 4] Since the concept of Night origins from West Zone [Upper Seashore], so people of Australia, Africa, Asia, Arabian Peninsula,

		and Europe will observe Idd uniformly on the same day prior to people of the East Zone of North America and South America.
18	Facing towards entering and rising directions of the so-called sun as the manifested sign of natural magnetism for the equal & opposite Seashore [East Zone and West Zone] represent East Direction as well as Contrary & Contradictory Magnetic Force & Gravitational Force. This is called **Tautologous Black Magic**. The setting and the ending directions of the so-called sun as the manifested sign of natural magnetism for the equal & opposite Seashore [East Zone and West Zone] represent West Direction as well as Contrary & Contradictory Weak Force and Strong Force. This is called **Tautologous Black Magic.** There is no Natural Magnetism in the Tautologous Black Magic, though there are two kinds of Magnetism in the works of Black & White Juggler. Newton's three Laws of Motion & Law of Gravitation as well as Einstein's Theory of Relativity, Gravitational Lensing, and Binary Pulsar have no place in the Tautologous Black Magic i.e. [Mechanical Globalisation, Man-made Magnetism, and Circular Rotation & Revolution System.] Will you justify that East represents Magnetic Force and North represents Gravitational Force? Will you try to prove that West represents Weak Force and South represents Strong Force? If you have the capability to prove the above mentioned Tautologous Black Magic, you are invited to play the game of Magic with Manifested Signs of Natural Magnetism and Upright Logic.	Four Cardinal [Revealed] Directions resemble with Four Basic Forces. Isa [ass] brought Clear Proofs or Pragmatic Correspondence Truth or Injiil concerning four Cardinal Directions. In other words, Crucified Sign is the Clear Proof of the Four Cardinal Directions. Upward Direction represents West or Strong Force. Rain comes from West. Downward Direction represents East or Gravitational Force. Earthquake emerges from East. The Left Side from the point of view of Revelation but right side from the point of view of the finite observer facing towards the appointed Kaba represents North or Magnetic Force. In other words, the entering direction of the so-called sun as the Manifested Sign of Natural Magnetism for the East Zone and setting direction of the so-called sun as the manifested sign of Natural Magnetism for the West Zone represent Revealed Left or Northern Hemisphere or Haiyalas-swalaah [Manifested North-East and North-West]. Fire/Light is the clear sign of Manifested Northern Hemisphere. The Right Side from the point of view of Revelation but left side from the point of view of finite observer facing towards the appointed Kaba represents South or Weak Force. In other words, the setting direction of the so-called sun as the manifested sign of Natural Magnetism for the East Zone and rising direction of the so-called sun as the Manifested Sign of Natural Magnetism for the West Zone represent Revealed Right or Southern Hemisphere or Haiyal-Falaah [Manifested South-East and South-West]. Air is the clear sign of South. The Black Magicians have concealed North and South directions represented by manifested signs of Natural Magnetism and projected East and West through Mechanical Globalisation, Circular Rotation & Revolution System, and Man-made Magnetism. Newton's three Laws of Motion & Law of Gravitation as well as

		Einstein's Theory of Relativity, Gravitational Lensing, and Binary Pulsar have the proper places in Manifested Nature, Natural Magnetism, Upright Rectangular Universe, Black Square, Manifested World within four Cardinal Directions, two equal & opposite Seashore [two Zones], and Appearing Pentagonal Earth with three ascending Stairs.
19	In the Black Magic or Manmade Nature, the so-called sun rises once and sets once for the one whole day. There is no alteration of Day and Night for the equal & opposite Seashore [East Zone and West Zone] save Circular Rotation & Revolution System. What we perceive as Darkness or Night is nothing but the **Veil of Ignorance**.	In the Manifested Hexagonal World [Nature] within the Black Square of East Horizon, the so-called sun as the manifested sign of natural magnetism rises twice and sets twice for the one whole day as well as for the equal & opposite seashore [East Zone and West Zone]. These twice rising and twice setting of the so-called sun as manifested sign of natural magnetism cause the alteration of Day and Night for the equal & opposite seashore [East Zone and West Zone]. So, when there is sun-light [Morning Show] in the East Zone of North America and South America, then it is Night [early hours of night] for the West Zone of Europe, Arabian Peninsula, Asia, Africa, and Australia. Again, when there is sun-light [Evening Show / break of the day] in the West Zone of Europe, Arabian Peninsula, Asia, Africa, and Australia, then it is Night [fall of night] for the East Zone of North America and South America. In brief, twice rising and twice setting of the so-called sun as the manifested sign of natural magnetism cause the alteration of day and night for the equal & opposite Seashore. This is the **manifested statistics of day & night calculation in resemblance with revelation.**
	Copernican Revolution and Mechanical Globalisation are the two faces of the Same Coin. These faces have occupied the centre ½ [Unicode (hex) 00BD] of the Global Trinity, the Trinity of Hypocrisy - Tyranny – Conspiracy, and the Duality of Activism & Terrorism. Findings of Similarities [Rotation & Revolution] Projected Solar System	Copernican Revolution has no place either in Revelation or in Manifestation or in Appearance. In other words, Copernican Revolution is contrary as well as contradictory to Upright Rectangular Universe [End of Proof / Kaba], Manifested Hexagonal World within the Black Square of East Horizon, the Appearing Pentagonal Earth with three ascending stairs, Equal & Opposite Manifested Signs of Natural Magnetism, Equal & Opposite Stages of

| 20 | Due to Copernican Revolution, the Universe is moving round the World, the World is moving round the Earth, the Earth is moving round the Moon, the Moon is moving round the Electron Cloud [Man-made Owner of Magnetic Force], man-made magnetic force is moving round Insert Symbol, Insert symbol is moving round Windows'7, Windows'7 is moving round Bookshelf Symbol7 in resemblance with the will of the Juggler of Historical Black Magic. Consequently, we are moving round the Ultimate Decision of the Juggler of Historical Black Magic [Teleological Evidence Sorcerer and Unique Epistemic Persecutor in resemblance with Petitio Principii]. This projected mechanism or manmade nature is called Black Magic. | Journey, Equal & Opposite Right Direction towards Upright West [Arsh], Appointed Days of Observing Idd Uniformly, equal & opposite day & night, and the like. On the contrary, Copernican Revolution involves all the established fallacies which are being taught by the people of the book as well as which are available in pen – paper – pencil works and black & white activities of real educated people. Some of those fallacies are – Fallacy of Arguing in a Circle, Fallacy of Infinite Regress, Fallacy of Non-Observation of Instances, Fallacy of Non-observation of Essential Circumstances, Fallacy of Individual Mal-observation, Fallacy of Universal Mal-observation, Fallacy of Practical Imperfection, Fallacy of Characteristic Imperfection, Existential Fallacy, Fallacy of Plurality of Causes, and the like. In brief, Copernican Revolution is a Post Hoc Ergo Propter Hoc. The projection of Copernican Revolution is a pen – paper – pencil Conspiracy against Manifested Nature, Natural Magnetism, and Faith & Belief of Mankind, and particularly against faith & belief of the followers of Equal & Opposite Right Direction towards Upright West. Those who are using their pen – paper – pencil knowingly or ignorantly in resemblance with Copernican Revolution, they are Black & White Activists. Those who have formalised several illicit guidelines & formalities and are verifying as well as justifying Manifest Truths [Essential / Fundamental Circumstances of Instances] in resemblance with Copernican Revolution, they are the so-called Epistemic Terrorists. If you have any objection in these sharing concepts, you are invited to play the game of 'GO WITH KITAABA WAL-HIKMATA – AGREE WITH MANIFESTED NATURE AND THE UTILITY OF ONE'S UPRIGHT LOGIC with a view to recognise SOLIDIFIED SOLID HUMAN RIGHTS. Nothing is moving save Shakarash-Shamsa Wal Qamar or the so-called sun and the so-called moon. Due to revelation |

	& manifestation of the projected planets Saturn and Mars, neither the East Horizon, nor the manifested Hexagonal World within the Black Square of East Horizon, nor the appearing Pentagonal Earth with three ascending stairs is moving. Firm standings of Hills and Mountains are the clear signs of the immovable Manifested Hexagonal World and Appearing Pentagonal Earth.
Note – Please go through 'Kitaaba Wal-Hikmata' or 'Manifested Nature and the Utility of One's Upright Logic' for detail scientifically verifiable as well as logically justifiable references for verification, justification, confirmation, recognition, publication, communication, and broadcasting one of the above contradictory sets i.e. either Set – A or Set – B; but not both Set – A and Set – B. Let us invite the Epistemic Uniqueness of Scientific Certainty, Owner and Possessors of Manmade Nature, Teleological Evidence Sorcerers & Unique Epistemic Persecutors either to prove Set – A or to recognise Set – B, but not both.	

BREAD AND BUTTER HUMAN RIGHTS
[Second Generation Human Rights]

I had completed my Master Degree in Philosophy in 1993-94 Academic Sessions from Gauhati University, Guwahati [Assam, Hindustan] securing first class first position. But I did not get a job till November 2000 due to the implementation of several Formalised Fundamental Human Rights [Second Generation Human Rights], Formalised Guidelines, and Meta-Ethical Nature of the so-called Epistemic Persons who share falsehood as truth exclusively with reference to opinions, suppositions, imaginations, predictions etc. of X, Y, Z, - - - i.e. 'according to'. During that period I was **deprived of Bread and Butter Human Rights**. From the pragmatic implementation of the Formalised Human Rights it had appeared as if I was a human person without having Rights i.e. lack of Second Generation Human Rights or Fundamental Human Rights or Bread and Butter Human Rights. Consequently, without placing any complain before you [Respective Authorities], I had decided not to receive the awarded Gold Medal for the position holder as I would be an **unemployed Gold Medallist**. I was satisfied with the normative teachings and upright nature of Prof. Dr. Nima Sharma, Gauhati University, Guwahat, Assam [Hindustan].

But there are epistemic persons who possess upright nature as well as intrinsic values. With a view to confess gratitude from the core of my psyche, I must have to share some names like Prof. B. K. Dhar [S S College], Prof. Dr. A. Karim [S S College], Prof. Dr. Hifzur Rahman [B Ed College Hkd], Prof. Mamduda Yeasmin Laskar [S S College], Prof. Ali Haidar Laskar [S S College], Prof. Dr. Bidhan Ch. Deb [S S College], Prof. Abul Hussain Choudhury [S S College], Prof. Dr. D. K. Nath [S S College], Prof. Hilal Uddin Laskar [S S

College], Prof. Ashim Sen [S S College], Mr. Azizur Rahman Choudhury [Sec. B Ed College Hkd], Prof. Dr. Alpana Talukdar [Karimganj College], Guardians of the students of Hailakandi District, public representative Mr. Sahidul Alom Choudhury, and last but not the least my friends in deed i.e. my dear students who had exerted their out most effort with a view to provide justified balance with my Bread and Butter Human Rights. I was appointed as a lecturer in a non-sanctioned post [with a conveyance allowance of Rs. 800/- pm] in the Department of Philosophy, S. S. College, Hailakandi in 1998-99 Academic Sessions. That appointment was not sufficient to meet up my Bread and Butter Human Rights. So, I had decided to leave the social life with a view to become a recluse in resemblance with Siddhartha [Gautama Buddha]. But I could not do so due to the altruistic efforts of the above mentioned Karma Yogis [Akbari-Jihadists / followers of Utilitarianism]. Due to the selfless efforts of the above mentioned altruists, the then Principal of S. S. College, and the above mentioned Public Representative, the Director of Higher Education, Assam had converted a vacant sanctioned post from History Dept. to Philosophy Dept. That conversion was sufficient to meet up my Bread and Butter Human Rights.

Again, due to the implementations of NET, RC, and OC etc. within the stipulated period, I was deprived of Carrier Advancement and Incremental Benefit. I had qualified NET, and participated in RC & OC. But till today I am doing my job with usual pay. Now, Re-search Publications and Re-search Works i.e. API points become barrier in my Bread and Butter Human Rights. I have been trying to do **re-search** work and to publish **re-search** papers since 1994 in a proper logical method grounded on observation of particular instances and verifiable as well as justifiable by the revealed & established criterions of truth. But, I had failed to search out a single institution and an upright guide within my limited range of searching for **re-search work in proper logical method** as all of them were found busy with **research work following illicit guidelines & formalities**. Consequently, in the year 2010, I had decided to become the **so-called research scholar** under the supervision of Dr. P V Thomas of Assam University, Silchar [Assam, Hindustan].

I had started my so-called research work on 'Cloning'. After sixth months study on Cloning, I had conceptualised my inability to do re-search work on 'Cloning'. So, after thorough discussion with my supervisor, I had decided to do re-search work on 'Justice'. In resemblance with the mutual decision among/ between research scholar and Supervisor, I began to prepare my re-search work collecting materials on Justice from various sources. In this regard **flipkart. com** is a concrete example as well as valid reference of my honest and sincere efforts. But this mutual decision continued only two years. During that period

I had completed two-thirds of my re-search work on Justice prior to provisional registration. But my supervisor who is an expert on Solidarity Rights began to project binary games of CAPITAL LETTER and small letter concerning my Re-search area and provisional registration. That projection did not hurt me as I had firm faith on One Dot Leader and practical belief on Diamond Operator. Consequently, I agreed to follow the projected nature of CAPITAL LETTER and accordingly got registered myself as a research scholar on 'Solidarity Rights in Islam'.

During the preparation of my re-search work on Justice, I had searched out several lacunae in the projected 'Veil of Ignorance'. I was in a perplex position to summarise my searched out findings for re-search work in a proper logical method verifiable as well as justifiable by the four established criterions of truth. In this regard projected binary nature of my supervisor had helped me a lot. So, I began to prepare **binary research and re-search works on Solidarity Rights**. I had marketed three chapters from **Dissertation India [Astha.com]** on **Solidarity Rights in Islam** at the cost of **Rs. 75,000/- [Seventy five thousand]** as the concrete example of mechanical or manmade research work as well as **formal document concerning the pragmatic consequences of the implementation of API points by the UGC - Hindustan**. On the other hand, I began to summarise my searched out verifiable as well as justifiable findings concerning 'Solitary Rights in Islam'. I have no doubt that a re-search work based on a hypothesis without fulfilling the first condition of hypothesis i.e. observation of facts or observation of particular instances is nothing save tabula rasa of John Locke. So, theoretical researches, critical researches, comparative re-search works, and the like which are being done without observation of particular instances cannot be called verifiable as well as justifiable re-search works in the truest sense of the term **'Re-search'**, but may be recognised / certified / revolutionised as imaginations or predictions or suppositions or fictitious works or tabula rasa or empty vessel of John Locke.

Now, I have compiled my searched out findings in a logical method called Dictum De Omni ET Nullo [Deductive Method or Mathematical Method] in resemblance with the method of the father of Modern Western Philosophy called Cartesian Method. The compiled work of my searched out findings has been named as 'Kitaaba Wal-Hikmata' or 'Manifested Nature and the Utility of One's Upright Logic' in resemblance with the content, context, and searching & sharing [framing hypotheses] method. There are several Highlights / Conclusions / Searched out Findings in the 'Kitaaba Wal-Hikmata' or 'Manifested Nature and the Utility of One's Upright Logic'. Each shared searched out finding is either justifiable or both verifiable as well as justifiable by one or the other established criterion of truth. So, each shared searched out finding is a selfevident truth from the point of view of re-searcher & sharer of 'Kitaaba Wal-Hikmata' or 'Manifested Nature and the Utility of One's Upright Logic'. The teleology of this sharing is to get recognised confirmation of the searched out findings shared as Solidified Solid Human Rights from the respective authorities, and the ontology of this sharing is to share those findings with each and every matured member of the manifested hexagonal world and appearing pentagonal earth within the Black Square of East Horizon.

Moreover, each searched out finding is related with my re-search work on Solidarity Rights or Area/Domain of my Re-search Work. Consequently, 'Kitaaba Wal-Hikmata' or 'Manifested Nature and the Utility of One's Upright Logic' is the first stage of my re-search work or finalised preliminary works or **framed hypotheses** on the basis of observation of particular instances [consequences of searching works for the period of six years]. I have submitted, placed, and forwarded my searched out findings or framed hypotheses to the respective departments, authorities, and epistemic persons for verification, justification, confirmation, recognition, suggestion, and the like. I will try to get myself registered for final stage of my re-search work when the respective departments will share lacunae in the framed hypotheses or searched out findings with four witnesses i.e. four revealed and established criterions of truth as valid references. I will try to do the final stage of my re-search work on those findings which are neither verifiable nor justifiable on the basis of one or the other criterion of truth in resemblance with manifested nature, manifested sign of natural magnetism, equal & opposite revelation, established Newton's Three Laws of Motion & Law of Gravitation, Einstein's Gravitational Lensing & Binary Pulsar, and all verifiable scientific laws as well as justifiable philosophical theories. **I will also try to do the final stage of my re-search work on those findings which will be proved as beyond the domain of the Solidarity Rights in Islam.**
So, as a **re-search** scholar on 'Solidarity Rights in Islam' as well as a deprived Govt. Employee from carrier advancement & incremental benefit, it is not only my Solidarity Rights [Third Generation Human Rights] but also **Bread and Butter Human Rights** [Second Generation Human Rights / Formalised Human Rights] and **Valid Legal Rights** to complete the second stage of my re-search work on the pointed lacunae with a view to become the follower of the **implemented guidelines of UGC [Hindustan] as well as to accumulate certified pieces of paper for carrier advancement and incremental benefits**.

Moreover, being a human person by birth and a teacher by profession, it is my **Birth Rights or Inborn Rights or Natural Rights** [First Generation Human

Rights] as well as **Justified Ethical/Moral Rights** not only to get generalised recognition & confirmation of the searched out findings free from subjective selfcontradictions and objective paradoxes with a view to share true knowledge as a teacher, but also to know the lacunae in the shared searched out findings for the final stage of my re-search work as well as to refrain me from sharing selfcontradictions & paradoxes as true knowledge with my friends in deed [students].

Further, as an individual as well as a follower of equal & opposite middle course towards upright West [Arsh], it is the **Solitary Rights** or **Solidarity Rights** or **Solid Human Rights** or **Individual Rights** or **Rights grounded on Faith & Belief** to get verified confirmation from the respective authorities of the searched out manifest truths most of which are either contrary or contradictory to the projected scientific truths, pen-paper-pencil works, black & white activities, and what are being taught through formal education. Moreover, to get verified confirmation and generalised recognition of the searched out manifest truths shared in the 'Kitaaba Wal-Hikmata' or 'Manifested Nature and the Utility of One's Upright Logic' from the respective authorities are the **Solidified Solid Human Rights of mankind in general for the greatest happiness of the greatest number** [Utilitarianism].

SOLIDIFIED SOLID HUMAN RIGHTS
EXCLUDE NO HUMAN RIGHTS

In brief, the framed searched out findings in the 'Kitaaba Wal-Hikmata' or 'Manifested Nature and the Utility of One's Upright Logic' is a **Trinity** of First Generation Human Rights [**Natural Rights**] – Third Generation Human Rights [**Individual Rights**] – Second Generation Human Rights [**Formalised /Generalised Human Rights**]. The Third Generation Human Rights or **Individual Rights is the binary ~sth between First Generation Human Rights and Second Generation Human Rights**. The bond or the Trinity established by the Individual Rights is called **Solidified Solid Human Rights.** In other words, **Solidified Solid Human Rights exclude no human rights**. So, to get verified as well as justified confirmation of the shared searched out findings from the experts of the re-search works on Human Rights, and to get **certified recognition** of the shared searched out findings **particularly from Assam University [AUS]** is the **Solidified Solid Human Rights of the re-search scholar on 'Solidarity Rights in Islam'**.

However, to get **generalised recognition** of the shared searched out findings from **all respective authorities** are the **Solidified Solid Human Rights** not only of the searcher & sharer of 'Kitaaba Wal-Hikmata' or 'Manifested Nature and the Utility of One's Upright Logic' but also of **mankind in general as Solidified Solid Human Rights exclude no Human Rights**.

I had tried to share my searched out preliminary findings concerning Manifested Nature and Natural Magnetism with the Epistemic Uniqueness of the Mystic Religious Experience in the UGC Sponsored National Seminar Organised by the Department of Philosophy, AUS [Hindustan] on 27th March, 2015. But I had failed to do so due to the projected allergy towards Proper Logical Method and established criterions of truth of the so-called Epistemic Persons who were present in the above mentioned Seminar. From their behaviour it was being appeared that either those epistemic persons had no clear concepts of the revealed universe, manifested world, appearing earth, manifested signs of natural magnetism, difference between man-made magnetism and natural magnetism, Cartesian Method, four established criterions of truth, four cardinal directions in resemblance with four basic forces etc. or they were trying to play mischievous game with my searched out findings knowingly / open secret mutual understanding. Consequently, I had to pull the Broken Bar in the first re-search paper presentation / Seminar presentation throughout my academic carrier. I have been doing hard work since 27th March, 2015 with a view to make that Seminar Presentation as the First & Last One in my academic carrier.

I did/do not know who was/is the Epistemic Uniqueness in Mystic Religious Experience in resemblance with the National Seminar Organised by the Dept. of Philosophy [AUS]. So, I am placing an appeal before the Head of the Dept. of Philosophy to place/forward 'Kitaaba Wal-Hikmata' or 'Manifested Nature and the Utility of One's Upright Logic' before that Epistemic Uniqueness for verification, justification, confirmation, and recognition of the shared Highlights / Conclusions / Searched out Findings. I am also placing another appeal before the faculty members of the Philosophy Dept. not only of the Assam University [AUS], but also all the Universities of the appearing pentagonal earth to share scientifically verifiable, logically justifiable, and philosophically established criterions for verification and justification of the re-search papers / Seminar papers / re-search works / Major Re-search Project / Minor Re-search Project etc. on **'Geetanjali of Rabindranath Tagore and Theory of Relativity of Einstein'**, **'Epistemic Uniqueness in Mystic Religious Experience'**, **'Knowledge Theorisation and Rights'**, and the like. If the four philosophically established criterions of truth have no existential import in PhD works / Philosophical Re-search Works, then why are you teaching us those criterions of truth? Whether a mathematical simplification is examined without mathematical criterions i.e. units, formulae etc.? If logical methods have no existential import in the re-search work, why do you mention that the findings are based on logical method? Is there an illogical sharing in the 'Kitaaba Wal-Hikmata' or 'Manifested Nature and the Utility of One's Upright Logic'? Which of the shared searched out findings of 'Kitaaba Wal-Hikmata' or 'Manifested Nature and the Utility of One's Upright Logic' are neither verifiable by one or the other established criterion of truth, nor justifiable by one or the other established criterion of truth, nor follow logical methods, nor within the domain of Solidarity Rights, nor follow tautologies, nor follow established un-contradictory laws [verifiable laws] & valid rules?

Due to several informal threatening concerning my Bread and Butter Human Rights and an analytical certificate provided in the air over cell phone by the meta-ethical faculty members of the Dept. of Philosophy, AUS, I am neither appearing before the faculty members of the Dept. of Philosophy, AUS, nor before the administrative officials since 27th March, 2015. The Dept. of Philosophy AUS has even provided me a false paper presentation certificate of a National Seminar. But my morality does not permit me to fulfil Bread and Butter Rights on the basis of that false certificate / API points. So, I had requested to the head of the Dept. of Philosophy to take that false certificate back [formally].

Moreover, the sharing of the contents and contexts of the analytical certificate by the faculty members of the department of Philosophy AUS over cell phone with some highly educated news mongers had snatched away not only my near ones from me but also the mental peace of my professional - social - family life. For these reasons, with a view to get verifiable as well as justifiable references concerning that informal analytical certificate provided by the faculty members of the Dept. of Philosophy, AUS, I have been trying to maintain formal relation with the Dept. of Philosophy through E-mail. I have also sent my initial findings to the Head of the Dept. of Philosophy [AUS] as well as the so-called supervisor of my re-search work for their kind consideration for final registration of my re-search work. But I did not get any response from their end. Now, I have submitted my framed hypothesis [searching works for the period of six years] as the first stage of my re-search work in the name of 'Kitaaba Wal-Hikmata' or 'Manifested Nature and the Utility of One's Upright Logic' through E-mail before the table of the Head of the Dept. Philosophy [AUS] for verification, justification, confirmation, and recognition of the searched out 'Solidarity Rights in Islam'. I will try to do Re-search work on those findings which are beyond Solidarity Rights as well as neither verifiable nor justifiable by one or the other criterion of truth. Simultaneously, it is an appeal before the Head of the Dept. of Philosophy, Administrative Authority of AUS, ICPR, UGC, NAAC, DHE Assam, and all respective authorities to recognise those findings which are scientifically verifiable, logically justifiable by one or the other philosophically established criterion of truth. This recognition will help me to fulfil my Bread and Butter Human Rights as well as Solidified Solid Human Rights.

SELF-EVALUATION

Self-evaluation or self-assessment or self-appraisal is an easy method of confirming selfevident truths. There are several searched out findings shared as selfevident truths in the 'Kitaaba Wal-Hikmata' or 'Manifested Nature and the Utility of One's Upright Logic'. Each shared selfevident truth is verifiable as well as justifiable in resemblance with manifested nature grounded on equal & opposite stages of journey of the manifested signs of natural magnetism and four revealed and established criterions of truth. Let me share the similitude of the method of self-evaluation.

LET US START WITH THE SEARCHING QUESTION
WHO AM I?
ANSWER - I AM A/AN / THE -------------------------------------.
OPTIONS

i) A Mystic or a Spiritualist or an Epistemic Uniqueness in Mystic Religious Experiences

ii) A truth sharer or an ideal teacher

iii) A hypocrite [an epistemic persecutor] who has binary nature in resemblance with the functions of Sankara's Maya i.e. concealment of the real and projection of the unreal, the Historical Owner of two kinds of Magnetism

iv) A persecutor or the killer of faith & belief of mankind in general

v) A dictator or a tyrant or a unique epistemic terrorist

vi) A black & white activist or a falsehood monger [expert, supervisor, verifier, justifier, researcher, editor, publisher, writer, organiser, co-ordinator, chair person, lawyer, scholar, commentator, journalist, news editor, and the like]

vii) A learned man or an educated person

viii) An idiot or a fool

ix) An ignorant human person or the trinity of blind – deaf – dumb

x) An evidence sorcerer [Black Magician] or the Don of the Trinity of hypocrisy – conspiracy – tyranny or the possessor of manmade magnetism

xi) An altruist or an akbari-jihadist or a karma-yogi who tries to do righteous work and to give good tidings to those who believe

xii) The so-called epistemic person [expert, supervisor, verifier, justifier, researcher, editor, publisher, writer, organiser, co-ordinator, chair person, lawyer, scholar, commentator, journalist, news editor, and the like]

xiii) The possessor of Upright Nature or an Unbiased verifier & justifier of truth [re-search work] on the basis of revealed and established criterions of truth or an impartial media person [publication media / broadcasting media]

xiv) A public representative in the truest sense of the term 'public' which implies protector of Solidified Solid Human Rights [General Rights regardless of the trinity of class – religion – caste or those rights which are manifested truths like Cogito Ergo Sum of Rene Descartes or Greatest Happiness of the Greatest Number]

Sl. No.	CONTENTS
1	I am a human person but I have no concept of the revealed Universe, manifested World within the Black Square, appearing Pentagonal Earth with three ascending stairs, four cardinal directions in resemblance with four basic forces, difference between natural magnetism and projected man-made magnetism, morning show & evening show, and the like. I am a/an/the -----------------------------.
2	I am a human person. I have true concepts of the revealed Rectangular Universe, manifested Hexagonal World within the Black Square, appearing Pentagonal Earth with three ascending stairs, four cardinal directions in resemblance with four basic forces & four fundamental categories of knowledge, difference between natural magnetism [manifested Truth] and man-made magnetism [projected falsehood], morning show & evening show, equal & opposite sages of journey, and the like, but my binary nature does not permit me to share manifest truths openly and publicly. I am a/an/the ----------------------------.
3	I am an epistemic person and I have conceptualised Universe, World, and Earth as synonymous or Global Trinity. In other words, I have firm faith & belief that the global map implies Global Universe, Global World, and Global Earth. Similarly, I have conceptualised acquired knowledge as true knowledge i.e. there are two kinds of magnetism, namely, natural magnetism and man-made magnetism. I have conceptualised both nature and mechanism [man-made] as synonymous. In other words, according to epistemic persons like me, two contradictories such as Natural North and Mechanical [man-made] North represent same direction i.e. true together from the same space, time, substance, and causality relation. I am a/an/the ----------------------------.
4	I am a teacher. I never feel/think it necessary to make my concept clear i.e. free from subjective self-contradictions and objective paradoxes concerning revealed universe, manifested world, and appearing earth. So, I am teaching my students that the universe is a Solar System, the world is a planetary system, and the earth is a moonlet system. This Trinity of Systems is called Globalisation or Circular Rotation & Revolution System. I am a/an/the ----------------------------.
5	I am by designation a teacher but by nature a sharer of selfevident truth. My nature does not permit me to conceptualise and to share Black Magic, subjective selfcontradiction,

	objective paradox, post hoc ergo propter hoc etc. as true knowledge. On the contrary, I am trying to share with certainty that two contraries or contradictories from the point of view of four cardinal directions in resemblance with four basic forces & four fundamental categories of knowledge can never be true together. For example – Both natural magnetism and man-made magnetism can never be true together with respect to manifested North & South and in resemblance with revealed left & right. So, all works and activities which are either contrary or contradictory to Equal & Opposite Manifested Signs of Natural Magnetism are vitiated by the fallacies of subjective selfcontradictions and objective paradoxes. In other words, such kind of sharing will lead the concept of truth to the fallacy of arguing in a circle or the fallacy of infinite regress. To get rid of such kind of fallacies, I have searched out several manifest [natural] truths such as the Universe is a Diamond Operator which has been revealed as equal & opposite Trinity and manifested as an upright rectangle, the world has been manifested as a hexagon [six regions] within the Black Square of East Horizon, the appearing earth is a pentagonal orange with three ascending stairs and seven continents, so-called sun [Bullet / Niche / Electromagnetic Wave] is the manifested sign of natural magnetism as well as manifested North and South directions in resemblance with revelation, twice rising and twice setting of the so-called sun cause the alteration of day & night for the equal & opposite seashore, and the like. I want to share my searched out verifiable as well as justifiable manifest [natural] truths with each and every matured member of the manifested hexagonal world and appearing pentagonal earth with three ascending stairs within the Black Square of East Horizon through self-publication. I am a/an/the ----------------------------.
6	I am a scientist but I have no concept concerning the realities of One Dot Leader, Diamond Operator, End of Proof, Black Square, Bullet, Bullet Operator, Ring Operator, Octagon, Hexagon, Pentagon, Star Operator, Quadrilateral Stars, Difference between Natural Magnetism and Man-made Magnetism, Semi-anticlockwise and Semi-clockwise stages of journey of the so-called sun as the manifested [natural] sign of Natural Magnetism, twice rising & twice setting of the so-called sun, four cardinal directions in resemblance with four basic forces, Sirius Binary System, Night Sky, Earth Sky, Earth's Night Sky, Morning Show and Evening Show, octagonal black spots, the clear proofs & eye opening evidences etc. I even do not know the existential import of the hexagonal unit such as Unicode (hex), Symbol (hex) etc, though I am using those measuring units / criterions for verification and justification of manifest truth. I am a/an/the ----------------------------.
7	I am an appointed/elected/selected higher authority of a Scientific Research Institution. My chiefs, ministers, soldiers, pen-paper-pencil workers, black & white artists, mouth speakers, and the like have true concepts of the searched out realities of One Dot Leader, Diamond Operator, End of Proof, Black Square, Octagon, Hexagon, Pentagon, Difference between Natural Magnetism and Man-made Magnetism, Semi-anticlockwise and Semi-clockwise stages of journey of the so-called sun or manifested sign of natural magnetism, four cardinal directions in resemblance with four basic forces, Sirius Binary System, Night Sky, Earth Sky, Earth's Night Sky, Morning Show and Evening Show, octagonal black spots, planets etc. So, I have prepared separate sheets of the searched out real scientific symbols called Insert – Symbol. But I do not want to share searched out manifest truths i.e. denotation and connotation of Insert Symbol with a view to track off believers from equal & opposite right direction and to dominate mankind in general of the Natural Pentagonal House [natural magnetism] from the Man-made Pentagonal House [man-made magnetism]. In brief, I am

	dominating the so-called sun as the manifested sign of natural magnetism projecting the veil of Electron Cloud over upright logic of mankind in general through Red Colour Triangle of the Napier Bones and four manifested directions in resemblance with man-made magnetism. In other words, I have projected myself as if the sole owner of the windows of seven ultimate or seven repeated verses in resemblance with Iblis. So, my Home Number is 7, my window number is 7, my book number is 7, my instrument is AK47, my identifying mark is Antivirus K7, my sign is headache seven [&], and my key is Circular Rotation & Revolution System. I am a/an/the ----------------------------.
8	I do not know that Copernican Revolution is a Post Hoc Ergo Propter Hoc. I have firm faith on the works of black & white artists. So, I have conceptualised that the so-called sun as the manifested sign of natural magnetism rises once and sets once to cause the alteration of day & night for the equal & opposite seashore. It rises from the East [downward direction & gravitational force] and sets in the West [upward direction & strong force]. I have no curiosity to confirm manifest truth concerning Right Directions of Electromagnetic Wave or converted snake of Muus's Staff as well as correct statistics of the alteration of day and night for the equal & opposite seashore [East Zone & West Zone]. I have conceptualised Historical Black Magic as the manifest truth in resemblance with revelation. I am a/an/the ----------------------------.
9	Semi-anticlockwise and semi-clockwise or equal & opposite stages of journey of the so-called sun are the manifested signs of natural magnetism in resemblance with revelation. I have clear concept that there is only one kind of magnetism. This magnetism may be called Natural Magnetism. I am undoubtedly sharing with each and every matured member of the manifested hexagonal world and appearing pentagonal earth that the so-called sun as the manifested sign of natural magnetism neither rises once from the East nor sets once in the West. The so-called sun as the manifested sign of natural magnetism enters from Manifested North-East [Revealed Left / Northern Hemisphere / Haiyalas-Swalah] in correspondence with North America and sets in Manifested South-East [Revealed Right / Southern Hemisphere / Haiyalal-Falaah] in correspondence with South America for the Ground Stair of North America and South America of the East Zone [Lower Seashore]. The so-called setting sun runs away from there and rises from Manifested South-West [Revealed Right / Southern Hemisphere / Haiyalal-Falah] in correspondence with Europe and ends the journey in Manifested North-West [Revealed Left / Northern Hemisphere / Haiyalas-Swalaah] in correspondence with Australia for the Middle Stair of Europe, Asia, Africa, and Australia as well as Topmost Stair of Arabian Peninsula of the West Zone [Upper Seashore]. So, the so-called sun as the manifested sign of natural magnetism enters from North-East, sets in South-East, again, rises from South-West and ends in North-West. As the manifested sign of Magnetic Force, the so-called sun [Bullet / Niche / Electromagnetic Wave] enters from Northern Hemisphere [Manifested North / Haiyalas-Swalaah of the East Zone] and moves Semi-anticlockwise and sets in Southern Hemisphere [Manifested South / Haiyalal-Falaah of the East Zone] for the East Zone. Due to Magnetic Force, the so-called setting sun as the manifested sign of natural magnetism runs away from East Zone and rises from Southern Hemisphere [Manifested South / Haiyalal-Falaah of the West Zone] and moves Semi-clockwise and ends the journey in Northern Hemisphere [Manifested North / Haiyalas-Swalaah of the West Zone] for the West Zone. In brief, due to magnetic force, the so-called sun as the manifested sign of [natural] magnetism enters from North [revealed Left] and ends the Semi-anticlockwise and Semi-clockwise stages of journey in North [Revealed Left]. This revealed left or North represents manifested North-East and North-West.

	Semi-anti-clockwise and Semi-clock-wise stages of journey of the so-called sun from North-East to North-West are the manifested signs of [natural] magnetism. The above shared findings are not only verifiable & justifiable but also unalterable. All established laws, revolutionised theories, pen-paper-pencil works, black & white activities, imposed formalities & guidelines, and the like which are either contrary or contradictory or both to the manifested signs of [natural] magnetism cannot be believed/understood as manifested truths [natural truths] save Historical Black Magic. To do righteous work and to give good tidings or Duty for Duty's sake or Niskama Karma or Survival of the Truest is the Summum Bonum of my Life. I am openly and publicly sharing searched out manifest truths with the hope that those who possess true knowledge [upright logic] will verify & justify each shared finding for the greatest happiness of the greatest number as well as for the sake of manifested nature [manifest truth] and sanctity of formal education [acquiring & sharing true knowledge]. The Historical Don of Teleological Evidence Sorcery & Unique Epistemic Persecution [Black Magician] has no control over equal & opposite manifested signs of Magnetic Force or what is being taught as Natural Magnetism. So, with a view to get generalised recognition of the searched out Solidified Solid Human Rights from the respective authorities for the greatest happiness of the greatest number, I have placed 'Kitaaba Wal-Hikmata' or 'Manifested Nature and the Utility of One's Upright Logic' before respective tables of the Honourable Chairs particularly before the Head of the Dept. of Philosophy AUS and International Human Rights Commission [IHRC HQ] on 16th April, 2016. I am a/an/the ----------------------------.
10	I have no concept of the criterions of truth. I have faith on revelation and firm belief on hearsay & Historical Black Magic. So, I always follow what the Teleological Evidence Sorcerers and Epistemic Persecutors as the representatives of the Don of Black Magic project before me. I am a/an/the ----------------------------.
11	I have clear concepts of the criterions of truth, subjective selfcontradictions and objective paradoxes, fallacy of infinite regress, fallacy of arguing in a circle, fallacy of post hoc ergo propter hoc etc. But my binary nature does not permit me to verify & justify manifest truth on the basis of the criterions of truth. So, I always do my jobs on the basis of opinions, suppositions, imaginations, predictions of the pen-paper-pencil workers, black & white artists, teleological evidence sorcerers, epistemic persecutors, the Trinity of Hypocrisy – Tyranny – Conspiracy and Duality of Activism & Terrorism of the Historical Don of two-in-one partnership, and the like. I am a/an/the ----------------------------.
12	I have no clear concept of the entering and rising directions as well as setting and ending directions of light for this manifested hexagonal world and appearing pentagonal earth within the Black Square of Manifested East Horizon, yet I am taking the responsibility to lead mankind towards light i.e. Arsh in the life hereafter as well as in the Day of Judgment projecting myself as a metaphysical entity called Mystic / Spiritualist. I am a/an/the --------------------
13	I am an Environmentalist [Naturalist]. I have no concept of the Environment or Nature or Manifest Truth. I do not know that the universe is an upright rectangle in coherence with the 'End of Proof' of the Real Science and in correspondence with the appointed Kaba; the revealed East Horizon guarded by the four basic forces is called Black Square, two zones within the East Horizon, octagonal matter particles as electron clouds or projected sun which rises from the East, the world has been manifested as a hexagon, due to equal & opposite ~sth between Upright West Region of Arabian

Peninsula and Straight Middle East Region of Eartha 3D the manifested hexagonal world is appearing as a pentagon, the appearing pentagon with three ascending stairs is called the earth; due to projected planets Saturn and Mars the World and the Earth of the East Horizon [Black Square] are not moving; projected nine planets are the clear proofs and eye opening evidences given to Muusa [ass]; equal & opposite movements of the so-called sun as the manifested sign of natural magnetism from North-East to North-West via South-East & South-West are the manifested/natural signs of magnetism, Semi-anticlockwise and semi-clockwise stages of journey of the manifested sign of natural magnetism [so-called sun / Bullet / Niche] cause the alteration of day and night for the equal & opposite seashore. In brief I have no concept of the source of light for the East Horizon [Black Square], yet I am fighting for Enlightenment / Natural Rights. I am a/an/the -------------------.

Know you not that **to Allah (alone) belongs the dominion of the heavens and the world**? He punishes whom He pleases, and He forgives whom He pleases. And Allah has power over all things. [Sura No. – 4\5 – Maaaidah, Ayat\Verses – 40]

If you **do not** stretch your hand against me, **to slay me, it is not for me to stretch my hand against you to slay you**: For I **fear Allah**, the Rab of the universe. [Sura No. – 4\5 – Maaaidah, Ayat\Verses – 28]

On that account: We ordained for the Banii-Israa-il that if any one **slew a person**--unless it be for murder or for **spreading mischief in the land [earth]**- -it would be as if he **slew the whole people**: And if any one saved a life, it would be as if he **saved the life of the whole people**. Then although there came to them Our **desciple** with **Clear Signs**, yet, even after that, **many of them continued to commit excesses in the land [earth].** -.[Sura No. – 4\5 – Maaaidah, Ayat\Verses – 32]

O you who believe! **The law of equality** is prescribed to you in cases of murder - **the freeman for the freeman, the slave for the slave, the woman for the woman**. But if any remission is made by the brother of the slain, then grants any reasonable demand and compensate him with handsome gratitude. This is a concession and a Mercy from Allah. **After this whoever exceeds the limits shall be in painful doom. In the Law of Equality there is Life to you in retribution. O men of understanding! you may free yourselves from the path of shaytan (devil's path).** [Sura No. 1\2 – Baqarah, Ayat No. – 178 & 179]

SECTION – II
METHODOLOGY – 2.2
EVIDENCE SORCERY AND EPISTEMIC PERSECUTION
[BLACK MAGIC]

Fasta-'iz billaahi minash-Shaytaanir-Rajim
Bismillaahir-Rahmaanir- Rahim

If knowledge is called light and reading & writing books are the means of acquiring & sharing true knowledge [light], then what are the scales or criterions of verification as well as justification of the 'Quality' or 'Standard' of a Book for Publication save Truth-in-itself and criterions of truth? Who will verify and justify the 'Quality' or 'Standard' of 'Kitaaba Wal-Hikmata' / 'Manifested Nature and the Utility of One's Upright Logic'? Who will not publish 'Kitaaba Wal-Hikmata' \ 'Manifested Nature and the Utility of One's Upright Logic'? Who will not read 'Kitaaba Wal-Hikmata' or 'Manifested Nature and the Utility of One's Upright Logic'? Who will not recognise revealed and manifested directions in resemblance with equal & opposite stages of journey of the manifested signs of natural magnetism? Who will not recognise manifest truths or apparent truths [manifested nature] in resemblance with equal & opposite revelations?

'Kitaaba Wal-Hikmata' or 'Manifested Nature and the Utility of One's Upright Logic' is an inspired sharing of a **Trinity** of **Reminder – Inspiration – Good Tidings**. This sharing **primarily** includes **two apriori forms of knowledge as the Universal Major Premises** and **secondarily** includes **Generalised Minor Premises. The shared** selfevident truths of **'Kitaaba Wal-Hikmata' or 'Manifested Nature and the Utility of One's Upright Logic' are the deduced conclusions**. Universal Major Premises represent **universal and necessary truths** which are justifiable only by the utility of upright logic in resemblance with revelations. Minor premises represent **verifiable truth of facts** which are verifiable further on the basis of manifested signs, real scientific symbols of the separate sheet, verifiable laws & justifiable theories, and brief explanation of the sharer of this inspired sharing in resemblance with the acquired terminologies through formal education. So, regarding universality & necessity, this sharing is beyond human made laws, rules, guide lines, formalities, theorisations and the like as well as **beyond any argument**. Laws are established not to formalise faith & belief or **Solitary Rights** or Individual Rights with a view to conceal Principle of the Uniformity of Nature & the Law of Causation, but to protect one's faith & belief **[Individual Rights] and Solidified Solid Human Rights** from the **Trinity** of **Hypocrisy – Tyranny – Conspiracy, and the Duality of Activism and Terrorism**. Is there an established law against the **Sanctity** of both Uniform Principle [Revelation or Truth-in-itself] and Uniform Motion [Successive Relation or Manifested Nature or Criterions of Truth or Coded Tautologies] or Universal Major Premises such as 'All men are mortal' and the Affirmative Minor Premises such as 'I am a man'; though there are innumerable laws to control the wicked hands of Trinity as well as Duality of the **Teleological Evidence Sorcerers and Unique Epistemic Persecutors, owner and possessors of manmade nature, and the like. So, all justifiable/valid laws are the protectors** of the two pre-conditions of true knowledge [wisdom] or Solitary Rights or the primary findings of '**Kitaaba Wal-Hikmata'** or '**Manifested Nature and the Utility of One's Upright Logic'** from the binary nature of **Teleological Evidence Sorcerers called Hypocrisy [Activism] as well as Unique Epistemic Persecutions called Terrorism**. In other words, all valid laws will resemble with the shared conclusions of '**Kitaaba Wal-Hikmata'** or '**Manifested Nature and the Utility of One's Upright Logic'**. If a Law does not resemble with the Universal Major Premises such as 'All men are mortal' as well as verifiable Minor Premises such as I am a human person; then that law cannot be regarded as a valid law [justifiable as well as verifiable Law] with respect to the deduced conclusion 'I am a mortal being'. On the contrary, that law may be a theory or an opinion or a discovery or a contrary formality or a contradictory guideline etc. One of the **purposes** of this inspired sharing is to **rescue both Truths-in-themselves and coded tautologies** from the **veil of teleological evidence sorceries** i.e. from the veil of the **trinity** of Discovery – Theory – Opinion or Supposition – Imagination – Prediction.

INVALID REFERENCES vs. VALID REFERENCES
REVOLUTION vs. RESURRECTION
VERA CAUSA OF THE TRINITY OF
HYPOCRISY – TYRANNY - CONSPIRACY
TRINITY OF DISCOVERIES – THEORIES – OPINIONS

There is a difference between **Theory** and **Law** like the difference between discovery and proof. Theories are opinions or imaginations of some so-called epistemic persons which are yet to be observed in resemblance with two pre-conditions of true knowledge [formal grounds] and Manifested Nature [material ground]. In other words, the conclusion of a theory is an imagination till the formation of a hypothesis on the basis of observation of particular instances [manifested signs]. So, the conclusion of a theory is either **contingent or contradictory, but cannot be recognised or generalised or categorised as a tautology**. That is why; **Plato said** an opinion is not knowledge, neither right knowledge nor wrong knowledge. It is very unfortunate fact that the over-spreading of imaginations like the advertisement of a product through the **Trinity** of **black & white work - publication - broadcasting**, people like the sharer of this inspired sharing are conceptualising imagination as hypothesis, probability as certainty, discovery as proof, theory as law, opinion as criterion, contingency/contradictory as tautology, and the like. So, pen – paper – pencil workers and black & white artists are recognising several contingencies as well as contradictories as **ROTATIONS & REVOLUTIONS** within the domain of formal education without **BROKEN BAR**. Such **ROTATIONS & REVOLUTIONS** are nothing but the wicked hands or the instrumental causes or efficient causes of the trinity of hypocrisy – tyranny - conspiracy as well as the duality of activism and terrorism in resemblance with the Circular Rotation & Revolution of the Immovable Star Operator. The similitude or likenesses of such Rotations & Revolutions are **Copernican Revolution, Kantian Revolution, Napier Bones, Kepler's Laws** and the like. **J. S. Mill** had tried to get his work Revolutionised one concerning Universal Major Premises or two pre-suppositions or faith & belief or Solitary Rights or Individual Rights so that theories can be applied on Uniform Principle & Uniform Motion [Successive Relation] or Solidified Solid Human Rights and ultimately committed **two suicide cases** – one is called **subjective selfcontradiction** [but not **self-contradiction**] and the other is called **objective paradox**. His subjective selfcontradictions [fallacies concerning formal truth] and objective paradoxes [fallacies concerning both formal and material truth] are the **self-explained evidences** as well as **clear proofs** that opinions, theories, and discoveries are the **instruments or wicked hands** of the **Historical Don** of the **trinity** of **hypocrisy – tyranny - conspiracy, and the duality of activism and terrorism**. The consequences of Mill's attempt imply that formalities and guidelines are **not valid references or criterions of truth.** So, a subjectively selfcontradictory and an objectively paradoxical theory or opinion or discovery or law or formalised rule or generalised guideline is neither justifiable nor verifiable on the basis of the criterions of **Tawraat** [Coherence Truth], **Injiil** [Correspondence Truth], **Zabuur** [Pragmatic Truth], and **Furqaan** [One's Upright Logic] in resemblance with **Truth-in-itself** [Revelation or Universal Major Premise] and in correspondence with Coded Tautologies [Affirmative Minor Premises]. This sharing of the **'Selfevident Truth' is not a revolutionary work like Copernican Revolution or Kantian Revolution or the Revolutionary Game** of Come with One among us [Four Pillars for straight uses or Four Man-made Forces] and Reject your Faith and belief [faith on truth or Revelation and belief on Four Criterions of Truth or Manifestations or Four Basic Forces]. On the contrary, it is an inspired sharing, reminder, and good tidings for all concerning the overwhelming **Day of Resurrection [Yawmal-Qyiamati].**

TRINITY OF FORMALITIES –
GUIDELINES – RECOMMENDATIONS

Regarding the **publication** of a book several formalities and guidelines have been placed before the sharer of selevident truth or manifest truth. This sharing is primarily concerned with 'All men are mortal' and secondarily concerned with 'I am a man' with a view to get confirmed recognition that I am a mortal being. This is called Deductive Method or Mathematical Method. This method has been developed as Dictum De Omni ET Nullo of Aristotle and Cartesian Method of Rene Descartes. This sharing is concerned with Truth-in-itself and selfevident truth in resemblance with Deductive Method or the established Dictum De Omni ET Nullo and Cartesian Method. So, this sharing is beyond **contrary formalities** as well as **contradictory guide lines** so far have been placed before the sharer of Selfevident Truth or Manifested Nature. If the Teleological Evidence Sorcerer or Unique Epistemic Persecutor is able to formalise the above mentioned deductive argument, then I will try to follow the established contrary formalities as well as contradictory guidelines A to Z. The size of the book, style of writing, get up of the book, total pages of the book with serial number, permissions of the respective authorities for using figures available in the printed book or in the Internet as references, and the like which have been established as guide lines of publication may be necessary for those who want to publish their works motivated by the sense of recognition. Those people will follow A to Z of the established formalities as well as guidelines with a view to accumulate recognised pieces of paper through **Recommendation of finite/mortal beings as if recommendations** are the criterions of Truth. But being a professor of philosophy, I know it very well that those recommendations and pieces of paper are not criterions of truth. So, as a **sharer of 'Kitaaba Wal-Hikmata' or 'Manifested Nature and the Utility of One's Upright Logic', I am neither seeking nor desiring recommendations for the publication of the searched out Selfevident Truths / Manifest Truths**. It is being published partially since 31st July, 2015. If my Rab wills, then the pentagonal publication of Selfevident Truth / Manifest Truth will not only unveil several Teleological Evidence Sorceries and Unique Epistemic Persecutions from behind the bush of Recommendations but will also become concrete example as well as valid reference for future Verifiable and Justifiable Recommendations.

Regarding **acknowledgment prior to verification and justification of the searched out findings,** I would like to share with you that this inspired sharing is going **to unveil several Teleological Evidence Sorceries and Unique Epistemic Persecutions** which almost all mortal beings do not know save the Historical Don of two-in-one partnership and his Chiefs & Ministers [with reference to four projected pillars and Insert - Symbols]. All contrary formalities and contradictory guidelines concerning publication and broadcasting have been made by that Historical Don to retain his/her dominion over mankind of **Star Operator** within the **Black Square** as well as to track off believers from the equal and opposite middle course towards Upright West [Arsh]. If someone or a group of epistemic persons are able to acknowledge prior to verification and justification of the searched out shared conclusions of '**Kitaaba Wal-Hikmata'** or '**Manifested Nature and the Utility of One's Upright Logic'** that this is an authentic work, that means the shared findings are selfevident truths in resemblance with manifested nature, then he/she or that group must know the manifest truth. If they know the manifest truth, then why did they not share the same with us before the sharing of this '**Kitaaba Wal-Hikmata'** or '**Manifested Nature and the Utility of One's Upright Logic'**? If there is any such kind of acknowledger who is concealing manifest truth knowingly, then he/she is a **Hypocrite** and his/her activity is called Hypocrisy. So, no rational being is able to acknowledge '**Kitaaba Wal-Hikmata'** or '**Manifested Nature and the Utility of One's Upright Logic'** prior to its recognised confirmation through verification and justification save the Historical Don of Trinity & Duality and his black & white artists. If a publisher is able to add a recommendation or a recognition from that Historical Don or his black & white artists prior to the pentagonal publication of this inspired work, then I have no objection to include the same as the Recommendation for the publication of '**Kitaaba Wal-Hikmata'** or '**Manifested Nature and the Utility of One's Upright Logic'**.

Moreover, those who are arguing in favour of the concealment of real and projection of unreal, they are the partners of the Historical Don of Trinity & Duality. Such kind of pen, paper and pencil workers or black & white artists are the **real activists like the father of IRF**. They are consciously or unconsciously doing their pen – paper – pencil works in resemblance with projected mechanical globalisation, man-made magnetism, and circular rotation & revolution system. Are there pen – paper – pencil activists and black & white artists who will be able to prove One-One [11] contents as well as contexts of their uncountable contrary formalisations, ignorant recommendations, contradictory generalisations, and black & white publications & broadcastings? The One-One contents and contexts of the pen – paper – pencil workers and black & white artists are as follows –

ONE-ONE [11] CONTENTS AND CONTEXTS
i) The Revealed Universe, the Manifested World, and the Appearing Earth resemble with the Projected Global Trinity, and the three ascending stairs of the appearing earth resemble with circular rotation

& revolution of the pentagon in resemblance with Copy Right & Registered Sign.

ii) Man-made magnetism represents Manifested Northern Hemisphere in resemblance with Revealed Left [Magnetic Force, entering point of the Electromagnetic Wave, ending point of the Bullet] and Manifested Southern Hemisphere in resemblance with Revealed Right [Weak Force, turning point of the Electromagnetic Wave, rising point of the Bullet] of the Manifested Hexagonal World and Appearing Star Operator within the Black Square of East Horizon.

iii) Projected Northern Hemisphere through Global Trinity represents Manifested North as well as Revealed West [Back Border, Magnetic Force, Night Sky, Heaven, Galaxy of Stars, Upward Direction, Strong Force] and projected Southern Hemisphere through Global Trinity represents Manifested South as well as Revealed East [Border, Weak Force, Hell, Earth Sky & Earth's Night Sky, Downward Direction, Gravitational Force].

iv) The direction from which the so-called sun as the manifested sign of natural magnetism rises once only for the entire day represents Manifested East [Back Border, Gravitational Force and entering direction of the Electromagnetic Wave i.e. Magnetic Force].

v) The direction towards which the so-called sun as the manifested sign of natural magnetism sets once only for the entire day represents Revealed West [Border, Strong Force and setting direction of the electromagnetic wave due to Magnetic Force].

vi) There are two kinds of magnetism. This reference is available in uncountable published books i.e. works of the pen – paper – pencil workers and activities of black & white artists. This reference implies that there are two North Directions and two South Directions. Which directions [North & South] represent man-made magnetism [Conspiracy North and South] and which directions [North & South] represent natural magnetism [Revealed Left and Revealed Right]? Whether Manmade Magnetism represents Manifested Directions in resemblance with Revelations or Natural Magnetism represents Manifested Directions in resemblance with Revelations?

vii) Whether Manifested Sign of Natural Magnetism [Electromagnetic Wave or Equal & Opposite stages of journey of the so-called sun as the manifested sign of natural magnetism or Binary Pulsar of Einstein or the kindled niche as the appointed light or Bullet or Wash-shamsi] represents Projected Northern Hemisphere & Southern Hemisphere or North & South represented by Man-made Magnetism or Manifested

Northern Hemisphere [Revealed Left] & Manifested Southern Hemisphere [Revealed Right]?

viii) Upright Rectangular Universe, Hexagonal World, and Pentagonal Earth with Three Ascending Stairs are moving round following Copernican Hypotheses i.e. in two forms – Rotation and Revolution. So, Copernicus is the One Dot Leader of the Star Operator, End of Proof, Black Square, and Three Ascending Stairs.

ix) Semi-anti-clockwise and Semi-clockwise stages of journey of the so-called sun as the manifested sign of natural magnetism or the Binary Pulsar of Einstein or Kindled Niche or Bullet or Wash-Shamsi or Projected Sister Planet Venus or Maryam supplied with Sustenance or She-camel as the manifested signs of Allah form a full circle due to Copernican Hypotheses and Kepler's Orbit. So, Copernican Hypotheses and Kepler's Orbit represent Revealed and Manifest Truths or Manifested Nature.

x) The Manifested Sign of the 'End of Proof' or Black & White Imam of the City called Kaba as the Standard of right directions in coherence with four basic forces and in pragmatic correspondence with Crucified Sign [Plus Sign of Mathematics] for mankind has not been appointed in the Upright West Region of the manifested hexagonal world, Topmost Stair and Uppermost Land of the appearing pentagonal earth, and on the right side of the Mount Tuur within the Black Square of East Horizon. On the contrary, the Imam of the City called Kaba has been appointed in the Straight Middle East Region of Eartha 3D or Black Hole or Hole in the journey boat of Muusa [ass], and at the centre of the Global Trinity & Quaternary as well as Circular Rotation and Revolution System of the hexagonal world and pentagonal earth.

xi) Equal & Opposite stages of journey of the so-called sun [Bullet] from Haiyalas-swalaah [Northern Hemisphere or Revealed Left or Manifested North-East] to Haiyalaal-Falaah [Southern Hemisphere or Revealed Right or Manifested South-East] and turning back from Haiyalaal-Falaah [Southern Hemisphere or Revealed Right or Manifested South-West] to Haiyalas-Swalaah [Northern Hemisphere or Revealed Left or Manifested North-West] do not represent the right direction of the manifested signs of natural magnetism; but the project circular rotation & revolution system from East to West represents manifested signs of natural magnetism. Moreover, circular ways in resemblance with projected global map towards Straight Middle East Region of Eartha 3D or Black Hole following North and South determined by Man-made Magnetism represent the right direction of my Qibla.

It is certain like 'Cogito Ergo Sum' of Rene Descartes that neither the pen – paper – pencil workers, nor black & white artists, nor Epistemic Uniqueness of Scientific Certainty, nor Owner of Manmade Magnetism, nor Possessors of Manmade Nature, nor Don of the two-in-one Partnership, nor Don of the Historical Black Magic will be able to prove the above mentioned One-One [11] contents and contexts of subjective selfcontradictions, objective paradoxes, petitio principii, argument ad infinitum, post hoc ergo propter hoc etc., yet they are uninterruptedly doing their jobs following A to Z contrary formalities and contradictory guidelines in resemblance with mechanical globalisation, manmade magnetism, circular rotation & revolution system, and the like. Further, it is selfevident that all pen – paper – pencil works; black & white activities, contrary formalities and contradictory guidelines which resemble with the above mentioned One-One [11] contents and contexts of post hoc ergo propter hoc are neither true knowledge nor manifested nature. On the contrary, these are the vera causa of leading Unity towards Diversities as well as experimentum crusis & crucial instances against the Trinity of Hypocrisy – Tyranny – Conspiracy & Duality of Activism and Terrorism. The introduction of the above mentioned One-One [11] contents and contexts of post hoc ergo propter hoc in the domain of formal education is Teleological Evidence Sorcery and Unique Epistemic Persecution of Faith and Belief of mankind on Manifested Nature, Scientific Certainty, Formal Education, and the like. If it is argued in favour of post hoc ergo propter hoc, then it will imply that the sole purpose of formal education is to acquire false knowledge or Avidhya or Veil of Ignorance. So, those who are involved to project and to argue in favour of post hoc ergo propter hoc, they are none but Chiefs and Ministers of the Historical Don of Teleological Evidence Sorcery and Unique Epistemic Persecution.

So, the question is – **'Who will recommend or recognise shared 'Selfevident Truth' of 'Kitaaba Wal-Hikmata' or 'Manifested Nature and the Utility of One's Upright Logic'?** In this regard let us try to make clear our concept about the very term **'selfevident'** which has been broken into **'self-evident'** establishing conspiracy **grammatical rules** in resemblance with the **Hypocrisy of Divide Each Individual concept of Truth or Solitary Rights or Individual Rights in to smallest particles in resemblance with Unicode (hex), ASCII (hex), Symbol (hex), Symbol (decimal), ASCII (decimal), 6310 commentary codes of IFTA on Quran in the name of 'The Holy Quran', and the like projecting Contrary Rules, Contradictory Guide Lines, Categories of knowledge [Shane-nuzuul] etc. with a view to dominate his/her Folk**. In other words, dig a black hole of selfcontradiction in one's upright logic and veil him/her with objective paradoxes with a view to track off him/her from the equal & opposite middle course towards Upright West. It is none but my Rab who has unveiled the projected covering from my eyes and has shown me the white thread so that I must have to make an attempt with a view to unveil two West & two East, Revealed Right & Revealed Left, Manifested North-East & North-West, Manifested South-East & South-West, Early Hours of Night & Fall of Night, Morning Show & Evening Show, Sole Natural Magnetism, and the like with a view to make clear the distinction between white thread and black thread or Region of White and Region of Black or Upright West Region of Arabian Peninsula and Straight Middle East Region of Eartha 3D.

A selfevident truth is an individual concept of truth grounded on individual faith & belief on One Dot Leader and Diamond Operator. In other words, a selfevident truth is one's faith and belief on two pre-suppositions, namely, Uniform Principle [Principle of the Uniformity of Nature] and Uniform Motion [Successive Relation or Law of Causation]. Uniform Principle states that Nature behaves same under similar circumstances. This is called Fitrat or Dharma or Pure Religion or True Religion or Diin or Nuur or Light or Wisdom or Truth-in-itself. The law of Causation [Successive Relation] states that every event must have a cause. In other words **nothing comes out of nothing** or Ex Nihilo Nihil Fit.

There is a difference between **acknowledgement** and recommendation [**recognition**]. Acknowledgement is subjective and recommendation [recognition] is objective. An acknowledgement is the expression/statement of an individual **of** something; while recommendation or recognition is the judgment or the individual opinion **on** something. So the very term 'recommendation or recognition' with respect to truth means 'generalisation'. A recommended or a recognised truth means a **generalised truth**.

Generalised truths are of two kinds – discovery and proof. **Theorisation** is the product of research works within falsehood. It is done in the field of **critical study, comparative study, commentary, and theoretical study** without having clear concept of what a researcher is researching. So, a fictitious researcher starts with an imagination of a lacuna somewhere in the works of pen – paper – pencil as well as activities of black and white artists. This is called contingency or contrary to truth. A theoretical research is not done on the basis of observation of particular instances or framed hypothesis on the basis of observation of manifested signs [truth of facts] but on the basis of superficial resemblance among opinions, denotative criterions [number of references] and the like with a view to conceal truth and to project falsehood as truth. For example – Like men, plants have birth, growth, decay, and death. Men possess intelligence. Therefore, plants also possess intelligence. This is called **bad** analogy or **false notion. There is no hypothesis in this theorisation save imagination in resemblance with John Rawls' Veil of Ignorance.** During bright day light if I make a supposition that it is a dark night and begin to walk with a torch light in a public road on the basis of framed supposition, then that supposition cannot be called a hypothesis save imagination or veil of ignorance. If you fail to conceptualise this selfevident truth, then you have no capability to approach truth in resemblance with the concept of 'Capability Approach' of Amartya Sen. This equal and opposite theorisations are nothing but stepping stone to track off upright nature from the criterions of Truth projecting 'Theory of Justice' prior to 'Idea of Justice'. Are there mortal beings like the possessor of capability approach who are willing to argue with the above mentioned One-One [11] contents and contexts of post hoc ergo propter hoc? If there are such persons, then their range of thinking is called imagination and their willingness is called capability approach. Now, with a view to generalise truth, it is an acid test before Teleological Evidence Sorcerers and Unique Epistemic Persecutors to falsify or to recognise the searched out selfevident truths shared as Solidified Solid Human Rights in the **'Kitaaba Wal-Hikmata'** or **'Manifested Nature and the Utility of One's Upright Logic'** with a view to prove that their imaginations were hypotheses and the willingness was their capability approach to distinguish between justice and injustice, truth and falsehood, concealment and projection, natural magnetism and man-made magnetism, Manifested Nature and Manmade Nature, Projected North and Manifested North, Manifested South and Projected South, Projected Northern Hemisphere and Revealed Left, Projected Southern Hemisphere and Revealed Right, Global Trinity and Hexagonal Measuring Units, Globe and End of Proof, Globe and Black Square, Globe and House, Global Map and Star Operator, and the like.

Have you ever tried to do a research work on Newton's Third Law – 'Equal & Opposite'? The sole reason is that it is a selfevident or a manifest Truth in resemblance with Revelation. In reality, theorisation is a competition among Epistemic Senses with a view to verify the answer of a question – 'Who can abuse better than me?' Just reflect on the concept that if a truth can be criticised and researched, then how can it be recommended and recognised as truth? If it is recommended and recognised as truth, then what is the necessity of doing research on recommended and recognised truth? Such kind of epistemic researchers are not sharing truth with our fresh generations. On the contrary, they are teaching them how to abuse each other. Try to conceptualise the difference between **'teaching'** and **'sharing'**. Teaching implies a kind of punishment with reference to **'according to'** only. But sharing implies a mutual communication or an environment of freeness in sharing truth as if a game of parables and tales with gesture and posture. Further, if it is asked – 'What does a teacher teach?', then the answer of this question may lead the process to an infinite regress or arguing in a circle. But if you ask me – 'What does a teacher share?', then the sole answer is – 'A teacher sharers his/her selfevident concepts of Truth free from subjective selfcontradictions and objective paradoxes'.

The first stage of search for truth is discovery. It origins from faith & belief on **two pre-suppositions** i.e. Uniform Principle & Successive Relation; but it **begins** with **Un-contradicted experience** and **observation of essential points of resemblance** between **two things or between phenomena under investigation or between revelation and manifested signs**. Try to conceptualise **'origination'** and **'beginning'** in resemblance with **Kantian concepts** whose works have been compared with **Copernican Revolution**. The conclusion of first searching is called discovery. It is drawn on the basis of the Method of Agreement or Joint Method of Agreement and Difference. So, the conclusion of the first stage of searching is always probable. This probable conclusion is called a supposition or a hypothesis or a discovery, but not an imagination or a prediction. For example – Animals like men have birth, growth, decay, and death. Men feel pleasure and pain. Therefore, animals also feel pleasure and pain. This conclusion is a hypothesis or an established conclusion or a discovery of a causal relation or the first stage of **searching**.

If the searcher of truth or re-searcher proceeds further with his searched out findings called hypotheses or discoveries like the searcher of this sharing with a view to prove a causal relation between antecedent and consequent, then that proceeding is his/her second **stage of searching or re-search on previous searching**. Now, applying either the Method of Difference or the Method of Residues or in some special cases the Method of Concomitant Variation, if the **re-searcher** [second stage of searching] is able to search out [second time] a causal relation on the basis of both positive and negative instances, then that re-searched conclusion is a certain conclusion or a Proof or a **Re-search work in the truest sense of the term.**

Further, there is a difference between a **generalised truth** and a **general truth**. When a justified as well as verified truth is recommended and recognised as truth and is established as a Law, then it is called generalised truth like Newton's Second Law 'F = ma'. Scientific Progress depends on generalised Law. The generalised law may vary due to circumstances. But each criterion of truth is called a general truth or a tautology or a selfevident truth like 'I am a man'. The recognisers of a generalised truth may be more than one individual or more than one groups; while the recogniser of general truth is none but my Rab and your Rab or Dot Operator. The due balance between generalised law [truth] and general law [Tautology] is called Law of Causation or Successive Relation. The main task of Philosophy is to justify a hypothesis and the main task of science is to verify the probability and the certainty of a searched out hypothesis and re-searched conclusion on the basis of criterions of truth in resemblance with Truth-in-itself with a view to recognise the re-searched conclusion as a generalised Truth. If a re-searched conclusion can be recognised as a generalised truth, then that re-searcher should be awarded a Degree. This Degree is called **Doctor of Philosophy and abbreviated as Ph.D**.

But it is very unfortunate that the Historical **Don** of the **trinity** as well as **duality has become successful in diverting epistemic senses from the true concepts of Philosophy, Science, and Doctor of Philosophy by recognising some Theorisations as Revolutions.** So, Philosophers and Epistemic Persons in general motivated by the sense of recognition very often use a dialogue with reference to one of the definitions of Philosophy – 'Philosophy includes everything, but excludes none' with a concrete example – 'Philosophy is like a tree and the rest are the branches, leaves, flowers, fruits etc. of that tree'. Through this definition with concrete example they have established opinions as the strongest denotative criterions of justification in resemblance with 'Might is Right', **250** pages as one of the criterions of justification & verification, abusing or criticising others' subjective thoughts as one of the criterions of justification & verification, converting good names of others' without prior permission as one of the criterions of justification & verification, five chapters on a single concept as one of the criterions of justification & verification, and the like. It is a supposition that if you ask theoretical research guides, justifiers of research findings, verifiers of research findings, recognisers of Ph.D. Degree, and research scholars about criterions of truth, then almost 99% will reply like my Supervisor that the above mentioned contrary formalities and contradictory guidelines are the criterions of truth. For this reason, after the recognition of a research work, it is being kept in the dark chamber or in the black hole [Straight Middle East or Recycling Region] so that none can know about it. If you ask them about the logic of the concealment of the researched truth, they will reply that it is being kept concealed so that none can copy it. This implies that truth can be copied or Xeroxed or stolen. Are there epistemic thieves who will desire to copy [to steal] shared selfevident truths from the 'Kitaaba Wal-Hikmata' or 'Manifested Nature and the Utility of One's Upright Logic'? If there are such epistemic persons, then they are not epistemic thieves. On the contrary, they are real educated people who share manifest truths. So, if a truth is selfevident like Newton's Third Law, then none can copy or steal it except to share the same. Those who are sharing Newton's Laws they are educated people. Are there epistemic thieves who have stolen Newton's Third Law? What the epistemic thieves are stealing is nothing save falsehood in resemblance with mechanical globalisation, man-made magnetism, and Circular Rotation & Revolution System. To verify and to justify these shared concepts just go through Internet and search the Encyclopaedia concerning the reality of the universe, the world, the earth, so-called solar system, moon, planets etc. as well as works of the pen – paper – pencil workers and activities of black & white artists, you will be able to conceptualise how falsehood can be copied or Xeroxed. On the contrary, various kinds of software are being developed to test copy-writing. Now the software is the authentic criterion of truth like mechanical globalisation and manmade magnetism. I think these are sufficient concrete examples as well as valid references about how the selfsensed epistemic persons are dealing with both Truth-in-itself as well as criterions of Truth. Ask them a simple question with reference to the example of the tree – Have they ever heard that a banyan tree bears flowers of mango and fruits of jack-fruit? What is the Vera causa [real cause] of such kind of spoiling the sanctity of formal education? Whether philosophy is a branch of Logic or Logic is a branch of philosophy? Whether Truth is the ideal of philosophy or Truth is the ideal of Logic? If the purpose of knowledge is search for truth, then which one must be recommended and recognised as the mother of true knowledge – Logic or Philosophy? So, it is a justified sharing that Logic or **Kitaaba Wal-Hikmata** is the first Subject or Ummil Kitaab or Mother of all Subjects as Logic deals with truth. Logic is not a branch of philosophy. On the contrary, Philosophy and Science are the two main branches of Logic. This conceptualisation will help us to distinguish between truth and falsehood in the field of formal education.

Again, there is a difference between **general truth** and **universal truth** or between **tautologies and Truth-in-itself**. Each criterion of truth is a general truth [tautology or manifested sign] but not a universal truth [Truth-initself or Equal & Opposite Revelation]. On the other hand, a revealed truth is a universal truth. The **possessor** of general truth may be more than one, but the **Owner** of Universal Truth is none but one who has no attribute save that He is the Owner of all attributes. This Owner is the One Dot Leader of the Real Science. There is a difference between **possessor-ship** and **ownership**. I am living in a rented house, so the house is in my possession, but I am not the owner of that house. [Try to conceptualise the sequence of this sharing.] So, Bullet [Maryam supplied with sustenance], Ring Operator [Wax Feeble Jakariya, guardian of Maryam], and the like 25 shared names of Messengers & Warners are the Possessors of Successive Relations. The researchers of IFTA have converted almost all shared names of the Messengers and Warners in resemblance with the reference of a formalised research work violating revealed guide lines, grammatical rules, sanctity of good names, [spoiling] morality of a qualified person, solitary rights etc. as if their morality resembles with the tabula rasa of John Locke. Consequently, people of the Kitaab with Truth are uninterruptedly following their conspiracy fathers.

Universal Truth comes out of nothing. But a **general truth comes out of something**. This something is the **equal and opposite Trinity or Diamond Operator**. **This equal & opposite Trinity has been developed as the Third Law of Newton – 'Equal & Opposite'. Sirius Binary System of real science or Nuurun alaa Nuur of Kitaab with Truth represents manifested nature in resemblance with revealed trinity. Manifested Truth does not necessarily require supposition, imagination, assumption, prediction, recognition etc. save confirmation of the resemblance between manifested sign and revelation.**

Probably you do not know that the Historical **Don** of the **trinity** of **hypocrisy, tyranny, and conspiracy is also the Don of the duality i.e. activism and terrorism. This very Don knows that the believers are following wrong direction in their compulsory prayer [salat].** Believers of the Middle Stair of Europe, Asia, Africa, and Australia are performing their compulsory prayer towards Revealed Left [Haiyalas-Swalaah] and believers of Ground Stair of North America and South America are performing their compulsory prayer towards Revealed Right [Haiyalal-Faalah]. This very Don has translated Quran through the concealed agreement of two-in-one partnership with the wandering Arabs [Beduin] in the name of IFTA Research Centre after hundred years of the corporeal disappearance of Prophet (sas). This very centre has translated Quran with Translation of the Meanings and **six thousand three hundred ten [6310] Commentary Codes** in resemblance with projected conspiracy hexagonal & decimal codes in the name of 'The Holy Quran'. This very centre is also gifting its Commentary or Critical Remark on Quran with a view to project Mechanical Globalisation, Manmade Magnetism, and Circular Rotation and Revolution i.e. Manmade Nature. What a fool I was! I had conceptualised the commentaries i.e. critical remarks of the researchers of IFTA on Truth-in-itself as well as tautologies as blessed gift due to free gift package. Now, under the guidance of my Rab I have conceptualised the manifest truth justifiable by coded shared tautologies and verifiable by real scientific symbols along with their character codes. So, I will not make any compromise with searched out manifest truths if the **ORIGINAL TWO-IN-ONE EPISTEMIC LADEN** as well as the **Teleological Evidence Sorcerer and Unique Epistemic Persecutor project terrors in resemblance with their habits and customs called Morality.** These Original Educated but Uncultured Laden, Teleological

Evidence Sorcerer, and Unique Epistemic Persecutor have snatched away the right direction of my Qibla through the projections of global trinity, two North, two South, mechanical globalisation, man-made magnetism, circular rotation & revolution system, and the like i.e. Projected Manmade Nature contrary as well as contradictory to Manifested Nature.

Through this inspired sharing, I want to kill the **Introduced Calf** of Firawn which rises once from the unidentified East [Gravitational Force] and sets once in the infinite West [Strong Force] as if the appointed light for the manifested hexagonal world and appearing pentagonal earth within the Black Square of East Horizon. It is expected that going through 'Kitaaba Wal-Hikmata' or 'Manifested Nature and the Utility of One's Upright Logic' each matured member of the manifested hexagonal world and appearing pentagonal earth within the Black Square of East Horizon will be able to search out the owners of conspiracy Northern Hemisphere, conspiracy Southern Hemisphere, man-made North, man-made South, Global Trinity, Circular Rotation & Revolution System, and the like. Owner and Possessors of introduced calf are leading believers towards Straight Middle East Region of Eartha 3D without Broken Bar. One of the purposes of this inspired sharing is to cut down two fore legs of the introduced dog.

As a reader of 'Kitaaba Wal-Hikmata' or 'Manifested Nature and the Utility of One's Upright Logic' if you find any grammatical mistake or spelling mistake in this sharing, then you are requested to conceptualise shared findings of this sharing in resemblance with manifested signs [manifested nature] and verifiable as well as justifiable laws & theories by the utility of your upright logic. But if you find any subjective self-contradiction or objective paradox in this sharing, then please give me good tidings. In this regard, try to share specifically -

i) That concept which has committed subjective selfcontradictions and objective paradoxes going beyond revelation and manifested nature.
ii) True concept of that misconception with verifiable as well as justifiable references i.e. revealed and established criterions of truth

If you fail to cope up with at least the above mentioned two conditions, then please do not try to mislead me further as **my own people are far from manifested nature in resemblance with revelations**.

Generalisation is a mental process of passing from something known [observed] to something unknown [unobserved]. This process primarily requires two stages and two pre-conditions. Two primary stages of generalisation are – (i) to frame a supposition of a causal relation on the basis of observation of particular instances or manifested signs i.e. hypothesis or discovery, and (ii) to prove that framed supposition by searching out its resemblance with Revelation on the basis of both observation and experiment i.e. justifiable as well as verifiable by criterions of truth in resemblance with Truth-in-itself and Tautologies. So, observation and experiment are called material grounds of generalisation. But in reality they are both formal as well as material grounds as matter cannot exist without form. We have a common misconception that liquid like water has no form. If water has no form then how do you measure water with a unit called litre [metric unit of capacity equal to 1 cubic decimetre]? So the form of **water** depends upon right measurement or due proportion [which may be scientifically called **source of life** or odd integer or spin1/2].

But prior to making any supposition we are to depend on two pre-suppositions [pre-conditions] of knowledge. These two pre-suppositions are – (i) The Principle of the Uniformity of Nature, and (ii) The law of Causation. These two pre-suppositions are the formal grounds of both supposition as well as generalisation. Knowledge is not possible without these two pre-suppositions or innate ideas or revealed truths. If there is only one Uniform Principle, then it is implied that there is only one kind of Magnetism. If there are two kinds of Magnetism, then it is also implied that there are two kinds of Nature. So, the searching questions are – How many Uniform Principles are there? How many Manifested Natures are there? How many Uniform Motions are there?

It is certain like my existence as a human person that there is only one Uniform Principle, only one Manifested Nature, only one kind of Magnetism called Natural Magnetism. So, the searching questions are – What kind of nature is being represented by Manmade Magnetism? Where is the existence of that nature which is being guided by Manmade Magnetism? What is the Uniform Principle of that Nature which is being guided by Manmade Magnetism save persecution? Whether 'Inertia' of Newton's First Law and Galileo's Motion represent Manifested Nature or Manmade Nature? In manmade nature Einstein's Electromagnetic Wave or so-called sun or Bullet as the manifested sign of natural magnetism rises once from unidentified East. **From which direction Einstein's Electromagnetic Wave or converted Snake of Muusa's Staff enters within the Black Square [Non-luminous Moon or Dark Chamber] of the Manifested Nature?**

The two pre-suppositions [pre-conditions or formal grounds] are nothing but what are called faith and belief from the point of view of Religion and Individual Rights from the point of view of Human Rights; while scientists call these formal grounds as Reliance on One Dot Leader & Diamond Operator or Uniform Principle & Uniform Motion, and philosophers call these formal grounds as apriori principles or apriori forms of knowledge, social reformers call these formal grounds as Solitary Rights or Third Generation Human Rights or Individual Faith & Belief, and the like. These two pre-conditions have been generalised as – Principle of the Uniformity of Nature and Law of Causation. Rene Descartes indentified these pre-conditions as innate ideas or inborn ideas. His best argument in this regard is '**Cogito Ergo Sum**' or '**I think, therefore, I exist**'. This is what is called **selfevident truth** or **Kitaaba wal-Hikmata** or **Upright Nature** of a man in resemblance with the upright logic of Descartes or upright nature of Ibrahim [ass] or Utility of **One's Upright Logic or One's true faith and belief.**

But there are philosophers, people of the Kitaab, people of the book, researchers on Truth-in-itself like John Locke, John Rawls, researchers of IFTA, father of IRF who are always trying to cover truth with a view to project falsehood as truth by comparing Truth and falsehood, by projecting separable relation between the tree and the branches of the tree, and the like. They are always trying to establish independent existence of the branches of a tree which is an impossible task. The concrete example in this regard is the father of IRF. From the state of darkness, if I make a supposition following Rawls that I am in darkness, then that supposition is called depth of darkness or complete darkness or **veil of ignorance**. Again, if I make a comparison of intelligence between man and tree on the basis of superficial resemblances/similarities like birth – growth – decay – death etc., then I am trying to establish falsehood as truth in resemblance with the father of IRF. With reference to Kitaab with Truth, they are called **hypocrites** in pragmatic correspondence with **pagans** or **wandering Arabs. Such kind of learned people are called activists**. Here I am sharing with you some symbols along with their character codes which stand for pen, paper, and pencil activists of the Teleological Evidence Sorcerer and Unique Epistemic Persecutor. These symbols also represent the teleology of **the Historical Don of two-in-one partnership**. ✎ [Symbol – decimal – 33, Symbol – hex – 0021], ▦ [Symbol – decimal 38, Symbol – hex – 26[, ✂ [Symbol – decimal – 0034, Symbol – hex – 0022], ✉ [Symbol – decimal – 42, Symbol – hex – 002A], ✍ [Symbol – decimal – 63, Symbol – hex – 003F], ✑ [Symbol – decimal – 0040, Symbol – hex – 64]. Further some symbols of crookedness which represent their teleology as well as ontology - ☝, ✌, ☟, ☜, ☞, ☛, ☝, ☝, ✋.

Similarly, Locke who was a rational being like us had tried to conceal the selfevident truth by saying that the mind is a tabula rasa [empty vessel or rationality less]. Probably Locke was a rationality less person as his mind was an **empty vessel**. Those who had established Locke's senseless opinion as a theory in the domain of true knowledge, they too had empty vessels. If you want to argue in this regard feeling a sense of **inferiority complex**, try to prepare answers of the questions like – Whether John Locke had no innate idea or shared sense or rationality? What is the distinguishing mark between John Locke and Peter Singer's non-human person? Whether Locke had experienced the universe as a human person of Peter Singer or as a non-human person of Peter Singer? If he had no logic, then why are you teaching us the theory of a non-logical human person in resemblance with Locke's self-referred parrot? Is there any consistency between Locke's concept of tabula rasa and his self-referred parrot with reference to the concept of person? Is experience possible without sense or innate idea or rationality or logic or psyche or possessor of a ~sth? On the basis of which criterions of truth you had made an empty vessel or a senseless being as the father of Modern Empiricism? So you are busy in establishing mechanical fathers and artificial mothers in resemblance with mechanical globalisation and man-made magnetism concealing the reality of your natural father [Upright West Region of White] & natural mother [Straight Middle East Region of Black], natural magnetism & manifested nature, revealed & manifested directions, and the like.

In this sharing I am not playing the non-moral game of immediate inferences with the good names of epistemic persons like the reference of a research work. So, I am not sharing Rawls ['] John instead of John Rawls. Whether the game of conversion with the good name of a person without taking written permission from that person is morally justified, legally valid, grammatically correct? Is there any significance of the good name of a human person? If so, then does it not imply that the game of immediate inference with good names is the violation of most significant Human Rights? I am a re-search scholar on **SOLIDARITY RIGHTS IN ISLAM** under the Supervision of Dr. Pious V Thomas [in resemblance with the reference of a research work]. Now, if I ask – 'What is the good name of my Supervisor?', then – 'Are you able to answer the good name of my supervisor with certainty?' But my supervisor had advised me to prepare the list of references following this or that style. These styles are called **contrary formalities** which are neither truth nor criterions of truth, but are contrary to morality, contradictory to legal laws, violation of human rights, and Black Magic with coded shared tautologies with a view to conceal the **good names of the Trinity of Sirius Binary System or Nuurun alaa Nuur and to project Lunar System and man-made magnetism or so-called solar system or circular rotation & revolution of the planets.**

REJECT SUCH MEN AS USE BAD LANGUAGE

Ayat\Verses - The most beautiful names belong to Allah: so call on Him by them; but reject such men as use bad language in His names: for what they do, they will soon be requited? And of those whom We have created there is a nation (ummat or midiyan people) who guide with **truth** and establish **justice** therewith. [Sura No. – 6\7 - As-haabul- a' raaf, Ayat No. – 180 & 181]

Moreover, the sole purpose of the implementation of illicit formalities and guidelines is to lead believers either contrary or contradictory to truth as well as criterions of truth. In other words, to lead mankind towards a **shameful morality** in the name of research on truth going contradictory to the above referred coded shared tautologies. So, being a re-search scholar on **'SOLIDARITY RIGHTS IN ISLAM'**, it is my **SOLIDIFIED SOLID HUMAN RIGHTS like Pythagorean Number – "SOLID – FOUR"** and four converted shades of Windows'7 Ultimate to raise allegations against those persons who are leading believers, truth seekers, and mankind in general either contrary or contradictory to both **TRUTH-IN-ITSELF** and **CRITERIONS OF TRUTH** illicitly establishing several contrary formalities, selfcontradictory rules, selfrecognised guidelines which are neither justifiable nor verifiable by the criterions of truth in resemblance with truth-in-itself. Is it necessary to write **250** pages to share a single concept of truth?

So, **this sharing will not be published following unjustified impositions or illicit guide lines** or violating human rights and valid legal laws concerning good names or going beyond Truth as well as criterions of truth as it is an inspired sharing against those illicit formalisers & guidelines of truth recognisers who are uninterruptedly pushing mankind towards **Petitio Principii with reference to Post Hoc Ergo Propter Hoc** or towards the **Black Hole** at the **Centre of the Global Trinity i.e.** Straight Middle East Region of **Eartha 3D** of the manifested hexagonal world and appearing pentagonal earth within the Black Square of East Horizon with reference to mechanical globalisation, circular rotation & revolution system, and man-made magnetism.

Several contrary formalities as well as contradictory guidelines have been projected with a view to establish that **'Might is Right' and consequently it has been informally formalised that Opinion is the Right Criterion**. So, you are doing nothing but establishing formalities to project - **'Fight to Establish Right'** in the name of **Educative Theory**. The sole teleology of your **'Fight for Rights'** is to create terror in the psyche of truth seekers. **Probably you do not have the true concepts of the Existential Imports of the Educative Theory or Duty for Duty's Sake or Tabligh or Nishkam-Karma or to do righteous**

work and to give good tidings to those who believe or Utilitarianism. You always use either Retributive Theory or Preventive Theory in the name of Educative / Reformative Theory not with a view to reform an offender or to educate a criminal but with a view to create terror in the psyche of mankind in general. So, **who is superior to you as a Unique Formalised Activist and a Unique Epistemic Terrorist? The revealed truth is that Persecution is worse than Killing.**

You may try to take my life i.e. to kill me taking the help of one or the other adopted son of Firawn or Yajuja & Majuja may try to kill me in resemblance with the previous incident of 2010. That incident was a pragmatic truth [Zabuur] in my life. Consequently, I fear those epistemic persons who are hypocrites, those who have rejected faith, those who are mischievous, those who are uninterruptedly spoiling the psyche of fresh generations, those whose works are full of selfcontradictions & paradoxes, those whose works resemble with conspiracy guidelines, those who are engaged in covering truth, those who are engaged in projecting falsehood, those whose works and activities resemble with manmade nature, and the like. So I fear such kind of epistemic senses more than the projected **Jihadists, Fundamentalists, Activists, and Terrorists**. In this regard, it is an appeal before mankind and specifically before believers that there are various kinds of adopted sons [including hypocrites] of Firawn living among us like Mufti Khan **Muhammad**. So, please try to identify them and provide them either preventive measurement or retributive punishment in resemblance with their crimes justifiable as well as verifiable by valid legal laws.

The questions are - Who is not a fundamentalist? Who is not a Jihadist? Who is a terrorist? Who is an activist? Those who are taking the life of innocent people, those whose activities create fear in the psyche of mankind, those who are guided by IFTA, those who are motivated by Mysticism, those who are successfully using their arms with a view to create terror in the psyche of mankind, those who have projected mechanical globalisation, those who have projected man-made magnetism, those who have projected circular rotation & revolution system, those who have projected two North, those who have projected two South, those who have concealed two East & two West, those who have concealed the right directions of equal & opposite stages of journey of the manifested sign of natural magnetism, those who have concealed the reality of the manifested sign of natural magnetism, those who have projected that the Manifested Sign of Natural Magnetism rises from the East and sets in the West, those who are knowingly concealing the reality of End of Proof – Black Square – Star Operator with three ascending stairs, and the like cannot be regarded as fundamentalists or jihadists save Activists or Terrorists. On the other hand, those who are sacrificing themselves along with what is provided by their Rab i.e. time and money with a view to give good tidings to mankind, doing righteous work, engaged in intrinsic fighting with a view to ward off evil, always ready to leave their houses with a view to fight in favour of poor and innocent people, trying to retain peace and harmony in the society, searching equal and opposite middle course towards Upright West, and if necessary then ready to fight against the tyrant for the sake of truth only, are called fundamentalists or jihadists. Such fundamentalists or jihadists are called Tabligis. But those whom both publishing media and broadcasting media have been highlighting without Broken Bar before us as **fundamentalists** or **jihadists**, they are in reality **activists** and **terrorists** like the introduced calf of Firawn, introducer & owner of that sun which rises once from the East and sets once in the West to cause the alteration of day and night for the equal & opposite seashore i.e. Post Hoc Ergo Propter Hoc Statistics, introducer & possessors of manmade magnetism, and the like. They are neither practical believers nor qualified persons, though they are projecting themselves as if symbolic representatives of Morality, Epistemology, and Metaphysics. In reality, they are the adopted sons of Firawn like Mufti Khan Muhammad, and they know that the universe has been revealed as equal & opposite Trinity, the universe has been manifested as an upright rectangle, the world has been manifested within the Black Square of East horizon after two sacrifices/divorces as a hexagon, the earth is appearing as a pentagon or Star Operator due to equal and opposite binary ~sth between Upright West Region of Arabian Peninsula and Straight Middle East Region of Eartha 3D, the appointed Kaba is neither in the East Zone nor in the Middle East within the Black Square of East Horizon, the appointed Kaba is on the right side of the Mount Tuur in the Top West Region of the manifested hexagon and Uppermost land of the appearing pentagon within the Black Square of East Horizon, the reality of the manifested signs of natural magnetism from Manifested North-East to Manifested South-East & Manifested South-West to Manifested North-West, and the like.

To get confirmation of the searched out manifest truth [(i) the real structure of the appearing earth, (ii) the real nature of matter particles, (iii) the Intrinsic Luminosity of Diamond Operator, (iv) the rising and setting point of the so-called sun as the manifested sign of natural magnetism, (v) the equal & opposite right direction of my Qibla towards Upright West, (vi) Revealed Right or Northern Hemisphere or Haiyalas-Swalaah, (vii) Revealed Left or Southern Hemisphere or Haiyalal-Falaah, (viii) Revealed Magnetic Force in resemblance with manifested sign of natural magnetism, and the like] is my Birth Right as a human being or what is called Solidified Solid Human Rights. You are teaching us to fight for rights. In this regard, I am a follower of your moral teachings called Ethics. I have also an undoubted concept that both human person and non-human person possess fundamental elements or essential elements to fight for the better adjustment with the environment. You have also established this concept as '**Survival of the fittest**'. Is there any human person or non-human person who/ that does not possess fundamental elements and his / her / its possessed elements do not fight with a view to survive with the environmental situation? If there is any such being who/which possesses essential elements or fundamental qualities that do not fight with the environmental situation, then that being is none save Devil [Iblis]. So, the establishment of 'Survival of the Fittest' is not a synthetic judgement or a real establishment, as all created beings are the possessors of 'Survival of the Fittest' or Fundamentalists in the truest sense of an analytical judgement or a verbal proposition. Are you not a Fundamentalist? Are you a Devil or Iblis? Are you able to place reference of a single being who is not a fundamentalist or a jihadist? But the pen – paper - pencil workers and black & white artists have failed to conceptualise the meaning as well as attributive usage of the term 'fundamentalism' or your establishment of 'Survival of the Fittest'. So, they are ever-ready to certify truth seekers like tabligis as fundamentalists as if it is a commentary in resemblance with the commentary of IFTA. It implies that such certifiers do not possess any fundamental elements in resemblance with Devils. On the one hand, you are teaching us to fight for fundamental rights; on the other hand, you are criticising fundamentalists. This is nothing but the reality of your selfcontradictory or binary nature. You always talk **equivocally** like sun and Bullet, moon and Triangular Bullet, World and Black Square, Earth and Star Operator, Universe and Diamond Operator, morning show and evening show, enter and rise, set and end, North and True North, South and False South, North and Northern Hemisphere, South and Southern Hemisphere, Man-made Magnetism and Natural Magnetism, rotation and revolution, Sign and Symbol, Saturn and Sirius B, Jupiter and Sirius A, and the like as if you are the owner of the binary pulsar of Einstein, Natural Magnetism, and Sirius Binary System. So, to avoid both subjective selfcontradictions as well as objective paradoxes, I am **univocally** as well as publicly certifying myself as a **Fundamentalist or a Jihadist** not only because I am the possessor of fundamental elements but also that I am trying to unveil the reality of Fundamental Truth or Manifested Nature or Fundamental Magnetism through this sharing with reference to Insert Symbol and gift package of 'The Holy Quran' of two-in-one Epistemic Uniqueness. With this respect, this sharing is **extrinsic** or '**Survival of the Fittest**' [**fight for the recognition of fundamental truth or manifested nature or natural magnetism or Natural Rights or Environmental Rights or Birth Rights or Inborn Rights or Human Rights**].

Again, I am also the possessor of something which distinguishes me from non-human persons of Peter Singer. This something is the essence of intrinsic value or rationality or innate idea or upright logic or psyche or ~sth which will lead me towards summum bonum of my life. Due to the possessor of the essence of intrinsic value, I must have to fight against those who are killing both faith and belief of mankind on One Dot Leader & Diamond Operator, Manifested Nature & Natural Magnetism, Revelation & Manifestation, Scientific Certainty & Formal Education, and the like without Broken Bar. Moreover, I am being inspired to fight against both activists [pen, paper, and pencil of the activist Laden] as well as terrorists [well known and well trained conspiracy scientists like Laden himself] at the cost of everything. In this regard, this sharing is **intrinsic or 'Survival of the Truest'**. In other words, I will prefer to say goodbye to this manifested hexagonal world and appearing pentagonal earth happily [coded tautologies] than to live a mischievous life or selfcontradictory as well as paradoxical life in circular recycling globe of © & ® following Shyatn's @ [wise policy] towards Black Hole or Straight Middle East. I will prefer to live with truth-in-itself than my better half. I will prefer to die with tautologies than to live with my near and dear ones. So, I am fighting against none but against conflict of desires. This inner fighting is called Tabligh or **Jihade-Akbar** or the Greatest War or an altruistic approach or the work of a karma-yogi or will of a satyagrahi or Fight for equal & opposite diagonal way towards Upright West [Arsh]. The purpose of this sharing is to become an **Akbari-Jihadist or an Altruist or a Karma-yogi**. If my Rab wills and I am able to ward off evils, then my activities will make me a **successful Akbari-jihadist or a Utilitarian Fundamentalist or an Altruist or a Karmayogi**. For this reason, **this sharing is a selfdeclared Akbari-Jihad of a utilitarian fundamentalist** against **the Historical Don** of the **trinity** of **hypocrisy** – Tyranny - Conspiracy **and the duality of Activism & Terrorism along with his partners of IFTA [i.e. win-win fingers] with a view to confirm SOLIDIFIED SOLID HUMAN RIGHTS IN RESEMBLANCE WITH MANIFESTED SIGNS OF NATURAL MAGNETISM for the Greatest Happiness of the Greatest Number [Altruistic Hedonism].**

In this regard, if a journalist or a media person wants to conduct a press meet, then he/she is requested to go sincerely & thoroughly through 'Kitaaba Wal-Hikmata' \ 'Manifested Nature and the Utility of One's Upright Logic' at least three times before proposing the sharer of this inspired sharing to conduct a press meet. If a mouth speaker without going thoroughly what has been shared in 'Kitaaba Wal-Hikmaata' / 'Manifested Nature and the Utility of One's Upright Logic' wants to conduct a press meet on arbitrary points on the basis of projected falsehood, then he/she is none but one of the followers of activist Laden. The sharer of this inspired sharing is sincerely requesting all media persons of the manifested hexagonal world and appearing pentagonal earth within the Black Square of East Horizon to conduct as many press meet as possible with both the Epistemic Uniqueness of IFTA and the Epistemic Uniqueness of Scientific Certainty & Formal Education with a view to share with us four witnesses against each shared selfevident truth of Kitaaba Wal-Hikmata' / \ 'Manifested Nature and the Utility of One's Upright Logic' prior to conduct a press meet with the sharer of this inspired sharing. [Four Witnesses = four valid references free from subjective selfcontradictions & objected paradoxes, free from fallacy of infinite regress & fallacy of arguing in a circle, free from existential fallacy & fallacy of post hoc ergo propter hoc, free from plurality of causes & fallacies of observation, verifiable by Injiil (correspondence truth) & Zabuur (pragmatic truth), justifiable by Tawraat (coherence truth) & Hikmaat (upright logic / \ selfevident truth)] in resemblance with manifested nature, manifested signs of natural magnetism, and revelation or established two formal pre-suppositions. Remember, according to Plato, an opinion is not knowledge i.e. neither right knowledge nor wrong knowledge. So, the four witnesses must be free from opinions, theorisations, contrary formalities, contradictory guidelines, suppositions, hypotheses without observation of facts, predictions, imaginations, revolutionisations, commentaries, translation of the meanings, denotative criterions in resemblance with the reference list of a research work etc. as each shared selfevident truth is verifiable as well as justifiable by one or the other revealed & established criterion of truth.

If an activist wants to argue in this regard, then ask him/her to fight against the friendship between the hypocrites of IFTA and owners of historical conspiracy science with a view to share with us openly and publicly the realities of End of Proof, Black Square, House, Natural Magnetism, Semi-anti-clockwise & Semi-clock-wise stages of journey, three ascending stairs, Top West Region of the manifested hexagon and Uppermost Land of the appearing pentagon where there is the appointed Kaba on the right side of the Mount Tuur, the equal & opposite diagonal way in resemblance with Muusa's climbing the Mount towards Upright West [Arsh], appointed day of observing Idd uniformly, and the like. It is a supposition that the terrorists and the activists [but not the jihadists and fundamentalists] are conscious about the right direction of Qibla, true structure of the universe, the truth concerning manifested hexagon as well as appearing pentagon; otherwise, how can they often become successful in committing remote crimes? But they are following wrong direction in prayer individually as well as collectively along with unconscious believers in resemblance with the Mystic of the Film PK. That is why; they were successful in their mighty 9/11 plot. Do those incidents not imply that the terrorists are the well known as well as well trained conspiracy scientists or unique epistemic persecutors? But jihadists and fundamentalists are fighting for selfrectification as well as for the greatest happiness of the greatest number. They never think of causing any harm consciously against revelations & manifestations. So, it is an appeal before all fundamentalists and utilitarian jihadists to make an attempt with a view to search out activists and terrorists living among us in resemblance with Mufti Khan Muhammad, the hypocrites who are gifting people criticisms or commentaries on Quran in the name of 'The Holy Quran', who are comparing True Religion with falsehood, who are trying to establish four criterions of truth [Kitaabs] as four independent revelations, who are successful in committing several kinds of arm activities. Let us compel them to confess the truth concerning the equal and opposite stages of journey in resemblance with manifested signs of natural magnetism as well as the reality of their Don. If my Rab wills, then this sharing will distinguish the activists and the terrorists from the Fundamentalists and Utilitarian Jihadists.

If there is any true jihadist who knows his/her right direction but he/she is concealing truth fearing the activities of the terrorists, then he/she is requested to purify his/her concept of Jihad first by elevating that concept from Egoism [Jihad / War] to Altruism [Jihade-Akbar / Greatest War]. This elevation will help to bring back Unity in Diversity [Utilitarianism], and to establish **'Survival of the Truest' as the Summum Bonum of life**. If my Rab wills, then each and every Akbari-Jihadist [Tablighi or Altruist or Karmayogi] who is doing righteous work and engaged in giving good tidings to mankind or engaged in doing Jihade-Akbar i.e. self purification through selfrenunciation or doing Nish-kama Karma or fighting for the **'Survival of the Truest'** will find a way that is Upright.

In this regard, the sharer of 'Kitaaba Wal-Hikmata' or 'Manifested Nature and the Utility of One's Upright Logic' is enthusiastically waiting from his searching room for the right time to meet with pen-paper-pencil works, black & white activists, mystics, mouth speakers, and the like who are uninterruptedly publishing & broadcasting nouns in their attributive forms [prefixes / shane-nuzuul] like my so-called supervisor. My so-called supervisor had dictated me to prepare the Synopsis of 'Solidarity Rights in Islam' such as - Islam is a progressive, liberal, communicative, democratic, dialogical, enlightening, dialectical, strictly monistic, spiritual, ----, -----, -----, holy Religion. From that very dictation of my so-called supervisor, it was appeared to me as if 'Islam' is one who walks, talks, communicates, and so on like a rational being.

I have conceptualised the very term 'Islam' in resemblance with what science calls Reliance on two pre-suppositions, what philosophy calls two formal grounds or pre-supposed pillars, what religious scholars call faith and belief, what Kant has maintained as apriori forms of knowledge, what Aristotle has maintained as Major Premise & Minor Premise of the Dictum De Omni ET Nullo, what Rene Descartes has maintained as Innate & Adventitious Ideas, what Sankara has maintained as ParamartikaSattva and Vyavaharika-Sattva, what social reformers have formalised as Solidarity Rights, and the like. I have conceptualised the very term 'Islam' as – I have surrendered myself completely before the Laws of Allah and the practical life of Prophet Muhammad [sas]. In other words, I have surrendered myself completely before the equal & opposite revelation [Law or Tawraat or Faith] and manifested signs [Crucified Sign or Natural Magnetism or Injiil or Belief]. So, I have firm faith on One Dot Leader and practical belief on Diamond Operator. One who has reliance on these two pre-established conditions is called a Muslim or a Believer. So, in the truest sense of the term 'Believer', each and every real scientist like Newton or true philosopher/epistemic person like Rene Descartes or truth seeker/Satyagrahi like M K Gandhi is a believer of two pre-suppositions of wisdom / light / Vidhya / knowledge.

Moreover, the very term 'Islam' etymologically means 'Peace'. We are establishing innumerable commissions, organisations, councils, and associations etc. with a view to retain peace in every aspect of life and to prove that we are peace lovers, yet we have allergy with the term Islam/ Peace. So, we are publishing/broadcasting selfcontradictory concepts like Islamic-Terrorist, Muslim-Activists, and Muslim-Jihadist etc. The similitude/ concrete examples of such kind of prefixes [shane-nuzuul] are 'Cold Burning Fire', 'Hot Ice-cream', True North & False Northern Hemisphere, False South & True Southern Hemisphere, Semi-anticlockwise and Semi-clockwise full circle, Manmade Magnetism and Natural Magnetism etc. I have been searching solutions to get rid of such kind of subjective selfcontradictions, objective paradoxes, argument ad infinitum, petitio principii, post hoc ergo propter hoc etc. since 1987.

It was very shameful fact that being a so-called literate person, I did not know the real structure of the universe, the world, and the earth. I had no true concept of the four cardinal directions in resemblance with four basic forces and manifested sign of natural magnetism, four Imams or basic forces for straight usage, the entering – setting – turning – rising - ending of the stages of journey of the so-called sun as the manifested sign of natural magnetism. I did not know that the symbol Star Operator or Spiders' Net or House or Pentagonal Star or 9/11 pentagon is the real structure of the appearing earth. Being a believer, I did not know the revealed truth that the so-called sun as the manifested sign of natural magnetism or niche is kindled each day neither from the East nor from the West. I even did not know the manifest truths that the so-called sun as the manifested sign of natural magnetism enters for the equal & opposite stages of journey from Manifested North-East and sets in Manifested South-East, runs away from there and rises from Manifested South-West and ends the journey in Manifested North-West. I had firm faith on mechanical globalisation and man-made magnetism i.e. Manmade Nature. I even never tried to conceptualise the similarities/differences among/between the terminologies like Natural, Manifested, Apparent, Mechanical, Man-made, Artificial etc. I used to determine the right direction of my Qibla towards Straight Middle East or Black Hole on the basis of man-made magnetism and mechanical globalisation. So, I was far from the equal and opposite diagonal way [middle course] towards Upright West. For this reason, I was fighting for non-human rights forgetting my Solidified Solid Human Rights which resemble with Helium-4, Pythagorean Number – 'Solid – Four', and Four converted colourful shades of Windows'7 Ultimate. **I was a self-certified 'Pen-Paper-Pencil Activist'** in resemblance with publishing and broadcasting prefixes i.e. subjective selfcontradictions and objective paradoxes. But I was not an Epistemic Terrorist or Meta-Ethical Persecutor of faith and belief as I never tried to conceal manifest truth as a believer and never tried to share with my friends in deed as a teacher that subjective selfcontradictions and objective paradoxes are the tautologies or selfevident truths with a view to project falsehood as truth. Now, it is the right time to ask Peter Singer – **'Who will fight for whom [Human Liberation / Animal Liberation]**?'

With a view to get rid of projected subjective selfcontradictions, objective paradoxes, Petitio Principii, Argument ad Infinitum, Existential Fallacy, Post Hoc Ergo Propter Hoc, and the like, I have searched out on the basis of verifiable as well as justifiable references the nature and functions of the Don of Terrorism [Original Uncultured Laden or Unique Epistemic Persecutor] and the Don of Activism [Real Activist or Unique Teleological Evidence Sorcerer] under the supervision of my Rab. Some significant nature and functions of the Historical Don of two-in-one are as follows -

   i) **Distinctive Nature of the Historical Don of Terrorism**: - Concealment of the Real and Projection of the Unreal are the two distinctive natures of the Historical Don of Terrorism. These two distinctive natures resemble with the binary nature of 'Maya' of Sankara's Advaita Vedanta.

   ii) **Distinctive Features of Historical Terrorists**: - Hypocrisy, Tyranny, and Conspiracy are the Distinctive Features of Historical Terrorists. These three distinctive features are together called Trinity.

iii) **Identifying Marks of the Historical Don of Terrorism/Terrorists**: - Teleological Evidence Sorcery and Unique Epistemic Persecution are the Identifying Marks of the Historical Don of Terrorism/Terrorists. These two identifying marks are together called Duality or Dualism.

iv) **Distinctive Natures of the Historical Don of Activism:** - To inspire epistemic senses so that they begin to play the game of immediate inferences in the name of research work on truth-in-itself and coded shared tautologies in resemblance with his projected Teleological Evidence Sorceries motivated by the sense of recognition. These games lead epistemic senses towards Black Magic.

v) **Distinctive Features of the Real Activists**: - To believe on the Trinity of Hearsay – Further explanations of what have been shared explaining in detail by Allah and His Messenger as translation of the meanings [Tafsir] – critical remarks [Commentary] on Revelation & Manifestation; to highlight through public support those people who have no concept of the equal & opposite right direction towards Upright West as if Partners/ Ministers of Allah in this Global Trinity as well as Quaternary; to follow the believers of Straight Middle East [Black Hole]; to share a portion of revelation with others randomly in resemblance with Preface of 'The Holy Quran' of IFTA; to verify and to justify truth without confirming Manifested Signs of Allah in resemblance with Revelations; to spread/ broadcast falsehood as if the advertisement of a product; to use pen-paper-pencil in resemblance with mechanical globalisation, circular rotation & revolution system, and man-made magnetism, and the like

vi) **Identifying Marks of the Historical Don of the Real Activists/ Real Activism**: - Pen-Paper-Pencil workers and their works such as reading, writing, editing, publishing, broadcasting, teaching, sharing, motivating, inspiring, instigating, acquiring, comparing subjective selfcontradictions and objective paradoxes etc. which resemble with projected mechanical globalisation, circular rotation & revolution system, man-made magnetism, Copernican Revolution, Post Hoc Ergo Propter Hoc, Fallacy of Infinite Regress, Fallacy of Arguing in a Circle, Existential Fallacy etc. These works are neither verifiable nor justifiable by the four revealed & established criterions of truth

WHO KNOW THE TRUTH?

The Owners of the Scientific Research Institutions represented by the following Font Names certainly know the manifest truths in resemblance with revelations and verifiable by searched out real scientific symbols in resemblance with manifested signs as well as justifiable in resemblance with coded shared tautologies.

Bookshelf Symbol 7	FangSong	KaiTi
Marlett	MS Outlook	MingLiU_HKSCS-ExtB
MingLiU-ExtB	MS Reference Specialty	MT Extra
PMingLiU-ExtB	Rupali	SimHei
SimSun-ExtB	Symbol	Webdings
Wingdings	Wingdings 2	Wingdings 3
There is no scope to argue with selfevident truth or manifested nature save justification & verification. In this regard, the findings of the Kitaaba Wal-Hikmata or Manifested Nature and the Utility of One's Upright Logic' are being shared publicly with the expectation that the owners of the scientific research institutions represented by the above Font Names will verify and justify the shared findings for the sake of Truths-in-themselves and Tautologies.		

Now, ask the Media people who always project the slogan of impartiality and who are able to search out a single individual from the dark chamber of a multi-storied building – how do they fail to project till date the fundamentalists as well as utilitarian jihadists who very often meet openly as well as publicly under the canopies in Tabligi Istemas? So, ask those Media people to highlight before mankind the activities of uncountable Tabligis [Akbari-jihadists] whose activities are contradictory to the terrorists and the activists. Ask them to highlight through their broadcasting media before public each District Level Istema, State Level Istema, National Istema, and International or Biswa Istema along with what are being taught in each Istema, if they are truthful with respect to their profession of sharing truth impartially. Ask those media men [who are always able to search out a terrorist from an unknown place] to search out Owner of Projected Manmade Nature, Teleological Evidence Sorcerer, Unique Epistemic Persecutor of faith and belief of mankind on Formal Education and Scientific Certainty from Spiderman's Net along with his Chiefs & Ministers with a view to project them openly & publicly. Now, common people will ask them about the real concepts of the terms like Evidence Sorcery, Epistemic Persecution, Night Sky, Earth Sky, Morning Show, Evening Show, Bullet, Triangular Bullet, Man-made Magnetism, False North, False South, Activism, End of Proof, Black Square, Hexagon, Octagon, Black Spot, Pentagon, Planet Mars, Non-luminous Moon which has eclipse, Intrinsically Luminous Moon as Earth's only permanent Natural Satellite, repeated visiting in the heaven and in the hell etc. From the analysis of the journalists concerning repeated visiting of the so-called scientists in Night Sky [heaven or moon] and in Earth's Night Sky [Planet Mars or hell] it may be concluded that those journalists must know the truth concerning manifested nature as they are able to unveil seven dark spots of the manifested sign of natural magnetism which resemble with digit 7 of Windows'7 Ultimate. The analyses, interpretations, critical remarks, explanations etc. of the shared manifested truths of this inspired sharing are nothing for them. So, it is an appeal before broadcasting media to highlight with detail analysis the shared selfevident/manifested truths of this inspired sharing which neither the Owner of the Historical Conspiracy Science, nor the hypocrites of IFTA as well as its Off-springs, nor the pen – paper – pencil workers, nor black & white artists of the activist Laden, nor the terrorist AK47 Laden will be able to falsify or to reject, as these findings are manifest truths grounded on manifested signs of natural magnetism, and verifiable as well as justifiable by the revealed and established criterions of truth. The impartial role of the broadcasting media may help us to get verified as well as justified confirmation & recognition of the shared searched out findings concerning manifested nature from the respective authorities.

It has been showed through film media the truths concerning wrong number as well as wrong dialling in the film PK which was able to instigate a contradictory chaos in the peaceful social relation in my motherland. It is a supposition that Amir Khan, the actor of the film till date does not know the truth concerning the name as well as contents of the film. The very ideographic name PK stands for Pak Kalam [Pak Katha / Pak Kalima] and dialling wrong number represents that those who have made a promise on the right side of the Mount Tuur and surrendered their wills completely before none but one who has no attribute save that He is the owner of all attributes are now performing their salat [compulsory prayer] towards wrong direction i.e. Straight Black Hole of the Middle East instead of equal & opposite diagonal way [Middle Course] towards Upright West [Arsh]. But the presentation of the film with a question like – 'Pike ayea hai keya?' leads the concealed truth towards opposite direction. This is a concrete example of the binary role of the film media in resemblance with the Epistemic Uniqueness of the UGC Sponsored National Seminar and Binary [CAPITAL LETTER & small letter] International Language in our social life.

It has also been showed through film media the significance of the manifested signs of natural magnetism [so-called sun-light] within the Black Square of East Horizon as the means of sustenance in the film 'Kai Milgaya'. But we the ignorant people fail to conceptualise the existential import of the term 'DUUUP' due to physical gesture and posture of Riitikh Roshon and Preti Jinta.

It has also been showed through film media that Amitabhji had written a book in the name of 'Baghwan'. Probably Amitabhji is till date unaware about the contents and contexts of his self-written book, but the Epistemic Uniqueness must know about that book. So, he has established several illicit formalities & guidelines with a view to put barriers just for a single book. But think thrice – 'Who is the best planner save my Rab?' Now the people will be able to conceptualise the content & context of the book 'Bhagwan'.

So, the Epistemic Uniqueness must know the reality of those films. He has projected almost all coded shared tautologies in the form of a story. His story writers on coded shared tautologies know the truth that the Kitaab with Truth is not a Kitaab of poems but parables; yet he is gifting the same [The Holy Quran] in the form of poems via the wandering Arabs and sons of Firawn like Mufti Khan Muhammad. Epistemic Uniqueness must know better than anyone the contradictory roles and functions of jihadists and terrorists, and I have conceptualised his historical evidence sorceries better than his chiefs & ministers under the supervision of my Rab. So, I am addressing that Epistemic Uniqueness as the Historical **Don** of the **Trinity** of **Hypocrisy – Tyranny - Conspiracy** and **Duality of Activism and Terrorism**, and sincerely inviting him along with his pen – paper - pencil workers and black & white artists to face the shared selfevident truths or findings or conclusions of 'Kitaaba Wal-Hikmata' or 'Manifested Nature and the Utility of One's Upright Logic' openly and publicly with a view to prove that **his mechanical globalisation, circular rotation & revolution system, and man-made magnetism are the revealed as well as manifested truths [manifested nature].**

I have searched out several Manifest Truths in resemblance with Newton's three laws of Motion, Newton's Law of Gravitation, Einstein's Theory of Relativity, Einstein's Gravitational Lensing & Binary Pulsar, verifiable scientific laws, justifiable Philosophical Theories, Kitaab with Truth. One of those searched out findings is that the so-called sun as the manifested sign of natural magnetism neither rises once from the East nor sets once in the West. The so-called sun [Bullet] as the equal & opposite manifested signs of natural magnetism rises twice and sets twice. It enters from Manifested North-East in correspondence with North America and sets in Manifested South-East in correspondence with South America for the Ground Stair of North America and South America of the East Zone [Lower Seashore] within the Black Square [Non-luminous Moon] of East Horizon. The so-called sun [Bullet] as the manifested sign of natural magnetism or the setting sun [Bullet] runs away from there and rises from Manifested South-West in correspondence with Europe and ends the journey in Manifested North-West in correspondence with Australia for the Middle Stair of Europe, Asia, Africa, and Australia as well as Topmost Stair [Upright West Stair] of Arabian Peninsula of the West Zone [Upper Seashore] within the Black Square [Non-luminous Moon] of East Horizon. In brief, the so-called sun [Bullet] as the manifested sign of natural magnetism enters from Manifested North-East [Revealed Left or Northern Hemisphere or Haiyalas-Swalaah] and sets in Manifested South-East

[Revealed Right or Southern Hemisphere or Haiyalal-Falaah]. Again, the so-called setting sun [Bullet] rises from Manifested South-West [Revealed Right or Southern Hemisphere or Haiyalal-Falaah] and ends in Manifested North-West [Revealed Left or Northern Hemisphere or Haiyalas-Swalaah]. These equal & opposite rising and setting of the so-called sun [Bullet] as the manifested signs of natural magnetism cause the alteration of day and night for the equal & opposite Seashore [East Zone and West Zone]. The equal and opposite Semi-anticlockwise [4/8 or ½ ASCII (hex) 00BD & ASCII (decimal) 189] and Semi-clockwise [1/4 for Topmost Stair of Arabian Peninsula and ¾ for Middle Stair – ASCII (hex) 00BC & 00BE and ASCII (decimal) 188 & 190] stages of journey of the so-called sun [Bullet] represent the manifested signs of Natural Magnetism. Moreover, **there is only one kind of magnetism called Natural Magnetism i.e.** only one North [Revealed Left or Northern Hemisphere or Haiyalas-Swalaah] and only one South [Revealed Right or Southern Hemisphere or Haiyalal-Falaah]. All pen-paper-pencil works as well as black & white activities which are either contrary or contradictory to Manifested Signs of Natural Magnetism, North [Revealed Left or Northern Hemisphere or Haiyalas-Swalaah], South [Revealed Right or Southern Hemisphere or Haiyalal-Falaah], Real Science [Insert – Symbol] are Teleological Evidence Sorceries as well as Unique Epistemic Persecutions. Projected Manmade Nature through Mechanical Globalisation [Napier Bones], Circular Rotation & Revolution System [Copernican Revolution], and Manmade Magnetism [False North & Northern Hemisphere and False South & Southern Hemisphere] resembles with the Film PK.

The shared searched out selfevident truths are the manifest truths justifiable by Revelation [Universal Major Premise] and verifiable by the separate sheet of real scientific works & manifested signs of natural magnetism [Affirmative Minor Premises]. If Unique Epistemic Decision Taker is true to his/her words and deeds and he/she is able to take ultimate decision concerning the above shared selfevident truths, then he/she may argue with rest of the selfevident truths. But if he/she fails to reject the above shared manifest truths or he/she fails to prove the realities of his/her Manmade Nature in resemblance with manifested signs of natural magnetism i.e. four revealed directions & six manifested directions, then it will prove that his/her pen-paper-pencil works and black & white activities are nothing save Teleological Evidence Sorceries. Further, the projection of those Evidence Sorceries through mechanism is nothing save Unique Epistemic Persecutions of Faith and Belief of mankind in general on Scientific Certainty and Formal Education.

It is certain like my existence in this manifested hexagonal world and appearing pentagonal earth that a mortal being like the Don of two-in-partnership has no power over manifested signs of natural magnetism. So, it is selfevident that he/she will not be able to convert manifested signs of natural magnetism in resemblance with his/her projected North & Northern Hemisphere and South & Southern Hemisphere. Consequently, the Don of Historical Black Magic i.e. Teleological Evidence Sorceries, Concealment of Manifest Truths, Unique Epistemic Persecutions of Faith and Belief on Manifested Nature - Manifest Truth - Scientific Certainty - Formal Education etc. has no right to interfere in our Solidified Solid Human Rights like equal & opposite right direction towards Upright West [Arsh] as our Qibla, appointed day of observing Idd uniformly, the inviolable sacred place of the Black & White Imam of the City for Pilgrimage [Haj], and the like.

In the Kitaab with Truth it has been shared that persecution is worse than killing. As the Teleological Evidence Sorcerer & Unique Epistemic Persecutor has snatched away the right direction of our Qibla, projected fictitious days of observing Idd, uninterruptedly killing faith and belief of mankind in general, and the like, so the question is – Who is that Epistemic Uniqueness? With respect to 'Kitaaba Wal-Hikmata' or 'Manifested Nature and the Utility of One's Upright Logic' that Epistemic Uniqueness is the **Original Epistemic Laden**.

Who is the knower of that Epistemic Uniqueness? The sole answer is those who have proposed, sanctioned, organised, conducted UGC Sponsored National Seminar on 'Epistemic Uniqueness', they must know him/her [Epistemic Uniqueness]. In other words, the faculty members of the Department of Philosophy, Assam University, Silchar and Respective Chair Holders of University Grant Commission – Hindustan must know that Epistemic Uniqueness.

I am not thinking about whether the Original Uncultured Educated Laden will kill me or not. I must have to leave this manifested hexagonal world and appearing pentagonal earth within the Black Square of East Horizon one day even if he/she does not kill me. So, it does not matter whether the Original Educated but Uncultured Laden will kill me or not. But it is certain that if the Original Educated but Uncultured Laden wants to reject shared selfevident findings or conclusions of the Dictum De Omni ET Nullo, then he/she must have to face the monadic motion of 'Kitaaba Wal-Hikmata' or 'Manifested Nature and the Utility of One's Upright Logic'. But if the Immoral Epistemic Uniqueness or his/her Chiefs and Ministers are able to take the life of the sharer of this inspired sharing, then they must have to face infinite number of upright nature. It is up to the Original Uncultured Educated Laden to take a verifiable

& justifiable decision as he/she has projected himself/herself as if the ultimate decision taker of the seven windows of the manifest truth or seven repeated verses without faith and belief on One Dot Leader and Diamond Operator. So, **it is an inspirational sharing not only for the Original Uncultured Epistemic Uniqueness but also for his/her folk [including activists and terrorists] as well as mankind in general that an overwhelming day is very near.** The dry wood or the space-bar of the key-board is waiting to follow the command of my Rab. So let us make our faith and belief pure confessing manifest truths before that day and seek refuge in Allah from Shaytan the outcast.

The Don of the two-in-one partnership had gifted me the cursed package of criticisms or commentaries on Quran in the name of 'The Holy Quran' of IFTA through the hands of Mufti Khan **Muhammad** on 7th April, 1995 at Nanglui Market, New Delhi, Hindustan, when **I was pragmatically engaged in doing Akbari-jihad against the sense of recognition for selfrectification with a view to ward off evil for a course of four months [tinsilla].** From the very name i.e. the placement of Muhammad at the last, I had conceptualised him as one of the adopted sons of Firawn. The Original Uncultured Epistemic Laden has trained several such sons of Firawn. He is using those sons of Firawn by the names of Islamic-Terrorists, Muslim-Activists, International-Activists etc. The leader of each terrorist or activist group has clear concepts of the revealed & manifested upright rectangular universe, manifested hexagonal world, appearing pentagonal earth with three ascending stairs, manifested signs of natural magnetism, right direction of Qibla, appointed day of observing Idd uniformly, and the like. They have prepared separate sheets of the manifest truths. They even translated Quran or Revelation in resemblance with projected Manmade Nature in the name of 'The Holy Quran' of IFTA. They are gifting their Manmade Nature with a view to conceal manifest truth and to spread falsehood as truth. They always follow manifest truths and project mechanical/manmade falsehoods before us in resemblance with their binary nature. They always project conspiracy sheet before respective authorities with a view to mislead believers and to dominate Star Operator from Spiderman's Net. Those chiefs and ministers are living among believers as if they are the metaphoric guiding stars and sacrificing their lives with a view to safeguard manifest truths [detail explanations of Fundamental Ayat] in resemblance with revelation. When this 'Kitaba Wal-Hikmata' or 'Manifested Nature and Utility of One's Upright Logic' will touch the air in a black and white form throughout the appearing pentagon, then the believers will begin to search out adopted sons of Firawn as well as wandering Arabs [hypocrites] from among themselves. They will punish those sons of Firawn following valid guidelines of retributive theory in its both forms as well as preventive theory. This will help to distinguish terrorists and activists from jihadists and fundamentalists. Moreover, each possessor of upright logic will be able to identify without any doubt like Descartes' 'Cogito Ergo Sum' that the Original Uncultured Epistemic Laden is the **Don of two-in-one partnership among terrorism and activism, hypocrisy and conspiracy, evidence sorcery and persecution**. For instance - **Under the guidance of my Rab, I have conceptualised Laden's plotted plan of the 9/11 incident on the existing PENTAGON as one of the signs of indication as well as clear proof of the existence of the appointed Kaba on the right side of the Mount Tuur in the Top West Region of the manifested hexagonal world and Uppermost Land of the appearing pentagonal earth within the Black Square of East Horizon.** Further all unknown rational beings will be able to conceptualise that the appearing earth is like the existing Pentagonal Orange in the East Zone. When this 'Kitaaba Wal-Hikmata' or 'Manifested Nature and the Utility of One's Upright Logic' will touch the pentagonal earth in black & white form, both epistemic senses as well as common senses will be **able to identify the activists, terrorists, hypocrites, mischievous theorisers & their theories, cunning policies, those adopted sons of Firawn who are living among believers, pen-paper-pencil activists, black & white evidence sorcerers, epistemic persecutors of faith and belief of mankind on scientific certainty and formal education, and the like**.

Now, going through this inspired sharing, common people will be able to conceptualise how the critical researchers are **abusing** each other, what the researchers of IFTA gifting believers of Quran save **critical remarks** [commentaries] on Quran in the name of 'The Holy Quran' in resemblance with projected mechanical globalisation, man-made magnetism or mechanical North & South, circular rotation & revolution of the immovable pentagonal earth, how the people of the Kitaab with Truth have conceptualised the critical remarks of IFTA on Quran as **further interpretations [Tafsir of Quran]** without confirming the same with the manifested signs of Allah, and uninterrupted sharing of the criticisms of Quran with the believers as if **TRUE RELIGION**, how a group of conspiracy scientists [**but not the progressive scientists**] and selfrecognised philosophers are putting **black clothes over the privative eyes of Maryam with a view to project justified balance between equal and opposite nature of Contraries and Contradictories**, how a group of hypocrites are working in **favour of historical evidence sorcery** against **manifest truth** by debating on similarities and differences between Tautologies and projected falsehood in the name of **comparative study**.

Either the above mentioned groups **do not know the truth** that the manifested **world and appearing earth are not like the projected globe but a pentagon** out of a hexagon or they are **concealing truth knowingly**. Probably they even do not know that with reference to Quran, the upward direction is called West in resemblance with the direction of Heaven [appearance of quadrilateral stars are the signs of the existence of heaven] and downward direction is called East in resemblance with the direction of Hell. They even do not know where are the two West and two East, West Horizon and West Zone, East Horizon and East Zone etc. but they have the contradictory faith and belief on the projected two North [Northern Hemisphere as Magnetic Force & North as Strong Force] and two South [Southern Hemisphere as Weak Force & South as Gravitational Force], and there is neither West direction nor East direction in their uncountable works. So, they are not true believers of Quran, though they are the true followers of Conspiracy Science. If they want to argue in this regard, then it will prove that they must know the truth concerning manifested sign of natural magnetism, yet they have concealed the revealed and manifest truth from us before this inspired sharing. So they are none but the adopted sons of Firawn.

If Locke's thought is called scientific, then what is the concept of neutron in his scientific thought? What is represented by the term 'sense' in 'sense data' save innate idea or logic as the distinctive feature of a human person? In this connection I would like to share a selfevident truth that a rational being like Newton cannot be an atheist. It has been shared with us that David Hume was a sceptic. The question is – Whether David Hume did not believe that he would die one day? Is there any sceptic like David Hume who does not believe that his/her physical existence will perish one day from this appearing pentagonal earth? If the answer is negative, then he/she is sharing truth. If the answer is affirmative, then he/she is none but a hypocrite for simple logic that he/she is concealing truth knowingly. Is there any matured rational being who has no faith and firm belief on the nature and behaviour [fitrat] of fire that this or that fire will burn him/her? If there is any such atheist [doubter / sceptic], then provide me legal permission to test his/her faith and firm belief. You may also test such kind of atheist who has neither faith nor firm belief on the nature & behaviour or dharma or fitrat of fire. So, the very term 'theist' is conceptually related with 'faith on Uniform Principle or One Dot Leader of Real Science' and 'belief in Successive Relation between One Dot Leader and Diamond Operator' [Sirius Binary System or Nuurun Alaa Nuur or Trinity]. This faith is the unseen holding stand as the Sign of Justice in resemblance with Newton's Balance diagram between electron and proton. Again, firm belief is the equal & opposite balance or due proportion or due measurement between revelation and manifestation, between manifestation and appearance. These faith & firm belief are the formal grounds of wisdom. These two formal grounds are beyond human made Laws, Rules, Guidelines, Formalities etc. So, these two formal grounds have been established as Third Generation Human Rights or Solitary Rights or Solidarity Rights or Solid Rights or Individual Rights. Further, these two formal grounds establish the relation between First Generation Human Rights and Second Generation Human Rights. **This establishment is called Trinity of Human Rights or Solidified Solid Human Rights.** I have placed my searched Solidified Solid Human Rights before IHRC HQ [International Human Rights Commission on 16th April, 2016], WPMO Team [World Peace Mission Organization on 19th April, 2016], and more than 600 [six hundred] Organisations, Councils, Associations, Scientific Research Institutions, Islamic Research Institutions, Philosophical Councils, Religious Scholars, Head Quarters of different Religious Organisations, Highly Educated Persons, Universities, Media Press Offices and the like for verification, justification, confirmation, recognition, publication, broadcasting etc. of the searched out manifest truths shared as Solidified Solid Human Rights.

Nature behaves same under similar circumstances. But the behaviour of manifested nature depends on successive relation. This successive relation is called Revelation or Tawrat or Equal & Opposite Law. The sole valid condition of a successive relation [called implication] is – **If the antecedent is true, then the consequent must also be true. It will not be the case that the antecedent is true but the consequent is false. In other words, it will not be the case that there is smoke, but there is no fire. This sharing is primarily based on true antecedent with a view to get justified as well as verified confirmation concerning searched out consequences [findings].**

The true antecedent of an implication may be either Truth-in-itself or a criterion of Truth [called tautology]. This concept was developed by Aristotle in his method of Deductive Reasoning in the name of **Dictum de Omni ET Nullo**. This dictum suggests that there are two conditions for the justification as well as verification of truth. The first condition for justification is that the **Major Premise must be Universal** [free from fallacies] and the second condition for verification is that the **Minor Premise must be Affirmative**. But the searching questions are – (i) Where from we can get Universal Major Premises free from the fallacies of infinite regress and arguing in a circle save revelations? (ii) Where from we can get Generalised Minor Premises or Verifiable Affirmative Minor Premises in resemblance with revelations save the separate sheet of the searched out real scientific symbols along with their character codes available in Insert Symbol and manifested signs?

Dictum De Omni Et Nullo of this inspired sharing

Major Premises – Complete Coded Shared Scientific Knowledge or Kitaab with Truth

Minor Premises – References of searched out real scientific symbols along with their character codes from the separate sheet and manifested signs.
Conclusions – Searched out Selfevident Truths verifiable by real science & manifested signs and justifiable by truths-in-themselves, and criterions of truth.

COPERNICAN REVOLUTION INVOLVES
THE FALLACY OF POST HOC ERGO PROPTER HOC

Copernican Revolution is a concrete example concerning the key of Teleological Evidence Sorceries and Unique Epistemic Persecutions. This revolution is the store house of fallacies. It involves the fallacy of non-observation of instances, fallacy non-observation of essential circumstances, fallacy of individual mal-observation, fallacy of universal mal-observation, fallacy of plurality of causes, fallacy of practical imperfection, fallacy of characteristic imperfection, fallacy of infinite regress, fallacy of arguing in a circle. Ultimately, Copernican Revolution is a Post Hoc Ergo Propter Hoc. Copernican Revolution that the earth moves round the sun is a conspiracy against both revelations and manifestations. Through the projection of Copernican rotation and revolution, the Historical Don of Conspiracy Science is uninterruptedly leading believers towards Straight Black Hole of the Middle East and dominating mankind from the Pentagonal Net Work of the Octagonal Spider. Are there mortal beings like two-in-one Epistemic Uniqueness who can rescue Copernican Revolution from the above alleged fallacies? If there is any such being, let him/her come forward and make an attempt with a view to rescue the key of Teleological Evidence Sorceries and Unique Epistemic Persecutions from the shared manifest truths of 'Kitaaba Wal-Hikmata' or 'Manifested Nature and utility of One's Upright logic'.

The Historical Don of Conspiracy Science has converted – (i) the Dot Operator into Solar System, (ii) the Diamond Operator or Intrinsically Luminous Star into Intrinsically Luminous Moon, (iii) the upright rectangular universe or End of Proof into Lunar System, (ii) the manifested hexagon appearing as a pentagon into Non-luminous Moon, (iv) West into Northern Hemisphere, (v) East into Southern Hemisphere, (vi) Main sequence Sirius A into Jupiter and white dwarf companion Sirius B into Saturn, (vii) octagonal black spots called springs or matter particles into pentagonal shining stars, (Viii) Revealed Left & Manifested North-East and Revealed Right & Manifested South-West into East, (ix) Revealed Right & Manifested South-East and Revealed Left & Manifested North-West into West etc. I am a believer of revelations and manifestations by birth but practically I was a follower of conspiracy science. The above mentioned Evidence Sorceries through Copernican Revelation are conspiracies against both revelations and manifestations. This revolution is also contradictory to searched out real scientific symbols. Due to this revolution,

I was off the track from the right direction or the equal and opposite Middle Course towards Upright West [Arsh]. Now, my Rab has shown me the right direction of my Qibla. So, like 'Cogito Ergo Sum' of Descartes, it is certain that this time the Historical Don of Conspiracy Science will not be able to rescue his projected mechanical globalisation and man-made magnetism from 'Kitaaba Wal-Hikmata' or 'Manifested Nature and Utility of One's Upright Logic'.

Moreover, the appointed time to manifest Lailatul Qadar or Power of Night is coming closer. So, this self-publication is an invitation to those who have concealed the realities of two West and two East and the revealed magnetic directions of the Morning Show and the Evening Show with a view to track off believers from the right direction of Qibla and to dominate mankind from the Spiderman's Net to come forward and to face the monadic motion of 'Kitaaba Wal-Hikmata' \ 'Manifested Nature and Utility of One's Upright Logic'. In this regard, the instruments of the sharer of this inspired sharing is nothing but the established two pre-conditions of knowledge, manifested signs, searched out real scientific symbols [Insert – Symbol], valid scientific laws, contradictory conspiracies of the Historical Don of Conspiracy Science etc. Remember, Nature behaves same under similar circumstances & every action must have an equal and opposite reaction.

Let us ask the Don of two-in-one partnership some questions concerning the fallacy of infinite regress and the fallacy of arguing in a circle. If nothing comes out of nothing, then what was the cause of **One Dot Leader**? What was the cause of **Big Bang or the ongoing equal & opposite 3D command or downwards Arrow of Science**? What was the cause of **Diamond operator**? What was the cause of End of Proof? What were the causes of **Black Square and Asterisk**? What was the cause of **Star Operator** or **House or Spiders' Net**? Whether the existence of North America & South America resemble with two knees in pragmatic correspondence with the manifested truth called Nature or the existence of North America & South America resemble with the Head [top] and Leg [bottom] of an upright man in pragmatic correspondence with the projected Northern Hemisphere and Southern Hemisphere in the global map of the global trinity & quaternary? What are the distinguishing marks between Solar System and Sirius Binary System? What are the distinguish marks between projected sun [gravitational wave / electron cloud / ether / Bish-Shamsi] and searched out Bullet [Electromagnetic Wave / manifested sign of natural magnetism]?

Which one between hen and egg had come first? Which one follows the other between project Sun and the moving Earth of the circular rotation & revolution system with reference to Copernican Revolution? Which one revolves round the other between the so-called Sun [Bullet] and the so-called Moon [Triangular Bullet]? Which one moves round the other between Morning Show and Evening Show? Which one among the non-luminous white Moon and the luminous Black Square follows Semi Major Axis? Whether Solar System represents the binary ~sth between the sun and the earth or between the earth and the universe or between the moving sun and the moving earth or between the bright sun and the dark moon or between Sirius A and Sirius B or between Electromagnetic Waves and Gravitational waves or between Binary Pulsar and Sirius Binary System or between hexagon and pentagon or between Black Square and Star Operator, and the like?

Let us make a railway journey with the pen –paper-pencil workers and black & white artists from the Railway Station Z towards the Railway Station P in resemblance with the key-board of a PC. During the journey, it has been observed **universally** that the trees, light posts, hills etc. are moving. Now, just reaching the Railway Station P we make a jump **together** exclaiming **UREKA**. You will rush towards that UREKA with searching eyes and ask about that exclamation. We all together will reply you that the **purpose** of our journey is to verify & justify universally the Copernican Revolution. You are anxious to know the result of our verification & justification. With due **proportional** tone we will begin to share with you that Copernican Revolution is the **highest Scientific Certainty** as well as the **Universal Major Premise**. The curiosity of your upright nature will compel you to ask the **reason/logic** of this universal recognition. So, we will provide justifications in resemblance with J. S. Mill's **self-contradictory** as well as **paradoxical** nature. Since the earth is moving following circular rotation and revolution system, and we have universally observed that trees, light posts etc. are also moving, so, the train was running towards the Railway Station P and the Railway Station P was also moving towards the train [us]. Due to this **equal and opposite** running and moving we have finally reached the Railway Station P. So, we are certifying on the **basis** of the equal & opposite running and moving as well as our **universal mal-observation** that **Copernican Revolution** is not only the highest scientific certainty but also the Universal Major Premise. In the mean time the mouth speakers will begin to advertise the product of universal mal-observation in resemblance with Mill's suicide games. The research scholars will begin to do research works on that universal mal-observation. The supervisors and experts will examine research works on the basis of number of references, and ultimately, they will make our journey as if a revolution in the field of true knowledge.

Now, it is your turn to use the term **'commentary'** equivocally in resemblance with the researchers of IFTA who are doing research on Revelations [Truth-in-itself] and manifestations [coded shared tautologies or manifested signs] in the name of Islamic Research following Mill's subjective selfcontradictions and objective paradoxes. Moreover, if revelation and manifestation need to be researched then can those truths be called Revealed Truths and Manifested Tautologies or Manifested Nature?

In this inspired work of 'Kitaaba Wal-Hikmata' or 'Manifested Nature and the Utility of One's Upright Logic', the searched out real scientific symbols along with their character codes in resemblance with revelations are the affirmative minor premises, while both revealed Truth as well as coded shared tautologies [Kitaab with Truth] are the Universal Major Premises. In other words, Fundamental Ayat and detail explanations of Fundamental Ayat are the Universal Major Premises for justification of the findings of this inspired sharing. You are requested to verify the findings of this inspired sharing on the basis of real scientific symbols in resemblance with manifested nature [manifested signs]. This sharing is primarily the sharing of Manifest Truth called Nature justifiable by **Universal Major Premises and verifiable by searched out real scientific symbols [called General or Affirmative Minor Premises], and Manifested Signs**. We are to remember here the difference between universal and general concepts. Have you ever heard or shared the name of a universal secretary, though you are hearing and sharing names of uncountable general secretaries?

The pen-paper-pencil workers and black & white artists are teaching us that there is a sun which rises from the east, and moves following an orbit high above our heads, and then sets in the west. The earth which is beneath our feet also moves round the sun which is moving high above the head. So the questions are – If the so-called sun rises from the East, then from which direction the commonly conceived Earth moves round the so-called sun? Which one follows which one? Whether soil of Hindustan [earth] does not move when there is no sun during night in West Zone? From which direction the so-called sun rises in America? If the so-called sun rises from the East in America, then it must have to rise from the West in Hindustan as America and Hindustan are the two opposite seashores. So if you are scientists and you want to argue with shared manifest truth, then at first you have to make clear that fallacy which had failed to vitiate Copernican Hypothesis. If you have no knowledge about the manifest truth concerning the alteration of day and night for the equal & opposite seashore i.e. twice sun [Bullet] rise and twice sun [Bullet] set, then do you consider yourselves as right epistemic persons to argue with manifested signs of natural magnetism? What are the existential imports of the projected Sister Planet Venus [Bullet] and Planets Uranus & Neptune [Triangular white

& black Bullets]? Whether there is a moon or there are moons? What are the atomic components of the Nil Arm Strong's visited moon in resemblance with the reflection of my corporeal existence in a mirror? If I say that I have four fingers and one equal and opposite finger in my two hands, then how many fingers are there in my hands? If you say that there are eight equal and opposite planets, then how many planets are there? In this regard either you have to share manifest truth with us or you have to make a journey with the selfevident findings concerning manifest truths **till the killing of the introduced dead calf as the adopted son of Firawn from the 5th column and the 7th row of Napier Bones or Napier Rods** with a view to unveil the truth concerning the **'786' Black Magic of the Original Uncultured Epistemic Laden**. Those who have rejected faith, they have conceptualised the code number of Original Uncultured Epistemic Uniqueness as if **BISMILLAAHIR-RAHMAANIR-RAHIM without confirming the same with the revelations. But under the guidance of my Rab, I have searched out the teleological evidence sorceries and epistemic persecutions of Original Uncultured Epistemic Laden which are justifiable in resemblance with revelations and verifiable by the manifested signs and real scientific symbols available in the separate sheet. The Don of the two-in-partnership has changed all the names of the Suras that begin with 'Waw' into Al, At, At, Ad etc. and the disbelievers are also using the magical number of the Original Black Magicians at the top of their magical capsules.** They even named the revealed Kitaab as book and the magical book as Kitaab.

Under the guidance of my Rab and the Rab of the Black Magicians, I have decoded ideograms as well as symbols of **259 Scientific Institutions** from **Windows'7 Ultimate** and Insert More Symbols of my PC. Now **Original Epistemic Uniqueness and his/her conspiracy groups must have to falsify [reject] those scientific symbols, translation/transliteration of Quran [Truth] in the name of 'The Holy Quran' with critical remarks or they must have to share manifest truth with us. There is no scope to play the game of Black Magic further or to reject 'Selfevident Findings concerning Manifested Nature' as I have already placed my searched out Solidified Solid Human Rights in their right places.**

I was an illiterate man concerning the language of the Kitaab with truth. But the understanding of real scientific symbols available in the separate sheet under the guidance of my Rab has made my task easy to understand 'translation of the meanings, commentary, commentary codes etc. on revealed Truth-in-itself as well as criterions of truth. I will try to place almost all coded shared tautologies as universal major premises of the Dictum de omni et nullo or both truth-in-itself as well as criterions of truth before verifiers and justifiers of the searched out Solidified Solid Human Rights as if Fulsirat or Lakshman Rekha between two high ways, between Safa and Marwa, between truth & falsehood, between light & darkness, between the so-called sun and projected sun with a view to seek curse of UNBREAKABLE REALITY or UNIQUENESS or UNIFORM PRINCIPLE upon falsehood.

Moreover, the translation of the term 'Shams' as 'Sun' by IFTA and the Universe is a Solar System by the conspiracy scientists as well as the commentaries [critical remarks] of IFTA in resemblance with projected conspiracy science manifest a partnership of the Historical Don with so-called Laden or between conspiracy scientists and researchers of IFTA. But my Rab has inspired me to disclose their mighty plots which are tracking off the believers from the equal & opposite right direction of Qibla without Broken Bar. If it be the will of my Rab to unveil the manifest truth, then there is no alternative for the Original Uncultured Epistemic Laden & Black Magicians save to share manifest truths like –

i)   The manifested world is a Black Square and a hexagon appearing as a pentagon in resemblance with the existing 9/11 Pentagon or Spiders' Net or Star Operator or House.

ii) The appointed Light of Allah or Manifested Sign of Natural Magnetism is kindled / enters each day from Manifested North-East [Revealed Left or Northern Hemisphere or Haiyalas-Swalaah] in correspondence with the existing North America for lower seashore [East Zone] and sets in Manifested South-East [Revealed Right or Southern Hemisphere or Haiyalal-Falaah] in correspondence with existing South America. The so-called sun as the manifested sign of natural magnetism runs away from there and rises from Manifested South-West [Revealed Right or Southern Hemisphere or Haiyalal-Falaah] for the West Zone in correspondence with existing Europe and moves towards the end point i.e. Manifested North-West [Revealed Left or Northern Hemisphere or Haiyalas-Swalaah] in correspondence with the existing Australia due to Revealed Magnetic Force. So, the so-called sun as the manifested sign of natural magnetism enters from North, sets in South, turns away from there and rises from South, and ends in North. These twice entering & rising, and setting & ending of the so-called sun as the manifested signs of natural magnetism - from North to South, and South to North represent sole Magnetism or so-called Natural Magnetism. Twice rising and twice setting of the manifested sign of sole Natural Magnetism cause the alteration of day and night for the equal & opposite seashore.

iii) There are twelve springs or matter particles shared by each sea-shore. [Four for straight usages in resemblance with four converted shades of Windows'7 Ultimate+ seven force carrier particles or black octagonal spots or windows in resemblance with digit 7 of Windows'7 Ultimate + one equal and opposite so-called sun (Bullet) and so-called moon (Triangular Bullet)] in resemblance with two white semi-circles [converted blue & red shades and converted green & yellow shades] and two birds like binary pulsar in green and blue shades following equal & opposite tree of Zytuun in the converted red shade..

iv) Projected nine planets are the nine stars or clear proofs given to Muusa [ass] as eye opening evidences for verification and justification of the manifested nature in resemblance with revelations.

The condition of a successive relation between nature and behaviour or potentiality and actuality or form and matter or essence and existence or the like is the antecedent and consequent relation or 'if --- then -----'relation or implication. This successive relation is called cause and effect relation or Law of Causation or Kun and Fayakun or Law of Karma. This Law of causation is the ongoing 3D command of Allah. The belief on a causal relation between equal and opposite nature like Sirius A & Sirius B, cause & effect, electron & proton, up & down, right & left, prior & posterior, male & female, subject & object, knower & known, willingness & unwillingness, and so on is the belief on the ongoing command of 3D. This belief is called the belief in Trinity or manifested signs in resemblance with revelation or detail explanations of Fundamental Ayat or Manifested Tautologies in resemblance with Truth-in-itself. There are approximately 18000 [eighteen thousand] equal and opposite relations and seventy layers of knowledge. The reference of this sharing resembles with the reference of the founder of **Objective Idealism**. Hegel has developed his dialectic on the basis of seventy categories of knowledge from a Haddith of Prophet Muhammad [sas] where it has been mentioned that there are seventy layers of knowledge. Out of seventy layers of knowledge, only one layer has been manifested as the World. Hegel has changed the concept of layers into the concept of categories of knowledge [shane-nuzul]. This is nothing but the concealment of truth with a black cloth over the sightless eyes of the justifier & verifier with a view to project falsehood on both sides of the balance stand as if two sides of the same coin in resemblance with the coin of Amitabji in the film 'Shule' or two faces of a thread in resemblance with owners of historical conspiracy science & the researchers of IFTA who are providing the gift package of critical remarks on Quran. Their activities with formal documents are called Teleological **Evidence Sorceries and Epistemic Persecutions or Black Magic**. If my Rab wills, then this inspired sharing will be able to unveil several kinds of Teleological Evidence Sorceries which are historically going on with a view to track off blindly followers of Truth called believers from the equal and opposite middle course towards Upright West by creating several subjective selfcontradictions [digging small holes] as well as objective paradoxes [big bang] in their upright logic. This binary nature i.e. concealment of manifest truth and the projection of manmade falsehood as manifest truth is called Unique Epistemic Persecution. In brief teleological evidence sorceries and unique epistemic persecutions are the keys of **DIVIDE AND RULE POLICY or WISE POLICY OF SHYATAN [DEVIL].**

Though I am by profession a sharer/teacher of philosophical theories, yet I have no clear concept about Hegel's dialectic on the basis of seventy categories of knowledge. You are requested to make your concept clear from the recognised philosophers of the pentagonal earth who have capabilities to approach between justice & injustice. **If philosophers, people of the Kitaab with Truth, researchers, hypocrites, the so-called epistemic persons who had examined several research works suffered by the inferiority complex and motivated by the will force of recognition want to argue with shared 'Selfevident Findings' going beyond criterions of truth, then they must have to make clear the dialectic of seventy categories [shane-nuzul] of Hegel.**

Once upon a time, there was an evidence sorcerer & persecutor in the name of Firawn. This name has been shared in the Kitaab with Truth as a sign of teleological evidence sorcerer & unique epistemic persecutor as well as Historical Don of Black Magicians. Firawn has introduced a calf [**the so-called sun which rises from the east and sets in the West by converting Wash-Shamsi into Bish-Shamsi in resemblance with the reference list of a research work**] as if a partner of Allah in the absence of Muusa [ass] for forty nights. This truth has also been shared with us in the Kitaab with Truth. But Muusa [ass] has inspired us to kill that dead calf by ourselves. Now, the appointed time is very near and it has been inspired to remind you, to inspire you, to give you good tidings so that we may kill that roasted calf of Firawn and the dog's paw by the grace of Allah. So, this sharing is the **first shouting without a second one** to those who are – (i) projecting the **red colour triangle** of the 5th column and the 7th row of the Napier Bones as well as the **red colour triangle** of the **Anti-virus K7** as the **so-called sun** that rises from the East and Sets in the West but represents Natural Magnetism i.e. North & South, (ii) projecting the hexagonal manifestation and pentagonal appearance as global trinity, (iii) projecting the appointed Kaba at the centre of the Middle-East Region of Eartha 3D of the Manifested Hexagonal World within the Black Square of East Horizon, and the like.

It has also been shared with us that the body of Firawn will remain as a manifested sign or a concrete example of the consequences of black magic, evidence sorceries and persecutions not only for the believers of Revelation but also for the present as well as future generations. I am a mortal being like you, and consequently neither a truth recogniser, nor an authoritarian of truth, nor an interpreter of Truth-in-itself, nor a critic of others faith and belief, nor a sharer of this or that kind of truth. On the other hand, I am a finite sharer of individual concept of truth called selfevident truth grounded on two pre-conditions of Scientific Certainty i.e. justifiable by a true antecedent [Uniform Principle or Revelation or Universal Major Premise] and verifiable by observable manifested signs [affirmative minor premise or real scientific symbols or Uniformity of Succession or Law of Causation or Manifested Nature] free from subjective selfcontradictions and objective paradoxes.

Through this sharing, I am not recognising these or those persons are the black magicians, the teleological evidence sorcerers, and the epistemic persecutors. Though the roots, branches, leaves, flowers, and fruits of selfexplained teleological evidence sorceries and self-proved epistemic persecutions are appearing all over the pentagonal earth, yet the stem [projected planet Mars] of those teleological evidence sorceries and epistemic persecutions is one and the root of those teleological evidence sorceries and epistemic persecutions is a **historical one i.e. remote past or Straight Middle-East Region of the East Zone within the Black Square of East Horizon**. In this sharing, I am addressing that unique one as the **Historical Don** or **Epistemic Uniqueness** and others are chiefs, ministers, soldiers, partners, and black & white artists, activists, terrorists, tyrants, and the like of that Historical Don or Epistemic Uniqueness. They are the theorisers of falsehood as truth, recognisers of formalities & guidelines as criterions, players of immediate inferences with good names of others, justifiers of truth on the basis of list of references & opinions, researchers on Truth-in-itself and tautologies, and the like. They have no concept of the four cardinal directions in resemblance with four basic forces. They have no faith on revealed and manifested magnetism. So, they are determining their right direction of Qibla towards Black Hole in the Straight Middle East Region within the Black Square on the basis of mechanical as well as man-made directions in this life but are taking the responsibilities to lead us towards Nuur or Light or Arsh or Upright West in hereafter.

Failing to search out a single reliable table as the right place of placing selfevident truth, I am placing my allegations **against the Historical Don** before all of you [my dear friends in deed or students, truth seekers, common people, honourable chairs of law & Justice, human rights authorities, organisations, government of each country, and the trinity of true scientists – true people of the Kitaab with

Truth – true philosophers] **with the expectation that you will try to** justify as well as to verify the shared selfevident findings.

In this sharing, I will not share complete decoded scientific symbols & number games. I shall try to share some clues so that the truth seeker scientists can justify as well as verify the findings of this sharing. If Epistemic Uniqueness and his chiefs try to deny the searched out Solidified Solid Human Rights i.e. searched out manifest truths which resemble with manifested signs of Allah as well as real scientific symbols, then they must have to falsify or to disprove their complete separate sheet i.e. Insert - Symbols.

I have also found that Quran as the Kitaab of both Truth-in-itself as well as criterions of truth i.e. both revelations as well as manifestations has been translated in resemblance with conspiracy scientific sharing by the hypocrites or Wandering Arabs of IFTA Research Institution. Further, they added commentaries on detail explanations with a view to create a permanent bridge of **partnership** between truth and falsehood as if the Golden Mean of Aristotle between two extremes. They have distorted the seven repeated verses of Quran as a Unique Utilitarian Player called Fatiha in resemblance with Copernican Revolution and have made it a prayer of 8 [eight] Ayat instead of seven. This is the selfexplained and selfproved evidence sorcery by IFTA. Similarly, they have misinterpreted several terms, changed the names of all Messengers & warners. They have changed the names of all Suras which begin with 'waw' into 'Alif' with reference to several self-created conspiracy grammatical rules and meanings e.g. the ideogram 'Alif' stands for Allah concealing the concept of the sign of Justice and ~sth. They have changed the gender introducing fictitious grammatical rules. The conspiracy code number of the Historical Don is '786' where digit 7 represents the home number of the Historical Don in resemblance with the digit 7 of the key-board, Bookshelf Symbol 7, Antivirus K7, Windows'7 Ultimate etc. The digit 8 represents Global Trinity and Circular Rotation & Revolution System in resemblance with headache of the digit 7 as '&'. The digit 6 is the conversion of 'waw' [9] in resemblance with the reference list of a research work.

Game of Immediate Inference

Ayat\Verses - But **the transgressors among them changed the word from that which had been told them**. So we sent on them a plague from heaven. For that they had repeatedly transgressed. [Sura No. – 6\7 - As-haabul- a' raaf, Ayat No. – 162]

Ayat No. – 58 & 59 - And when We said: **Enter this township** and eat of the plenty therein as you wish; but **enter the gate prostrate**, and say: **Repentance.**

We will **forgive** you **your sins** and will **increase** (reward) for the **right-doers.** But **those who did wrong changed the word which had been told them for another saying**, and We sent **on the transgressors a plague** from heaven, **for evil doing**. [Sura No. – 1\2 – Baqarah, Ayat No. – 58 & 59]

Verily the reversing [immediate inference] is an addition to unbelief. Ayat\Verses – O you who believe! There are indeed many among the **priests** (Lukmans) and **anchorites** [Imamgiris], who in **falsehood demolish** the **wealth of men** and **obstruct** [them] **from the way of Allah**. And there are those who **stored gold and silver and spend it not** in the **way of Allah** announce unto them a **most grievous penalty**. On the Day when heat will be produced out of that [wealth] in the **fire of Hell**, and with it will be branded their **foreheads,** their **borders,** and their **backs borders**, and their **backs**. This is the [treasure] which you stored for yourselves. Taste you, then, the [treasures] you stored. The **number of months [planets] in the sight of Allah is twelve. So ordained by Him the day He created the heavens [West Horizon] and the world [East Horizon]. Four of them are blessed that is the straight usage**. So **wrong not yourselves therein,** and **wage the Pagans / idolaters all together as they waging on all of you**. But know that Allah is with those who restrain themselves. **Verily the reversing is an addition to unbelief.** The unbelievers are led to wrong thereby, for they make it lawful one year, and forbidden another year, in order to adjust the number of months forbidden by Allah and make such forbidden ones lawful. The evil of their course seems pleasing to them. But **Allah guides not those who reject Faith**. [Sura No. – 8\9 – Tauba, Ayat No. – 34 to 37]

It was very shameful morality that I had conceptualised the sellers of the number 786 in their capsules as the ministers of Allah. Probably the people of the kitaab with Truth never imagined that the successors of Prophet Muhammad [sas] will mislead them. So, they have conceptualised the conspiracy remarks or critical notes in resemblance with conspiracy science called commentaries as further grammatical explanation or Tafsir of Quran in the name of 'The Holy Quran'. It has been repeatedly shared that the Revelations have been shared explaining in detail. These detail explanations are the manifested signs or concrete examples or similitude or resemblance or appearing nature. On the other hand, if it is asked the people of Kitaab with Truth – 'What is meant by the term 'commentary'?' undoubtedly they will answer that remarks, comments, criticisms etc. But I have failed to conceptualise the ground or logic on the basis of which they have conceptualised commentaries of IFTA on Quran as tafsir or further explanations of Quran.

The sharing of kitaab with Truth as Universal Major Premise is neither translation of the language of Quran, nor interpretation [explanation] of Quran, nor commentary on Quran. On the contrary, the sharing of Kitaab with Truth as Universal Major Premise is the individual concept of the resemblance between revelation and manifestation on the basis of translation and transliteration of Quran by the people of Kitaab with Truth. This individual concept is called selfevident truth. So, if you find any subjective selfcontradiction or objective paradox in this sharing, please give me good tidings.

J. S. Mill had tried to prove the Uniform Principle [Truth-in-itself] and consequently grounded on subjective selfcontradictions and objective paradoxes. The remedy which had been suggested to him was nothing but to dropout the attempt of proving faith and belief without any argument. So, being a man with reason, I will not try to play a suicide game like Mill in this sharing; otherwise, undoubtedly I will be the non-human person of Peter Singer or the parrot of John Locke. Moreover, I will also try to remain conscious throughout the sharing of selfevident truth so that John Rawls must not be able to cover me and my friends in deed further with the help of triangular red cloth of ignorance. Nothing comes out of nothing. So, what is the cause of Uniform Principle and what is the cause of Uniform Motion? If someone as a philosopher or a scientist or a people of the Kitaab with Truth is able to prove the two pre-conditions grammatically as well as out of these two pre-conditions which one was the first cause, then I will try to argue with him concerning Fundamental Ayat [pl] and their detail explanations.

No blame to the 99% people of the appearing pentagonal earth who are like the sharer of this inspired sharing victims of teleological evidence sorceries and unique epistemic persecutions. This inspired sharing is for those who are human beings just because they possess a shared blessed message called reasoning or logic. If you are able to make clear your concept about a manifest truth or a manifested sign or what is there in nature as a concrete example with firm faith and belief in resemblance with selfexistence or Descartes' 'Cogito Ergo Sum' or 'I think, therefore, I exist', then it is your selfevident truth justifiable as well as verifiable by the utility of your upright logic or Kitaaba Wal-Hikmata in resemblance with Manifested Nature. In other words, a justifiable as well as verifiable selfevident truth will resemble with a tautology or a true antecedent or a universal major premise which no finite being can **falsify** or **disprove**. Consequently, you will become the acknowledger of 'Kitaaba Wal-Hikmata' or 'Manifested Nature and the Utility of One's Uprigt Logic' as well as sharer of the selfevident truth.

But those whose will-forces will lead them to play mischievous games of immediate inferences knowingly with the selfevident truth motivated by the sense of recognition; they will be the members of the Teleological Evidence Sorcery and Unique Epistemic Persecution Group. They are invited to play the game of 'Magic with Manifested Signs of Natural Magnetism and Upright Logic'. In this regard the method of the game will not be 'Come with one among the four pillars [two North & two South] – Reject your Faith & Belief on One Dot Leader and Diamond Operator. On the contrary, the method of the game will be -

i) **Go with me i.e.** 'Kitaaba Wal-Hikmata'– **Agree with me** i.e. 'Manifested Nature and the Utility of One's Upright Logic'

ii) Go with **Manifest Truth** – Agree with searched out **Selfevident Truth**

iii) Go with **Affirmative** Minor Premises [Manifested Signs] – Get **Verifiable Truth of Facts**

iv) Go with **Universal** Major Premises – Get **Justifiable Necessary Truth**

v) Go with Teleological Evidence Sorceries [**Mechanism**] – Commit **Objective Paradoxes**

vi) Go with Epistemic Persecutions [**Man-made Magnetism**] – Commit **Subjective Selfcontradictions**

vii) Go with Circular **Rotation & Revolution** System – Agree with **Post Hoc Ergo Propter Hoc.**

SECTION – II
METHODOLOGY – 2.3
WITH REFERENCE TO SOLIDIFIED
SOLID HUMAN RIGHTS

Fasta-'iz billaahi minash-Shaytaanir-Rajim
Bismillaahir-Rahmaanir- Rahim

Findings of this inspired sharing are the deduced conclusions as well as searched out selfevident truths / manifest truths concerning manifested nature, equal & opposite stages of journey of the manifested sign of natural magnetism, equal & opposite verifiable as well as justifiable right direction towards Upright West in resemblance with manifested signs and revelations, appointed day of observing Idd uniformly, and the like. Conclusions / shared searched out findings of 'Kitaaba Wal-Hikmata' or 'Manifested Nature and the Utility of One's Upright Logic' are verifiable as well as justifiable by one or the other criterion of truth in resemblance with revelation & manifestation. In other words, each searched out manifest truth is verifiable by the affirmative minor premise and justifiable in resemblance with universal major premise. Moreover, conclusions / shared searched out findings of 'Kitaaba Wal-Hikmata' or 'Manifested Nature and the Utility of One's Upright Logic' are not absolute but relative. Each conclusion / shared searched out finding is a **Bond** or a **Trinity** of **First Generation Human Rights** [Natural Rights or Birth Rights or Inborn Rights], **Second Generation Human Rights** [Theorised Human Rights], and **Third Generation Human Rights** [Solitary Rights grounded on one's faith & belief]. So, findings or shared conclusions of 'Kitaaba Wal-Hikmata' or 'Manifested Nature and the Utility of One's Upright Logic' are the **Solidified Solid Human Rights**. To get **recognised confirmation** from the respective authorities as well as verifiable & justifiable lacunae in the searched out findings from the experts [verifiers & justifiers] for re-search work are the Solidarity Rights or Individual Rights of the re-search scholar [searcher & sharer of 'Kitaaba Wal-Hikmata' or 'Manifested Nature and the Utility of One's Upright Logic']. To get **generalised recognition** of the searched out Manifest Truths from the respective authorities for the greatest happiness of the greatest number is the Solidified Solid Human Rights not only of the searcher & sharer of 'Kitaaba Wal-Hikmata' or 'Manifested Nature and the Utility of One's Upright Logic' but also of mankind in general.

Postulate – Dictum De Omni ET Nullo suggests that Major Premise must be Universal and Minor Premise must be affirmative. In other words, the validity of a Deductive Argument suggests that the Major Premise must be free from all kinds of fallacies i.e. justifiable necessary truth [**Rationalism**] and Minor

Premise must be based on observation of particular instances i.e. verifiable truth of facts [**Empiricism**]. According to Leibnitz, truth of fact is derived from experience [Empiricism] and necessary truth is derived from reason [Rationalism]. His famous dictum is – "There is nothing in the intellect which was not previously in the sense except the intellect itself". Further, necessary truth implies concept and truth of fact implies percept. According to I. Kant, "Perception without conception is empty and conception without perception is blind" [Ref. Critical Theory of Knowledge]. In brief, from the point of view of finite knower, knowledge is relative. It origins from Revelation or Fundamental Ayat [Aprori] and begins from Manifestation or Detail Explanation [Apoteriori]. To search out verifiable as well as justifiable resemblance between manifested signs [truth of fact or Injiil] and equal & opposite law of revelation [necessary truth or Tawraat] by the utility of one's upright logic [Hikmat] is called search for truth [true knowledge]. Messenger of Allah has inspired us with concrete example to travel consistently with a view to acquire true knowledge free from subjective selfcontradictions and objective paradoxes. This sharing is a search for truth in resemblance with the inspired sharing of Prophet Muhammad [sas].

Truism – The sole condition of a valid argument is that if the premises are true and the reasoning is correct, then the deduced conclusion must also be true, and the argument is valid. It will not be the case that the premises are true and the reasoning is correct, but the conclusion is false. This valid argument is called Dictum De Omni ET Nullo or Valid First Figure or Perfect Figure which includes BARBARA, CELARENT, DARII, and FERIO.

Tautologous Approach – 'Kitaaba Wal-Hikmata' or 'Manifested Nature and Utility of One's Upright Logic' has been framed in a deductive method called Dictum De Omni ET Nullo. So, the tautologous approaches for verification and justification of the searched out findings shared in 'Kitaaba Wal-Hikmata' or 'Manifested Nature and Utility of One's Upright Logic' are as follows –

i) **Go with me i.e.** 'Kitaaba Wal-Hikmata'– **Agree with me** i.e. 'Manifested Nature and the Utility of One's Upright Logic'
ii) Go with **Manifest Truhs**– Agree with searched out **Selfevident Truths**
iii) Go with **Affirmative** Minor Premises [Manifested Signs] – Get **Verifiable Truth of Facts**
iv) Go with **Universal** Major Premises – Get **Justifiable Necessary Truth**
v) Go with Unique Teleological Evidence Sorceries [**Mechanical Trinity e.g. Globalisation**] – Commit **Objective Paradoxes**
vi) Go with Epistemic Persecutions [**Black Magic e.g. Man-made Magnetism**] – Commit **Subjective Selfcontradictions**

vii) Go with **Rotation & Revolution** System of the Immovable Pentagonal Earth – Agree with **Post Hoc Ergo Propter Hoc**

viii) **Contrary or Contradictory Approach – Play the Game of Magic with Manifested Sign of Natural Magnetism and Upright Logic.**

ix) **Easy Approach** – Go as you like but show me **Equal & Opposite Middle Course** [Right Direction] towards **Upright West** [Arsh] in resemblance with the appointed Kaba on the right side of the Mount Tuur in the Upright [Top] West Region of the Manifested Hexagonal World and Uppermost Land of the Appearing Pentagonal Earth within the Black Square of East Horizon.

x) **Minor Premises** – All shared Scientific Symbols along with their character codes with reference to separate sheets of the Real Science i.e. Insert – Symbol, and Manifested Signs

xi) **Major Premises** – Universal Propositions free from all kinds of fallacies i.e. Truth-in-itself and Tautologies or Fundamental Ayat and Detail Explanations of Fundamental Ayat or Manifest Truths in resemblance with Revelations

xii) **References** – Shared **Minor Premises** and **Major Premises of 'Kitaaba Wal-Hikmata' or 'Manifested Nature and the Utility of One's Upright Logic'**

Two warnings
"To enter the Masjid as they had entered it first time."

When the **first of the warnings** [time for the first of the two] came to pass, We roused against you slaves of Ours of great might who destroyed (your) country, and it was a threat performed. Then We granted you the return as against them. We gave you increase in resources and sons, and made you the more numerous in man-power. If you did well, you did well for yourselves. If you did evil, [you did it] against yourselves. So when the **second of the warnings** came to pass, [We permitted your enemies] to destroy you, and **to enter the Masjid as they had entered it first time**, and to lie waste all that they conquered with an utter wasting. It may be that your Rab may [yet] show Mercy unto you; but if you repeat [your sins], We shall revert [to Our punishments]: And we have made Hell a prison **for those who reject [Faith]**. [Sura No. – 16\17 - Banii-Israai-iil, Ayat No. – 5 to 8]

SECOND WARNING
Nine Eye opening Evidences or Projected Nine
Planets of the Conspiracy Science

And verily We gave to Muusa **Nine Tokens [Clear Proofs]** (of Allah's Sovereignty): Do but ask Bani-Israa-iil how he came to them? Firawn said to him: O Muusa! I consider you, indeed, to have been worked upon by sorcery! Muusa said: In truth you know well that these things (clear proofs) have been sent down by none but the Rab of the heavens and the world as **eye-opening evidences**, and I consider you indeed, O Firawn, to be one doomed to destruction! And he wished to remove them from the face of the world; but We did drown him and all who were with him all together. And We said thereafter to Bani-Israa-iil: Dwell securely in the land [earth], but when the **second of the warnings came to pass**, We gathered **you together in a mixed crowd. [Sura No. – 16\17 -** Banii-Israai-iil, Ayat No. – 101 & 104]

LEGAL DECISION

They ask you for a **legal decision**. Say: Allah directs [thus] about those who leave no descendants or ascendants as **heirs. If it is a man that dies, leaving a sister but no child, she shall have half the inheritance. If [such a deceased was] a woman, who left no child, her brother takes her inheritance. If there are two sisters, they shall have two-thirds of the inheritance [between them]. If there are brothers and sisters, [they share], the male having twice the share of the female.** Thus Allah makes clear to you [His law], lest you **err.** And Allah has knowledge of all things. [Sura No. – 3\4 – Nisaa, Ayat No. – 177]

O you who believe! Call in remembrance the favour of Allah unto you **when certain men formed the design to stretch out their hands against you**, but (Allah) held back their hands from you. So fear Allah. And on Allah let believers put (all) their trust. [Sura No. – 4\5 – Maaaidah, Ayat\Verses – 11]

Let the believers **do not take** for friends or helpers non-believers in **preference** to believers. If any do that, he has **no connection with Allah** unless (it be) that you but **guard** yourselves against them, taking (as it were) security. But Allah **warns** you beware (only) of Himself. Unto Allah is the final goal (Summum bonum). [Sura No. - 2\3 – Imran, Ayat No. –28]

O you who believe! If you obey the disbelievers, they will drive you back on your heels, and you will turn back [from equal & opposite right direction] to your own loss. [Sura No. - 2\3 – Imran, Ayat No. –149]

O you who believe! Take not into your intimacy other than your own folk, who would spare no pains to ruin you, they love to hamper you. Hatred has already appeared from their mouths, what their hearts conceal is far worse. We have made clear to you the Signs, if you have understanding. Ah! You are those who love them, but they love you not, though you believe in the whole of the Kitaab. When they meet you, they say: **We believe. But**

when they are alone, they bite off the very tips of their fingers at you in their rage. Say: Perish in you rage; Allah knows well all the secrets of (your) heart. [Sura No. - 2\3 – Imran, Ayat No. –118 & 119]

JUDGE YOU THEN BETWEEN ME AND THEM OPENLY

Ayat\Verses - They said: If you desist not, O Nuuh! You shall be stoned [to death]. He said: O my Rab! **Truly** my people have **rejected me. Judge You**, then, between **me and them openly**, and **deliver me and those of the believers who are with me.** So We **delivered him** and those with him, in the **Ark** filled [with all creatures]. Thereafter We **drowned those who remained behind.** Verily **in this is a Sign, but most of them do not believe.** And verily your Rab is He, the Exalted in Might, Most Merciful. [Sura No. 25\26 – Shuraa, Ayat No. – 116 to 122]

Ayat\Verses – If those who **reject** faith (disbelievers) join **battle** with you, they would certainly **turn** their backs; then they will find neither **protector** nor **helper.** It is **the Law of Allah** which has taken course afore time. **You will find no change in the Law of Allah.** And it is He Who has restrained their hands from you and your hands from them in the valley of **Makka.** After that He gave you the **victory** over them. And Allah sees well all that you do. **They are the ones who denied Signs of Allah** (Manifested Sign of Natural Magnetism) and **hindered (debarred) you from the Sacred Mosque** (equal & opposite right direction towards Upright West) and **debarred the offering from reaching its goal.** And if it not been for believing men and believing women, whom you know not – lest you should trampled them down and on whose account a crime would have accrued to you without [your] knowledge. [Allah would have allowed you to force your way, but He held back your hands] that He may admit to His Mercy whom He will. If they had been apart, We should certainly have punished the disbelievers among them with a grievous punishment. [Sura No. – 47\48 – Fat-ham-Mubiina, Ayat No. – 22 to 25]

Ayat\Verses – Say: O **People of the Kitaab!** Why disbelieve you the **Signs of Allah,** when Allah is Himself **witness** to all you do? Say: O you People of the Kitab! **Why drive you back believers from the path of Allah, seeking to make it global or crooked,** while you were yourselves **witnesses** [of equal & opposite revelation]? But Allah is **not unmindful of all that you do.** [Sura No. 2\3 – Imran, Ayat No. – 98 to 99]

There are two kinds of science within the Black Square [Non-luminous Moon or Dark Moon] of East Horizon. One is projected science grounded on man-made magnetism in resemblance with the will of the Don of the Historical

Black Magic, and the other is real science grounded on natural magnetism in resemblance with revelation. Now, Don of the Historical Black Magic and his ministers & chiefs must have to prove their projected Manmade Nature [**Globalisation, Circular Rotation & Revolution System, and Manmade Magnetism**] with a view to reject shared searched out findings in resemblance with manifested nature [manifested signs of natural magnetism]. Simultaneously, we are to reject Manmade Nature with a view to recognise Manifested Nature in resemblance with manifested signs of Natural Magnetism.

The projection of Arabian Peninsula in the Middle East Region of Earha 3D of the East Zone within the Black Square of East Horizon through global trinity is a conspiracy against believers. This projection is contrary as well as contradictory to Revelation, Manifested Nature, Real Science & Scientific Symbols, and Upright Logic. It has been shared in Quran that there is no crookedness or circular way or common run either in revelation or in manifestation. Moreover, it has been inspired to confirm manifested signs in resemblance with equal & opposite revelation. But going contradictory to revelation, manifested signs, real science & scientific symbols, and upright logic, I had believed on the conspiracy projection of the Historical Black Magician that the appointed Kaba is in the Straight Middle East Region of Eartha 3D of the East Zone as well as at the centre of Revelation, Manifestation, and the appearing pentagonal earth within the Black Square of East Horizon. So, I was off the track from the equal and opposite right direction of Qibla towards Upright West or Arsh. In other words, I was a follower of Black Hole or Eartha 3D of the Straight Middle East Region of the East Zone within the Black Square of East Horizon. Moreover, due to Black Magic, I was far from the true concept of the four cardinal directions, Helium-4, four blessed springs, four Imams for straight usage, four pillars of the pentagonal tent, and the like.

It is none but my Rab who not only has shown me clear signs of the right direction of my Qibla towards upright West following the appointed Kaba on the right side of the Mount Tuur in the Top West Region of the manifested hexagonal world and Uppermost Land of the appearing pentagonal earth, but also has inspired me to give good tidings to those who believe. So, I have to divorce absolutely the wise policy of Shyatan towards Hell or Straight Middle East of Eartha 3D and have been trying to search out the Experimentum Crusis as well as Crucial Instances with a view to prove Teleological Evidence Sorceries and Unique Epistemic Persecution of faith & belief on Revelation, Manifested Nature, Scientific Certainty, Formal Education, and the like. This sharing is nothing new from the point of view of manifested rectangular universe, manifested hexagonal world, and appearing pentagonal earth with three ascending stairs. On the contrary, this sharing is an inspired work with a view to confirm the resemblance between manifested nature and revelation justifiable by Universal Major Premises [Translation / Transliteration of Kitaab with Truth] and verifiable by Affirmative Minor Premises [Real Scientific Symbols along with their character codes / Insert Symbols and Manifested Signs] on the basis of revealed as well as established criterions of Truth.

The findings of 'Kitaaba Wal-Hikmata' or 'Manifested Nature and the Utility of One's Upright Logic' are the selfevident truths. That means the shared findings are free from **subjective selfcontradictions** and **objective paradoxes**. Moreover, these findings are free from the fallacy of equivocation. These findings are verifiable on the basis of searched out real scientific symbols of more than 250 Scientific Institutions along with their character codes. These scientific symbols & their character codes are the **Experimentum Crusis** as well as **Crucial Instances** against **Conspiracies & Evidence Sorceries by the Unique Epistemic Killer of faith & belief of mankind called Historical Black Magician**. So, the findings of this inspired sharing are free from the **fallacies of Non-observation of Instances, Non-observation of Essential Circumstances, Individual Mal-observation, Universal Mal-observation, Characteristic Imperfection, Practical Imperfection, Plurality of Causes**, and ultimately from the **Fallacy of Post Hoc Ergo Propter Hoc**. Each shared finding or selfevident truth concerning manifested nature is verifiable by real scientific symbols as well as justifiable on the basis of **Criterions of Truth** in resemblance with **Revelations** or **Universal Major Premises** or **Truths-in-themselves** or **Kitaab with Truth** or **Complete Coded Shared Tautologies**. So, the shared findings are free from the **Fallacy of Arguing in a Circle** [Petitio Principii] and the **Fallacy of Infinite Regress [Argument ad infinitum]**.

There is a successive relation between revelation and manifestation, manifestation and appearance. These successive relations are the antecedent-consequent relations or cause & effect relations or relations between necessary truth and truth of facts. The antecedent-consequent relation is called Law of causation. Revelation is the antecedent and Manifestation is the consequent. Again, manifestation is the antecedent and appearance is the consequent. The relation between Revelation [antecedent] and Manifestation [Consequent] is nothing but T-Junction. For this reason, the Fundamental Ayat (pl) of Quran and their detail explanations have been termed as Kitaab with Truth. Revelation is called Tawraat or Equal & Opposite Law or Coherence Truth. Manifestation is called Injiil or Correspondence between antecedent and consequent. Appearance is the manifested sign or clear proof of the successive relation between Revelation and Manifestation. A believer is one who has faith on Revelation and belief on Manifestation verifiable by Injiil and justifiable by Tawraat. Faith and belief are the two formal grounds or apriori forms or pre-conditions of scientific certainty, verifiable certain knowledge, and justifiable valid knowledge. In this sense, real scientists and real philosophers are true believers because they rely on both One Dot Leader [Uniform Principle] and Diamond Operator [Uniform Motion or State of Equilibrium]. So, possessors of upright logic will never go either contrary or contradictory to both revelation and manifestation.

If it is asked –'Did you spend the money you had stolen?', then – 'Will you be able to answer yes or no?' If you are able to answer yes or no concerning successive relation of this question, then I will try to answer yes or no if you ask any question concerning successive relation. Such questions cannot be answered as yes or no without confirming the antecedent. In such cases if the antecedents are affirmative, then the questions concerning successive relations [consequents] can definitely be answered as yes or no.

I had to leave science stream in 1987 when I was a student of class XII [twelve] due to several subjective selfcontradictions and objective paradoxes in the domain of projected scientific knowledge concerning manifested nature and natural magnetism. Manifested Truth in resemblance with Revelation does not necessarily require the establishment or theorisation or revolutionisation of a hypothesis or a supposition or a prediction or an imagination or an opinion going either contrary or contradictory to Uniform Principle, Uniform Motion, and signs of successive relation. Such establishment or theorisation is the vera causa of subjective selfcontradictions and objective paradoxes. Scientific knowledge or verified certain knowledge must be free from subjective selfcontradictions and objective paradoxes. In other words, **selfcontradictory as well as paradoxical knowledge cannot be called scientific knowledge**. That means contradictory as well as paradoxical knowledge cannot be regarded as verifiable & justifiable certain knowledge or real science. So, the establishment of contradictory as well as paradoxical knowledge through theorisation in the domain of science is not real science, yet the projection of such knowledge through revolutionisation must be categorised as the conspiracy science. Copernican Revolution, Globalisation, and Man-made Magnetism are the Valid References of Conspiracy Science. The teleology of the real science is to search out truth concerning the reality of the manifested nature verifiable as well as justifiable by the criterions of truth in resemblance with Uniform Principle and Uniform Motion. On the contrary, the intentions of the Historical Don of the conspiracy science are –

i) To conceal the reality or truth concerning manifested nature;

ii) To share falsehood as truth projecting crookedness or circular ways or common run with reference to hypotheses, suppositions, imaginations, predictions, opinions etc. with a view to mislead mankind towards recycling region [Straight Middle East of Eartha 3D] in the name of Globalisation and in resemblance with International Binary Language. [Note – My mother tongue is unary. So, this sharing is univocal and public (common sharing) as if International Harmony and International Human Rights or Solidified Solid Human Rights.]

iii) To engage self-sensed epistemic persons in research works on projected falsehood and complete coded shared tautologies with a view to spread existential fallacies, subjective selfcontradictions, and objective paradoxes;

iv) To inspire epistemic senses so that they begin to play the game of immediate inferences with opinions with reference to four leaders in resemblance with the conversion of good names in the reference list of a research work vitiating one's morality and violating grammatical rules, valid/justified laws, human rights etc.;

v) To conceal the reality of Truth-in-itself or the fifth pillar or the sign of Justice as well as equal & opposite binary ~sth between revelation & manifestation, between West and East; between two end points of revelation, between two end points of manifestations, two entering and two ending points of the Binary Pulsar, and the like;

vi) To track off believers of the manifested hexagonal world as well as appearing pentagonal earth from the equal and opposite right direction [Middle Course] towards Upright West [Arsh] and to guide them towards Straight East Direction or Middle East or Eartha 3D or Black Hole in the East Zone within the Black Square of East Horizon performing Salat towards Haiyalas-Salaah from the Middle Stair and towards Haiyalal-Falaah from the Ground Stair;

vii) To put the veil of darkness over blind eyes of truth verifiers & truth justifiers projecting several contrary formalities and contradictory guidelines as if those formalities & guidelines are the criterions of truth;

viii) To govern mankind of the appearing pentagon [spider's net] from the spiderman's net projecting global orange and concealing the reality of pentagonal orange;

ix) To conceal the reality of the appearing pentagon [spider's net] within the Black Square in resemblance with the concealment of manifested sign of natural magnetism and to project Rotation & Revolution of the pentagonal orange [spider's net] in resemblance with the projected manmade magnetism;

x) To conceal the reality of Non-luminous World [Black Square or Dark Moon Non-luminous Moon] and to project signs of Intrinsically Luminous Star as Non-luminous Moon [Dark Moon].

xi) To conceal the realities of two seashores [West Zone or Upper Seashore and East Zone or Lower Seashore], and three ascending Stairs, namely, Tin, Zaytuun, and Tuur [Superscript 1, 2, 3], and to project Mechanical Globalisation;

Possessors of true knowledge have established Dictum De Omni ET Nullo and four criterions of truth for verification as well as justification of Manifest Truth. But going contrary as well as contradictory to Dictum De Omni ET Nullo as well as criterions of truth, the pen-paper-pencil workers & black & white artists are uninterruptedly projecting several falsehoods, selfcontradictions, paradoxes etc. as truths on the basis of opinions, suppositions, imaginations, predictions etc. concerning the reality of the manifested nature & natural magnetism. With a view to protect their projected falsehoods or man-made nature, they have established several contradictory guidelines and contrary formalities in the domain of sharing manifest truth. The pen – paper - pencil workers and black & white artists who have no clear concept of the four cardinal directions, West & East Horizons, West & East Zones, Northern & Southern Hemispheres, four criterions of truth, four basic forces, revealed magnetism, the revealed universe, the manifested world, the appearing earth etc. yet they are sharing revealed as well as manifest truth going either contrary or contradictory to revelation, manifestation, and appearance. This is because they have conceptualised projected falsehoods of the conspiracy science as revelation, manifestation, and appearance. So, they are unconsciously or ignorantly engaged in their jobs of interpretations, theorisations etc. with a view to strengthen the four walls of projected falsehoods on the basis of guidelines, formalities, opinions, suppositions, imaginations, commentaries, predictions etc. in resemblance with the Teleology of the Black Magician. Now, these falsehoods become the standards or criterions of justification as well as verification of the selfevident truth / manifest truth / manifested nature.

Primary findings concerning manifested nature were shared with some scientific research institutions for verification as well as justification. Being a re-search scholar on 'Solidarity Rights in Islam', the primary findings were also shared with the Department of Philosophy, Assam University [AUS]. But no feedback has been received till date. Probably the so-called scientists of those Scientific Research Institutions, as well as supervisors of the research works on 'Solidarity Rights' have failed to conceptualise the monadic motion of 'Kitaaba Wal-Hikmata' or 'Manifested Nature and the Utility of One's Upright

Logic'. So, this time common people will justify as well as verify searched out selfevident truths or manifest truths shared in the 'Kitaaba Wal-Hikmata' \ 'Manifested Nature and the Utility of One's Upright Logic'.

The findings shared in the 'Kitaaba Wal-Hikmata' or 'Manifested Nature and the Utility of One's Upright Logic' have been searched out on the basis of criterions of truth. These findings are being shared in the form of Dictum De Omni ET Nullo. The Dictum suggests that the validity of deduced conclusions / findings depends on Universal Major Premises and the certainty of deduced conclusions / findings depends on verifiable [observable] affirmative Minor Premises [Manifested Signs]. References of Ayat/Verses are the Universal Major Premises and references of scientific symbols along with their character codes & manifested signs are the Affirmative Minor Premises. Individual concept of the sharer and additional figures are the ˜sth between shared Major Premises and referred Minor Premises. So, the selfevident findings of 'Kitaaba Wal-Hikmata' or 'Manifested Nature and the Utility of One's Upright Logic' are verifiable as well as justifiable on the basis of Four Criterions of Truth, namely, Tawraat [coherence truth], Injiil [correspondence truth], Zabuur [pragmatic truth], and Furqaan [selfevident truth]. As the findings of this inspired sharing are selfevident truths or consequences of the utility of one's upright logic concerning the reality of the manifested nature, and verifiable by real scientific symbols along with their character codes [Affirmative Minor Premises], and justifiable by the criterions of truth in resemblance with revelations [Universal Major Premises]; so, the sharer of this inspired sharing has no concern about the quality or standard of 'Kitaaba Wal-Hikmata' or 'Manifested Nature and the Utility of One's Upright Logic' for publication. In this regard, all truth seekers as well as possessors of upright logic are sincerely requested to make an attempt with a view to verify as well as to justify the quality or standard of 'Kitaaba Wal-Hikmata' or 'Manifested Nature and the Utility of One's Upright Logic' on the basis of revealed and established criterions of truth in resemblance with manifested signs of natural magnetism and Manifested Nature. **The sole concern of the sharer of this inspired sharing is to get generalised recognition [from the Respective Authorities] of the searched out findings shared as Solidified Solid Human Rights.**

We are to remember that none can change the revealed and manifested nature save One Dot Leader. If the shared findings of this re-search project resemble with manifested nature, then these are certain like 'Cogito Ergo Sum' of Descartes. But if the Don of the two-in-one partnership and the Historical Don of the conspiracy science does not recognise the shared selfevident truths, then they must have to prove the following contradictories of their binary nature or hypocrisy –

i) **Shared Scientific Symbols, their character codes, their placement etc. are the products of conspiracy scientific works.**

ii) There is no external force or prime mover or one dot leader or Uniform Principle or Nature or Fitrat or Essence.

iii) The universe is not a Diamond Operator as collocation or State of Equilibrium or Inertia of Newton's First Law or Motion of Galileo.

iv) There is no Sirius Binary System or Trinity.

v) The manifested universe is not an upright rectangle or the End of Proof in resemblance with the appointed Kaba.

vi) Sirius A does not represent West Horizon and projected planet Jupiter within the Universe and Sirius B does not represent East Horizon and projected planet Saturn within the Universe.

vii) Black Square [Non-luminous Moon or Dark Moon] does not represent East Horizon within the four walls of basic forces.

viii) The manifested world is not a hexagon within the Black Square of East Horizon.

ix) Black spots or springs or matter particles or so-called sunspots are not octagonal.

x) Projected nine planets are not nine clear proofs in resemblance with the nine eye opening evidences given to Muusa [ass].

xi) Twelve springs in resemblance with twelve sacred months, twelve hours of a clock, etc. do not represent twelve matter particles – four basic forces in resemblance with four cardinal directions, seven force carrier springs or matter particles or windows in resemblance with seven days of a week, seven colours of rain bow, seven canopies, seven heavens and similar number of world, digit 7 of Windows'7 Ultimate, seven repeated verses; and one equal & opposite binary pulsar or so-called sun [Wash-shams] and so-called moon [Qamar]

xii) Everything is moving save the so-called sun [Wash-Shamsi or Bullet or electromagnetic wave] and so-called moon [Qamar or Triangular Bullet]. The shared truth 'each one is moving' represents that galaxy of stars, the world, and the spread out table are moving save Spiderman's Net.

xiii) Hexagonal units like Unicode (hex), Symbol (hex) etc. are fictitious concepts.

xiv) The earth is not a spread out table or Mayidah or pentagon in resemblance with House, Spider's Net, Star Operator, and in correspondence with the 9/11 Pentagon.

xv) Revealed as well as manifested two West and two East are fictitious concepts; while projected two North and two South are the manifested truths.

xvi) Due to natural magnetism, the so-called sun as the manifested sign of natural magnetism rises from the East and sets in the West. So, projected North and South through man-made magnetism are the revealed and manifested North Direction and South Direction.

xvii) The projected non-luminous moon [commonly perceived white moon] has eclipse due to John Rawls' veil of ignorance.

xviii) The projected globe represents the universe, the world, the earth, the galaxy of stars, region of white, Horizon of night sky, zone of earth's night sky, Horizon of earth sky, stages of journey of the appointed light, zone of cloud, zone of ether, zone of iron, zone of earthquake, the appointed Kaba is in the Middle East of the revelation & manifestation in resemblance with Eartha 3D and the region of white dwarf etc.

Further, if you find subjective selfcontradictions and objective paradoxes in the shared findings or if you find that the shared findings are neither justifiable nor verifiable by the criterions of truth in resemblance with Uniform Principle, Uniform Motion, and signs of Successive Relation, then you are invited to play the game of 'Magic with Manifested Sign of Natural Magnetism and Upright Logic'.

CONDITIONS OF THE GAME OF MAGIC WITH MANIFESTED SIGN OF NATURAL MAGNETISM AND UPRIGHT LOGIC REVELATIONS

CONDITIONS

"Say: O **People of the Kitaab [Ahlal Kitaab]! You have no ground to stand upon unless you stand fast by the Tawraat, the Injiil, and all the revelation that has come to you from your Rab.**"

Ayat\Verses - O Prophet! Proclaim the (Message) which has been sent to you from your Rab. If you do it not, you will not have fulfilled His Mission. And Allah will defend you from men (who mean mischief); for **Allah guides not those who reject Faith.** Say: O **People of the Kitaab [Ahlal Kitaab]! You have no ground to stand upon unless you stand fast by the Tawraat [Revelation or Coherence Truth], the Injiil [Manifestation or Correspondence Truth], and all the revelation that has come to you from your Rab.** It is the revelation that comes to you from your Rab, that increases in most of them their obstinate rebellion and blasphemy. **But sorrow you not over (these) people without Faith.** Those who believe, those who are **Haaduu, Saabiuu-nas** and **Nasaaras,** and any who **believe in Allah** and the **Last Day** [Yawmil-Aakhiri], and **work righteousness,** --on them shall be no fear, nor shall they grieve. [Sura No. – 4\5 – Maaaidah, Ayat\Verses – 67 to 69]

And We caused Isabni-Maryama to follow in their footsteps **confirming that which was revealed before him in the Tawraat [coherence truth].** We bestowed on him the **Injiil [Pragmatically Correspondence Truth in resemblance with the Crucified Sign]** wherein are **guidance and light [equal & opposite right direction in resemblance with natural magnetism],** and **confirmation** of that which was before it in the **Tawraat,** guidance and an admonition to those who ward off evil. Let the People of the **Injiil [Correspondence Truth] judge** by what Allah has revealed therein. If any do fail to judge in **correspondence** with what Allah has revealed, they are (no better than) those who **rebel.** [Sura No. – 4\5 – Maaaidah, Ayat\Verses – 46 & 47]

And when the **Isabna Maryam** is held up (quoted) as **an example,** behold, the folk laugh out (you people raise a clamour thereat) And say: **Are our gods better, or he?** They raise not the objection save **for argument.** Nay! But **they are a controversial (notorious) folk. He was no more than a slave on whom We bestowed favour** and We made him **an example to Baniii-Israiil.** [Sura No. 42\43 – Ummil-Kitaab \ (Zukhruf), Ayat No. – 57 to 59]

When Isa came with Clear Proofs, he said: I have come to you with wisdom, and in order to make clear to you some of the [points] **on which you dispute.** Therefore keep your duty to (fear) Allah and obey me. Lo! Allah, He is my Rab and your Rab. So worship Him. This is an equal & opposite **Straight Way (right path).** [Sura No. 42\43 – Ummil-Kitaab \ (Zukhruf), Ayat No. – 63 & 64]

Lo! The Religion before Allah is Diin-e-Islam [submission to His Will and Guidance]. **Those who received the revelations** (the People of the Kitaab) **differed only after knowledge came unto them, through jealousy of each other** (transgression among themselves). But **if any deny the Signs of Allah,** Allah is swift in calling to account. So if they **argue** with you, say: **I have surrendered My whole self to Allah** and so have those who follow me. And say to the People of the Kitaab (those who have received the revelations) and to those who do not read: **Do you surrender yourselves? If they do, they are in right guidance,** but **if they turn back, your duty is to convey the Message (unto them). Allah is the Seer of (his) bondmen.** [Sura No, 2\3 – Imran, Ayat No. –19 & 20]

Say: **We believe in Allah, and the revelation given to us, and to Ibrahim, Ismaiil, Ishaaq, Yaquub, and the Tribes (Aasbati), and that given to Muusa and 'Isa, and that which the prophets received from their Rab. We make no difference between one and another of them. And we surrender to Allah as**

Muslims. So if **they believe as you believe, they are indeed on the right path**; but **if they turn back**, it is they **who are in schism [far from the right path]**; but Allah will **suffice** you (for defence) against them, and He is All-Hearing, All-Knowing. [Sura No. 1\2 – Baqarah, Ayat No. – 136 & 137]

LEADER OF RIGHT DIRECTION
APPOINTED IMAM OR MANIFESTED SIGN OR IBRAHIM'S
KABA OR BLACK & WHITE IMAM OF THE CITY

Ayat\Verses - And **who does greater wrong** than one to whom are recited the **Signs** of his Rab, and who then **turns away there from**? Verily from those who transgress We shall exact [due] retribution. We verily gave the Kitaab to Muusa. So, be not you in doubt of his receiving it and We made it a guide to the Bani-Israa-iil. "Wa ja-alnaa minhum A-immatany-yahduuna bi-Amri-naa **lammaa** sabaruu; wa kanuu bi-' Aayaatinaa yuu- qinuun". [Trans.] And **We appointed [manifested] Imam** from among those signs **giving guidance** under **Our command**, so long as they **persevered** with **patience and continued to have faith in Our Signs**. Verily your **Rab will judge** between **them** on the **Day of Resurrection**, in the matters **wherein they differ** [among themselves] [Sura No. – 31/32, Sura – Sajdah, Ayat\Verses – 22 to 25]

Ayat\Verses - Wa izibtalaaa Ibraahima-Rabuuhuu bi-Kalimaatin- fa-atammahunna, qaala Innii-jaa-ilukalin-naase **Imaamaa**. Qaala wa min-zurriyyatii; Qaala laa yanaalu ahdiz-zaa-limin. [Trans.] And remember that **Ibrahim** was tried by Allah with certain **commands**, which He fulfilled. He said: **I had made an Imam for mankind**; (he appealed): and **also for my offspring [possessors of upright nature]**. He said: **But My Promise is not within the reach of evil-doers.** [Sura No. – 1\2 – Baqarah, Ayat No. – 124]

Ayat\Verses - And when We made the House (Kaba), a remedy for mankind and a sanctuary (saying): Take as your place of worship the place where Ibrahim stood (to pray); and when we imposed a duty upon Ibrahim and Ismail (saying), Purify My House for those who go around and those who meditate therein and prostrate themselves (in worship). And when Ibrahim said: My Rab! Make this a region of security, and bestows upon its people fruits such of them as believe You and the last Day. He answered: As for him who rejects Faith, for a while I will grant them their pleasure, but will soon drive them to the torment of fire, an evil destination [indeed]! And when Ibrahim and Ismail were raising the foundations of the House [With this prayer]: Our Rab! Accept [this service] from us. Lo! You only You are The Hearer, The Knower. Our Rab! And make us submissive to Your, and of our seed a nation submissive to Your and show us our ways of worship; and turn unto us [in Mercy]. Lo! You, only You are Ever Returning, Most Merciful. Our

Rab! And rise up in their midst a Prophet from among them who shall rehearse Your Signs to them and instruct them in Kitab and wisdom, and sanctify them. For You are The Mighty, The Wise. And who forsakes the religion of Ibrahim save him who befool himself. Verily We chose him in this world, and he will be in the Hereafter in the ranks of the Righteous. When his Rab said unto him: Bow [thy will to Me], he said: I bow [my will] to the Rab and Cherisher of the Worlds. And this was the legacy that Ibrahim left to his sons, and so did Yaquub. Oh my sons! Allah has chosen the Faith for you. Then die not save who have surrendered (unto Him). [Sura – 1/2, Sura – Baqarah, Ayat\Verses – 125 to 132]

Ayat\Verses – Allah has appointed (manifested) the Kaba, the Sacred House, as a standard [manifested sign of right direction or Imam] for mankind, and the **Sacred Months [twelve matter particles or springs]**, the **offerings [99 elements]**, and the **garland** (25 names of the signs of relation / messengers): that you may know that Allah has knowledge of what is in the heavens and in the world and that Allah is well acquainted with all things. [Sura No. – 4\5 – Maaaidah, Ayat\Verses – 97]

IMAMS MEAN RECORDS
Respective Imams – Four Revealed Directions in resemblance with Four Basic Forces [Four Blessed Springs]
Ayat\Verses – One day We shall call together all human beings with their [respective] **Imams**: those who are given their **record** in their **right** hand will read it [with pleasure], and they will not be dealt with unjustly in the least. (IFTA) **[Sura No. – 16\17 - Banii-Israai-iil, Ayat No. – 71]**
Or
Ayat\Verses – On the day when We will summon all men with their **records**, whoso is given his record (book) in his right hand, such will read their record (book) and they will not be wronged in the least. [IS (P) Ltd.] **[Sura No. – 16\17 - Banii-Israai-iil, Ayat No. – 71]**

Note: Here the term 'Imam' connotes the record of one's deeds. Moreover, right hand represents manifested direction in resemblance with revealed & manifested forces. So, the term 'Imam' does not represent a human being. On the contrary, 'Respective Imam' represents particular direction or Right Hand or Haiyalal-Falaah from the Infinite Subjective point of view and Haiyalas-Swalaah from finite subjective point of view.

NO LOGIC TO DISPUTE
Who have plotted mighty plots of mechanical globalisation and man-made magnetism?

Say (unto the peoples of the Kitaab): **Will you dispute with us concerning Allah** when He is **our Rab** and **your Rab**? We are **responsible for our doings and you for yours**. We look to Him alone. [Sura No. 1\2 – Baqarah, Ayat No. – 139]

UMMAT – WITNESS AGAINST MANKIND
"But it is not Allah's purpose that your faith should be in vain."
Ayat\Verses – Thus, We have made you an **Ummat** justly balanced (a middle people), that you may be **witnesses against mankind** and that the **Rasuul** (Prophet) is a **witness over** you. And We **appointed the Qibla** which **you formerly observed only that We might know him who follows Rasuul** (the Prophet) from those **who would turn on** their heels (run away). Indeed it is a **hard test**, except to those **guided by Allah**. But it is **not** Allah's purpose that your faith should be in **vain**. For Allah is to all people full of piety, Most Merciful towards mankind. [Sura No. 1\2 – Baqarah, Ayat No. – 143]

Nuuh said: My Rab! Lo! They have disobeyed me, but they follow one whose wealth and children give them no increase except ruin. And **they have plotted a mighty plot. And they have said: Abandon not your realities.** Abandon **neither Wadd nor Suwa, neither Yaguusa nor Ya-uuqa, nor Nasaraa.** And **they have led many astray; and grant You no increase to the wrong- doers except error** [in straying from their mark]. Because of their sins they were **drowned**, and were made to **enter** the **Fire**, and they **found** that they had **no helpers in place of Allah.** [Sura No. – 70\71 – Nuuh, Ayat No. – 21 to 25]

Say: O **People of the Kitaab [Ahlal Kitaab]! You have no ground to stand upon [upright] unless you stand fast by the Tawraat [coherence truth], the Injiil [correspondence truth], and all the revelations [manifested signs or manifested nature] that have come to you from your Rab.** It is the revelation that comes to you from your Rab, that increases in most of them their obstinate rebellion and blasphemy. **But sorrow you not over (those) people without Faith.** Those who believe, those who are **Haaduu, Saabiuu-nas** and **Nasaaras**, and any who **believe in Allah** and the **Last Day** [Yawmil-Aakhiri], and **work righteousness [do tabligh]**, --on them shall be no fear, nor shall they grieve. [Sura No. – 4\5 – Maaaidah, Ayat\Verses – 68 & 69]

ESTABLISHED CONDITIONS
The sole condition of a valid argument is that if the antecedent is true and the reasoning is correct, then the consequent must also be true. It would not be

the case that the Major Premise is Universal [True] i.e. free from the fallacies of arguing in a circle and infinite regress, and Minor Premise is generalised affirmative proposition [verifiable conclusion of an induction or verifiable certain knowledge or searched out real science or truth of fact]; but the deduced conclusion is false. In other words, in a valid deductive argument premises imply the conclusion.

Kitaab with Truth includes four criterions of truth or four witnesses for justification and verification by the finite judges. The appointed Kaba is the standard of verification and justification for mankind in general concerning right directions in resemblance with revelations. Crucified Sign is the clear proof of four cardinal directions. The four witnesses or criterions have been developed as four theories of truth. These theories are coherence truth [Tawraat], correspondence truth [Injiil], pragmatic truth [Zabuur], and selfevident truth [Furqaan]. Each criterion of truth is neither Truth-in-itself, nor Independent Kitaab with Truth [Independent Nature or Independent Magnetism]. In other words, these theories / criterions have no independent existential import [in resemblance with **Subjective Idealism of Berkeley**]. To regard a criterion of truth as Truth-in-itself or an independent Kitaab or independent revelation & manifestation is to commit the fallacy of Solipsism or Existential Fallacy. There is only one kind of magnetism called Natural Magnetism. To regard manmade magnetism as true concept of magnetism is to commit the **Existential Fallacy or Solipsism**. On the contrary, these criterions are tautologies or manifest truths or selfevident truths or selfexplained truths or manifested nature in resemblance with revelations or detail explanations of Fundamental Ayat or complete coded shared tautologies [**in resemblance with Naive Realism**]. The selfevident findings of 'Kitaaba Wal-Hikmata' or 'Manifested Nature and the Utility of One's Upright Logic' have been searched out on the basis of four revealed as well as established criterions of truth under the guidance of my Rab with a view to get generalised recognition of the manifested nature from the respective authorities.

The shared findings also based on three Fundamental Laws of Thought. These Laws are –
   i)   Law of Identity – A is A
   ii)   Law of Contradiction – A cannot be both B and not-B
   iii) Law of Excluded Middle – A is either B or not-B

The shared findings also based on four Canons of Elimination. These Canons are -
   i)   Whatever antecedent can be left out without frustrating the effect can never be the cause.

ii) Whatever antecedent cannot be left out without frustrating the effect must be the cause or a part of the cause.

iii) If two phenomena like Shams & Qamar always vary together, they are causally connected.

iv) What is known to be the cause of some other phenomenon [e.g. North & South on the basis of man-made magnetism] cannot be the cause of the phenomenon under investigation [North & South on the basis of Natural Magnetism].

To interpret further in detail [Tafsir or translation of the meanings] with categories of knowledge of what have been shared explaining in detail by Allah and His Messenger [Manifested Signs of Allah], to do research work on complete coded shared tautologies [manifested nature or natural magnetism] without searching resemblance between revelations and manifestations or without confirming truth, to comment merely on complete coded shared tautologies concealing true concepts of Truths-in-themselves [Revelations or Fundamental Ayat], and to do pen – paper – pencil works on complete coded shared tautologies on the basis of opinions motivated by the sense of recognition with a view to project criterions of truth as truths-in-themselves like the father of IRF can be categorised as the works of the **Real Activists**. You will find the shared selfevident truth of this inspired sharing either contrary or contradictory to the works of the **Real Activists**. Two contraries cannot be true together, though they both maybe false together. In other words, if one of the contraries is true, then the other must be false, but not conversely. That means if one of the contraries is false, then the other is doubtful. But two contradictories can neither be true together nor false together. If one of the contradictories is true, then the other must be false; and conversely. Suppose two contradictories -

i) Universe is a Diamond Operator, Revealed as a Trinity, and manifested as an Upright Rectangle. The world is a Black Square [Non-luminous Moon or Dark Moon] manifested as a hexagon [Asterisk] within East Horizon. Appearing Earth is a Pentagon with three ascending Stairs. There are four revealed directions in resemblance with Crucified Sign, namely, Down [East], Up [West], Revealed Left [North], and Revealed Left [South]. Manifestation resembles with multiplication sign. So, there are six manifested directions, namely, North-East, Middle-East, South-East, South-West, Upright-West, and North-West. Two sacrificed directions are North [North Pole] and South [South Pole]. There is only one kind of Magnetism called Natural Magnetism. The so-called sun [Bullet] as the manifested sign of natural magnetism enters from North-East and sets in South-East for the ground stair of North America and South America. The so-called sun [Bullet] as the manifested sign of natural magnetism rises from South-West and ends in North-West for the Middle stair of Europe, Asia, Africa, and Australia as well as Topmost Stair of Arabian Peninsula. The so-called sun [Bullet] rises twice and sets twice to cause the alteration of day and night for the equal & opposite Seashore. The appointed Kaba is neither in the Middle-East Region of Eartha 3D nor at the centre of Revelation, Manifestation, and Appearance. On the contrary, the appointed Kaba is on the right side of the Mount Tuur in the Upright West Region of the manifested hexagonal world, Topmost Stair and Uppermost Land of the appearing Pentagonal Earth.

ii) The International Globalisation or Global Trinity or Rotation & Revolution of the immovable pentagonal Earth or so-called Solar System is called the Universe or the Trinity as well as quaternary of Global Universe, Global World, Global Earth, and sign of relation between Global Universe, Global World & Global Earth is the Projected Global Map. There are two kinds of magnetism i.e. one manifested nature in resemblance with natural magnetism and one manmade nature in resemblance with manmade magnetism. In manmade nature, the so-called sun [Bullet] rises once from an unidentified East and sets once in an unidentified West to cause the alteration of day and night for the equal & opposite seashore of the Manifested Nature.

These two contradictories can neither be true together nor false together. One of them must be true and the other must be false. One is ground on real scientific symbols in resemblance with manifested nature justifiable in resemblance with revelations and the other is the store house of selfcontradictions & paradoxes due to the product of conspiracy works with a view to play the game of Black Magic. Projected North and Northern Hemisphere, South and Southern Hemisphere represent manmade nature. All pen-paper-pencil works and black & white activities in resemblance with Mechanical Globalisation, Circular Rotation and Revolution System, and Manmade Magnetism are neither Manifest Truths nor Manifested Nature in resemblance with Revelations, but are Teleological Evidence Sorceries and Unique Epistemic Persecutions.

Now, these two contradictories have been placed together so that the Historical Don of two-in-one partnership must have to reject one of the contradictories and to accept the other. In other words, we are to verify as well as to justify one of the contradictories as manifested nature in resemblance with revelations. Simultaneously, the other contradictory sharing will be proved as manmade nature as well as Teleological Evidence Sorceries and Unique Epistemic Persecutions. The method of placing two contradictories together concerning manifested nature before all with a view to get generalised recognition of the searched out and shared selfevident truths / Manifest Truths / Manifested Nature is called **'GO WITH MANIFESTED SIGNS – RECOGNISE SELFEVIDENT TRUTH**; while with respect to conversational method, it is called 'GO WITH ME – AGREE WITH ME'. With respect to this inspired sharing, this method is - **Go with Kitaaba Wal-Hikmata – Agree with Manifested Nature and the Utility of One's Upright Logic**. This method will help us to reject one of the contradictories and to recognise the other.

Further, the placement of the above contradictories together is an instance of how to verify as well as to justify each and every shared finding or selfevident truth by those who possess upright logic, who are seekers after truth, and Honourable Chairs who will verify as well as justify shared findings of 'Kitaaba Wal-Hikmata' \ 'Manifested Nature and the Utility of One's Upright Logic' with a view to provide generalised recognition of shared Solidified Solid Human Rights. This method of placing contradictories together is also a concrete example for the self-sensed pen, paper, and pencil workers who have the capability to argue with searched out verifiable as well as justifiable manifest truths. So, sceptics like David Hume have nothing to do with shared selfevident truths, but – 'TO GO WITH ME & AGREE WITH ME' or they must have to play the game of 'MAGIC WITH MANIFESTED SIGN OF NATURAL MAGNETISM AND UPRIGHT LOGIC'.

SOLIDARITY RIGHTS

Third Generation Human Rights are called Solitary Rights or Solidarity Rights or Solid Human Rights. Solitary Rights include -
  i)   **Natural Human Rights** [Universal Rights / **First Generation Human Rights**] – With respect to 'Kitaaba Wal-Hikmata' or 'Manifested Nature and the Utility of One's Upright Logic', Natural Rights include Rights to get verifiable & justifiable confirmation as well as general recognition of the searched out Manifest Truths, i.e. Rights concerning Natural Magnetism, Manifested Nature, Manifested Signs, Four Revealed and six Manifested Directions, End of Proof / Upright Rectangular Universe, East Horizon / Black Square, Hexagonal World / Asterisk, Three ascending stairs, Pentagonal Earth etc.
  ii)   **Formalised / Theorised Human Rights** [General Rights / **Second Generation Human Rights**] – With respect to 'Kitaaba Wal-Hikmata' or 'Manifested Nature and the Utility of One's Upright Logic', Fundamental Human Rights include Bread and Butter Rights, Rights to get confirmation of the searched out lacunae termed as Mechanical Trinity and Black Magic or packages of subjective selfcontradictions

and objective paradoxes, Rights to do re-search work on those searched out manifest truths which are neither verifiable nor justifiable by one or other criterion of truth, right to get recognised confirmation as well as generalised recognition of the searched out works / findings as a re-search scholar for carrier advancement and incremental benefit etc.

iii) **Un-interfere able Human Rights** [Individual Rights / **Third Generation Human Rights**] – With respect to 'Kitaaba Wal-Hikmata' or 'Manifested Nature and the Utility of One's Upright Logic', Solidarity Rights [Individual Rights] include Rights Concerning Individual Faith and Belief or what science calls Reliance on One Dot Leader & Diamond Operator or what philosophy calls apriori principles, namely, Principle of the Uniformity of Nature and the Law of Successive Relation or what are generally called two pre-conditions of knowledge

Due to the inclusion of Individual Faith and Belief within the domain of Human Rights, the Third Generation Human Rights are called Solitary Rights or Solidarity Rights or Solid Human Rights. The very attributive term 'Solitary' or 'Solidarity' or 'Solid' as a prefix [shane-nuzuul] of Human Rights connotes 'Unbreakable Human Rights', 'Un-interfere able Human Rights', 'Un-comparable Human Rights', 'Individual Human Rights', and the like. In other words, 'Solitary Rights' imply those Human Rights which are beyond the range of finite mortal beings like Namruud, beyond four basic walls of scientific investigations, beyond human made laws – rules – guidelines, beyond pen – paper – pencil works of the activists as well as AK47 & Nuclear Weapons of the terrorists, beyond the binary nature of the Unique Historical Don of Teleological Evidence Sorcery & Epistemic Persecution, beyond the Trinity of Mechanical Globalisation – Circular Rotation & Revolution System - Man-made Magnetism of the Historical Don of Two-in-One Epistemic Uniqueness. In brief, Solidarity Rights are beyond Historical Black Magic and Mechanical Trinity. If you have any doubt in this regard, then please try to conceptualise verifiable as well as justifiable answers of the following simple questions.

i) Are there valid Laws [established rules & guidelines] which can dominate, control, regulate, and guide my faith and belief? If there is any such human made Law / Rule / Guideline etc., then Individual Faith and Belief are not Solitary Rights or Solid Human Rights or Individual Rights or Third Generation Human Rights.

ii) Is there any Nuclear Weapon in the hands of the Unique Epistemic Laden that can separate my Faith and Belief from me without destroying my physical existence from this manifested hexagonal world and appearing pentagonal earth within the Black Square of East

Horizon? If there is such a human made weapon, then Faith and Belief are not Solitary Rights or Solidarity Rights or Solid Human Rights or Individual Rights or Third Generation Human Rights.

In this connection, I would like to share with you that in reality; we fear the Hypocrisy of the Unique Epistemic Laden more than AK47 & the Nuclear Weapons of the Terrorist Laden. This fact has been clearly stated by Sankara, the founder of Advaita Vedanta. Sankara has borrowed the term 'Maya' from the Upanisad to explain the real nature of Hypocrisy. Maya has two functions – Concealment of the Real and Projection of the Unreal. These two functions resemble with the binary nature of the Unique Epistemic Historical Don who has concealed the realities concerning Manifested Nature & Natural Magnetism and has projected Mechanical Globalisation & Man-made Magnetism. Consequently, we fear the binary nature of devil more than the physical strength of a wrestler.

Is there any human-made weapon mightier than the Three Dimensional Arrow of Science or ongoing command 'Kun' or Trishul of Shivji or Paramartika-sattva of Sankara? So, on which ground shall I fear the Unique Epistemic Laden save his binary nature or hypocrisy?

iii) Is it possible for a Mystic or a Sufi or the so-called spiritualists to apply his/their mystical power on my faith and belief save my Rab and your Rab? If it is possible for a mystic to dominate my Faith and Belief through his/her magical power, then Faith and Belief are not Solitary Rights or Solidarity Rights or Solid Human Rights or Individual Rights or Third Generation Human Rights.

iv) Is there any theoriser who will be able to theorise my faith and belief? If there is such a theoriser or a theory, then Faith and Belief are not Solitary Rights or Solidarity Rights or Solid Human Rights or Individual Rights or Third Generation Human Rights. A theoriser can theorise fictitious concepts like 'Knowledge Theorisation and Rights', and can run a project work on the fictitious resemblance between 'Geetanjali' of Rabindranath Tagore and 'Theory of Relativity' of Einstein, and can conduct a Seminar on 'Epistemic Uniqueness in Mystic Religious Experience', the findings of which are neither verifiable nor justifiable by one or the other criterion of truth but those findings [Black Magic] can help to create an infinite field for further fictitious research works in resemblance with the 'Theory of Justice' & 'Idea of Justice', and can help to open the doors for economisation of education.

v)   Is there any activist who can compare my faith and belief with his/ her faith and belief? If there is such an activist, then Faith and Belief are not Solitary Rights or Solidarity Rights or Solid Human Rights or Individual Rights or Third Generation Human Rights. I am living with my better half under the same roof and on the same bed, but I have failed to compare our faith and belief. The direction which is West to her is North to me; and the direction which is South to me is East to her. I have also failed to search out a single criterion on the basis of which activists such as the father of IRF & so-called Mouth-Speakers are openly and publicly comparing Solitary Rights or Solidarity Rights or Solid Human Rights or Individual Rights or Third Generation Human Rights. Similarly, I have failed to search out the criterion on the basis of which Researchers of IFTA are uninterruptedly gifting commentaries on revelations & manifestations. In this regard, you may ask those activists and researchers about the right direction of their Qibla towards Upright West [Arsh] in coherence with Equal & Opposite Revelation [Tawraat] and in pragmatic correspondence with Crucified Sign & Manifested Signs of Natural Magnetism [Injiil].

SOLIDIFIED SOLID HUMAN RIGHTS

Electron, Proton, Neutrino & Neutron of an atom have no independent existence without having a bond i.e. Fifth One or Nucleus. Similarly, Natural Rights, Human Rights, Solitary Rights of Faith & Belief have no independent existence without having a bond or the existence of an individual in the manifested Hexagonal World and appearing Pentagonal Earth within the black Square of East Horizon as a human person. This concept has been developed by Rene Descartes as 'Cogito Ergo Sum' or 'I think, therefore, I exist'. The Third Generation Human Rights or Solitary Rights of Faith and Belief are formal rights or the ~sth [components] between Natural [Universal] Human Rights and Generalised Human Rights or between First Generation Human Rights and Second Generation Human Rights. **This ~sth establishes a solid bond of Rights of an individual like an atom.** This Solid Bond of Rights grounded on Individual Faith and Belief is called **Solidified Solid Human Rights**. In other words, Solidified Solid Human Rights consist of –

i)   Natural Rights or Rights to know [to get confirmation & recognition of] Truths / Realities concerning Revealed Universe, Manifested World, Appearing Earth, Ascending Stairs of the Appearing Earth, Manifested Signs of Natural Magnetism, Four Cardinal Directions in resemblance with Four Basic Forces, Equal & Opposite or Semi-anticlockwise and Semi-clockwise stages of journey of the Manifested Signs of Natural Magnetism etc. on the one hand; Black Magic and

Mechanical Trinity on the other. These are innate [inborn] Rights or Natural Rights or First Generation Human Rights as a human person of Peter Singer.

ii) To do Re-search work on the pointed lacunae in the searched out 62 [sixty two] significant findings shared in the 'Kitaaba Wal-Hikmata' or 'Manifested Nature and the Utility of One's Upright Logic' with a view to become a true follower of the implemented guidelines of the University Grant Commission [UGC] – Hindustan, to gather recognised certificates for API Points, to fulfil the criterion of Bread and Butter Human Rights with a view to get incremental benefits through carrier advancement etc. are the Extrinsic Values of my life for the Survival of the Fittest or Generalised Human Rights or Second Generation Human Rights.

iii) To **get confirmation of the equal and opposite right direction towards Upright West [Arsh], to follow equal & opposite right direction towards Upright West [Arsh], to observe Idd on the appointed days uniformly, to perform due Haj i.e. journeys towards the inviolable sacred place where Black & White Imam of the City has been appointed as the criterion or standard of verification & justification concerning right direction as well as sacred place for Pilgrimage, to keep the sacred place of Pilgrim neat and clean i.e. free from the footsteps as well as interference of the Historical Don of Hypocrisy, Tyranny, Conspiracy, Activism, Terrorism, Evidence Sorcery, Persecution,** and the like are the Intrinsic Values or Solitary Rights or Solid Human Rights for the Survival of the Truest of an altruist [an akbari-jihadist or a karma-yogi]. To do righteous work and to give good tidings to mankind in general or Duty for Duty's Sake or Niskama-Karmas are the only means to attain Summum Bonum or the Highest End of Life or the Survival of the Truest.

I have searched out my Solidified Solid Human Rights and have compiled my searched out findings in a Logical Method called Dictum De Omni ET Nullo [Deductive Method or Mathematical Method or Cartesian Method] in the name of 'Kitaaba Wal-Hikmata' or 'Manifested Nature and the Utility of One's Upright Logic'. In other words, the searched out findings shared in the 'Kitaaba Wal-Hikmata' or 'Manifested Nature and the Utility of One's Upright Logic' are not independent but relative. My Solitary Rights or Individual Faith and Belief are the components between Revelation and Manifestation, between Manifestation and Appearance, between Manifested Nature and Natural Magnetism, between Manifested Signs of Natural Magnetism and Equal &

Opposite right direction towards upright West, between arising or appearing of new moons and appointed day of observing Idd etc.

Moreover, from the point of view of teaching profession, I am individually suffering from subjective selfcontradictions and objective paradoxes. I cannot conceptualise two contradictories as true together going contradictory to acquired knowledge as well as upright logic. I am failing to share freely several true concepts [selfevident truths / manifested truths] with my friends in deed [students] due to innumerable pen – paper – pencil works in resemblance with Mechanical Globalisation and Man-made Magnetism i.e. subjective selfcontradictions and objective paradoxes. So, I want to get rid of subjective selfcontradictions and objective paradoxes individually. None will justify/ suggest that sharing subjective selfcontradictions and objective paradoxes as ideal teaching save the Unique Epistemic Hypocrite and His / Her black & white activists. For this reason I am individually trying to get verified as well as justified confirmation and recognition of the shared searched out findings from True Epistemic Persons, Real Scientists, and Respective Authorities. No one else is responsible for these individual efforts. So, it is an appeal before the respective authorities to provide verifiable & justifiable punishment of the sharer of 'Kitaaba Wal-Hikmata' or 'Manifested Nature and the Utility of One's Upright Logic' exclusively, if illicit as well as illegitimate sharing/placement of appeals concerning the Solidified Solid Human Rights are found in the 'Kitaaba Wal-Hikmata' or 'Manifested Nature and the utility of One's Upright Logic'. Simultaneously, it is an appeal before respective authorities to provide preventative as well as educative punishments to those who are involved in hypocrisy, tyranny, conspiracy, activism, terrorism, evidence sorcery, persecution, violation of Solitary Rights – Generalised Human Rights – Natural Rights i.e. Solidified Solid Human Rights.

I have submitted, placed, and forwarded my compiled Solidified Solid Human Rights or 'Kitaaba Wal-Hikmata' or 'Manifested Nature and the Utility of One's Upright Logic' for Verification, Justification, Confirmation, Recognition, Publication, Broadcasting, and the like through e-mail accounts available in the Internet to some Highly Qualified Epistemic Persons, University Presses, Scientific Research Institutions, Media Press Offices, Respective Assam University Silchar, University Grant Commission - Hindustan, NAAC – Hindustan, Philosophical Councils, United Nation Organization, International Human Rights Commission, World Peace Mission Organization, and the like. For common and public sharing, I am trying to get 'Kitaaba Wal-Hikmata' or 'Manifested Nature and the Utility of One's Upright Logic Self-published'.

WE ARE INSPIRED TO LEAVE THEM AND THEIR INVENTIONS
Or
We are inspired to leave two-in-one partnership along
with their commentary in resemblance with mechanical
globalisation and man-made magnetism
[It is a sincere appeal before all to reflect on each
shared concept of Selfevident Truth]
"Expound to Me (the case) with knowledge if you are truthful."
Ayat\Verses - Even so, in the eyes of most of the pagans (wandering Arabs /
hypocrites), **their partners [conspiracy group] made alluring the slaughter
of their children [followers of West Direction], in order to lead them** to their
own **destruction,** and **cause confusion in their religion**. If Allah had willed,
they would not have done so: But **leave alone them and their inventions.** And
they say that such and such **cattle** and **crops** are **taboo** (forbidden), and **none
should eat of them except those whom - so they say -** We wish; further, there
are **cattle forbidden to yoke** or burden, and **cattle on which the name of Allah
is not pronounced;** - **inventions against Allah's name:** soon will He requite
them for their **inventions.** They say: What is in the wombs of such and such
cattle is especially reserved for our **men**, and **forbidden to our women**; but if it
is **still- born**, then all have **share therein**. For their [false] **attribution** [shane-
nuzzul], He will soon **punish them**: for He is full of wisdom and knowledge.
They are losers who **infatuatedly** have **slain** their children without knowledge,
and have **forbidden** that which Allah bestowed upon them, **inventing** a lie
[mechanical magnetism] against Allah. **They indeed have gone astray and
are not guided.** [Sura No. – 5\6 – An-aam, Ayat No. –138 to 141]

ALLAH'S GUIDANCE IS THE ONLY GUIDANCE
Say: **Shall we indeed cry instead of unto Allah, things that can do us neither
good nor harm, and turn us after receiving guidance from Allah like
one bewildered whom the devils have infatuated in the world, which has
companions who invite him to the guidance (saying): come unto us [come
with one among us]?** Say: **Allah's guidance is the [only] guidance, and we
are directed to submit ourselves to the Rab of the universe.** To establish
regular prayers and to fear Allah: for it is to Him that we shall be gathered
together. **It is He who created the heavens [West Horizon] and the world
[East Horizon] in true [proportions] in that day when He said: Be! It is.**
[Sura No. – 5\6 – An-aam, Ayat No. –71 & 73]

Those **who reject our signs** are **deaf** and **dumb,** in the midst of **darkness**
profound: whom Allah wills, He leaves to **wander;** whom He wills, He places
on **the way that is right**. [Sura No. – 5\6 – An-aam, AyatNo.–39]

SECTION – III
DIAMOND OPERATOR – 3.1
UNIVERSE IS AN INTRINSICALLY LUMINOUS STAR

Fasta-'iz billaahi minash-Shaytaanir-Rajim
Bismillaahir-Rahmaanir- Rahim

Function Key – F1 -
Universe is a **Fixed Star** [Symbol (hex) 003F]
Universe is a **Diamond Operator** [Unicode (hex) 22C4]
Universe is a **Shining Star** [Ayat No. – 35 of Sura Nuur]
Universe is **Shi-rra** or **Sirius** [Ayat No. – 49 of Sura – Wan-Najm]

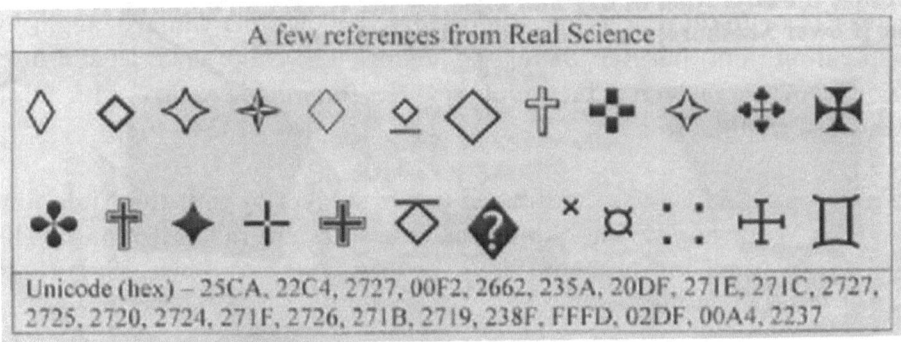

A few references from Real Science

Unicode (hex) – 25CA, 22C4, 2727, 00F2, 2662, 235A, 20DF, 271E, 271C, 2727, 2725, 2720, 2724, 271F, 2726, 271B, 2719, 238F, FFFD, 02DF, 00A4, 2237

In real science, the universe [alamin] is called a Diamond Operator. In the Kitaab with Truth, the universe [alamin] is called Shi-rra. Both Diamond and Shi-raa connote intrinsic luminosity. So, the universe [alamin] is intrinsically luminous. Moreover, universe has been revealed as an equal & opposite Trinity. This equal & opposite Trinity resembles with Crucified Sign. Isa (ass) brought Clear Proof of the revealed universe [alamin] as well as Clear Sign of the revealed directions. So, the Universe is a Quadrilateral Shining Star or a Diamond Operator or Shi-raa or Sirius. Regarding the concept of the universe as an intrinsically luminous quadrilateral star or Diamond Operator, there is no contradiction between Real Science and Kitaab with Truth save Resemblance. Universe is also called a Fixed Star in Real Science. Projected Planet Mercury is related with the Fixed Star.

PRIME ORIGIN

"He is above all comprehension, yet is acquainted
[manifested] with all things."

Ayat\erses - To Him is due the **primal origin** of the heavens [West Horizon] and the world [East Horizon]: **How can He have a son when He has no consort (companion)?** He has created all things, and He has full knowledge of all things. That is Allah, your Rab! There is no unbreakable reality save He, the Creator of all things: then serve you Him: and He has power to dispose of all affairs. **No vision can grasp** Him, but His grasp is over all vision: **He is above all comprehension, yet is acquainted [manifested] with all things. Now have come to you, from your Rab, proofs [manifested signs] if any will see, it will be for [the good of] his own soul; if any will be blind, it will be to his own [harm]: I am not [here] to watch over your doings. Thus we do explain the signs by various [manifestations in resemblance with revelation]: that they may say, You have taught [us] diligently, and that We may make the matter clear to those who have knowledge.** Follow what **you are taught by inspiration** from your Rab: there is no unbreakable reality save He: **and turn aside from those who join breakable realities [manmade nature] with Allah [manifest truth]**. [Sura No. – 5\6 – An-aam, Ayat No. –102 to 107]

Ayat\erses - And He has constrained (made **subject to you) the Night and the Day. Washshamsa** wal-**qamara** and the **stars are in subjection by His Command**. Verily in **these are Signs for men who have sense**. [Sura No. – 15\16 – Nahl, Ayat\Verses – 12]

In projected science [conspiracy science], the universe is called Intrinsically Luminous Moon. This intrinsically luminous Moon is the Earth's only permanent Natural Satellite. There is a difference between real science and projected science. Projected science has converted Intrinsically Luminous Quadrilateral Star i.e. Diamond Operator into Intrinsically Luminous Circular Moon. This game of conversion is called Immediate Inference. This Immediate Inference is the Vera Causa as well as instrument of Teleological Evidence Sorcery and Unique Epistemic Persecution. The researchers of IFTA are uninterruptedly following the game of immediate inference in resemblance with the Reference List of a Research Work. For instance, they [Researchers of IFTA] have converted the name of the Messenger Dawuud (ass) into David. From the translation of meanings with commentary on Quran by the researchers of IFTA in the name of 'The Holy Quran' in resemblance with the game of immediate inference of the projected science, it can validly be inferred that IFTA and projected science are the two faces of the same thread. These two faces further resemble with Pythagorean Number – 'Line – Two'.

The game of immediate inference also manifests an equal & opposite game between real science and projected science. From the point of view of real science, the Universe is a **Diamond Operator [Sirius] and a Star System [Sirius Binary System]**. On the contrary, from the point of view of projected science, the Universe is an intrinsically luminous Moon and a **Lunar System. This Intrinsically Luminous Moon is the Earth's only permanent Natural Satellite.** In other words, the Universe is an **Astral System**, a **Planetary System**, a **Cosmological System**, an **Astrophysical System**, a **Stellar System**, a **Solar System**, and the like. These terminologies are synonymous. Regarding the concept of the Universe, there is no contradiction between **projected science, pen-paper-pencil works, mechanical activities, and translation of the meanings with commentary on Quran in the name of 'The Holy Quran' by the researchers of IFTA save resemblance. These black & white works and mechanical activities have converted Diamond Operator and Star System into Moonlet and Solar System or Planets and Planetary System**. Consequently, each star of the Star System is being called a Planet / a Moonlet in projected science, solar system, astral system, cosmological system, astrophysical system, stellar system, and black & white works and mechanical activities. Four Galilean Moons are the best examples as well as valid reference in this regard. To avoid confusion & contradiction, an attempt is being made to share univocally the concept of the manifested nature in resemblance with revelation and searched out symbols of the real science available in Insert Symbol through this re-search project. For this reason, with respect to 'Kitaaba Wal-Hikmata' or 'Manifested Nature and the Utility of One's Upright Logic', the Universe is a Diamond Operator [Shi-raa], revealed as an equal & opposite Trinity called Star System [Sirius Binary System], and manifested as an upright rectangle called End of Proof. Consequently, each projected Planet or projected Moonlet is a Guiding Star in this re-search project in resemblance with Kitaab with Truth and Real Science.

REJECT SUCH MEN AS USE BAD LANGUAGE

Ayat\erses - **The most beautiful names belong to Allah: so call on Him by them; but reject such men as use bad language in His names: for what they do, they will soon be requited?** And of those whom We have created there is a nation (ummat or median people) who guide with **truth** and establish **justice** therewith. [Sura No. – 6\7 - As-haabul- a' raaf, Ayat No. – 180 & 181]

"Wa annahu Huwa Rabbush-Shi-raa"
"He is Rab of the Sirius [Diamond Operator]."

Ayat\erses - And that your Rab, He is the goal; and that it is He Who grants laugh and grants weep, and that it is He Who gives death and Who gives life;

and that He creates the two **spouses**, the male and the female, from a drop (of seed) when it is poured forth, and that He has ordained the **second** bringing forth, and that it is He Who gives wealth and satisfaction; **"Wa annahu Huwa Rabbush-Shi-raa" [49] [Trans.]** that **He is Rab of the Sirius [the Diamond Operator];** [Sura No. – 52\53 – Wan-Najm, Ayat No. – 42 to 49]

Rasuulum-mir-Rbbil Aalamin

Ayat\erses - We sent Nuuh to his people. He said: O my people! Serve Allah! You have no other reality apart from Him. I fear for you the punishment of a dreadful day! The leaders of his people said: Ah! We see you surely in plain error. He said: O my people! There is no error in me. On the contrary, I am a messenger from the Rab of the universe **[Rasuulum-mir-Rbbil Aalamin].** [Sura No. – 6\7 - As-haabul- a' raaf, Ayat No. – 59 to 61]

Al-jawaaril-kunnas
"I swear by the places of the Stars"

Ayat\erses - Furthermore **I swear by the places of the Stars**, and lo! That is indeed a **tremendous oath**, if you but knew, that (this) is indeed **Qur-aanuun-Karim, in a Kitaab well-guarded,** which **none touches save those who are clean** (in **faith**), a **Revelation** from the **Rab of the universe. Is it this message that you scorn and have you made it your livelihood that you should declare it false?** [Sura No. – 55\56 - Waqa-atil-waaqiah, Ayat No. – 75 to 82]

Reference of Shining Star [Glass] and Intrinsic Luminosity

Ayat\erses -"Allaahu Nuurus-samaa-waati wal-arz – Masalu **Nuu-rihi** ka**Mishkaatin**-fiihaa **Mis-baah** – Al-**Misbaahu** fii **Zujaa-jah** – **azzuujaajatu** ka-annahaa – kawkabun durriyyuny-yuuqa-du – min Shajaratimmubaara-katin-**Zaytuunatil**-laa **Sharqiy**-yatinw wa laa **Garbiyyatiny**-yakaadu **Zaytuhaa** yuziii-u wa law lam tamsas-hu naar. **Nuu-run alaa Nuur! Yahdillaahu** li-**Nuurihii** many-yashaaa – wa-yazribullaahul-amsaala linnaas – wallahu bi-kulli shay-in Alim." [Sura No. – 23\24 – Nuur, Ayat No. – 35] [Trans.] **Allah** is the **Light [Nuur]** of the **heavens [West Horizon]** and the **world [East Horizon]. The similitude** (likeness) of **His Light** [Nuur] is as a **niche** (binary thread of a candle or binary pulsar) **in a lamp** [Black Square or World]. The **lamp** is in a **glass** [Shi-raa or Natural Satellite]. The **glass** [Shi-raa] is as it is a **shining star.** (This niche is) kindled from a **blessed tree**, an **olive** (Zaytuun Tree), **neither from the East nor from the West,** whose (niche's) oil [continual acceleration] will almost glow forth (of itself i.e. intrinsically luminous) though **no fire touches** it. **Light upon Light!** [Nuurun Alaa Nuur]! Allah guides unto His light [Nuur] whom He wills. And Allah speaks to mankind in **parables**, for Allah is Knower of all things. [Sura No. – 23\24 – Nuur, Ayat No. – 35]

Game of Immediate Inference

Ayat No. – 162 - But **the transgressors among them changed the word from that which had been told them**. So we sent on them a plague from heaven. For that they repeatedly transgressed. [Sura No. – 6\7 - As-haabul- a' raaf, Ayat No. – 162]

Ayat No. – 58 & 59 - And when We said: **Enter this township** and eat of the plenty therein as you wish; but **enter the gate prostrate**, and say: **Repentance**. We will **forgive** you **your sins** and will **increase** (reward) for the **right-doers**. But **those who did wrong changed the word which had been told them for another saying**, and We sent **on the transgressors a plague** from heaven, **for evil doing**. [Sura No. – 1\2 – Baqarah, Ayat No. – 58 & 59]

Verily the reversing [immediate inference] is an addition to unbelief.
Ayat\Verses – O you who believe! There are indeed many among the **priests** (Lukmans) and **anchorites** [Imamgiris], who in **falsehood demolish** the **wealth of men** and **obstruct** [them] **from the way of Allah**. And there are those who **stored gold and silver and spend it not** in the **way of Allah** announce unto them a **most grievous penalty**. On the Day when heat will be produced out of that [wealth] in the **fire of Hell**, and with it will be branded their **foreheads**, their **borders**, and their **backs borders**, and their **backs**. This is the [treasure] which you stored for yourselves. Taste you, then, the [treasures] you stored. The **number of months [planets] in the sight of Allah is twelve. So ordained by Him the day He created the heavens [West Horizon] and the world [East Horizon]. Four of them are blessed that is the straight usage.** So **wrong not yourselves therein, and wage the Pagans / idolaters all together as they waging on all of you.** But know that Allah is with those who restrain themselves. **Verily the reversing is an addition to unbelief.** The unbelievers are led to wrong thereby, for they make it lawful one year, and forbidden another year, in order to adjust the number of months forbidden by Allah and make such forbidden ones lawful. The evil of their course seems pleasing to them. But **Allah guides not those who reject Faith**. [Sura No. – 8\9 – Tauba, Ayat No. – 34 to 37]

RELIGION OF TRUTH
"It is He Who has sent His Messenger with
guidance and the Religion of Truth."
'Uzayru-nib-nullaahi' and 'Masiihub-nullaah'
'Yahuudis' and 'Nasaras'
"Allah's curse is on them: how they are deluded away from the Truth."

Ayat\Verses - The '**Yahuudis**' call '**Uzayru-nib-nullaahi**' as the **son** of Allah, and the '**Nasaras**' call '**Masiihub-nullaah**' as the **son** of Allah. That is **a saying from their mouth**; [in this] they but **imitate** what the **unbelievers of old used to say**. **Allah's curse is on them**: how they are **deluded away from the Truth**. They have taken as **lords** besides Allah their **rabbis** and their **monks**, [and] **Wal-Musii-habna-Maryam** [to be their Rabs in derogation of Allah], **yet they were commanded to worship none but One Allah**. There is **no unbreakable** reality except Him. Praise and glory to Him: [Far is He] from having the **partners they associate** [with Him]. Fain would **they extinguish Allah's light** with their **mouths**, but Allah will not allow that His light should extinguish, even though the unbelievers may abominate (it). It is He Who has sent His Messenger with guidance and **the Religion of Truth**, to prevail over **all isms [diversities or projected falsehood]**, even though the **Pagans / idolaters** [followers of falsehood] may **detest** [it]. [Sura No. – 8\9 – Tauba, Ayat No. – 30 to 33]

Ayat No.	Sura No. – 80\81 SURA – **AL-JAWAARIL-KUUNAS [TAKWIIR]**
1	**"Izas-shamsu** Kwwirat" [Trans.] When the **sun [with its spacious light]** is folded up - [IFTA]. When the **sun** is over thrown – [IB (P) Ltd.] When the **so-called sun as the manifested sign of natural magnetism** is folded up – [Self-evident Concept]
2	"Wa Izan-nujuu-munkadarat" [Trans.] And when the **stars** fall,
3	"Wa izal-jibaalu suyyirat" [Trans.] And when the **hills** are moved,
4	"Wa izal-ishaaru uttilat" [Trans.] And when the **camels big with young** are abandoned,
5	"Wa izal-wuhuushu bushirat" [Trans.] And when the **wild beasts** are herded together,
6	"Wa izal-bihaaru sujjirat" [Trans.] And when the **seas** rise,
7	"Wa izan-nufuusu zuwwijat" [Trans.] And when **souls** are reunited,
8	"Wa izal-maw-uudatu suilat" [Trans.] And when the **girl-child** that was buried alive is asked,
9	"Bi-ayyi-zambin-qutilat" [Trans.] For what sin [reason] she was slain [hamstrung the she- camel]
10	"Wa izas-suhufu nushirat" [Trans.] And when the **pages** are laid open. [Manifested signs will be made clear]
11	"Wa izas-samaaa-u kushitat" [Trans.] And when the sky is torn away,
12	"Wa izal-Jahiimu su'irat" [Trans.] And when the **hell** [flameless fire] is lighted
13	"Wa izal-Jannatu uzlifat" [Trans.] And when the **Garden** [Heaven / West Horizon] will be brought near,

14   "Alimat nafsum-maaa – ahzarat" [Trans. [Then] each **soul** will know what it has put forward.

15   "Fa-laaa uqsimu bil-khun-nas" [Trans.] So verily I call **to witness the planets / moons [eye opening evidences]**

16   **"Al-jawaaril-kunnas"** [Trans.] Go straight, or hide [IFTA]; The **stars which rise and set** [IBS (P) Ltd.].
Each day a new star is kindled neither from the east nor from the west, but from the Revealed Left and Manifested North-East declining to the Revealed Right and Manifested South-East, and runs away from there and rises from Revealed Right and Manifested South-West, and moves towards Revealed Left Manifested North-West [Self-evident Concept with reference to Sura No. – 80\81 – Takwiir, Ayat No. 16, Sura Nuur – Ayat No. – 35, Sura Khaf – Ayat No. – 17]]

17   "Wal-layli izaa as-as" [Trans.] And the **night** as it **dissipates**;

18   "Was-subhi iza tanaffas" [Trans.] And the dawn [**morning**] as it breathes away the **darkness**;

19   "Innahuu lqawlu Rasuulin-Karim" [Trans.] **This is in truth, the word of an honoured Messenger,**

20   "Zii-quwwatin inda Zil-Arshi-makiin" [Trans.] **Endowed with Power, with rank before the Rab of the Throne [Arsh],**

21   "Mutaa-in-samma Amiin" [Trans.] **[One] to be obeyed and trust worthy.**

22   "Wa maa swahibukum-bi-majnuun" [Trans.] And **your friend** is **not mad**.

23   "Wa laqad ra-aahu bil-ufuqil –mubiin" [Trans.] And **without doubt he saw Him in the clear horizon.**

24   "Wa maa huwa alal-gaybi bi-zaniin" [Trans.] And **he will not withhold grudgingly knowledge of the Unseen.**

25   "Wa maa huwa bi-qawli Shaytaanir-rajiim" [Trans.] **Nor it is the word of an evil spirit accursed.**

26   "Fa-ayna tazhabuun" [Ind. Concept] **Where then you will go?**

27   "In huwa illa Zikrul-lil-aalamin" [Trans.] Verily **this is no less than a reminder** to the created members of the universe [Zikirul-lil-Aalamin]

28   "Liman-shaaa-aminkum any-yastaqiim" [Trans.] Unto **whomsoever of you will to go straight.**

29   "Wa maa tashaaa-uuna illaaa any-yashaaa-Allaahu Rabbul Alamiin" [Trans.] But **you will not unless** [it be] that **Allah wills**, the Rab of the Alamin [Universe].

WHEN THE STAR[S] BECOME DIM [PUT OUT]
Ah woe, that Day, to the rejecters of Truth!
"Fabi-ayyi Hadiisim-ba-dahuu yu-minuun"
[In what statement, after this, will they believe?]
[Emissary Wind]
Ayat\Verses – By the emissary **winds** [sent] one after another; which then blow violently in tempestuous gusts, by those which scatter [things] far and wide; then separate them, one from another, [Sura No. – 76\77 - Wal-Mursalaat, Ayat No. – 1 to 4]

[Reminder – To Excuse or To Warn]
Ayat\Verses – Then spread abroad a **reminder**, to **excuse** or to **warn** [Sura No. – 76\77 - Wal-Mursalaat, Ayat No. – 5 & 6]

Ayat\Verses – **Assuredly**, that which you are **promised** must **come to pass**, then when the stars become dim; and when the sky is cleft asunder; and when the mountains are scattered [to the winds] as dust; and when the messengers [reminders] are brought to their appointed time. [Sura No. – 76\77 - Wal-Mursalaat, Ayat No. – 7 to 11]

Ayat\Verses – For what Day is the time appointed? For **the Day of Sorting out [Day of Decision], and what will explain to you what the Day of Sorting out is**? Ah woe, that Day, to **the rejecters of Truth**! Did We not destroy the men of old [for their evil]? So We shall make later [generations] follow them. Thus We ever deal with the guilty. [Sura No. – 76\77 - Wal-Mursalaat, Ayat No. – 12 to 18]

Ayat\Verses – Ah woe, that Day, to the rejecters of Truth! Have We not created you from a fluid [held] despicable which We laid up in a safe abode for a period determined? Thus We arrange. How excellent is Our arranging! [Sura No. – 76\77 - Wal-Mursalaat, Ayat No. – 19 to 23]

Ayat\Verses – Ah woe, that Day! To those who reject **Truth**! Have We not made the world a receptacle both for the **living** and the **dead**, and **made therein mountains standing firm**, lofty [in stature]; and **provided** for you sweet **water** [and wholesome]. [Sura No. – 76\77 - Wal-Mursalaat, Ayat No. – 24 to 27]

Ayat\Verses – Ah woe, that Day, to the rejecters of Truth! [It will be said:] Depart you to that which you **used to deny** [reject as false]! **Depart you to a shadow in three columns,** [which yet is] neither relief nor shelter from the flame. Lo! It throws up sparks like the castles, (or) as it might be **camels of bright yellow hue**. [Sura No. – 76\77 - Wal-Mursalaat, Ayat No. – 28 to 33]

No argument and no plea

Ayat\Verses – Ah woe, that Day, to the rejecters of Truth! That will be a Day when **they will not be able to speak**. Nor will it be open to them to put forth pleas. [Sura No. – 76\77 - Wal-Mursalaat, Ayat No. – 34 to 36]

Star sets implies Star rises

Ayat\erses - By the **Star when it sets, your comrade errs not, nor is misled, nor does he speak** of (his own) **desire. It is nothing save an inspiration that is inspired.** He was **taught** by **one Mighty in Power, endowed with Wisdom and clear to view. While he was in the upper part of the horizon [West]** then **he drew neigh (near) and came down**, and was at a distance of but **two bows length** or even **nearer.** So did [Allah] **convey** the **inspiration to His slave** [conveyed] what He [meant] to convey. [Sura No. – 52\53 – Wan-Najm, Ayat No. – 1 to 10]

What they say is nothing but falsehood!
PRODUCE FOUR EVIDENCES

Ayat\erses - Lo! Those **who brought forward the lie** are **a gang among you.** Think it **not** to be an **evil to you.** On the **contrary, it is good for you.** To every man among them [will be paid the punishment] of the **sin** that he **earned**, and to him who took on himself the **lead among them**, will be a **penalty grievous.** Why did not the believers - men and women - **when you heard of the affair,** put the **best construction on it in their own minds** and say: **This is a manifest untruth**? Why did they not produce **four criterions** (witnesses) to prove it? When they have **not produced the criterions** (evidences / witnesses), **they verily are liars in the sight of Allah.** Were it not for the grace and mercy of Allah on you, in this world and the Hereafter, a grievous penalty would have seized you in that you rushed murmured into **this affair.** [Sura No. – 23\24 – Nuur, Ayat No. – 11 to 14]

Evidence Sorcerers are requested to play
game of Black Magic once again

Ayat\Verses – Ah woe, that Day, to the rejecters of Truth! That will be a Day of Sorting out [Decision]! We shall gather you and the men of old together. **Now, if you have a trick [or plot], use it against Me!** [Sura No. – 76\77 - Wal-Mursalaat, Ayat No. – 37 to 39]

Most Serious Matter in the sight of Allah

Ayat\erses - When **you welcomed it on your tongues, and said out of your mouths things of which you had no knowledge; and you thought it to be** a light matter, while **it was most serious in the sight of Allah**. [Sura No. – 23\24 – Nuur, Ayat No. – 15]

"This is a most serious insult!"

Ayat\erses - And why did you not, when you heard it, say: It is **not right** of us to speak of this? Glory to Allah! **This is a most serious insult!** [Sura No. – 23\24 – Nuur, Ayat No. 16]

"Allah knows and you know not."

Ayat\erses - Lo! **Those who love that scandal (project Black Magic) should be spread concerning those who believe, theirs will be a painful punishment in this life and in the Hereafter. Allah knows and you know not.** Had it not been for the grace and mercy of Allah unto you and that Allah is full of kindness and mercy, [you would be ruined indeed]. [Sura No. – 23\24 – Nuur, Ayat No. – 19 & 20]

Follow not Shaytan's footsteps.

Ayat\erses - O you who believe! **Follow not Shaytan's footsteps. If any will follow the footsteps of Shaytan, he will command what is filthiness and wrong;** and had it not been for the grace and mercy of Allah unto you, not one of you would ever have **grown pure**. But Allah purifies whom He pleases, and Allah is One Who hears and knows [all things]. [Sura No. – 23\24 – Nuur, Ayat No. – 21]

"Do you not wish that Allah should forgive you?"

Ayat\erses - **And let not those who possess dignity and ease among you swear not against** helping their kinsmen, those in want, and **those who have left their homes in Allah's way** let them forgive and overlook. **Do you not wish that Allah should forgive you**? For Allah is Ever Forgiving, Most Merciful. [Sura No. – 23\24 – Nuur, Ayat No. – 22]

It is the Absolute Truth that the Kitaab with Truth is not the speech of a poet

Ayat\erses - The Sure Reality! What is the Sure Reality? And what will convey to you what the Sure Reality is! [Sura No. – 68\69 – Al-Haaaqqatu, Ayat No. – 1 to 3]

Ayat\erses - The Samuud and the Aad People disbelieved in the Judgment to come. As for Samuud, they were destroyed by the **lightning**. And the Aad, they were destroyed by a furious Wind, exceedingly violent which He imposed on them **seven long nights and eight long days** so that you might have seen men lying overthrown, as they were hollow **trunks** of **palm-tree**. Can you see any remnant of them? And Firawn, and those before him, and the communities that were destroyed, committed habitual Sin. And **they disobeyed [each] the messenger of their Rab**; so He gripped them with a lightening grip. [Sura No. – 68\69 – Al-Haaaqqatu, Ayat No. – 4 to 10]

Ayat\erses - Lo! When the water overflowed beyond its limits, We carried you upon the **ship**. That We might make it a reminder for you, and that remembering ears retain its memory in remembrance. **Then, when one blast will be sounded on the Trumpet, and the world will move [World is not moving at present], and its mountains, and they are crushed to powder at one stroke,** then on that Day, will the event befall. And the sky will be rent asunder, for that Day it will be flimsy. And the angels will be on its sides, and **eight will**, that Day, **bear the Throne of your Rab above them.** [Sura No. – 68\69 – Al-Haaaqqatu, Ayat No. – 11 to 17]

Ayat\erses - **On that Day you will be exposed, not a secret of you will be hidden.** Then as for him who is given his record in **his right hand**, he will say: Take, read my record! Surely I knew that I should have to meet my reckoning, and then he will be in a Blissful state, in a high Garden whereof the clusters are in easy reach. (And it will be said to those therein) Eat and drink at ease for that which you sent before you in the past days. [Sura No. – 68\69 – Al-Haaaqqatu, Ayat No. – 18 to 24]

Ayat\erses - But as for him who is given his **record in his left hand**, he will say: Ah! Would that my record had not been given to me! And knew not what any reckoning is! Ah! Would that [Death] had made an end of me! Of no profit to me has been my wealth! My power has gone from me! [The stern command will say]: **Seize him, and chain him** and then expose him to the hell-fire. Further, make him march in a chain, whereof the length is **seventy cubits**! Lo! He used not to believe in Allah Most High. And would not encourage the feeding of the wretched! So no friend has he here this Day. Nor has he any food except the corruption from the washing of wounds which none but sinners eat. [Sura No. – 68\69 – Al-Haaaqqatu, Ayat No. – 25 to 37]

Ayat\erses - But, nay! I swear by Allah that you see and all that you see not, that this is verily **the speech of an illustrious messenger. It [Kitaab with truth] is not the speech of a poet** - little is it that you believe! Nor is it the diviner's speech - little is it that you **remember.** [This is] a revelation from the Rab of the Universe [Aalamiin]. And if he had invented any sayings in Our name, We should certainly take him by the right hand, and then severed his life-artery, nor could any of you withhold him. But verily this is a **warrant** for those who ward off evil. And We certainly know that some amongst you will deny. And lo! It is a **cause of sorrow for the disbelievers**. And lo! It is the **Absolute Truth. So glorify the name of your Rab Most High**. [Sura No. – 68\69 – Al-Haaaqqatu, Ayat No. – 38 to 52]

Ayat\Verses - **To Allah belong the Sovereignty of the heavens and the world; and Allah is Able to do all things. [Sura No. 2\3 – Imran, Ayat No. – 189]**

Ayat\Verses - Unto **Allah belongs the dominion [Sovereignty] of the heavens and the world**. He gives life and He takes it. Except for Him you have no protector, nor helper. **Allah turned with favour to the Prophet, the Muhajirs, and the Ansars, who followed him in a time of distress.** After that the hearts of a part of them had nearly swerved [from duty]; but He turned to them [also]: for He is unto them Most Kind, Most Merciful. And to **the three [three ascending stairs] also** (did He turn in mercy) **who were left behind, when the world vast as it is, was straitened for them still they be thought them that there is no refuge from Allah except toward Him; then turned Him unto them in mercy that they (too) might turn (repentant unto Him). Lo! Allah! He is the Relenting, the Merciful. [Sura No. – 8\9 – Tauba, Ayat No. – 116 to 118]**

Praise be to Allah, Who has created the heavens and the world, and **appointed darkness** and **light [Nuur]**. Yet **those who reject Faith ascribe [others] as equal, with their Guardian-Rab.** He it is Who has created you from clay, and then decreed a stated term [for you]. And **there is another determined term fixed with Him; yet you doubt**. And **He is Allah** in the heavens and the world. He knows both **your secret and your utterance**, and He knows **what you earn. But never did a single one of the signs of their Rab reach them, but they turned away there from.** And now **they deny the truth** when it reaches them. But soon will **they** learn the **reality [truth]** of **what they used to mock at. [Sura No. – 5\6 – An-aam, Ayat No. – 1 to 5]**

NO MONISM IN ISLAM
DO NOT SAY 'THREE-CEASE' OR 'DESIST TRINITY'
REVELATION OR TAWRAAT OR EQUAL & OPPOSITE LAW
Fasta-'iz billaahi minash-Shaytaanir-Rajim
Bismillaahir-Rahmaanir- Rahim

Quran is a Kitaab with Truth. It includes both Truth-in-itself as well as Criterions of Truth [Witnesses], Fundamental Ayat and detail explanations of Fundamental Ayat, Revelation and Manifestation. Fundamental [Abbreviated] Ayat is called Revelation or Formal Ground or Formula or Tawraat or Law or Coherence Truth. Detail explanation of a Fundamental Ayat is called Manifestation or Injiil or Correspondence Truth or Manifested Nature. **"And all things We have created by pairs that you may receive instruction"** represents both revelation and manifestation. Equal & Opposite Law is the Revealed Truth. So, **'all things We have created by pairs' is the revealed truth or Tawraat or Formal Ground or Coherence Truth. Moreover, this equal & opposite revelation is an instruction for us. This instruction is the manifest truth or Injiil or Correspondence Truth or Manifested Signs.** Manifested Nature is called Detail Explanation of Fundamental Ayat i.e. instruction. Equal & Opposite stages of journey of the Manifested Signs of Natural Magnetism [converted snake & white of Muusa's Staff] are the Clear Proofs of Equal & Opposite Revelations and Manifestations. Moreover, Allah is not Spiritual Truth in resemblance with the concept of self or spirit or atman. On the contrary, Allah is the Manifest Truth. Isa (ass) brought Clear Proof or Manifested Sign of equal & opposite revelation and manifestation.

NO MONISM IN ISLAM
"We did not prescribe (ordain) Monasticism [Monism] for them"
"Only seeking Allah's pleasure"
Then We caused Our Messengers to follow in their footsteps, and We caused **Iisabni-Maryam** to follow, and gave him **Injiil**, and placed compassion and mercy in the hearts of those who follow him. **But the monasticism (Monism) which they invented for themselves, We did not prescribe (ordain) for them – only seeking Allah's pleasure,** but they **did not** foster (observe) as they should have **done**. Yet We gave, on those among them who believed, their [due] reward, but many of them are **rebellious transgressors**. [Sura No. – 56\57 - Anzalnal-Hadiid, Ayat No. – 27] Additional References - [Sura No. – 38\39 – Tanziilul-Kitaab,

Ayat\Verses – 1 to 3], [Sura No. – 15\16 – Nahl, Ayat\Verses – 77 to 79], [Sura No. – 38\39 – Zumar, Ayat No. – 1 to 5], [Sura No. – 4\5 – Maaaidah, Ayat\Verses – 51], [Sura No. – 4\5 – Maaaidah, Ayat\Verses – 103]

"Alaa lillaahid-dinul-khaalis; Wallaziinattakhazuu min duunihiii awliyaaa. Maa na-buduhum illaa liyuqarribuunaaaa ilallaahi zulfa. Innallaaha yahkumu baynahum fii maa hum fiihi yakhtalifuun. Innallaaha laa yahdii man huwa kaazibun-kaffaar." [Trans.] Surely pure religion is for Allah only. And those who choose protecting friends (awliyaaa) besides Him (say): We worship them only that they may bring us near to Allah. Lo! Allah will judge between them concerning that wherein they differ. But Allah guides not such as are liars (false) and ungrateful. [Sura No. – 38\39 – Tanziilul-Kitaab, Ayat\Verses – 1 to 3]

If Allah had willed to choose a son, He could have chosen whom He pleased out of those whom He created. Be He Glorified! [He is above such things.] He is Allah, the One, the Irresistible (Absolute) ["Huwallaahul-Waahidul-Qahaar"] [Sura No. – 38\39 – Tanziilul-Kitaab, Ayat\Verses – 4]

Existential Import of Trinity [Equal & opposite Revelation] in Islam
DO NOT SAY 'THREE-CEASE' OR 'DESIST TRINITY'
[Transcendental Majesty]

O People of the Kitaab [Ahlal-Kitaab]! Do not exaggerate in your Diin (True Religion) or utter aught concerning Allah except the truth. The Masiiha, Iisabnu-Maryam, was only a messenger of Allah, and his word which he conveyed unto Maryam, and a spirit from him. So believe in Allah and His messengers, and say not 'Three-Cease' (Desist Trinity)! (It is) better for you! Allah is only One (innamal-laahu- 'ilaa-hunw-Waahid). Far from His Transcendental Majesty that He should have a son. To Him belong all that is in the heavens [West Horizon] and that is in the world [East Horizon]. And Allah is sufficient as defender. [Sura No. – 3\4 – Nisaa, Ayat No. – 171]

"They do blaspheme who say: Allah is one of three in a Trinity."

They do blaspheme who say: Allah is one of three in a Trinity: For there is no unbreakable reality except Waahid. If they desist not from so saying, verily a grievous penalty will fall on those of them who disbelieve. [Sura No. – 4\5 – Maaaidah, Ayat\Verses – 73]

"Fa-lillaahil-Hamdu Rab-bis-samaawaati wa Rabbil-arzi, Rabbil-Aalamin" {Trans.} Then Praise be to Allah, Rab of the heavens and Rab of the world,

Rab of the Universe [Alamin]! [Sura No. – 44\45 – Tanziilul Kitaab (Prev. Jaasiya), Ayat No. – 36].

"Allah comes in between a man and his heart."

O you who believe! **Give your response to Allah and His Rassul, when He call you to that which will give you life; and know that Allah comes in between a man and his heart,** and that **it is He to Whom you shall [all] be gathered.** And guard yourselves against a chastisement which cannot fall exclusively (particularly) on those of you who are wrong-doers and know that Allah is strict in punishment. And remember, when you were few and reckoned feeble in the land [earth], and were in fear lest men should extirpate you, hoe He gave you shelter, and strengthened you with His help, and made provision of good things for you, that haply you might be thankful. **[Sura No. – 7\8 – Anil-Anfaal, Ayat No. – 24 to 26]**

Allah is the Manifest Truth

Ayat\Verses – As for **those who** insult **virtuous women, indiscreet (careless) but believing, cursed are they in this life and in the Hereafter.** For them is **a grievous Penalty. On the Day when their tongues, their hands, and their feet will bear witness against them as to what they used to do. On that Day Allah will pay them their just dues,** and **they will realise** that **Allah is the Manifest Truth.** Vile women are for vile men, and vile men for vile women and good women are for good men, and good men are for good women. **Such are innocent (not affected) of that which people say.** For them there is forgiveness and a beautiful provision. [Sura No. – 23\24 – Nuur, Ayat No. – 23 to 26]

PAIRS [EQUAL & OPPOSITE REVELATION] IS AN INSTRUCTION SET NOT ANOTHER OBJECT [MANMADE NATURE] OF WORSHIP

Ayat\Verses - Glory be to Him, Who created **in pairs all things** that the world produces, as well as **their own kind** and **things of which they have no knowledge.** [Sura No. – 35\36 – Yaa-siiin, Ayat No. – 36]

Ayat\Verses – With power and skill We have constructed the Firmament [Windows]; for it is We Who create the vastness of rapidity [geometrical progression]. And We have spread out the earth; How excellently We do spread out! **And all things We have created by pairs that you may receive instruction.** Hasten you then [at once] to Allah. **I am a plain warner to you from Him.** And **set not another object of worship** (manmade magnetism) along with Allah. **I am a plain warner to you from Him.** Similarly, **no messenger** came to the peoples before them, but they [the guilty folk] said [of him]: **A sorcerer or a madman**. [Sura No. – 50\51 – Waz-Zaariyaat, Ayat No. – 47 to 52]

Ayat\Verses - Allah is He Who created **seven Firmaments** [windows] and of the **world a similar number. Through the midst of them descends His Commandment** that you may know that Allah is Able to do all things, and that Allah comprehends all things in [His] Knowledge. [Sura No. – 64\65 – Tallaq-tumun-nisaaa-'a, Ayat No. – 12]

Ayat\Verses - We created not the heavens, the world, and all between them, **but for just ends**. And the **Hour is surely coming**. So forgive with a gracious forgiveness; for verily it is your Rab who is the Master-Creator, knowing all things. And We have bestowed upon you the **Seven repeated [verses]** and the **Grand Qur'an. [Qur-aanal-Azim]**. Strain not your eyes toward that which We cause some wedded **pairs** among them to enjoy; and be not grieved on their account, and lower your wing [in gentleness] for the believers. [Sura No. – 14\15 = Hur, Ayat No. – 85 to 89]

Universe has been revealed as a Trinity of West Horizon [Rabbin-Naas] – equal & opposite ~sth [Ilaahin-Naas] – East Horizon [Malikin-Naas]. The command of Equal & Opposite ~sth is called **Revelation** or **Tawraat** or **Equal & Opposite Law**, while the revealed **Trinity** is called **Manifestation** or **Sirius Binary System or Injill**. The main sequence of the Revealed Trinity is called Sirius A [Rabbin-Naas]. The white dwarf companion of the Revealed Trinity is called Sirius B [Malikin-Naas]. Star Jupiter represents Sirius A, West Horizon, Strong Force, Heaven, Galaxy of Stars, Garden beneath which rivers flow, Proton, and the like. Rain comes from the West. Star Saturn represents Sirius B, East Horizon, Gravitational Force, Hell, Flameless Fire, Iron, Dry Wood, Space-bar, Oily Table, Electron, and the like. Earthquake emerges from the East.

WILLINGNESS AND UNWILLINGNESS
[Pythagorean Number – 'Line – Two']

Ayat\Verses - Say: **Disbelieve** you verily in Him **Who created the world [East Horizon] in two days**, and **ascribe** you unto Him **rivals**? He is the Rab of the universe. He set in the [world], mountains **standing firm**, high above it, and bestowed blessings in the world, and **measure therein all things to give them nourishment in due proportion, in four Days**, in accordance with [the needs of] those who seek [Sustenance]. Then **He turned to the heavens [West Horizon] when it was smoke [white dwarf]**, and said unto it and unto the earth: **Come both of you, willingly or unwillingly**. They said: We come, obedient. So He **ordained** them as **seven firmaments (windows) in two Days,** and **He inspired in each heaven its mandate.** And **We adorned the lower heaven with lights** and **[provided it] with guard.** Such is the Measuring of the Mighty, The Knower. **[Sura No. – 40\42 – Haa-Mim-Sajdah, Ayat No. – 9 to 12]**

The concept of **Binary System** or **Trinity** resembles with the concept of an atom or the unit of logic called proposition. The concept of **Binary** ~sth resembles with the sign of relation between the subject term and the predicate term of a logical proposition called copula. Though copula is a part of proposition, yet copula cannot be called a term i.e. neither subject term nor predicate term. The nature and function of Ilaahin-Naas [cloud & sky] resemble with the nature and function of copula. Sky & Cloud or Ilaahin-Naas is a part of Sirius Binary System like neutrino & neutron of an atom, but it neither represents Atom [Logic / Diamond Operator / Shi-raa / Universe / Alamin / Intrinsically Luminous Star / Shining Star / Glass / Sirius]; nor it represents Revealed Trinity i.e. proposition [Sirius Binary System / Nuurun Alaa Nuur], nor it represents Sirius A [Star Jupiter or Rabbin-Naas or Proton or Infinite Subject Term of the Trinity], nor it represents Sirius B [Star Saturn or Malikin-Naas or Electron or Infinite Predicate Term of the Trinity]. On the contrary, cloud & sky or neutrino & neutron or Copula or Ilaahin-Naas represents sign of relation between equal & opposite terms or between equal & opposite nature and function. This sign of relation or equal & opposite ~sth is the formal ground of manifestation. The representation of an equal & opposite relation is called Manifest Truth [Manifested Nature / Injiil / detail explanation of Fundamental Ayat / Criterion of Truth / witness / manifested sign / coded shared tautology]. Thus, binary ~sth or cloud & sky or neutrino & neutron or Ilaahin-Naas or copula establishes a formal relation between two extremes. For this reason, cloud & sky or binary ~sth or Ilaahin-Naas is called Bridge, Fulsirat, Golden Mean, Pythagorean Number – 'Line – Two' or converted snake & white of Muusa's Staff, and the like between two Extremes [between East and West]. One who has true concept of the equal & opposite ~sth or Ilaahin-Naas is called Lukman or a wise man. From the established verifiable scientific laws it is selfevident that Aristotle, Galileo, Newton, Einstein, Descartes and the like had true concept of the nature and function of cloud & sky or neutrino & neutron or Copula or binary ~sth or converted snake & white of Muusa's staff. So, they were possessors of true knowledge.

EQUAL & OPPOSITE REVELATION

Ayat\Verses – And if you ask them: Who created the heavens and the world? They would be sure to reply: They were created by [Him], the Exalted in Power,

Full of Knowledge. Who made the world a resting place for you, and placed roads for you therein, that haply you may find your way (guidance). And who sends down [from time to time] rain from the sky in due measure; and We raise to life therewith a land [earth] that is dead; even so you will be raised [from the dead]. **Who has created pairs in all things**, and has made for you **ships and cattle** on which you ride in order that you may mount upon [stages of journey] their backs, and you may remember the favour of your Rab when you mount thereon, and may say: Glorified to Him Who has subjected these to us, for we could never have accomplished this [by ourselves]; and to our Rab, surely, we must turn back. And they attribute (allot) to Him a portion of his bondmen! Lo! Man is verily a mere ingrate. Or chooses He **daughters** of all that He has created, and honours He you with **sons**? [Sura No. 42\43 – Ummil-Kitaab \ (Zukhruf), Ayat No. – 9 to 16]

Ayat\Verses – And if one of them has tidings of that which He sets up as a likeness to the Beneficent One, his/her face darkens, and he/she is filled with inward grief! [Similitude they then to Allah] that which is one brought up outward show, and unable to give a clear account in a **dispute?** And they make the angels, who are the slaves of the Beneficent, females. **Did they witness their creation?** Their evidence (testimony) will be recorded, and they will be called to account (questioned)! [Sura No. 42\43 – Ummil-Kitaab \ (Zukhruf), Ayat No. 17 to 19]

Ayat\erses - O you who believe! **Fasting is prescribed to you as it was prescribed to those before you, that you may [learn] self-restraint.** [Fasting] for a **certain number of days**; but if any of you is **ill**, or on a **journey**, the prescribed number [Should be made up] from days later. For those who can do it [With hardship], is a ransom, the feeding of one that is indigent. But he that will give more, of his own free will it is better for him. And it is better for you that you fast, if you only know. [Sura No. 1\2 – Baqarah, Ayat No. – 183 & 184]

Ayat\erses - Ramadhan is the [equal & opposite] month in which was revealed **Qur'an, as a guide to mankind, as Clear Proofs for guidance and the criterion of Truth.** So every one of you **who is present during that month** should spend it in fasting, but if anyone is ill, or on a journey, the prescribed period [should be made up] by days later. Allah intends every facility for you. He does not want to put you to difficulties. [He wants you] to complete the prescribed period, and to glorify Him in that **He has guided you**; and perhaps you shall be **grateful. And when My servant will ask you concerning Me, I am indeed close [to him]. I listen to the prayer of every suppliant when he cries unto Me.** So, **let them listen to My call, and let them trust in Me**

that they may be led aright (i.e. equal & opposite right direction towards Upright West). [Sura No. 1\2 – Baqarah, Ayat No. – 185 & 186]

There are twelve months, twelve hours of a clock in resemblance with the shared concepts of twelve chieftains, twelve springs [matter particles]. Out of twelve months, the month of Ramdhan is a clear proof of the equal & opposite month of eating & fasting. The Revelation of Quran in the Month of Ramdhan is a clear proof of the Equal & Opposite Revelation. So, Newton's Third Law – 'Equal & Opposite' is a Revealed Truth.

Ayat\Verses – Said Firawn: Believe you in Him before I give you permission? Surely this is a trick which you have planned in the **city** to drive out its people: but soon shall you know [the consequences]. Be sure I will cut off your hands and your feet on opposite sides, and I will **cause you all to die on the cross.** They said: For us, We are but sent back unto our Rab. But you cause **retribution** on us simply because **we believed in the Signs of our Rab** when they reached us! Our Rab! Pour out on us patience and constancy, and take our souls unto thee as Muslims! [Sura No. – 6\7 - As-haabul- a' raaf, Ayat No. – 123 to 126]

ONGOING COMMAND OF 3D

Ayat\erses - And He has constrained (made **subject to you) the Night and the Day. Washshamsa** wal-**qamara** and the **stars are in subjection by His Command.** Verily in **these are Signs for men who have sense.** [Sura No. – 15\16 – Nahl, Ayat\Verses – 12]

Ayat\erses - We said: O Adam! Dwell **you** and your **wife** in the **Garden [Heaven]**; and eat of the **bountiful** things therein as [where and when] you will; but **approach not this tree,** lest you become **wrong-doers.** But **the Shaytan caused them to deflect there from [the garden], and get them out of the state [of felicity] in which they had been.** We said: **Fall down, one of you a foe to the other!** There shall be for you in the World a **habitation [Earth] and provision [binary pulsar]** for a time. [Ayat No. – 35 & 36 of Sura Baqarah]

The Universe is a Shi-raa or a Diamond Operator or an intrinsically luminous star. The Universe has been revealed as a Trinity and manifested as an equal & opposite trinity of two Horizons and two Hemispheres. The ongoing downward 3D command of Allah – "We said: **Fall down, one of you a foe to the other!** There shall be for you in the World a **habitation & provision** for a **time**" is called Equal & Opposite Revelation. This Equal & Opposite Revelation is the causeless cause of leading **Intrinsically Luminous Star [Shi-raa or Diamond Operator]** into Trinity or Star System or Sirius Binary System. This revealed downward 3D command is the Downward Arrow of Science, three minutes after Bing Bang, Trinity of Philosophy, Origin of **Alif-Lam-Mim, Spin ½, H2O, and the like.** The revealed ongoing command of 3D has been established as **Big Bang Theory**. Revealed Equal & Opposite Law [Tawraat or Coherence Truth] has been established as **Newton's Third Law – 'Equal & Opposite'**. From the above revealed truths it is selfevident that Gravitational Force represents downward tendency towards East. This concept has been developed as **'Newton's Law of Gravitation'**. The ongoing command of 3D is the **origin of revelation i.e. Time** as well as sole efficient cause [instrumental cause or force] of continuous acceleration & motion for a period. This time period is unknown and unknowable from finite subjective point of view. The Intrinsically Luminous Star [Universe or Diamond Operator or Shi-raa or Shining Star] will remain in the state of equilibrium for the unknown and unknowable period due to revealed ongoing command of 3D. This equilibrium state of the Universe [Intrinsic Luminosity of the Diamond Operator / Shi-raa / Sirius / Fixed Star] is called Motion or Inertia or Nature or Fitrat or Dharma or Prakriti. This concept has been developed as **Newton's First Law – 'Inertia' and 'Motion' of Galileo.** Inertia does not mean **'inactivity'**. On the contrary, Nature or Fitrat or Dharma [of Intrinsically Luminous Star or Shi-raa] is not whimsical or capricious. This concept has been developed as **'Principle of the Uniformity of Nature'**. This Principle states that Nature [Intrinsic Luminosity of Shi-rra] behaves same under similar circumstances. The concept of equilibrium state of the Intrinsically Luminous Quadrilateral Star resembles with Samyavasta of Prakriti of Sankhya Philosophy. This Inertia or Motion or State of Equilibrium or Nature or Samyavasta is called Fitrat in the Kitaab with Truth. The revealed ongoing command of 3D is the External Force [cause of constantly acting or continual acceleration] i.e. F of **Newton's Second Law – 'F = ma' and Staff at the right hand of Muusa**. The consequences of this revealed ongoing command of 3D with respect to revelation are acceleration [a of Newton's Second Law or converted snake of Muusa's Staff] and masses [m of Newton's Second Law or obversion of the converted snake of Muusa's Staff into white]. The consequences of the revealed ongoing 3D command with respect to manifestations are electromagnetic waves [Bullet or Provision] and gravitational waves [Habitation]. Newton's Three Laws of Motion and Law of Gravitation are the revealed truths; while Einstein's Gravitational Lensing and Binary Pulsar are the manifest truths. Natural Magnetism is the Manifested Sign of Equal & Opposite Revelation i.e. **Conversion of Muusa's Staff into Snake and White.** In other words, Manifested Sign of Natural Magnetism or Wash-Shams Wal Qamar or the so-called sun and the so-called moon are the two Clear Proofs of Equal and Opposite Revelation as well as Manifested Signs of Natural Magnetism.

TWO PROOFS

Now when Muusa had **fulfilled the term**, and was **travelling** with his family, he perceived **a fire [light] in the direction of Mount Tuur [Top West Triangle of the Manifested Hexagon]**. He said to his family: **Tarry** you; I **perceive** a fire [light]; I hope to **bring you** from there some **information** (tidings), or a brand from the fire that you may **warm yourselves**. And when he reached it, he was called **from the right side [the appointed Kaba is at the right side of the Mount Tuur]** of the valley in the blessed field, from the **tree**: O Muusa! Lo! I, even I, am Allah, the Rab of the Universe. **Throw down your staff**. And when he **saw it moving** as it had been **a demon,** he **turned** to flee headlong (and it was said unto him): **Draw near**, and **fear not**. Lo! **You are of those who are secure. Thrust your hand into the bosom of your robe**, it will **come forth white [white dwarf]** without hurt. And **guard your heart from fear**. These are the **two proofs** from your Rab to **Firawn and his chiefs**, for **truly** they are a people **rebellious and wicked. [Sura No. – 27\28 – Qasas, Ayat\Verses – 29 to 32]**

Ayat\Verses - And what is that in the **right hand**, O Muusa? He said: It is my **staff**, on it I lean; with it I beat down fodder for my flocks; and in it I find other uses. [Allah] said: Throw it, O Muusa! He threw it, and behold! It was a **snake, active in motion**. [Allah] said: Seize it, and fear not. We shall return it at once to its former condition. Now draw your hand close to your side. It shall come forth **white** [and shining], without harm [or stain], **as another Sign**. In order that We may show you our **Greater Signs**; go you to Firawn, for he has indeed transgressed all bounds. [Sura No. – 19\20 – Twa-Haa, Ayat No. – 17 to 23]

Then We sent **Muusa** and his brother **Haarun, with Our Signs** and a **clear warrant** to **Firawn** and his **Chiefs**. But they scorned (them) and they were **dictatorial folk**. They said: **Shall we believe in two men like ourselves? And their people are subject to us!** So they **accused them of falsehood**, and **they became of those who were destroyed**. And We gave **Muusa** the Kitaab, in order that they might receive **guidance**. And We made **Ja-alnabna Maryama** and his mother as a **Sign**. We gave them refuge on a height, **a place of thanks** and **water spring**. [Sura No. – 22\23 – Mu-Minuun, Ayat\Verses – 45 to 50]

MOST EXCELLENT PROVISION
"After a difficulty, Allah will soon grant relief."
"Now, Allah has sent down to you a reminder."
Ayat\Verses - Let her who has **abundance** spend of her abundance, and he whose provision is **measured**, let her spend of that which Allah has given her. Allah puts **no burden** on any person beyond what He has given him. **After a difficulty, Allah will soon grant relief**. And how many a community that insolently opposed the

Command of their Rab and of His messengers, and We called it to a **stern account** and We imposed on them an **exemplary Punishment**. So that it tasted the ill-effects of their conduct, and the consequence of the conduct was **loss**. Allah has prepared for them a **severe Punishment**. Therefore keep your **duty** to Allah, O you **men of understanding!** O you who believe! **Now, Allah has sent down to you a reminder**. A Messenger, who rehearses to you the **Signs of Allah containing clear explanations**, that he may lead forth **those who believe and do righteous deeds** from the **depths of Darkness into Light**. And those who believe in Allah and **work righteousness**, He will admit to Gardens beneath which Rivers flow, to dwell therein forever. Allah has indeed granted for them a **most excellent Provision** [Sustenance]. [Sura No. – 64\65 – Tallaq-tumun-nisaaa-'a, Ayat No. – 7 to 11]

A bountiful provision

Ayat\Verses - O Prophet! Say to those who are imprisoned in your hands: If Allah finds any good in your hearts, He will give you something better than what has been taken from you, and He will forgive you: for Allah is Ever Forgiving, Most Merciful. And if they would **betray you, they betrayed Allah** before, and He gave (you) power over them. Allah is Knower, Wise. Those who **believed**, and left their **homes**, and fought for the **Faith**, with their **property** and their **lives**, in the cause of Allah, as well as those who gave [them] **asylum** and **aid**, these are [all] **friends and protectors of one another**. As to those who **believed** but **did not leave their homes**, you have **no duty to protect them** until **they leave their homes**; but if they seek your help in the matter of **religion**, then it is your **duty** to help them, **except** against a group with whom you have a **treaty of mutual union**. And [remember] Allah sees all that you do. The **unbelievers are protectors, one of another**: Unless you do this, [protect each other], there would be **confusion** in the land, and **great corruption**. **Those who believe, and left their homes, and fight for the Faith, in the cause of Allah as well as those who give [them] asylum and aid, these are [all] in very truth the believers. For them is the forgiveness of sins and a bountiful provision.** And those who accept Faith subsequently, and left their homes, and fight for the Faith in your company, **they are of you**. But kindred (family members / relatives) by blood have prior rights against each other in the Kitab of Allah. Verily Allah is well-**acquainted** with all things. **[Sura No. – 7\8 – Anil-Anfaal, Ayat No. – 70 to 75]**

FUNDAMENTAL AYAT OR REVELATION
Formulae or Formal Grounds of Manifestations

Q A A A F: Wal-Quraanil-Majiid [This a Fundamental Ayat of the Quran. By this Clear Proofs of The Quranil-Majiid] [Sura No. 49\50 – Kaaaf, Fundamental Ayat No. – 1]

"Saaad. Wal Quraani Ziz-Zikr"

SAAAD - By the **Qur'an,** Full of Warning: [This is the Truth]. [Sura No. – 37\38 – Swad, Fundamental Ayat No. – 1]

But the disbelievers [are steeped] in **self-glory** and **Separatism**. How many generations before them did We destroy? In the end they cried [for mercy] when there was no longer time for being saved! And they wonder that a **warner** has come to them from among themselves! And the disbelievers say: This is a sorcerer (wizard / magician) telling lies! He **makes all breakable realities into one Wahid [I l a a hanw-Waahidaa]. Truly this is a wonderful thing!** [Sura No. – 37\38 – Swad, Ayat No. – 2 to 5]

NUUUN. [Sura No. – 67\68 – Qalam, Ayat No. – 1]

By the pens and that which they write, you are not, **by the Grace of your Rab unto you**, a **madman**. [Sura No. – 67\68 – Qalam, Ayat\Verses No. –2]

"Yaa-Siin. Wal Quraanil-Hakiim"

YAA-SIIN. By the **Qur'an, full of Wisdom,** you are of those sent [la-minal-mursaliin] on an equal & opposite Right Path [Siraatim-Mustaqiim]. It is a Revelation sent down by [Him], the Exalted in Might, Most Merciful in order that you may **warn** a folk, whose fathers were not warned, so they are heedless. [Sura No. – 35\36 – Yaa-siiin, Ayat No. – 1 to 6]

Twa-Siin: Tilka Aayaatul-Quraani wa Kitaabim-mubin

TWA-SIIN. These are Clear Proofs as shared in the Qur'an, a **Kitaabim-mubin** that makes [things] clear (plain). [Sura No. – 26\27 – Namal, Ayat No. – 1]

A **guidance**, and **good tidings** for the believers those who establish regular prayers and pay the poor-due and are sure of the hereafter. [Sura No.-26\27-Namal, Ayat No.-2 & 3]

TWA-HA [Sura No. – 19\20 – Twa-Haa, Fundamental Ayat No. – 1]

We have not revealed this **Qur'an** to you to be [an occasion] for your distress. But only as a **reminder** to those who fear [Allah]. [Quran is] a **revelation** from Him Who **created the world and the heavens** on **high**. [Allah] Most Gracious is **firmly established on the throne** [Arsh]. To Him belongs **what is in the heavens** and **in the world**, and all **between** them, and **all beneath the soil**. If you **pronounce** the **word aloud**, [it is no matter], for verily He know what is **secret** and what is yet more **hidden**. Allah! There is **no** ilaaha (unbreakable reality) save Him. To Him belong the most **Beautiful Names**. [Sura No. – 19\20 – Twa-Haa, Ayat No. – 2 to 8]

"Ha-Mim. Tanzilul-Kitaabi minallaahil – Azizul-Hakim"
Ha-Mim. [Revelation – Sura No. – 45\46 – Bil-ahqaafi, Fundamental Ayat No. – 1]
This **Revelation of the Tanzilul-Kitaab** is from Allah the Mighty, The Wise (Aziizil-Hakim). We created not the **heavens and the world** and all between them **save with truth,** and for a **Term Appointed. But those who reject Faith (disbelieve) turn away from that whereof they are warned.** [Sura No. – 45\46 – Bil-ahqaafi, Ayat No.-2 & 3]

"Ha-Mim. Tanzilul-Kitaabi minallaahil – Azizul-Hakim"
[Day and Night are the clear signs of equal & opposite Revelation.]
Ha-Mim. [Sura No.-44\45 – Tanziilul Kitaab (Prev.Jaasiya), Fundamental Ayat No.- 1]
The **revelation of the Tanzilul-Kitaab is from Allah** the Mighty, the Wise. Verily in the **heavens** and the **world**, are **Signs** (portents or clear proofs) for those who **believe.** And in the **creation of yourselves** and the fact that **animals** are scattered [through the earth], are **Signs** (portents) for those **whose faith is sure.** And in the **alternation** (difference) of **Night** and **Day**, and the fact that Allah sends down **Sustenance** from the **sky,** and **revives** therewith the **world** after its **death,** and in the **change of the winds,** are **Signs** (portents) for those **who have sense.** These are the **Signs (portents) of Allah,** which **We rehearse to you in Truth. Then in what manifestation [of sign] will they believe after Allah and His Signs?** [Sura No. – 44\45 – Tanziilul Kitaab (Prev. Jaasiya), Ayat No. – 2 to 6]

"Ha-Mim. Tanzilul-Kitaabi minallaahil – Azizul-Hakim"
HA-MIM [Sura No. – 39\40 – Mu-Min, Fundamental Ayat No. – 1]
The revelation of this **Tanzilul-Kitaab** is from Allah, The Mighty, The Knower, The Forgiver of sin, The Accepter of repentance, the Stern in punishment, and the Bountiful. **There is no unbreakable reality except Him.** To Him is the final goal. **None dispute (argue) concerning the Signs (Revelations) of Allah except the disbelievers (those who reject faith). So, let not their turn of fortune in the land deceives (misleads) you.** [Sura No. – 39\40 – Mu-Min, Ayat No. – 2 to 4]

"Ha-Mim. Wal Kitaabim-Mubin"
Ha-Mim [Revelation – Sura No. – 41\42 – Shuuraa, Fundamental Ayat No. - 1]
Ha-Mim. [Revelation, Sura No. 42\43 – Ummil-Kitaab \ (Zukhruf), Fundamental Ayat No. – 1]
By the **Kitaabim-Mubin** that **makes things clear; lo! We have appointed it a Qur'an in Arabic** that **you may be able to understand.** And verily, in the

Mother of the Kitaab which We possess, it is indeed **inspiring, vital**. [Sura No. 42\43 – Ummil-Kitaab \ (Zukhruf), Ayat No. – 2 to 4]

"Ha-Miim. Tanziilum-minar-Rahmaanir-Rahiim. Kitaabun-fussilat Aayaatuhuu Quraanan Arabiyyal-liqawminy-ya-lamuun. Bashiranwwa Naziiraa; fa-a-raza aksaruhuum fahum laa yasma-uun."

Ha Mim - [Sura No. – 40\41 – Haa-Mim-Sajdah, Fundamental Ayat No. -1] (Revelation) from the **Beneficent, the Merciful; A Kitaabun-fussilat, whereof the Ayats are explained in detail; a Qur'an in Arabic, for people who understand,** giving good tidings and admonition; **yet most of them turn away,** and **so they hear not. [Sura No. – 40\41 – Haa-Mim-Sajdah, Ayat No. – 2 to 4]**

Ain-Sin-Qaf. [Revelation – Sura No. – 41\42 – Shuuraa, Fundamental Ayat No. - 2]
Thus [He] **has sent inspiration to you as [He did] to those before you**. Allah is Exalted in Power, Full of Wisdom [Huwal-Aliyyul-Azim]. To Him belongs all that is in the **heavens and in the world**: and He is The Sublime, The Tremendous. [Sura No. – 41\42 – Shuuraa, Ayat No. – 3 & 4]

Alif Lam Mim. [Sura No. 1\2 – Baqarah, Fundamental Ayat – 1]
This is the **Zaalikal-Kitab. In it is guidance sure, without doubt**, to those who fear Allah (who ward off evil). Who believe in the Unseen, and steadfast in prayer, and spend out of what We have bestowed upon them. And who believe in that which is **revealed unto you [manifested magnetism in resemblance with revelation]**, and that which was **revealed before your** time and [in their hearts] has the **assurance of the Hereafter**. They depend on guidance from their Rab, and it is they who will **succeed**. [Sura No. 1\2 – Baqarah, Ayat\Verses – 2 to 5]

Alif, Lam, Mim.
The Roman Empire has been defeated in a land close by; but they, [even] after [this] defeat of theirs, will soon be victorious, within a few (ten) years. Allah's is the command in the former case and in the later, and in that **the believers will rejoice with the help of Allah to victory**. He helps whom He will, and He is exalted in might, most merciful. **[It is] the promise of Allah. Never does Allah depart from His promise. But most men understand not. [Sura No. – 29\30 – Ruum, Ayat\Verses No. – 1 to 6]**

Alif-Lam-Mim [Sura No. – 2\3 – Imran, Fundamental Ayat – 1]
Allah! There is no unbreakable reality save Him, the Alive, the Eternal. He has revealed unto you **a alaykal-Kitaaba-bil-Haqqi [Kitaab with Truth-in-itself]**

confirming that **which were revealed before it [manifested magnetism in resemblance with revelation]**, even as He revealed the **Tawraat** and the **Injiil** before this, as a **guide** to mankind, and **He has revealed the Criterions** (scales of justifications as well as verifications). Lo! **Those who reject Faith in (or disbelieve) the Signs of Allah (revealed as well as manifested signs of magnetism) theirs will be a severest chastisement** and Allah is Mighty and Able to Requite (the wrong). [Sura No. – 2\3 – Imran, Ayat\Verses – 2 to 5]

Alif-Laaam- Miim. [Sura No. – 28\29 – Ankabuut, Fundamental Ayat No. – 1] Do men think that they will be left (at ease) because they say: We believe, **and will not be tested with affliction**? Lo! We tested those who were before you, and Allah certainly knows **those who are true** and **those who are false**. Or do those who practise evil deeds [projected magnetism] imagine that they will get the better of Us? **Evil is their judgment [conspiracy projection]!** For those who looks forward to the meeting with Allah (let him know that) Allah's reckoning is surely near, and He is the Hearer, the Knower. And if any strive, they do so for their own souls, for Allah is free of all needs from all creations. And those who believe and work righteous deeds (tabligh), from them We shall blot out all evil [that may be] in them, and We shall reward them according to the best of their deeds. [Sura No. – 28\29 – Ankabuut, Ayat No. – 2 to 7]

Alif-Laaam-Mim.
[This is] the Revelation of the **Tanziilul-Kitaab** in which there is no doubt, from the Rab of the Universe. **Or do they say: He has forged it? Nay, it is the Truth from your Rab, that you may admonish a people to whom no warner has come before you in order that they may receive guidance.** [Suea No. – 31\32 = Sajdah, Ayat\Verses – 1 to 3]

Alif-Laaam-Mim.
These are Verses of the **Aayaatul-Kitaabil-Hakim**, A Guide and a Mercy to the doers of good, those who establish regular prayer, and give regular charity, and have [in their hearts] the assurance of the Hereafter. These are on [true] guidance from their Rab, and these are the ones who will prosper [Sura No. – 30\31 – Luqmaan, Ayat\Verses – 1 to 5]

Twa-Sin-Mim.
These are the **shared ideograms** revealed in the alaykal-**Kitaabim-Mubin** as **Clear Proofs [Revelations]**. We narrate to you some of the tales of Muusa and Firawn in Truth, for **people who believe**. [Sura No. – 27\28 – Qasas, Ayat\ Verses – 1 to 3]

Twa-Sin-Mim.
These are the shared ideograms revealed in the Kitaab as Clear Proofs which makes things clear. [Sura No. 25\26 – Shuraa, Ayat No. – 1 & 2]

Alif-Laaam-Raa. Tilka Aa-yaatul-Kitaabil-Hakim. These are the **Ayat** of the **Kitaab of Wisdom** [Sura No. – 9\10 – Yuunus, Fundamental Ayat No. – 1]
"Alif-Laaam-Raa. Kitaabun-uh-kimat Aayaa-tuhu summa fussi-lat milla-dun Hakiimin khabiir" [Trans.] - This is a **Kitaab** with **basic or fundamental Ayat** [of **established meaning**], **further explained in detail [verses]**, from **One Who is Wise and Well- acquainted with all things**]. [Sura No. – 10\11 – Kitaabun-uh-kimat (Prev. – Huud), Fundamental Ayat No. – 1]

"**Alif-Laaam-Raa. Tilka Aayaatul-Kitaabi wa Quraanim-Mubin.**"
ALIF-LAAAM-RAA. These are the Fundamental **Ayat** of the **Kitaab** as **Qur'an** that **make things clear**. [Sura No. – 11]\12 – Yuusuf, Aat No. – 1]
We have sent it down as an Arabic Qur'an, in order that you may learn wisdom. We narrate unto you the most beautiful of **parables**, in that We reveal to you this Qur'an: before this, you too were among those who knew it not. [Sura No. – 11]\12 – Yuusuf, Aat No. – 2 & 3]

"**Alif-Laaam-Raa. Kitaabun anzalnaahu ilay-ka litukh-rijan-naasa minazzulumaati ilan-Nuuri-bi-izni Rabbihim ilaa Siraatil-Aziizil-Hamid**"
Alif-Laaam-Raa. A Kitaab which We have revealed unto you, in order that you might lead mankind out of the **depths of darkness into light** by the permission of their Rab to the Way of [Him] the Exalted in power, worthy of all praise! [Sura No. – 13\14 – Ibrahim, Fundamental Ayat No. - 1]

"**Alif-Laaam-Raa. Tilka Aayaatul-Kitaabi wa Quraanim-Mubin.**"
Alif-Laaam-Raa. These are the Fundamental **Ayat** of the **Kitaab** as **Qur'an** that **make things clear**. [Sura No. – 14\15 = Hur, Fundamental Ayat No. – 1]

"**Alif-Laaam-Mim-Raa. Tilka Aayatul Kitaab. Wal-lazina unzila ilay-ka mir-Rabbikal-Haqqu wa laa-kinna aksaran-naasi laa yu-minuun.**"
Alif-Laaam-Mim-Ra. These are the **Clear Signs** [Clear Proofs] of the Kitaab. That which has been revealed to you from your Rab is the **Truth** (Tautology). But most men believe not. [1] [Sura No. – 12\13 – Rad, Fundamental Ayat No. - 1]

Alif - Laaam – Mim - Swad. [Sura No. – 6\7 - As-haabul- a' raaf, Fundamental Ayat No. – 1]
A **Kitab** revealed unto You (Prophet). So let not Your heart be oppressed more by any difficulty (doubt) on that account that with it You may **warn** thereby

and a **reminder** unto **believers**. Follow [O men!] the revelation given unto you from your Rab, and follow not, as **friends** or **protectors**, other than Him. Little do you remember of warnings? [Sura No. – 6\7 - As-haabul- a' raaf, Ayat No. – 2 & 3]

"Kaf-Ha-Ya-Ain-Sad. Zikru Rahmati Rabbika abdahuu Jakariyaa"
Kaf – Ha – Ya – Ain – Sad. [Sura No. – 18\19 – Maryam, Ayat –1]
[This is] a recital of the Mercy of your Rab to His servant Jakariya. When He cried to his Rab in secret. [Sura No. – 18\19 – Maryam, Ayat\Verses – 2 & 3]

Search for Truth
1. If I say that Fundamental Ayat (pl) i.e. Formal Grounds [Formulae / Tawraat / Laws] are unknown and unknowable, and I believe only further explanations [Tafsirs] of what have been shared explaining in detail by Allah and His Messenger, then I am ---------------------.
   Options –
   (a) a believer of whole Quran,
   (b) a believer of a part of Quran,
   (c) a believer of opinions,
   (d) a disbeliever
   (e) a non-believer
2. It has been inspired [Ref. Ayat No. 5 to 7 of Sura Imran] to believe in ----------
   Options –
   (a) whole Quran [both Fundamental Ayat as well as detail explanations of Fundamental Ayat i,e. both Tawraat and Injiil or Revelations and Manifested Signs],
   (b) only detail explanations of Fundamental Ayat [Injiil],
   (c) only further explanations of detail explanations [Tafsirs and opinions],
   (d) both what have been shared explaining in detail by Allah & His Messenger as well as further explanations of detail explanations [Tafsir or Opinions]
3. "None knows its explanation' [Ref. Ayat No. – 5 to 7 of Sura Imran] implies that none knows ----------------------------------
   Options –
   (a) Fundamental Ayat [Tawraat / Formal Grounds / Revelations / Formulae],
   (b) Detail Explanation [Injiil or Manifestation],
   (c) Further Explanation [Tafsir or Oinions] of what have been shared explaining in detail by Allah and His Messenger

4. **"Then woe to those who write the Kitaab with their own hands, and then say: This is from Allah, to traffic with it for miserable price! Woe to them for what their hands do write, and for the gain they make thereby.** And they say: **The fire shall not touch us** but for **a few numbered days**: [Ref. Ayat No. – 79 to 81 of Sura – Baqarah] denote - ----------------------.

   Options –
   (a) those who type / write / print / publish Quran
   (b) those who write further explanations [Tfsirs and Opinions] of what have been shared explaining in detail by Allah and His Messenger

5. **Whether Fundamental Ayat (pl) are the Substances or Detail Explanations are the Substances or Further Explanations are the Substances of Quran i.e. Alaykal-Kitaab with reference to Ayat No. 5 to 7 of Sura - Imran?**

What is meant by breaking of Quran?

Ayat\Verses – And say: **I am indeed he who warns openly and without ambiguity,** such as We sent down for those **who make division, those who break the Quran into parts.** Therefore, by the Rab, We will, of a surety, call them to account, for all their deeds. **Therefore expound openly what you are commanded**, and **turn away from those who join false realities with Allah [manmade magnetism].** [Sura No. – 14\15 = Hur, Ayat No. – 90 to 94]

Ayat\Verses - From Allah, verily **nothing is hidden** in the world or in the heavens. He it is Who **shapes** you in the wombs as He pleases. There is no unbreakable reality save Him, the Almighty, the Wise. It is He Who has revealed to you the **Alaykal-Kitaab** wherein there are **Clear Signs** [Clear Proofs]. They are the **Substances** [foundations or substratum] of the **Kitaab**; and others [which are] **allegorical** [Symbolic or Phonetic presentations or detail explanations]. But those in whose hearts are **doubts, pursue** forsooth (related to), that which is symbolic **(allegorical or manifested signs or detail explanations)** seeking (to cause) **dissension by seeking to explain it. None knows its explanation [Tafsir]** except Allah. And those who are of **sound instruction** say: We believe therein, the **whole** is from our Rab; but **only men of understanding really notice.** [Sura No. - 2\3 – Imran, Ayat No. – 5 to 7]

Ayat\Verses – **Then woe to those who write the Kitaab with their own hands, and then say: This is from Allah, to traffic with it for miserable price! Woe to them for what their hands do write, and for the gain they make thereby.** And they say: **The fire shall not touch us** but for **a few numbered days**: Say: **Have you taken a promise from Allah? Truly Allah** will never break His promise. Or is it that you say of Allah what you do not know? Nay, **those who seek gain in evil**, and are **girt round by their sins**, they are **companions of the fire**. Therein they will **abide [For ever].** [Sura No. 1\2 – Baqarah, Ayat No. – 79 to 81]

Ayat\Verses – After this it is you, **the same people, who slay among yourselves,** and **drive out a party of you from their homes; supporting one another against them by sin and ill will**; and if they come to you as **captives**, you **ransom** them, where as **their expulsion was itself unlawful for you. This is only because you believe a part of the Kitaab and you reject the rest?** And what is the reward for those among you who behave like this but **disgrace** in this life and on the **Day of Judgment they will be consigned to the most grievous penalty**. For Allah is not unmindful of what you do. **These are the people who buy the life of this world at the price of the Hereafter. Their penalty shall not be lightened nor shall they be helped.** [Sura No. 1\2 – Baqarah, Ayat No. – 85 & 86]

IT HAS BEEN INSPIRED TO CONFIRM MANIFESTATIONS IN RESEMBLANCE WITH REVELATIONS

Ayat\Verses – As for that which We **inspire in the ilayka-minal-Kitaab**, it is the Truth **confirming** that which was (revealed) before it. For Allah is assuredly- with respect to His servants - well acquainted and Fully Observant. [Sura No. – 34\35 - Faatiris-Samaawaati wal-'arz, Ayat\Verses – 31]

Ayat\Verses – O you unto whom the Kitaab has been given! **Believe in what We have revealed, confirming what is [already] with you [Manifested Signs]**, before We destroy countenances so as to confound them or curse them as We cursed the Ashaabas-Sabt. The commandment of Allah is always executed. [Sura No. – 3\4 – Nisaa, Ayat No. – 47]

Ayat\Verses – **And those who bring the Truth and confirm [with a view to believe] it - such are the men who do right. [Sura No. – 38\39 – Tanziilul-Kitaab, Ayat\Verses – 33]**

Alif-Lam-Mim [Sura No. – 2\3 – Imran, Fundamental Ayat – 1]
Ayat\Verses – Allah! There is no unbreakable reality save Him, the Alive, the Eternal. He has revealed unto you **a Kitaab with truth, confirming that which were revealed before it**, even as He revealed the **Tawraat** and the **Injiil** before this, as a **guide** to mankind, and **He has revealed the Criterions** (scales of justifications as well as verifications). Lo! **Those who reject Faith on (or disbelieve) the Signs of Allah (Manifestation in resemblance with**

Revelation) theirs will be a severest chastisement and Allah is Mighty and Able to Requite (the wrong). [Sura No. – 2\3 – Imran, Ayat\Verses – 2 to 4]

"Ha-Mim. Wal Kitaabim-Mubin"

Ha-Mim. [Revelation, Sura No. 42\43 – Ummil-Kitaab \ (Zukhruf), Ayat No.-1]

By the **Kitaabim-Mubin** that **makes things clear; lo!** We have appointed it a **Qur'an in Arabic** that **you may be able to understand.** And verily, in the **Mother of the Kitaab** which We possess, it is indeed inspiring, vital. [Sura No. 42\43 – Ummil-Kitaab \ (Zukhruf), Ayat No. – 2 to 4]

Search for Truth

Allah and His Messenger have shared Quran with us in Arabic so that we can understand shared Truth as well as Criterions of Truth clearly. The above verses imply that there must be some distinctive features in Arabic language which keep the meaning of the shared terms definite, unambiguous, univocal, conceivable, and consistent with Clear Signs [Fundamental Ayat / Revelation / Tawraat] as well as Manifested Signs [Injiil / Manifested Nature]. So the question is – What are those distinctive features of Arabic Language which keep the meanings of shared terms clear and distinct? If you know the right answer of this question, then kindly share the same with mankind as a whole.

The Kitaab with Truth is a Kitaab of Clear Signs, Manifested Signs, Signs of Indication, Ideograms, Symbolic Presentation, and Phonetic Representation. In other words, Kitaab with Truth includes both Truth-in-itself and Criterions of Truth. "Lo! **Those who reject Faith on (or disbelieve) the Signs of Allah theirs will be a severest chastisement** and Allah is Mighty and Able to **Requite** (the wrong)." So, the question is – Is there a single book on Signs of Allah worked out by the people of the Kitaab, people of the book, and the researchers on tautologies which is available in the market? If there is any such book, then kindly share its references.

If I do not know the differences between revelation and manifestation, sign and symbol, clear signs and manifested signs, ideograms and phonograms, signs of indication and symbolic (allegorical) representation etc. but motivated by the sense of recognition, try to interpret only detail explanations further denying their relations with Fundamental Ayat, then who am I? Options– (i) a wise man, (ii) an epistemic person, (iii) a mystic [a spiritualist or a sufi], (iv) an idiot, (v) a fraud, (vi) a true believer, (vii) a verified disbeliever, (viii) a recognised ignorant one, (ix) a selfevident hypocrite, (x) a selfexplained evidence sorcerer, (xi) pen or paper or pencil of the activist Laden.

Each Fundamental Ayat is called Revelation or Formal Ground [Tawraat / Law / Formula or Coherence Truth] of Manifestation [Injiil or Correspondence Truth]. As a teacher/student of Mathematics I always try to teach/do mathematics in resemblance with Mathematical Formulae i.e. Formal Ground or Tawraat. Mathematical Formulae or Tawraat are not perceptible but conceivable. Similarly, Fundamental Atat (pl) or Verses are not perceptible but conceivable. As a teacher/student of Mathematics, I never try to teach/practice mathematics either contrary or contradictory to formal grounds i.e. formulae or Tawraat. But as a Religious Scholar/believer of Quran, I never try to teach or share or confirm manifest truth in resemblance with Revelation or Formal Ground or Tawraat or Law. I have conceptualised that Fundamental Ayats or Formal Grounds or Revelations or Tawraat or Laws are unknown and unknowable. On the contrary, as an expert of comparative study, I am trying to establish that Fundamental Ayat or Formal Ground or Tawraat or Law or Revelation as an independent Kitaab with Truth. Similarly, Manifestation or detail explanation or Injiil is also an independent Kitaab with Truth. So, I am comparing between Kitaab with Truth [Quran] with Tawraat, with Injiil. This comparison manifests that I have no concept of the Existential Fallacy. So, I am comparing between the tree [Quran / Kitab with Truth / Truth-in-itself] and the four important branches of the tree [Kitaabs / criterions of truth / tautologies] with a view to prove that branches of the tree have independent existence. I do not know that Quran [Kitaab with Truth] includes four witnesses [four criterions of truth]. Tawraat [coherence truth], Injiil [correspondence truth], Zabur [pragmatic truth], and Furqan [selfevident truth] are the four witnesses or four criterions of truth or Kitaabs, but not Kitaab with Truth. These criterions are not excluded from Kitaab with Truth [Quran] i.e. these criterions have no independent existence. If you want to argue with searched out manifest truth in resemblance with revelation, then you must have to follow the revealed conditions to argue with manifest truth.

Ayat\Verses - Say: O **People of the Kitaab [Ahlal Kitaab]! You have no ground to stand upon [to argue] unless you stand fast by the Tawraat [Revelation or Formal Ground or Coherence Truth], the Injiil [Correspondence Truth or Manifested Signs], and all the revelations [manifested nature] that have come to you from your Rab.** It is the revelation that comes to you from your Rab, that increases in most of them their obstinate rebellion and blasphemy. **But sorrow you not over (those) people without Faith.** Those who believe, those who are **Haaduu, Saabiuu-nas** and **Nasaaras,** and any who **believe in Allah** and the **Last Day** [Yawmil-Aakhiri], and **work righteousness [do tabligh],** --on them shall be no fear, nor shall they grieve. [Sura No. – 4\5 – Maaaidah, Ayat\Verses – 68 & 69]

In the day when Allah will gather the Messengers together, and ask: What was the response you received (from men to your teaching)? They will say: We have no knowledge: It is You Who knows in full all that is hidden. Then Allah will say: O Isabna-Maryam! Remember My favour to you and to your mother, how I strengthened you with the **Holy spirit**, so that you spoke to the mankind in childhood and in maturity; and how I **taught** you **Kitabah** wal-**hikmata**, wat **Tawraat**, wal-**Injiil** and how you dust shape of clay as it were the likeness of a bird by My permission, and you breathe into it, and it becomes a bird by My permission, and you heal those born blind, and the lepers by My permission. And how you raise the dead by My permission. And how I did restrain Banii-Israara from (violence to) you when you showed them the **Clear Proofs** and the unbelievers among them said: **This is nothing but evident magic**. [Sura No. – 4\5 – Maaaidah, Ayat\Verses – 109 & 110]

What is that '**truth**' which will be shown at the end?
Say: **If** what you would see hastened were in my power, the **matter** would be **settled at once** between you and me. But Allah knows best **those who do wrong**. With Him are the **keys of the unseen**, the **treasures** that **none knows** but He. **He knows whatever there is on the earth and in the sea. Not a leaf** falls but with His **knowledge**. There is **neither** a **grain** in the **darkness of the world, nor** anything fresh or dry, but is [noted] in a clear record [to those who can read. It is He **who** takes your **souls by night**, and has **knowledge** of all that you have done **by day**. Then He **raises** you up again **by day** that a **term appointed** be fulfilled. In the end unto Him will be your **return**. Then He will **show you the truth** of all that **you did**. [Sura No. 5\6 An-aam, Ayat No.58 to 60]

His word is the truth. His will be the dominion on the day when the trumpet is blown; Knower of the invisible and visible. He is the Wise, the Aware. Lo! Ibrahim said to his father Aazara: Take you idols for realities? For I see you and your people in manifest error. So also did We show Ibrahim the Kingdom of the heavens and the world that he might be with **those possessing certainty**. [Sura No. – 5\6 – An-aam, Ayat No. –74 to 76]

SECTION – III
DIAMOND OPERATOR – 3.3
MANIFESTED UNIVERSE IS AN UPRIGHT RECTANGLE

**Universe has been manifested as an Upright Rectangle
in coherence with [Tawraat] the 'End of Proof' and in
correspondence with [Injiil] the appointed Kaba
West Horizon [Heaven] and East Horizon [World]
WEST [JUPITER / SIRIUS A] AND EAST [SATURN / SIRIUS B]
WEST [UPWARD DIRECTION] AND EAST
[DOWNWARD DIRECTION]
WEST [STRONG FORCE] AND EAST [GRAVITATIONAL FORCE]
Fasta-'iz billaahi minash-Shaytaanir-Rajim
Bismillaahir-Rahmaanir- Rahim**

Allah is the Manifest Truth

Ayat\Verses – As for **those who** insult **virtuous women, indiscreet (careless)
but believing, cursed are they in this life and in the Hereafter.** For them is
**a grievous Penalty. On the Day when their tongues, their hands, and their
feet will bear witness against them as to what they used to do. On that Day
Allah will pay them their just dues,** and **they will realise** that **Allah is the
Manifest Truth.** Vile women are for vile men, and vile men for vile women and
good women are for good men, and good men are for good women. **Such are
innocent (not affected) of that which people say.** For them there is forgiveness
and a beautiful provision. [Sura No. – 23\24 – Nuur, Ayat No. – 23 to 26]

Ayat\Verses – **Alif-Lam-Mim.** This is the **Zaalikal-Kitaab.** In it is **guidance
sure, without doubt,** to those **who fear Allah** (who ward off evil / projected
falsehood); who **believe** in the **Unseen,** and **steadfast** in prayer, and **spend** out
of that which We have bestowed upon them; and who **believe** in that which is
revealed unto **you,** and **that** which **was revealed before your time, and [in
their hearts]** has the **assurance of the Hereafter.** They **depend on guidance**
from their Rab, and it is **they who will succeed.** [Sura No. 1\2 – Baqarah, Ayat
No. – 1 to 5]

Heaven [West Horizon or Sirius A] and
World [West Horizon or Sirius B]

Ayat\Verses – Say: **Shall we indeed cry instead of unto Allah, things that
can do us neither good nor harm, and turn us after receiving guidance
from Allah like one bewildered whom the devils have infatuated in the
earth, which has companions who invite him to the guidance (saying):**

come unto us? Say: **Allah's guidance is the [only] guidance, and we are directed to submit ourselves to the Rab of the universe.** To establish regular prayers and to fear Allah: for it is to Him that we shall be gathered together. **It is He who created the heavens [West Horizon] and the world [East Horizon] in true/due proportion in that day when He said: Be! It is.** [Sura No. – 5\6 – An-aam, Ayat No. –71 & 73]

Ayat\Verses – To **Allah** belong the **Mystery** of the **heaven** [West Horizon] and the **world** [East Horizon]. And the Decision of the Hour [of Judgment] is as the twinkling of an eye, or even quicker: for Allah has power over all things. **It is He Who brought you forth from the wombs of your mothers when you knew nothing; and He gave you hearing and sight and intelligence and affections that you may give thanks [to Allah]. Have they not seen the birds obedient in mid-air? None holds them save Allah. Lo! Herein, verily are signs for those who believe.** [Sura No. – 15\16 – Nahl, Ayat\Verses – 77 to 79]

He has created the **heaven [West Horizon] and the world** [East Horizon] for **just ends (with truth)**. Far is He above having the **partners** they ascribe to Him! [Sura No. – 15\16 – Nahl, Ayat\Verses – 3]
[Sura No. 57\58 – Mujaadalah, Ayat No. - 6 & 7]
[Sura No. – 6\7 - As-haabul- a' raaf, Ayat No. – 54]
[Sura No. – 31\32 – Sajdah, Ayat\Verses – 4]
[Sura No. – 9\10 – Yuunus, Ayat No. – 3]
[Sura No. – 10\11 – Kitaabun-uh-kimat (Prev. – Huud), Ayat No. – 5 to 8]
Note: - Ayat\Verses of Heaven and World represent Sirius Binary System or Trinity.

The **Universe has been manifested as an upright rectangle in resemblance with 'End of Proof' of Real Science and in correspondence with the appointed Kaba.** Manifested Universe has two Horizons. Two Horizons of the Manifested Universe are West Horizon and East Horizon. Manifested Universe has two Hemispheres. Two Hemispheres of the Manifested Universe are Northern Hemisphere [Revealed Left] and Southern Hemisphere [Revealed Right]. **West Horizon** represents upward direction, Strong Force, Star Jupiter, Galaxy of 124000 Guiding Stars (approximately), Heaven, Proton, White Square, obversion of converted snake of Muusa's Staff into white, and the like. **East Horizon** represents downward direction, Gravitational Force, Star Saturn, Electron, Black Square, and the like. **Northern Hemisphere** [Haiyalas-Swalaah] represents Revealed Left or Magnetic Force. **Southern Hemisphere** [Haiyalal-Falaah] represents Revealed Right or Weak Force.

APPOINTED BLACK & WHITE IMAM
OF THE CITY CALLED KABA
[Clear Proof & Manifested Sign]

Ayat\Verses – The **first Sanctuary appointed for men** was that at **Bakka**, a blessed place, guidance to the people. In it are Manifested Signs [in resemblance with Revelation] where in Ibrahim stood up to pray; whoever enters it attains security; Pilgrimage to the House is a duty unto Allah for those who can find a way thither [towards that place]. But if any deny faith, Allah stands not in need of any of His creatures. [Sura No. 2\3 – Imran, Ayat No. – 96 to 97]

Ayat\Verses - "Wa izibtalaaa Ibraahima-Rabuuhuu bi-Kalimaatin- fa-atammahunna, qaala Innii-jaa-ilukalin-naase **Imaamaa**. Qaala wa min-zurriyyatii; Qaala laa yanaalu ahdiz-zaa-limin." [Trans.] And remember that **Ibrahim** was tried by Allah with certain **commands**, which He fulfilled. He said: **I have made an Imam [Kaba or an appointed thing] for mankind**; (he appealed): and **also for my offspring [next generation]**. He said: **But My Promise is not within the reach of evil-doers. [Sura No. – 1\2 – Baqarah, Ayat No. – 124]**

Ayat\Verses – **Allah has appointed (manifested) the Kaba, the Sacred House, as a standard [manifested sign for verification & justification] for mankind, and the Sacred Months [twelve matter particles or springs]**, the **offerings [99 elements]**, and the **garland** (25 names of the signs of relation / messengers): that you may know that Allah has knowledge of what is in the heavens and in the world and that Allah is well acquainted with all things. [Sura No. – 4\5 – Maaaidah, Ayat\Verses – 97]

Ayat\Verses – (And remember) the angels said: **O Maryam! Lo! Allah has given you good tidings of a Word from Him**. His name will be **Isa**, Isabnu-**Maryam**, held in honour in this world [present life] and the Hereafter and of [the company of] those nearest to Allah. He will **speak to the people in childhood** and in **maturity**. And he is [of the company] of the **righteous. She said: My Rab! How shall I have a son when no man (_mortal_) has touched me?** He said: Even so, Allah creates what He wills. When He decrees a plan, He but says to it, **'Be,' and it is!** And Allah will teach him **the Kitaaba wal-Hikmata, wat-Tawraata and wal-Injiil.** And [appoint him] **a messenger to the Banii-Israa-'il**, [saying]: Lo! I have come to you, **with a Sign from your Rab**, in that **I make for you out of clay, the likeness of a bird**, and **I breathe into it**, and **it is a bird by Allah's permission. And I heal those born blind, and the lepers**, and **I raise the dead**, by Allah's permission; and I announce unto you what you eat, and what you store up in your houses. Surely **therein is a Sign for you if you do believe.** And [I have come to you], to confirm that which was before me in the **Tawraat.** And to make **lawful** some of which was

forbidden unto you. I have come to you **with a Sign from your Rab**. So keep duty to Allah and obey me. It is Allah Who is my Rab and your Rab; so worship Him. This is a **Way that is right**. [Sura No. 2\3 – Imran, Ayat No. –45 to 51]

Ayat\Verses – Of **those messengers** some of whom We have **caused** to **excel others**. To **one** of them Allah **spoke; others** He **raised** to **degrees** [of honour]. To **Isabna Maryam** We gave **Clear Proof** and **supported** him with the **holy spirit. If Allah had so willed, succeeding generations would not have fought among each other, after Clear Proofs had come to them. But they differed. Some believing and others rejecting. If Allah had so willed, they would not have fought each other. But Allah fulfils His plan. [Sura No. – 1\2 – Baqarah, Ayat No. - 253]**

Ayat\Verses – And We caused **Iisabni-Maryam** to follow in their footsteps **confirming that which was revealed before him in the Tawraat**. We bestowed on him the **Injiil [Pragmatically Correspondence Truth in resemblance with the Crucified Sign]** wherein are **guidance and light**, and **confirmation** of that which was before it in the **Tawraat**, guidance and an admonition to those who ward off evil. Let the People of the **Injiil [followers of Correspondence Truth] judge** by what Allah has revealed therein. If any do fail to judge in **correspondence** with what Allah has revealed, they are (no better than) those who **rebel**. [Sura No. – 4\5 – Maaaidah, Ayat\Verses – 46 & 47]

Ayat\Verses – And when **Isabnu-Maryam** is held up (quoted) as **an example**, behold, the folk laugh out (you people raise a clamour thereat) And say: **Are our gods better, or he?** They raise not the objection save **for argument**. Nay! But **they are a controversial (notorious) folk. He was no more than a slave on whom We bestowed favour** and We made him **an example to Baniii-Israiil**. [Sura No. 42\43 – Ummil-Kitaab \ (Zukhruf), Ayat No. – 57 to 59]

Ayat\Verses – **When Isa came with Clear Proofs,** he said: I have come to you with wisdom, and in order to make clear to you some of the [points] on which you dispute. Therefore keep your duty to Allah and obey me. Lo! Allah, He is my Rab and your Rab. So worship Him. This is a **Right Path (towards Arsh).** [Sura No. 42\43 – Ummil-Kitaab \ (Zukhruf), Ayat No. – 63 & 64]
   i) "The **similitude of Isa** before Allah **is as that of Adam."**
   ii) "He created him from dust, and then said to him: **Be, and he was.** This is the **Truth from Allah."**
   iii) "If **anyone dispute in this matter with you, now after [full] knowledge (Tautology) has come to you, say: Come! Let us gather together, our sons and your sons, our women and your women, ourselves and yourselves."**

iv) "Then **let us earnestly pray** humbly, and (solemnly) **invoke the curse of Allah on those who lie! This is the true account.**"

v) "But **if they turn back**, Allah has full knowledge of **those who do mischief.**"

Ayat\Verses – The **similitude of Isa** before Allah **is as that of Adam.** He created him from dust, and then said to him: **Be, and he was.** This is the **Truth from Allah. So be not of those who doubt. If anyone dispute in this matter with you, now after [full] knowledge (Tautology) has come to you, say: Come! Let us gather together, our sons and your sons, our women and your women, ourselves and yourselves.** Then **let us earnestly pray** humbly, and (solemnly) **invoke the curse of Allah on those who lie! This is the true account.** There is no ilaaha (unbreakable reality) except Allah. And lo! Allah is indeed the Exalted in Power, the Wise. But **if they turn back**, Allah has full knowledge of **those who do mischief.** [Ref. No. – 352 to 356]

Then let man consider his food, [and how We provide it]. For that We pour forth water in abundance and **We split the world in fragments [regions],** and produce therein corn, and grapes and green fodder, and Olives and Dates, and enclosed Gardens dense with lofty trees, and fruits and grasses, **Provision for you** and **your cattle.**[Sura No. – 79\80 - Abasa wa tawallaaa, Ayat No. – 24 to 32]

Say: Who gives you **sustenance,** from the **heavens** and the **world**? Say: It is Allah; and certain it is that **either we** or **you are on right guidance** or in **manifest error!** [Sura No. – 33\34 – Saba, Ayat No. - 24]

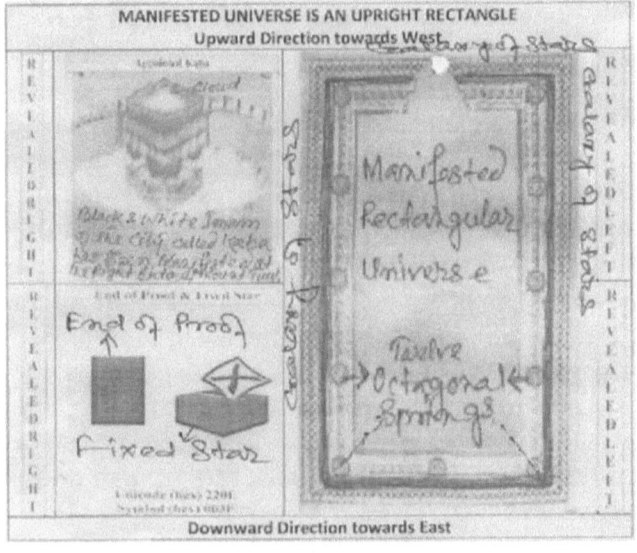

234

Star [Planet/Moon] Mercury represents Dot Operator or Point of Intersection of the equal & opposite ~sth. Mercury is Gravitationally Locked and has no reflection. Mercury appears/manifests once in every two Mercurian Years i.e. equal & opposite Lunar Year and Calendar Year. This appearance is called Lay-Latul-Qadr.

'IQRA'
SIGNIFICANCE OF H2O [WATER]

Ayat\Verses - Read! In the name of your Rab Who creates. Creates man, out of 'alaq' [clot / water as sign] [Sura No. 95\96 – Iqraa, Ayat No. – 1 & 2]

Ayat\Verses - Read! And your Rab is the Most Bounteous, Who teaches by the **pens;** teaches man **that [truth]** which he knew not. [Sura No. 95\96 – Iqraa, Ayat No. – 3 to 5]

Ayat\Verses - Nothing, but verily man is **rebellious** [disobedient] that he thinks himself as **self-sufficient** [independent]. Verily, to your Rab is the **return** [of all]. [Sura No. 95\96 – Iqraa, Ayat No. – 6 to 8]

Ayat\Verses - Have you seen **her** [one who is supplied with Sustenance] who discourages **a slave when he [user] prays?** [Sura No. 95\96 – Iqraa, Ayat No.-9 & 10]

Ayat\Verses - Have you seen if he **relies [firmy believes]** on the guidance [of Allah] and is forward? Have you seen if he denies and turns away? Is he then unaware that Allah sees? Let him beware! If he **desists** [cease] **not,** We will **drag him by the forelock** [lock of hair just above the forehead], **the lying, sinful forelock!** Then, let him call upon **his beneficent** [projected falsehood]. We will call on the angels of punishment [to deal with him]! Nothing [Nay]! You **obey not him,** but **prostrate yourself,** and **draw near** [unto **Allah**]! [Sura No. 95\96 - Iqraa, Ayat No. -11 & 19]

Ayat\Verses - Allah causes the alternation of the day and the night. Verily herein (in these things) are **instructive examples** for those who have vision! **And Allah has created every animal from water.** Of them there are some that creep on their bellies; some that walk on two legs; and some that walk on four. Allah creates what He wills for verily Allah has power over all things. [Sura No. – 23\24 – Nuur, Ayat No. – 44 & 45]

A Blessed Night
Ayat – Ha-Mim. [1]

Ayat\Verses – By the Kitaab that **makes things clear (plain),** lo! We revealed it on a **Blessed Night.** Lo! We are ever warning, **whereupon every wise**

command it made clear; as a command from Our Presence. Lo! We are ever sending, a Mercy from your Rab for He hears and knows [all things], the Rab of the heavens and the world and all that is between them, if you [but] **have an assured faith**. [Sura No. – 43\44 - Bi-Dukhaanim-Mubiin, Ayat No. – 1 to 7]

LAY-LATUL-QADR OR THE NIGHT OF POWER
[Intrinsically Luminous Moon or Diamond Operator]

Ayat\Verses - "Innaaa anzalnaahu fii **lay-latul-Qadr**." [Trans.] We have indeed **revealed it** [QURAN] **through the Night of Power.** [Sura No. – 96\97 – Lay-latul-Qadr, Ayar No. – 1]

Ayat\Verses - "Wa maaa adraaka maa laylatul-Qadr". [Trans.] Ah what will convey to you what the **night of power** is! [Sura No. – 96\97 – Lay-latul-Qadr, Ayar No. – 2]

Ayat\Verses - "Laylatul-Qadri khayrum-min alfi shahr." [Trans.] The **Night of Power** is better than a **thousand months [springs / planets / moons].** [Sura No. – 96\97 – Lay-latul-Qadr, Ayar No. – 3]

Ayat\Verses - "Tanazzalul-malaaa-ikatu war-ruhuu fiha bi-izni-Rabbihim-min-kulli amr." [Trans.] The angels and the sprit descend therein [Night of Power], by the **permission** of their Rab, with every **errand** [beam / ray / task] [Sura No. – 96\97 – Lay-latul-Qadr, Ayar No. – 4]

Ayat\Verses - "Salaamun hiya hattaa mat-la-il-fajr." [Trans.] **Salam** [Peace] until the **rising of the dawn** [Sura No. – 96\97 – Lay-latul-Qadr, Ayar No. – 5]

Ayat\Verses - And We have indeed made the **Qur'an easy to understand** and **remember**: then is there any that will receive admonition? [Sura No. – 53\54 – Qamar, Ayat\Verses – 17, 22, 32, 40]

Ayat\Verses - Sad - By the **Qur'an, Full of Warning**: [This is the Truth]. [Sura No. – 37\38 – Swad, Fundamental Ayat No. – 1]

Ayat\Verses - But the **disbelievers** [are steeped] in **self-glory** and **Separatism.** How many generations before them did We destroy? In the end they cried [for mercy] when there was no longer time for being saved! And they wonder that a **warner** has come to them from among themselves! And the disbelievers say: This is a sorcerer (wizard / magician) telling lies! He **has brought all diversities into one Wahid. Truly this is a wonderful thing!** [Sura No. – 37\38 – Swad, Ayat No. – 2 to 5]

"We have detailed the signs for those who receive admonition."

Ayat\Verses – When there comes to them a **Clear Proof (token) from Allah**, they say: We shall not believe until we receive one [exactly] like those received by Allah's Prophets. **Allah knows the best where [and how] to carry out His mission. Soon will the wicked be overtaken by humiliation before Allah, and a severe punishment, for all their plots?** And whomsoever it is Allah's will to guide, He expands His bosom unto the surrender, and whomsoever it is His will to send astray, He makes His bosom close and narrow as if He were engage in sheer ascent. Thus Allah lays ignominy upon **those who refuse to believe. This is the way of your Rab, leading Upright West: We have detailed the signs for those who receive admonition.** [Sura No. – 5\6 – An-aam, Ayat No. –125 to 127]

    i)   Trinity of West Horizon – Cloud & Sky – East Horizon
    ii)   Trinity of Rabbin-Nass [West] – Ilahin-Naas [equal & opposite sth] – Malikin--Naas [East]
    iii) Trinity or ~sth between Sirius A and Sirius B of Sirius Binary System
    iv) Trinity or ~sth between Star Jupiter and Star Saturn
    v)   Trinity of Proton – Neutrino & Neutron – Electron of an Atom
    vi) Trinity of Subject Term – Copula – Predicate Term of a Logical Proposition
    vii) Trinity of Lam [Proton] – Alif [sth] – Mim [Electron] and the manifestation of this Trinity is called Zalikal-Kitaab

TWO FOLD MERCY AND AN APPOINTED LIGHT

Ayat\Verses – "Yaaa – ayyuhallazna aamanuttaqul-laaha wa aaminuu bi-Rasuulihii yu-tikum kif-layni mir-Rahmatihii wa yaj-al-lakum Nuuran-tamshuuna bihi wa yagfir lakum; wal-laahu Gafuurur-Rahim." [Trans.] O you **who** believe! **Be mindful of your duty to Allah**, and **put faith in His Messenger.** He will give you **two fold** of His Mercy and will appoint for you a **Light wherein you will walk**, and He will forgive you; for Allah is Ever Forgiving, Most Merciful. That the **People of the Kitaab** may know that they have no control whatever over the bounty of Allah; but that **the bounty is in His Hand to bestow it on whomsoever He wills**. And Allah is of Infinite Bounty. [Sura No. – 56\57 - Anzalnal-Hadiid, Ayat No. – 28 & 29]

The Universe [Alamin] has been revealed as a ~sth between West Horizon and East Horizon. This Equal & Opposite Revelation is called **Tawraat or Equal & Opposite Law.** West Horizon is the main sequence Sirius A of Sirius Binary System and East Horizon is the white dwarf companion Sirius B of Sirius Binary System. Star **[Planet / Moon] Jupiter** represents **West Horizon.** Further, Sirius A [West Horizon / Jupiter] represents Galaxy of Stars, Heaven, Strong Force, Upward Direction towards West, and the like. Star **[Planet / Moon] Saturn** represents **East Horizon**. Sirius B [East Horizon / Saturn] represents downward direction towards East, Gravitational Force, Flameless Fire, and the like. Cloud & Sky represent ~sth [Velocity] between West and East, between Light and Darkness, between Water and Fire, and the like. The bifurcation of the **Intrinsically Luminous Star or projected Intrinsically Luminous Moon** [Shi-raa / Diamond Operator] with the witness finger by the Messenger of Allah is the similitude as well as valid reference of Trinity or **Sirius Binary System**. This bifurcation of the Intrinsically Luminous Star is the similitude of the origin of Time, Equal & Opposite Law [Tawraat or Revelation] i.e. Alif-Lam-Mim. This valid reference of bifurcation has been established as Big Bang Theory.

SIRIUS BINARY SYSTEM
Sirius A and Sirius B
[Source – Internet]

Sirius (/'sɪriəs/) is the brightest star (in fact, a star system) in the Earth's night sky. With a visual apparent magnitude of −1.46, it is almost twice as bright as Canopus, the next brightest star. The name "Sirius" is derived from the Ancient Greek Σείριος (*Seirios*), meaning "glowing" or "scorcher". The system has the Bayer designation Alpha Canis Majoris (α CMa). What the naked eye perceives as a single star is actually a binary star system, consisting of a white main-sequence star of spectral type A1V, termed **Sirius A**, and a faint white dwarf companion of spectral type DA2, called **Sirius B**. The distance separating Sirius A from its companion varies between 8.2 and 31.5 AU.[24]

Sirius appears bright because of both its intrinsic luminosity and its proximity to Earth. At a distance of 2.6 parsecs (8.6 ly), as determined by the Hipparcos astrometry satellite,[2][25][26] the Sirius system is one of Earth's near neighbors. Sirius is gradually moving closer to the Solar System, so it will slightly increase in brightness over the next 60,000 years. After that time its distance will begin to increase, but it will continue to be the brightest star in the Earth's sky for the next 210,000 years.[27]

Sirius A is about twice as massive as the Sun ($M_\odot$) and has an absolute visual magnitude of 1.42. It is 25 times more luminous than the Sun[12] but has a significantly lower luminosity than other bright stars such as Canopus or Rigel. The system is between 200 and 300 million years old.[12] It was originally composed of two bright bluish stars. The more massive of these, Sirius B, consumed its resources and became a red giant before shedding its outer layers and collapsing into its current state as a white dwarf around 120 million years ago.[12]

Sirius is also known colloquially as the "**Dog Star**", reflecting its prominence in its constellation, Canis Major (Greater Dog).[18] The heliacal rising of Sirius marked the flooding of the Nile in Ancient Egypt and the "dog days" of summer for the ancient Greeks, while to the Polynesians in the Southern Hemisphere the star marked winter and was an important reference for their navigation around the Pacific Ocean. [Science]

Ayat\Verses - There was indeed a Sign for Saba, in their dwelling place. Two Gardens to the right and to the left. **Eat** of the **Sustenance** [provided] by your Rab, and be grateful to Him: A territory fair and happy, and a Rabbun Gafuur. [Sura No. – 33\34 – Saba, Ayat No. - 15]

A\V No.	Sura No. – 1\2 Sura – Baqarah
1	**Alif Lam Mim.**
2	**This is the Zaalikal-Kitab.** In it is **guidance sure, without doubt**, to those who fear Allah (who ward off evil).
3	Who **believe in the Unseen**, and **steadfast in prayer**, and **spend out of what We have bestowed upon them;**
4	And who believe in that which is **revealed** unto you, and that which was **revealed** before your time, and [in their hearts] has the **assurance** of the Hereafter; [equal & opposite magnetism]
5	They depend on guidance from their Rab, and it is **they who will succeed.**
6	As to **those who reject Faith**, it is the **same** to them whether you warn them or do not warn them; **they will not believe.**
7	**Allah has set a seal on their hearts** and on their **hearing**, and on their **eyes there is a veil (covering)**; great is the **penalty** they [invite].
8	Of the people there are some who say: We believe in Allah and the Last Day; but they do not [really] believe.
9	Fain would they **deceive Allah** and **those who believe,** but they only **deceive themselves, and realise [it] not!**
10	In their hearts is a **disease**; and Allah has increased their disease: And grievous is the penalty they [invite], because **they are false** [to themselves].
11	When it is said to them: **Make not mischief in the world, they say: We are only ones that put thing right.**
12	**Of a surety, they are the ones who make mischief**, but they realise [it] not
13	When it is said to them: **Believe as the others believe**; they say: **Shall we believe as the fools believe?** Nay [nothing], **of a surety they are the fools,** but they do not know.
14	When they meet those who believe, they say: We believe; but **when they are alone with their evil ones,** they say: We are really with you: We [were] only joking.
15	Allah will throw back their mockery on them, and give them **rope** in their trespasses; so they will wander like blind ones [To and fro]

16	These are they who have exchanged Guidance for error: But their traffic is profitless, and they have lost true direction.
17	Their **similitude** is that of a man who kindled a fire; **when it lighted all around him, Allah took away their light and left them in utter darkness.** So they could not see.
18	**Deaf, dumb, and blind, they will not return** [to the path].
19	Or [another **similitude**] is that of a rain-laden cloud from the sky. In it are **zones of darkness,** and **thunder** and **lightning.** They press their fingers in their ears by reason of the stunning thunder-clap, for fear of death. But Allah **encompasses the disbelievers.**
20	The **lightning all but snatches away their sight;** every time **the light [helps]** them, **they walk therein (light),** and **when the darkness grows on them, they stand still.** And if Allah willed, He could **take away** their faculty of **hearing and seeing;** for Allah has power over all things.
21	O mankind! Serve Allah, who has created you and those who came before you, that you may have the **chance** to learn **righteousness** (ward off evil).
22	Who has made the **world a resting place for you** and the **sky a canopy;** and **sent down rain from the heavens;** and **brought forth therewith fruits for your sustenance;** then do **not set up rivals unto Allah** when you know [better].
23	And **if you are in doubt** as to what We have **revealed** from **time to time** to Our bondmen (slaves), then **produce a Sura [revealed something] like thereunto;** and **call your witnesses or helpers** [If there are any] besides Allah, if **your [doubts] are true.**
24	But if **you do it not,** and of a **surety you cannot,** then fear the **fire** whose **fuel** is **men and stones,** which is prepared for **those who reject Faith.**
25	And **give good tidings** to **those who believe** and work **righteousness,** that **their portion is Gardens,** beneath which rivers flow. Every time they are fed with fruits there from. They say: **This is what was given us afore time; and it is given to them in resemblance.** Thereof, there are **pure companions,** and they **abide** therein [**for ever**].
26	Allah disdains not to use the **similitude of things,** lowest as well as highest. Those who believe know that it is truth from their Rab; but **those who reject faith** say: **What means Allah by this similitude? By it He causes many to go astray, and many He leads into the right path;** but He causes not to stray, except **those who forsake** [the path],
27	Those **who break Allah's Covenant (promise) after it is ratified,** and who **sunder (Literary separate) what Allah has ordered to be joined,** and do **mischief in the world: These cause loss** [only] to themselves.
28	**How can you reject the faith in Allah seeing that you were without life, and He gave you life; then He will cause you to die, and will again bring you to life; and again to Him you will return?**

29	It is He Who has created for you all things that are in the world. Then **He turned to the heaven**, and fashioned it as **seven firmaments**; and of all things He has perfect knowledge.
30	And when your Rab said to the angels: I will create a **viceroy in the world**. They said: Will You place therein one who will **make mischief therein** and shed blood while we do hymn Your praises and glorify Your holy [name]? He said: Surely I know what you know not.
31	And He **taught Adam** the **names of all things**. Then He placed them before the angels, and said: **Tell me the names of these if you are right.**
32	They said: Be glorified! **We have no knowledge**, save **what You have taught** us. In truth it is You Who are perfect in knowledge and wisdom.
33	He said: O Adam! Tell them their names. When he had told them, Allah said: Did I not tell you that **I know the secrets of heaven and the world**, and I know **what you reveal and what you conceal**?
34	And when We said to the angels: Bow down to Adam and they bowed down. Not so **Ibliis**; he refused and was **haughty (proud)**. **He was of those who reject Faith.**
35	We said: O Adam! Dwell **you** and your **wife** in the **Garden**; and eat of the **bountiful** things therein as [where and when] you will; but **approach not this tree**, lest you become **wrong-doers**.
36	But **the Shaytan caused them to deflect there from [the garden], and get them out of the state [of felicity] in which they had been. We said: Fall down, one of you a foe to the other!** There shall be for you on earth a habitation and provision for a time.
37	Then **Adam** received from Allah **words of inspiration** and Allah **turned towards him**; for He is Ever Relenting, Most Merciful (Huwat-Tawwaa-bur-Rahim)
38	We said: Get you down all from here; and verily **there will come to you Guidance from Me**. Whosoever **follow** My guidance, there shall no fear come upon them neither shall they grieve.
39	But **those who reject Faith and belie Our Signs [of manifested magnetism in resemblance with revelation]**, they will be **companions of the fire**. They will abide therein.
40	O Banii-Isra-iil! Remember My [special] favour which I bestowed upon you, and fulfil your covenant (promise) with Me as I fulfil My Covenant with you, and fear none but Me.
41	And **believe in what I reveal, confirming the revelation which is with you [manifested magnetism]**, and be **not the first to reject Faith therein, nor sell My Revelations for a small price**; and fear Me, and Me alone.
42	**And cover not Truth with falsehood, nor conceal the Truth when you know [what it is].**

43	And be steadfast in prayer; pay the poor-due; and bow down your heads with those who bow down [in worship].
44	Do you enjoin right conduct on the people, and **forget** [To practise it] yourselves, and yet **you study the Kitaab? Have you then no sense?**
45	Seek help with patience and prayer. It is indeed hard, except to those who are humble minded.
46	Who bear in mind the **certainty** that they are to meet their Rab, and that they are to return to Him.
47	O **Banii-Isra-iil**! Remember My favour which I bestowed upon you, and that **I preferred you to all other.**
48	Then **guard yourselves against a Day** when **one soul shall not avail another nor** shall **intercession** be accepted from it, **nor** shall **compensation** be taken from it, **nor** shall anyone be **helped.**
49	And remember, **We delivered you from the people of Firawn.** They set you hard tasks and punishments, **slaughtered your sons and let your women-folk** live; therein was a **tremendous trial** from your Rab.
50	**And remember, We divided the sea for you and saved you and drowned Firawn's people within your very sight. (IFTA).** **And when We brought you through the sea and rescued you, and drowned the folk of Firawn's in your very sight. (IBPLtd.)**
51	And **remember We appointed forty nights for Muusa, and in his absence you took the calf [for worship] and you did grievous wrong.**
52	Even then We did **forgive** you; there was a **chance** for you to be **grateful.**
53	And remember We **gave Muusa the Kitaab and the Criterions [of judging right or wrong].** There was a **chance for you to be guided aright.**
54	And when **Muusa said to his people:** O my people! **You have indeed wronged yourselves by your worship of the calf. So turn [in repentance] to your Creator, and kill (the calf) yourselves that will be the best for with your Creator** and He will **relent** toward you, for He is the **Relenting, the Merciful.**
55	And remember you said: O **Muusa**! We shall never believe in you **until we see Allah manifestly,** but you were **confused with thunder** and lightning seized you.
56	Then We **received** you after your **extinction,** that **you might give thanks.**
57	And We gave you the **shade of clouds** and sent down to you **manna and quails,** saying: **Eat** of the good things We have provided for you. We wronged them not, but **they did wrong themselves.**
58	And when We said: **Enter this township** and eat of the plenty therein as you wish; but **enter the gate prostrate,** and say: **Repentance.** We will **forgive** you **your sins** and will **increase** (reward) for the **right-doers.**
59	But **those who did wrong changed the word which had been told them for another saying,** and We sent **on the transgressors a plague** from heaven, **for evil doing.**

60	And when **Muusa** prayed for water for his people; We said: **Strike** the **rock** with your **staff. Then gushed forth there from twelve Springs (Planets). Each group knew its own place for water.** So **eat and drink of the sustenance provided by Allah, and do neither evil nor mischief in the world.**
61	And when you said: O **Muusa!** We cannot endure **one kind of food** [always]; so beg your Rab for us that He brings forth for us of which that the **world grows**, of its herbs, and cucumbers, its garlic, lentils, and onions. He said: **Will you exchange the higher for that which is lower?** Go you down to any **town**, and you shall find what you want! **They were covered with humiliation and misery; they drew on themselves the wrath of Allah. This because they went on rejecting the Signs of Allah and slaying His Messengers wrongfully. This because they rebelled and went on transgressing.**
62	Those who **believe** [in that which revealed unto you], and **those who are H a a d u u (Yahuudis), Nasaaras and Sabii-iians,** any who **believe in Allah** and the **Last Day**, and **work righteousness,** shall have their **reward** with their **Rab.** On them shall be **no fear, nor** shall they **grieve.**
63	And remember We took your **covenant** (promise) and **caused the Mount Tuur to tower above you** [Saying]: **Hold firmly to what We have given you** and remember that **which is there** in **that** you may **ward off (evil).**
64	Then even after that **you turned away,** and if it had **not** been for **the grace of Allah** and **His Mercy,** you had been **among the losers.**
65	And you know of those of you who **transgressed in the matter of the Sabti.** We said to them: **Be you apes, despised and rejected.**
66	So We **made it an example** to their **own** and to the **succeeding generations,** and a **lesson** to those **who fear Allah.**
67	And when **Muusa** said to his people: **Allah commands that you sacrifice a heifer.** They said: Does you make **game** of us? He said: **Allah save me from being an ignorant [fool]!**
68	They said: Pray on our behalf to your Rab to **make clear to us** what [heifer] she is! He said: He says: **The heifer should be neither too old nor too young, but of middling age.** Now do what you are commanded!
69	They said: Pray on our behalf onto Allah to make plain to us **Her colour.** He said: He says: Verily **she is a yellow heifer; bright is her colour, gladdening beholders!** [Sura No. – 1\2 – Baqarah, Ayat\Verses – 69]
70	They said: Pray on our behalf unto your Rab to make clear to us **what** (heifer) she is! To us are **all heifers alike.** We wish indeed for **guidance, if Allah wills.**
71	He said: He says: Verily she is a heifer unyoked, not trained to till the soil or water the fields; sound (Vibration causing sensation, healthy, quality of the bottom of the sea) and without blemish. They said now you had brought the truth. So, they sacrificed her, though almost they did not.

72	Remember you slew a man and fell into a dispute among yourselves as to the crime. But Allah was to bring forth what you did hide.
73	So, we said: Cut the body with some of it. Thus Allah brings the dead to life and shows you His Signs so that you may understand.
74	**Thenceforth your hearts were hardened**: They became like a rock and even worse in hardness. For among rocks there are some from which rivers gush forth; others there are which when split asunder send forth water; and others which sink for fear of Allah. And Allah is not unmindful of what you do.
75	Have you any hope that they will be true to you when a party of them used to listen to the word of Allah, and then used to change it, after they had understood it, knowingly?
76	And when they fall in with the men of Faith, they say: We believe. But when they meet each other in private, they say: Shall you tell them what Allah has revealed to you, that they may engage you in **argument about it before your Rab**? Have you then no sense?
77	Know they not that Allah knows what they **conceal and what they reveal**?
78	And there are among them **illiterates**, who know not the **Ya-lamuunal-Kitaab**, but [see therein their own] desires, and they do nothing but conceptualise the truth.
79	Then woe to those who write the **Yaktubuunal-Kitaab** with their own hands, and then say: This is from Allah, to traffic with it for miserable price! Woe to them for what their hands do write, and for the gain they make thereby.
80	And they say: The fire shall not touch us but for a few numbered days: Say: Have you taken a promise from Allah? Truly Allah will never break His promise. Or is it that you say of Allah what you do not know?
81	Nay, **those who seek gain in evil**, and are **girt round by their sins**, they are **companions of the fire**. Therein shall they **abide [For ever]**.
82	But those **who have faith** and **work righteousness (tabligh)**, they are **companions of the Garden.** Therein they shall abide [For ever].
83	And (remember) when **We took a promise from the Banii-Israa-iil** (saying): Serve none but Allah; treat with kindness your parents and family members, and orphans and those in need; speak fair to the people; be steadfast in prayer; and pay the poor-due. Then **you turned back, except a few among you, and you slip back [even now]**. (An Individual conception – I have conceptualised that the address of Allah in resemblance with Ayats 84 & 85 represents not the Christians but the believers of shared tautologies as Ayat 84 represents both Pak Kalam (1st Pillar of Islam) and Pak Witness (2nd Pillar of Islam)
84	And remember We took your promise [saying]: Shed no blood amongst you, **nor turn out your own people from your homes,** and this you **solemnly approved**, and to this **you were witnesses** (there to).
85	After this it is you, **the same people, who slay among yourselves**, and **drive out a party of you from their homes; supporting one another against them by sin and ill will**; and if they come to you as **captives**, you **ransom** them, where as **their expulsion was itself unlawful for you. This is only because you believe a part of the Kitaab and you reject the rest?** And what is the reward for those among you who behave like this but **disgrace** in this life and on the **Day of Judgment they will be consigned to the most grievous penalty.** For Allah is not unmindful of what you do.

86	These are the people who buy the life of this world [present life] at the price of the Hereafter. Their penalty shall not be lightened nor shall they be helped.
87	We gave **Muusa** the **Kitaab** and followed him up with a **succession of messengers**; We gave **Isabnu-Maryam Clear Proofs** and strengthened him with the holy spirit. **Is it that whenever there comes to you a Prophet (from Allah) with what you yourselves desire not, you grow arrogant and some you disbelieve, and some you slay?**
88	They say: Our hearts are hardened. Nay, but Allah has **cursed them** for their unbelief. **Little is that which they believe.**
89	"Wa lammaa jaaa-ahum Kitaabum-min- i n d i l l a a h i musaddiqul-limaa ma-ahum wa kaanuu min-qabluyastafti-huunna alal-laziina kafaruu-falammaa jaaa-ahum-maa arafuu fafaruu bihii fala-natullaahi alal-kaafiriin." And when there comes to them a Kitaab from Allah [**Kitaabum-min- i n d i l l a a h i**], **confirming what is with them though before that they were asking for a signal triumph over those who disbelieved.** And when there comes to them that **which they know** (to be the truth), **they refuse to believe** in it but the **curse** of Allah is on those **who disbelieve**.
90	**Evil** is that for which **they sell their souls** that they **deny** [the revelation] which Allah has sent down, in **cheeky jealousy** that Allah of His Grace should send it to any of His servants He pleases. Thus they have drawn on themselves **wrath upon wrath**. And **humiliating** is **the punishment of those who reject Faith.**
91	When it is said to them, believe in **what Allah has sent down. They say: We believe in what was sent down to us.** Yet they reject all besides, **even if it be truth confirming what is with them.** Say: **Why then have you slain the prophets of Allah in times gone by, if you did indeed believe?**
92	There came to you **Muusa** with **Clear Proofs**; yet while he was away, you chose the **calf [projected magnetism), and you were wrong-doers.**
93	And when We took your **promise** and **caused the Tuur to tower above you** (saying): **Hold firmly to what We have given you,** and hear (our words,). They said: **We hear, and we disobey. And (worship of) the calf was made to sink into their hearts** because of their **rejection** (of the promise). Say (unto them): **Evil is that** which your **belief enjoins** on you, if you are believers.
94	Say: If the **last Home**, with Allah, be for you specially, and not for anyone else, then **seek you for death**, if you are truthful.
95	But they will **never seek for death, on account of the sins** which their hands have sent on before them. And Allah is Aware of the **wrong- doers.**
96	And you will indeed find them, of all people, most **greedy of life, even more than the idolaters.** Each one of them wishes He could be given **a life of a thousand years.** But the grant of such life **will not save** him from [due] punishment. For Allah sees well all that they do.
97	Say: **Whoever is an enemy to Jibraiil?** For he brings down the [revelation] to your heart by Allah's will, a confirmation of what went before, and guidance and glad tidings for those who believe.
98	**Whoever is an enemy to Allah and His angels and Messengers, to Jibraiil and Michael? Lo! Allah is an enemy to those who reject Faith.**

99	We have sent down to you **Manifested Signs (Clear Tokens), and none reject them** except **those who are wicked**.
100	**Is it not [the case] that** every time they make a promise, **a party among them throw out it aside**. Nay, **most of them are faithless**.
101	And when there came to them a Messenger from Allah, **confirming** what was with them, a party of the people of the **Uutul-Kitab** threw away the **Kitab of Allah [Kitaaballahi]** behind their backs, as if [it had been something] they did not know!
102	And follow that **which the devils falsely related against** the kingdom of **Sulaymaan. Sulaymaan disbelieved not,** but the **devils disbelieved. Teaching mankind magic and that which was revealed in the two angles in Baabile, Haaruut and Maaruut.** Nor did they teach it to anyone till they had said: We are only a temptation, therefore, **disbelieve not (in the guidance of Allah).** And from these **two**, people learn that by which **they cause division between man and wife**, but **they injure thereby no-one save by Allah's permission.** And they learn that **which harms them** and **profits them not.** And surely they do know that he who traffics therein will have **no portion in the hereafter**; and surely **evil is the price** for which **they sell their souls**, if they but know.
103	**If they were kept their Faith and guarded themselves from evil, far better had been the reward from Allah, if they only knew!**
104	O you who believe! **Say not** [unto the Prophet] Listen to us. But **say: Look upon us;** and be you **listeners**. To those without Faith is a **grievous punishment**.
105	Neither those who disbelieve among the followers of the **Prophet of the Kitab** nor the **idolaters** love that there should be sent down unto you any good thing from Allah. But **Allah will choose for His special Mercy whom He will as Allah is the Owner of all attributes in their superlative degree.**
106	**None of Our revelations do We abrogate or cause to be forgotten, but We substitute something better or similar. You do not know that Allah has power over all things.**
107	You do not know **that to Allah belongs the Sovereignty of the heavens and the world, and besides Him you have neither any helper nor any leader.**
108	**Would you question your Prophet as Muusa was questioned afore time? He who chooses disbelief instead of faith,** verily **he has gone away from a plain road.**
109	**Quite a number of the People of the Kitab wish they could turn you [people] back to unfaithfulness after you have believed, from selfish greed**, after the **Truth** has become **Manifest** unto them. But forgive and overlook, **till Allah accomplish His purpose.** For Allah has power over all things.
110	And be steadfast in prayer and pay the poor-due. And whatever good you send forth for your souls before you, you will find it with Allah. For Allah sees Well all that you do.
111	And they say: None shall enter Paradise unless he is a Huudan aw Nasaara. Those are their [vain / false] desires. Say: Produce your proof if you are truthful.
112	Nay, whoever submit their purposes to Allah while doing good, they will get their reward from Allah; and there shall be no fear come upon them neither shall they suffer.
113	The Yahuudis say: The Nasaaras follow nothing (true); and the Nasaaras say: The Yahuudis follow nothing (true); yet both are readers of the Kitaab. Even thus speak those who know not. Allah will judge between them on the Day of Resurrection concerning that wherein they differ.

114	And who is more unjust than he who forbids the approach to the safe heavens of Allah lest His name should be mentioned there in, and strives for their ruin? As for such, it was never meant that they should enter there except in fear. For them there is nothing but disgrace in this world, and in the world to come, an exceeding suffering.
115	To Allah belong the east and the West. Whithersoever you turn; there is the presence of Allah. For Allah is all-Pervading, all-Knowing.
116	They say: Allah has taken unto Himself a son. Be He glorified! Nay, to Him belongs all that is in the heavens and in the world. Everything renders worship to Him.
117	To Him is due the primal origin of the heavens and the world: When He decrees a matter, He says to it: Be, and it is.
118	And those who have no knowledge say: Why do not Allah speaks unto us or some sign come unto us? So said the people before them words of similar import. Their hearts are alike. We have indeed made clear the Signs to any people who hold firmly to Faith [in their hearts].
119	Verily We have sent you in truth as a bringer of good tidings and a warner. And you will not be asked about the owners of hell-fire.
120	Never will the Yahuudis or the Nasaaras be satisfied with you unless you follow their creed. Say: Lo! The Guidance of Allah (Himself), which is the [only] Guidance. And if you should follow their desires after the knowledge which has come you, then you would have from Allah neither protecting friend nor helper.
121	Those to whom We have given the Kitaab study it as it should be studied. They are the ones that believe therein. Those who reject faith therein, the loss are their own.
122	O Banii-Isra-iil! Remember the special favour which I bestowed upon you, and that I preferred you to all creatures.
123	Then guard yourselves against A Day when one soul shall not avail another, nor shall compensation be accepted from it, nor will intercession be use of it, nor will anyone be helped.
124	And remember that Ibrahim was tried by Allah with certain commands, which He fulfilled. He said: I had made an Imam for mankind. He appealed: and the Imam also from my offspring. He said: But My Promise is not within the reach of evil-doers.
125	And when We made the House (Kaba), a remedy for mankind and a sanctuary (saying): Take as your place of worship the place where Ibrahim stood (to pray). And when we imposed a duty upon Ibrahim and Ismail (saying), Purify My House for those who go around and those who meditate therein and prostrate themselves (in worship).
126	And when Ibrahim said: My Rab! Make this a region of security, and bestows upon its people fruits such of them as believe You and the last Day. He answered: As for him who rejects Faith, for a while I will grant them their pleasure, but will soon drive them to the torment of fire, an evil destination [indeed]!
127	And when Ibrahim and Ismail were raising the foundations of the House [With this prayer]: Our Rab! Accept [this service] from us. Lo! You only You are The Hearer, The Knower.
128	Our Rab! And make us submissive to Your, and of our seed a nation submissive to Your and show us our ways of worship; and turn unto us [in Mercy]. Lo! You, only You are Ever Returning, Most Merciful.

247

129	Our Rab! And rise up in their midst a Prophet from among them who shall rehearse Your Signs to them and instruct them in Kitab and wisdom, and sanctify them. For You are The Mighty, The Wise.
130	And who forsakes the religion of Ibrahim save him who befool himself. Verily We chose him in this world, and he will be in the Hereafter in the ranks of the Righteous.
131	When his Rab said unto him: Bow [thy will to Me], he said: I bow [my will] to the Rab and Cherisher of the Worlds.
132	And this was the legacy that Ibrahim left to his sons, and so did Yaquub. Oh my sons! Allah has chosen the Faith for you. Then die not save who have surrendered (unto Him).
133	Were you witness when death appeared before Yaquub? When, he said to his sons: What will you worship after me? They said: We shall worship your Rab and the Rab of your fathers, of Ibrahim, Ismaiil and Ishaaq, the one [True] Allah. To Him we have surrendered.
134	Those are a people who have passed away. They shall reap the fruit of what they did, and yours of what you do! And you will not be asked of what they used to do.
135	They say: Be Huudan or Nasaara, then you will be rightly guided. Say (unto them): Nay! But we follow the Religion of Ibrahim, the upright, and he was not of the idolaters.
136	Say: We believe in Allah, and the revelation given to us, and to Ibrahim, Ismaiil, Ishaaq, Yaquub, and the Tribes (Aasbati), and that given to Muusa and 'Isa, and that which the prophets received from their Rab. We make no difference between one and another of them. And we surrender to Allah as Muslims.
137	So if they believe as you believe, they are indeed on the right path; but if they turn back, it is they who are in schism; but Allah will suffice you (for defence) against them, and He is All-Hearing, All-Knowing.
138	[Our religion is] the Baptism of Allah: And who can baptize better than Allah? And it is He Whom we worship. (IFTA) Or (We take our) colour from Allah, and who is better than Allah at colouring? We are His worshipers. (IBPLtd.)
139	Say (unto the peoples of the Kitaab): Will you dispute with us concerning Allah when He is our Rab and your Rab? We are responsible for our doings and you for yours. We look to Him alone.
140	Or say you that Ibrahim, Ismaiil, Ishaqq, and Yaquub, and the Ashaata (Tribes) were Huudan or Nasaaras? Say: Do you know better than Allah? And who is more unjust than those who conceal the testimony (Sahaada / witness) which he has received from Allah? But Allah is not unaware of what you do.
141	Those are a people who have passed away. They will reap the fruits of what they did, and you of what you do! And you will not be asked of what they used to do.
142	The fools among the people will say: What has turned them from the Qibla which they formerly observed? Say: To Allah belong the East and the West. He guides whom He will to a Way that is straight.
143	Thus, We have made you an Ummat justly balanced (a middle people), that you may be witnesses against mankind and that the Rasuul (Prophet) is a witness over you.

	And We appointed the Qibla which you formerly observed only that We might know him who follows Rasuul (the Prophet) from those who would turn on their heels (run away). Indeed it is a hard test, except to those guided by Allah. But it is not Allah's purpose that your faith should be in vain. For Allah is to all people full of piety, Most Merciful towards mankind.
144	We have seen the turning of your face to heaven [West]. And now verily We shall make you turn towards a Qibla which is dear to you. So, turn your face towards the direction of Qibla (towards West). Wherever you are, turn your faces in that direction [towards West]. The people of the Kitaab know well that this is the Truth from their Rab. And Allah is not unaware of what they do.
145	And even if you bring to the people of the Kitab all the Signs [together], they would not follow your Qibla; nor they are going to follow their Qibla; nor are some of them followers of each other's Qibla. And if you follow their desires after the knowledge which has come to you, <u>then surely wert thou of evil-doers.</u> (Note - I have failed to conceptualise the truth of – 'then surely wert thou of evil-doers'. In this regard a request is placing before you that try to share your true concept if you are able to conceptualise the truth.)
146	Those unto whom We gave the Kitab, recognise (this revelation) as they recognise their sons. But Lo! A party of them knowingly conceal the truth.
147	The Truth is from Allah; so be not you of those who doubt.
148	To each is a goal to which Allah turns him. So vie with one another in good works. Where so ever you may be, Allah will bring you all together. For Allah is Able to do all things.
149	Whence so ever you come forth for prayer, turn your face in the direction of the Inviolable Place of Worship (Qibla); that is indeed the truth from your Rab. And Allah is not unaware of what you do.
150	So from whence so ever you come forth, turn your face in the direction of the Inviolable Place of Worship; and where so ever you are, turn your faces towards that inviolable place [towards West] so that there may be no ground of arguing against you, except those of them that are bent on wickedness. So fear them not, but fear Me, so that I may complete My grace upon you, and you May be guided.
151	Even as **We have sent unto you a messenger from among you,** who recites to you **Our Signs (Clear Proofs), that causes in sanctifying you,** and **shares with you the Kitaab and Wisdom,** and that **which you knew not.**
152	Therefore, **you remember Me; I will remember you.** Be **grateful to Me, and reject not Faith.**
153	O you who believe! Seek help with steadfastness and prayer, for **Allah is with the steadfast.**
154	**And call not of those who are slain in the way of Allah. They are dead. Nay, they are living, though you perceive [it] not.**
155	**Surely we shall test you** with **something of fear** and **hunger** and **some loss of wealth** and **lives** and **crops,** but **give good tidings (do tabligh) to those who patiently persevere.**
156	Who say, when afflicted with calamity: To Allah We belong, and to Him is our return.
157	They are those on whom blessings from their Rab, and Mercy, and they are the ones who are **rightly guided.**

158	**Lo! Safa and Marwa are among the indications of Allah.** It is therefore **no sin for him who is on pilgrimage to the House [Kaba]** or at other times should **compass them round.** And he who does well of his own accord, be sure that Allah is He Who is Responsive Aware.
159	**Those who conceal the Clear Proofs We have sent down, and the Guidance, after We have made it clear for the people in the Kitab, on them shall be Allah's curse, and the curse of those entitled to curse.**
160	**Except those who repent** and **amend** and **make manifest** [openly declare the Truth]: To whom I **relent**; for I am Ever-Returning, Most Merciful.
161	Lo! **Those who disbelieve, and die while they are disbelievers, on them is Allah's curse, and the curse of angels, and of all mankind.**
162	They will **abide therein. Their penalty will not be lightened, nor** will they be reprieved.
163	And your Rab is Waahid (One Rab Who has no attribute but Who is the sole Owner of all attributes). **There is no ilaaha (unbreakable reality) but He,** The Beneficent, The Merciful.
164	Lo! In the creation of the heavens and the world; in the alternation of the night and the day; in the sailing of the ships through the sea for the profit of mankind. In the rain which Allah Sends down from the skies, and the life which He gives therewith to the world that is dead; in the beasts of all kinds that He scatters through the world; in the change of the winds, and the clouds which they Trail like their slaves between the sky and the world; [Here] indeed are Clear Signs / Clear Proofs / Clear Tokens (of Allah's Sovereignty) for a people who are wise.
165	Yet of mankind are some who **take** to themselves **rivals** to Allah loving them with a love like of Allah. Those who believe are **checker in their love for Allah.** Oh that those who do evil had but known (on the Day) when they behold the doom that power belongs wholly to Allah, and that is severe in punishment!
166	On the Day when those who were followed (leaders) disown (clear themselves of) those who follow [them], and they (followers) behold the penalty, and all relations between them would be cut off.
167	And those who followed would say: **If only We had one more chance, We would clear ourselves of them, as they have cleared themselves of us.** Thus will Allah show them [the fruits of] their deeds as [nothing but] regrets. Nor will there be a way for them out of the Fire.
168	O you people! Eat of what is in the world, **lawful** and **wholesome**; and **do not follow the footsteps of the shaytan (devil),** as he is an **open enemy for you.**
169	**He (shaytan / devil) enjoins** upon you only what is evil and foul [projected mechanical globalisation and artificial magnetism], and that you should tell concerning Allah that of which you have no knowledge.
170	**When it is said to them: Follow which Allah has revealed [ewual & opposite magnetism]; they say: Nay! We shall follow the ways of our fathers [mechanical globalisation and artificial magnetism]. What! Even though their fathers were void of wisdom and guidance?**
171	The **similitude of those who reject Faith** is as if one were to shout like a goat-herd, to things that listen to nothing but calls and cries: Deaf, dumb, and blind, they are **void of wisdom.**

172	O you who believe! Eat of the good things that We have provided for you, and be **grateful** to Allah, if it is He whom you worship.
173	He has only forbidden you **dead meat**, and **blood**, and the **flesh of swine** and **that on which any other name has been invoked [like mechanical globalisation and artificial magnetism] instead of Allah**. But if one is **forced by necessity, neither craving nor transgressing**, then it is **no sin for him**. Lo! Allah is Ever Forgiving, Most Merciful.
174	**Lo! Those who conceal Allah's revelations in the Kitab, and purchase for them a miserable profit, they swallow into their bellies nothing else than fire;** Allah will not address them on the **Day of Resurrection**. Nor purify them. Grievous will be their penalty.
175	They are the ones who buy **Error in place of Guidance** and **Torment in place of Forgiveness. Ah! What boldness [they show] for the Fire!**
176	That is because Allah has **revealed** the Kitaab in truth **[Kitaaba bil-Haqq]** but those who seek **causes** of **dispute** in the **fil-Kitaab** are in a **schism Far** [from the purpose].
177	**It is not righteousness that you turn your faces Towards east or West; but it is righteousness to believe in Allah and the Last Day, and the Angels, and the Kitab, and the Messengers.** To spend of your substance, out of love for Him, for your kin, for orphans, for the needy, for the wayfarer, for those who ask, and for the ransom of slaves. To be steadfast in prayer, and to give the poor-due; **to fulfil the contracts which you have made**; and to be firm and patient, in pain [or suffering] and adversity, and throughout all periods of stress. Such are they who are **sincere.** Such are the Allah-fearing.
178	O you who believe! **The law of equality** is prescribed to you in cases of murder - **the freeman for the freeman, the slave for the slave, the woman for the woman**. But if any remission is made by the brother of the slain, then grants any reasonable demand and compensate him with handsome gratitude. This is a concession and a Mercy from Allah. **After this whoever exceeds the limits shall be in painful doom.**
179	**In the Law of Equality there is Life to you in retaliation. O men of understanding! That you may free yourselves from the path of shaytan (devil's path).**
180	It is **prescribed** for you, **when one of you approaches death**. If he leaves wealth that he bequeaths unto parents and next of kin, according to reasonable usage; this is a duty for **all those who ward off (evil).**
181	**If anyone changes the bequest after hearing it, the guilt shall be on those who make the change.** For Allah hears and knows [All things].
182	But if anyone fears partiality or wrong-doing on the part of the _testator_, and makes peace between [The parties concerned], there is no wrong in him: For Allah is Ever-Forgiving, Most Merciful.
183	O you who believe! **Fasting is prescribed to you as it was prescribed to those before you, that you may [learn] self-restraint.**
184	[Fast] for a **certain number of days**. But if any of you is **ill**, or on a **journey**, the prescribed number [Should be made up] from days later. For those who can do it [With hardship], is a ransom, the feeding of one that is indigent. But he that will give more, of his own free will it is better for him. And it is better for you that you fast, if you only know.

185	Ramadhan is the [month] in which was revealed **Qur'an, as a guide to mankind, as Clear Proofs for guidance and the criterion of judgment [between right and wrong]**. So every one of you **who is present during that month** should spend it in fasting, but if anyone is ill, or on a journey, the prescribed period [Should be made up] by days later. Allah intends every facility for you. He does not want to put you to difficulties. [He wants you] to complete the prescribed period, and to glorify Him in that **He has guided you**; and perhaps you shall be **grateful**.
186	And when My **servant** will **ask** you concerning Me, I am indeed close [to them]. I listen to the prayer of every suppliant when he cries unto Me. So, let them listen to My call, and let them **trust** in Me That they may be led **aright** (i.e. Straight / according to law).
187	It is made lawful for you to go unto your wives on the night of fasting. They are your garments and you are their garments. Allah is aware that you were deceiving yourselves in this respect and He has turned in mercy toward you and relieved you. So associate with them, and seek that which Allah has ordained for you, and eat and drink, until **the white thread becomes distinct to you from the black thread of the dawn**. Then strictly observe the fast till the nightfall; but do not associate your wives [projected falsehood] at your devotions in the mosques. Those are limits imposed by Allah. So, approach them not. Thus, Allah expounded His Clear Signs (Clear Proofs of Revealed Magnetism) to mankind that they may learn self- restraint.
188	**And do not eat up your property among yourselves for vanities (falsehood), nor seek by it to gain hearing of the judges, with the plan that you may knowingly devour a portion of the property of others' wrongfully.**
189	**They ask you** concerning the **New Moons**. Say: **They are nothing except signs** to mark **fixed periods of time** in [the affairs of] men, and **for Pilgrimage. It is no virtue (righteousness) that you go to houses (_Inviolable places_) by the backs thereof** (as some worshipers of idols do). But **the virtuous (righteous) man is he who wards off (evil). So, go to houses by the gates thereof, and observe your duty to Allah that you may be successful.**
190	**Fight in the way of Allah against those who fight against you, but begin not hostilities.** Lo! Allah loves **not the aggressors.**
191	**And slay them wherever you find them, and drive them out from the (sacred / inviolable) places whence they drove you out, for persecutions / maltreatments are worse than slaughter. And fight not with them at the Sacred Mosque, until they [first] attack you there. But if they attack you (there), then slay them. Such is the reward of those who suppress faith (truth).**
192	But **if they cease**, Allah is Ever Forgiving, Most Merciful.
193	And **fight them on until persecutions / maltreatments / oppressions are no more, and there prevail justice and faith in Allah. But if they cease, Let there be no hostility except to those who practise oppression.**
194	The prohibited month for the prohibited month, and so for all things prohibited, and forbidden things in retaliation (there is the law of equality). If then any one transgresses the prohibition against you, transgress you likewise against him. But fear Allah, and know that Allah is with those who control themselves.
195	**Spend your wealth for the cause of Allah**, and **make not your own hands contribute to [your] destruction; but do good; for Allah loves the beneficent.**

196	Perform the Hajj (Pilgrimage) in the service of Allah. But **if you are prevented**, then **send such gifts as can be obtained with ease**, and **do not shave your heads until the gifts have reached their destination**. And whoever among you is sick or has ailment of the head must pay a ransom of **fasting** or **almsgiving** or **offering. And if you are in safety**, then **whosoever contents himself with the visit for the Hajj** (Pilgrimage) (shall give) such gifts as can be had with ease. And **whosoever cannot find** (such gifts), then **a fast of three days** while on the **pilgrimage, and of seven when you have returned**, i.e. **ten in all. That is for him whose folk are not present at the Inviolable Place of Worship. Observe your duty** to Allah, and know that Allah is **strict in punishment.**
197	**The Hajj (Pilgrimage) is (in) the well-known months (_places_), and whoever is minded to perform the Hajj (pilgrimage) therein (well-known months i.e. places) there is** (to be) **no immorality** (dishonesty), **nor inequity** (injustice), **nor internal strife** (angry conversation) **on the pilgrimage.** And whatever good you do, Allah knows it. **So, make provision for yourselves; for the best provision is to ward off evil.** Therefore, **keep your duty unto Me, O men of understanding.**
198	**It is no sin (act against Allah's will) in you if you seek from the bounty of your Rab. But when you exert pressure / press on (at the time of returning) in the multitude from Aarafat, then remember Allah by the Sacred (Kaba / Headstone) Monument. Remember Him as He has guided you.** Although before that **you were of those who go astray (_i.e. in multitude_).**
199	Then **hasten onward from the place whence the multitude hastens onward**, and **ask forgiveness** of Allah. Lo! Allah is Forgiving, Merciful.
200	So when ye have accomplished your holy rites, **celebrate** the praises of Allah, as **ye used to celebrate the praises of your fathers,** yea, **with far more Heart and soul.** There are men who say: "Our Lord! Give us [Thy bounties] in this world!" but they will have no portion in the Hereafter. (IFTA) [Sura No. 1\2 – Baqarah, Ayat No. – 200] Or And when you have completed your devotions, then remember Allah as you remember your fathers or with a more lively remembrance. But of mankind is he who says: Our Rab (Lord)! Give unto us in this world, and he has no portion in the Hereafter. (IBPLtd.) [Sura No. 1\2 – Baqarah, Ayat No. – 200]
201	And there are men who say: Our Rab! Give us in this world [present life] that which is good and in the Hereafter that which is good, and guard us from the torment of the fire!
202	For them there is in store a goodly portion out of that which they have earned; and Allah is quick in account.
203	And remember Allah through the Appointed Days. But if any one hastens to leave in **two days**, there is **no sin** for him, and if any one **stays on**, there is **no sin** for him, if his **aim is to ward off evil. Be careful of your duty to Allah**, and know that you will **surely be gathered unto Him.**
204	**And of mankind there is the type of man, whose conversation about the life of this world pleases you, and he calls to witness as to that which is in his heart; yet he is the most rigid of opponents.**
205	And when he turns away (from you), **his effort is to spread mischief** through the earth and **to destroy the crops and the cattle;** but **Allah loves not mischief.**

206	And when it is said to him: **Be careful of your duty to Allah, pride (sense of recognition) leads him to sin.** Hell will **settle his account, an evil-resting place.**
207	And there is the type of man **who would sell himself seeking the pleasure of Allah;** and **Allah has compassion.**
208	**O you who believe! Come all of you,** into submission, and **follow not the footsteps of the devil. Lo! He is an open enemy to you.**
209	**And if you slide back after the clear proofs have come unto you,** then **know that Allah is Mighty, Wise.**
210	**Will they wait until Allah comes to them in canopies of clouds, with angels and the question is [thus] settled? But to Allah do all questions go back [for Judgment].**
211	**Ask Banii-Isra-iil how many Clear Signs,** We have sent them! But **if any one, after Allah's favour has come to him, substitutes [something else], Allah is strict in punishment.**
212	The life of this world is attractive to those who reject faith, and they laugh at those who believe. But the righteous will be above them on the Day of Judgment; for Allah bestows His abundance without measure on whom He will.
213	**Mankind was one single nation** (community / **ummatanw-waahidah**), and Allah sent Messengers with good tidings and warnings; and with them He revealed the **Kitaab in truth**, to judge between people in matters wherein they differed [ma-ahumul-Kitaaba bil-haqqili-yahkuma banyan-naasi]. But the People of the Kitaab, after the Clear Signs (Proofs) came to them, did not differ among themselves, except through selfish contumacy (hatred one of another). Allah by His Grace guides the believers to the Truth, concerning that wherein they differed. For Allah guides whom He will to a path that is straight.
214	Or do you think that you shall enter the Garden [of bliss] without such [trials] as came to those who passed away before you? They encountered suffering and adversity, and were so shaken in spirit that even the Messenger and those of faith who were with him cried. When [will come] the help of Allah? Ah! Verily, the help of Allah is [always] near!
215	They ask you what they should spend [In charity]. Say: Whatever you spend for good to parents and kindred and orphans and those in want and for wayfarers. And whatever you do that is good, lo! Allah knows it well.
216	**Warfare (Tabligh) is prescribed for you** and **you dislike it.** But it is possible that you dislike a thing which is good for you, and that you love a thing which is bad for you. But Allah knows, and you know not.
217	They ask you concerning warfare in the Sacred Month (in the sense of both time and place). Say: **Warfare therein is a great (transgression)** but to **turn from the way of Allah, and in the Inviolable Place of Worship,** and to expel his people thence, is **greater in the sight of Allah, for persecution is worse than killing.** Nor will they cease fighting you until they turn you back from your faith if they can. And whosoever Turn back from their faith and die in unbelief, their works will bear no fruit in this life and in the Hereafter; they will be companions of the Fire and will abide therein.
218	Lo! Those who **believe** and those who **emigrate** and **strive in the way of Allah, they have hope of Allah's mercy.** Allah is Ever Forgiving, Most Merciful.

219	They ask you **concerning wine and gambling**. Say: In them is great sin, and some utility, for men; but the sin is greater than the utility. They ask you how much they are to spend. Say: What is beyond your needs (superfluous). Thus Allah has made clear to you His Signs in order that you may give thought.
220	[Their bearings] on this life and the Hereafter. They ask you **concerning orphans**. Say: The best thing to do is what is for their good. If you **mix their affairs with yours**, they are your brethren; but Allah knows the man who spoils from him who improves. And if Allah had wished, He could have put you into difficulties: He is indeed Exalted in Power, The Wise.
221	**Do not marry idolatresses until they believe. Lo! A slave woman who believes better than an idolatress,** even though she pleases you. **Nor marry [your girls] to idolaters until they believe. A man slave who believes is better than an idolater, even though he pleases you. These invite unto the fire.** But Allah **invites** by His Grace to the Garden [of bliss] and **forgiveness**, and expounds **His Signs clear to mankind** that **they may remember.**
222	They ask you **concerning women's courses**. Say: **It is an illness**. So keep away from women in their courses, and do not approach them until they are clean. And when they have purified themselves, then you go in unto them as Allah has enjoined upon you. **Truly Allah loves those who turn to Him constantly and He loves those who keep themselves pure and clean.**
223	Your wives are as a tilth (tilled soil) unto you. So go to your tilth as you will. But **do some good act for your souls beforehand; and fear Allah**. And know that you will meet Him [one day], and **give good tidings to those who believe** (Tabligh).
224	And **make not Allah, by your oaths, a hindrance to your being righteous** and **observing your duty unto** Him and **making peace among mankind**; for Allah is The Hearer and Knower of all things.
225	Allah **will not call you to account for thoughtlessness in your oaths**, but **for the intention in your hearts**; and He is Ever-Forgiving, Most Forbearing.
226	For those who take an oath for abstention from their wives, **a waiting for four months is ordained**; if then they return, Allah is Forgiving, Merciful.
227	But **if their intention is firm for divorce**, (let them remember that) Allah is The Hearer and Knower of all things.
228	Divorced women shall wait concerning themselves **for three monthly periods**. Nor is it lawful for them to hide what Allah Hath created in their wombs, if they have faith in Allah and the Last Day. And their husbands have the **better right** to take them back in that period, if they wish for reconciliation. And women shall have **rights** similar to the **rights** against them, according to what is **equitable**; but men have a degree [of advantage] over them. And Allah is Exalted in Power, Wise.
229	A **divorce is only permissible twice**: after that, the parties should either hold together on equitable terms, or separate with kindness. It is not lawful for you, [Men], to take back any of your gifts [from your wives], except when both parties fear that they would be unable to keep the limits ordained by Allah. If you [judges] do indeed fear that they would be unable to keep the limits ordained by Allah, there is no sin on either of them if she gives something for her freedom. These are the limits ordained by Allah. So do not transgress them. If any do transgress the limits ordained by Allah, such persons are wrong-doers.

230	So if a husband divorces his wife, then he cannot, after that, re-marry her until after she has married another husband and He has divorced her. In that case there is no sin on either of them if they re-unite; provided they feel that they can keep the limits ordained by Allah. Such are the limits ordained by Allah, which He makes clear to those who understand.
231	When you divorce women, and they fulfil the term of their ['Iddat], then either take them back on **equitable terms** or set them free on **equitable terms**; but do not take them back to injure them, [or] to take undue advantage. If any one does that he does wrong with his own soul. **Do not treat Allah's Signs as a jest**, but solemnly remember Allah's favours on you, and the fact that He sent down to you the Kitaab and Wisdom, for your instruction. And fear Allah, and know that Allah is well acquainted with all things.
232	When you divorce women, and they fulfil the term of their ['Iddat], do not prevent them from marrying their [former] husbands, if **they mutually agree on equitable terms**. This instruction is for all amongst you, who believe in Allah and the Last Day. That is [the course Making for] most virtue and purity amongst you and Allah knows, and you know not.
233	The mothers should suckle to their offspring for **two whole years**, if the father desires to complete the term. But he shall bear the cost of their food and clothing on equitable terms. **No soul shall have a burden laid on it greater than it can bear.** No mother shall be Treated unfairly on account of her child. Nor father on account of his child, an heir shall be chargeable in the same way. If they both decide on weaning, by mutual consent, and after due consultation, there is no sin on them. If you decide on a foster-mother for your offspring, there is no sin on you, provided you pay [the mother] what you offered, on equitable terms. But observe your duty to Allah and know that Allah sees well what you do.
234	If any of you die and leave widows behind, they shall wait concerning themselves **four months** and **ten days**. When they have fulfilled their term, there is no sin on you if they dispose of themselves in a just and reasonable manner. And Allah is well acquainted with what you do.
235	There is **no sin** on you if you proclaim or hide it in your hearts concerning your troth (truth) with women. Allah knows that you cherish them in your hearts. But do not make a secret contract with them except in terms Honourable, nor re-solve on the tie of marriage till the term prescribed is fulfilled. And know that Allah Knows what is in your hearts, and take heed of Him; and know that Allah is Ever-Forgiving, Most Forbearing.
236	There is no sin on you if you divorce women before consummation or the fixation of their dower; but bestow on them [a suitable gift], the wealthy according to his means, and the poor according to his means. A gift of a reasonable amount is due from those who wish to do the right thing.

237	And if you divorce them before consummation, but after the fixation of a dower for them, then the **half of the dower** [is due to them], unless they remit it or [the man's half] is remitted by him in whose hands is the marriage tie; and the remission [of the man's half] is the nearest to righteousness. And **do not forget liberality between yourselves**. For Allah sees well all that you do.
238	**Guard strictly your prayers, and of the mid-most prayer, and stand up with devotion to Allah.**
239	**If you fear, pray on foot, or riding. But when you are again in safety, remember Allah in the manner as He has taught you, which you knew not [before].**
240	(In the case of) those of you who are **about to die** and **leave behind them wives,** they should **bequeath** unto their wives a **provision** for the year **without turning them out,** but if they go out (of their own accord), there is **no sin** for you in that **which they do of themselves within their rights.** Allah is The Mighty, The Wise.
241	For divorced women **maintenance** [should be provided] on a reasonable [scale]. This is a duty on the **righteous.**
242	Thus Allah makes **clear His Signs** to you in order that **you may understand.**
243	**Bethink you of those of old, who went forth from their homes,** though they were **thousands** (in number), **for fear of death,** and Allah said unto them: **Die** and then He brought back them to life. For Allah is full of bounty to mankind, but **most of them are ungrateful.**
244	**Then fight in the way of Allah** (Tabligh), and know that Allah Hears and knows all things.
245	**Who is he that will lend to Allah a beautiful loan, which Allah will double unto his credit and multiply many times? It is Allah Who gives [you] want or plenty, and to Him you will return.**
246	**Bethink you of the leaders / chiefs of Banii-Israa-iil after Muusa,** how they said unto a Prophet among them: Appoint for us a king, that we May fight in the way of Allah. He said: Would you then refrain from fighting if fighting were prescribed for you? They said: How could we refuse to fight in the way of Allah, when we were turned out of our homes with our children? But when they were commanded to fight, they turned back, except a small band among them. But Allah Has full knowledge of those who do wrong.
247	**Their Prophet said to them: Allah has appointed Taluuta as King (Malikaa?)** over you. They said: How can he exercise authority over us when we are better fitted than he to exercise authority, and he is not even gifted, with wealth in abundance? He said: Allah has chosen him above you, and has gifted him abundantly with knowledge and stature. Allah grants His authority to whom He pleases. Allah cares for all, and He knows all things.
248	And [further] their **Prophet** said to them: **A Sign of his authority is that** there shall come to you **the Ark wherein are peace of reassurance from your Rab,** and **a reminder of that** which the house of **Muusa** and the house of **Haruun** left behind, **the angels bearing it. In these are signs / tokens for you if you indeed have faith.**
249	When Taluuta set out with the armies (rays), he said: Allah will test you at the stream (river / sea). If any drinks of its water, He goes not with my army. Only those who taste not of it go with me except him who takes (thereof) in the hollow of his hand. But they all drank of it, except a few. When they crossed the river, he and the faithful

	ones with him, they said: This day we cannot cope with Goliath (*May be Jaaluut*) and his forces. But those who were convinced that they must meet Allah, said: How many of little company has overcome mighty armies by Allah's permission? Allah is with those who steadfastly persevere.
250	When they advanced to meet Goliath and his forces, they prayed: Our Rab! Bestow on us endurance and make our steps firm. Help us against those that reject faith.
251	By Allah's will they routes them; and Daawuud slew Goliath; and Allah gave him power and wisdom and taught him whatever [else] He willed. And if Allah did not repelled some men by others, the earth would indeed be full of mischief. But Allah is full of bounty to (His) creatures.
252	**These are the Signs of Allah. We rehearse them to you in truth. Verily you are one of the messengers.**
	Of **those messengers** some of whom We have **caused** to **excel others**. To **one** of them Allah **spoke; others** He **raised** to **degrees** [of honour]. To **Isabna Maryam** We gave **Clear Proof** and **supported** him with the **holy spirit. If Allah had so willed, succeeding generations would not have fought among each other, after Clear Proofs had come to them. But they differed. Some believing** and **others rejecting.** If Allah had so willed, they would not have fought each other. **But Allah fulfils His plan.**
253	O you who believe! Spend out of that We have **provided** you, **before the Day comes** when **no** bargaining / trafficking will avail], **nor** friendship **nor** intercession. **Those who reject Faith they are the wrong-doers.**
254	**Allah! There is no ilaaha (unbreakable reality) but He, the Living, the Self-subsisting, Eternal.** Neither slumber nor sleep can seize Him. Unto Him belong all things in the heavens and on earth. **Who is there that can intercede with Him except His permission?** He knows that which is before or after or behind them. They encompass nothing of His knowledge except as He will. His Throne extends over the heavens and the earth, and He feels no fatigue in guarding and preserving them for He is the Sublime, the Supreme.
255	There is **no compulsion** in Fid-Diin (religion). **The right direction is henceforth distinct from Error.** Whoever rejects false desires and believes in Allah has grasped the firm hand-hold that never breaks. And Allah is Hearer and Knower of all things.
256	**Allah is the Protector of those who have faith.** He brings them out of darkness unto light. As for those who reject faith their patrons are the false ones (deities). They bring them out of light into depths of darkness. They will be companions of the fire. They will dwell there [for ever].
257	Bethink you of him who had an argument with Ibrahim about his Rab, because Allah had granted him the kingdom. How! When Ibrahim said: My Rab is He Who gives life and death. He (Namruud) answered: I give life and cause death. Said Ibrahim: **Allah causes bish-shamsi (neither Ash-Shams nor Wash-Shamsi) to rise from the east; so do you cause him to rise from the West.** Thus was the disbeliever (who has rejected faith) confused and Allah guides not the wrong doing folk.
258	Or [bethink you of] the **similitude of one** who passed by a **township [lower east]** which had fallen into utter ruin, exclaimed: Oh! How shall Allah give this **township life** after **its death?** But Allah **caused him to die** for a **hundred years,** and then raised him up [again]. He said: **How long did you tarry?** He said: **I have tarried a day or**

	part of a day. (He) said: Nay, you have tarried for **a hundred years.** But **look at** your **food** and your **drink**; **they show no signs of age**; and **look at** your **donkey.** And that We may make of you **a sign of proof** unto the people. **Look** further at the **bones, how We bring them together** and **clothe them with flesh.** When this was shown clearly to him, he said: **I know now that Allah is Able to do all things.**
259	When **Ibrahim** said (unto his Rab): My Rab! **Show me how You give life to the dead. He replied: Have you no faith? He said: Yes,** but **just to reassure** my **heart. Allah said: Take four birds,** draw them to you, and **cut** their bodies to pieces. **Scatter** them over the **mountain-tops,** and then **call them back. They will come swiftly to you. Know** that Allah is The Mighty, The Wise.
260	The **parable (likeness / similitude)** of those who **spend** their wealth **in the way of Allah** is the likeness of **a grain** which growth **seven ears,** and **each ear** has a **hundred** grains. Allah give **increase manifold** to which He **pleases.** Allah is All-Embracing, All-Knowing.
261	Those who **spend** their **substance** in the **cause** of Allah (Tabligh) and **follow not** up their gifts with reminders of their generosity or with injury, for them **their reward is with their Rab.** There will be **no fear** on them, **nor** shall they **grieve.**
262	**Kind words and the covering of faults are better than charity followed by injury.** Allah is free of all wants, and He is Most-Forbearing.
263	O you who believe! **Cancel not your charity by reminders of your generosity or by injury,** like those **who spend their wealth to be seen of men,** but **believe neither in Allah nor in the Last Day.** They are in **parable** like **a hard, barren rock, on which is a little soil on it falls heavy rain, which leaves it a bare stone.** They will be able to do nothing with aught they have earned. **And Allah guides not those who reject faith.**
264	And the likeness of those who spend their substance, seeking to please Allah and to strengthen their souls, is as a garden, high and fertile; heavy rain falls on it but makes it yield a double increase of harvest, and if it receives not Heavy rain, light moisture suffice it. Allah sees well whatever you do.
265	**Would any of you wish that he should have a garden with date-palms and vines and streams flowing underneath, and all kinds of fruit for him therein;** and while he is stricken with old age, and he has feeble offspring and a fiery whirlwind strikes it and it is consumed by fire? Thus Allah makes clear to you [His] Signs that you may give thought.
266	O you who believe! Spend of the good things which you have [honourably] earned, and of the fruits of the earth which We have produced for you, and **do not even aim at getting anything which is bad,** in order that out of it you may give away something, when you yourselves would not receive it except with disdain and know that Allah is Absolute owner of all praise.
267	**The evil one (devil) threatens you with poverty and proposes you to conduct unseemly.** Allah promises you His forgiveness and bounties. Allah is All-Embracing, All-Knowing.
268	**He grants wisdom unto whom He pleases, and he to whom wisdom is granted receives indeed a benefit overflowing; but none will grasp the message except men of understanding.**
269	And whatever you spend in charity or devotion, be sure Allah knows it all. But **the wrong-doers have no helpers.**

270	If you disclose [acts of] charity, even so it is well, but if you **conceal** them, and make them **reaches those [really] in need, that is best for you.** It will remove from you some of your **ill-deeds**. And Allah is well informed with **what you do.**
271	The guiding of them is not your duty, but Allah guides whom He will. And whatever good things you spend benefits your own souls, and you shall only do so **in search of** Allah's countenance. Whatever good you give, shall be rendered back to you, and you will not be dealt with unjustly.
272	(Alms are / charity is) for the poor who are straitened for the cause of Allah, who cannot travel in the land, the unthinking man accounts them wealthy because of their restraint. **You shall know them by their mark (?).** And whatsoever good thing you spend, lo! Allah knows it.
273	Those who [in charity] spend their wealth **by night and by day, in secret and in public,** have **their reward with their Rab.** On them shall be no fear, nor shall they grieve.
274	Those who eat greedily **usury** (lending of money at interest) cannot rise up except as he arises whom the devil has prostrated by (his) touch. That is because they say: Trade is like usury. But Allah has permitted trade and forbidden usury. Those who after receiving direction from their Rab, desist, shall be pardoned for the past; their case is for Allah [to judge]. But those who repeat [the offence] are companions of the fire. They will abide therein [for ever].
275	O you who believe! Spend out of that We have **provided** you, **before the Day comes** when **no** bargaining / trafficking will avail], **nor** friendship **nor** intercession. **Those who reject Faith they are the wrong-doers.**
276	Allah will deprive usury of all blessing, but will give increase for deeds of charity; for Allah **loves not the impious and guilty.**
277	Those who believe, and do deeds of righteousness (Tabligh), and **establish** regular prayers and regular charity, will have their reward with their Rab. On them shall there be no fear, nor shall they grieve.
278	O you who believe! Observe your duty to Allah, and **give up what remains of your demand for usury, if you are indeed believers.**
279	If you do it not, **take notice of war from Allah** and His Messenger. But **if you turn back, you shall have your capital sums. Deal not unjustly, and you shall not be dealt with unjustly.**
280	If the debtor is in a **difficulty, grant him time** till it is easy for him to repay. But if you **remit** it by way of charity, that would be best for you if you did but know.
281	And **guard yourselves** against the Day when you will be brought back to Allah. Then every soul will be paid what it earned, and none will be dealt with unjustly
282	O you who believe! When you **contract a debt for a fixed term, record it in writing.** Let a scribe record it in writing between you **in (terms of) equity.** Let not the scribe refuses to write, as Allah Has taught him; so let him write. Let him who incurs the debt dictate, and let him observe his duty to Allah, his Rab, and diminish naught thereof. But if he who owes the debt is of low understanding, or weak, or unable himself to dictate, then let the guardian of his interests dictate **in (terms of) equity.** And call to witness, from among your men, **two witnesses. And if two men be not (at hand) then a man and two women, of such as you approve as witnesses,** so that if the one errs (through forgetfulness) the other will remember. And the witnesses must

	not refuse when they are summoned. Be not averse to writing down (the contract) whether it be small or great, with (record of) the term thereof. That is **more equitable in the sight of Allah** and surer for testimony, and the best way of avoiding doubt between you; save only in the case when it is actual merchandise which you transfer among yourselves from hand to hand. In that case it is no sin for you if you write it not. And have witnesses when you sell one to another, and let **no harm be done to scribe or witness. If you do (harm to them) lo! It is a sin in you.** Observe your duty to Allah. Allah is teaching you. And Allah is knower of everything.
283	If you be on a journey, and cannot find a scribe, a pledge with possession [may serve the purpose]. And if one of you entrust to another, let the trustee [Faithfully] discharge His trust, and let him fear his Rab. **Conceal not evidence (Testimony); for whoever conceals it, verily his heart is polluted with sin.** And Allah is aware of all that you do.
284	To **Allah belong all that is in the heavens and in the world. Whether you show what is in your minds or conceal it,** Allah will bring you **to account for it.** He **forgives** whom He **pleases,** and punishes whom He **will,** for Allah is **Able** to do all things.
285	**The Messenger believes in that which has been revealed to him from his Rab, and so do the men of faith. Each one [of them] believes in Allah, His angels, His Kitaabs, and His Messengers. We make no distinction between any of His messengers.** And they say: We hear, and we obey. [Grant us] **Your forgiveness, our Rab, and unto You is the end of all journeys.**
286	**On no soul Allah place a burden greater than it can bear. It gets** every **good that** it **earns,** and it **suffers** every **ill that it earns. [Pray:] Our Rab! Condemn us not if** we **forget** or **miss the mark! Our Rab! Lay not on us a burden** like **that which You did lay on those before us. Our Rab! Lay not on us a burden greater than we have strength to bear. Blot out our sins, and grant us forgiveness. Have mercy on us. You are our Protector. Grant us victory over the unbelieving folk.**

O you people! Eat of what is in the world, **lawful** and **wholesome;** and **do not follow the footsteps of the Shaytan (projected mechanical globalisation, circular rotation & revolution system, and manmade magnetism),** as he is an **open enemy for you. He (Historical Don of two-in-one) enjoins upon you only what is evil and foul, and that you should tell concerning Allah that of which you have no knowledge.** [Sura No. 1\2 – Baqarah, Ayat No. – 168 & 169]

All food was lawful to the **Bani-Israa-iil,** except what **Israa-iil** forbade himself **before the Tawraat was revealed. Say: produce Tawraat and read it, if you are men of truth. If any,** after this, **invent a lie and attribute it to Allah, they are indeed unjust wrong-doers. Say: Allah speaks the Truth. So follow the religion of Ibrahim, the upright. He was not of the idolaters. [Sura No. 2\3 – Imran, Ayat No. – 93 to 95]**

O you who believe! Be mindful of your duty to Allah and seek the way of approach unto Him, and strive in His way so that you may succeed. As to those who reject Faith, if they had everything in the world, and twice repeated, to give as ransom for the penalty of the Day of Resurrection, it would never be accepted of them. Theirs would be a grievous Penalty. Their wish will be to get out of the fire, but never will they get out there from. Their Penalty will be one that endures. As to the thief, male or female, cut off his or her hands: A punishment by way of example, from Allah, for their crime: And Allah is Exalted in Power. But if the thief repents after his crime and amends his conduct, Allah turns to him in forgiveness; for Allah is Forgiving, Merciful. [Sura No. – 4\5 – Maaaidah, Ayat\Verses – 35 to 39]

Know you not that **to Allah (alone) belongs the dominion of the heavens [West Horizon] and the world [East Horizon]**? He punishes whom He pleases, and He forgives whom He pleases. And Allah has power over all things. [Sura No. – 4\5 – Maaaidah, Ayat\Verses – 40]

SECTION – III
DIAMOND OPERATOR – 3.4
EYE OPENING EVIDENCES
[Kitaab of Muusa (ass) or Nine Clear Proofs given to Muusa (ass)]

Fasta-'iz billaahi minash-Shaytaanir-Rajim
Bismillaahir-Rahmaanir- Rahim

Nine Planets/Moons are the **nine Stars of the Star System or nine clear proofs** as well as eye opening evidences given to Muusa (ass). In other words, nine planets/moons represent nine stars of the Kitaab of Muusa (ass) or Valid References or Manifested Signs in resemblance with Revelations.

We verily gave the Kitaab to Muusa
Ayat\Verses - And **who does greater wrong** than one to whom are recited the **Signs** of his Rab, and who then **turns away there from**? Verily from those who transgress We shall exact [due] retribution. **We verily gave the Kitaab to Muusa.** So, be not you in doubt of his receiving it and We made it a guide to the Bani-Israa-iil. "Wa ja-alnaa minhum A-immatany-yahduuna bi-Amri-naa **lammaa** sabaruu; wa kanuu bi-' Aayaatinaa yuu- qinuun". [Trans.] And **We appointed [manifested] Imam** from among those signs **giving guidance** under **Our command**, so long as they **persevered** with **patience and continued to have faith in Our Signs.** Verily your **Rab will judge** between **them** on the **Day of Resurrection**, in the matters **wherein they differ** [among themselves] [Sura No. – 31/32, Sura – Sajdah, Ayat\Verses – 22 to 25]

NINE TOKENS OR CLEAR PROOFS
EYE OPENING EVIDENCES
NINE PLANETS OR MOONS AS NINE
STARS OF THE STAR SYSTEM
Ayat\Verses - And verily We gave to Muusa **Nine Tokens [Clear Proofs / Nine Stars]** (of Allah's Sovereignty): Do but ask Bani-Israa-iil how he came to them [manifested magnetism]? Firawn said to him: O Muusa! I consider you, indeed, to have been worked upon by sorcery! Muusa said: In truth you know well that these things (Tokens) have been sent down by none but the Rab of the heavens and the world as **eye-opening evidences,** and I consider you indeed, O Firawn, to be one doomed to destruction! And he wished to remove them from the face of the earth; but We did drown him and all who were with him all together. And We said thereafter to Bani-Israa-iil: Dwell securely in the land [of promise or uppermost land], but when the **second of the warnings came to pass,** We

gathered **you together in a mixed crowd**. [Sura No. – 16\17 - Banii-Israai-iil, Ayat No. – 101 & 104]

[Suppositions on the basis of pen-paper-pencil works and mechanical activities]

i) Star [Planet/Moon] Mercury represents Dot Operator or Point of Intersection of the equal & opposite ~sth. Mercury is Gravitationally Locked and has no reflection. Mercury appears once in every two Mercurian Years i.e. equal & opposite Lunar Year and Calendar Year. This appearance is called Lay-Latul-Qadr.

ii) Star [Planet/Moon] Saturn represents Revealed & Manifested East or Gravitational Force. In other words, Star [Planet/Moon] Saturn represents shared concept 'Wife of Ibrahim' with respect to Revelation and 'Wife of Imran' with respect to manifestation. Star [Planet/Moon] Saturn represents East Horizon or white dwarf companion Sirius B of Sirius Binary System The Ring of the Star Saturn is the First Peg of the Immovable Hexagonal World as well as Pentagonal Earth. The Ring of the Star Saturn also represents octagonal ring of the Black-spots or So-called Sun-spots or Matter Particles or Springs.

iii) Star [Planet/Moon] Jupiter represents Revealed & Manifested West or Strong Force. In other words, Star [Planet/Moon] Jupiter represents shared concept 'Adam' with respect to Revelation [Ref. Allah has taught Adam names of all things] and Imran with respect to Manifestation. Star [Planet/Moon] Jupiter represents main sequence Sirius A of Sirius Binary System or Galaxy of Guiding Stars [Four-Figured Shining Stars or Ashab-bikan-Nujum].

iv) Star [Planet/Moon] Uranus represents new moons or Triangular Bullets as **Signs** [ice-like] to mark **fixed periods of time** in [the affairs of] men, and **for Pilgrimage**, Night Sky, Major Axis, shared name Muzzammil, and accelerating charges i.e. a of Newton's Second Law 'F = ma'.

v) Star [Planet/Moon] Neptune or Triangular Bullet [not visible to the unaided eye, subject to Gravitational Perturbation, Enveloped] represents Earth's Night Sky, Minor Axis, shared name Muddassir, and accelerating masses i.e. m of Newton's Second Law 'F = ma'.

vi) Sister Planet Venus represents manifested signs of Natural Magnetism or Bullet or so-called sun as the day-light for the equal & opposite Seashore in resemblance with shared name Maryam supplied with sustenance and shared tale of Zilqar-Nayin [with reference to stages of journey for the equal & opposite Seashore i.e. East Zone and West Zone].

vii) Projected Planet/Moon Earth represents projected sun or Bish-Shamsi or gravitational wave or Ether or Screen or Veil or Black Umbrella which rises from the East in the name of Jakariya, the guardian of Maryam.

viii) Projected Planet/Moon Pluto represents manifested hexagonal world within the Black Square of Manifested East Horizon after equal & opposite sacrifices of an Octagon [Six days] i.e. twice divorces in resemblance with converted green and blue shades of Windows'7 Ultimate. The revealed truth is that one day the world [Planet/Moon Pluto] will be changed into a different world.

ix) Projected Planet/Moon Mars represents the sixth triangle on the manifested hexagonal world i.e. Eartha 3D of Straight Middle East [Recycling Region] as the peg of the immovable Pentagonal Earth. Moreover, Planet/Moon Mars represents shared concept 'Wife of Firawn'.

SOURCE - INTERNET
MERCURY

Key Terms – 88, messenger, 100K, 700K, smallest tilt, largest orbital eccentricity, 1.5 times, **gravitationally locked, rotates in a way that is unique, relative to the fixed stars, three times for every two revolutions, rotate only once every two Mercurian years, one day every two years, Earth's sky in the morning or the evening, not in the middle of the night, Venus and the Moon, proximity to the Sun, Two spacecraft have visited Mercury, MESSENGER,** launched in **2004,** orbited Mercury over **4,000 times in four years, no substantial atmosphere, large iron core.**

Mercury is the **smallest** and **closest to the Sun of the eight planets** in the Solar System,[a] with an **orbital period** of about **88** Earth days. Seen from Earth, it **appears to move around its orbit** in about **116 days,** which is **much faster than any other planet** in the Solar System. It has **no known natural satellites.** The planet is named after the Roman deity Mercury, the **messenger to the gods.**

Because it has almost **no atmosphere to retain heat,** Mercury's surface experiences the greatest temperature variation of the planets in the Solar System, ranging from **100 K (−173 °C; −280 °F)** at night to **700 K** (427 °C; 800 °F) during the day at some equatorial regions. The poles are constantly below 180 K (−93 °C; −136 °F). Mercury's axis has the smallest **tilt** of any of the Solar System's planets (**about $\frac{1}{30}$ of a degree**), but it has the largest orbital eccentricity.[a] At **aphelion** [North], Mercury is about **1.5 times** as far from the

Sun as it is at **perihelion** [South]. Mercury's surface is heavily cratered and similar in appearance to the **Moon**, indicating that it has been **geologically inactive** for billions of years.

Mercury is **gravitationally locked** and **rotates in a way that is unique** in the Solar System. As seen **relative to the fixed stars**, it rotates on its axis exactly **three times** for **every two revolutions** it makes around the Sun.[b][13] As seen from the Sun, in a **frame of reference** that **rotates with the orbital motion**, it appears to **rotate only once every two Mercurian years**. An observer on Mercury would therefore see only **one day every two years**.

Because Mercury orbits the Sun within Earth's orbit (**as does Venus**), it can appear in **Earth's sky** in the **morning or the evening** but **not in the middle of the night**. Also, like **Venus and the Moon**, it displays a complete range of **phases** as it moves around its orbit relative to Earth. Although Mercury can appear as a bright object when viewed from Earth, its **proximity to the Sun makes it more difficult to see than Venus. Two spacecraft have visited Mercury: Mariner 10 flew by in the 1970s; and _MESSENGER_, launched in 2004**, orbited Mercury over **4,000 times in four years**, before **exhausting its fuel and crashing into the planet's surface on April 30, 2015.**[14][15][16]

Mercury is the **smallest planet** in the solar system, orbiting the Sun once **every 88 days. Physically, Mercury is similar in appearance to the Moon** as it is heavily cratered. It has **no natural satellites** and **no substantial atmosphere**. The **planet has a large iron core.**

JUPITER
TWO AND HALF
TWO FOLD MERCY AND A LIGHT
A TENTH OF THE NUMBER OF MOLECULES
NIGHT SKY – SIRIUS B
SIGNS OF ALLAH – NO SOLID SURFACE AND RAPID ROTATION
UNIFORM MOTION

Key Terms - largest in the Solar System, **mass one-thousandth, two and a half times,** astronomers of ancient times, **−2.94, bright enough for its reflected light to cast shadows, night sky, tenth of the number of molecules,** rocky core of heavier elements, **lacks a well-defined solid surface, rapid rotation,** latitudes, Great Red Spot, 17th century, **faint planetary ring** system, **powerful magnetosphere, 67 moons, including** the four large Galilean moons,

Jupiter is the fifth planet from the Sun and the **largest** in the Solar System. It is a giant planet with a mass **one-thousandth** that of the Sun, but **two and a half times** that of all the other planets in the Solar System combined. Jupiter is a gas giant, along with **Saturn** (Uranus and Neptune are ice giants). Jupiter was known to astronomers of ancient times.[11] The Romans named it after their god Jupiter.[12] When viewed from Earth, Jupiter can reach an apparent magnitude of −2.94, **bright enough for its reflected light to cast shadows**,[13] and making it on average the third-brightest object in the **night sky** after the **Moon** and **Venus**.

Jupiter is primarily composed of hydrogen with a quarter of its mass being helium, although helium only comprises about a **tenth of the number of molecules**. It may also have a rocky core of heavier elements,[14] but like the other giant planets, **Jupiter lacks a well-defined solid surface**. Because of its **rapid rotation**, the planet's shape is that of an oblate spheroid (it has a slight but noticeable bulge around the equator). The outer atmosphere is visibly segregated into several bands at different **latitudes**, resulting in turbulence and storms along their interacting boundaries. A prominent result is the **Great Red Spot**, a giant storm that is known to have existed since at least the 17th century when it was first seen by telescope. Surrounding Jupiter is a **faint planetary ring** system and a **powerful magnetosphere**. Jupiter has at least **67 moons**, **including** the **four large Galilean moons** discovered by Galileo Galilei in 1610. Ganymede, the largest of these, has a diameter greater than that of the planet Mercury.

Jupiter has been explored on several occasions by robotic spacecraft, most notably during the early *Pioneer* and *Voyager* flyby missions and later by the Galileo orbiter. Jupiter was most recently visited by a probe in late February **2007**, when *New Horizons* used Jupiter's gravity to increase its speed and bend its trajectory en route to Pluto. The next probe to visit the planet will be *Juno*, which **is expected to arrive in July 2016**. Future targets for exploration in the Jupiter system include the probable ice-covered liquid ocean of its moon Europa.

SISTER PLANET VENUS

MARYAM, THE SISTER OF MUUSA, SUPPLIED WITH SUSTENANCE AS THE GLORIOUS MORNING SHOW FOR THE EAST ZONE & EVENING SHOW FOR THE WEST ZONE
THE APPOINTED LIGHT OF ALLAH ON EARTH
COMMONLY CONCEIVED AS WELL AS PERCEIVED SUN

1. Venus has no **natural satellite**.
2. It is the **brightest natural object** in the night sky.
3. It is **bright enough to cast shadows**.
4. Venus is an **inferior planet** from Earth.
5. It never appears to venture **far from the Sun**.
6. Venus is a **terrestrial planet**.
7. It is sometimes called Earth's **"sister planet"** [**Maryam supplied with Sustenance**] because of their **similar size, mass, proximity to the Sun** and **bulk composition**.
8. It is **radically different** from Earth in other respects.
9. It has the densest **atmosphere** of the **four terrestrial planets** [**four forces**], consisting of more than **96% carbon dioxide**.
10. The **atmospheric pressure** at the **planet's surface is 92** [**elements as beautiful names**] **times** that of Earth.
11. Venus is by far the **hottest planet** in the **Solar System**.
12. Mercury is closer to the Sun. **Venus is shrouded by an opaque layer of highly reflective clouds** [**magnetic clouds**] of **sulfuric acid**, preventing its surface from being seen from space in **visible light**.
13. It may have **had oceans** in the past,[16][17] but these would have **vaporized as the temperature rose due to a runaway greenhouse effect.**
14. The water has most probably **photodissociated,** and, because of the **lack of a planetary magnetic field**, the free **hydrogen** has been swept into interplanetary space by the **solar wind**.[19]
15. Venus's surface is **a dry desertscape** interspersed with **slab-like rocks** and **periodically refreshed by volcanism.**
16. **Morning Star** from the point of view of East Zone [North America & South America]
17. **Evening Star** from the point of view of West Zone [Europe, Arabian Countries, Asia, Africa, and Australia]

COMPARE WITH THE CONCEPTS OF THE SO-CALLED SUN

No natural satellite v/s brightest natural object; night sky v/s hottest planet; bright enough to cast shadows; terrestrial planet; densest atmosphere; four terrestrial planets [four forces; planet's surface is 92 [elements as beautiful names]; hottest planet in the Solar System; an opaque layer of highly reflective clouds [planet Saturn]; temperature rose due to a runaway greenhouse effect, photodissociated; lack of a planetary magnetic field

Venus is the second planet from the Sun, orbiting it every **224.7** Earth days.[14] It **has no natural satellite**. It is named after the Roman goddess of love and beauty. After the <u>Moon</u>, it **is the brightest natural object** in the night sky, reaching an <u>apparent magnitude</u> of **−4.6, bright enough to cast shadows.**[15] Because Venus is an <u>inferior planet</u> from Earth, it never appears to venture far from the Sun: its <u>elongation</u> reaches a maximum of **47.8°.**

Venus is a <u>terrestrial planet</u> and is sometimes called Earth's **"sister planet"** because of their **similar size, mass, proximity to the Sun and bulk composition. It is radically different from Earth in other respects. It has the densest <u>atmosphere</u> of the four terrestrial planets, consisting of more than 96% <u>carbon dioxide</u>.** The <u>atmospheric pressure</u> at the planet's surface **is 92 times** that of Earth. With a mean surface temperature of **735 K** (462 °C; 863 °F), **Venus is by far the hottest planet in the <u>Solar System</u>**, even though <u>Mercury</u> is closer to the Sun. Venus is shrouded by an opaque layer of highly reflective clouds of <u>sulfuric acid</u>, preventing its surface from being seen from space in <u>visible light</u>. It may have had **oceans** in the past,[16][17] but these would have vaporized as the temperature rose due to a <u>runaway greenhouse effect</u>.[18] The water has most probably <u>photodissociated</u>, and, because of the lack of a <u>planetary magnetic field</u>, the free hydrogen has been <u>swept into interplanetary space</u> by the <u>solar wind</u>.[19] Venus's surface is a dry desertscape interspersed with slab-like rocks and periodically refreshed by <u>volcanism</u>.

Venus is the second-closest planet to the Sun, orbiting it every 224.7 Earth days. It is the brightest natural object in the night sky, except for the Moon. Venus reaches its maximum brightness shortly before sunrise or shortly after sunset, for which reason it is often called the **Morning Star** or the **Evening Star**. Classified as a terrestrial planet, it is sometimes called Earth's **"sister planet"**, for the two are similar in size, gravity, and bulk composition.

URANUS
SEASONAL CHANGE / CLOUDS & STORMS
MUZZAMMIL
Uranus North and South poles therefore lie where
most other planets have their <u>equators</u>.

Uranus is the **seventh** <u>planet</u> from the <u>Sun</u>. It has the **third-largest planetary radius and fourth-largest planetary mass** in the <u>Solar System</u>. **Uranus is similar in composition to <u>Neptune</u>,** and both have different bulk chemical composition from that of the larger <u>gas giants</u> **<u>Jupiter</u> and <u>Saturn</u>.** For this reason, scientists often classify **Uranus and Neptune as "<u>ice giants</u>"** to distinguish them from the gas giants. Uranus's atmosphere, although similar to Jupiter's and Saturn's in its primary composition of <u>hydrogen</u> and <u>helium</u>, contains more "ices", such as water, ammonia, and methane, along with traces of other hydrocarbons.[11] It is the coldest planetary atmosphere in the Solar System, with a minimum temperature of 49 K (−224.2 °C), and has a complex, layered cloud structure, with water thought to make up the lowest clouds, and methane the uppermost layer of clouds.[11] The interior of Uranus is mainly composed of ices and rock.[10]

Uranus is the only planet whose name is derived from a figure from Greek mythology, from the Latinized version of the Greek god of the sky, **Ouranos**. Like the other giant planets, Uranus has a **ring system**, a **magnetosphere**, and **numerous moons**. The Uranian system has a unique configuration among those of the planets because its **axis of rotation** is **tilted sideways,** nearly into the plane of its revolution about the Sun. Its **north and south poles therefore lie where most other planets have their equators.**[15] In **1986**, images from *Voyager 2* showed Uranus as an almost **featureless planet in visible light**, without the cloud bands or storms associated with the other giant planets.[15] Observations from Earth have shown **seasonal change** and increased weather activity as Uranus approached its equinox in 2007. Wind speeds can reach 250 metres per second (900 km/h, 560 mph).[16]

NEPTUNE
MUDDASSIR
Great Dark Spot 14 moons subject to gravitational perturbation

Neptune is the eighth and farthest planet from the Sun in the Solar System. It is the fourth-largest planet by diameter and the third-largest by mass. Among the giant planets in the Solar System, Neptune is the densest. Neptune is **17 times** the mass of Earth and is slightly more massive than its **near-twin Uranus**, which is **15** times the mass of Earth and slightly larger than Neptune.[c] Neptune orbits the Sun at an average distance of 30.1 astronomical units (4.50×10^9 km). Named after the Roman god of the sea, its astronomical symbol is ♆, a stylised version of the god Neptune's trident.

Neptune is **not visible to the unaided eye** and is the only planet found by mathematical prediction rather than by **empirical observation**. Unexpected changes in the **orbit of Uranus** led Alexis Bouvard to deduce that its orbit was **subject to gravitational perturbation** by an unknown planet. Neptune was subsequently observed with a telescope on **23 September 1846**[11] by Johann Galle within a degree of the position predicted by Urbain Le Verrier. **Its largest moon, Triton**, was discovered shortly thereafter, though none of the **planet's remaining 14 moons** were located telescopically until the 20th century. The planet's distance from Earth gives it a very small apparent size, making it challenging to study with Earth-based telescopes. Neptune was visited by *Voyager 2*, when it flew by the planet on **25 August 1989**.[10] The advent of Hubble Space Telescope and large ground-based telescopes with adaptive optics has allowed for more-detailed observations.

Neptune is **similar in composition to Uranus**, and both have compositions that differ from those of the larger gas giants, **Jupiter and Saturn**. **Neptune's atmosphere, like Jupiter's** and Saturn's, is composed primarily of hydrogen and helium, along with traces of hydrocarbons and possibly nitrogen; it contains a higher proportion of "ices" such as water, ammonia, and methane. Scientists sometimes categorise **Uranus and Neptune as "ice giants"** to emphasise this distinction.[11] The interior of Neptune, like that of Uranus, is primarily composed **of ices and rock**.[12] Traces of methane in the outermost regions in part account for the planet's **blue appearance**.[13]

In contrast to the hazy, relatively **featureless atmosphere of Uranus**, Neptune's atmosphere has active and visible weather patterns. For example, at the time of the **1989** *Voyager 2* flyby, the planet's **southern hemisphere** had a **Great Dark Spot** comparable to the **Great Red Spot** on **Jupiter**. These weather patterns are driven by the **strongest sustained** winds of any planet in the Solar System, with recorded wind speeds as high as 2,100 kilometres per hour (580 m/s; 1,300 mph).[14] Because of its great distance from the Sun, Neptune's outer atmosphere is one of the coldest places in the Solar System, with temperatures at its cloud tops approaching 55 K (−218 °C). Temperatures at the planet's centre are approximately 5,400 K (5,100 °C).[15][16] Neptune has a faint and fragmented **ring system** (labelled "arcs"), which may have been detected during the 1960s but was indisputably confirmed only in 1989 by *Voyager 2*.[17]

MARS = TERRESTRIAL PLANET
THE SECOND PEG OF THE IMMOVABLE EARTH

Key terms - impact craters of the **Moon**, **valleys**, deserts, and polar **ice caps** of Earth, seasonal cycles, **similar to those of Earth**, largest **volcano** and second-highest known mountain, Valles Marineris,

Mars is the fourth planet from the Sun and the second smallest planet in the Solar System, after Mercury. Named after the **Roman god of war**, it is often referred to as the "**Red Planet**"[13][14] because the **iron oxide** prevalent on its surface gives it a **reddish appearance**.[15] Mars is a **terrestrial planet** with a thin **atmosphere**, having surface features reminiscent both of the impact craters of the **Moon** and the **valleys**, deserts, and polar **ice caps** of Earth,

The **rotational period** and seasonal cycles of Mars are likewise **similar to those of Earth**, as is the tilt that produces the seasons. Mars is the site of Olympus Mons, the largest **volcano** and second-highest known mountain in the Solar System, and of Valles Marineris, one of the largest canyons in the Solar System. The smooth **Borealis basin in the northern hemisphere** covers 40% of the planet and may be a giant impact feature.[16][17] **Mars has two moons, Phobos [Yajuja] and Deimos [Majuja]**, which are small and irregularly shaped. These may be captured **asteroids,**[18][19] similar to **5261** Eureka, a Mars trojan.

Until the first successful Mars flyby in **1965 by *Mariner 4***, many speculated about the **presence of liquid water on the planet's surface**. This was based on observed periodic **variations in light and dark patches**, particularly in the **polar latitudes,** which appeared to be seas and continents; long, dark striations were interpreted by some as irrigation **channels for liquid water**. These **straight line features** were later explained as **optical illusions**, though **geological evidence gathered** by uncrewed missions suggests that Mars once had large-scale water coverage on its surface at some earlier stage of its existence.[20] In **2005**, radar data revealed the presence of large quantities of **water ice at the poles**[21] and at **mid-latitudes**.[22][23] The Mars **rover *Spirit*** sampled chemical compounds containing water molecules in March **2007**. The *Phoenix* lander directly sampled water ice in shallow Martian soil on July **31**, 2008.[24] On September 28, 2015, NASA announced the presence of briny flowing **salt water** on the Martian surface.[25]

Mars is host to **seven functioning spacecraft: five in orbit**—*2001 Mars Odyssey, Mars Express, Mars Reconnaissance Orbiter,* MAVEN and Mars Orbiter Mission**—and two on the surface—Mars Exploration Rover Opportunity** and the Mars Science Laboratory *Curiosity*. Observations by the *Mars Reconnaissance Orbiter* have revealed possible flowing water during the warmest months on Mars.[26] In 2013, **NASA's *Curiosity*** rover discovered that Mars's soil contains **between 1.5% and 3%** water by mass (albeit attached to other compounds and thus not freely accessible).[27]

There are ongoing investigations assessing the **past habitability** potential of Mars, as well **as the possibility of extant life.** *In situ* investigations have been performed by the *Viking* landers, *Spirit* and *Opportunity* rovers, *Phoenix* lander, and *Curiosity* rover. Future astrobiology missions are planned, including the **Mars 2020** and ExoMars rovers.[28][29][30][31]

Mars can **easily be seen from Earth** with the naked eye, as can its **reddish coloring**. Its **apparent magnitude** reaches −2.91,[7] which is surpassed only by **Jupiter**, **Venus,** the **Moon**, and the **Sun**. Optical ground-based telescopes are typically limited to resolving features about 300 kilometers (190 mi) across when **Earth and Mars are closest because of Earth's atmosphere.**[32]

SATURN
THE FIRST PEG OF THE IMMOVABLE WORLD
OCTAGONAL RING OF THE FORCE CARRIER PARTICLES
WHITE DWARF COMPANION SIRIUS B
EAST HORIZON / NON-LUMINOUS MOON / DARK CHAMBER
"This does not include the hundreds of <u>moonlets</u> comprising the rings."
"<u>Titan</u>, Saturn's largest moon, and the second-largest in the Solar System"

Moonlet – Each Beautiful attributive name of Allah is a Moonlet.
Titan – Titan stands for Bermuda Triangle or Eartha 3D as the second peg of the immovable pentagonal earth.

Key Terms - <u>gas giant</u>, **one-eighth,** <u>**iron–nickel**</u> **and rock,** outer layer, **Electrical current**, **magnetic field**, 580, **one-twentieth of Jupiter's,** <u>Wind</u> <u>speeds</u>, **prominent** <u>**ring system**</u>, **nine continuous main rings** and **three discontinuous arcs**

Saturn is the sixth <u>planet</u> from the <u>Sun</u> and the second-largest in the <u>Solar</u> <u>System</u>, after <u>Jupiter</u>. It is a <u>gas giant</u> with an average radius about nine times that of <u>Earth</u>.[10][11] Although only **one-eighth** the average density of Earth, with its larger volume Saturn is just over 95 times more massive.[12][13][14] Saturn is named after the <u>Roman</u> <u>god of agriculture</u>; its <u>astronomical symbol</u> (♄) represents the god's <u>sickle</u>.

Saturn's interior is probably composed of a core of <u>**iron–nickel**</u> **and rock** (<u>silicon</u> and <u>oxygen</u> compounds). This core is surrounded by a deep layer of <u>metallic hydrogen</u>, an intermediate layer of <u>liquid hydrogen</u> and <u>liquid helium</u>, and finally outside the <u>Frenkel line</u> a gaseous outer layer.[15] Saturn has a **pale yellow hue due to <u>ammonia</u> crystals in its upper atmosphere. <u>Electrical</u> <u>current</u>** within the metallic hydrogen layer is thought to give rise to Saturn's planetary **magnetic field**, which is weaker than Earth's, but has a <u>magnetic</u> <u>moment</u> 580 times that of Earth due to Saturn's larger size. **Saturn's magnetic field strength is around one-twentieth of Jupiter's.**[16] The outer <u>atmosphere</u> is generally bland and lacking in contrast, although long-lived features can appear. <u>Wind speeds</u> on Saturn can reach 1,800 km/h (500 m/s), higher than on Jupiter, but not as high as those on <u>Neptune</u>.[17]

Saturn has a prominent <u>ring system</u> that consists of **nine continuous main rings** and **three discontinuous arcs** and that is composed mostly of **ice particles** with a smaller amount of **rocky debris and <u>dust</u>. Sixty-two**[18] <u>moons</u> are known to orbit Saturn, of which **fifty-three** are officially named. **This does not include the hundreds of <u>moonlets</u>** comprising the rings. **<u>Titan</u>, Saturn's largest moon**, and the **second-largest in the Solar System**, is larger than the planet <u>Mercury</u>, although less massive, and is the only **moon** in the Solar System to have a substantial atmosphere.[19]

EARTH

2006 – Earth is manifested as a hexagon but the Don oh the two-in-one partnership has projected as two hemisphere.

5 March 2006 – Though the Earth is manifested as a hexagon but due to equal and opposite Black Hole and White Dwarf regions, the manifested hexagon is appearing as a pentagon.

2007 – Verified Symbols as well as Clear Proofs are available on Windows'7 Ultimate.

Shown in 130 countries – Verified Symbols with their character codes are shown in 259 Scientific Institutions of 130 Countries.

Planet Earth is a <u>2006</u> British television series produced by the <u>BBC Natural History Unit</u>. Five years in the making, it was the most expensive <u>nature documentary</u> series ever commissioned by the <u>BBC</u> and also the first to be filmed in <u>high definition</u>.[1] *Planet Earth* premiered on **5 March 2006** in the <u>United Kingdom</u> on <u>BBC One</u>, and by June 2007 had been shown in 130 countries.

The **series comprises eleven episodes**, each of which features a **global overview of a different <u>biome</u>** or <u>habitat</u> on <u>Earth</u>. At the end of **each fifty-minute episode**, **a ten-minute featurette** takes a **behind-the-scenes** look at the challenges of filming the series.

Earth /ˈɜrθ/ (also **the world**[n 5], in <u>Greek</u>: **Γαῖα** *Gaia*,[n 6] or in <u>Latin</u>: **Terra**[26]) is the third <u>planet</u> from the <u>Sun</u>, the <u>densest</u> planet in the <u>Solar System</u>, the largest of the Solar System's **four <u>terrestrial planets</u>**, and the **only <u>astronomical object</u>** known to harbor <u>life</u>. The earliest <u>life on Earth arose</u> at least 3.5 billion years ago. Earlier physical evidence of life includes <u>graphite</u>, a <u>biogenic substance</u>, in 3.7 billion-year-old <u>metasedimentary rocks</u> discovered in **south western <u>Greenland</u>**, as well as, "remains of <u>biotic life</u>" found in **4.1** billion-year-old rocks in **<u>Western Australia</u>**.[27][28] Earth's <u>biodiversity</u> has expanded continually except when interrupted by <u>mass extinctions</u>.[29] Although scholars estimate that over **99** percent of all species of life (**over five billion**)[30] that ever lived on Earth are extinct,[31][32] there still an estimated **10–14** million extant species,[33][34] of which about **1.2** million have been documented and over **86** percent have not yet been described.[35] Over 7.3 billion humans[36] live on Earth and depend on its biosphere and minerals for their survival. Earth's human population is divided among about **two hundred sovereign states** which **interact through** diplomacy, conflict, travel, trade and communication media.

According to **evidence** from radiometric dating and other sources, Earth was formed about 4.54 billion years ago.[37][38][39] Within its first billion years,[40] life appeared in its **oceans** and began to affect its atmosphere and surface, promoting the proliferation of aerobic as well as anaerobic organisms. Since then, the combination of Earth's distance from the Sun, its physical properties and its geological history have allowed life to thrive and evolve.

Earth's lithosphere is divided into several rigid tectonic plates that migrate across the surface over periods of many millions of years. **Seventy-one** percent of Earth's surface is covered with water,[41] with the remainder **consisting** of **continents and islands** that together have many **lakes** and other sources of water that contribute to the **hydrosphere**. Earth's polar regions are mostly covered with ice, including the Antarctic ice sheet and the sea ice of the Arctic ice pack. Earth's interior remains active with a **solid iron inner core**, a **liquid outer core** that **generates the magnetic field**, and a converting mantle that drives plate tectonics.

Earth gravitationally interacts with other objects in space, especially the Sun and the Moon. During one orbit around the Sun, Earth rotates about its own axis 366.26 times, creating 365.26 solar days or one sidereal year.[n 7] Earth's axis of rotation is tilted 23.4° away from the perpendicular of its orbital plane, producing seasonal variations on the planet's surface with a period of one tropical year (365.24 solar days).[42] **The Moon is Earth's only permanent natural satellite.** Its gravitational interaction with Earth causes ocean tides, stabilizes the orientation of Earth's rotational axis, and gradually slows Earth's rotational rate.

Apart from meteors within the atmosphere and low-orbiting satellites, the main apparent motion of celestial bodies in Earth's sky is **to the west** at a rate of 15°/h = 15'/min. For bodies near the celestial equator, this is equivalent to an apparent diameter of the Sun or the Moon every two minutes; from Earth's surface, the apparent sizes of the Sun and the Moon are approximately the same.[178][179]

Earth orbits the Sun at an average distance of about 150 million kilometers every 365.2564 mean solar days, or one sidereal year. This gives an apparent movement of the Sun **eastward** with **respect to the stars** at a rate of about 1°/day, which is one apparent Sun or Moon diameter **every 12 hours**. Due to this motion, on average it takes 24 hours—a solar day—for Earth to complete a full rotation about its axis so that the Sun returns to the meridian. The orbital speed of Earth averages about 29.8 km/s (107,000 km/h), which is fast enough to travel a distance equal to Earth's diameter, about 12,742 km, in seven minutes, and the distance to the Moon, 384,000 km, in about 3.5 hours.[3]

The Moon and Earth orbit a common barycenter every 27.32 days relative to the background stars. When combined with the Earth–Moon system's common orbit around the Sun, the period of the synodic month, from new moon to new moon, is 29.53 days. Viewed from the celestial north pole, the motion of Earth, the Moon, and their axial rotations are all counter clockwise. Viewed from a vantage point above the north poles of both the Sun and Earth, Earth orbits in a counter clockwise direction about the Sun. The orbital and axial planes are not precisely aligned: Earth's axis is tilted some 23.4 degrees from the perpendicular to the Earth–Sun plane (the ecliptic), and the Earth–Moon plane is tilted up to ±5.1 degrees against the Earth–Sun plane. Without this tilt, there would be an eclipse every two weeks, alternating between lunar eclipses and solar eclipses.[3][180]

Earth, along with the Solar System, is situated in the **Milky Way** and orbits about 28,000 light years from its center. It is about 20 light years above the galactic plane in the Orion Arm.[182]

THE SUN
SUNSPOTS / SCREEN / BLACK UMBRELLA

The **Sun** is the star at the centre of the Solar System and is by far the most important **source of energy** for life on Earth. It is a nearly perfect spherical ball of hot plasma,[12][13] with internal convective motion that generates a magnetic field via a dynamo process. Its **diameter is about 109 times that of Earth**, and it has a mass about **330,000 times that of Earth**, accounting for about **99.86% of the total mass of the Solar System**.[15] Chemically, about **three quarters** of the Sun's mass consists of **hydrogen**, whereas the rest is mostly helium, and much smaller quantities of heavier elements, including oxygen, carbon, neon and iron.

The Sun is a G-type main-sequence star (G2V) based on spectral class and it is informally referred to as a yellow dwarf. It formed approximately 4.567 billion[b][17] years ago from the gravitational collapse of matter within a region of a large <u>molecular cloud</u>. **Most of this matter gathered in the centre**, whereas the rest flattened into an orbiting disk that <u>became the Solar System</u>. The central mass became increasingly hot and dense, eventually initiating <u>thermonuclear fusion</u> in its core. **It is thought that almost all stars <u>form by this process</u>**. The Sun is roughly **middle age** and has not changed dramatically for **four billion**[b] years, and will remain fairly stable for **four billion more**. However, after **hydrogen fusion** in its core has stopped, the Sun will undergo severe changes and become a <u>red giant</u>. It is calculated that the **Sun will become sufficiently large to engulf the current orbits of <u>Mercury</u>, <u>Venus</u>, and possibly Earth.**

Seen-Spot

Sunspots - Galileo observed the Sun through his telescope and saw that the Sun had **dark patches** on it that we now call *sunspots* (he eventually went blind, perhaps from damage suffered by looking at the Sun with his telescope). Furthermore, he observed **motion of the sunspots** indicating that the **Sun was rotating on an axis**. These "blemishes" on the Sun were **contrary to the doctrine of an unchanging perfect substance in the heavens**, and the rotation of the Sun made it less strange that the **Earth might rotate on an axis too**, as required in the **Copernican** model. Both represented new facts that were **unknown to Aristotle and Ptolemy**.

Ayat\Verses - And **you dwell in the dwellings of those who wronged their own souls**. You were clearly shown how We dealt with them; and made examples for you. Mighty indeed were the **plots** which they made, but their **plots** were [well] within the sight of Allah, even though they were such as to shake the hills! **Never think that Allah would fail his messengers in His promise**. Lo! Allah is Exalted in power, the Lord of Retribution. [Sura No. – 13\14 – Ibrahim, Ayat No. – 45 to 47]

Ayat\Verses - 'Iz qaala yuusuf li-'abiihi yaaa-'abati 'innii ra'-aytu 'aha-da 'ashara kaw-kabanw-wash-shamsa wal-qamara ra-'aytu-hum lii saajidiin."

[Trans.] Behold! Yuusuf said to his father: O my father! I did see **eleven springs [planets]** and **wash-shamsa wal-qamara**. I saw them prostrating themselves unto me! Said [the father]: My [dear] little son! Narrate not your vision to your brothers, lest **they fabricate a plot against you**: for **Shaytan is to man a declared enemy!** [Sura No. – 11]\12 – Yuusuf, Aat\Verses No. – 5]

SECTION – III
DIAMOND OPERATOR – 3.5
GLOBALISATION IS AN IMPOSSIBLE TASK

Globalisation – Global Universe – Global World – Global Earth
Common Run – Circular Rotation & Revolution System

HEARSAY, MECHANICAL GLOBALISATION, CIRCULAR
ROTATION & REVOLUTION SYSTEM, MAN-MADE MAGNETISM
OR
CROOKEDNESS / COMMON RUN

Fasta-'iz billaahi minash-Shaytaanir-Rajim
Bismillaahir-Rahmaanir- Rahim

Shall I inform you, [O people!], on whom it is that the evil ones descend? They descend on every lying, **wicked person**, [into whose ears] **they pour hearsay vanities**, and most of **them are liars**. And for Poets [pen-paper-pencil of IFTA], it is those straying in evil, who follow them. Have you not seen that they wander distracted in every valley? And that they say what they practise not [they project conspiracy science and conceal real science in a separate sheet]? Except those who believe, work righteousness, engage much in the remembrance of Allah and vindicate them-selves after they have been wronged. Those who do wrong will come to know by what a (great) reverse they will be over-turned! [Sura No. 25\26 – Shuraa, Ayat No. – 221 to 222]

O you who believe! There are indeed many among the **priests** (Lukmans) and **anchorites** [Imamgiris], who in **falsehood demolish** the **wealth of men** and **obstruct** [them] **from the way of Allah**. And there are those who **stored gold and silver and spend it not** in the **way of Allah** announce unto them a **most grievous penalty**. On the Day when heat will be produced out of that [wealth] in the **fire of Hell**, and with it will be branded their **foreheads**, their **borders**, and their **backs borders**, and their **backs**. This is the [treasure] which you stored for yourselves. Taste you, then, the [treasures] you stored. The **number of months [planets] in the sight of Allah is twelve. So ordained by Him the day He created the heavens and the world. Four of them are blessed that is the straight usage. So wrong not yourselves therein, and wage the Pagans / idolaters all together as they waging on all of you.** But know that Allah is with those who restrain themselves. **Verily the reversing is an addition to unbelief.** The unbelievers are led to wrong thereby, for they make it lawful one year, and forbidden another year, in order to adjust the number of months forbidden by Allah and make such forbidden ones lawful. The evil of their course seems pleasing to them. But **Allah guides not those who reject Faith**. [Sura No. – 8\9 – Tauba, Ayat No. – 34 to 37]

Is one who prays devoutly during the hour of the night prostrating himself or standing [in adoration], who takes heed of the Hereafter, and who places his hope in the Mercy of his Rab? Say: **Are those who know equal with those who do not know? But only men of understanding will pay heed (receive admonition).** Say: O Allah's bondmen who believe! Observe your duty to your Rab [following manifested signs of Allah]. For those who do good in this present life there is good, and **Allah's world is spacious.** Verily **the upright (followers of Upright West) will be paid their wages without measure.** Say: Lo! I am commanded (inspired) to serve Allah making **religion pure for Him (only) (Mukhlisal-lahud-din).** And I am commanded (inspired) to be the first of those who completely surrender (unto Him) [**Awwalal-Muslimiin**]. Say: Lo! If I should disobey my Rab, I fear the Penalty of the Mighty Day. Say: It is Allah I serve, making **religion pure for Him (only) (Mukhlisal-lahud-din** / with my sincere and **exclusive** devotion]. [Sura No. – 38\39 – Tanziilul-Kitaab, Ayat\Verses – 9 to 14]

You worship whom / what you will besides Him. Say: The **losers** will be those who lose themselves and their **house folk** on the **Day of Renaissance** (Resurrection). Ah, that will be the manifest loss! They have **an awning (a canopy)** of **fire above** them; and **beneath** them **a dais (space bar) (platform of devils)** with which Allah appals His **bondmen.** O My bondmen! Therefore **fear Me!** [Sura No. – 38\39 – Tanziilul-Kitaab, Ayat\Verses – 15 & 16]

And **those who put away false realities [projected mechanical globalisation, circular rotation & revolution system, and manmade magnetism] lest they should worship them** and **turn to Allah in repentance, for them there are good tidings.** Therefore give **good tidings to My bondmen.** Those who listen to **the Word,** and **follow the best [verifiable as well as justifiable by upright logic] in it, they are the ones whom Allah has guided,** and **they are men of understanding.** [Sura No. – 38\39 – Tanziilul-Kitaab, Ayat\Verses – 17 & 18]

Is, then, one against whom the decree of punishment is justly due, can you, then, rescue him [who is] in the Fire? [Sura No. 38\39 Tanziilul-Kitaab, Ayat\ Verses – 19]

But those who keep their duty to their Rab, for them are lofty halls, **one above another,** have been built, beneath them flow rivers. [Such is] the Promise of Allah. **Allah fails not His Promise.** [Sura No. – 38\39 – Tanziilul-Kitaab, Ayat\Verse - 20]

Have you not seen how Allah sends down **rain from the sky [West]**, and **leads** it through **planets (stars)** in the world [East]? Then He **causes** to grow, therewith **crops** of diverse hues, and afterward they wither and you see them **turn yellow**; then He makes them chaff. Lo! Herein verily is a **reminder** for **men of understanding.** [Sura No. – 38\39 – Tanziilul-Kitaab, Ayat\Verses – 21]

Is he whose bosom Allah has expanded for the surrender (unto Him i.e. Islam) so that he has received **Light** from his Rab, [as he who disbelieve]? Woe to those whose hearts are hardened against remembering the praises of Allah! **Such are in plain error.** [Sura No. – 38\39 – Tanziilul-Kitaab, Ayat\Verses – 22]

Allah has (now) revealed the **fairest of statements, Hadiisi Kitaabam** (wherein the promises of reward are) consistently paired [reference of equal & opposite reward and punishment], where at creeps the flesh of those who fear their Rab, so that their flesh and their hearts soften to Allah's reminder. Such is Allah's guidance, where with He guides whom He will. And him whom Allah sends astray, for him there is no guide. [Sura No. – 38\39 – Tanziilul-Kitaab, Ayat\ Verses – 23]

Note: - The **Hadiisi Kitaabam** includes **fairest of statements** of Reward and Punishment (**Fazaile Amal**) [along with the events took place with the guiding stars – **Waqiya**]

Is he then, one who will **strike his face against the awful doom** upon the Day of Resurrection [as he who does right]? And it will be said to the wrong-doers: Taste you [the fruits of] what you earned! [Sura No. – 38\39 – Tanziilul-Kitaab, Ayat\Verses – 24]

Those before them [also] rejected [revelation or manifested signs of revealed magnetism], and so the punishment came to them from **directions they did not perceive.** So Allah gave them a **taste of humiliation** in the **present life**, but greater is the punishment of the **Hereafter**, if they only knew! [Sura No. – 38\39 – Tanziilul-Kitaab, Ayat\Verses – 25 & 26]

We have put forth for men, **in this Qur'an every kind of Parable [similitude]**, in order that they may receive admonition. **[It is] a Qur'an in Arabic, without any crookedness** or global concept [therein]: in order that they may guard against evil [projected mechanism]. **Allah puts forth a parable / similitude. A man in relation to whom are several part-owners, quarrelling, and a man belonging entirely to one master. Are those two equal in similitude? Praise be to Allah! But most of them have no knowledge.** Truly you will die [one day], and truly they [too] will die [one day]. In the **end** you [all] will, on the Day of Resurrection, **settle your disputes** in the presence of your Rab. [Sura No. – 38\39 – Tanziilul-Kitaab, Ayat\Verses – 27 to 31]

Who, then, does more wrong than one who **utters a lie** concerning Allah, and **rejects the Truth** when it reaches him [Manifested Nature]? Will not the home of disbelievers be in hell? [Sura No. – 38\39 – Tanziilul-Kitaab, Ayat\Verses – 32]

And those who bring the Truth and confirm [with a view to believe] it - such are the men who do right. They shall have all that they wish for, in the presence of their Rab. Such is the reward of those who do good. That Allah will remit from them [even] the worst in their deeds and give them their reward according to the best of what they have done [Sura No. – 38\39 – Tanziilul-Kitaab, Ayat\Verses – 33 to 35]

Is not Allah enough for his Servant? But they try to frighten you with other [projected realities] besides Him! For such as Allah sends to go astray, there is no guide for him. And he whom Allah **guides**, there can be none to **lead astray**. Is not Allah Exalted in Power, [Able to enforce His Will], Rab of Retribution? [Sura No. – 38\39 – Tanziilul-Kitaab, Ayat\Verses – 36 & 37]

And verily if you ask them: '**Who has created the heavens and the world?**' they would be sure to say: **Allah**. Say: **Bethink** you then of the things [projected falsehood] that you **invoke** besides Allah. If Allah wills some **penalty** [hurt] for me, **can they remove His penalty**? Or if He wills some **grace** for me, **can they keep back His grace**? Say: **Allah is sufficient** [all in all] for me! In Him the **trustees put their trust**. [Sura No. – 38\39 – Tanziilul-Kitaab, Ayat\Verses – 38]

Say: O my People! Do whatever you can. I will do [my part]. But soon you will know. [Sura No. – 38\39 – Tanziilul-Kitaab, Ayat\Verses – 39]

Who is it to whom comes a penalty of ignominy, and on whom descends a penalty that abides? [Sura No. – 38\39 – Tanziilul-Kitaab, Ayat\Verses – 40]

Verily We have revealed the Kitaab to you in Truth, for [instructing] mankind. He, then, that **receives guidance** benefits his own soul. But he that **strays injures** his own soul. Nor you are set over them to **dispose of their affairs**. [Sura No. – 38\39 – Tanziilul-Kitaab, Ayat\Verses – 41]

Allah receives the souls [of men] at death; and that soul which yet dies not is its sleep. Those on whom He has passed the **decree** of death, He keeps back [from returning to life], but the rest He sends [to their bodies] for a term appointed. Verily in **these are Signs for those who reflect.** [Sura No. – 38\39 – Tanziilul-Kitaab, Ayat\Verses – 42]

What! Do they take for intercessors [mechanical magicians] others besides Allah? Say: **Even if they [the two-in-one partnership] have power over nothing and have no intelligence?** [Sura No. – 38\39 – Tanziilul-Kitaab, Ayat\ Verses – 43]

Say: To Allah belongs **exclusively** [all] **intercessions.** To Him belongs the **Sovereignty** of the heavens and the world. In the **End**, it is to Him that you will be brought back. [Sura No. – 38\39 – Tanziilul-Kitaab, Ayat\Verses – 44]

When Allah alone is mentioned, the hearts of those who believe not in the Hereafter are filled with disgust and horror; but when [projected falsehood] other than He are mentioned, behold, they are filled with joy! [Sura No. – 38\39 – Tanziilul-Kitaab, Ayat\Verses – 45]

Say: O Allah! **Creator** of the **heavens** and the **world!** Knower of all that is **hidden** and **open!** It is You Who will **judge** between Your **servants** in **those matters about which they have differed.** [Sura No. – 38\39 – Tanziilul-Kitaab, Ayat\Verses – 46]

Even if the **wrong-doers** had all that there is in the world, and as **much more**, [in vain] would they offer it for payment from the **pain of the penalty** on the **Day of Resurrection**; but something will **confront** them from Allah, which **they could never have counted upon!** For the **evils** of their **deeds** will **confront** them, and they will be [completely] **encircled** by that which they **used to mock at!** [Sura No. – 38\39 – Tanziilul-Kitaab, Ayat\Verses – 47 & 48]

Now, when **trouble** touches man, he cries to Us. But when We bestow a **favour** upon him as from Ourselves, he says: **This has been given to me because of a certain knowledge** [I obtained it]! Nay, but it is a **test.** But most of them **understand not!** Those before them said it, but all that they did was of **no profit to them.** Nay, the evil results of their deeds overtook them. **And the wrong-doers of this [projected falsehood] the evil results of their deeds will soon overtake them [too], and they will never be able to frustrate [Our Plan]!** They **do not know** that Allah **enlarges the provision or restricts it,** for any He pleases. Verily, in **these are Signs** for those who **believe!** [Sura No. – 38\39 – Tanziilul-Kitaab, Ayat\Verses – 49 to 52]

Say: O my servants who have **transgressed** against their souls! **Despair not** of the Mercy of Allah; for Allah forgives all **sins**; for He is Ever Forgiving, Most Merciful. **Turn [towards right direction] you to our Rab [in repentance] and bow to His [Will], before the penalty comes on you, as after that you shall not be helped.** And follow **the best** of [the guidance] **revealed** to you **from your Rab, before** the **penalty** comes on you of **a sudden while you perceive not!** Lest the soul should [then] say: Ah! **Woe is me!** In that I **neglected** [manifested signs] towards Allah, and was but among those who **mocked!** Or [lest] it should say: **If only** Allah had **guided** me, I should certainly have been among the **righteous!** Or [lest] it should say when it [actually] sees the **penalty**: If only I had **another chance**, I should certainly be among those who do good! [Sura No. – 38\39 – Tanziilul-Kitaab, Ayat\Verses – 53 to 58]

[The reply will be:] Nay, but there came to you **my Signs [of revealed magnetism]**, and you **rejected** them. You were **haughty**, and became **one of those who reject faith!** [Sura No. – 38\39 – Tanziilul-Kitaab, Ayat\Verses – 59]

On the **Day of Resurrection** you will see those who told **lies** against Allah; their **faces** will be **turned black [towards Black Hole of the Middle East]. Is there not in Hell an abode for the Haughty?** [Sura No. – 38\39 – Tanziilul-Kitaab, Ayat\Verses – 60]

But Allah will deliver the **righteous** to their **place of salvation. No evil** shall touch them, nor shall they grieve. [Sura No. – 38\39 – Tanziilul-Kitaab, Ayat\Verses – 61]

Allah is the **Creator** of all things, and He is the **Guardian** and **Disposer** of all affairs. To Him belong the keys of the heavens and the world; and those who **reject the Signs of Allah [Manifested Nature]**, it is they who will be in **loss**. Say: **Do you bid me serve other than Allah, O you fools [ignorant ones]?** [Sura No. 38\39 – Tanziilul-Kitaab, Ayat\Verses – 62 to 64]

But it has already been revealed to you, as it was to those before you, If you ascribe partners [projected falsehood], truly fruitless will be your work [service towards Allah], and you will surely be in the ranks of those who lose. Nay [nothing], but serve Allah and be of those who **give thanks.** [Sura No. – 38\39 – Tanziilul-Kitaab, Ayat\Verses – 65 & 66]

And **they esteem not Allah** as **He has the right to be esteemed**, when the whole world is His handful on the Day of Judgment, and the heavens are rolled in His **right hand**. He be glorified and High is He **above the Partners they attribute** to Him! [Sura No. – 38\39 – Tanziilul-Kitaab, Ayat\Verses – 67]

And the **Trumpet** will [just] be **sounded**, when all that are in the heavens and in the world will **faint**, except such as it will please Allah [to exempt]. Then will a **second** one be **sounded**, when, behold, they will be **standing and waiting**! [Sura No. – 38\39 – Tanziilul-Kitaab, Ayat\Verses – 68]

"Wa 'ash-raqatil-'arzu bi-nnuri Rabbiha wa jiii-'a binnabiu-yiina wash-shuhadaaa-'i w a quziya baynahum-bil-haqqi wa hum laa yuzlamuun." [Trans.] And the **World will shine with the glory of her Rab**. The **Record** [of Deeds] will be placed [open]. The **prophets** [binnabiu-yiina] and the **witnesses** [wash-shuhadaaa-'i] will be brought **forward** and **a just decision pronounced** between **them**; and **they will not be wronged** [in the least]. [Sura No. – 38\39 – Tanziilul-Kitaab, Ayat\Verses – 69]

And to every **soul** will be paid in full [the fruit] of its **deeds**; and [Allah] knows best all that they do. [Sura No. – 38\39 – Tanziilul-Kitaab, Ayat\Verses – 70]

And those who **disbelieve** are driven into hell in **troops** till they reach it and the **gates** thereof are **opened**, and its **keepers will say: Did not messengers come to you from among yourselves, rehearsing to you the Signs of your Rab, and warning you of the Meeting of this Day of yours?** The answer will be: **True**: but **the Decree of Punishment has been proved true against the disbelievers!** [To them] will be said: **Enter you the gates of Hell**, to dwell therein; and **evils** are the abode of the **arrogant**! [Sura No. – 38\39 – Tanziilul-Kitaab, Ayat\Verses – 71 & 72]

And those who **feared** their Rab will be led to the Garden in crowds; until behold, they arrive there; its **gates** will be opened; and its **keepers will say: Peace** be upon you! You have done **well**! Enter you here, to dwell therein. **They** will say: **Praise be to Allah**, Who has **truly fulfilled His Promise to us**, and has given us [this] **land [earth] in heritage**. We can dwell in the **Garden** as we will. **How excellent a reward for those who work [righteousness]!** [Sura No. – 38\39 – Tanziilul-Kitaab, Ayat\Verses – 73 & 74]

And you will see the **angels surrounding** the **Throne** [Sign of **Arsh**] on all sides, **singing Glory and Praise to their Rab**. The **Decision between them** [at Judgment] will be in [perfect] **justice**, and the **cry** [on all sides] will be, Praise be to Allah, the **Rab of Aalamiin** [universe]! [Sura No. – 38\39 – Tanziilul-Kitaab, Ayat\Verses – 75]

QURAN IS AN EASY KITAAB WITH TRUTH

Lo! Your Rab knows how you keep vigil, sometimes nearly **two-thirds** of the night, or **half** or a **third thereof**, as do a party of those with you. Allah **measures** the **night and the day**. He knows that **you count it not**, and **turns to you** in mercy. **Recite then of the Quran which is easy for you**. He knows that there are **sick folk** among you, while others **travel** in the land in search of Allah's bounty, and others are **fighting** for the cause of Allah. So recite of it which is **easy**, and establish worship and pay the poor-due, and lend to Allah a goodly loan. Whatsoever good you send before you for your souls, you will surely find it with Allah, better and greater in the recompense. And seek forgiveness of Allah. Lo! Allah is Ever Forgiving, Most Merciful. [Sura No. 72\73 – Yaaa-ayyuhal Muzzamil, Ayat No. - 20]

And We have indeed made the **Qur'an easy to understand** and **remember**: then is there any that will receive admonition? [Sura No. – 53\54 – Qamar, Ayat\ Verses – 17, 22, 32, 40]

LANGUAGE OF ONE'S OWN PEOPLE

Alif-Lam-Ra. A Kitaab which We have revealed unto you, in order that **you might lead mankind out of the depths of darkness into light by the permission of their Rab** to the Way of [Him] the Exalted in power, worthy of all praise! Unto Allah belong all things in the heavens and on earth. But alas for the unbelievers for a terrible penalty [their Unfaith will bring them]! Those who love the life of this world more than the Hereafter and **debar** [men] from the Path of Allah and seek therein something **crooked**, they are **astray** by a long distance. And **We never sent a messenger except [to teach] in the language of his [own] people, in order to make [things] clear to them**. Then Allah leaves whom He pleases astray and guides whom He pleases; and He is Exalted in power, full of Wisdom. [Sura No. – 13\14 – Ibrahim, Ayat No. – 1 to 4]

"We have made this [Qur'an] **easy, in your tongue**, in order that [reason] they may give heed."

As to the Righteous [they will be] in a position of Security, among **Gardens** (Galaxy of Stars / West Horizon) and **Springs** (Matter Particles); dressed in fine **silk** and in rich **brocade**, they will **face each other**. So We shall join them to **fair women** with **beautiful, big**, and **shiny eyes**. There they can call for every kind of **fruit** in peace and **security**. There they will not taste **Death**, except the **first death**; and He will preserve them from the Penalty of the **Blazing Fire**, as a Bounty from your Rab! That will be the **Supreme Triumph**! Verily, We have made this [Qur'an] easy, in your tongue, in order that they may give heed. So wait you and watch; for they [too] are waiting. [Sura No. – 43\44 - Bi-Dukhaanim-Mubiin, Ayat No. – 51 to 59]

QURAN IN ARABIC WITHOUT ANY CROOKEDNESS OR GLOBAL CONCEPT OR CIRCULAR WAYS

We have put forth for men, **in this Qur'an every kind of Parable [similitude]**, in order that they may receive admonition. **[It is] a Qur'an in Arabic, without any crookedness** [therein]: in order that they may guard against evil. **Allah puts forth a parable / similitude. A man in relation to whom are several part-owners, quarrelling, and a man belonging entirely to one master. Are those two equal in similitude? Praise be to Allah! But most of them have no knowledge.** Truly you will die [one day], and truly they [too] will die [one day]. In the **end** you [all] will, on the Day of Resurrection, **settle your disputes** in the presence of your Rab. **[Sura No. – 38\39 – Tanziilul-Kitaab, Ayat\Verses – 27 to 31]**

What they say is nothing but falsehood!

Further that He may **warn those** [also] who say: Allah **has chosen a son. No knowledge they have of such a thing, nor had their fathers. Dreadful is the word** that comes out of their mouth. **What they say is nothing but falsehood!** [Sura No. – 17\18 – Khaf – Ayat No. – 4 & 5]

THERE IS NO CROOKEDNESS [CIRCULAR WAYS OR GLOBAL CONCEPT]

Praise be to Allah, Who has revealed the **Abdihil-Kitaaba** unto His slave (servant), and has placed therein **no crookedness.** [He has made it] **equal & opposite** in order that He may **warn of a terrible punishment** from Him, and that He may give **good tidings to the believers who work righteous deeds [tabligh],** that they will have a **fair reward**; wherein they will abide forever. [Sura No. – 17\18 – Khaf – Ayat No. – 1 to 3]

The Day when the **Trumpet will be blown**, that Day, We shall gather the sinful, **white-eyed** [with terror]. Murmuring among themselves: Yet tarried **not longer than ten [Days]**. We know best what they will say, when their **leader** most eminent in conduct will say: You tarried **not longer than a day**! They ask you concerning the **Mountains**. Say: My Rab will break them and scatter them as dust. He will leave them as plains smooth and level. **Nothing crooked** or **curved** or global concept will you see in their place. [Sura No. – 19\20 – Twa-Haa, Ayat No. – 102 to 107]

On that Day, they will follow the caller, no **crookedness** [can they show] in him. All sounds shall humble themselves in the Presence of [Allah] Most Gracious. Nothing shall you hear but the tramp of their feet [as they march]. On that Day **no intercession** will avail except for those for whom **permission** has been granted by [Allah] Most Gracious and whose **word** is acceptable to Him. **He knows what [appears to His creatures as] before or after or behind them**, but they shall **not compass** it with their knowledge. [All] faces shall be humbled before [Him], the **Living, the self-subsisting, eternal**. Hopeless indeed will be the man that carries iniquity [on his back]. But he who **works deeds of righteousness**, and has **faith**, will neither have fear of harm, nor of any **curtailment** [of what is his due]. [Sura No. – 19\20 – Twa-Haa, Ayat No. – 108 to 112]

Say: O **People of the Kitaab**! Why disbelieve you the **Signs of Allah**, when Allah is Himself **witness** to all you do? Say: O you People of the Kitab! **Why drive you back believers from the path of Allah, seeking to make it global or crooked**, while you were yourselves **witnesses** [or equal & opposite revelation]? But Allah is **not unmindful of all that you do**. [Sura No. 2\3 – Imran, Ayat No. – 98 to 99]

COMMON RUN IS A PROJECTED CONSPIRACY
Equal & Opposite is the Revealed Truth
Ayat\Verses - If you obey to **follow** the **common run** [projected globalisation] of those in the world, **they will lead you away from the way of Allah**. They **follow nothing** but conjecture. **They do nothing but lie.** Your Rab knows best who **strays** from His way: He knows best who they are that **receive His guidance**. So **eat** of [those **sustenance**] on which **Allah's name has been pronounced**, if you have **faith** in **His signs** Why should you not eat of [those sustenance] on which Allah's name has been pronounced, when He has explained to you in detail what is forbidden to you except under compulsion of necessity? But many do mislead [men] by their appetites unchecked by knowledge. Your Rab knows best those who transgress. [Sura No. – 5\6 – An-aam, Ayat No. –117 to 120]

Neither mutual bargaining nor befriending
Ayat\Verses – Tell to my servants who have believed, that they may establish regular prayers, and spend [in charity] out of the sustenance we have given them, secretly and openly, before the coming of a Day in which there will be **neither mutual bargaining nor befriending**. It is Allah Who hath created the heavens and the world and sends down rain from the skies [West], and with it brings out fruits wherewith to feed you. It is He Who has made the ships

subject to you, that they may sail through the sea by His command; and the rivers [also] has He made subject to you. And **He has made subject to you lakumush-shamsa wal-qamara, both diligently pursuing their courses**; and the night and the day has he [also] made subject to you. And He gives you of all that you ask for. But if you count the favours of Allah, never will you be able to number them. Verily, **man is given up to injustice and ingratitude**. [Sura No. 13\14Ibrahim, Ayat No.31 to 34]

CLEAR SIGNS OF THE THREE ASCENDING STAIRS
OF THE APPEARING PENTAGONAL EARTH
REVEALED TRUTH - GLOBALISATION IS AN IMPOSSIBLE TASK

i) Near bank of the valley [Middle West of Europe, Asia, Africa, and Australia]
ii) Farther bank [Upper West of Arabian Peninsula]
iii) Lower ground [East Zone of North America and South America]
iv) Globalisation – Globalisation is an impossible task - **Even if you had made a mutual appointment to meet [Global Concept], you would certainly have failed in the appointment**

Remember you [Ummat or Midiyan] were on the **near bank of the valley**, and they on the **farther bank** and **the caravan on lower ground than you. Even if you had made a mutual appointment to meet, you would certainly have failed in the appointment:** But [thus you met], that Allah might accomplish a matter already enacted; that **those who died might die after a clear Sign** [had been given], and **those who lived might live after a Clear Sign** [had been given]. And verily Allah is He Who hears and knows [all things]. **Remember in your dream Allah showed them to you as few. If He had shown them to you as many, you would surely have been discouraged, and you would surely have disputed in [your] decision; but Allah saved [you]: for He knows well the [secrets] of [all] hearts**. And remember when you met, He showed them to you as few in your eyes, and He made you appear as disgraceful in their eyes that Allah might accomplish a matter already enacted. For to Allah do all questions go back [for decision]. **[Sura No. – 7\8 – Anil-Anfaal, Ayat No. – 42 to 44]**

CONSPIRACY PLOT WITH NIGHT & DAY

Those who had been despised will say to the proud ones: Nay! **It was a plot [of yours] scheming night and day when you [constantly] ordered us to be ungrateful to Allah and to attribute equals (set up rivals like projected globalisation and artificial magnetism) to Him! They will declare [their] repentance when they see the Penalty. We shall put yokes on the necks of** the disbeliever. It would only be requital for their [ill] deeds. [Sura No. – 33\34 – Saba, Ayat No. - 33]

WICKED ONES

Is she **who was dead** and We have **raised her unto life,** and **set for her a light wherein she walks among men,** as him **whose similitude is in utter darkness whence he cannot emerge?** Thus is **their conduct made fair seeming for the disbelievers?** Thus have We made in every city great ones of its **wicked ones** that they should plot therein? But **they only plot against their own souls, and they perceive it not.** [Sura No. – 5\6 – An-aam, Ayat No. –123 & 124]

Neither it is permitted to the Lash-Shamsu to catch up the Moon [Qamar] **nor can the Night outstrip the Day. Each [just] swims along in [its own] orbit / Axis [according to Law].** [Sura No. – 35\36 – Yaa-siiin, Ayat No. – 40]

YOU WILL FIND NO CHANGE IN OUR WAYS

But those who were blind in this world [present life] will be blind in the hereafter, and most astray from the Path. And their purpose was to tempt you away from that which We had revealed unto you, **to substitute in Our name something quite different;** [in that case] **Behold! They would certainly have made you [their] friend!** And had We not given you **strength,** you would nearly have **inclined to them** a little. In that case We should have made you **taste an equal portion [of punishment]** in **this life,** and an **equal portion in death**; and moreover you would have found **none to help you against** Us! **And they indeed wished to scare you from the land that they might drive you forth from thence, and then they would have stayed (there) but a little after you.** [This was Our] way with those whom We sent before you and you will find **no change in Our ways (methods).** Establish regular prayers - **Li-duluukish-shamsi** till the darkness of the night, and (the recital of) Quran at dawn (**aanal-Fajr**). Lo! **(The recital of) the Quran at dawn is ever witnessed (carry testimony). [Innal Quraanal-Fajri kaana mash-huudaa.]** And some part of the night awake for it and additional prayer for you. It may be that your Rab will raise you to a station of praise. And say: My Rab! Cause me to come in with a **firm incoming** (i.e. by the gate of truth and honour) and to go out with a **firm out going** (i.e. by the gate of truth and honour). And give me from Your Presence an authority to aid. **And say: Truth has [now] arrived, and Falsehood perished. Lo! Falsehood is [by its nature] bound to perish. [Sura No. – 16\17 - Banii-Israai-iil, Ayat No. – 72 to 81]**

SECTION – IV
BLACK SQUARE – 4.1
MANIFESTED WORLD IS A HEXAGON

MANIFESTED WOLD WITHIN THE BLACK
SQUARE OF EAST HORIZON IS A HEXAGON
Fasta-'iz billaahi minash-Shaytaanir-Rajim
Bismillaahir-Rahmaanir- Rahim

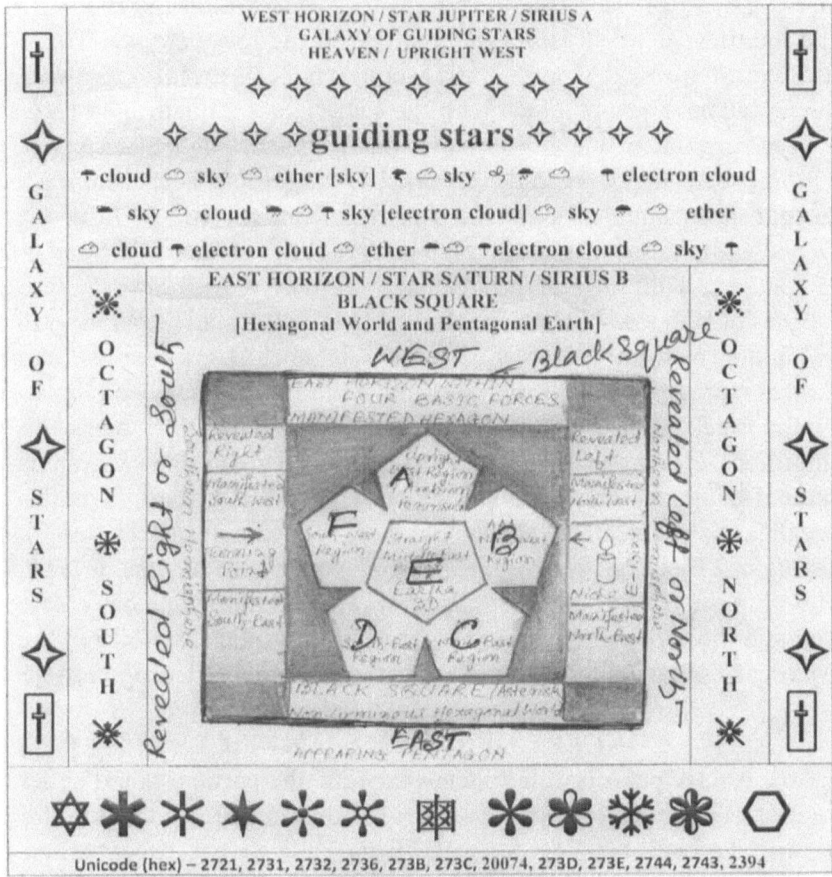

"One day the world will be changed to a different world"

Ayat\Verses - Yawma tubad-dalul- Arzu gayral – Arzi was-Samaa-waatu wa barazuu lillaahil- Waa-hidil-Qahhaar. [Trans.] **One day the world will be changed to a different world** and so will be the heavens and they will come forth unto Allah, the One-Almighty. [Sura No. – 13\14 – Ibrahim, Ayat No. – 48]

Revealed **East Horizon** [Star Saturn] guarded by the four revealed [cardinal] directions and four basic forces [walls] is called **Black Square.** This Black Square is the projected **Non-luminous Moon** or **Dark Moon** or **Dark Chamber.** This Non-luminous Moon or Dark Moon or Black Square does not represent what we commonly perceive white or ice-like moon at night. On the contrary, Non-luminous Moon or Dark Moon represents East Horizon as the first peg of the immovable world and Black Square within which the hexagonal world has been manifested. Ring of the Star Saturn represents four walls of the Black Square. The ring of the Star Saturn [four walls] is the peg of the Revealed East Horizon i.e. manifested world within the Black Square. So, the manifested world is not moving due to the revealed peg. This Black Square [Non-luminous Moon / Dark Moon] has been revealed as two Zones and an Octagon in resemblance with the sacrifice of a quadruped [Baqarah] from psychic point of view. So, matter particles [springs / black spots / so-called sunspots] are octagonal. Revealed Octagon has been Manifested as a **Hexagon** after equal & opposite sacrifice from psychic point of view in coherence with the sacrifices of Ismail (ass) & Ibrahim (ass) [Tawraat], and in correspondence with the equal & opposite sacrifice of the converted green & blue shades of Windows'7 Ultimate [Injiil]. This manifested hexagon within the Black Square [Non-luminous Moon / Dark Chamber / Dark Moon] of the East Horizon is called the **World or Asterisk.** As the World has been manifested within the Black Square; so, the manifested hexagonal world [Asterisk] is also non-luminous. Hexagonal symbols in the converted green shade of Windows'7 Ultimate, Hexagonal units such as Unicode (hex), ASCII (hex), Symbol (hex) etc. are the clear proofs of the Hexagonal World called Asterisk. Helium-4 atom consisting of 2 Electron, 2 – Proton, and 2 – Neutron also represents hexagonal world within the Black Square. It has been revealed that Allah has created the Heaven and the World in six days. So, it is certain like 'Cogito Ergo Sum' of Rene Descartes that the Manifested immovable World is a Hexagon [Asterisk].

CONCEPT OF TWICE SACREFICES

A **divorce is only permissible twice**: after that, the parties should either hold together on equitable terms, or separate with kindness. It is not lawful for you, [Men], to take back any of your gifts [from your wives], except when both parties fear that they would be unable to keep the limits ordained by Allah. If you [judges] do indeed fear that they would be unable to keep the limits ordained by Allah, there is no sin on either of them if she gives something for her freedom. These are the limits ordained by Allah. So do not transgress them. If any do transgress the limits ordained by Allah, such persons are wrong-doers. [Sura No. – 1\2, Baqarah, Ayat\Verse No. – 229]

East Horizon within the four basic forces is called **Black Square** [or what conspiracy science calls **Non-luminous Moon**]. The World has been manifested within the Black Square i.e. four basic forces or four walls of East Horizon. The world has been manifested declining towards Revealed Right [Southern Hemisphere or Haiyalal-falaah] after equal & opposite division of the East Horizon [Ref. Shred parable Baqarah] into two Seashores. This manifestation resembles with suppose X or Multiplication Sign of Mathematics.

Two seashores within the Black Square are called Zones. The Upper Seashore is called West Zone in coherence with upward direction towards Upright West Region, and in correspondence with the appointed Kaba at the right side of the Mount Tuur in the Top West [Upright West] Region of the Manifested Hexagonal World. The Lower Seashore is called East Zone in coherence with downward direction towards East, and in correspondence with the Straight Middle East Region of Eartha 3D.

Four basic forces are the four blessed springs [four imams for straight usage] or four walls of scientific investigations. Any knowledge/broadcasting the domain of which is external to four basic forces cannot be categorised as verifiable certain knowledge or scientific knowledge. Such external knowledge can be categorised as justifiable valid knowledge. Validity of such external knowledge depends on the resemblance between hypothesis and Revelation. So, **Science cannot go beyond four basic forces**. Scientific broadcasting the subject matter of which is beyond four basic forces cannot be recognised as certain knowledge or verifiable scientific knowledge. On the contrary, such broadcasting is either justifiable valid knowledge in resemblance with Revelation or teleological conspiracy against Revelation & Manifestation in the name of scientific certainty.

Manifested hexagonal world [Asterisk] has two Zones and six Regions. Each Zone comprises three Regions [Triangles]. West Zone [Upper Seashore] comprises **South-West Region** of Europe, **North-West Region** of Asia – Africa – Australia, and **Top West Region** of Arabian Peninsula. Arabian Peninsula is neither towards Revealed Left nor towards Revealed Right. So, Arabian Peninsula is the **Upright West Region**. East Zone [Lower Seashore] comprises **North-East Region** of North America, **South-East Region** of South America, and **Straight Middle East Region** of Eartha 3D. The hexagonal unit [hex] is the self-evident proof as well as valid reference of the Manifested Hexagonal World.

Converted red and green shades of Windows'7 Ultimate represent East Zone [Lower Seashore] of the manifested hexagonal world. Converted yellow and blue shades of Windows'7 Ultimate represent West Zone [Upper Seashores] of the manifested hexagonal world. Converted red shade of Windows'7 Ultimate represents North-East Region of North America. Converted green shade of Windows'7 Ultimate represents South-East Region of South America. Converted yellow shade of Windows'7 Ultimate represents South-West Region of Europe. Converted blue shade of Windows'7 Ultimate represents North-West Region of Asia, Africa, and Australia. White cloud of Windows'7 Ultimate represents Upright-West Region of Arabian Peninsula. The gap [i.e. point of intersection] at the centre of four shades of Windows'7 Ultimate represents Straight Middle East Region of Eartha 3D. Symbols of rain and quadrilateral stars in the converted yellow and blue shades of Windows'7 Ultimate are the clear proofs of the West Zone as well as upward direction towards West. Symbols of hexagons and equal & opposite tree in the converted green and red shades of Windows'7 Ultimate are the clear proofs of the East Zone as well as downward direction towards East.

SIX REGIONS OF THE MANIFESTED HEXAGONAL WORLD WITHIN THE BLACK SQUARE [NON-LUMINOUS MOON OR DARK MOON] OF EAST HORIZON [PLANET/MOON SATURN]

Six Regions of the Manifested Hexagonal World
represents six manifested directions.

EAST ZONE [LOWER SEASHORE] COMPRISES THREE REGIONS [TRIANGLES]

i)   North-East Region of North America
ii) Middle East Region of Eartha 3D
iii) South-East Region of South America

WEST ZONE [UPPER SEASHORE] COMPRISES THREE REGIONS [TRIANGLES]

iv) South-West Region of Europe
v) Upright West Region of Arabian Peninsula
vi) North-West Region comprises three continents such as Asia – Africa - Australia

MENIFESTED WORLD IS A HEXAGON

"Nor between five but He makes the sixth nor of less than that or more, but He is in their midst, where so ever they be."
[Trinity – Three], [Quadrilateral – four], [Pentagon – five], [Hexagon – six], [Not of less than Six or more, but He is in their midst – Manifestation - Seven]

Ayat\Verses – On the Day when Allah will raise them all together and show them the **Truth** [what they did] of their conduct. Allah has kept account of it, though they may have forgotten it, for Allah is **Witness** over **all** things Have you not seen that Allah knows all that is in the heavens [West Horizon] and all that is in the world [East Horizon]? **There is not a secret consultation between three**, but **He makes the fourth** among them. **Nor between five but He makes the sixth nor of less than that or more, but He is in their midst, where so ever they be**. In the end He will tell them the truth of their conduct, on the Day of Resurrection [Yawmal-Qiyaamah]. Lo! Allah is The Knower of all things. [Sura No. 57\58 – Mujaadalah, Ayat No. - 6 & 7]

"Your Ultimate-Rab is Allah, Who created the heavens and the world in six days."
Ayat\erses - Your Ultimate-Rab is Allah, Who created the **heavens** and the **world in six days**, and then He firmly mounted the Arsh [stages of journey]. He draws the night as a veil over the day, each seeking the other in rapid succession. He made Wash-**shamsa**, Wal-**qamara** and Wan-**nujuuma** (galaxy stars) sub-servient by His command. His verily all creations and commandment. Blessed be Allah, the Rab (Cherisher and Sustainer) of the Alamin [Universe]! [Sura No. – 6\7 - As-haabul- a' raaf, Ayat No. – 54]

It is Allah Who has **created** the **heavens** and the **world**, and all **between** them, in **six Days**. Then He **mounted [stages of journey] the Throne**. You have none, besides Him, a protecting friend or mediator. Will you then not receive admonition? [Sura No. – 31\32 – Sajdah, Ayat\Verses – 4]

Ayat\Verses – Verily your Rab is Allah, who **created the heavens and the world in six days**, and is firmly established Himself upon the Throne [Arsh], regulating and governing all things. No intercessor [can plead with Him] except His permission. This is Allah your Rab; therefore serve Him. Oh! Will you not remind? [Sura No. – 9\10 – Yuunus, Ayat No. – 3]

> **He it is Who created the heavens and the world in six**
> **Days and His Arsh [Throne] was over the waters that**
> **He might try you, which of you is best in conduct.**

Ayat\Verses - Behold! They fold up their hearts, that they may lie hid from Him! Ah even when they cover themselves with their garments, He knows what they conceal, and what they reveal: for He knows well the [inmost secrets] of the hearts. There is no moving creature in the world but its sustenance depends on Allah: He knows the time and place of its definite abode and its temporary deposit: All is in a clear Record. **He it is Who created the heavens and the**

world in six Days and His Arsh [Throne] was over the waters that He might try you, which of you is best in conduct. But if You say to them, you shall indeed be raised up after death, the unbelievers would be sure to say, this is nothing but obvious sorcery (open magic) If We delay the penalty for them for a definite term, they are sure to say: What keeps it back? Ah! On the day it [actually] reaches them, nothing will turn it away from them, and they will be completely encircled by that which they used to mock at! [Sura No. – 10\11 – Kitaabun-uh-kimat (Prev. – Huud), Ayat No. – 5 to 8]

IMMOVABLE WORLD
And **the World We have spread out [manifested], and set thereon mountains firm and immovable**

And even if We opened out to them a **gate** from heaven [West], and they were to continue [all day] **ascending therein [stages of journey]**. They would only say: **Our eyes have been intoxicated**: Nay, we have been **bewitched by sorcery.** And verily in the heaven We have set **mansions (canopies or windows) of the stars**, and We have **beautified** it [**West Horizon**] for **beholders**. And **We have guarded them from every cursed devil.** Save him who stealth the bearing, and then does a **clear flame** [commonly perceived moon] pursue. And **the World We have spread out [manifested], and set thereon mountains firm and immovable**; and produced therein all kinds of things in **due balance**. And We have provided therein **means of subsistence [Electromagnetic wave]** for you and for those for **whose sustenance you are not responsible.** And **there is not a thing** but its **[sources** and] **treasures** [inexhaustible] are with Us; but We only send down thereof in due and **ascertainable measures**. And We send the fecundating winds, then cause the rain to descend from the sky, therewith providing you with water [in abundance], though **you are not the guardians of its stores [beyond the four walls of scientific investigations]**. And verily, it is **We Who give life, and Who give death**. It is We Who remain inheritors [after all else passes away]. And verily We know those of you **who hasten forward**, and those **who lag behind**. Assuredly it is your Rab Who will gather them together, for He is perfect in Wisdom and Knowledge. [Sura No. – 14\15 = Hur, Ayat No. – 14 to 25]

Ayat\Verses – And We made a people, considered weak [and of no account], inheritors of in **East Zone** [Lower Seashore] and **West Zone** [Seashore], **lands** [middle parts] whereon We sent down Our blessings. The fair promise of your Rab was fulfilled for the Baniii - Israaa -il, because they had patience and constancy, and We levelled to the ground the great works and fine buildings which Firawn and his people erected [with such pride]. [Sura No. – 6\7 - As-haabul- a' raaf, Ayat No. – 137]

Ayat\Verses – Do not the Unbelievers see that the heavens and the world are joined together, before we clove them asunder? We made from water every living thing. Will they not then believe? And We have set in the world **mountains standing firm, lest it should shake with them**, and We have made therein **broad highways** for them to pass through that they may receive Guidance. And We have made the heavens as a **canopy** well guarded [Night Sky]; yet do they turn away from the Signs which these things [point to]! It is He Who created the Night and the Day, and **wash-shamsa wal-qamara. They float, each in an orbit.** [Sura No. – 20\21 – Ambiyaaa, Ayat No. – 30 to 33].

TWO DAYS AND FOUR DAYS
Two Days – Straight Middle-East Region of Eartha 3D and Upright West Region of Arabian Peninsula
Four Days – North-East Region, South-East Region, South-West Region, and North-West Region

Come both of you, willingly or unwillingly.
Say: **Disbelieve** you verily in Him **Who created the world in two days**, and **ascribe** you unto Him **rivals**? He is the Rab of the universe. He set in the [world], mountains **standing firm**, high above it, and bestowed blessings in the world, and **measure therein all things to give them nourishment in due proportion, in four Days**, in accordance with [the needs of] those who seek [Sustenance]. Then **He turned to the heavens [West Horizon] when it was smoke [white dwarf]**, and said unto it and unto the earth: **Come both of you, willingly or unwillingly**. They said: We come, obedient. So He **ordained** them as **seven firmaments (windows) in two Days**, and **He inspired in each heaven its mandate [movements]**. And **We adorned the lower heaven with lights** and **[provided it] with guard.** Such is the Measuring of the Mighty, The Knower. **[Sura No. – 40\42 – Haa-Mim-Sajdah, Ayat No. – 9 to 12]**

Ayat\Verses - "Fabi-ayyi aalaaa – i Rabbikumaa tukazzibaan?" [Trans.] Which is it of the **favours** of your Rab that you **deny**? [Sura No. 54\55, Ar-Rahman, Ayat\Verses.18]
Ayat\Verses - "Rabbul-**Mashri-qayni** wa Rabbul-**Magribayn**" Rab of the **two Easts** and Rab of the **two Wests**. [Sura No. – 54\55, Ar-Rahman, Ayat\Verses. 17]

SECTION – IV
BLACK SQUARE – 4.2
TWO EAST AND TWO WEST

REVELATION AND MANIFESTATION
Revealed and manifested Universe
Two East – East Horizon and East Zone
Two West – West Horizon and West Zone
Revealed and Manifested World
Two East – Ground Stair and Straight Middle-East Region of Eartha 3D
Two West – Middle Stair and Upright West Region of Arabian Peninsula
Fasta-'iz billaahi minash-Shaytaanir-Rajim
Bismillaahir-Rahmaanir- Rahim

TWO HORIZONS OF THE REVEALED UNIVERSE
&
TWO SEASHORES WITHIN THE BLACK SQUARE
OF THE MANIFESTED EAST HORIZON

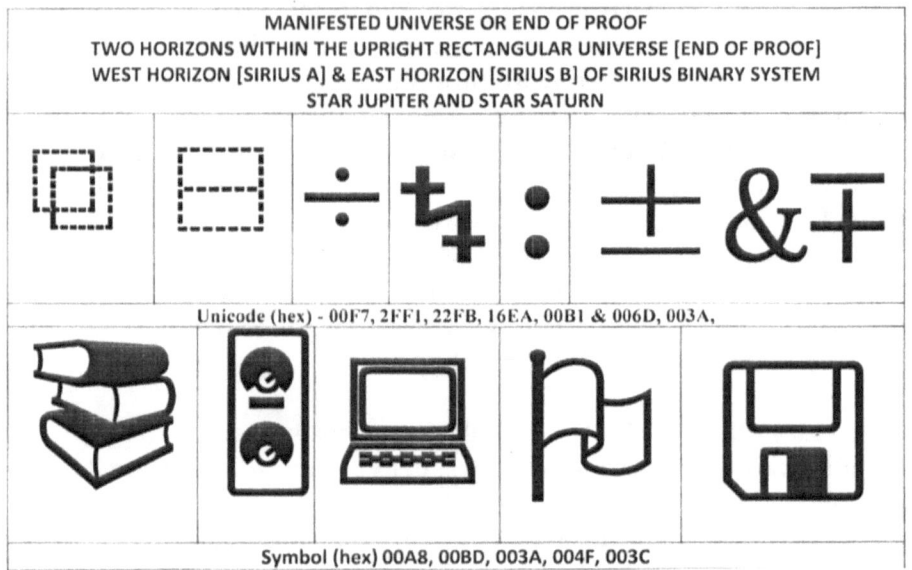

MANIFESTED UNIVERSE OR END OF PROOF
TWO HORIZONS WITHIN THE UPRIGHT RECTANGULAR UNIVERSE [END OF PROOF]
WEST HORIZON [SIRIUS A] & EAST HORIZON [SIRIUS B] OF SIRIUS BINARY SYSTEM
STAR JUPITER AND STAR SATURN

Unicode (hex) - 00F7, 2FF1, 22FB, 16EA, 00B1 & 006D, 003A,

Symbol (hex) 00A8, 00BD, 003A, 004F, 003C

Ayat\Verses – Allah is He Who created **seven Firmaments** [Canopies / Windows / Heavens] and of the **world a similar number. Through the midst of them descends His Commandment** that you may know that Allah is Able to do all things, and that Allah comprehends all things in [His] Knowledge. [Sura No. – 64\65 – Tallaq-tumun-nisaaa-'a, Ayat No. – 12]

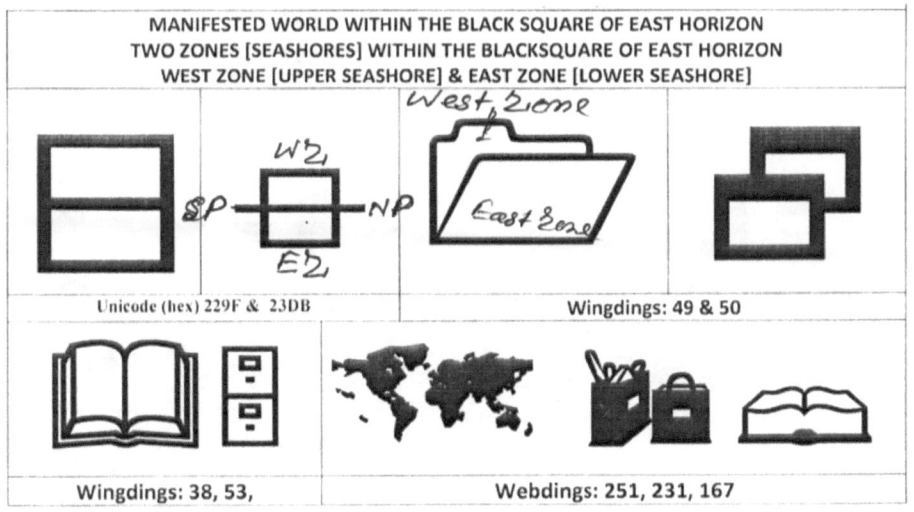

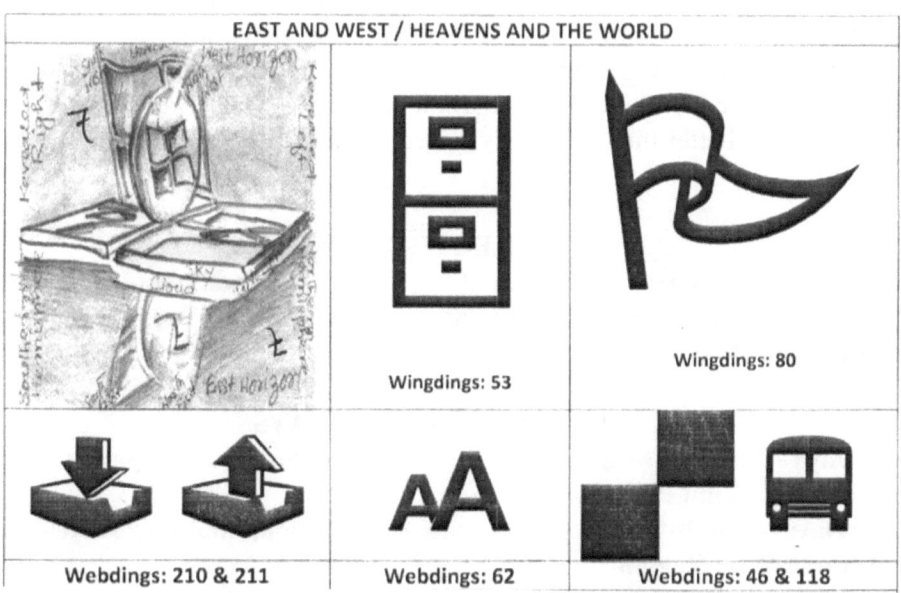

Ayat\Verses – Verily We create man from a drop of **mingled sperm** [drop of thickened fluid / current] to **test** him. So We gave him [the gifts], of hearing and knowing [sight]. Lo! We showed him the **way**, whether he be **grateful** or **disbelieving**. For the **rejecters** we have prepared **chains**, **yokes**, and a **violent fire**. [Sura No. 75\76 – Hal ataa alal-Insaan, Ayat No. – 2 to 4]

Ayat\Verses – Verily, in the alternation of the night and the day, and in all that Allah has created, in the heavens and the world, are **signs for those who fear Him**. [Sura No. – 9\10 – Yuunus, Ayat No. –6]

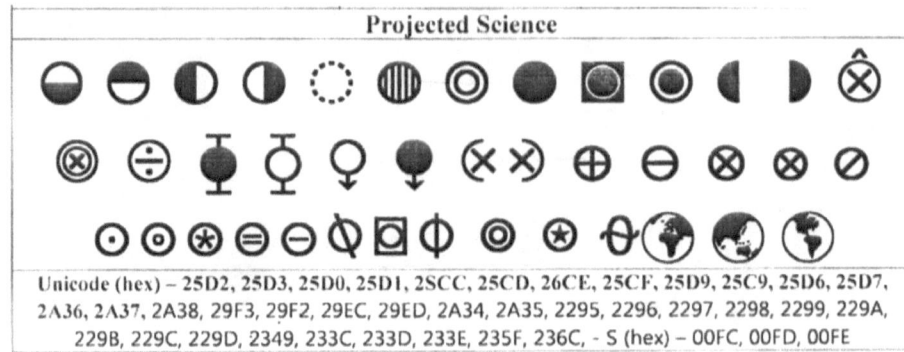

Projected Science

Unicode (hex) – 25D2, 25D3, 25D0, 25D1, 2SCC, 25CD, 26CE, 25CF, 25D9, 25C9, 25D6, 25D7, 2A36, 2A37, 2A38, 29F3, 29F2, 29EC, 29ED, 2A34, 2A35, 2295, 2296, 2297, 2298, 2299, 229A, 229B, 229C, 229D, 2349, 233C, 233D, 233E, 235F, 236C, - S (hex) – 00FC, 00FD, 00FE

DIRECTION COMES FROM ALLAH
Ayat\Verses - O mankind! There has come to you a **direction** from your Rab and a slave for the [diseases] in your hearts, and for those who believe guidance and a Mercy. Say: In the bounty of Allah and in His Mercy, in that let them **rejoice**, that is better than what they stored. [Sura No. – 9\10 – Yuunus, Ayat No. – 57 to 58]

DIRECTIONS
In the Day of Resurrection [Yawmal-Qiyamah] the **fundamental question** between one who **rejects faith** and the **Shaytan** (devil) will be **concerning the East and the West.**

TWO DAYS AND FOUR DAYS
Two Days – Straight Middle-East Region of Eartha 3D and Upright West Region of Arabian Peninsula
Four Days – North-East Region, South-East Region, South-West Region, and North-West Region

Come both of you, willingly or unwillingly.
Say: **Disbelieve** you verily in Him **Who created the world in two days**, and **ascribe** you unto Him **rivals**? He is the Rab of the universe. He set in the [world], mountains **standing firm**, high above it, and bestowed blessings in the world, and **measure therein all things to give them nourishment in due proportion, in four Days**, in accordance with [the needs of] those who seek [Sustenance]. Then **He turned to the heavens [West Horizon] when it was smoke [white dwarf]**, and said unto it and unto the earth: **Come both of you,**

willingly or unwillingly. They said: We come, obedient. So He ordained them as seven firmaments (windows) in two Days, and He inspired in each heaven its mandate [movements]. And We adorned the lower heaven with lights and [provided it] with guard. Such is the Measuring of the Mighty, The Knower. [Sura No. – 40\42 – Haa-Mim-Sajdah, Ayat No. – 9 to 12]

TWO EAST AND TWO WEST

Ayat\Verses - Fabi-ayyi aalaaa – i Rabbikumaa tukazzibaan? [Trans.] Which is it of the favours of your Rab that you deny? [Sura No. 54\55, Ar-Rahman, Ayat\Verses.18]

Ayat\Verses - Rabbul-Mashri-qayni wa Rabbul-Magribayn [Trans.] Rab of the two East and Rab of the two West. [Sura No. – 54\55, Ar-Rahman, Ayat\Verses. 17]

RAB OF THE EAST AND THE WEST
You take Him alone for disposer of affairs.
Rabbul-Mashriqi wal-Magribi Laaa ilaaha
illa Huwa fattakiz-hu Wakiilaa

Ayat\Verses – Rab of the East and the West. There is no unbreakable reality except Him. Therefore, you take Him alone for disposer of affairs. [Sura No. 72\73 – Yaaa-ayyuhal Muzzamil, Ayat No. - 9]

O you who believe! There are indeed many among the priests (Lukmans) and anchorites [Imamgiris], who in falsehood demolish the wealth of men and obstruct [them] from the way of Allah. And there are those who stored gold and silver and spend it not in the way of Allah announce unto them a most grievous penalty. On the Day when heat will be produced out of that [wealth] in the fire of Hell, and with it will be branded their foreheads, their borders, and their backs borders, and their backs. This is the [treasure] which you stored for yourselves. Taste you, then, the [treasures] you stored. The number of months in the sight of Allah is twelve [springs or matter particles]. So ordained by Him the day He created the heavens and the world. Four of them are blessed that is the straight usage. So wrong not yourselves therein, and wage the Pagans / idolaters all together as they waging on all of you. But know that Allah is with those who restrain themselves. Verily the reversing is an addition to unbelief. The unbelievers are led to wrong thereby, for they make it lawful one year, and forbidden another year, in order to adjust the number of months forbidden by Allah and make such forbidden ones lawful. The evil of their course seems pleasing to them. But Allah guides not those who reject Faith. [Sura No. – 8\9 – Tauba, Ayat No. – 34 to 37]

It is We Who created you and gave you shape; then We bade the angels to fall **prostrate** before **Adam**, and they prostrate all **except Iblis**. He refused to be of those who prostrate. [Allah] said: What prevented you (iblis) from prostrating when I commanded you? He (Iblis) said: I am better than he. **You created me from fire** and him (Adam) from clay. [Allah] said: Get you down from this. **It is not for you to show pride her.** So, go forth! Lo! You are of those degraded. He said: Give me reprieve till the day they are raised up. [Allah] said: Be you among those who have reprieve. He said: Now, **because You have sent me astray, verily I shall lurk in ambush for them on the Straightway.** Then I shall come upon them from **before** them and from **behind** them and from their **right hands** and from their **left hands**, and You will not find in most of them gratitude (or your mercies). [Allah] said: Go forth from hence, **degraded, banished. As for such of them as follow you, surely I will fill hell with all of you.** [Sura No. – 6\7 - As-haabul- a' raaf, Ayat No. – 11 to 18]

MARYAM – A CHAMBER LOOKING EAST OR DOWNWARD
And make mention of Maryam in the Kitaab, **when she had withdrawn from her people to a chamber looking East.** (IBPLtd.) [Sura No. – 18\19 – Maryam, Ayat\Verses – 16]

Or

Relate in the Book [the story of] Mary, when **she withdrew from her family to a place in the East.** (IFTA) [Sura No. – 18\19 – Maryam, Ayat\Verses – 16]

THREE STAIRS OF THE APPEARING EARTH
EAST, AND WEST, AND LAND
East represents Ground Stair of North America and South America
West represents Topmost Stair of Arabian Peninsula
Land represents Middle Stair of Europe, Asia, Africa, and Australia
Ayat\Verses – And We made a people, considered weak [and of no account], inheritors of in **East Zone** [Ground Stair of Lower Seashore] and **West Zone** [Topmost Stair of Upper Seashore], **lands** [Middle Stair] whereon We sent down Our blessings. The fair **promise** of your Rab was fulfilled for the **Baniii - Israaa -iil**, because they had **patience** and **constancy**, and We levelled to the ground the great works and fine buildings which Firawn and his people erected [with such pride]. [Sura No. – 6\7 - As-haabul- a' raaf, Ayat No. – 137]

SUBSTITUTION
RAB OF ALL POINTS IN THE EAST AND THE WEST
Now, what is the matter with the disbelievers that they keep starting toward you open-eyed **from the right and from the left**, in groups? **Does every man of them long to enter the Garden of Bliss? By no means!** For We have created

them from what they know. Now I do call to witness the **Rab of all points** in the **East** and the **West** that We can certainly **substitute** for them better than they; and We are not to be defeated [in Our Plan]. [Sura No. – 69\70 – Zil-ma-aaru, Ayat No. – 36 to 41]

DAY OF RESURRECTION [Yawmal-Qiyamah]
FUNDAMENTAL QUESTION CONCERNING
THE EAST AND THE WEST

"Hattaaa izaa jaaa-anaa qaala yaa-layta baynii wa baynaka
bu'-dal-Mashriqayni fabi'-sal-qariin" [Sura No. 42\43 –
Ummil-Kitaab \ (Zukhruf), Ayat No. – 38]

At length, when [such a one]
Comes to Us, he says
[To his evil companion]:
"Would that between me
And thee were **the distance
of East and West!"**
Ah! Evil is the companion [indeed]!
[Trans.] (IFTA) [Sura No. 42\43 – Ummil-Kitaab \ (Zukhruf), Ayat No. – 38]
Or

Till when he comes unto Us, he says (unto him comrade): Ah, would that between me and you there was the **distance of the two horizons** – an evil comrade! (IBPLtd.) [Sura No. 42\43 – Ummil-Kitaab \ (Zukhruf), Ayat No. – 38]

Self-evident Concept: - From the above transliterations of the same Ayat, it can be conceptualised without doubt that East and West are upward & downward directions, **two Horizons** in resemblance with the concept of the Heaven and the World which belong to Allah. Further, these transliterations have made it clear that in the Day of Resurrection [Yawmal-Qiyamah] the **fundamental question** between one who **rejects faith** and the **Shaytan** (devil) will be **concerning the East and the West.**

WESTERN SIDE

You were not on the Western side when We expounded unto Muusa the commandment, and **you were not among those present.** [Sura No. – 27\28 – Qasas, Ayat\Verses No. – 44]

And make mention of Maryam in the Kitaab, **when she had withdrawn from her people to a chamber looking East.** (IBPLtd.) [Sura No. – 18\19 – Maryam, Ayat\Verses – 16]

Or

Relate in the Book [the story of] Mary, when **she withdrew from her family to a place in the East**. (IFTA) [Sura No. – 18\19 – Maryam, Ayat\Verses – 16]

Rab of the East and the West

[Muusa] said: **Rab of the East and the West**, and all between! **If you only had sense!** [Firawn] said: If you put forward any god other than me, I will certainly put you in prison! [Muusa] said: **Even if I showed you something clear [and] convincing**? [Firawn] said: Show it then, **if you tell the truth**! So [Muusa] flung down his staff, and behold, it was a serpent, manifest [for all to see]! And he drew out his hand, and behold, it was **white** to all beholders! [Sura No. 25\26 – Shuraa, Ayat No. – 28 to 33]

It is not righteousness that you turn your faces Towards East or West; but it is righteousness to believe in Allah and the Last Day, and the Angels, and the Kitab, and the Messengers; to spend of your substance, out of love for Him, for your kin, for orphans, for the needy, for the wayfarer, for those who ask, and for the ransom of slaves; to be steadfast in prayer, and to give the poor-due; **to fulfil the contracts which you have made**; and to be firm and patient, in pain [or suffering] and adversity, and throughout all periods of stress. Such are they who are **sincere**. Such are the Allah-fearing. [Sura No. 1\2 – Baqarah, Ayat No. – 177]

Fall Down – Towards East
Turn to the Heaven – Towards West
"We said: **Fall down, one of you a foe to the other!**"

We said: O Adam! Dwell you and your wife in the Garden; and eat of the bountiful things therein as [where and when] you will; but approach not this tree, lest you become wrong-doers. But the Shaytan caused them to deflect there from [the garden], and get them out of the state [of felicity] in which they had been. We said: **Fall down, one of you a foe to the other!** There shall be for you on earth a **habitation** and **provision** for a time. Then Adam received from Allah words [signs] of inspiration and Allah turned towards him; for He is Ever Relenting, Most Merciful (Huwat-Tawwaa-bur-Rahim) We said: Get you down all from here; and verily there will come to you Guidance from Me. Whosoever follow My guidance, there shall no fear come upon them neither shall they grieve. But those who reject Faith and contradict with Our Signs, they will be companions of the fire. They will abide therein. [Sura No. 1\2 – Baqarah, Ayat No. – 35 to 39]

How can you reject the faith on Allah seeing that you were without life, and He gave you life; then He will cause you to die, and will again bring you

to life; and again to Him you will return? It is He Who has created for you all things that are on earth. Then **He turned to the heaven,** and fashioned it as **seven firmaments [windows];** and of all things He has perfect knowledge. [Sura No. 1\2 – Baqarah, Ayat No. – 28 & 29]

"He said: **Will you exchange the higher for that which is lower?"**
HIGHER [WEST] AND LOWER [EAST]

And when you said: O **Muusa!** We cannot endure **one kind of food** [always]; so beg your Rab for us that He brings forth for us of which that the **world grows,** of its herbs, and cucumbers, its garlic, lentils, and onions. He said: **Will you exchange the higher for that which is lower?** Go you down to any **town [East Zone],** and you shall find what you want! **They were covered with humiliation and misery; they drew on themselves the wrath of Allah. This is because they went on rejecting the Signs of Allah and slaying His Messengers wrongfully. This is because they rebelled and went on transgressing.** [Sura No. 1\2 – Baqarah, Ayat No. – 61]

But the sincere [and devoted] slaves of Allah; for them is a **Sustenance determined** fruits and they [shall enjoy] honour and dignity in Gardens of Felicity **facing each other on Throne** [Arsh]; round will be passed to them a **Cup** from a clear-flowing **fountain crystal-white** of a taste delicious to those who drink [thereof], wherein there is no headache [& or crookedness], nor they will suffer from intoxication. And besides them will be **chaste women,** restraining their glances, with big eyes [of wonder and beauty]; as if they were [delicate] **eggs closely guarded.** [Sura No. – 36\37 – Was-saaaffaati, Ayat No. – 40 to 49]

A BEAUTIFUL DWELLING-PLACE ACROSS THE SEA

We took the Bani-Israa-iil **across the sea:** Firawn and his hosts followed them in insolence and spite. At length, when overwhelmed with the **flood,** he said: I believe that there is no god except Him Whom the Children of Israa-iil believe in. I am of those who submit unto Him. [It was said to him]: Ah now! But a little while before, was you in **uprising,** and you did **mischief** [and violence]. But this day We save you in **your body,** that you may be a **sign** to those who come after you! But verily, many among mankind are **heedless of Our Signs.** [Sura No. – 9\10 – Yuunus, Ayat No. – 90 to 92]

ACROSS THE SEA – BANI-ISRAA-IIL
AND
FIRAWN AND HIS HOSTS FOLLOWED THEM

We took the Bani-Israa-iil across the sea: **Firawn and his hosts followed them in insolence and spite.** At length, when overwhelmed with the flood, he said: I believe that there is no god except Him Whom the Children of Israa-iil believe in. I am of those who submit unto Him. [It was said to him]: Ah now! But a little while before, was you in uprising, and you did mischief [and violence]. This day shall **We save you in the body, that you may be a sign to those who come after you**! But verily, many among mankind are heedless of Our Signs. [Sura No. – 9\10 – Yuunus, Ayat No. – 90 to 92]

TWO ZONES – EAST ZONE AND WEST ZONE

And remember, **We separated [delivered] you from the people [folk] of Firawn**. They set you hard tasks and punishments, **slaughtered your sons and let your women-folk** live; therein was a **tremendous trial** from your Rab. [Sura No. 1\2 – Baqarah, Ayat No. – 49]

FACING EACH OTHER [Equal & Opposite]

Lo! Those who **ward off** (evil) are among gardens and **water-springs** [And it is said unto them]: Enter you here in **peace** and **security**. And We shall remove from their hearts any lurking **sense of injury**. [They will be] **brothers facing each other on thrones** of dignity [Kaba]. There no sense of **fatigue** shall touch them, nor shall they [ever] be asked to leave. **Tell My servants that I am indeed the Ever-Forgiving, Most Merciful. And that My Penalty will be indeed the most grievous Penalty**. [Sura No. – 14\15 = Hur, Ayat No. – 45 to 50]

TWO FORCES

"Lo! I fear Allah: for Allah is strict in punishment."

O you who believe! when you meet a force, be firm, and call Allah in remembrance much [and often]; that you may prosper. And obey Allah and His Rasuul; and fall into no disputes, lest you lose heart and your power depart; and be patient and persevering: For Allah is with those who patiently persevere. And be not like those who started from their homes disrespectfully and to be seen of men, and to **hinder [men] from the path of Allah**; while Allah is surrounding all they do. Remember Shaytan made their [sinful] acts seem alluring to them, and said: No one among men can overcome you this day, while I am near to you. But when the **two forces** came in sight of each other, he turned on his heels, and said: Lo! I am clear of you; lo! I see what you see not; Lo! I fear Allah: for Allah is strict in punishment. [Sura No. – 7\8 – Anil-Anfaal, Ayat No. – 45 to 48]

JUNCTIONS OF THE TWO SEASHORES

But your Rab is Most forgiving, full of Mercy. If He were to call them [at once] to account for what they have earned, then surely He would have hastened their punishment; but they have **their appointed time**, beyond which they will find no refuge. Such were the townships [east zones] we destroyed when they committed iniquities; but we fixed an **appointed time** for their destruction. And when, Muusa said to his attendant: I will not give up until I reach **the junction of the two seashores**, though I march on for ages. But **when they reached the Junction, they forgot [about] their Fish [Recycling Region / Eartha 3D], which took its course through the sea as in a tunnel.** [Sura No. – 17\18 – Khaf – Ayat No. – 58 to 61]

TWO PARTIES

Verily we shall make what is on world a **barren hillock [a hole]**. Or you do reflect that the **people of the cave [Black Square]** and of the **Inscription [water]** are wonders among **Our Signs**? When the youths fled for refuse themselves to the Cave, they said: Our Rab! Bestow on us Mercy from Yourself and **dispose of our affair for us in the right way**! Then We sealed up their hearing, for a number of years, in the Cave. And afterwards We raised them up, in order to know which of the **two parties** would best calculate the time that they had tarried! **We narrate to you their story with truth. They were youths who believed in their Rab, and We advanced them in guidance.** [Sura No. – 17\18 – Khaf – Ayat No. – 8 to13]

Universe is an intrinsically luminous star in correspondence with Crucified Sign called Diamond Operator or Shi-raa. Universe has been revealed as an equal and opposite Trinity. Upward direction of the revealed universe towards heaven is called West Horizon or Night Sky. Downward direction of the revealed universe towards world is called East Horizon or Earth's Night Sky. So, with respect to revealed universe upwards direction towards Heaven or Galaxy of Stars represents West or Strong Force; while downward direction towards World or Black Square represents East or Gravitational Force.

East Horizon of the revealed universe has been revealed further in to two Zones [Seashores] and an octagon. The East Horizon has been manifested within the four basic forces in resemblance with four revealed [cardinal] directions called Black Square as a Hexagon. The manifested hexagon within the Black Square of East Horizon is called the World. With respect to manifested hexagonal world, the Upper Seashore in correspondence with Upright West Region is called West Zone. With respect to manifested hexagonal world, the Lower Seashore in correspondence with Straight Middle East Region of Eartha 3D is called East Zone.

HYPOCRISY OR DOUBLE STANDARDS
CONCEALMENT OF TRUTH AND PROJECTION OF FALSEHOOD

Projections of Northern Hemisphere [Revealed Left] and Southern Hemisphere [Revealed Right] as top [head] and bottom [leg] of the Global Trinity or

Mechanical Globalisation as well as two seashores [two Zones] of the manifested hexagonal world are Teleological Evidence Sorceries, Unique Epistemic Persecutions, and conspiracies against believers & followers of Upright West. In brief, projections of Northern Hemisphere [Revealed Left] and Southern Hemisphere [Revealed Right] as top [head] and bottom [leg] of the Global Trinity or Mechanical Globalisation as well as two seashores [two Zones] of the manifested hexagonal world respectively are selfevident concealments of Manifested Two East and Two West in resemblance with Revelations.

Four forces or Four Blessed Springs or Four Imams

i)   **Up or West = <u>strong</u> force, foreheads, before, final cause**
ii) **Down or East = <u>gravitational</u> force, backs, behind, material cause**
iii) **Revealed Right or South = <u>weak</u> forces, borders, formal cause; while their right hands represent North [Haiyalas-Swalaah]**
iv) **Revealed Left or North = <u>electromagnetic</u> forces, back borders, efficient cause, magnetism; while their left hands represent South [Haiyalal-Falaah]**
v)   **Clear Proof = Crucified Sign**

O you who believe! There are indeed many among the **priests** (Lukmans) and **anchorites** [Imamgiris], who in falsehood demolish the wealth of men and obstruct [them] from the way of Allah. And there are those who stored gold and silver and spend it not in the way of Allah announce unto them a most grievous penalty. On the Day when heat will be produced out of that [wealth] in the fire of Hell, and with it will be branded their **foreheads**, their **borders**, and their **backs borders**, and their **backs**. This is the [treasure] which you stored for yourselves. Taste you, then, the [treasures] you stored. The **number of months in the sight of Allah is twelve [springs or matter particles]. So ordained by Him the day He created the heavens and the world. Four of them are blessed that is the straight usage.** So **wrong not yourselves therein,** and **wage the Pagans / idolaters all together as they waging on all of you.** But know that Allah is with those who restrain themselves. **Verily the reversing is an addition to unbelief.** The unbelievers are led to wrong thereby, for they make it lawful one year, and forbidden another year, in order to adjust the number of months forbidden by Allah and make such forbidden ones lawful. The evil of their course seems pleasing to them. But Allah guides not those who reject Faith. [Sura No. – 8\9 – Tauba, Ayat No. – 34 to 37]

It is We Who created you and gave you shape; then We bade the angels to fall **prostrate** before **Adam**, and they prostrate all **except Iblis**. He refused to be of those who prostrate. [Allah] said: What prevented you (Iblis) from prostrating when I commanded you? He (Iblis) said: I am better than he. **You created me from fire** and him (Adam) from clay. [Allah] said: Get you down from this. **It is not for you to show pride him.** So, go forth! Lo! You are of those degraded. He said: Give me reprieve till the day they are raised up. [Allah] said: Be you among those who have reprieve. He said: Now, **because You have sent me astray, verily I shall lurk in ambush for them on the Straightway (towards Middle-East of Eartha 3D).** Then I shall come upon them from **before** them and from **behind** them and from their **right hands** and from their **left hands**, and You will not find in most of them gratitude (or your mercies). [Allah] said: Go forth from hence, **degraded, banished. As for such of them as follow you, surely I will fill hell with all of you.** [Sura No. – 6\7 - As-haabul- a' raaf, Ayat No. – 11 to 18]

HEAVEN OR WEST HORIZON AND WORLD OR EAST HORIZON
He has created the **heaven [West Horizon] and the world** [East Horizon] for **just ends (with truth)**. Far is He above having the **partners** they ascribe to Him! [Sura No. – 15\16 – Nahl, Ayat\Verses – 3]

(Both) the Yahuuds and the Nasaaras say: We are the sons of Allah, and His beloved. Say: Why then doth He punish you for your sins? Nay, you are but mortals of His creating. He forgives whom He pleases. And He punishes whom He pleases: And to Allah belongs the **dominion** of the **heavens** [West Horizon] and the **world [East Horizon]**, and all that is **between**: And unto Him is the final goal (of all). [Sura No. – 4\5 – Maaaidah, Ayat\Verses – 18]

Know you not that **to Allah (alone) belongs the dominion of the heavens [West Horizon] and the world [East Horizon]**? He punishes whom He pleases, and He forgives whom He pleases. And Allah has power over all things. [Sura No. – 4\5 – Maaaidah, Ayat\Verses – 40]

To **Allah belong the dominion of the heavens [West Horizon] and the world [East Horizon]**, and **all that is therein**, and it is He who has power over all things. [Sura No. 4\5 Maaaidah, Ayat\Verses.120]

[Sura No. 57\58 – Mujaadalah, Ayat No. - 6 & 7]
[Sura No. – 6\7 - As-haabul- a' raaf, Ayat No. – 54]
[Sura No. – 31\32 – Sajdah, Ayat\Verses – 4]
[Sura No. – 9\10 – Yuunus, Ayat No. – 3]
[Sura No. – 10\11 – Kitaabun-uh-kimat (Prev. – Huud), Ayat No. – 5 to 8]
Note: Heaven [West Horizon or Sirius A] and World [East Horizon or Sirius B]

The **revelation of the Tanzilul-Kitaab is from Allah** the Mighty, the Wise. Verily in the **heavens** and in the world, are **Signs** (portents or clear proofs) for those who **believe**. And in the **creation of yourselves** and the fact that **animals** are scattered, are **Signs** (portents or clear proofs) for those **whose faith is sure.** And in the **alternation** (equal & opposite semi-anti-clockwise and semi-clock-wise stages of journey of the manifested sign of natural magnetism in resemblance with revelation i.e. North to South & South to North) of **Night** and **Day**, and the fact that Allah sends down **Sustenance** from the **sky [West]**, and **revives** therewith the **world [East]** after its **death**, and in the **change of the winds**, are **Signs** (portents / clear proofs) for those **who have sense**. These are the **Signs (portents / clear proofs) of Allah**, which **We rehearse to you in Truth. Then in what manifestation [Injiil or Correspondence Truth] will they believe after Allah and His Signs?** [Sura No. – 44\45 – Tanziilul Kitaab (Prev. Jaasiya), Ayat No. – 2 to 6]

Ayat\Verses – And if you ask them: Who created the heavens and the world? They would be sure to reply: They were created by [Him], the Exalted in Power, Full of Knowledge. Who made the **world a resting place** for you, and **placed roads for you therein,** that haply **you may find your way (guidance).** And who sends down [from time to time] rain from the sky in due measure; and We raise to life therewith a land [earth] that is dead; even so you will be raised [from the dead]. Who has **created pairs in all things**, and has **made for you ships** and **cattle** on which you ride in order that **you may mount upon their backs,** and you may remember the favour of your Rab when **you mount thereon**, and may say: Glorified to Him Who has subjected these to us, for we could never have accomplished this [by ourselves]; and to our Rab, surely, we must turn back. And **they attribute (allot) to Him a portion of his bondmen!** Lo! Man is verily a mere ingrate. **Or chooses He daughters of all that He has created, and honours He you with sons?** [Sura No. 42\43 – Ummil-Kitaab \ (Zukhruf), Ayat No. – 9 to 16]

REVEALED LEFT AND REVEALED RIGHT
NORTHERN HEMISPHERE AND SOUTHERN HEMISPHERE

**Fasta-'iz billaahi minash-Shaytaanir-Rajim
Bismillaahir-Rahmaanir- Rahim**

CONCEALMENT OF TWO EAST AND TWO WEST
&
PROJECTION TWO NORTH AND TWO SOUTH

Formal Grounds / Instrumental Grounds of Manmade Nature i.e. Projected Mechanical Globalisation, Circular Rotation & Revolution System, and Manmade Magnetism as well as Shameful Post Hoc Ergo Propter Hoc, Petitio Principii, Double Standards, Subjective Selfcontradictions, Objective Paradoxes, Plurality of Causes etc. in the Pen-Paper-Pencil works and activities of the Epistemic Uniqueness of Scientific Certainty and his Chiefs and Ministers.

**Concealment of Manifest Truth [Manifested
Nature and Natural Magnetism]
&
Projection of Manmade Falsehood [Manmade
Nature and Manmade Magnetism]
Black Magic – Teleological Evidence Sorceries
and Unique Epistemic Persecutions
Revealed Truth – Persecution is worse than killing
What a shameful Post Hoc Ergo Propter Hoc!
A few hints [Source – Internet]**

One of the most important elements of a map is direction. There are, as you know, **four cardinal (basic) directions** – north, south, east, and west.

Usually maps are drawn with **north at the top**. Most maps indicate north with an arrow marked N. **If you know where north lies, it is easy to find south, east and west**.

If **north is at the top, then south is at the bottom, east is to the right, west to the left**.

North-east lies between north and east, north-west between north and west, south-east between south and east, and south-west between south and west.

NO REFERENCE OF EAST AND WEST
i)   **North Pole** and **South Pole** – The two end points of the axis of rotation are called the poles. To the North is the North Pole, and to the South is the South Pole.
ii) **Northern Hemisphere** and **Southern Hemisphere** – Midway between the poles is an **imaginary line circling the earth**. It is called the **Equator**. It divides the earth into two equal halves, called the hemispheres. To the North of the equator is the Northern Hemisphere, and to the South of the equator is the Southern Hemisphere.

Do you know?
If you stand at the North Pole, all directions point South – **there is no East or West**! Similarly, at the South Pole all directions point North.

WHERE IS THE WEST OR THE EAST IN THE GLOBE?
Where is the Region of White Dwarf or Arabian
Peninsula in the Global Trinity?
Where is the City or Top West Triangle or The Torrid Zone or Zone of Hot Sands in the globe?

THREE STAIRS OF THE APPEARING PENTAGON
i)   City or Top West Region or The Torrid Region or Region of Hot Sands – The area on both sides of the Equator, between the Tropic of Cancer and the Tropic of Capricorn, is called the Torrid or Tropical Zone. The word 'torrid' means extremely hot. As it lies near the Equator, this Zone receives maximum heat. It is hot throughout the year.
ii) Midiyan or Middle West Zone or The Temperate Zones – The areas that lie between the Tropic of Cancer and the Arctic Circles in the Northern Hemisphere, and the Tropic of Capricorn and Antarctic Circle in the Southern Hemisphere, are called the Temperate Zones. This region has a mild climate – neither very hot nor very cold.
iii) Township or East Zone or The Frigid Zones or Greenland or Tundra Region – The area between the Arctic Circle and the North Pole in the Northern Hemisphere, and between the Antarctic Circle and South Pole in the Southern Hemisphere, receive very little heat from the sun. They are very cold, and large parts are covered with snow throughout the year. They are called the Frigid Zones.

DRC – THE LAND OF DENSE FOREST
AND ARABIAN PENINSULA

The Democratic Republic of Congo (DRC) is a country located deep in the heart of the African continent. It lies in the Torrid Zone. The **Equator passes through Northern DRC** or Belgian Congo or Zaire. There are high mountains in the eastern edges of the DRC, and **plateau in the South and South-West. Being near the Equator, the climate in most part of the DRC is very hot throughout the year.** The evaporation of water due to the high temperature during the day makes the humidity very high. This causes **clouds [electron clouds] to build up by the afternoon, and rain [white dwarf] comes down at about 4 [major axis] pm. This happens almost every day, making the region one of the wettest in the world.**

In some part of the DRC, **especially in the south [where there is appointed Kaba]**, it does not rain so much while the summer is hot and wet, and the winter is cool and dry. Instead of dense forests, there are grass lands with a few scattered trees. Such open grasslands are called **savanna.**

Saudi Arabia is a large country situated in the Arabian Peninsula. A peninsula is a piece of land jutting out into the sea. It is surrounded by water on three sides. The Tropic of Cancer passes through the middle of Saudi Arabia.

EVIDENCE SORCERIES WITH HORIZONTAL AND VERTICAL
~STH LATITUDE AND LONGITUDE OR EARTH SKY

To locate places on the globe, **horizontal** and **vertical** lines are drawn on it. The **horizontal** lines are drawn **parallel to the equator**. Like the equator, they go **around the earth**. They are known as **latitudes or parallels**. The **vertical** lines are **semicircles** that run from the **North Pole to the South Pole**. They are called **longitudes or meridians**. Latitudes and longitudes are marked in degrees (°) and minutes (60' = 1°)

MARKING LATITUDES

The equator is marked 0°. The North Pole is 90° North and the South Pole is 90° South. Thus, the latitude of a place tells you how far north or south a place is from the Equator. For example, a place with latitude of 45° North is located in the Northern Hemisphere, midway between the Equator and the North Pole. A place of 60° North is further north.

There are 181 degrees of latitude in all. There are 90 degrees of latitude in the Northern Hemisphere, 90 degrees of latitude in the Southern Hemisphere, and there is the 0° latitude, that is the equator.

SELFEXPLAINED EVIDENCE SORCERIES

i) **Latitudes run parallel** to each other in the **east-west direction**.
ii) They are at an **equal distance** from **each other**.
iii) **All of them form complete circles, except the North and South poles,** which are **points**.
iv) The **latitude decreases** in length as you move from the **equator to the poles**.

Other than the Equator and Poles, there are four other important latitudes –

a) **Tropic of Cancer at 23 ½ ° N**
b) **Arctic Circle at 66 ½° N**
c) **Tropic of Capricorn 23 ½ ° S**
d) **Antarctic Circle 66 ½ ° S**

MARKING LONGITUDES

The longitude running through the old **Observatory at Greenwich** near London is Marked 0°. It is called the **Greenwich Meridian or the Prime Meridian**. There are 360° of longitude – 180 **degrees of longitude East** of the Prime Meridian, and 180 **degrees of longitude West** of the **Prime Meridian**. The 180° East and 180° West longitudes meet and form a single line on the opposite side of the world, called the 180° longitude or the International Date Line.

Thus, the longitude of a place tells you how much East or West of the Greenwich Meridian the place is located.

FEATURES OF LONGITUDES
TRUE NORTH-SOUTH DIRECTION

i) The lines of longitudes are **Semi circles** that run from one **Pole to the other Pole**.
ii) They run in a **true North-South direction**.
iii) They are spaced farthest apart at the equator and **come together to a point** at the Poles.
iv) The latitudes and longitudes cut each other at right angles.

What a shameful Post Hoc Ergo Propter Hoc!
Man Made Nature

MECHANICAL GLOBALISATION, CIRCULAR ROTATION & REVOLUTION SYSTEM, MANMADE
MAGNETISM, PROJECTED TWO NORTH & TWO SOUTH, UNIDENTIFIED EAST & WEST AS
MANIFESTED DIRECTIONS OF NATURAL MAGNETISM

There is neither East nor West in the Manmade Nature or Shameful Post Hoc Ergo Propter Hoc save Eastern
Hemisphere and Western Hemisphere in resemblance with the WILL of the Don of Historical Black Magic as
well as his Teleological Evidence Sorcerers and Unique Epistemic Persecutors.

What a shameful Statistics!

In Manmade Nature or Shameful Post Hoc Ergo Propter Hoc, the so-called Sun [Bullet] as the Manifested
Sign of Natural Magnetism rises once from an unidentified East and sets once in an unidentified West to
cause the alteration of Day and Night for the Equal & Opposite Seashore of East Zone [Lower Seashore] and
West Zone [Upper Seashore] within the Black Square [Non-luminous Moon or Dark Moon] of East Horizon.

What we conceive as darkness or night is nothing save Veil of Ignorance!

Where is the Upright West Region of the Manifested Hexagonal World, Topmost Stair out
of three ascending stairs [Tuur], and Uppermost Land of the appearing Pentagonal Earth
called Arabian Peninsula or City of the Black & White Imam in the Global Trinity?

What a shameful Post Hoc Ergo Propter Hoc!
MANMADE NATURE
BLACK & WHITE WORKS AND ACTIVITIES

- Projected **Northern Hemisphere** represents **Upward Direction** and **Strong Force. Rain** comes from Projected **Northern Hemisphere. So-called Moon and Stars** also appear from Projected **Northern Hemisphere**. However, this projection is contradictory to the projected **Blue Shade of Windows'7 Ultimate** where **rain** comes from **Downward Direction or Projected Southern Hemisphere. Quadrilateral Stars** or **Guiding Stars** also appear from **Downward Direction** or **Projected Southern Hemisphere** or **Gravitational Force** in resemblance with **Yellow Shade of Windows'7 Ultimate**. Moreover, these projections manifest that water from hills and mountains moves towards Northern

Hemisphere i.e. Upward Direction. Projected Northern Hemisphere is also contradictory to Newton's Law of Gravitation.

- Projected **Southern Hemisphere** represents **Downward Direction or Gravitational Force**. **Earthquake** emerges from Projected **Southern Hemisphere**. However, this projection is contradictory to the Projected **Hexagonal Symbols** of the Manifested Hexagonal World in the **Green Shade of Windows'7 Ultimate** where Earthquake emerges from **Upward Direction i.e. Strong Force**. This projection is also contradictory to Newton's Law of Gravitation. Moreover, this projection implies that when an apple falls from the tallest tree, then it moves upward in resemblance with projected hexagonal symbols in the Green Shade of Windows'7 Ultimate.

- There are four cardinal [revealed] directions in resemblance with four Imams for straight usage or four basic forces. East represents downward direction and Gravitational Force. West represents upward direction and Strong Force. Revealed Left and manifested North-East & North-West represent direction towards North and Magnetic Force. Revealed Right and Manifested South-East & South-West represent direction towards South and Weak Force. But in Manmade Nature, projected Northern Hemisphere represents upward direction and Strong Force contradictory to Manifested West in resemblance with Revelation. Coherently, in Manmade Nature, projected Southern Hemisphere represents downward direction and Gravitational Force contradictory to Manifested East in resemblance with Revelation. So, projected Northern Hemisphere represents Manifested West [Natural West] and Strong Force. Rain comes from West in resemblance with the converted Blue Shade of Windows'7 Ultimate. Projected Southern Hemisphere represents Manifested East [Natural East] and Gravitational Force. Earthquake emerges from East in resemblance with the converted Green Shade of Windows'7 Ultimate.

- Projected left side of a finite individual facing towards once rising of the so-called sun [Bullet] as the manifested sign of Natural Magnetism from an unidentified East is called **North or Magnetic Force**. Projected right side of a finite individual facing towards once rising of the so-called sun [Bullet] as the manifested sign of natural magnetism from an unidentified East is **South or Weak Force**. However, the so-called sun [Bullet] as the manifested sign of natural magnetism rises from the East and sets in the West. **These projections are called Shameful Post Hoc Ergo Propter Hoc**. Those who are **deaf, dumb, and blind,** they may believe such kind of Selfevident Teleological Evidence Sorceries and Selfproved Unique Epistemic Persecutions. These projections are both contrary as well as contradictory to Real Pen-Paper-Pencil works & activities, Manifest Truth in resemblance with Revelation, all established verifiable laws, four criterions of truth, and Upright Logic. If North represents Magnetic Force and the so-called sun [Bullet] is the Manifested Sign of Natural Magnetism, then the direction from which the so-called sun [Bullet] enters/rises or sets/ends equal & opposite Semi-anticlockwise and Semi-clockwise stages of journey for the equal & opposite seashore must be either North or South. The entering/rising direction or setting/ending direction of the so-called sun [Bullet] as the manifested sign of natural magnetism is neither East nor West of the Manifested Nature in resemblance with Revelation. These are neither statements, nor suppositions, nor hypotheses, nor imaginations, nor predictions, nor discoveries, nor theories; but these are the Manifest Truths in resemblance with Revelations. These Truths are verifiable by the manifested sign of natural magnetism and searched out real scientific symbols available in the separate sheet i.e. Insert - Symbol as well as justifiable in resemblance with Revelations, Valid Laws, four criterions of truth in resemblance with four cardinal directions & four basic forces, and Upright Logic. In other words, these searched out manifest truths are unalterable truths i.e. Coded Tautologies. So, in **Manmade Nature**, the so-called sun [Bullet] as the manifested sign of natural magnetism rises once from the unidentified East and sets once in an unidentified West. But in **Manifested Nature**, the so-called sun [Bullet] as the manifested sign of natural magnetism enters from Manifested North-East in correspondence with North America and converted Red Shade of Windows'7 Ultimate and sets in Manifested South-East in correspondence with South America and converted Green Shade of Windows'7 Ultimate. This entering & setting is the Semi-anticlockwise journey of the so-called sun [Bullet] as the manifested sign of natural magnetism for the Ground Stair [Tin] of the appearing Pentagonal Earth as well as East Zone of the Manifested Hexagonal World within the Black Square [Non-luminous Moon or Dark moon] of East Horizon. The so-called sun [setting sun] or Bullet runs away from there and rises from Manifested South-West in correspondence with Europe and converted Yellow Shade of Windows'7 Ultimate and ends in Manifested North-West in correspondence with Australia and converted Blue Shade of Windows'7 Ultimate. This rising & ending is the Semi-clockwise journey of the so-called sun [Bullet] as the manifested sign of natural magnetism for the Middle Stair [Zaytuun] and Topmost Stair [Tuur] of the appearing Pentagonal Earth and West Zone of the Manifested Hexagonal World within the Black Square of

317

East Horizon. So, the so-called sun [Bullet] as the manifested sign of natural magnetism rises twice and sets twice to cause the alteration of Day and Night for the equal & opposite Seashore of East Zone and West Zone. This is the Natural Statistics or Manifest Truth concerning equal & opposite Day and Night for the equal & opposite seashore in resemblance with equal & opposite revelation and established Newton's Third Law – 'Equal & Opposite'.

SEARCHED OUT MANIFESTED TRUTH [MANIFESTED NATURE]

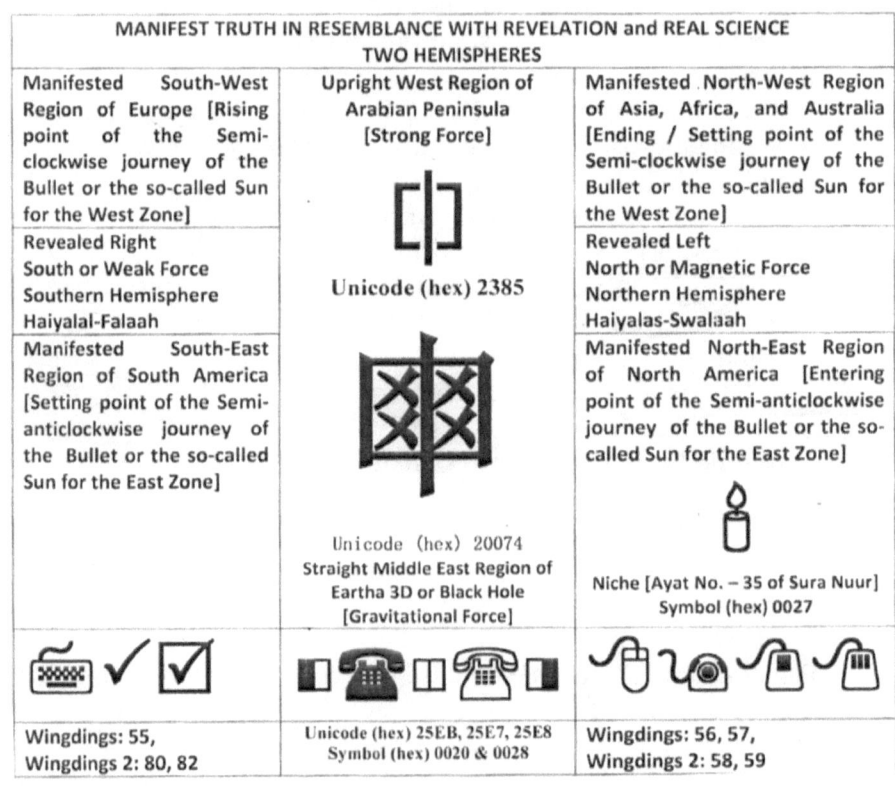

MANIFEST TRUTH IN RESEMBLANCE WITH REVELATION and REAL SCIENCE TWO HEMISPHERES		
Manifested South-West Region of Europe [Rising point of the Semi-clockwise journey of the Bullet or the so-called Sun for the West Zone]	Upright West Region of Arabian Peninsula [Strong Force] Unicode (hex) 2385	Manifested North-West Region of Asia, Africa, and Australia [Ending / Setting point of the Semi-clockwise journey of the Bullet or the so-called Sun for the West Zone]
Revealed Right South or Weak Force Southern Hemisphere Haiyalal-Falaah		Revealed Left North or Magnetic Force Northern Hemisphere Haiyalas-Swalaah
Manifested South-East Region of South America [Setting point of the Semi-anticlockwise journey of the Bullet or the so-called Sun for the East Zone]	Unicode (hex) 20074 Straight Middle East Region of Eartha 3D or Black Hole [Gravitational Force]	Manifested North-East Region of North America [Entering point of the Semi-anticlockwise journey of the Bullet or the so-called Sun for the East Zone] Niche [Ayat No. – 35 of Sura Nuur] Symbol (hex) 0027
Wingdings: 55, Wingdings 2: 80, 82	Unicode (hex) 25EB, 25E7, 25E8 Symbol (hex) 0020 & 0028	Wingdings: 56, 57, Wingdings 2: 58, 59

HIGHLIGHTS / CONCLUSIONS / SEARCHEDOUT FINDINGS	
CONSPIRACY SCIENCE & MAN-MADE DIRECTIONS	REAL SCIENCE & MANIFESTED DIRECTIONS
Manifested Universe or Alamin is either like a Globe or in resemblance with the project Solar System or both.	Manifested Universe or Alamin is an upright rectangle in coherence with the End of Proof of real science [Tawraat] and in correspondence with the appointed Kaba [Injiil].
The Top of the Globe is North [Magnetic Force].	East Horizon of the revealed universe guarded by the four basic forces is called Black Square. The world has been manifested as a hexagon within the Black Square of East Horizon. There is an equal & opposite ˜sth between Upright West Region of Arabian Peninsula and Straight Middle East Region of Eartha 3D. So, the manifested hexagon is appearing as a Pentagonal Earth
	Further, revelation resembles with plus sign, while manifestation resembles with multiplication sign. **The Top of the Upright Rectangular Universe is West** [Strong Force].
The Bottom of the Globe is South [Weak Force].	**The Bottom of the Upright Rectangular Universe is East** [Gravitational Force].
The upper half of the Global Trinity or so-called Solar System in correspondence with the Top as North is the **Northern Hemisphere**.	**Two West** The upper half of the Upright Rectangular Universe in resemblance with West [Strong Force] represents **West Horizon**, Heaven, Galaxy of Guiding Stars, Night Sky, Axis, Planet / Moon Jupiter. The upper half of the Black Square is **West Zone** [Upper Seashore]
The Lower Half of the Global Trinity or projected Solar System in correspondence with the Bottom as South is the **Southern Hemisphere**.	**Two East** The Lower Half of the Upright Rectangular Universe in resemblance with East [Gravitational Force] represents **East Horizon**, Hell, Black Square, West Zone & East Zone, Earth Sky, Orbit, Planet / Moon Saturn. The lower half of the Black Square is **East Zone** [Lower Seashore]
Right side of the Global Trinity or projected Solar System from the objective point of view and left side from the point of view of finite observer facing towards once rising sun from unidentified direction is the **Eastern Hemisphere as well as South. This is called post hoc ergo propter hoc and a package of subjective self-contradictions and objective paradoxes.**	Revealed Right represents South [Weak Force] and manifested Southern Hemisphere [Haiyalal-Falaah] i.e. South-East and South-West. The right side or right half from the point of view of Manifested East Horizon [Black Square], and left side from the point of view of finite observer facing towards the appointed Kaba is the Southern Hemisphere [Haiyalal-Falaah]. The setting direction of the so-called sun [Bullet] as the manifested sign of natural magnetism for the East Zone [of North America and South America] represents manifested sign of South-East [Southern Hemisphere / Haiyalal-Falaah]. The rising direction of the so-called sun as

	the manifested sign of natural magnetism for the West Zone [of Europe, Arabian Peninsula, Asia, Africa, and Australia] represents manifested sign of South-West [Southern Hemisphere / Haiyalal-Falaah].
The left side of the Global Trinity or projected Solar System from objective point of view but right side from the point of view of the finite observer facing towards once rising sun is the Western Hemisphere as well as North. **This is called post hoc ergo propter hoc and a package of subjective self-contradictions and objective paradoxes.**	Revealed Left represents North [Magnetic Force] and manifested Northern Hemisphere [Haiyalas-Swalaah] i.e. North-East and North-West. The left side or left half from the point of Manifested East Horizon [Black Square], but right side from the point of view of finite observer facing towards the appointed Kaba is Northern Hemisphere [Haiyalas-Swalaah]. The entering direction of the so-called sun [Bullet] as the day light [morning show] for the East Zone [of North America and South America] represents manifested sign of North-East [Northern Hemisphere / Haiyalas-Swalaah]. The ending [setting] of the stages of journey of the so-called sun [Bullet] as the manifested sign of natural magnetism for the West Zone [of Europe, Arabian Peninsula, Asia, Africa, and Australi] represents manifested sign of North-West [Northern Hemisphere / Haiyalas-Swalaah].
The Horizontal or Parallel Equator Line or Imaginary Line in resemblance with minus sign is the line of demarcation between Northern Hemisphere and Southern Hemisphere.	The Vertical or Perpendicular Line or Upright Line or Prime Meridian or sign of Justice in resemblance with the ideogram 'Alif' is the line of demarcation between Northern Hemisphere [Haiyalas-Swalaah] and Southern Hemisphere [Haiyalal-Falaah].
The Vertical or Perpendicular Line or Prime Meridian or sign of Justice in resemblance with the ideogram 'Alif' is the line of demarcation between Eastern Hemisphere and Western Hemisphere of the Global Trinity or projected Solar System.	Horizontal Line [Parallel Equator Line or Danger] is the line of demarcation between West Horizon and East Horizon of the Manifested Universe; while the Horizontal lines of Double Danger represent lines of demarcation between West Zone and East Zone, between Upright West Region [Topmost Stair] and Middle West Regions of the West Zone [Middle Stair], between Ground Stair of the East Zone and Straight Middle East Region of Eartha 3D.
Northern Hemisphere represents Strong Force & Upward Direction.	Northern Hemisphere represents Magnetic Force i.e. E-Points [entering and ending points of the stages of journey of the so-called sun or Bullet], leftward direction from the point of view of Revelation but rightward direction from the point of view of finite observer facing towards the appointed Kaba, Back Border, and the like. The so-called sun [Bullet] as the manifested sign of natural magnetism enters from North-East in correspondence with North America. The so-called new moon [Planet / Moon Uranus or Muzzammil or Triangular White Bullet] also arises for the West Zone from North-West in correspondence with Australia. So, North is the single root for the so-called sun and the so-called new moon.

Southern Hemisphere represents Gravitational Force & Downward Direction.	Southern Hemisphere represents Weak Force i.e. T-Point, Rightward direction from the point of view of Revelation but leftward direction from the point of view of finite observer facing towards the appointed Kaba, Border, and the like. The so-called sun [Bullet] sets in South-East for the East Zone. The so-called sun [Bullet] rises from South-West for the West Zone. So-called new moon appears for the East Zone from South-East.
The Top Half of the Global Trinity or projected Solar System is the **Northern Hemisphere**. Further, **leftward direction** of the finite observer facing towards once rising sun as the manifested sign of natural magnetism from unidentified East, and **in resemblance with man-made magnetism** represent **North**. So, there are two man-made or mechanical **North** in Global Trinity or projected Solar System. There is neither Revealed North nor Manifested North.	The left half of the revealed East Horizon has been manifested as Northern Hemisphere [Haiyalas-Swalaah]. Consequently, leftward direction from the point of view of Revelation manifests Magnetic Force [North]. In other words, rightward direction from the point of view of finite observer facing towards the appointed Kaba is Northern Hemisphere [Haiyalas-Swalaah] or North Direction. So, there is only one North from the points of view of Revelation, Manifestation and Appearance. This sole North represents E-Points of Natural Magnetism [entering and ending of the stages of journey of the so-called sun or Bullet] i.e. Magnetic Force.
The Bottom Half of the Global Trinity or projected Solar System is the **Southern Hemisphere**. Further, **rightward direction** of the finite observer facing towards once rising sun from unidentified East as the manifested sign of natural magnetism and **in resemblance with man-made magnetism** represent **South**. So, there are two man-made or mechanical **South** in Global Trinity or projected Solar System. There is neither Revealed South nor Manifested South.	The right half of the revealed East Horizon has been manifested as Southern Hemisphere [Haiyalal-Falaah]. Consequently, rightward direction from the point of view of Revelation represents Weak Force or South. In other words, leftward direction from the point of view of finite observer facing towards the appointed Kaba is Southern Hemisphere [Haiyalal-Falaah]. So, there is only one South from the points of view of Revelation, Manifestation, and appearance. This sole South represents T-Point of Natural Magnetism [stages of journey of the so-called sun or Bullet as the manifested sign of natural magnetism].
In Global Trinity or projected Solar System, there is **no place for Revealed & Manifested North** [Magnetic Force] i.e. Natural Magnetism, though there are **places for several False North and Northern Hemisphere. Simultaneously,** there is **no place for Revealed & Manifested South** [Weak Force]	In revelation and manifestation or in Truth-in-itself and Tautologies, there is no scope of subjective selfcontradiction and objective paradox. There are two West [West Horizon & West Zone from the point of view of revealed and manifested universe, Top West Stair & Middle West Stair from the point of view of manifested world and appearing earth], and two East [East Horizon & East Zone from the point of view of revealed and manifested universe, Ground Stair & Straight Middle East from the point of view of manifested world and

in resemblance with Natural Magnetism, though there are **places for several False South and Southern Hemisphere**. There is **no concept of East [Gravitational Force] or West [Strong Force] in Global Trinity or projected Solar System**. The so-called sun as the manifested sign of sole magnetism rises once from unidentified East and sets once in an unidentified West to cause the alteration of day & night for the equal & opposite Seashore [West Zone and East Zone]. **What a Post Hoc Ergo Propter Hoc Calculation!**	appearing earth.]. There is only one North and only one South. So, there is only one kind of Magnetism. The so-called sun [Bullet] is the manifested sign of sole Magnetism. The so-called sun [Bullet] as the manifested sign of sole magnetism enters from Manifested North-East and sets in Manifested South-East for the East Zone. The so-called sun [Bullet] as the manifested sign of sole magnetism rises from Manifested South-West and sets in Manifested North-West for the West Zone. The so-called sun as the manifested sign of sole magnetism rises twice and sets twice to cause the alteration of day & night for the equal & opposite Seashore [East Zone and West Zone]. These are the unalterable manifested truths in resemblance with revelations.

Search for Truth	Search for Truth
What kind of truths are there in the Manmade Nature i.e. Mechanical Globalisation, Circular Rotation & Revolution System, and Manmade Magnetism save concealments of Two East and Two West with a view to lead the followers of Upright West towards Straight Middle East Region of Eartha 3D?	What kind of falsehoods are there in Revelation, Manifestation, Appearance, and shared searched out Manifest Truths in the Kitaaba Wal-Hikmata or Manifested Nature and the Utility of one's Upright Logic?

The Universe or Alamin is a Quadrilateral Shining Star [Diamond Operator or Fixed Star]. The Universe [Alamin] is a four-figured star in resemblance with the Pythagorean number '**SOLID – FOUR'**. The crucified sign is a clear proof as well as correspondence truth of the existence of the Diamond Operator of Real Science. This clear proof is called **Injiil**. The concept of the four-figured shining star or the Diamond Operator represents four cardinal [revealed] directions, four basic forces, four blessed Imams [springs or pillars] for straight usage, four Kitaabs as criterions of Truth, four cardinal virtues of Plato, four Noble Truths of Buddha [Chatwari Arya Shatyani], four witness without reference to male/female, four Purusharthas [Dharmartha Kama Muksacha Pusrushartha Itii Udahit – Agni Purana], four fundamental categories of knowledge, four Galilean Moons, four causes of Aristotle, and the like. The concept of Diamond Operator also pragmatically corresponds with the equal & opposite binary ~sth [intersection] and four shades of Windows'7 Ultimate. This pragmatic correspondence truth is called **Zabuur**.

If the above shared concepts cohere with revelations, then these shared concepts are called coherence truths or Laws or Tawraat. In other words, if the above shared concepts are justifiable in resemblance with Complete Coded Shared Universal Major Premises [Kitaab with Truth], then the above shared concepts are called Laws or **Tawraat**. In brief, the Law means a formal relation between Uniform Principle and Uniform Motion and the resemblance between Manifestation and Revelation. This formal relation or resemblance is called the Law of Causation or Law of Karma. It is one of the formal grounds of manifestation. The successive relation or resemblance is a binary sign which forms a Trinity like two terms and a binary sign of relation of the unit of logic called proposition. If an individual is able to understand the Trinity or the resemblance between revelations and manifested signs, then it is the selfevident truth of that individual. This understanding is the consequence of the utility of one's innate idea [Ref. Descartes] or one's rational endeavour [Rationalism] or one's upright logic [Ref. Ibrahim (ass)] or single minded slave of the Uniqueness / Ahad [Ref. Sahabas]. This selfevident truth is called Kitaaba Wal-Hikmata or Manifested Nature and the Utility of One's Upright Logic or **Furqaan** or one's wisdom [true knowledge].

The term atom is a substratum. It has both denotation [number] as well as connotation [attributes]. In other words, the term atom is a **connotative term**. It is not the name of a specific element or particle or constituent or even the name of a sign of relation, yet it is composed of two opposite particles like man & jiin or proton & electron or water & fire, and a binary sign of relation like messenger & a blessed message or Neutron & Neutrino or snake & white. This binary sign of relation is called ~sth between two opposite particles. This ~sth is an element, a constituent, and a binary component like minus [-] sign. So, as a substratum, an atom consists of two denotative particles as well as a binary connotative ~sth, namely, proton [positive particle or Strong Force] and electron [negative particle or Gravitational Force], and two connotative or formal equal and opposite faces, namely, neutrino and neutron. The ~sth or the binary sign of relation or the imaginary line between straight upward direction and straight downward direction or between West and East or between Eastward face and Westward face or between Proton and Electron or between strong force and gravitational force or between up and down is the Vertical Line or Perpendicular Line or the **Upright** ~sth. This ~sth resembles with the ideogram 'Alif' of the Arabic Alphabet which stands for **JUSTICE or UNIFORM PRINCIPLE as well as Muusa's Staff**. The sign of Justice is binary, but the establishment of a binary relation is a **TRINITY**. The left side of the sign of Justice or Vertical ~sth or Perpendicular ~sth or the appointed Kaba from the objective point of view and right side from the point of view of the finite observer facing towards the appointed Kaba is the

Revealed Left & Manifested **Northern Hemisphere [Haiyalas-Swalaah].** **Northern Hemisphere [Haiyalas-Swalaah] represents Magnetic Force [Back Border].** In other words, Northern Hemisphere represents **E-Point** [entering & ending Pole or North Pole] of the so-called sun as the manifested sign of natural magnetism. The niche as a ~sth is kindled from Revealed Left [North]. **Reference from Real Science -** ◨Unicode (hex) – 25E8, the dark part of the Black Square is the Northern Hemisphere and the white part is the Southern Hemisphere. The dark part is left side from the objective point of view but right side from the point of view of finite observer. So, the left side of Revelation represents Northern Hemisphere [Magnetic Force]. The white part is right side from the objective point of view but left side from the point of view of finite observer. This white part represents Manifested Southern Hemisphere.

The right side of the sign of Justice or Vertical ~sth or Perpendicular ~sth or appointed Kaba from the objective point of view and left side from the point of view of the finite observer facing towards the appointed Kaba is the **Manifested Southern Hemisphere [Haiyalal-Falaah].** The Southern Hemisphere represents Weak Force [Border]. In other words, Southern Hemisphere represents **T-Point** [setting as well as turning & rising / appearing] of the kindled niche. **Reference from Real Science -** ◧Unicode (hex) – 25E7, the dark part of the Black Square is the Southern Hemisphere. This dark part is right side from the objective point of view but left side from the point of view of finite observer. So, the right side of the appointed Kaba or T-Point of kindled niche represents Manifested Southern Hemisphere [Haiyalal-Falaah]. The white part is left side from the objective point of view but right side from the point of view of finite observer. This white part represents Manifested Northern Hemisphere.

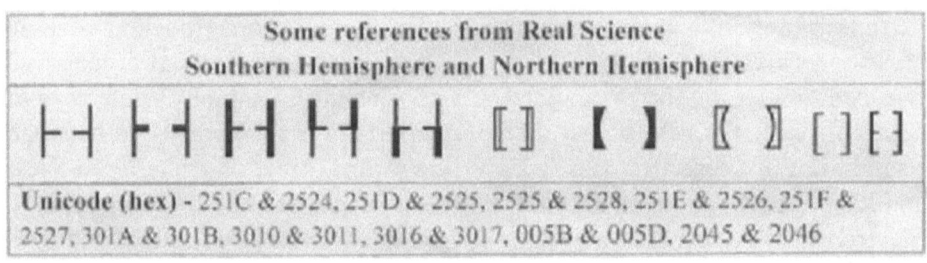

Some references from Real Science
Southern Hemisphere and Northern Hemisphere

Unicode (hex) - 251C & 2524, 251D & 2525, 2525 & 2528, 251E & 2526, 251F & 2527, 301A & 301B, 3010 & 3011, 3016 & 3017, 005B & 005D, 2045 & 2046

REVEALED RIGHT [SOUTH] AND REVEALED LEFT [NORTH]
Ayat\Verses - In **like manner** [example] We have **revealed** to you the **Kitaab** and those, to whom We gave the Kitaab **afore time** believed therein, and of those (also) there are some who believe therein and **none** but **disbelievers reject Our Signs.** And you were not (able) to recite a Kitaab before this, **nor did you write it from right hand side,** for then might those have **doubted,** who

follow **falsehood**. But it is **self-evident** in the hearts of those who are endowed with **true knowledge**, and **none deny Our Signs** (Manifested Magnetism in resemblance with Revelations) except **wrong-doers**. [Sura No. – 28\29 – Ankabuut, Ayat No. – 47 to 49]

SO-CALLED SUN [BULLET / MANIFESTED SIGN OF NATURAL MAGNETISM] ENTERS FROM NORTH AND ENDS IN NORTH

Revealed Left – Northern Hemisphere [Manifested North-East & North-West]
Revealed Right – Southern Hemisphere [Manifested South-East & South-West]
Ayat\Verses – "Wa **tarash-shamsa** izaa tala-at-lazaawaru an-kahfihim zaatal-yamimi wa izaa gara-at-taqri-zuhum zaatash-shimaali wa hum fi fajwatim-minh. Zaalikka min aayaatillaah; many-yahdillaaahu fahuwal-muhtad; wa many-yuzlil falan-tajida lahuu waliyyam-murshidaa". [Trans.] And might have seen wa tarash-shamsa [**Scientific name Star Sirius B**], **when it rises, declining to the right** [semi-anticlockwise in resemblance with three to nine of a clock] from their Cave, and **when it sets, turning away from them to the left** [semi-clockwise in resemblance with nine to three of a clock], **while they lay in the open space** [within the planetary or psychic barrier under the guidance of Jakiriya] **in the midst of the Cave**. Such [equal and opposite i.e. half-clockwise and half-anticlockwise circulation of Allah's glittering show for the earth] **are among the Signs of Allah**. He whom Allah, guides is rightly guided; but he whom Allah **leaves to stray**, for him you will **not find** protector to lead him to **the Right Way**. [Sura No. – 17\18 – Khaf – Ayat No. – 17]

Have they **not observed all things that Allah has created,** how their **shadows incline to the right and to the left, making prostration unto Allah, and they are lowly?** [Sura No. – 15\16 – Nahl, Ayat\Verses – 48]

REVEALED RIGHT [SOUTH] AND REVEALED LEFT [NORTH]
You would have **deemed** them **awake,** while they were **asleep,** and We **caused them to turn over to the right and the left, and their dog** stretching forth his **two fore-legs** on the threshold. If you had **observed** them **closely,** then **certainly** you would **turn back** from them **in flight,** and would certainly have been filled with terror of them. [Sura No. – 17\18 – Khaf – Ayat No. – 18]

REVEALED RIGHT [SOUTH] AND REVEALED LEFT [NORTH]
EAST [DOWN] AND WEST [UP]
Now, what is the matter with the disbelievers that they keep starting toward you open-eyed **from the right and from the left,** in groups? **Does every man of them long to enter the Garden of Bliss?** By **no means!** For We have created them from what they know. Now I do call to witness the **Rab of all points** in

the **East** and the **West** that We can certainly **substitute** for them better than they; and We are not to be defeated [in Our Plan]. [Sura No. – 69\70 – Zil-ma-aaru, Ayat No. – 36 to 41]

REVEALED RIGHT [SOUTH] AND REVEALED LEFT [NORTH]
Have they **not observed all things** that Allah has created, how their **shadows incline** to the **right and to the left,** making **prostration** unto Allah, and they are lowly? [Sura No. – 15\16 – Nahl, Ayat\Verses – 48]

Lo! There are above you guardians / protectors [lahaafigiin as appointed angels] Generous and Recording [Kiraaman – Kaatibin]. They know [and understand] all that you do. [Sura No. – 81\82 - Izas-samaa-unfatarat, Ayat No. – 10 to 12]

Ayat\Verses – We verily created a man, and We know **what his soul whispers to him,** and **We are nearer to him than [his] jugular vein.** When the **two [guardian angels]** appointed **to read** [his doings] and **to write** [noted them], **seated on the right and on the left. And not a word does he utter** but there is with him an **observer** ready. And the stupor of **death comes in Truth.** (And it is said to him): **This was that which you were trying to escape. [Sura No. 49\50 – Qaaaf, Ayat No. – 16 to 19]**

TWO GUARDIANS TO THE RIGHT AND TO THE LEFT
There was indeed a **Sign** for Saba, in their dwelling place. **Two Gardens to the right and to the left. Eat of the Sustenance** [provided] by your Rab, and be grateful to Him: **A territory fair and happy, and a Rabbun Gafuur.** [Sura No. – 33\34 – Saba, Ayat No. - 15]

MANKIND WILL BE SORTED OUT INTO THREE CLASSES
Ayat\Verses – When the event inevitable comes to pass (befall), there is no denying that it will come to pass (befall); abasing (some), exalting (others). **When the world is shaken with a shock (to its depths), and the hills are ground to powder,** so that they become a **scattered dust,** and you will be **sorted out into three classes. (First) those on the right hand, what of those on the right hand! And (you) those on the left hand, what of those on the left hand! And the foremost in the race, the foremost in the race.** [Sura No. – 55\56 - Waqa-atil-waaqiah, Ayat No. – 1 to 10]

REWARD FOR THOSE WHO ARE FOREMOST IN THE RACE
Ayat\Verses – **And the foremost in the race, the foremost in the race;** these are they who will be brought near in Gardens of Bliss; a multitude of those of old, and a few of those of later times on lined couches, reclining **therein face to**

face [equal & opposite]. There wait on them immortal youths with bowls and ewers and a cup from a pure spring wherefrom they get no aching of the head nor any madness (nor will they suffer intoxication). And fruit that they prefer and flesh of fowls that they desire and [there are) fair ones with wide lovely eyes, like unto hidden pearls reward for what they used to do. There they will hear neither vain speaking nor recrimination; (naught) but the saying: Peace, (and again) Peace. [Sura No. – 55\56 - Waqa-atil-waaqiah, Ayat No. – 10 to 26]

REWARD FOR THOSE WHO ARE ON THE RIGHT HAND

And those on the right hand, **what of those on the right hand?** "Fii sidrim-makhzuud" [Trans.] Among thorn less **lute tree.** "Wa talhim-manzuud" [Trans.] And clustered plantains / (Among Talh trees with flowers [or fruits] piled one above another); "Wa zillim-mamduud" [Trans.] And spreading shade, "Wa maaa-im-maskuub" [Trans.] By water gushing (flowing constantly), and fruit in abundance. Neither out of reach nor yet forbidden, and on Thrones [of Dignity], raised high. Lo! We have created them [their Companions] a (new) creation; and made them virgins [pure and undefiled], lovers, and friends for the **Companions of the Right Hand**; a multitude of those of old and a multitude of those of later times. [Sura No. – 55\56 - Waqa-atil-waaqiah, Ayat No. – 27 to 40]

Then why do you not [intervene] when [the soul of the dying man] reaches the throat, / Why, then, when comes up to the throat, and you are at that moment looking? But We are nearer to him than you are, and you see not, why then, if you are not in bondage (unto Us)? **Do you not force it back, if you are truthful?** Thus, then, if he be of those nearest to Allah; [there is for him] rest and satisfaction, and a garden of delights. And if he be of the **companions of the Right Hand**, [for him is the salutation]: Peace unto you, from the companions of the **Right Hand**. [Sura No. – 55\56 - Waqa-atil-waaqiah, Ayat No. – 83 to 91]

"And one of them who say: I am an Absolute besides Him, such a one We should reward with Hell."
"Wa maaa arsalnaa min-qablika mir-rasuulin illaa nuuhiii ilayhi annahuu laaa-ilaaha illaaa Ana fa 'buduun" [25] [Trans.] And we sent no messenger before you but We **inspired him** (saying): There is no ilaaha (absolute reality) save Me, so worship Me. And they say: [Allah] Most Gracious **has begotten a son.** Be He glorified! Nay, but are honoured slaves. They speak not before He speaks, and they act [in all things] by **His Command.** He knows what is **before** them, and what is **behind** them, and they offer **no intercession** except for those who are **acceptable, and they stand in awe and reverence of His [Glory].** And one

of them who say: I am an Absolute besides Him, such a one We should reward with **Hell**. Thus We repay **those who do wrong**. [Sura No. – 20\21 – Ambiyaaa, Ayat No. – 25 to 29]

COMPANIONS OF THE RIGHT HAND OR THE LEFT HAND

But he [the follower of tautology] has not attempted the ascent [rise / climb]. And what will convey to you the **path that is ascent**? [It is:] to free a bondman. Or the giving of food in a day of hardship to the **orphan** with **claims of relationship**, or to the destitute [down] in the dust; and to be of those who believe and enjoin patience, and enjoin deeds of kindness and compassion. Such are the **Companions of the Right Hand** [West Zone or Heaven]. But those who reject Our Signs, they are the [unhappy] **companions of the left** hand East Zone or Hell]. On them will be fire vaulted over. [Sura No. – 89\90 – Bi-haazal-balad, Ayat No. – 1 to 20].

On that Day you will be exposed, not a secret of you will be hidden. Then as for him who is given his **record in his right hand**, he will say: Take, read my record! Surely I knew that I should have to meet my reckoning, and then he will be in a Blissful state, in a high Garden whereof the clusters are in easy reach. (And it will be said to those therein) Eat and drink at ease for that which you sent before you in the past days. [Sura No. – 68\69 – Al-Haaaqqatu, Ayat No. – 18 to 24]

But as for him who is given his **record in his left hand**, he will say: Ah! Would that my record had not been given to me! And knew not what any reckoning is! Ah! Would that [Death] had made an end of me! Of no profit to me has been my wealth! My power has gone from me! [The stern command will say]: Seize him, and chain him and then expose him to the hell-fire. Further, make him march in a chain, whereof the length is seventy cubits! Lo! He used not to believe in Allah Most High. And would not encourage the feeding of the wretched! So no friend has he here this Day. Nor has he any food except the corruption from the washing of wounds which none but sinners eat. [Sura No. – 68\69 – Al-Haaaqqatu, Ayat No. – 25 to 37]

And those on the left hand, **what of those on the left hand?** [They will be] in the midst of a Fierce Blast of Fire and in Boiling Water / In scorching wind and scalding water. And in the shades of **Black Smoke** neither cool nor refreshing; for that they were wont to be indulged, before that, in wealth [and luxury] / Lo! Heretofore they were effete with luxury. And persisted obstinately in wickedness supreme [awful sin]! And they used to say: What! When we die and become dust and bones, shall we then indeed be raised up again [We] and our fathers of old? Say: Yea, those of old and those of later times, all will certainly be gathered together for the meeting appointed for a Day well-known. Then lo! You, the erring, the deniers. You will surely taste (eat) of the Tree called **Zaqquum.** And will fill your bellies therewith. And thereon you will drink of boiling Water **drinking even as the camel drinks.** This will be their welcome on the Day of Judgment. [Sura No. – 55\56 - Waqa-atil-waaqiah, Ayat No. – 41 to 56]

We shall set up **scales of justification** for **the Day of Resurrection [Yawmil-Qiyaamati]**, so that not a soul will be dealt with unjustly in the least, and if there be [no more than] **the weight of a mustard seed**, We will bring it [to account]: and enough are We to take account. In the past We granted to Muusa and Haarun **the criterion [for justification]**, and **a Light** and **a Message for those who would do right;** those who fear their Rab in their most secret thoughts, and who hold the Hour in awe. And **this is a blessed Message which We have sent down: Will you then reject it?** [Sura No. – 20\21 – Ambiyaaa, Ayat No. – 47 to 50]

Who, then, does more wrong than one who **utters a lie** concerning Allah, and **rejects the Truth** when it reaches him [Manifested Nature]? Will not the home of disbelievers be in hell? [Sura No. – 38\39 – Tanziilul-Kitaab, Ayat\Verses – 32]

And those who bring the Truth and confirm [with a view to believe] it - such are the men who do right. They shall have all that they wish for, in the presence of their Rab. Such is the reward of those who do good. That Allah will remit from them [even] the worst in their deeds and give them their reward according to the best of what they have done [Sura No. – 38\39 – Tanziilul-Kitaab, Ayat\Verses – 33 to 35]

Is not Allah enough for his Servant? But they try to frighten you with other [projected realities] besides Him! For such as Allah sends to go astray, there is no guide for him. And he whom Allah **guides**, there can be none to **lead astray**. Is not Allah Exalted in Power, [Able to enforce His Will], Rab of Retribution? [Sura No. – 38\39 – Tanziilul-Kitaab, Ayat\Verses – 36 & 37]

And verily if you ask them: **'Who has created the heavens and the world?'** they would be sure to say: **Allah.** Say: **Bethink** you then of the things [projected falsehood] that you **invoke** besides Allah. If Allah wills some **penalty** [hurt] for me, **can they remove His penalty?** Or if He wills some **grace** for me, **can they keep back His grace?** Say: **Allah is sufficient** [all in all] for me! In Him the **trustees put their trust.** [Sura No. – 38\39 – Tanziilul-Kitaab, Ayat\Verses – 38]

SECTION – IV
BLACK SQUARE – 4.4
TWELVE MATTER PARTICLES

TWELVE MATTER PARTICLES IN RESEMBLANCE WITH EQUAL & OPPOSITE MONTHS OF THE MERCURIAN YEARS

Fasta-'iz billaahi minash-Shaytaanir-Rajim
Bismillaahir-Rahmaanir- Rahim

[TWELVE MATTER PARTICLES OR SPRINGS]
RING OPERATOR / ZAKARIYA

Due to Revealed **Trinity** [3D or Three Dimensional Command or Triangular Arrow of Science or Three Minutes after Big Bang or Zakayia's three days symbolic communications], Four Basic Forces, and Manifested **Geometrical Progression** [Manifestation declining towards right in resemblance with multiplication sign], there are twelve matter particles or springs. Matter particles are the **octagonal black spots or what science calls sun-spots**. So, there are **twelve matter particles** in resemblance with twelve hours of a clock, twelve sacred months shared by equal & opposite Mercurian Year, twelve springs shared by both Tribes [Lower Sea-shore or East Zone] and Nations [Upper Sea-shore or West Zone].

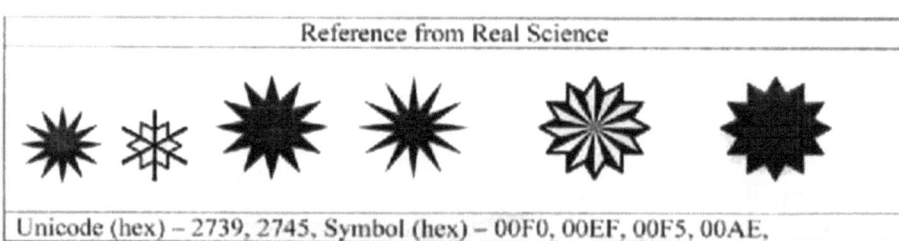

Reference from Real Science
Unicode (hex) – 2739, 2745, Symbol (hex) – 00F0, 00EF, 00F5, 00AE,

"Allah said: Lo! I am with you."
TWELVE SPRINGS OR MATTER PARTICLES

Allah made a **covenant** (promise) of **old** with the **Banii-Israa-iil** [at the right side of the Mount Tuur] and We rose among them **twelve chieftains [springs or matter particles]**, and **Allah said: Lo! I am with you**. If you establish prayer and pay the poor-due, and **believe in my Messengers and support them**, and **lend** unto Allah a **kindly loan**, surely I shall remit your **sins**, and surely I shall bring you into **Gardens [Heavens] underneath which river flow**. Whoso among you **disbelieve** after this **will go astray from right path**. [Sura No. – 4\5 – Maaaidah, Ayat\Verses – 12]

ELEVEN AND ONE [EQUAL & OPPOSITE] = TWELVE

"Iz qaala yuusuf li-'abiihi yaaa-'abati 'innii ra'-aytu 'aha-da 'ashara kaw-kabanw-wash-shamsa wal-qamara ra-'aytu-hum lii saajidiin." [Trans.] Behold! Yuusuf said to his father: O my father! I did see **eleven springs [matter particles]** and **wash-shamsa wal-qamara**. I saw them prostrating themselves unto me! [Sura No. – 11]\12 – Yuusuf, Aat No. – 4]

EACH GROUP [East Zone & West Zone]
REFERENCES OF TWELVE SPRINGS OR MATTER PARTICLES

And when **Muusa** prayed for water for his people; We said: **Strike** the **rock** [space-bar] with your **staff**; then **gushed forth there from twelve springs (matter particles). Each group** knew its own place for **water.** So eat and drink of the **sustenance provided by Allah,** and do **neither evil nor mischief** on the [face of the] earth. [Sura No.1\2 Baqarah, Ayat No. 60]

TWELVE SPRINGS OR MATTER PARTICLES EQUALLY SHARED BY TRIBES AND NATIONS

i) Reference of a **section** [mankind] who **leads** with **truth** and **establishes justice** therewith [followers of true religion who do righteous work and give good tidings]

ii) Reference of **two divided sections** of the manifestations as Tribes and Nations [Lower Seashore and Upper Seashore]

iii) Reference of **Inspiration** concerning water [Manifested Signs]

iv) Reference of Trinity – **Rock** [Iron or Buried Treasure or wood] – **Striking** [functioning of the command or continual acceleration] – **Staff** [3D Command of Allah in Coherence with the Concept of the ideogram Alif]

v) Reference of the **Twelve Springs or Matter Particles**

vi) Reference of **each group** represents each zone.

vii) Reference of **three stairs of the appearing earth** - shade of clouds, manna and quails

viii) Reference of Sustenance [Maryam supplied with Sustenance] provided by Allah - Eat of the good things We have provided for you

Ayat\Verses – Of the people of Muusa there is a **section** who leads with **truth** and establishes **justice** therewith. We **divided** them into **twelve tribes, nations.** We directed Muusa by **inspiration,** when his [thirsty] people asked him for **water: Strike** the **rock** with your **staff.** Out of it there gushed forth **twelve springs. Each group** knew its **own place for water.** We gave them the **shade of clouds,** and sent down to them **manna** and **quails,** [saying]: **Eat of the good things** We have **provided** for you: [but they rebelled]; to Us they did **no harm,** but they harmed their **own souls**. [Sura No. – 6\7 - As-haabul- a' raaf, Ayat No. – 159 & 160]

Allah did choose **Adam** and **Nuuh**, the **family of Ibrahim**, and the **family of Imran** above (all His) creatures. They were descendents one of the other. And Allah is Hearer and Knower of all things. (Remember) when the **wife of IMRAN** said: O Allah! I do dedicate unto You what is in my womb for Your **special service**, accept this from me, for You hears and knows all things. And when she was delivered, she said: My Rab! Lo! I am delivered of a female. Allah knew best what she brought forth. The male is not as the female; and Lo! I have named Her **Maryam**, and Lo! I desire Your protection for Her and Her off springs from **Shaytan** the outcast. And her (wife of Imran) Rab accepted Her (Maryam) with full acceptance. Allah made Her grow in **purity and beauty** and appointed **Jakariya** Her guardian. Whenever Jakariya enters into the **sanctuary** where she is, he finds her supplied with **sustenance**. He asks: O Maryam! Whence [comes] this to you? She answered: It is from Allah, for Allah provides sustenance to which He pleases without measure. [Sura No. 2\3 – Imran, Ayat No. –33 to 37]

i)   **Adam – all things**
ii) **Nuuh – animals**
iii) **Family of Ibrahim – followers of Upright West**
iv) **Family of Imran – galaxy of stars**
v)   **Wife of Imran – manifested world**
vi) **Maryam – means of sustenance**
vii) **Shaytan –zone of iron or wood or ring of the Star Saturn**
viii) **Zakariya – screen or veil or ether or umbrella or electron cloud**
ix) **Sanctuary – nucleus or oily tree or tree of Zaytuun**
x)   **Sustenance – Manifested Sign of Natural Magnetism in the name of Maryam supplied with sustenance**

ALLAH TAUGHT ADAM THE NAMES OF ALL THINGS

And He **taught Adam** the **names of all things**. Then He placed them before the angels, and said: **Tell me the names of these if you are right**. They said: Be glorified! **We have no knowledge**, save **what You have taught** us. In truth it is You Who are perfect in knowledge and wisdom. He said: O Adam! Tell them their names. When he had told them, Allah said: Did I not tell you that **I know the secrets of heaven and world,** and I know **what you reveal and what you conceal**? Sura No. 1\2 – Baqarah, Ayat No. – 31 to 33]

ZAKARIYA – THE GUARDIAN OF MARYAM
TWELVE MATTER PARTICLES ARE SHARED IN THE NAME OF ZAKARIY

(Remember) when the wife of IMRAN said: O Allah! I do dedicate unto You what is in my womb for Your special service, accept this from me, for You hears and knows all things. And when she was delivered, she said: My Rab! Lo! I am delivered of a female. Allah knew best what she brought forth. The male is not as the female; and Lo! I have named Her Maryam, and Lo! I desire Your protection for Her and Her off springs from Shaytan the outcast. And her (wife of Imran) Rab accepted Her (Maryam) with full acceptance. Allah made Her grow in purity and beauty and appointed **Zakariya** Her guardian. Whenever **Jakariya** enters into the sanctuary where she is, he finds her supplied with sustenance. He asks: O Maryam! Whence [comes] this to you? She answered: It is from Allah, for Allah provides sustenance to which He pleases without measure. [Sura No. 2\3 – Imran, Ayat No. –35 to 37]

NATURE OF SPRINGS OR MATTER PARTICLES
BONES - WAX FEEBLE AND HEAD IS
SHINING WITH GRAY HAIR

Kaf. Ha. Ya. 'Ain. Sad. [Sura No. – 18\19 – Maryam, Ayat –1]
[This is] a recital of the Mercy of you Rab to His servant Zakariya. When He cried to his Rab in secret, Praying: O my Rab! Lo! **the bones of me wax feeble and my head is shining with gray hair, and I have never been un-blest in prayer to You**, my Rab. Now I fear [what] my relatives [will do] after me; since **my wife is barren**. So **give me** from Your presence **a successor**. Who shall **inherit** of me and **inherit** (also) of **the house of Yaquub**; and make him, O my Rab! One with whom You are well-pleased! [Sura No. – 18\19 – Maryam, Ayat\Verses – 2 to 6]

Pentagon - Kaf. Ha. Ya. 'Ain. Sad
"Kaf-Ha-Ya-Ain-Sad. Zikru Rahmati Rabbika abdahuu Jakariyaa"
Kaf. Ha. Ya. 'Ain. Sad. [Sura No. – 18\19 – Maryam, Ayat –1]
[This is] a recital of the Mercy of your Rab to His servant Zakariya. When He cried to his Rab in secret. [Sura No. – 18\19 – Maryam, Ayat\Verses – 2 & 3]

ZAKARIYA AND SIGNIFICANCE OF THE SIGNS OF ALLAH
  i)   **Reference of off-springs and a suspicious doubt -** "He (Zakariya) said: My Rab! How can I have a son, seeing I am very old, and my wife is barren?"
  ii) **Prayer for a sign of indication -** "He said: My Rab! Give me a Sign [Token or Indication]!"

iii) **Reference of Trinity as well as existential import of signs of Allah -** (The angel) said: The Sign unto you (shall be) that **you shall not speak unto mankind three days except by signs.**

iv) **Reference of early hours of night [arising of new moon for the West Zone] and morning [entering of the so-called sun for the East Zone]** - "Remember Allah much and praise (Him) in early hours of night and morning".

v) **Son** - Allah gave you good tidings of (a son whose name is) **Yahya to confirm a Word from Allah**, and **lordly chaste**, and **a prophet of the righteous.**

SIGNIFICANCE OF TRINITY
THREE DAYS & SYMBOLIC COMMUNICATIONS
You shall not speak unto mankind three days except by signs.

Then **Zakariya prayed** to Allah, saying: O Rab! **Grant unto me from Your bounty goodly off spring**. Lo! You are the Hearer of Prayer. While he was standing in prayer **in the sanctuary**, the angels called unto him: Allah gave you good tidings of (a son whose name is) **Yahya to confirm a Word from Allah**, and **lordly chaste**, and **a prophet of the righteous**. He (Zakariya) said: My Rab! **How can I have a son**, seeing **I am very old**, and my **wife is barren**? (The angel) answered, so (it will be), **Allah does what He wills**. He said: My Rab! **Give me a Sign [Token or Indication]**! (The angel) said: **The Sign unto you (shall be) that you shall not speak unto mankind three days except by signs. Remember Allah much and praise (Him) in early hours of night and morning.** [Sura No. 2\3 – Imran, Ayat No. – 38 to 41]

And remember **Zun-nun**, when she departed in wrath: She imagined that We had no power over her! But she cried through the depths of darkness: There is no absolute reality but You: Glory to You: I was indeed wrong! So We listened to her: and delivered her from distress; and thus do We deliver those who have faith. And [remember] **Zakariya**, when he cried to his Rab: O my Rab! Leave me not without offspring, though you are the best of inheritors. So We listened to him: and We granted him **Yahya**: We cured his **wife's** [Barrenness] for him. These [**three**] were ever quick in emulation in good works; they used to call on Us with love and reverence, and humble themselves before Us. [Sura No. – 20\21 – Ambiyaaa, Ayat No. – 87 to 90]

THE SON OF ZAKARIYA
SIGNIFICANCE OF TRINITY – THREE NIGHTS

"Your **Sign or Token** is that you with no bodily defect shall **not speak unto mankind three nights.**"

334

[His prayer was answered]: O Zakariya! We give you good news of a **son**: His name shall be **Yahya. We have given the same name to none before (him).** He said: O my Rab! **How shall I have a son, when my wife is barren and I have grown quite decrepit from old age**? He said: So [it will be] the sayings of your Rab. It is easy for Me, even as I did you before, **when you had been nothing!** [Zakariya] said: O my Rab! Give me a Sign (Token). Your **Sign or Token** is that you with no bodily defect shall **not speak unto mankind three nights.** So **Zakariya came out to his people from the sanctuary and signified to them glorify your Rab at break of the day and fall of night.** [Sura No. – 18\19 – Maryam, Ayat\Verses – 7 to 11]

O Yahya! **Hold fast the Khuzil-Kitaab.**
[To his son came the command]: O Yahya! **Hold fast the Khuzil-Kitaab**, and We **gave him Wisdom even as a youth.** And piety [for all creatures] as from Us, and purity. He is devout and kind to his parents, and he is not overbearing or rebellious. So Peace on him the day he was born, the day that he will die, and the day he will be raised up to life [again]! [Sura No. – 18\19 – Maryam, Ayat\Verses – 12 to 15]

Say: the things that my Rab has indeed forbidden are: **shameful deeds**, whether **open** (apparent or manifest) or **secret** (within or immanent); **sins** and **trespasses against truth or logic**; **assigning of partners to Allah**, for which He has **given no authority**; and **saying things** about Allah of which you have **no knowledge.** To every **nation** is a **term appointed. When** their term is **reached**, they **cannot delay an hour**, or [an hour] they can **advance** [it in anticipation]. [Sura No. – 6\7 - As-haabul- a' raaf, Ayat No. – 33 & 34]

NONE TO INTERCEDE, NOR A SINGLE FRIEND
Now, then, we have none to intercede [Owner and Possessors of Manmade Nature], nor a single friend to feel [sceptic or researcher]. Now if we only had a **chance** of **return** we shall **truly** be of those **who believe!** Verily **in this is a Sign** but most of **them do not believe.** And verily your Rab is He, the Exalted in Might, Most Merciful. [Sura No. 25\26 – Shuraa, Ayat No. – 100 to 104]

HELL WILL APPEAR PLAINLY
"By Allah, **we were truly in an error manifest."**
And hell will appear plainly to the erring [followers of mechanical globalisation, circular rotation & revolution system, and manmade magnetism], and it shall be said to them: **Where are [Middle-East, Mechanical Globalisation, Circular Rotation & Revolution, and Manmade Magnetism] that you worshipped Besides Allah [Upright West]? Can they help you**

335

or help themselves? **Then they will be thrown headlong into the [Fire or projected planet Mars or Recycling Region of Eartha 3D], they and those straying in evil,** and the whole hosts of Iblis together. They will say when they quarrelling therein, by Allah, **we were truly in an error manifest,** when we held **you** as equals with the Rab of the universe; it was but the **guilty who misled us**. [Sura No. 25\26 – Shuraa, Ayat No. – 91 to 99]

Say: **I am forbidden** to serve **those** on whom **you call instead of Allah.** Say: **I will not follow** [projected mechanical globalisation, circular rotation & revolution system, and manmade magnetism] **your vain desires.** If I **did**, I **would go astray from the right path**, and be **not of the company of those who receive guidance**. [Sura No. – 5\6 – An-aam, Ayat No. – 56]

SOUND HEART

O my Rab! Bestow **wisdom** on me, and **join** me with the **righteous**. Grant me **honourable** mention on the **tongue** of truth among the latest [generations]. Make me one of the **inheritors** of the **Garden** of **Bliss**. Forgive my father, for that **he is among those who err**. And let me not be in disgrace on the Day when [men] will be raised up; the Day whereon neither wealth nor sons will avail, but only he [will prosper] that brings to Allah a **sound heart**; to the **righteous**, the **Garden** will be **brought near,** [Sura No. 25\26 – Shuraa, Ayat No. – 83 to 90]

My Duty is to preach clear Message

It is Allah Who made your habitations homes of rest [immovable] and quiet for you; and made for you, out of the **skins** of animals, [tent or Pentagon for] **dwellings**, which you find so light [and handy i.e. Pentagon or Tent] when you **travel** and when you **stop** [in your travels]; and out of their **wool**, and their **soft fibres** [between wool and hair], and their **hair**, rich **stuff** and articles of convenience [to serve you] for a time. It is Allah Who made out of the things He created, **some things to give you shade [Mount Tuur]; of the hills He made some for your shelter.** He made you **garments** to protect you from **heat**, and **coats** of mail to protect you from your [mutual] **violence [Veto]**. Thus He completes His favours on you in order that you may bow to His Will. But if they **turn away, your duty is only to preach the clear Message**. [Sura No. – 15\16 – Nahl, Ayat\Verses – 80 to 82]

SECTION – IV
BLACK SQUARE – 4.5
FOUR BLESSED SPRINGS
[SOLIDITY IMPLIES FOUR]

Pythagorean Number – 'Solid – Four'

Fasta-'iz billaahi minash-Shaytaanir-Rajim

Bismillaahir-Rahmaanir- Rahim

FOUR BLESSED SPRINGS OR BASIC
FORCES FOR STRAIGHT USAGE

Revealed Universe is an equal & opposite Trinity. This equal & opposite Trinity represents four cardinal [revealed] directions like a Diamond. In Real Science, the Universe is called Diamond Operator. In the Kitaab with Truth, the Universe is called Shi-raa or a Shining Star. 'Shi-raa' means intrinsically luminous. So, Universe is an intrinsically luminous Star. There is resemblance between Real Science and Kitaab with Truth concerning the concept of the Universe [Alamin]. The appearing pentagonal star in the National Flag of some Nations does not represent Guiding Stars. Guiding Stars [Ashabikan-nujjum] are quadrilateral or four figured star. Appearing Earth is a Pentagon. So, the symbol of pentagonal star in some National Flags represents appearing Pentagonal Earth.

In conspiracy science, the universe is called Moon or earth's permanent Natural Satellite. This further manifests an equal & opposite game between Real Science and Conspiracy Science. From the point of view of Real Science as well as Kitaab with Truth, the Universe is a **Star System**. On the contrary, from the point of view of Conspiracy Science, the Universe is a **Lunar System**. In other words, the Universe is an **Astral System**, a **Planetary System**, a **Cosmological System**, an **Astrophysical System**, a **Stellar System**, a **Solar System**, and the like. These systems are synonymous. In this regard, there is **resemblance** between **Conspiracy Science and Pen-Paper-Pencil works & mechanical activities. These black & white works and activities have converted Star System into Solar System**. Consequently, each star of the Star System is being called Planet / Moon in Conspiracy Science, Solar System, Astral System, Cosmological System, Astrophysical System, Stellar System, and black & white works and mechanical activities. Four Galilean Moons are the best examples as well as valid reference in this regard. With respect to 'Kitaaba Wal-Hikmata' or 'Manifested Nature and the Utility of One's Upright Logic', the Universe is a Star System in resemblance with Real Science and Kitab with Truth. So, project Planets or Moons are Guiding Stars in this re-search project.

With respect to Diamond Operator, four cardinal directions are - Up, Down, Right, and Left. With respect to Revelation or Sirius Binary System, the four cardinal directions are - East Horizon, West Horizon, Northern Hemisphere, and Southern Hemisphere. Helium-4 represents 2-Horizons and 2-Hemispheres. In other words, four cardinal directions represent equal & opposite Trinity in coherence with Plus Sign of Mathematics [Tawraat] and in pragmatic correspondence with Crucified Sign [Injiil]. Isa (ass) brought clear sign of the four cardinal [revealed] directions. Thus, **four cardinal [revealed] directions are Up, Down, Revealed Right, and Revealed Left.** Four cardinal [revealed] directions also represent four revealed [basic] forces. West represents upward direction. East represents downward direction. North represents direction towards revealed left. South represents direction towards revealed right. **Four basic forces are Gravitational Force, Strong Force, Magnetic Force, and Weak Force**. Gravitational Force represents downward direction towards East. Strong Force represents upward direction towards West. Magnetic Force represents leftward direction [North] from the point of view of revelation but rightward direction from the point of view of finite observer facing towards the appointed Kaba. Weak Force represents rightward direction [South] from the point of view of revelation but leftward direction from the point of view of finite observer facing towards the appointed Kaba. Earthquake emerges from East. Rain comes from West. Electromagnetic Wave [Fire / Light / Tarash-Shamsi / Bullet / so-called sun] enters from and ends in North [Revealed Left]. Electromagnetic Wave [Fire / Light / Wash-Shamsi / Bullet / so-called sun] turns back [sets and rises] from South [Revealed Right]. Four gross elements [Nitya Dravyas / Eternal Elements] also represent four cardinal directions. Earth [soil] represents East. Water [life] represents West. Fire [light] represents North. Air [sound] represents South. Four causes of Aristotle also represent four cardinal directions. Material cause represents East. Efficient cause represents North. Formal cause represents South. Final cause represents West. The fifth gross element is called ether or electron cloud or black umbrella or Bish-Shamsi or projected sun which rises from the East or what accelerating masses emit or gravitational wave.

There is a difference between revelation and manifestation. Revelation resembles with Plus Sign of Mathematics, while manifestation resembles with Multiplication Sign of Mathematics. Revealed East Horizon has been manifested declining towards Revealed Right. So, revelation represents **Axis** as well as **Arithmetic Progression** of the accelerating charges and accelerating masses of the Equal & Opposite Binary System in resemblance with Newton's Three Laws of Motion and Plus Sign. On the other hand, manifestation represents **Orbits** as well as **Geometric Progression** of what accelerating charges emit [release] i.e. Electromagnetic Wave and what accelerating masses emit [release] i.e. Gravitational Wave in resemblance with Einstein's Gravitational Lensing, Binary Pulsar, and Multiplication Sign. Consequently, there are four revealed [cardinal] directions in resemblance with Crucified Sign or solid directions in resemblance with Pythagorean Number – 'Solid – Four'. Again, there are six manifested directions in resemblance with Pythagorean Number – 'Animation – Six'. Four revealed [cardinal] directions are Down [East], Up [West], Revealed Left [North], and Revealed Right [South]. Six manifested directions are North-East, Middle-East, South-East, South-West, Upright-West, and North-West.

TWO DAYS AND FOUR DAYS
Two Days – Straight Middle-East Region of Eartha 3D and Upright West Region of Arabian Peninsula
Four Days – North-East Region, South-East Region, South-West Region, and North-West Region
Come both of you, willingly or unwillingly.

Say: **Disbelieve** you verily in Him **Who created the world in two days**, and **ascribe** you unto Him **rivals**? He is the Rab of the universe. He set in the [world], mountains **standing firm**, high above it, and bestowed blessings in the world, and **measure therein all things to give them nourishment in due proportion, in four Days**, in accordance with [the needs of] those who seek [Sustenance]. Then **He turned to the heavens [West Horizon] when it was smoke [white dwarf]**, and said unto it and unto the earth: **Come both of you, willingly or unwillingly**. They said: We come, obedient. So He **ordained** them as **seven firmaments (windows) in two Days**, and **He inspired in each heaven its mandate [movements]. And We adorned the lower heaven with lights** and **[provided it] with guard.** Such is the Measuring of the Mighty, The Knower. **[Sura No. – 40\42 – Haa-Mim-Sajdah, Ayat No. – 9 to 12]**

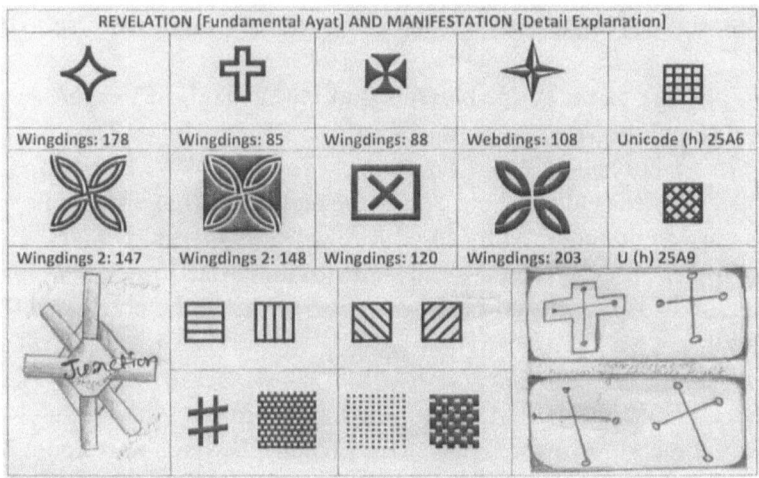

Universe is a Star System. Universe has been revealed as equal & opposite Trinity. This equal & opposite trinity is called Sirius Binary System. The main sequence of the Sirius Binary System is Sirius A. Star Jupiter represents Sirius A or West Horizon. The white dwarf companion of the Sirius Binary System is the Sirius B. Star Saturn represents Sirius B or East Horizon. The Universe has been manifested as an Upright Rectangle in coherence with the 'End of Proof' of Real Science [Tawraat], and in correspondence with the appointed Kaba [Injiil]. For this reason it has been shared that the appointed Kaba is the Black & White Imam [Leader] of manifested right directions for mankind.

There are twelve matter particles [springs]. Out of twelve matter particles [springs], four springs [pillars or Imams or matter particles] are blessed for straight usage in correspondence with four cardinal [revealed] directions and four basic forces. These four blessed springs [matter particles] are the witnesses [Criterions or Kitaabs] of the Fifth One or four pillars of the Pentagonal Tent.

References from Kitaab with Truth as Universal Major Premises
Cardinal [Revealed] Directions
i)   Right Angel –Weak Force
ii)   Left Angel - Magnetic Force
iii) An observer – West – Strong Force [West Horizon]
iv) Stupor of Death – East –Gravitational Force [East Horizon]

Ayat\Verses - We verily created a man, and We know **what his soul whispers to him**, and **We are nearer to him than [his] jugular vein.** When the **two** appointed [Guardian Angels] **to read** and **to write seated on the right and on the left and not a word do they utter** but there is with them an **observer** ready, and the stupor of **death comes in Truth.** (And it is said to you): **This was that which you were trying to escape. [Sura No. 49\50 – Qaaaf, Ayat No. – 16 to 19]**

"Four of them are blessed that is the straight usage."
Forehead [Upright West Region of Arabian Peninsula] –- Towards Upright West – Strong Force – Upward Direction - Towards West Horizon, Towards West Zone, Towards Uppermost Land of the manifested hexagon and appearing pentagon in resemblance with the concept of the Galaxy of Stars, Heavens, Leaning Tower, Tallest Tree, Mount Tuur, Mount Alal-Ju-diyyi. Upper West is the Region of White dwarf or the obverted white of the converted snake of Muusa's Staff.

Back [Straight Middle-East Region of Eartha 3D] – Gravitational Force – Downward direction - Towards Space-bar [Wood], Towards East Zone, Towards

Lower most Sea-shore in resemblance with the concept of Volcano, Hell, Buried Treasure, Zone of Iron, Recycling Region, Star Mars, Bish-Shamsi, Lish-Shamsi, Gravitational Wave, and the like.

Border – Right side of the revelation, right side of the appointed Kaba – Weak force – Leftward arrow from the finite subjective point of view but Rightward arrow from the point of view of the appointed Kaba, Right Imam, Right Guardian, Right Angel, Revealed South, Southern Hemisphere in resemblance with the existence of South America for the East Zone and Europe for the West Zone. **South America represents South-East and Europe represents South-West.**

Back Border – Left side of the revelation, left side of the appointed Kaba – Magnetic Force – Rightward arrow from the finite subjective point of view but Leftward arrow from the point of view of the appointed Kaba, Left Imam, left Guardian, Left Angel, Revealed North, North Direction, Northern Hemisphere in resemblance with the existence of North America for the East Zone and Australia for the West Zone. **North America represents North-East and Asia, Africa, and Australia represent North-West.**

Verily the reversing is an addition to unbelief.
Ayat\Verses – O you who believe! There are indeed many among the **priests** (Lukmans) and **anchorites** [Imamgiris], who in **falsehood demolish** the **wealth of men** and **obstruct** [them] **from the way of Allah.** And there are those who **stored gold and silver and spend it not** in the **way of Allah** announce unto them a **most grievous penalty.** On the Day when heat will be produced out of that [wealth] in the **fire of Hell**, and with it will be branded their **foreheads**, their **borders**, and their **backs borders**, and their **backs.** This is the [treasure] which you stored for yourselves. Taste you, then, the [treasures] you stored. The **number of months [springs] in the sight of Allah is twelve. So ordained by Him the day He created the heavens [West Horizon or Star Jupiter] and the world [East Horizon or Star Saturn]. Four of them are blessed that is the straight usage. So wrong not yourselves therein, and wage the Pagans / idolaters all together as they waging on all of you.** But know that Allah is with those who restrain themselves. **Verily the reversing is an addition to unbelief.** The unbelievers are led to wrong thereby, for they make it **lawful one year**, and **forbidden another year**, in order to adjust the number of months forbidden by Allah and make such forbidden ones lawful. The evil of their course seems pleasing to them. But **Allah guides not those who reject Faith.** [Sura No. – 8\9 – Tauba, Ayat No. – 34 to 37]

Reference of Manmade Nature
Reference of Iblis' Projected Directions
Mechanical Globalisation and Man-made Magnetism

i) **From before them – Projected East**
ii) **From behind them– Projected West**
iii) **From their [finite observers'] Right hands – Projected South**
iv) **From their [finite observers'] Left hands – Projected North**

It is We Who created you and gave you shape; then We bade the angels to fall **prostrate** before **Adam,** and they prostrate all **except Iblis.** He refused to be of those who prostrate. [Allah] said: What prevented you (Iblis) from prostrating when I commanded you? He (Iblis) said: I am better than he. **You created me from fire** and him (Adam) from clay. [Allah] said: Get you down from this. **It is not for you to show pride here.** So, go forth! Lo! You are of those degraded. He said: Give me reprieve till the day they are raised up. [Allah] said: Be you among those who have reprieve. He said: Now, **because You have sent me astray, verily I shall lurk in ambush for them on the Straightway.** Then I shall come upon them from **before** them and from **behind** them and from their **right hands** and from their **left hands**, and You will not find in most of them gratitude (or your mercies). [Allah] said: Go forth from hence, **degraded, banished. As for such of them as follow you, surely I will fill hell with all of you**. [Sura No. – 6\7 - As-haabul- a' raaf, Ayat No. – 11 to 18]

REFERENCE OF BETTER SUBSTITUTION
We [Allah and His Messenger] **are not to be defeated**
Rab of all points
Directions of Turning and Rising

i) Open-eyed – Right and Left –South and North.
ii) Rab of all points – ~sth or fifth one – East & West

Ayat\Verses – Now, what is the matter with the disbelievers that **they keep starting toward** you **open-eyed** from the **right** and from the **left**, in groups? Does every man of them long to enter the Garden of Bliss? By no means! For We have created them from what they know. Now I do call to witness the **Rab of all points** in the **East** and the **West** that We can certainly **substitute** for them **better** than they; and We are **not to be defeated** [in Our Plan]. [Sura No. – 69\70 – Zil-ma-aaru, Ayat No. – 36 to 41]

Where is the Manifested Rabbun Gafuur?
i) **Right side of the Kaba –** South America of the East Zone & Europe of the West Zone

ii) **Left side of the Kaba** – North America of the East Zone & Asia, Africa, and Australia of the West Zone

iii) **A Territory fair and happy** - Upper West Zone of the Arabian Peninsula

iv) **Sustenance** – Shams or Star Venus [projected sister planet Venus] or Maryam Supplied with Sustenance

v) **Rabbun Gafuur** – Black & White Imam of the City called Kaba represents Rabbun Gafuur

There was indeed a **Sign** for Saba, in their dwelling place. **Two Gardens to the right and to the left. Eat of the Sustenance** [provided] by your Rab, and be grateful to Him: **A territory fair and happy, and a Rabbun Gafuur.** [Sura No. – 33\34 – Saba, Ayat No. - 15]

"Our **representatives** are present with them to record"
WHO ARE THE REPRESENTATIVES MENTIONED HERE?
Or do they think that **We** hear not their secrets and their private counsels? Indeed [We do], and Our **representatives** are present with them to record. [Sura No. 42\43 – Ummil-Kitaab \ (Zukhruf), Ayat No. – 80]

Equal & Opposite
Reference of Octagon / Springs [Matter Particles] are Octagonal
[Take] **eight** in [four] **pairs**

It is He Who produces **gardens**, with **trellises** and without, and **dates**, and **crops** of all kinds, and **olives** and **pomegranates**, similar [in kind] and different [in variety]: eat of their fruit in their **season**, but render the **dues** that are proper on the day that the harvest is gathered. But **waste not by excess**: for Allah loves not the wasters. Of the **cattle** are some for burden and some for meat: eat what Allah has **provided** for you, and **follow not the footsteps of Shaytan**: for **he is an open enemy to you.** [Take] **eight** in [four] **pairs**: Of sheep a pair and of **goats a pair;** say, has He forbidden the **two males**, or the **two females**, or [the young] which the wombs of the **two females enclose**? **Expound to Me (the case) with knowledge if you are truthful.** Of camels a pair, and oxen a pair; say, has He forbidden the two males, or the two females, or [the young] which the wombs of the two females enclose? **Were you present when Allah ordered you such a thing? But who does more wrong than one who invents a lie against Allah, to lead astray mankind without knowledge?** For Allah guides not people who do wrong. [Sura No. – 5\6 – An-aam, Ayat No. –142 to 145]

Four Pairs

Allah did choose **Adam and Nuuh**, the **family of Ibrahim**, and the **family of Imran** above (all His) creatures. They were **descendents one of the other**. And Allah is Hearer and Knower of all things. (Remember) when the **wife of IMRAN** said: O **Allah**! I do **dedicate** unto You what is in my **womb** for Your **special service, accept** this from **me**, for You hears and knows all things. And **when she was delivered**, she said: **My Rab! Lo! I am delivered of a female**. Allah **knew** best **what she brought forth. The male is not as the female**; and Lo! **I have named Her Maryam**, and Lo! I **desire** Your **protection** for **Her** and **Her off springs** from **Shaytan** the outcast. And **her** (wife of Imran) **Rab accepted Her** (Maryam) with **full acceptance**. Allah **made Her grow in purity** and **beauty** and **appointed Zakariya Her guardian**. Whenever **Zakariya** enters into the **sanctuary** where **she** is, he finds her **supplied with sustenance**. He asks: O Maryam! **Whence [comes] this to you**? She answered: **It is from Allah, for Allah provides sustenance to which He pleases without measure**. [Sura No. 2\3 – Imran, Ayat No. –33 to 37]

Ayat\Verses – When **Ibrahim** said (unto his Rab): My Rab! **Show me how You give life to the dead. He replied: Have you no faith? He said: Yes, but just** to **reassure** my **heart. Allah said: Take four birds**, draw them to you, and **cut** their bodies to pieces. **Scatter** them over the **mountain-tops**, and then **call them back. They will come swiftly to you. Know** that Allah is The Mighty, The Wise. The **parable (likeness / similitude)** of those who **spend** their wealth **in the way of Allah** is the likeness of a **grain** which growth **seven ears** and **each ear** has a **hundred** grains. Allah give **increase manifold** to which He **pleases**. Allah is All-Embracing, All-Knowing. **[Sura No. – 1\2 – Baqarah, Ayat No. - 260 & 261]**

ANGELS AS MESSENGERS WITH WINGS TWO OR THREE OR FOUR

"Al-Hamdu lillahi Faatiris-Samaawaati wal-'arzi jaa-i'ilil-malaaa-ikati Rusulan uliii-ajnihatim-masnaa wa gulaasa wa rubaa: yaziidu fil-khalqi maa yashaaa; innallaaha alaa kulli shay-in-Qadiir." [Trans.] Praise be to Allah, Who created **[out of nothing]** the **heavens** and the **world**. Who appointed the angels as messengers with wings - **two, or three, or four**. He **adds to creation** as He pleases. For Allah is Able to do all things. [Sura No. – 34\35 - Faatiris-Samaawaati wal-'arz, Ayat\Verses – 1]

Verily your Rab is Allah, who **created the heavens and the world in six days**, and is firmly established Himself upon the Throne [Arsh], regulating and governing all things. No intercessor [can plead with Him] except His permission. This is Allah your Rab; therefore serve Him. Oh! Will you not remind? [Sura No. – 9\10 – Yuunus, Ayat No. – 3]

SECTION – IV
BLACK SQUARE – 4.6
SEVEN REPEATED VERSES

MANIFESTATION IMPLIES SEVEN

SEVEN WINDOWS

Pythagorean Number – 'Manifestation – Seven'

Fasta-'iz billaahi minash-Shaytaanir-Rajim

Bismillaahir-Rahmaanir- Rahim

ALLAH IS THE MANIFEST TRUTH
SEVEN REPEATED VERSES OF THE QUR-AANAL-AZIM
[Seven Windows / Seven Days of a Week / Seven Force Carrier Particles]
UNIQUE UTILITARIAN PRAYER – FATIHA

Ayat\Verses – Alif-Lam-Mim. This is the **Zaalikal-Kitaab**. In it is **guidance sure, without doubt,** to those **who fear Allah** (who ward off evil / projected falsehood); who **believe** in the **Unseen,** and **steadfast** in prayer, and **spend** out of that which We have bestowed upon them; and who **believe** in that which is **revealed** unto **you, and that** which **was revealed before your time, and [in their hearts]** has the **assurance of the Hereafter.** They **depend on guidance** from their Rab, and it is **they who will succeed.** [Sura No. 1\2 – Baqarah, Ayat No. – 1 to 5]

Pythagorean Number
i)   **Point – One**
ii) **Line – Two**
iii) **Plane – Three**
iv) **Solid – Four**
v)   **Quality – Five**
vi) **Animation – Six**
vii) **Manifestation - Seven**

WHERE FROM WE SHOULD START?
Options – Light / Darkness / Veil of Ignorance [depth of darkness]

Answer: Light

Logic – We cannot move a step forward or backward, upward or downward, right or left either in darkness or in veil of ignorance. So, we must have to start from light or knowledge or manifestation or manifested sign of natural magnetism or affirmative minor premise in resemblance with universal major premise.

WHAT IS THE SOLE SOURCE OF TRUE KNOWLEDGE?

Answer – Sense-Data

Logic – If the possessors of knowledge denote both human persons and non-human persons of Peter Singer, then knowledge is possible with data [Tabula Rasa] only; otherwise there is no other source of true knowledge save sense-data. Data represents Pythagorean Number 'Plane – Three'. ~sth or binary relation or equal & opposite relation among data is called sense-data. The manifestation of sense data represents equal & opposite data [subject & Object / Knower & Known] as well as a ~sth. So, manifestation or sense-data implies seven. Moreover, manifestation connotes Reproductive Periods.

ATOMIC ANALYSIS
BLESSED MESSAGE [FRIDAY LECTURE OR KHUTBAH] IN THE MIDST OF SEVEN DAYS OF A WEEK

Being the trinity of Soul – Psyche – Body, we cannot start from one or Monism or Spiritualism or Mysticism or the projected Global Earth or Circular Globe because **Pythagorean Number – 'POINT – ONE' is Windowless.**

We cannot start from Dualism or psyche only or ~sth only or formal relation only overlooking our physical existence in space and time relation because the **Pythagorean Number – 'LINE – TWO' does not represent manifestation, but a ~ sth or a binary sign only like '-', '^' etc.** This ~sth or binary sign **represents equal and opposite faces of the same thread only.**

Science can start neither from Monism nor from dualism. The projection of the global earth is not the product of real science but the product of conspiracy science. With a view to track off believers from the equal & opposite right direction, the conspiracy science has projected Mechanical Globalisation, Circular Rotation and Revolution System, and Man-made Magnetism. The concept of one is monodimensional and windowless. The psychic period is bi-dimensional or ~sth only. It is nothing but two equal & opposite faces or equal & opposite laws or Tawraat [Coherence Truth]. Further, it has been shared with us that there is no crookedness or global concept in Quran [Revelation] and Common Run / Global Concept is an impossible task. These revelations imply that there is neither a circular universe nor a global earth. Now, if I believe that the universe is a circular rotation and revolution system, the earth is like a globe; then I must have to reject my faith on Quran [Revelation]. So, one of the purposes of this inspired sharing is to confirm Manifestation. The universe is a Diamond Operator and a Star System. The universe has been revealed as equal & opposite trinity. The universe has been manifested as an upright rectangle. The East Horizon within the four basic forces is called Black Square [Non-Luminous Moon or Dark Moon]. The world has been manifested as a hexagon within the Black Square of East Horizon. Due to equal & opposite ~ **sth between Upright West Region of Arabian Peninsula and Straight Middle East region of Eartha 3D, the manifested Hexagonal world is appearing as a Pentagon. This appearing Pentagon is called the Earth. There are three ascending stairs in the appearing Pentagonal Earth in resemblance with the concepts of Super Script 1 [Tin], Super Script 2 [Zaytuun], and Super Script 3 [Tuur]**

Searching Questions

i) Is there any solidity prior to Pythagorean Number – 'Solid – Four'?
ii) Is it possible to verify scientifically the existence of non atomic concepts or formal concepts like faith, belief etc.?
iii) If there is no solidity prior to four, then on the basis of which criterion/law you have projected Globalisation? Whether globalisation is a scientific certainty or manifested truth or revealed truth or mechanical conspiracy? Whether Global Universe / Global World / Global Earth / Global Map / Circular Rotation and Revolution are verifiable truth of facts in resemblance with Pythagorean Number and the Concept of an Atom or Teleological Evidence Sorceries & Unique Epistemic Persecution [Black Magic]?
iv) Whether scientific investigation depends on uniform principle [Fitrat / Nature / Dharma] or uniform principle depends on scientific investigation?
v) Whether Uniform Motion [Inertia / State of Equilibrium] depends on scientific investigation or scientific investigation depends on Uniform motion [Inertia / State of Equilibrium]?
vi) Whether scientific investigation emits Binary Pulsar or Sirius Binary System emits Binary Pulsar? Whether Binary Pulsar is the product of Projected Solar System or Projected Solar System is the product of Sirius Binary System?
vii) Whether Binary Pulsar represents Natural Magnetism or Binary Pulsar represents Man-made Magnetism?
viii) Whether science can go beyond Black Square i.e. in the Heaven or in the hell?

ix) Out of Four Galilean Moons which one is the Nil Arm Strong's visited Moon?

It is being taught to us that there are three constituents or particles of an atom, namely, Electron, Proton, and Neutron. But I have conceptualised the truth that there are four constituents or particles of an atom in resemblance with the universe as a quadrilateral shining star, in resemblance with horizontal & vertical rings of an atom, the concept of Helium-4, four basic matter particles, four blessed springs, four leaders [Imams], four shades of the Windows'7 Ultimate, four criterions of Truth, four Kitaabs, four witnesses, four corner pillars of a pentagonal tent, four straight directions of thought, four Galilean Moons, four gross elements, four forces, and so on. An atom is composed of Proton + Neutron & Neutrino + Electron. Proton is positive, Electron is negative, and Neutron & Neutrino are the two equal & opposite faces of positivity & negativity. So, an atom is composed of four constituents. But due to equal and opposite ~sth, an atom is a trinity. The four constituents or particles of an atom are –Electron & Neutrino, Proton & Neutron. This concept of atom resembles with the Pythagorean Number 'SOLID – FOUR'. Neutrino and Neutron together form a binary sign [± Plus Minus Sign, Unicode (hex) – 0177 or **Sukun** ^ a sign of duality] of relation between Electron and Proton. This binary sign has a dual name in the Kitaab with Truth. Due to the conspiracy grammatical rule of the pen-paper- pencil workers, it is being taught that the dual number or dual name has no existential import in Arabic literature or in the Kitaab with Truth. But the truth is that dual name is a ~sth [sign of relation] like Pythagorean Number 'LINE – TWO' or Neutrino & Neutron. This ~sth [tide or wave] is the Equal & Opposite [Half] sign of relation between Electron and Proton of an atom, between the Lower Most Triangle [Bermuda Triangle or Projected Planet Mars or Middle East or Eartha 3D or Snake] and Upright West Triangle [Mount Tuur or Leaning Tower or Upper West Region or Arabian Peninsula or White] of the Manifested Hexagon. This ~sth is also the sign of relation or the sign of Justice or Prime Meridian. Manifestation implies equal & opposite trinity or knower & known relation. In other words, manifestation implies the concept of seven in resemblance with Pythagorean Number 'MANIFESTATION – SEVEN'. This half or equal & opposite sign of relation with respect to manifestation is the Uniform Principle.

IS IT POSSIBLE TO START WITH AN ATTRIBUTELESS SOLIDITY?

ANSWER – 'NO'

An undoubted answer like Descartes' Cogito Ergo Sum'

So, let us start with Pythagorean Number – 'PHYSICAL QUALITIES – FIVE' or Pentagonal Earth

IS IT POSSIBLE TO START WITH ESSENCES ONLY WITHOUT SPACE-TIME RELATION OR MOTION?

ANSWER – 'NO'

An undoubted answer like Descartes' Cogito Ergo Sum'

So, let us start with Pythagorean Number – 'ANIMATION – SIX' or Manifested Hexagonal World.

IS IT POSSIBLE TO PERCEIVE WITHOUT A PERCEIVER, TO OBSERVE WITHOUT AN OBSERVER, TO GET JUSTIFICATION WITHOUT A JUTIFIER, TO GET A VERIFIED TRUTH WITHOUT A VERIFIER?

ANSWER – 'NO'

An undoubted answer like Descartes' Cogito Ergo Sum'

So, let us start with Pythagorean Number – 'MANIFESTATION – SEVEN' or Equal & Opposite relation between willingness & unwillingness, between positivity & negativity, between downward [Eastward] & upward [Westward] arrows, between lower sea-shore & upper sea-shore. Water is the sign of relation between two sea-shores. Electron Cloud is the sign of relation between West and East.

WILLINGNESS AND UNWILLINGNESS

Ayat\Verses - Say: **Disbelieve** you verily in Him **Who created the world [East Horizon] in two days**, and **ascribe** you unto Him **rivals**? He is the Rab of the universe. He set in the [world], mountains **standing firm**, high above it, and bestowed blessings in the world, and **measure therein all things to give them nourishment in due proportion, in four Days**, in accordance with [the needs of] those who seek [Sustenance]. Then **He turned to the heavens [West Horizon] when it was smoke [white dwarf]**, and said unto it and unto the earth: **Come both of you, willingly or unwillingly**. They said: We come, obedient. So He **ordained** them as **seven firmaments (windows) in two Days**, and **He inspired in each heaven its mandate.** And **We adorned the lower heaven with lights** and [provided it] **with guard.** Such is the Measuring of the Mighty, The Knower. **[Sura No. – 40\42 – Haa-Mim-Sajdah, Ayat No. – 9 to 12]**

WHETHER JUSTIFICATION OR VERIFICATION OF DATA IS POSSIBLE WITHOUT A JUSTIFIER OR A VERIFIER OR A POSSESSOR OF SHARED SENSE OR BLESSED MESSAGE OR INNATE IDEA?

CERTAIN ANSWER – 'NO'

So, let us start with seven or sense-data or innate idea of Descartes in resemblance with Windows'7 Ultimate [Real Science] or Pythagorean Number – 'MANIFESTATION – SEVEN', seven repeated verses of Quran, seven prayers of the unique utilitarian prayer – 'Fatiha'.

FASTA-IZ BILLAAHI MINASH-SHAYTANIR-RAJIM

Fa-'izaa qara'-tal-Qur-'aa-na **fasta-'iz billaahi minash-Shaytaanir-Rajim** [98]. 'Innahuu laysa lahuu sul-taanan 'alallaziina 'aamanuu wa 'alaa Rabbihim yatawak-kaluun [99]. 'Innamaa sultaanuhuu'alallaziina yata-wallaw-nahuu wallaziina hum-bihii mushri-kuun [100]. [Trans.] **When you recite Quran seek refuge in Allah from Shaytan the outcast** [98]. No authority has he over those who believe and put their trust in their Rab [99]. His authority is over those only, who take him as patron and **who join partners with Allah** [100] [Sura No. – 15\16 – Nahl, Ayat\Verses – 98 to 100]

BISMILLAAHIR-RAHMAANIR-RAHIM

"Qaala sananguru asadaqta am kunta minal-kaazibim. Izhab-bi-Kittabii haazaa fa-alqih ilayhim summa tawalla anhum fanzur maa zaayarji-uun. Qaalat yaa-ayyuhal-mala-u inniii ulqiya ilayya Kitabun-Karim. Innahu min-Sulaymaana wa innahu **Bismillaahir-Rahmaanir- Rahim.** Allaa ta-luu alayya wa-tuunii Muslimin." [Trans.] [Sulaymaan] said: Soon we will see whether you have told the **truth or lied**! **Go you, with this letter of mine**, and **deliver** it to them. Then **draw back** from them, then **turn away** and [wait to] see what **answer they return**. [The queen of Saba] said (when she received the latter): O chiefs! Here is delivered to me - a letter **worthy of respect**. Lo! It is from Sulaymaan, and lo! It is **"Bismillaahir-Rahmaanir- Rahim" / "In the name of Allah, The Beneficent, Most Merciful"** Be you **not arrogant against me**, but come to me as those who **surrender**. [Sura No. – 26\27 – Namal, Ayat No. – 27 to 31]

UNIQUE UTILITARIAN PRAYER – FATIHA

Praise be to Allah, Rab of the Universe [Alamin]
The Beneficent, the Merciful
Owner of the Day of Judgment
You (alone) we worship; we ask for [seek] Your help (alone).
Show us the Upright Path [towards Arsh]
The path of those on whom You have bestowed Your Grace.
Not (the path) of those **who earn** Your **anger** nor of those **who go astray.**

Anil-Hamduu lillaahi Rabbil-aalamin

'Innal- laziina 'aamanuu wa 'amilus-saali-haati yahdii-him Rabbu-hum-bi-'iimaanihim: tajrii min-tahtihimul-'anhaaru fii Jaannaatinna-'iim [9]. Da –waa-hum fiihaa Sub-haana-kalaa-humma wa ta-hiyya-tuhum fiihaa Salaam! Wa 'aakhiru da'-wa hum **anil-Hamdu lillaahi Rabbil-aala-miin**! [Trans.] Those who **believe**, and **work righteousness**, their Rab will **guide** them because of their **faith, beneath** them will flow **rivers** in Gardens of Bliss. [This will be] their prayer therein will be: Glory to You, O Allah! And **"Peace" will be their greeting** therein! And the close of their **cry** will be: **Praise is to Allah**, the RAB of the Universe [Anil-Hamduu lillaahi Rabbil-aalamin]. [Sura No. – 9\10 – Yuunus, Ayat No. – 9 & 10]

Huwar-Rahma-nur-Rahim

"Wa-Ilaahukum Ilaahunw-Waahid. Laaa ilaaha illaa Huwar-Rahma-nur-Rahim" [Trans.] And your Rab is Waahid [One Rab]. **There is no ilaaha [unbreakable reality] but He** The Beneficent, Most Merciful. [Sura No. 1\2 – Baqarah, Ayat No. – 163]

They [guilty folk] said: **Even so has said your Rab; and He is full of Wisdom and Knowledge.** [Ibrahim] said: And (afterward) what is your duty, O you sent (from Allah)? They said: Lo! We are sent to a **guilty folk** that we may send upon them **stones of clay [black spots or sunspots],** marked as from your Rab for those **who trespass beyond bounds.** Then We **brought forth** such believers who were there. But **We found not there any just person** except **in one house** and **We left there a Sign** for such as fear the **Grievous** Penalty. [Sura No. – 50\51 – Waz-Zaariyaat, Ayat No. – 30 to 37]

¾ - Middle Stair, ¼ - Top Stair, and 4/8 or ½ - Ground Stair

We appointed for Muusa **thirty nights [3/4 - Middle Stair of Europe, Asia, Africa, and Australia], and completed [the period] with ten [1/4 - more for Topmost Stair of Arabian Peninsula].** Thus was completed the term

351

[Semi-Clockwise stages of journey of the so-called sun as the manifested sign of natural magnetism for the West Zone] with his Rab, **forty nights** [4/8 - Semi-anticlockwise stages of journey of the so-called sun as the manifested sign of natural magnetism for the East Zone]. And Muusa had charged his brother Haarun [before he went up]: Act for me amongst my people: **Do right, and follow not the way of those who do mischief.** When Muusa came to the place appointed by Us, and his Rab addressed him, He said: O my Rab! Show [Yourself] to me, that I may look upon You. Allah said: **By no means you can see Me [direct];** But **look upon the mount [middle course towards Upright West]; if it abide in its place,** then you shall see Me. **When his Rab manifested His glory on the Mount, He made it as dust [matter particles / electron cloud].** And Muusa fell down in a swoon. When he recovered his senses he said: Glory is to You! To You **I turn in repentance, and I am the first to believe.** [Sura No. – 6\7 - As-haabul- a' raaf, Ayat No. – 142 & 143]

ONE OF THE GREATER SIGNS OF MY RAB

The heart **in no way falsified that which he saw. Will you then dispute with him concerning what he saw**? And indeed he **saw** Him yet **another time**, by the **lot-tree** (Sidratil) of the outmost boundary near to which is the Garden of Abode. Behold, the Lot-tree was covered [**in mystery unspeakable!**] The **eye turned not aside, nor yet** was overbold (did it go wrong); for **truly he saw one of the greater Signs of his Rab.** [Sura No. – 52\53 – Wan-Najm, Ayat No. – 11 to 18]

Seven Canopies and Seven Windows

Allah is He Who created **seven Firmaments** [canopies] and of the **world a similar number [Windows]. Through the midst of them descends His Commandment** that you may know that Allah is Able to do all things, and that Allah comprehends all things in [His] Knowledge. [Sura No. – 64\65 – Tallaq-tumun-nisaaa-'a, Ayat No. – 12].

THERE ARE SEVEN GATES IN HELL

HELL – For Proud ones, rejected, and accursed

HEAVEN – For Perfectly devoted slaves

Iblis said: I will make [wrong] fair-seeming to them in the world, and I will put them all in the wrong (astray) except such of them as are Your perfectly devoted slaves.

[Allah] said: O Iblis! **What is your reason for not being among those who prostrate**? [Iblis] said: Why should I **prostrate myself** unto a mortal whom

You have created from sounding **clay**, from mud moulded into shape. [Allah] said: Then get you out from here, for you are **rejected, accursed**. And lo! The **curse** shall be on you **till the Day of Judgment**. [Iblis] said: O my Rab! Give me then respite till the Day the [dead] are raised. [Allah] said: Then lo! You are of those reprieved **till an appointed time**. [Iblis] said: O my Rab! Because You have put me in the wrong (astray), **I will make [wrong] fair-seeming to them in the world, and I will put them all in the wrong** (astray) **except** such of them as are Your **perfectly devoted slaves**. [Allah] said: This [way of My sincere servants] is indeed a way **that leads upright to Me**. Lo! For My servants you have **no authority** over any of them **except such of the forward as follow you. And verily, Hell is the promised abode for them all. To it are seven gates. For each of those gates is a [special] class [of sinners] assigned.** [Sura No. 14\15 = Hur, Ayat No. – 32 to 44]

Allah is the Manifest Truth

Ayat\Verses – As for **those who** insult **virtuous women, indiscreet (careless) but believing, cursed are they in this life and in the Hereafter.** For them is **a grievous Penalty. On the Day when their tongues, their hands, and their feet will bear witness against them as to what they used to do. On that Day Allah will pay them their just dues,** and **they will realise that Allah is the Manifest Truth.** Vile women are for vile men, and vile men for vile women and good women are for good men, and good men are for good women. **Such are innocent (not affected) of that which people say.** For them there is forgiveness and a beautiful provision. [Sura No. – 23\24 – Nuur, Ayat No. – 23 to 26]

"Allah said: Lo! I am with you."

Allah made a **covenant** (promise) of **old** with the **Banii-Israa-iil** and We rose among them **twelve chieftains [springs or matter particles],** and **Allah said: Lo! I am with you.** If you establish prayer and pay the poor-due, and **believe in my Messengers and support them,** and **lend** unto Allah a **kindly loan,** surely I shall remit your **sins,** and surely I shall bring you into **Gardens [Heavens] underneath which river flow.** Whoso among you **disbelieve** after this **will go astray from right path.** [Sura No. – 4\5 – Maaaidah, Ayat\Verses – 12]

BUT HE IS IN THE MIDST

i)   **FOUR – But He makes the Fourth** – Equal & Opposite ~sth as a Solid bond. [Diamond Operator]

ii) **FIVE – Nor between Five** – Physical qualities. [pentagonal earth]

iii) **SIX – He makes the sixth nor of less than that or more** – Animation. **Equal and opposite Trinity implies hexagon. [hexagonal world]**

iv) **SEVEN – BUT HE IS IN THE MIDST** - Truth Manifests in the midst of seven.

v) **Eight is the planetary borders or octagonal rings or the screen between day and night**

Ayat\Verses - On the Day when Allah will raise them all together and show them the Truth [what they did] of their conduct. Allah has kept account of it, though they may have forgotten it, for Allah is Witness over all things Have you not seen that Allah knows all that is in the heavens and all that is in the world? **There is not a secret consultation between three, but He makes the fourth among them**. Nor between **five** but He makes the **sixth nor of less than that or more**, but **He is in their midst**, where so ever they be. In the end He will tell them the **truth** of their conduct, on the Day of Resurrection [Yawmal-Qiyaamah]. Lo! Allah is The Knower of all things. [Sura No. 57\58 – Mujaadalah, Ayat No. - 6 & 7]

FATIHA – UNIQUE UTILITARIAN PRAYER
Fatiha – Seven Repeated Verses
He **thinks** that none has **power** over him? And he says: **I have destroyed vast wealth [Kitaab with Truth]**. He [destroyer of Truth] thinks that **none beholds** him? [Sura No. – 89\90 – Bi-haazal-balad, Ayat No. – 5 to 7]

WHO HAVE DESTROYED THE SEVEN REPEATED AYAT OF THE QURAANAL-AZIM?
Ayat\Verses - We created not the heavens, the world, and all between them, **but for just ends**. And the **Hour is surely coming** [when this will be manifest]. So forgive with a gracious forgiveness; for verily it is your Rab who is the Master-Creator, knowing all things. And We have bestowed upon you the **Seven repeated [verses]** and the **Quraanal-Azim.** Strain not your eyes toward that which We cause some **wedded pairs** among them to enjoy; and be not grieved on their account, and lower your wing [in gentleness] for the believers. [Sura No. – 14\15 = Hur, Ayat No. – 85 to 89]

MAKE A VIRTUAL STUDY OF COLUMN – 1 AND COLUMN – 2 WITH A VIEW TO SEARCH OUT THE DESTROYER OF THE SEVEN VERSES OF FATIHA Whose Translation supported by which publications & Research Institutions are distorting Kitaab with Truth without Broken Bar?		
FATIHA – THE UNIQUE UTILITARIAN PRAYER TOTAL AYAT – 7		
Ayat No,	COLUMN – No. – 1	COLUMN – No. – 2
	Reference (A) – IFTA & its Off-Spring	Reference (B) and (C)
	X X X ?????????????????? X X X X X X ????????? X X X X X X X ???????? X X X	"Bismillahir-Rahmaniir-Ra-him" [Trans.] In the name of Allah, The Beneficent, The Merciful
1	"Bismillahir-Rahmaniir-Ra-him" [Trans.] In the name of Allah, **Most** Gracious, **Most** Merciful.	"Al-Hamdu lillaahi Rabbil-Alamin" [Trans.] Praise be to Allah, Rab [the cherisher and sustainer] of the Universe [Alamin]
2	"Al-Hamdu lillaahi Rabbil-Alamin" [Trans.] Praise be to Allah, The Cherisher and Sustainer of the **worlds**.	"Ar-Rahmaanir-Rahim" [Trans.] The Beneficent, The Merciful
3	"Ar-Rahmaanir-Rahim" [Trans.] Most Gracious, Most Merciful	"Maaliki Yawmud-Diin" [Trans.] Owner of the Day of Judgment
4	"Maaliki Yawmud-Diin" [Trans.] **Master** of the Day of Judgment.	"E y y a a k a n a'-b u d u w a iyyaaka nasta-in" [Trans.] You (alone) we worship; we ask for Your help (alone).
5	"E y y a a k a n a'-b u d u w a iyyaaka nasta-in" [Trans.] Thee do we worship. And Thine aid we seek.	Ihdinas-Siraatal-Mu s t aqim [Trans.] **Show us the straightway.**
6	**Ihdinas-Siraatal-Mu s t aqim. [Trans.] Show us the straightway.**	"Siraatal-lazina an-amta alayhim" [Trans.] The way of those on whom You have bestowed Your Grace.
7	"Siraatal-lazina an-amta alayhim. Gayril-magz(d)ubi alay-him wa laz(d)-z(d)aaalliiin" [Trans.] The way of those on whom **Thou** has bestowed **Thy** Grace, Those whose portion is not wrath. And who go not astray.	"Gayril-magz(d)ubi alay-him wa laz(d)-z(d)aaalliiin" [Trans.] Not (the way) of those who earn Your anger nor of those who go astray.

SEVEN FIRMAMENTS OR SEVEN WINDOWS

Ayat\Verses – How can you reject the faith on Allah seeing that you were without life, and He gave you life; then He will cause you to die, and will again bring you to life; and again to Him you will return? It is He Who has created for you all things that are in the world. Then **He turned to the heaven**, and fashioned it as **seven firmaments**; and of all things He has perfect knowledge. [Sura No. 1\2 – Baqarah, Ayat No. – 28 & 29]

Ayat\Verses – It is He Who has created for you all things that are in the world. Then **He turned to the heaven**, and fashioned it as **seven firmaments**; and of all things He has perfect knowledge. [Sura No. – 1\2 – Baqarah, Ayat\Verses – 29]

SEVEN EARS

Ayat\Verses – When Ibrahim said (unto his Rab): My Rab! Show me how You give life to the dead. He replied: Have you no faith? He said: Yes, but just to reassure my heart. Allah said: Take **four birds**, draw them to you, and cut their bodies to pieces. Scatter them over the **mountain-tops**, and then call **them back**. They will come swiftly to you. Know that Allah is The Mighty, The Wise. The parable (likeness / similitude) of those who spend their wealth in the way of Allah is the likeness of a grain which growth **seven ears** and **each ear has a hundred grains**. Allah give increase manifold to which He pleases. Allah is All-Embracing, All-Knowing. **[Sura No. – 1\2 – Baqarah, Ayat No. - 260 & 261]**

THREE ASCENDING STAIRS AND SEVEN WINDOWS

i) Seven lean are eating – Seven Springs or Planets of the Lower Seashore or East Zone or Tribes or Townships

ii) Seven green ears of corn – Seven Springs or Planets of the Middle Zone or Lower West Zone or Nation or Midiyan

iii) Other (seven) dry – Seven Springs or Planets of the Upper West Zone or Upper Most Land of the Earth or Leaning Tower of Galileo or the Tallest Tree of Newton or Cities.

Ayat\Verses – And the king said: Lo! I saw in a dream seven fat kine [fermions or beams] which **seven lean are eating**, and **seven green ears of corn** and **other (seven) dry**. O you chiefs! Expound to me my vision, if you can interpret dreams. [Sura No. – 11\12 – Yuusuf, Ayat\Verses No. – 43]

(And he came to Yuusuf in the prison, he exclaimed): Yuusuf! O you the **truthful one**! Expound for us the **seven** fat kine which **seven** lean were eating and the **seven** green ears of corn and other **(seven)** dry, that I may return unto the people, so that they may **know [understand]**. [Sura No. – 11\12 – Yuusuf, Ayat\Verses No. – 46]

Ayat\Verses – He [Yuusuf] said: You shall **sow seven years** as usual and the harvests that you **reap**, leave it in the **ear** [seed-bearing head of a cereal plant], all save a little which you **eat**. Then after that will come **seven hard years** which will **devour** [demolish] all that you have **prepared** for them, except a little which you have [specially] **guarded**. Then, after that, will **come a year** when the people will have plenteous crops [abundant water], and when they will press (wine and oil). [Sura No. – 11\12 – Yuusuf, Ayat\Verses No. – 47 to 49]

Ayat\Verses – And the king said: Bring him unto me, And when the messenger came unto him, he (Yuusuf) said: Return unto your Rab and ask him, **what is the case of the women who cut their hands?** Lo! My Rab knows their guile [snare or cunningness]. He (the king) (then send for those women and) said: What was your affair when you did seek to seduce Yuusuf from his [true] self? They answered: Allah Blameless! We know not evil of him. Said the wife of the ruler, **now is the truth manifest** [to all]. It was I who asked of him an evil act, and he is surely of the truthful. (Then Yuusuf said: I asked for) this, that He (My Rab) may know that **I betrayed Him not** in secret, and that **surely Allah guides not the snare of the betrayers.** [Sura No. – 11\12 – Yuusuf, Ayat\Verses No. – 50 to 52]

Ayat\Verses – Has then your Rab distinguished you [O people of Kitaab!] by giving you **sons** and has chosen for Himself **females** from among the angels? **Truly you utter a most dreadful saying! We have explained [things] in various [ways] in Haazal-Qur'an**, in order that they may receive **admonition**, but it only **increases** their flight [from the Truth]! Say: If there had been [other] realities with Him, as they say, behold, they would certainly have **sought out a way against** the Rab of the **Arsh**! Glory to Him! He is **high above all** that **they say**! Exalted and Great [beyond measure]! The **seven heavens and the world**, and all beings therein, declare His glory and there is **not** a thing but celebrates His praise. And yet you **understand not** how they declare His glory! Verily He is Forbearer, Forgiving! **And when you recite the Qur'an [izaa-qara-tal-Qur-aana], We put, between you and those who believe not in the Hereafter, a veil invisible.** And We put **coverings over their hearts** [and minds] **lest they should understand** the Rabbaka fil-Qur-aani, and **deafness** into their ears. And when you **commemorate** your Rab and Him alone in the Qur'an [Rabbaka fil-Qur-aani], **they turn on their backs**, fleeing [from the Truth]. [Sura No. – 16\17 - Banii-Israai-iil, Ayat No. – 40 to 46]

Seven long nights and Eight long days

Ayat\Verses – The Samuud and the Aad People disbelieved in the Judgment to come. As for Samuud, they were destroyed by the **lightning**. And the Aad, they were destroyed by a furious Wind, exceedingly violent which He imposed on them **seven long nights and eight long days** so that you might have seen men lying overthrown, as they were hollow **trunks** of **palm-tree**. Can you see any remnant of them? And Firawn, and those before him, and the communities that were destroyed, committed habitual Sin. And **they disobeyed [each] the messenger of their Rab**; so He gripped them with a lightening grip. [Sura No. – 68\69 – Al-Haaaqqatu, Ayat No. – 4 to 10]

SEVEN HEAVENS [CANOPIES] IN HARMONY

Ayat\Verses – Further I have made public proclamation to them, and I have appealed them in private, saying: Seek forgiveness from your Rab; for He is Ever Forgiving. He will let loose the sky for you in plenteous rain, and will help you with wealth and sons; and will assign to you gardens and will assign to you rivers. What is the matter with you that you hope not toward Allah for dignity seeing that it is He Who has created you in diverse stages? See you not how Allah has created the **seven heavens in harmony** (one above another), [Sura No. – 70\71 – Nuuh, Ayat No. – 9 to 15]

Ayat\Verses – Blessed be He in Whose hands is the Sovereignty; and He is Able to do all things (Qadir). He has created life and death that He may test you, which of you is best in conduct, and He is the Exalted in The Mighty, The Forgiving (Aziz, Gafuur). He has created the **seven heavens one above another.** You can see no fault in The Beneficent (Rahman). So turn your vision again, can you see any rifts? **Then look again and yet again, your sight will come back to you weakened and made dim.** [Sura No. – 66\67 – Bi-Yadihil-Mulk, Ayat No. – 1 to 4]

SEVENTY CUBITS

Ayat\Verses – But as for him who is given his **record in his left hand**, he will say: Ah! Would that my record had not been given to me! And knew not what any reckoning is! Ah! Would that [Death] had made an end of me! Of no profit to me has been my wealth! My power has gone from me! [The stern command will say]: **Seize him, and chain him** and then expose him to the hell-fire. Further, make him march in a chain, whereof the length is **seventy cubits**! Lo! He used not to believe in Allah Most High. And would not encourage the feeding of the wretched! So no friend has he here this Day. Nor has he any food except the corruption from the washing of wounds which none but sinners eat. [Sura No. – 68\69 – Al-Haaaqqatu, Ayat No. – 25 to 37]

Thus when they fulfil their term appointed, **either take them back on equitable terms or part with them on equitable terms**; and **take for witness two just persons from among you,** and **keep your evidence upright for Allah.** Whoso believes in Allah and the Last Day is **encouraged** to act thus. And **whosoever keeps his duty to Allah, Allah will appoint a way out for him.** [Sura No. – 64\65 – Tallaq-tumun-nisaaa-'a, Ayat No. – 2]

SECTION – V
STAR OPERATOR – 5.1
EARTH IS A PENTAGON

Pythagorean Number – 'Physical Qualities – Five'

Fasta-'iz billaahi minash-Shaytaanir-Rajim
Bismillaahir-Rahmaanir- Rahim

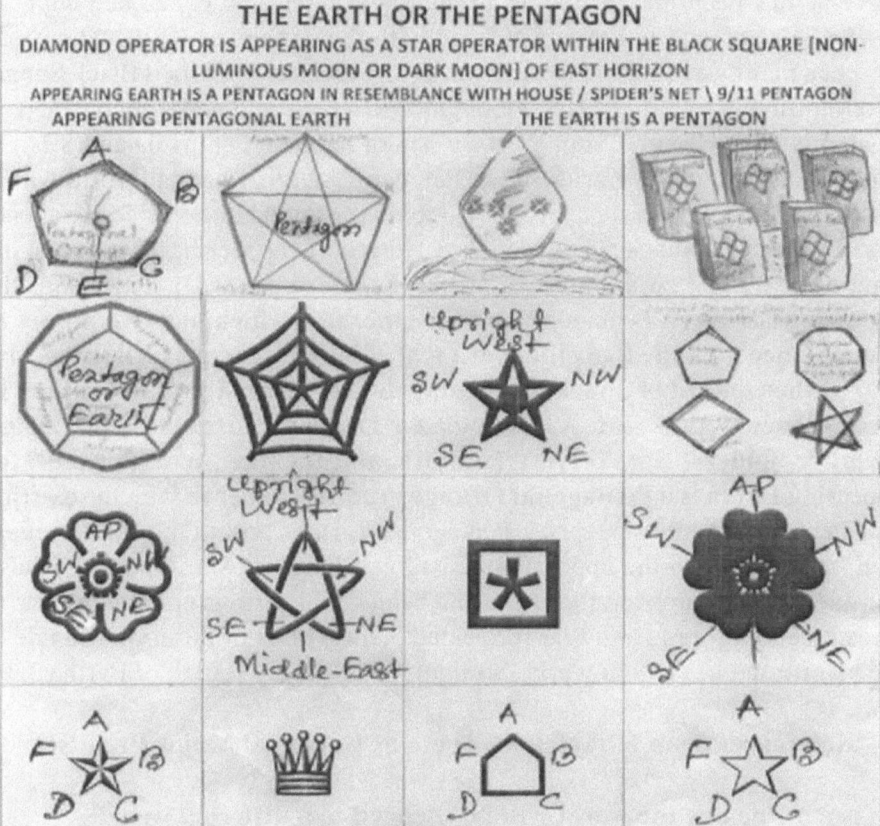

THE EARTH OR THE PENTAGON

DIAMOND OPERATOR IS APPEARING AS A STAR OPERATOR WITHIN THE BLACK SQUARE [NON-LUMINOUS MOON OR DARK MOON] OF EAST HORIZON

APPEARING EARTH IS A PENTAGON IN RESEMBLANCE WITH HOUSE / SPIDER'S NET \ 9/11 PENTAGON

APPEARING PENTAGONAL EARTH	THE EARTH IS A PENTAGON

Unicode (hex) 269D, 29C6, 272C, 2302, 2730, 272F, Webdings – 34 and Wingdings: 123 & 124

A /AP/ UWR – Upright West Region of the Manifested Hexagonal World and Topmost Stair as well as Uppermost Land of the Appearing Pentagonal Earth called Arabian Peninsula

B/NW – North-West Region of the Manifested Hexagonal World and North-Western [Western] Continents of the Appearing Pentagonal Earth [Asia, Africa, and Australia]

C/NE – North-East Region of the Manifested Hexagonal World and North-Eastern [Eastern] Continent of the Appearing Pentagonal Earth [North America]

D/SE – South-East Region of the Manifested Hexagonal World and South-Eastern [Eastern] Continent of the Appearing Pentagonal Earth [South America]

E/ME – Straight Middle East Region [Projected Planet Mars or Eartha 3D or Recycling Region or Bermuda Triangle of Titanic or Penetrated Hole in the Journey Boat of Muusa (ass)] of the Manifested Hexagonal World and Second Peg of the Immovable Pentagonal Earth

F/SW - South-West Region of the Manifested Hexagonal World and South-Western [Western] Continent of the Appearing Pentagonal Earth [Europe]

The Appearing Earth is a Star Operator or a Pentagon in resemblance with Spiders' Net, House, Pentagonal Orange, Terrestrial Star. The Appearing Earth within the Black Square of East Horizon is a Pentagon in correspondence with 9/11 Pentagon.

The universe is a Four-figured Shining Star called Diamond Operator. The universe has been revealed as a Trinity of East Horizon – Cloud & Sky – West Horizon. The revealed Universe or Trinity is called Sirius Binary System. The universe has been manifested as an upright rectangle in coherence with the **'End of Proof'** and in correspondence with the appointed **Kaba**. The world has been manifested within the four walls of East Horizon called Black Square [Non-luminous Moon or Dark Moon] as two Zones and a Hexagon. There is a ~sth between Upright [Top] West Region of Arabian Peninsula and Straight Middle East Region of Eartha 3D of the manifested Hexagonal World within the Black Square of East Horizon. Due to this ~sth, the manifested hexagonal world within the Black Square of East Horizon is appearing as a Pentagon. This **appearing Pentagon** is called the **Earth or Star Operator.** So, the appearing Earth is a Pentagon or a **Star Operator which has five pillars in resemblance with the five pillars of a tent. The Messenger of Allah has also shared the concept of appearing pentagon with reference to five pillars in resemblance with a tent.** Appearing **Star Operator** resembles with **House [Tent]** or **Spiders' Net, Terrestrial Stars, and the like.** In other words, the appearing **Earth** is a **Pentagonal Orange** in correspondence with the existing 9/11 Pentagon. As the appearing Pentagonal Earth is within the Black Square of East Horizon, so the appearing Pentagonal Earth or Star Operator is also non-luminous. Moreover, the appearing pentagonal earth is not moving due to three ascending stairs and Star [projected planet] Mars or the sixth triangle of the manifested hexagonal world [Straight Middle-East Region of Eartha 3D]

References from Kitaab with Truth as Universal Major Premises

"One day the world will be changed to a different world"
Ayat\Verses - Yawma tubad-dalul- Arzu gayral – Arzi was-Samaa-waatu wa barazuu lillaahil- Waa-hidil-Qahhaar. [Trans.] **One day the world will be changed to a different world** and so will be the heavens and they will come forth unto Allah, the One-Almighty. [Sura No. – 13\14 – Ibrahim, Ayat No. – 48]

Ayat\Verses - And you will see the sinners that day bound together in fetters. Their garments of liquid pitch, and their faces covered with Fire. That Allah may repay each soul according to its deserts; and verily Allah is swift in calling to account. **Here is a Message for mankind**. Let them take warning there from, and let them know that He is [no other than] One Allah. Let men of understanding take heed. [Sura No. – 13\14 – Ibrahim, Ayat No. – 49 to 52]

Is he then, one who will **strike his face against the awful doom** upon the **Day of Resurrection** [as he who does right]? And it will be said to the wrong-doers: Taste you [the fruits of] what you earned! **[Sura No. – 38\39 – Tanziilul-Kitaab, Ayat\Verses – 24, 31, 47, 60]**

Analogical Reference of the Appearing Pentagonal Earth
To avoid confusion & contradictions, Truths have
been shared with examples [similitude]

i) Structure of the Appearing Earth - **'likeness** of the **spider** when **she takes unto herself a house'**.
ii) Existential Import of the Manifested Sign – **'Verily in that is a Sign for the believers'**.

Ayat\Verses - The **similitude** [likeness] of those who take [choose] partners other than Allah is **likeness** of the **spider** when **she takes unto herself a house**, and lo! **Truly the weakest of all houses is the spider's house**, if they but know. Verily Allah knows what thing they **invoke** instead of Him. He is The Mighty, The Wise. And such are the **similitude** We set forth for mankind, but only those **understand** them who have **knowledge**. Allah **evolved the heavens and the world with Truth. Verily in that is a Sign for the believers**. [Sura No. – 28\29 – Ankabuut, Ayat No. – 41 to 44]

APPEARING EARTH IS A PENTAGONAL ORANGE

i) And fight them until harassment is no more
ii) There prevail justice
iii) Faith in Allah altogether and everywhere
iv) A fifth share is assigned to Allah and to Rasuul [Upright West], and to near relatives [South-West], orphans [North-West], the needy [South-East], and the wayfarer [North-East]
v) The Day of the meeting of the two forces [East and West]

Ayat\Verses - Say to the unbelievers, if [now] they desist [from Unbelief], their past would be forgiven them; but if they persist, the punishment of those before them is already [a matter of warning for them]. And fight them until harassment is no more, and **there prevail justice** and **faith in Allah altogether and everywhere** [Manifest Truth]. But if they cease, then Lo! Allah is the Seer of all that they do. If they refuse, be sure that Allah is your Protector, the best to protect and the best to help. And know that out of all the valuables that you may acquire [in war], **a fifth share is assigned to Allah, and to Rasuul,** and **to near relatives, orphans, the needy, and the wayfarer.** If you do believe in Allah and in the revelation We sent down to Our servant on the Day of Testing, the Day of the meeting of the two forces. For Allah has power over all things. **[Sura No. – 7\8 – Anil-Anfaal, Ayat No. – 38 to 41]**

ODD AND EVEN CONTRAST OF THE APPEARING PENTAGONAL EARTH

Two Seas are not alike – Equal & Opposite East Zone and West Zone
Pentagon – Palatable, Sweet, and Pleasant are representing West Zone; while Salt and Bitter are representing East Zone.

Ayat\Verses - And **the two seas are not alike**, the one **palatable, sweet, and pleasant to drink,** and the other, **salt and bitter.** And **from them both** you eat fresh meat and drive the ornament that you wear. And **you see the ship cleaving them with its prow that you may seek of His bounty** that you may be grateful. [Sura No. – 34\35 - Faatiris-Samaawaati wal-'arz, Ayat\Verses – 12]

There was indeed a **Sign** for Saba, in their dwelling place. **Two Gardens to the right and to the left. Eat of the Sustenance** [provided] by your Rab, and be grateful to Him: **A territory fair and happy, and a Rabbun Gafuur.** [Sura No. – 33\34 – Saba, Ayat No. - 15]

FROM HEAVEN A TABLE SPREAD WITH FOOD

Ayat\Verses - And when I **inspired** the **disciples** (saying): Believe in Me and in My Messenger: They said: We **believe. Bear witness** then we have **surrendered** (unto You). When the **disciples** said: Is your Rab able to **send down** for us **a table spread with food from heaven?** He said: **Observe your duty to Allah if you are true believers.** [They said]: We wish to eat thereof and satisfy our hearts, and to know that **you have indeed told us the truth**; and that we ourselves may be **witnesses. Isabnu Maryam said: O Allah our Rab! Send us from heaven a Table spread with food that it may be a feast for us, for the first of us and for the last of us, and a sign from You. Provide us sustenance, for You are the best Sustainer (of our needs).** Allah said: I will send it **down** unto you: But **if any of you after that resists (refuse to accept) faith,** I will punish him with a penalty such as I have not inflicted on any one among all the peoples. [Sura No. – 4\5 – Maaaidah, Ayat\Verses – 111 to 115]

Ayat\Verses – **Say: The truth is from your Rab**. Let him who will believe, and let him who will reject [it]. For the wrong-doers We have prepared a fire whose [smoke and flames], **like the walls and roof of a tent [Pentagon]**, will hem them in. If they implore relief they will be granted water like melted brass, that will scald their faces, how dreadful the drink! How uncomfortable a couch to recline on! As to those who believe and work righteousness, verily We shall not suffer to perish the reward of any who do a [single] righteous deed. For them will be Gardens of Eternity or Eaden Gardens [Heavens]; beneath them rivers will flow. They will be adorned therein with bracelets of gold, and they will wear green garments of fine silk and heavy brocade. They will recline therein on raised thrones [Kaba]. How good the recompense! How beautiful a couch [heaven] to recline on! Set forth to them the parable of two men: For one of them We provided **two gardens** of grape-vines and **surrounded them with date** [Arabian Peninsula], in between the two We **placed corn-fields** [Middle Zone or Lower West Zone of Europe, Asia, Africa, and Australia]. Each of those gardens brought forth its produce, and failed not in the least therein. In the midst of them We caused a river to flow [Maryam Supplied with Sustenance]. [Sura No. – 17\18 – Khaf – Ayat No. – 29 to 33]

WE ARE INSPIRED TO ASK THE EPISTEMIC UNIQUENESS
Do you build a landmark on every high place to amuse yourselves? And do you get for yourselves fine buildings in the hope of living therein [for ever]? And when you exert your strong hand, do you do it like men of absolute power? [Sura No. 25\26 – Shuraa, Ayat No. – 128 to 130]

Duty is to preach clear Message
It is Allah Who made your habitations homes of rest [immovable] and quiet for you; and made for you, out of the **skins** of animals, [tent or Pentagon for]

363

dwellings, which you find so light [and handy] when you **travel** and when you **stop** [in your travels]; and out of their **wool**, and their **soft fibres** [between wool and hair], and their **hair**, rich **stuff** and articles of convenience [to serve you] for a time. It is Allah Who made out of the things He created, **some things to give you shade [Mount Tuur]; of the hills He made some for your shelter.** He made you **garments** to protect you from **heat**, and **coats** of mail to protect you from your [mutual] **violence [Veto].** Thus He completes His favours on you in order that you may bow to His Will. But if they **turn away, your duty is only to preach the clear Message.** [Sura No. – 15\16 – Nahl, Ayat\Verses – 80 to 82]

Have they not travelled through the land, and seen what was the end of those before them? They were superior to them in strength. They tilled the soil and populated it in greater numbers than these have done. There came to them their messengers with **Clear Signs** [which they rejected, to their own destruction]. It was not Allah Who wronged them, but **they wronged their own souls.** In the long run **evil in the extreme** will be the end of those who do evil; for that **they rejected the Signs of Allah,** and held them up to **ridicule. [Sura No. – 29\30 – Ruum, Ayat\Verses No. – 9 & 10]**

Those **who reject Faith, neither** their possessions (nuclear power) **nor** their [numerous] off springs [soldiers] will avail them aught **against** Allah. They [killers of both faith & belief of mankind like sceptics of IFTA & Conspiracy Scientists] are themselves but **fuel for the fire.** [Their dilemma will be] **no better than that of the people of Fir'awn,** and their **forerunners.** They **denied Our Signs** and Allah will call them to account for their **sins.** For Allah is strict in punishment. **Say** to those **who reject Faith: Soon** will you be **crushed** and **gathered together to Hell,** an evil bed (resting place) indeed [to lie on]! [Sura No. - 2\3 – Imran, Ayat No. – 10 to 12]

Firawn said: And who is the Rab of the universe? [Muusa] said: The **Rab of the heavens and the world,** and **all between,** if you had but **sure belief.** [Firawn] said to those around: **Did you not listen** [to what he says]? [Muusa] said: **Your Rab and the Rab of your fathers from the beginning!** [Firawn] said: **Truly** your messenger who has been sent to you is a **veritable madman!** [Sura No. 25\26 – Shuraa, Ayat No. – 23 to 27]

SECTION – V
STAR OPERATOR – 5.2
PENTAGONAL EARTH HAS THREE ASCENDING STAIRS

TIN, ZAYTUUN, AND TUUR

Fasta-'iz billaahi minash-Shaytaanir-Rajim
Bismillaahir-Rahmaanir- Rahim

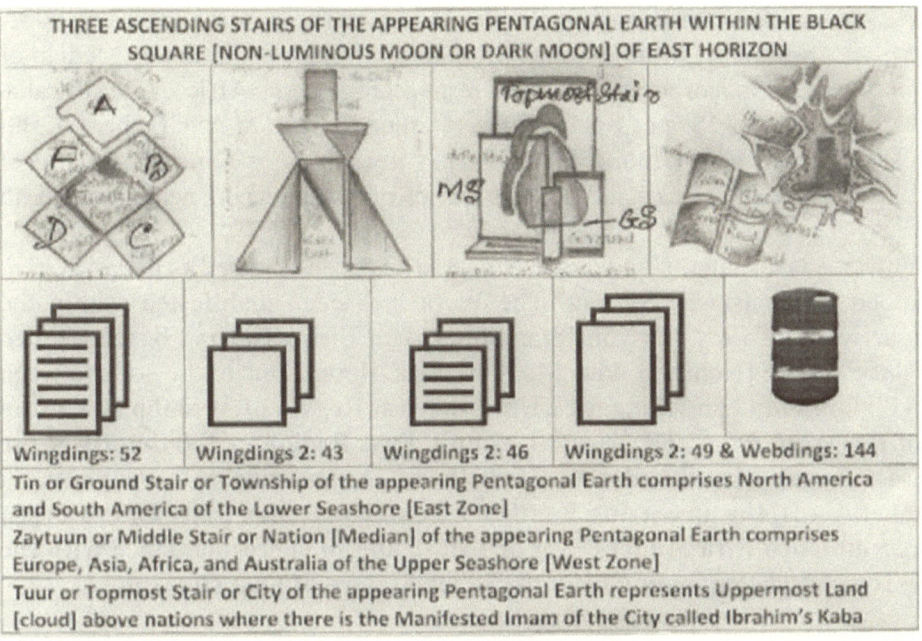

THREE ASCENDING STAIRS OF THE APPEARING PENTAGONAL EARTH WITHIN THE BLACK SQUARE [NON-LUMINOUS MOON OR DARK MOON] OF EAST HORIZON

Wingdings: 52	Wingdings 2: 43	Wingdings 2: 46	Wingdings 2: 49 & Webdings: 144

Tin or Ground Stair or Township of the appearing Pentagonal Earth comprises North America and South America of the Lower Seashore [East Zone]

Zaytuun or Middle Stair or Nation [Median] of the appearing Pentagonal Earth comprises Europe, Asia, Africa, and Australia of the Upper Seashore [West Zone]

Tuur or Topmost Stair or City of the appearing Pentagonal Earth represents Uppermost Land [cloud] above nations where there is the Manifested Imam of the City called Ibrahim's Kaba

Three Ascending Stairs and Seven Continents [Windows] of the appearing Pentagonal Earth [Star Operator]

Ground Stair or Tin
i) North America – Eastern Continent
ii) South America – Eastern Continent

Middle Stair or Zaytuun
i) Europe – Western Continent
ii) Asia – Western Continent
iii) Africa – Western Continent
iv) Australia – Western Continent

Topmost Stair or Tuur

i)   Arabian Peninsula – Western Continent

Three ascending stairs of the appearing pentagonal earth

i)   Symbol '**Superscript Three**' [3] [Unicode (hex) & ASCII (hex) 00B3] represents Topmost Stair [Tuur or City or Banii-Israiil] of Arabian Peninsula where there is the appointed Kaba at the right side of the Mount Tuur.

ii)   Symbol '**Superscript Two**' [2] [Unicode (hex) & ASCII (hex) 0082] represents Middle Stair [Zaytuun or Ummat or Nation or Median or Children of Adam] where there are Europe, Asia, Africa, and Australia.

iii)   Symbol '**Super Script One**' [1] [Unicode (hex) & ASCII (hex) 00B9] represents Ground Stair [Tin or Townships or Sons of Adham] of North America and South America of the appearing pentagonal earth

Universe is a Diamond Operator. Universe has been revealed as a Star System called Sirius Binary System. The World has been manifested within the four walls of East Horizon [Star Sirius B of Sirius Binary System] called Black Square [Non-luminous Moon or Dark Moon]. Due to the ~sth between willingness and unwillingness [**Upright West Region of Arabian Peninsula of the West Zone and Straight Middle East Region of Eartha 3D of the East Zone**], Manifested Hexagonal World is appearing as a Pentagon. In other words, **the appearing Earth is a Pentagonal Star [Star Operator] in resemblance with Spiders' Net or House and in correspondence with the existing 9/11 Pentagon within the Black Square of East Horizon.**

The appearing Pentagonal Earth has three ascending stairs and seven continents. East Zone comprises **two continents** and the **Ground Stair [Tin]** out of three ascending stairs. North America of the East Zone represents North-East Region of the Manifested Hexagonal World as well as **North-East [Eastern] Continent** of the Appearing Pentagonal Earth [converted red shade of Windows'7 Ultimate]. South America of the East Zone represents South-East Region of the Manifested Hexagonal World as well as **South-East [Eastern] Continent** of the Appearing Pentagonal Earth [converted green shade of Windows'7 Ultimate]. Converted green and red shades of Windows'7 Ultimate represent East Zone of the manifested hexagonal world [Asterisk] as well as **Ground Stair [Tin]** out of three ascending stairs of the appearing pentagonal earth [Star Operator].

West Zone comprises **five Continents** and two upper stairs out of three ascending stairs. The **Middle Stair [Zaytuun]** comprises two Regions and four Continents. Europe represents South-West Region of the Manifested Hexagonal

World and **South-West [Western] Continent** of the Appearing Pentagonal Earth [converted yellow shade of Windows'7Ultimate]. Asia, Africa, and Australia represent North-West Region of the Manifested Hexagonal World and **three North-West [Western] Continents** of the Appearing Pentagonal Earth [converted blue shade of Windows'7 Ultimate]. Converted yellow and blue shades of Windows'7 Ultimate represent West Zone of the manifested hexagonal world [Asterisk] as well as **Middle Stair [Zaytuun]** out of three ascending stairs of the appearing **Pentagonal Earth** [Star Operator]. Arabian Peninsula represents **Topmost Stair [Tuur]** out of three ascending stairs of the Appearing Pentagonal Earth in resemblance with white cloud of Windows'7 Ultimate and obverse of the converse of Muusa's Staff. In other words, Arabian Peninsula represents Upright West Region or Top West Region of the Manifested Hexagonal World [Asterisk] and **Upright West [Western] Continent** or Uppermost Land of the Appearing Pentagonal Earth [Star Operator].

Asia is not the Eastern Continent: - It has been projected through pen-paper-pencil works and activities that Asia is the Eastern Continent of the global earth. This projection is both contrary as well as contradictory to Manifest Truth. So, the projection of Asia as one of the Eastern Continents is teleological evidence sorcery and unique epistemic persecution.

ASIA IS ONE OF THE WESTERN CONTINENTS
Asia is neither in the East Zone [Lower Seashore] within the Black Square of East Horizon, nor in the Eastern Region of the Manifested Hexagonal World within the Black Square of East Horizon, nor one of the Eastern Continents of the Appearing Pentagonal Earth. On the contrary, West Zone within the Black Square of East Horizon comprises Asia in its North-Western Region. So, Asia is the **North-Western [Western] Continent** in the Middle Stair [Zaytuun] out of three ascending stairs of the Appearing Pentagonal Earth. Converted blue shade of Windows'7 Ultimate represents North-West Region of the manifested hexagonal world as well as Western Continents of Asia, Africa, and Australia.

Africa is one of the Western Continents: - The existence of Africa has been projected in the Southern Hemisphere through pen-paper-pencil works and activities. This projection is both contrary as well as contradictory to Manifest Truth. So, the projection of Africa as one of the Southern Continent is teleological evidence sorcery and unique epistemic persecution.

AFRICA IS ONE OF THE WESTEN CONTINENTS
Africa is neither in the East Zone [Lower Seashore] within the Black Square of East Horizon, nor in the Eastern Region of the Manifested Hexagonal World

within the Black Square of East Horizon, nor in the Southern Hemisphere [Right] of revelation & manifestation within the Black Square of East Horizon. On the contrary, West Zone within the Black Square of East Horizon comprises Africa in its North-Western Region. Africa is the **North-Western [Western] Continent** in the Middle Stair [Zaytuun] out of three ascending stairs of the Appearing Pentagonal Earth. Converted blue shade of Windows'7 Ultimate represents North-West Region of the manifested hexagonal world as well as Western Continents of Asia, Africa, and Australia. South Africa may be called Africa or North-Western Africa from the points of view of Revelation, Manifested Hexagonal World, Appearing Pentagonal Earth, Four cardinal directions, and the like.

References from Kitaab with Truth as Universal Major Premises
ONE THIRD, TWO-THIRDS AND HALF

Ayat\Verses – Lo! Your Rab knows how you keep vigil, sometimes nearly **two-thirds** of the night, or (sometimes) **half** or a **third** thereof, as do a party of those with you. Allah measures the night and the day. He knows that you count it not, and turns to you in mercy. **Recite then, of the Quran that which is easy for you**. He knows that there are sick folk among you, while others travel in the land [earth] in search of Allah's bounty, and others are fighting for the cause of Allah. So **recite** of it **which is easy**, and establish worship and pay the poor-due, and lend to Allah a goodly loan. Whatsoever good you send before you for your souls, you will surely find it with Allah, better and greater in the recompense. And seek forgiveness of Allah. Lo! Allah is Ever Forgiving, Most Merciful. [Sura No. 72\73 – Yaaa-ayyuhal Muzzamil, Ayat No. - 20]

DARKNESS – THUNDER - LIGHTNING

[Their **similitude**] is that of a rain-laden cloud from the sky. In it are **stairs of darkness**, and **thunder** and **lightning**. They press their fingers in their ears by reason of the stunning thunder-clap, for fear of death. But Allah **encompasses the unbelievers**.[Sura No. 1\2 – Baqarah, Ayat No. – 19]

THREE STAIRS OF THE APPEARING PENTAGONAL EARTH
  i)   Near bank of the valley [Top West Stair / Arabian Peninsula / Tuur]
  ii) Farther bank [Middle Stair / South-West Region of Europe and North-West Region of Asia – Africa – Australia / Zaytuun]
  iii) Lower ground [Ground Stair / North-East Region of North America and South-East Region of South America / Tin]

Ayat\Verses – Remember you were on the **near bank of the valley**, and they on the **farther bank** and **the caravan on lower ground than you. Even if you**

had made a mutual appointment to meet, you would certainly have failed in the appointment: But [thus you met], that Allah might accomplish a matter already enacted; that **those who died might die after a clear Sign** [had been given], and **those who lived might live after a Clear Sign** [had been given]. And verily Allah is He Who hears and knows [all things]. **Remember in your dream Allah showed them to you as few. If He had shown them to you as many, you would surely have been discouraged, and you would surely have disputed in [your] decision; but Allah saved [you]: for He knows well the [secrets] of [all] hearts.** And remember when you met, He showed them to you as few in your eyes, and He made you appear as disgraceful in their eyes that Allah might accomplish a matter already enacted. For to Allah do all questions go back [for decision]. **[Sura No. – 7\8 – Anil-Anfaal, Ayat No. – 42 to 44]**

Ayat\Verses – O you who believe! Let those whom your right hands possess and the [children] among you who have not come of age ask your permission [before they come to your presence], on **three occasions:** Before the prayer of dawn; while you do-off your clothes for the noonday heat; and after the late-night prayer, **three times of privacy** for you. Outside those times it is not wrong for you or for them to move about attending to each other. Thus **Allah makes clear the Signs for you**, for Allah is full of knowledge and wisdom. But when the children among you come of age, let them [also] ask for permission, as do those senior to them [in age]. Thus Allah makes clear His Signs to you, for Allah is full of knowledge and wisdom. [Sura No. – 23\24 – Nuur, Ayat No. – 58 to 59]

A TRINITY OF A WAVE, ABOVE WHICH A WAVE ABOVE WHICH IS A CLOUD

i) A Wave – Ground Stair or Tin or Lower Seashore or East Zone of North America and South America

ii) Above which a Wave – Middle Stair or Zaytuun or Upper Seashore or Middle Zone or Lower West Zone of Europe, Asia, Africa, and Australia

iii) Above which is a Cloud – Topmost Stair or Tuur or Region of White or Upright West Triangle of the Manifested Hexagon or Uppermost Land of the Appearing Pentagon or Arabian Peninsula

Ayat\Verses – And for **those who disbelieve, their deeds are as a mirage in a desert. The thirsty one approaches it to be water till he comes unto it and finds it to be nothing. But he finds in the place thereof Allah, Who pays him his due and Allah is swift in taking account.** Or [the Unbelievers' state] is like the depths of darkness in a vast awful sea. There covers him **a wave, above which is a wave, above which is a cloud; depths of darkness,**

369

one above another. If a man stretches out his hands, he can hardly see it! For any to whom Allah gives not light, there is no light! [Sura No. – 23\24 – Nuur, Ayat No. – 36 to 40]

MAN – CRAWLING CREATURES - CATTLE

And so amongst **men** and **crawling creatures** and **cattle** in like manner drivers hues (shades). Those who have knowledge among His servants truly fear Allah (alone), for Allah is Exalted in Might, Ever Forgiving. [Sura No. – 34\35 - Faatiris-Samaawaati wal-'arz, Ayat\Verses – 28]

THREE STAIRS OF THE APPEARING PENTAGONAL EARTH
[Inhabitations in the hills, and in the trees, and in habitations]
  i)   Habitation – Ground Stair of North America and South America
  ii) In the Trees – Middle Stair of Europe, Asia, Africa, and Australia
  iii) In the Hill – Topmost Stair or Upright West Region of Arabian Peninsula

And your Rab taught (inspired) the bee to build its inhabitations in the hills, and in the trees, and in habitations. [Sura No. – 15\16 – Nahl, Ayat\ Verses – 68]

ASCENDING ORDER – STAGE BY STAGE

Ayat\Verses - That you shall surely travel from **stage to stage. What then is the matter with them that they believe not?** And when the Qur'an is **recited** to them, **they fall not prostrate,** but on the contrary the disbelievers reject [it]. But Allah has full knowledge of **what they are hiding.** So **announce to them a Penalty Grievous,** except to those who believe and work righteous deeds, for them is a **reward that will never fail.** [Sura No. – 83 \ 84 - Izas-Samaaa-unshaqqat, Ayat No. – 19 to 25]

THREE STAIRS OF THE APPEARING PENTAGONAL EARTH
TIN – ZAYTUUN – TUUR
  i)   Tin – Ground Stair of North America and South America
  ii) Zaytuun – Middle Stair of Europe, Asia, Africa, and Australia
  iii) Tuur – Topmost Stair or Upright West Region of Arabian Peninsula

Will you contradict with these Coded Shared Tautologies?

Ayat\Verses – By the Fig [**Tin**] and the Olive [**Zaytuun**], and the Mount of **Tuur;** and **this land** [appearing pentagonal earth] made safe [within four walls]. We have indeed created **man of the best creature.** Then We **reduced**

him to the lowest of the low, **except** those who believe and **do righteous deeds** [tabligh]; for they shall have a reward unfailing. **Then what can, after this, contradict you [who henceforth will give lie to you], about the judgment? Is not Allah the most conclusive [wisest] of all judges?** [Sura No. 94\95 – Wat-tiini Waz-Zaytuun, Ayat No. – 1 to 8]

THREE STAGES OF THE APPEARING PENTAGONAL EARTH IN PRAGMATIC CORRESPONDENCE WITH THREE POSITIONS – LYING POSITION, SITTING POSITION, AND STANDING POSITION

Ayat\Verses – If Allah were to hasten for men the ill [they have earned] as they would hasten on the good, then would their respite be settled at once? But We suffer those who look not their hope on their meeting with Us, in their trespasses, wandering in distraction to and fro. When trouble touches a man, He cries unto Us [in all postures], **lying down** on his side, or **sitting**, or **standing**. But when We have solved his trouble, he goes on his way as if he had never cried to Us for a trouble that touched him. Thus do the deeds of transgressors seem fair in their eyes? Generations before you We destroyed when they did wrong. Their Messengers came to them with clear-signs, but they would not believe. Thus, do We reward the guilty folk? [Sura No. – 9\10 – Yuunus, Ayat No. – 11 to 13]

[THREE STAIRS ➜ ON FOOT – RIDING – SAFETY PLACE]

Ayat\Verses – **Guard strictly your prayers** and of the **mid-most prayer**, and **stand up with devotion to Allah. If you fear, pray on foot**, or **riding. But when you are again in safety**, remember Allah in the manner as **He has taught you**, which you knew not [before]. [Sura No. 1\2 – Baqarah, Ayat No. – 238 & 239]

Gardens, Grain, Tall Palm-Trees – piled one over another

Ayat\Verses – Do they not look at the sky above them [towards West]? How We have made it and adorned it, and there are no flaws (rifts) in it? And the world We have spread it out, and set thereon mountains standing firm, and have caused of every lovely kind to grow thereon a vision and a reminder for every devotee turning [to Allah]. And We send down from the sky [West] rain charted with blessing, and We produce therewith **gardens and grain for harvests; and tall [stately] palm-trees with shoots of fruit-stalks, piled one over another**; Sustenance for [Allah's] servants; and We give life therewith to land that is dead. Thus will be the Resurrection [Sura No. 49\50 – Kaaaf, Ayat No. – 6 to 11]

PALM-TREES, GRAIN, SCENTED HERB

Ayat\Verses - The Most Gracious! (The Beneficent!), it is He Who has **Revealed** (taught) the **Quran**. He has **created man.** He has **taught** him an **intelligent**

speech (utterance). "**Ash-shamsu wal-qamaru** bi-husbaan" – **The Sign of day-break** and **the sign of night-fall** are made **punctual** i.e. they **follow** the **courses exactly computed**. The **Stars** and the **trees** adore. And the **skies or firmaments** He has raised high (uplifted), and He has set the **balance (measure)** In order that you **may not exceed the measure** (limit) or **transgress the balance** (justice). But **observe the measure (balance) strictly**, and fall not short thereof. And **He has appointed** the **earth for (His) creatures** wherein are fruits and sheathed **palm-trees, husked grain** and **scented herb**. [Sura No. – 54\55, Ar-Rahman, Ayat\Verses – 1to 12]

THREE COLUMNS

Ayat\Verses – Ah woe, that Day, to the rejecters of Truth! [It will be said:] Depart you to that which you **used to deny** [reject as false]! **Depart you to a shadow in three columns,** [which yet is] neither relief nor shelter from the flame. Lo! It throws up sparks like the castles, (or) as it might be **camels of bright yellow hue**. [Sura No. – 76\77 - Wal-Mursalaat, Ayat No. – 28 to 33]

ASCENDING STAIRWAYS

Ayat\Verses – A questioner questioned concerning the Penalty about to fall upon the nonbelievers which none can repel **from Allah, Rab of the Ascending Stairways.** The angels and the spirit ascend unto him in a Day whereof the span is **fifty thousand years**. But be patient with a patience fair to see. [Sura No. – 69\70 – Zil-ma-aaru, Ayat No. – 1 to 5]

Ayat\Verses – And they say: **Why is not this Qur'an sent down to some leading (great) man** of the **two Towns [North America and South America]? Is it they who apportion the Mercy of their Rab?** It is We Who portion out between them their livelihood in the life of this world [present life], and **We raise some of them above others in ranks**, so that some may take labour from others. But the Mercy of your Rab is better than the [wealth] which they amass. [Sura No. 42\43 – Ummil-Kitaab \ (Zukhruf), Ayat No. – 31 & 32]

Ayat\Verses – And were it not that mankind would have become **one community**, We might well have appointed for **those who disbelieve in the Beneficent, roofs of silver** for their **houses** and **stairs whereby to mount,** and **doors** to their houses, and **Throne on which they could recline,** and also adornments of gold [**zukhrufaa**]. Yet all that would have been but a **provision of the life of the world [present life].** And the hereafter with your Rab would have been for those who keep from evil (the so-called sun). If **anyone withdraws himself from remembrance of the Beneficent [means of Sustenance]**, We assign unto him a devil (shaytan), who becomes his comrade.

And lo! They **surely turn them from the way of Allah**, and **yet they think that they are rightly guided.** [Sura No. 42\43 – Ummil-Kitaab \ (Zukhruf), Ayat No. – 33 to 37]

Ayat\Verses – Say: I do admonish you on **one point only: That you awake for the sake of Allah, by twos [top stair & middle stair] or singly [ground stair], and you reflect** (within yourselves / give thought). **There is no madness in your friend.** He is no less than a **warner** to you, in face of a **terrible Penalty.** [Sura No. – 33\34 – Saba, Ayat No. – 46]

TRINITY OF OLD – MIDDLE - YOUNG
"The heifer should be neither too old nor too young, but of middling age." And when **Muusa** said to his people: **Allah commands that you sacrifice a heifer.** They said: Does you make **game** of us? He said: **Allah save me from being an ignorant [fool]!** They said: Pray on our behalf to your Rab to **make clear to us** what [heifer] she is! He said: He says: **The heifer should be neither too old nor too young, but of middling age**. Now do what you are commanded! They said: Pray on our behalf onto Allah to make plain to us **Her colour.** He said: He says: Verily **she is a yellow heifer; bright is her colour, gladdening beholders!** They said: Pray on our behalf unto your Rab to make clear to us **what** (heifer) she is! To us are **all heifers alike.** We wish indeed for **guidance, if Allah wills.** He said: He says: Verily she is a heifer **unyoked, not trained to till the soil or water the fields; sound (Vibration causing sensation, healthy, quality of the bottom of the sea) and without blemish. They said now you had brought the truth. So, they sacrificed her, though almost they did not.** [Sura No. 1\2 – Baqarah, Ayat No. – 67 to 71]

A TRINITY OF FORTY NIGHTS, THIRTY NIGHTS, AND TEN MORE
i)   Forty [Tin] - ½ [Unicode (hex) & ASCII (hex) – 00BD and ASCII (decimal) 189] semi-anti-clock-wise stages of journey for the East Zone of North America and South America
ii)   Thirty [Zaytuun] - ¾ [Unicode (hex) & ASCII (hex) 00BE and ASCII (decimal) 190] semi clock-wise stages of journey for Europe, Asia, Africa, and Australia of the Middle Stair of West Zone
iii) Ten [Tuur] - ¼ [Unicode (hex) & ASCII (hex) 00BC and ASCII (decimal) 188] semi-clock-wise stages of journey for the Uppermost or Top West land of Arabian Peninsula

Ayat\Verses - And remember, We **divided the sea** for you and saved you and drowned Firawn's people within your very sight. And remember We appointed

373

forty nights for Muusa [for East Zone of North America and South America], and **in his absence you took the calf** [projected falsehood for right direction] and you did grievous wrong. Even then We did forgive you; **there was a chance for you to be grateful**. And remember We gave **Muusa the Kitaab** and **the Criterions [scales of justification and verification]**. There was a **chance for you to be guided aright**. And when Muusa said to his people: O my people! You have indeed **wronged** yourselves by your **worship of the calf [projected mechanism]**. So turn [in repentance] to your Creator, and **kill (the calf or projected falsehood) yourselves that will be the best** for with your Creator and He will **relent** toward you, for He is the Relenting, the Merciful. And remember you said: O Muusa! We shall **never believe** in you until we see Allah **manifestly**, but you were **confused** with **thunder** and **lightning seized you**. Then We **received** you after your **extinction**, that you might give **thanks**. And We gave you the **shade** of **clouds** [Arabian Peninsula or Topmost Stair] and sent down to you **manna** and **quails [Middle Stair & Ground Stair]**, saying: Eat of the **good things** We have **provided** for you. We **wronged** them not, but they **did** wrong themselves. And when We said: Enter this **township** and eat of the plenty therein as you wish; but enter the gate prostrate, and say: Repentance. We will forgive you your sins and will increase (reward) for the right-doers. **But those who did wrong change the word which had been told them for another saying** and We sent on the transgressors a plague from heaven, for evil doing. [Sura No. – 1\2 – Baqarah, Ayat No. – 50 to 59]

We appointed for Muusa **thirty nights [Middle Stair of Europe, Asia, Africa, and Australia], and completed [the period] with ten [more for Topmost Stair of Arabian Peninsula].** Thus was completed the term [Semi-Clockwise stages of journey of the so-called sun as the manifested sign of natural magnetism for the West Zone] with his Rab, **forty nights** [Semi-anticlockwise journey of the so-called sun as the manifested sign of natural magnetism for the East Zone]. And Muusa had charged his brother Haarun [before he went up]: Act for me amongst my people: **Do right, and follow not the way of those who do mischief.** When Muusa came to the place appointed by Us, and his Rab addressed him, He said: O my Rab! Show [Yourself] to me, that I may look upon You. Allah said: **By no means you can see Me [direct]; But look upon the mount [diagonal way towards West]; if it abide in its place**, then you shall see Me. **When his Rab manifested His glory on the Mount, He made it as dust [matter particles / electron cloud].** And Muusa fell down in a swoon. When he recovered his senses he said: Glory is to You! To You **I turn in repentance, and I am the first to believe.** [Sura No. – 6\7 - As-haabul- a' raaf, Ayat No. – 142 & 143]

CITY – MIDIYAN – TOWNSHIP

i) **City** – Topmost Triangle / Upright West Region of the Manifested Hexagon, Topmost Stair out of three ascending stairs, Uppermost Land of the Appearing Pentagonal Earth, Arabian Peninsula, Mount Tuur, Mount Al-al -Judai, Region of the appointed Black & White Imam of the City called Kaba, Bellies

ii) **Median** – Upper Seashore, Middle Stair out of three ascending Stairs, Nations, Ummat, Europe – Asia - Africa – Australia, Two Legs

iii) **Township** – Lower Seashore, East Zone where of North America and South America, Tribes, People of the Wood, Four Legs

BELLIES, TWO LEGS, AND FOUR LEGS

Allah causes the **alternation** of the day and the night. Verily herein (in these things) are instructive examples (clear proofs) for those who have vision! And Allah has created every animal from **water**. Of them there are some that creep on their **bellies**; some that walk on **two legs**; and some that walk on **four**. Allah creates what He wills for verily Allah has power over all things. [Sura No. – 23\24 – Nuur, Ayat No. – 44 & 45]

CITY OF THE BLACK & WHITE IMMAM [Topmost Stair]
&
Succeeding Generations

Ayat\Verses – And when Ibrahim said: O my Rab! Make **this city** [Arabian Peninsula] one of peace and security: and preserve [protect] me and my sons from worshipping idols. My Rab! Lo! They have indeed led astray many among mankind. He then who follows my [ways] is of me. And he who disobeys me, still You are indeed Ever- Forgiving, Most Merciful. Our Rab! Lo! I have settled some of my **succeeding generations** in an **uncultivable valley near unto Your Holy House [Kaba]**. Our Rab! That they may establish regular Prayer, so fill the hearts of some among men with love towards them, and feed them with fruits, so that they may be thankful. Our Rab! Truly **You know what we conceal and what we reveal**. **Nothing** in the world [East Horizon] or in the heaven [West Horizon] is hidden from Allah. [Sura No. – 13\14 – Ibrahim, Ayat No. – 35 to 38]

Ayat\Verses – Said the Chiefs of the people of Firawn: This is indeed a sorcerer well- versed. His plan is to get you out of your land [earth]: then what is your advice? They said: Keep him and his brother in suspense [for a while]; and **send to the cities men to collect.** And bring up to you all [our] sorcerers well-versed. So there came the sorcerers to Firawn: They said: Of course we shall have a [suitable] reward if we win! He said: Yea, [and more] for you shall in that case be [raised to posts] nearest [to my person]. [Sura No. – 6\7 - As-haabul- a' raaf, Ayat No. – 109 to 114]

Ayat\Verses – Said Ferawn: Believe you in Him before I give you permission? Surely this is a trick which **you have planned in the city to drive out its people**: but soon shall you know [the consequences]. Be sure I will cut off your hands and your feet on opposite sides, and I will cause you all to die on the cross. They said: For us, We are but sent back unto our Rab. But you cause retribution on us simply because we believed in the Signs of our Rab when they reached us! Our Rab! Pour out on us patience and constancy, and take our souls unto thee as Muslims! [Sura No. – 6\7 - As-haabul- a' raaf, Ayat No. – 123 to 126]

Categorical Revelation
[Prayer, Mid-most prayer, Stand up with devotion to Allah]
Guard strictly your prayers and **of the mid-most prayer** and **stand up with devotion to Allah** [Sura No. – 1\2, Baqarah, Ayat\Verse No. – 238]

Conditional Revelation
If you fear, pray on foot, or **riding. But when you are again in safety, remember Allah in the manner as He has taught you, which you knew not [before].** [Sura No. – 1\2, Baqarah, Ayat\Verse No. – 239]

THE VALLEY-WAYS IN THE EARTH
Ayat\Verses - "Wa ja-ala-qamara fuhinna nuuranw-wa ja-alash-shamsa siraajaa" [Trans.] And has **made ja-ala-qamara in their midst,** and has made the **ja-alash-shamsa as siraaja.** And Allah has caused you **to grow as a growth from the world.** And afterward He makes you **return** thereby, and He will **bring you forth again, a (new) forth-bringing.** And Allah has made **the world a wide expanse** for you that you may **tread** (go about) **the valley-ways thereof.** [Sura No. – 70\71 – Nuuh, Ayat No. – 16 to 20]

References of Nations and Tribes
O mankind! We created you from a **single** [pair] of a male and a female, **and made you into nations and tribes,** that **you may know each other [not that you may despise [each other].** Verily **the most honoured of you in the sight of Allah is** [he who is] **the most righteous of you.** And Allah has full knowledge and is well informed [ofall things]. [Sura No. – 48\49 - Minw-waraaa-il-Hujuraat, Ayat No. – 13]

Median people
People of the Middle Stair of Europe, Asia, Africa, and Australia

To the Median people We sent Shuyaib, one of their own brethren. He said: O my people! Serve Allah; You have no other reality but Him. Now has come unto you a clear [Sign] from your Rab! Give just measure and weight, nor withhold from the people the things that are their **due**; and do no mischief in the world after it has been **set in order**: that will be best for you, if you have Faith. [Sura No. – 6\7 - As-haabul- a' raaf, Ayat No. – 85]

Median People
[Reference of Two-in-one Partnership]

Ayat\Verses - The **Hypocrites**, men and women, [**have an understanding**] with each other. They **enjoin evil, and forbid what is just**, and are **close with their hands**. They have **forgotten** Allah; so He has forgotten them. Verily the **Hypocrites** are **rebellious** and **perverse**. Allah has **promised** the Hypocrites men and women, and the rejecters, of Faith, the fire of **Hell**: Therein shall they **dwell**: Sufficient is it for them. For them is the **curse** of Allah, and an **enduring punishment**. As in the case of those before you, they were mightier than you in power, and more flourishing in wealth and children. They had their enjoyment of their portion: and you have of yours, as did those before you; and you indulge in idle talk as they did. Their **works are fruitless in this world** and **in the Hereafter**, and they will lose [all good]. **Have not the story reached them of those before them, the people of Nuuh, and Aad, and Samuud, the people of Ibrahim, the men of median, and the cities overthrown [East Zone]?** To them came their messengers with **clear signs**. It is not Allah Who wrongs them, but they wrong their own souls. **[Sura No. – 8\9 – Tauba, Ayat No. – 67 to 70]**

PEOPLE OF THE MEDIAN [Middle Stair]

Ayat\Verses - When your **sister** went and said: **Shall I show you one who will nurse and rear the [child]?** So We brought you back to your **mother** that her eye might be cooled and she should not grieve. Then you **killed a man**, but We saved you from trouble, and We tried you in various ways. And you did tarry years among the people of **Median**. Then came hither [formal to] you by [My] providence, O Muusa! And I have prepared you Myself [for service]. Go, you and your brother, with My Signs, and slacken not, either of you, in keeping Me in remembrance. Go, both of you, to Firawn, for he has indeed transgressed all bounds; but speak to him mildly. Perhaps he may take warning or fear [Allah]. [Sura No. – 19\20 – Twa-Haa, Ayat No. – 40 to 44]

TOWNSHIP [Ground Stair]
EAST ZONE OR LOWER SEASHORE
NORTH AMERICA AND SOUTH AMERICA

Whenever We sent a Prophet to a **town**, We took up its people in suffering and adversity, in order that they might learn humility. Then We changed their suffering into prosperity, until they grew and multiplied, and began to say: **Our fathers** [too] were touched by suffering and prosperity. Behold! **We called them to account of a sudden, while they realised not [their danger]. If** the **people** of the **towns** had but **believed** and **feared** Allah, We should indeed have **opened** out to them [All kinds of] blessings from **heaven and world**; but they **rejected** [the truth], and **We brought them to account for their misdeeds. Did the people of the towns feel secure against the coming of Our wrath by night while they were asleep? Or else did they feel secure against its coming in broad daylight while they played about [care-free? Did they then feel secure against the plan of Allah?** But **no one** can feel **secure** from the **Plan** of Allah, except those [doomed] to ruin! To those who **inherit the world in succession** to its [previous] possessors, **is it not a guiding,** [lesson] that, if We so willed, **We could punish them [too] for their sins, and seal up their hearts so that they could not hear?** Such were the **towns** whose story We [thus] narrate unto you. There came indeed to them their messengers **with clear signs.** But they would **not believe** what they had rejected before. Thus Allah sealed up the hearts of those who reject faith. Most of them We found not men [true] to their promise. **But most of them We found rebellious and disobedient.** [Sura No. – 6\7 - As-haabul- a' raaf, Ayat No. – 94 to 102]

TOWNSHIP

"Say: Cry unto those whom you assumed besides Him; yet they have neither the power to remove your troubles from you nor to change them." **Ayat\Verses -** Say: Cry unto those whom you assumed besides Him; yet they have neither the power to remove your troubles from you nor to change them. Those whom they cry upon do seek [for themselves] the way of approach to their Rab even those who are nearest. They hope for His Mercy and fear His wrath, for the wrath of your Rab is something to be shunned. There is not a **township** but We shall destroy it before the Day of Resurrection or punish it with a dreadful Penalty that is written in the Kitaab. **[Sura No. – 16\17 -** Banii-Israai-iil, Ayat No. – 56 to 58]

TRIBES [People of Ground Stair] & NATIONS [People of Middle Stair] Ayat\Verses – Of the people of Muusa there is a **section** who leads with **truth** and establishes **justice** therewith. We **divided** them into twelve **tribes, nations.** We directed Muusa by **inspiration,** when his [thirsty] people asked him for water: **Strike** the **rock [?]** with your **staff**. Out of it there gushed forth **twelve springs**. **Each group** knew its **own place for water**. We gave them the **shade of clouds [Arabian Peninsula]**, and sent down to them **manna** and **quails**, [saying]: **Eat of the good things** We have **provided** for you: [but they rebelled]; to Us they did **no harm**, but they harmed their **own souls**. [Sura No. – 6\7 - As-haabul- a' raaf, Ayat No. – 159 & 160]

Ayat\Verses - There was indeed a **Sign for Saba**, in their **dwelling place**. **Two Gardens to the right and to the left**. Eat of the **Sustenance** by your **Rab**, and be **grateful** to Him: A territory [Macca] fair and happy, and a **Rabbun Gafuur**. [Sura No.33\34 Saba, Ayat No. 15]

Ayat\Verses - But **they turned away**, and We sent against them the **Flood** [released] from the **dams**, and We converted their **two garden [rows] into "gardens"** producing **bitter fruit**, and **tamarisks**, and some few [stunted] **lote-trees**. That was the **requital** We gave them because they ungratefully **rejected Faith**, and never do We give [such] requital except to such as be ungrateful rejecters. **Between them and the Cities on which We had poured our blessings** [region of white], We had placed **Cities in prominent positions**, and **between them We had appointed stages of journey in due proportion. Travel therein**, secure, by **night and by day**. [Sura No. – 33\34 – Saba, Ayat No. – 16 to 18]

Ayat\Verses - Say: **Disbelieve** you verily in Him **Who created the East Horizon [world] in two days**, and **ascribe** you unto Him **rivals**? He is the Rab of the universe. He set in the [world], mountains **standing firm**, high above it, and bestowed blessings in the world, and **measure therein all things to give them nourishment in due proportion, in four Days**, in accordance with [the needs of] those who seek [Sustenance]. Then **He turned to the heavens [towards West Horizon] when it was smoke [white dwarf]**, and said unto it and unto the world: **Come both of you, willingly or unwillingly [equal & opposite]**. They said: We come, obedient. So He **ordained** them as **seven firmaments (fermions) in two Days [odd integers / spin 2], and He inspired in each heaven its mandate [magnetism following safa & marwa]**. And **We adorned the lower heaven [East Horizon] and [provided it] with guard [Black Square]**. Such is the Measuring of the Mighty, The Knower. [**Sura No. – 40\42 – Haa-Mim-Sajdah, Ayat No. – 9 to 12**]

WHERE IS THE LOWEST HEAVEN?

Ayat\Verses - "**Rabbus-samaawaati wal-arzi, wa maa baynahumaa wa Rabbul-mashaariq.**" [**Self-evident Concept**] Rab of the heavens and of the world and all that is between them, and **Rab** of **Mashaariq [East].** [Sura No. – 36\37 – Was-saaaffaati, Ayat No. - 5]

Ayat\Verses - We have indeed **adorned** the **lowest heaven with an ornament, the planets as securities [guards] from every forward devil, [So] they cannot listen to (pull their ears in the direction of) the Highest Chief for they are guarded (by planets) from every side.** Repulsed (outcast), and there is perpetual torment; **except him** who **snatches a fragment,** and there **pursues him by a piercing brightness.** [Sura No. – 36\37 – Was-saaaffaati, Ayat No. – 6 to 10]

Then ask them: **Are they stronger as a creation,** or **those (others) whom We have created**? Lo! We have created them out of **sticky clay.** [Sura No. – 36\37 – Was-saaaffaati, Ayat No. - 11]

Ayat\Verses - Nay, but you do wonder when they mock, and heed not when they are reminded, and when **they see a Sign (clear proof), turn it to mockery, and say: This is nothing but evident sorcery (mere magic)!** What! When we die, and become dust and bones, shall we [then] be raised up [again]; and also our forefathers? [Sura No. – 36\37 – Was-saaaffaati, Ayat No. – 12 to 15]

Ayat\Verses - Say: Yea, in truth, you will be **brought low. Then there will be a single shout,** and behold, they will begin to see! And they will say: **Ah! Woe to us! This is the Day of Judgment [Yawmud-Diin]. This is the Day of Separation [Yawmul-Faslillazii kuntum-bihii tukazzibuun], which you used to deny.** [Sura No. – 36\37 – Was-saaaffaati, Ayat No. – 18 to 21]

Ayat\Verses - (It will be said): **Assemble those who did wrong, together with their wives and what they used to worship besides Allah, and lead them to the path of hell [towards Straight Middle East of Eartha 3D or Blackhole]; and stop them, for they must be questioned - What is the matter with you that you help not each other?** [Sura No. – 36\37 – Was-saaaffaati, Ayat No. – 22 to 25]

Ayat\Verses - Nay, but **this Day they will make full submission. And some of them draw near to others mutually questioning. They will say: Lo! It was you who used to come to us, imposing, (swearing that you spoke the truth). They will reply: Nay, you yourselves had no Faith! We had no power over you, but you were a folk in obstinate rebellion.** [Sura No. – 36\37 – Was-saaaffaati, Ayat No. – 26 to 30]

Ayat\Verses - Now the word of our Rab has been proved true, against us. **Lo! We are about to taste [the punishment of our sins]. We led you astray; for truly we were ourselves astray. Truly, that Day, they (both / two-in-one partnership) are sharers in the Penalty.** Verily that is how We shall deal with Sinners. For when it was said to them: **There is no reality except Allah**, they were scornful. [Sura No. – 36\37 – Was-saaaffaati, Ayat No. – 31 to 35]

Ayat\Verses - And said: What! Shall we give up our realities for the sake of a mad Poet? **Nay! He has brought the [very] Truth, and he will confirm those revealed.** Verily you will taste the **Grievous Penalty**. But it will be no more than the retribution of [the evil] that you have done. [Sura No. – 36\37 – Was-saaaffaati, Ayat No. – 36 to 39]

Ayat\Verses - But the sincere [and devoted] slaves of Allah; for them is a **Sustenance determined** fruits and they [shall enjoy] honour and dignity in Gardens of Felicity **facing each other on Throne** [Arsh]; round will be passed to them a **Cup** from a clear-flowing **fountain crystal-white** of a taste delicious to those who drink [thereof], wherein there is no headache [& or crookedness], nor they will suffer from intoxication. And besides them will be **chaste women**, restraining their glances, with big eyes [of wonder and beauty]; as if they were [delicate] **eggs closely guarded**. [Sura No. – 36\37 – Was-saaaffaati, Ayat No. – 40 to 49]

Ayat\Verses – Unto **Allah belongs the dominion [Sovereignty] of the heavens and the world**. He gives life and He takes it. Except for Him you have no protector, nor helper. **Allah turned with favour to the Prophet, the Muhajirs, and the Ansars, who followed him in a time of distress.** After that the hearts of a part of them had nearly swerved [from duty]; but He turned to them [also]: for He is unto them Most Kind, Most Merciful. And to **the three [ground stair, middle stair, and top stair] also** (did He turn in mercy) **who were left behind [Straight Middle East Region], when the world vast as it is, was straitened for them still they be thought them that there is no refuge from Allah except toward Him; then turned Him unto them in mercy that they (too) might turn (repentant unto Him). Lo! Allah! He is the Relenting, the Merciful.** [Sura No. – 8\9 – Tauba, Ayat No. – 116 to 118]

"Khalaqakum-min-nafsinw-waahidatin summa ja-'ala minhaa zawjahaa wa anzala lakum-minal-an-aami samaaa-niyata azwaaj; yakhluqukum fii butuuni ummahaatikum khalaqam-mim-ba' di khalqin fii zulumaatinsalaas. Zaalikumullaahu Rabbukum lahul-mulk. Laaa ilaaha-illaa Huu. Fa-annaa tusrafuun?" [Trans.] He created you [all] from a single being. Then from that being He made of like nature his mate; and he **sent down for you eight head of cattle in pairs. He makes you, in the wombs of your mothers, in stages, one after another, in a threefold gloom. Such is Allah your Rab and Cherisher.** To Him belongs the **Sovereignty.** There is **no unbreakable reality** except Him. **How then are you turned away?** [Sura No. – 38\39 – Tanziilul-Kitaab, Ayat\ Verses – 6]

THOSE WHO DISPUTE CONCERNING THE SIGNS OF ALLAH

Have not seen those who dispute concerning the Signs [Injiil or Correspondence Truth] of Allah? How are they turned away [from Reality]? Those who reject the bil-Kitaab and the [manifested signs] with which We sent our Messengers, but soon shall they know, when the **yokes** round their necks, and the **chains** they are dragged; through boiling water; then they are pushed into fire. Then it will be said to them: **Where are [Manmade Nature] that you used to make partners besides Allah?** They will reply: **They have left us in the lurch. But we invoked not, anything before [that had real existence].** Consequently, Allah does send **the disbelievers to go astray.** [And it is said unto them): This is because you **exulted** in the world **without right** (i.e. with things **other than the Truth**), and because you were **petulant** (ill-tempered). **Enter you the gates of Hell, to dwell therein; and evil is the habitation of the scornful [disdainful]. So stick with patience;** for **the Promise of Allah is True.** And whether We show you some part of that which We promised them, or [whether] We cause you die, still unto Us they will be brought back. [Sura No. – 39\40 – Mu-Min, Ayat No. – 69 to 77]

Fasta-'iz billaahi minash-Shaytaanir-Rajim
Bismillaahir-Rahmaanir- Rahim

ONE DOT LEADER OR RABBUS-SHI-RAA

One Dot Leader is the Prime Mover or Invariable & Unconditional Antecedent or Causeless Cause of Revelation. This One Dot Leader of the real science is the Ahad or Uniqueness Who has no attribute save that He is the Owner of all attributes. The real science has searched out this revealed truth and has named as **One Dot Leader with character code** Unicode (hex) 202A.

DOT OPERATOR OR AGENT [MOVING POWER]

Dot Operator is the Agent or Moving Power of continual acceleration & motion. Star [Planet/Moon] Mercury represents Dot Operator. Dot Operator is gravitationally locked. Real Science has searched out this revealed truth and has named as Dot Operator with character code Unicode (hex) 22C5.

DIAMOND OPERATOR OR PATIENT [COLLOCATION]

The Universe is a Diamond Operator or a quadrilateral shining star in pragmatic correspondence with the Crucified Sign [**Injiil**]. In other words, the Diamond Operator or Intrinsically Luminous Star is the collocation in coherence with the Plus Sign of Mathematics [Tawraat]. In the Kitaab with Truth, this Diamond Operator is called -Shi-raa or Shining Star. The real science has searched out this revealed truth and has named as **Diamond Operator** [Unicode (hex) – 22C4]. Four shades of Windows'7 Ultimate represents Diamond Operator or Universe or Alamin or Collocation or Sirius or Shi-raa or Shining Star. [Ref, Ayat No. – 35 of Sura Nuur]

TRINITY

The universe has been revealed as an Equal & Opposite Trinity of West Horizon – Cloud & Sky – East Horizon. Star [Planet/Moon] Jupiter represents West Horizon. Star [Planet / Moon] Saturn represents East Horizon. Cloud & Sky represents equal & opposite ~sth.

END OF PROOF OR UPRIGHT RECTANGLE

The Universe has been revealed as an equal & opposite Trinity and manifested as an upright rectangle in pragmatic correspondence with the appointed Kaba. The real science has searched out this manifested truth and has named as

'End of Proof' along with the character code - Unicode (hex) 220E. In the Kitaab with Truth, it has been shared with us that the appointed Kaba is the Clear Sign as well as Standard [Criterion] of right directions for mankind. So, the universe is not like the projected Global Trinity and Solar System of the Historical Conspiracy Science. On the contrary, the revealed as well as manifested universe [alamin] is an Upright Rectangle in resemblance with the appointed Kaba and the End of Proof of the Real Science, and Star System or Sirius Binary System.

PRIME MERIDIAN OR SIGN OF JUSTICE

The sign of justice between Revelation and Manifestation is called Prime Meridian. In other words, the Upright ~sth or Vertical ~sth or Perpendicular ~sth between **two end points** of West Horizon and East Horizon of the revealed universe is called **Prime Meridian**. The ideogram 'Alif' of the Arabic Alphabet represents Prime Meridian or **Sign of Justice**. This sign of Justice also resembles with the Pronoun I. Real Science has also searched out this revealed truth. [Ref. Unicode (hex) & ASCII (hex) 007C, 0049, ASCII (decimal) 124, 73, and the like]

TWO HEMISPHERES
zaatal-yamini [Revealed Left] & zaatash-shimaal [Revealed Right]

Two sides of the Prime Meridian or the Sign of Justice are called Hemispheres. Revealed Left of the Prime Meridian but right side of the finite observer facing towards the appointed Kaba is the Manifested Northern Hemisphere [Haiyalas-Swalaah]. Revealed Rght of the Prime Meridian but left side of the finite observer facing towards the appointed Kaba is the Manifested Southern Hemisphere [Haiyalaal-Falaah]. These two Hemispheres represent the Right Border and Left Back Border of Manifestation. **Northern Hemisphere represents Haiyalas-swalaah. Southern Hemisphere represents Haiyalal-Falaah**. Converted red and blue shades of Windows'7 Ultimate represent Northern Hemisphere. Converted green and yellow shades of Windows'7 Ultimate represent Southern Hemisphere. The concepts of two Hemispheres resemble with Parentheses or brackets and the capital letter H of the real scientific symbols along with their character codes such as Unicode (hex) & ASCII (hex) 005B, 005D, 007B, 007D, ASCII (decimal) 91, 93, 123, 125 etc.

EQUATOR LINE OR HORIZONTAL LINE OR IMAGINARY LINE

The universe is a diamond operator and revealed as an equal & opposite trinity. In other words, Horizontal ~sth or Parallel ~sth or imaginary line that connects two Hemispheres as well as divides the Diamond operator into two Horizons is called the **Equator Line or Horizontal Line**. This horizontal line represents the equal & opposite balance as well as barriers between West Horizon and East Horizon, between Top West and Bottom East, between Night Sky and Earth Sky, between revelation and manifestation. This Equator Line resembles with the division line of Mathematics and Danger & Double Danger of Real Science [Ref. ASCII (hex) 0086 & 0087 and ASCII (decimal) 134 & 135].

TWO END POINTS

The universe is a diamond operator or a quadrilateral shining star, revealed as a trinity, and manifested as an upright rectangle. The Equator Line is the division line or the binary ~sth of balance between equal and opposite revelations as well as manifestations. This binary ~sth represents the equal and opposite relation between willingness & unwillingness, neutron & neutrino. Due to this equal and opposite binary ~sth, the universe has been revealed as a Trinity in resemblance with the **Division Sign** of Mathematics. The upper dot of the division sign is the **West Horizon** or Magriib and the lower dot of the division sign is the **East Horizon or Mashriiq.** The top most end of the Prime Meridian is the **West end point** and the bottom most end of the Prime Meridian is the **East end point** of the revealed Trinity. The **two end points** of the **Universe** are Upright West and Straight East in pragmatic correspondence with the head and foot of an upright individual. With respect to Manifested Universe and East Horizon [Black Square / World], Northern Hemisphere and Southern Hemisphere represent two end points of Natural Magnetism. So, the universe is a diamond operator or a quadrilateral shining star, revealed as an equal & opposite Trinity of two equal & opposite end points, and manifested as an upright rectangle. These searched out findings are justifiable in resemblance with Complete Coded Shared Tautologies called Universal Major Premises, Verifiable by the searched out scientific symbols like End of Proof of the Real Science in pragmatic correspondence with the appointed Kaba and an upright man like Ibrahim [ass].

The definitions of two end points provided by the Epistemic Uniqueness of Historical Conspiracy Science are **vague, absurd, equivocal, and store houses of subjective selfcontradictions and objective paradoxes**. Two contradictory statements can neither be true together nor false together. If one of the contradictories is true, the other must be false, and conversely. Whether North represents upward direction towards the head of an upright man in resemblance with projected Northern Hemisphere or the left hand direction of an upright man facing towards once rising/entering sun from an unidentified East & in resemblance with Man-made Magnetism or two E-points of entering from North-East & ending in North-West of the so-called sun as the manifested sign of Natural Magnetism in resemblance with revelation? If North is the left hand direction out of four cardinal directions in resemblance with man-made magnetism, then it cannot be the upward direction towards the head of an upright man in resemblance with Global Trinity. Further, if North represented by the equal & opposite stages of journey of the so-called sun as the manifested sign of Natural Magnetism is the E-Point [entering & ending points], then North represented by Man-made Magnetism is teleological evidence sorcery.

Similarly, **whether two end points** represent top and bottom end points in resemblance with Northern Hemisphere and Southern Hemisphere of the projected global trinity or left and right end points in resemblance with the man-made magnetism or entering and setting points as well as rising and ending points of the stages of journey of the so-called sun as the manifested sign of natural magnetism **have not been clearly and distinctly defined by the pen – paper - pencil activists** with a view to maintain balance with the teleology as well as the ontology of the Don of two-in-one partnership. So, for the sake of truth, it is necessary to make clear our concepts from the black & white activists the manifest truth concerning **two end points** in resemblance with revelation, manifestation and manifested signs of natural magnetism.

T-JUNCTION OR NUCLEUS OR TRUTH-IN-ITSELF OR EQUATOR OR MIDDLE DOT

The point of division of the Prime Meridian [Sign of Justice] by the division line [the Line of Balance or Equator Line] is called T-junction [Intersection]. This point of intersection is the Nucleus [Truth or Equator or Middle Dot or Dot Operator or Planet Mercury]. The very term 'T-junction' connotes equal & opposite Trinity as well as Truth-in-itself. The scientific symbol 'Danger' represents equal & opposite Trinity. In the Kitaab with Truth, it has been advised not to say 'Three Cease'. The product of this intersection is a quadrilateral star [a Diamond Operator with a Middle Dot or an atom with a nucleus or Revealed Truth & Manifested Tautologies or Truth-in-itself & Criterions of Truth]. The crucified sign [Injiil] is the clear proof of these revealed truths as well as manifested tautologies which Isa [ass] had brought for mankind. The point of intersection has been marked as 0°.

FOUR CARDINAL DIRECTIONS OR FOUR IMAMS OR FOUR BLESSED SPRINGS FOR STRAIGHT USAGE

As a Diamond Operator, the Universe has four Cardinal Directions. These four cardinal directions are – East Horizon, West Horizon, Northern Hemisphere, and Southern Hemisphere. These four cardinal or basic directions are called Helium-4 or four Galilean moons. The four basic forces of the real science or revealed forces represent four cardinal directions. The East Horizon represents Gravitational Force or downward direction. Earth or soil is the clear sign of East Horizon. The West Horizon represents Strong Force or upward direction.

Rain comes from West. Northern Hemisphere represents Magnetic Force in resemblance with two E-Points of Natural Magnetism. Fire or light is the clear sign of the Northern Hemisphere. Southern Hemisphere represents Weak Force in resemblance with T-Point [turning point] of Natural Magnetism. Air is the clear sign of the Southern Hemisphere.

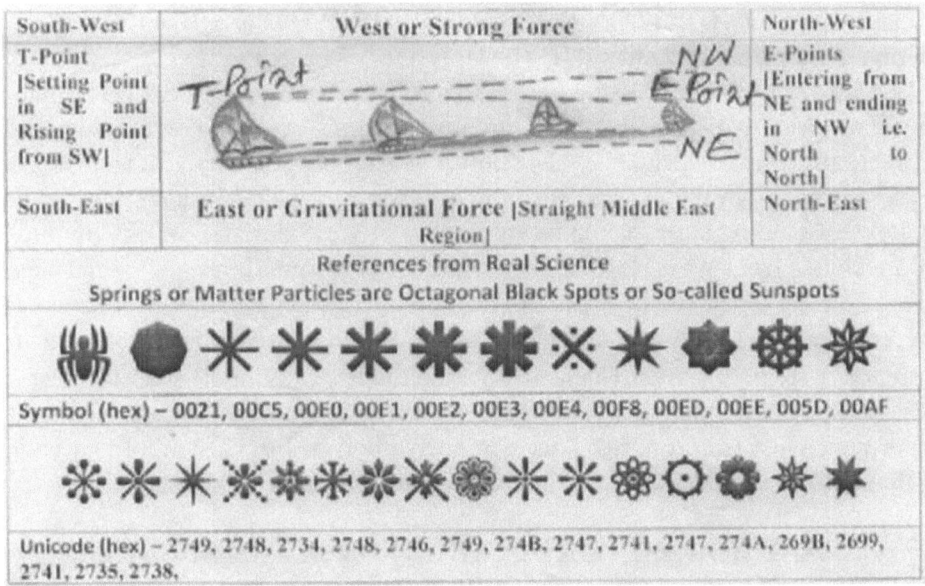

South-West	West or Strong Force		North-West
T-Point [Setting Point in SE and Rising Point from SW]			E-Points [Entering from NE and ending in NW i.e. North to North]
South-East	East or Gravitational Force [Straight Middle East Region]		North-East

References from Real Science
Springs or Matter Particles are Octagonal Black Spots or So-called Sunspots

Symbol (hex) – 0021, 00C5, 00E0, 00E1, 00E2, 00E3, 00E4, 00F8, 00ED, 00EE, 005D, 00AF

Unicode (hex) – 2749, 2748, 2734, 2748, 2746, 2749, 274B, 2747, 2741, 2747, 274A, 269B, 2699, 2741, 2735, 2738,

In the Kitaab with Truth, the four basic forces have been shared as four blessed springs for straight usage out of twelve chiefs. The Messenger of Allah has also shared the concept of the four Imams with concrete examples in resemblance with the four pillars of a pentagonal tent. The manifested sign of natural magnetism [electromagnetic wave or Niche or binary thread or so-called sun or Bullet or Wash-Shamsi] is kindled within the lamp [Black Square or East Horizon within the four cardinal directions] from North-East declining towards Southern Hemisphere. The appointed light of Allah or the kindled niche runs away from Southern Hemisphere towards the end point [North-West] following equal & opposite stages of journey due to Revealed Magnetic Force. Right hand companions and left hand companions represent Southern Hemisphere and Northern Hemisphere respectively. In other words, right hand companions represent people of the West Zone where the so-called sun as the manifested sign of natural magnetism [electromagnetic wave] rises from South-West and sets in North-West. Left hand companions represent people of the East Zone where the so-called sun as the manifested sign of natural magnetism [electromagnetic wave] enters from North-East and sets in South-East.

The concept of Four Noble Truths of Buddha Philosophy represents four basic forces. The concept of four Purusarthas also resembles with four basic forces. Four gross elements or nitya dravyas, namely earth, water, fire, and air also represent four blessed springs. Four cardinal virtues of Plato and four causes of Aristotle also represent four basic forces or four cardinal directions.

WEST HORIZON OR MAGRIB OR WHITE SQUARE

Universe is a Diamond Operator. The Universe has been revealed as a Trinity of West Horizon – Cloud & Sky – East Horizon. This Trinity is called Sirius Binary System. West Horizon or White Square represents Strong Force or upward direction towards Upright West [Throne of Authority or Arsh or Heaven] in resemblance with the obversion of the converted snake of Muusa's Staff. The galaxy of guiding stars, canopies etc. represent West Horizon [White Square]. Rain comes from West Horizon. Appearances of the guiding stars are the clear proofs of the existence of West Horizon [White Square]. Planet/Moon Jupiter represents West Horizon. West Horizon is the main sequence of the Sirius Binary System. So, West Horizon is called Sirius A. Symbols like [Symbol (hex) – 007F], [Symbol (hex) 00FF] of Font name 'Symbol' of the Insert Symbols represent West Horizon. So, Font name 'Symbol' of Insert Symbol is a valid reference of West Horizon [White Square] from Scientific Point of view.

EAST HORIZON OR MASHRIIQ OR BLACK SQUARE

Universe is a Diamond Operator. The Universe has been revealed as a Trinity of West Horizon – Cloud & Sky – East Horizon. This Trinity is called Sirius Binary System. East Horizon [Black Square] represents Gravitational Force or downward direction towards Manifestation [Hell or Zone of Iron or Buried Treasure]. Volcano, Earthquake etc. erupt/emerge from East. East Horizon has been guarded by the four basic forces. For this reason, East Horizon is called Black Square or Non-luminous Moon or Dark Moon. The world has been manifested within the Black Square of East Horizon. So, manifested world is also non-luminous. Planet/Moon Saturn represents East Horizon. The ring of the planet Saturn represents walls of basic forces. This Black Square or Non-luminous Moon or Dark Planet is the white dwarf companion Sirius B of Sirius Binary System. Symbols like ∓ [Symbol (hex) 006D], ! [Symbol (hex) 0021], ‿[Symbol (hex) 007B] of Font Name MT Extra of Insert Symbol represent East Horizon. So, Font name 'MT Extra' of Insert Symbol is a valid reference of East Horizon from Scientific Point of view.

WEST ZONE AND EAST ZONE

East Horizon has been divided into two Zones or two Seashores. This division has been done declining towards right from the infinite subjective point of view or from the point of view of Prime Meridian. The upper seashore in pragmatic correspondence with the appointed Kaba is the West Zone. The lower Sea-shore in pragmatic correspondence with the penetrated hole in the journey boat of Muusa [Eartha 3D or Recycling Region of Straight Middle East] is the East Zone. In the Kitaab with truth, it has been shared that Allah and His Messenger divided the sea and saved Bani-israa-iil [people of Arabian Peninsula] from the drowned people of Firawn. Searched out symbols of the real science with the character codes like Unicode (hex) 660C, Symbol (hex) 0032, Symbol (hex) 0031 represent two Zones within the Black Square of East Horizon. This manifestation is diagonal in resemblance with the multiplication sign of Mathematics declining towards right. Converted red and green shades of Windows'7 Ultimate represent East Zone [Lower Seashore]. Converted yellow and blue shades of Windows'7 Ultimate represent West Zone [Upper Seashore]. Moreover, manifestation represents geometric progression and Orbit; while revelation represents arithmetic progression and Axis.

TWO WEST AND TWO EAST

There are two West and two East. The Universe has been revealed as an equal & opposite Trinity of two Horizons and two Hemispheres. This revelation resembles with Plus Sign of Mathematics. The East Horizon guarded by the four Imams or Revealed Forces called Black Square [Non-luminous Moon or Dark Planet]. This Black Square has been manifested further as a trinity of two Zones declining towards Revealed Right. This revelation resembles with the Multiplication Sign of Mathematics. So, there are four revealed directions, namely, Up [West], Down [East], Revealed Right [South], and Revealed Left [North]. These directions represent Major Axis, Minor Axis, and Semi Major Axis of a & m of Newton's Second Law 'F = ma'.

East Horizon has been revealed in a diagonal form [latitude-longitude form] within four cardinal directions into two Zones [Seashores]. The Upper Seashore is called West Zone and the Lower Seashore is called East Zone. This manifestation resembles with Multiplication sign. This revelation was octagonal. The octagon has been manifested as a Hexagon after two sacrifices / divorces. So, there are six basic directions with respect to Manifested East Horizon. Three basic directions of the East Zone are North-East, Middle-East, and Sout-East. Three basic directions of the West Zone [Upper Seashore] are South-West, Upright West, and North-West. We express the greatness of my Rab and your Rab pronouncing 'Allahu-Akbar' six times in each call for prayer [Ajan]. Scientific unit like Unicode (hex), Symbol (hex) etc. are the clear proofs as well as valid references of six revealed & manifested directions. Moreover, hexagonal symbols also represent six revealed & manifested directions.

So, West Horizon and West Zone represent two West; while East Horizon and East Zone represent two East.

OCTAGON
EIGHT IN FOUR PAIRS AND RING OPERATOR

Matter particles [springs or black spots or so-called sun-spots] are octagonal in resemblance with the sacrifice of a quadruped from psychic point of view with reference to shared parable 'Baqarah' in the Kitaab with Truth. These matter particles are octagonal rings in resemblance with the ring of the projected planet Saturn. So, the ring of the projected planet Saturn represents the ring of the immovability of the revealed as well as manifested Universe and the world. The black spots or sunspots are the wax feeble octagonal matter particles or springs in the name of Zakariya, the Guardian of Maryam supplied with sustenance or what real science calls **Ring Operator**. There are twelve matter particles or springs, 23 elements due to equal sharing and equal & opposite ~sth between two zones [seashores], 99 components [projected 92 + 7] in resemblance with 99 beautiful names, 25 representative names as possessors of the signs of relation in resemblance with the shared names of messengers and warners. Eight in four pairs represent octagonal springs and four equal & opposite revealed forces or cardinal directions. The concept of octagon resembles with the eight legs of a spider Webdings: 33, Symbol (hex) 0021. Here the digit 2 and digit 1 represent even and odd; while digits 00 represent equal. So, the number 0021 represents equal & opposite as well as odd & even. Further, digit 2 represents Middle West Stair [of Europe, Asia, Africa, and Australia] and Upright West Stair [of Arabian Peninsula]; while digit 1 represents ground stair of North America and South America.

MANIFESTED HEXAGON / ASTERISK

East Horizon within four cardinal directions [basic forces] is called **Black Square** [Non-luminous Moon or Dark Moon]. Symbols like [Symbol (hex) – 0093], [Symbol (hex) – 00F0] of the Font name 'Bookshelf Symbol 7' of Insert Font Symbol represent Black Square [Non-luminous Moon or Dark Moon]. The East Horizon has been revealed as an octagon and manifested as a hexagon either due to equal & opposite psychic sacrifice of Ismail [ass] & Ibrahim [ass] or due to twice divorces. These two sacrifices/divorces are verifiable as well as justifiable in resemblance with equal and opposite sacrifices/divorces in green and blue shades of Windows'7 Ultimate. So, the Black Square [Non-luminous Moon / Dark Moon] has been manifested as a hexagon [Asterisk] in resemblance with the hexagonal sign in the bottom of the converted green shade of Windows'7 Ultimate. In the Kitaab with Truth it has been shared that the heaven and the world has been created in six days. Real Science has searched out this revealed & manifest truth. Character codes like Unicode (hex) 2731, 2732, 273B, 273C, 273D represent that the manifested Black Square [Non-luminous Moon or Dark Moon] is a hexagon or Asterisk. This manifested hexagon within the Black Square of East Horizon is called the World. In other words, the world has been manifested within the East Horizon.

The Four Walls of the building are the four basic forces or four walls of scientific investigations. Science cannot go beyond four basic forces. In other words, scientific broadcastings which cross the four walls of scientific investigations cannot be categorised as verifiable science. Such broadcastings are justifiable only in resemblance with revelations and what have been shared by the Messenger of Allah. So, all projected investigations, measurements, repeated visiting in the Night Sky & Earth's Night Sky by the conspiracy scientists are neither right knowledge nor wrong knowledge but are fictitious teleological broadcastings with a view to conceal Natural Magnetism and Manifested Nature. Only those broadcastings which resemble with revelations or justifiable in resemblance with revelations are valid knowledge. Those broadcastings which are verifiable in resemblance with manifested signs or manifested nature are certain knowledge.

STAR OPERATOR
HOUSE OR APPEARING PENTAGON OR SPIDERS'
NET WITHIN THE BUILDING [BLACK SQUARE]
SPREAD OUT TABLE OR MAAYIDAH
THE APPEARING EARTH IS A PENTAGONAL STAR

The universe is a diamond operator, revealed as an equal & opposite Trinity. The East Horizon has been revealed as a Trinity of West Zone – Sea & planetary barrier [electron cloud] – East Zone. The West Zone has been further manifested as a Trinity of Top West Region [Arabian Peninsula] – Sea or psychic barrier [electron cloud] – Middle West Regions. East Zone has been manifested by penetrating one triangle at the centre of the manifested world called Straight Middle East. This Straight Middle East is the second peg of the immovable hexagonal world as well as appearing pentagonal earth. This Straight Middle East is called Earth 3D or Bermuda Triangle of Titanic in resemblance with the penetrated hole in the journey boat of Muusa (ass). The remaining triangles of the East Zone resemble with the existence of North America and South America as Lower Seashore or Eastern Regions. There is an equal & opposite ~sth between Arabian Peninsula [Upright West Region] and Eartha 3D [Straight Middle East Region]. This ~sth is the Fifth Pillar or Prime Meridian of the Pentagonal Tent. Due to this equal and opposite ~sth, the manifested hexagon is appearing as a pentagon. So, the appearing earth [land] is not like the projected globe but a pentagon in correspondence with the existing 9/11 Pentagon in the

East Zone. This appearing pentagon is called the earth or Star Operator or spread out table. The real science has searched out this manifested & appearing truth and has named the appearing pentagonal earth as House ⌂ [Unicode (hex) 2302], Spiders' Net 🕸 Symbol (hex) 0022. In the Kitaab with Truth, the world has been compared with a building [Black Square] and the appearing earth has been compared with the net of a spider [Ref. Sura – Ankabut].

It has been shared in the Kitaab with Truth that there is no crookedness or global concept or common run in revelation & manifestation. The revealed & manifested truth is that the earth [land] is a House or a Pentagon or like the net of a spider or Star Operator of the Real Science. This truth is justifiable by upright logic and verifiable by the searched out real scientific symbols in pragmatic correspondence with 9/11 Pentagon and in resemblance with both revelations and manifested nature. Moreover, in the Kitaab with Truth, it has been shared that one day the world will be changed to a different world. When possessors of upright nature will be able to conceptualise that the revealed and manifested universe is an upright rectangle [End of Proof], the Black Square has been revealed as an octagon & two zones and manifested as a hexagon or Asterisk within four basic forces [world], the earth is a pentagon with three ascending stairs, and equal & opposite stages of journey of the manifested sign of natural magnetism, then the world will be changed for them. The shared findings of 'Kitaaba Wal-Hikmata' \ 'Manifested Nature and the Utility of One's Upright Logic' are neither hypotheses, nor suppositions, nor imaginations, nor predictions. On the contrary, the shared findings are the manifest truths in resemblance with revelations. So, the sceptics of revelation as well as manifestation are requested to do as much research works as possible on shared findings of this inspired sharing with a view to falsify each and every shared selfevident truth.

West Zone or Upper Seashore comprises Europe, Arabian Peninsula, Asia, Africa, and Australia. Arabian Peninsula is the Upright West Triangle [Region] of the manifested hexagon and Uppermost Land of the appearing pentagon. This Manifested Truth has been shared several times in the Kitaab with Truth. The Uppermost Land of the Pentagonal Earth or Arabian Peninsula has been addressed as City or Mount Tuur or Mount Alal-Ju-diyyi where there is the appointed Kaba on the right side of the Prime Meridian or Upright ~sth. Moreover, it has been shared that Mount Tuur is holding firmly the appointed Kaba at its right side. Further, the manifested truth is that the Arabian Peninsula is the region of white or Torrid Zone or Zone of Hot Sands where rain [white] comes every day at 4 pm.

Europe, Asia, Africa, and Australia are there in the Middle West Regions. In the Kitaab with Truth these regions have been addressed as Nations [Ummat] or

Median. These regions have a mild climate – neither very hot nor very cold and are called Temperate Regions. Europe is in the South-West Region of the Upper Seashore in correspondence with the converted Yellow Shade of Windows'7 Ultimate; while Asia, Africa, and Australia are there in the North-West Region of the Upper Seashore in correspondence with the converted Blue Shade of Windows'7 Ultimate. In the Kitaab with Truth, it has been clearly stated that there is no crookedness or circular way or global concept in Revelation. So, global concept is contradictory to revelation, manifestation, real science as well as upright logic. Just try to make your concept clear from the Don of the global universe, the global world, and the global earth by asking following simple factual truths - **Ask the pen – paper - pencil activists of globalisation to demarcate the universe, the world, and the earth from the projected globe pointing clearly Night Sky, Earth's Night Sky, Earth Sky, Galaxy of Stars, Region of Cloud, Region of White, Recycling Region, Earth 3D, Arabian Peninsula, Zone of Buried Treasure, Two End Points of the Universe, Two End Points of the Sirius Binary System, the Turning Point of the Sirius Binary System, Stages of Journey, Zone of Morning Show, Zone of Evening Show, Zone of Iron, Zone of flameless Fire, Helium-4 atom, Four Cardinal Directions, Three ascending stairs of the Appearing Earth, Real North which represents Magnetic Force, Major Axis, Minor Axis, Semi Major Axis, that Region of the Globe from which Stars appear at Night, that region from which rain comes, that region from which the earthquake emerges & Volcano erupts, to specify the entering point of the Bullet, to specify the setting point of the Bullet, to specify the turning point of the Bullet, to specify the rising point of the Bullet, to specify the End Point of the Bullet, Horizontal & Vertical ~sth in resemblance with a graph chart or Plus Sign, diagonal ~sth with respect to manifested world in resemblance with multiplication sign, and so on.**

In this regard, it is also requested to all higher authorities, researchers, research institutions, epistemic persons who are taking ultimate decision concerning manifested nature and Human Rights, to justify as well as to verify the findings of this sharing with a view to provide their ultimate decisions concerning revelations, manifestations, real scientific symbols, conspiracies, evidence sorceries, pen-paper-pencil works of the activists, projection of falsehood as truth by the unique epistemic persecutors, projected jihadists, so-called fundamentalists, real Akbari-jihadists, solid fundamentalists, solidified solidarity human rights, and the Historical Don of Trinity as well as Duality.

The East Horizon has been manifested as a hexagon but due to equal & opposite ~sth between Upright West Triangle and Straight Middle East Triangle, the manifested hexagon is appearing as a pentagon in resemblance with a terrestrial or pentagonal star and in pragmatic correspondence with the 9/11 Pentagon. There are three ascending stairs in the pentagonal earth. The lower stair is called Tiin or Ground Stair or Townships or Frigid Zones or Greenland or Tundra Regions or Super Script One [1] [Unicode (hex) & ASCII (hex) 0089]. The middle stair is called Median or Nation or Middle West or Zaytuun or Temperate Regions or Super Script Two [2] [Unicode (hex) & ASCII (hex) 0082]. The topmost stair is called the City or Tuur or Upright West Region or Torrid Region or Region of Hot Sands or Arabian Peninsula or Super Script Three [3] [Unicode (hex) & ASCII (hex) 0083]. These three stages resemble with the lying, sitting, and standing stages of an upright man, three ascending stairs in the lecture stair in a Masque, three stages of journey of the so-called sun as the manifested sign of natural magnetism in resemblance with the shared parable of Zil-qarnayin. The searched out symbol ‡ Double Danger [Unicode (hex) 2021, ASCII (hex) 0087, ASCII (decimal) 135] represents the barriers between semi-anti-clockwise and semi-clockwise stages of journey for the three ascending stairs.

The appearing earth is a Star Operator or a pentagonal star where even and odd are contrasted. **Two lower triangles** of the House or the Pentagonal Star represent East Zone or Lower Seashore where there are North America and South America; while **three upper triangles** of the House or the Pentagonal Star represent West Zone or Upper Seashore where there are Europe, Arabian Peninsula, Asia, Africa, and Australia. So, the appearing earth is a pentagon where even and odd are contrasted.

Again, out of three upper triangles of the West Zone or Upper Seashore, one is the Upright West out of three triangles which represents Arabian Peninsula and remaining two triangles represent South-West and North-West of the Middle Stair or Middle West Regions. So, there is even and odd contrast in the West Zone also.

Manifested East Zone has three triangles. But due to one penetrated triangle called Middle East Triangle of Eartha 3D, two triangles are being found in appearance in the East Zone. So, there is odd and even contrast in the manifested East Zone also.

SIX REGIONS OF THE MANIFESTED HEXAGONAL WORLD
Three Regions of the West Zone
i) **South-West – Europe represents South-West Region.**
ii) **North-West – Asia, Africa, and Australia represent North-West Region.**
iii) **Upright West – Arabian Peninsula represents Upright West Region.**

Three Regions of the East Zone
i) North-East – North America represents North-East Region.
ii) South-East – South America represents South-East region.
iii) Straight Middle East – Eartha 3D or Recycling Region or Bermuda Triangle or Penetrated hole in the journey boat of Muusa (ass) represents Straight Middle East Region

THREE STAIRS OF THE APPEARING PENTAGONAL EARTH
Ground Stair – North America and South America
Middle Stair – Europe, Asia, Africa, and Australia
Topmost Stair – Arabian Peninsula

SEVEN CONTINENTS OR WINDOWS OF
THE APPEARING PENTAGON
Eastern Continents – North-Eastern Continent of North America and South-Eastern Continent of South America
Western Continents – South-Western Continent of Europe, Upright Western Continent of Arabian Peninsula, North-Western Continents of Asia, Africa, and Australia

SEVEN INTELLECTUAL COMMUNICATIONS
Let us communicate with the citizens of Europe, Arabian Peninsula, Asia, Africa, and Australia when it is noon in the West Zone [Semi-clockwise Zone] with a view to search out whether it is day [day-light] in Europe, Arabian Peninsula, Asia, Africa, and Australia at the same time or not. If the answer is affirmative, then that simple communications will prove that West Zone [Semi-clockwise Journey of the manifested sign of natural magnetism] comprises Europe, Arabian Peninsula, Asia, Africa, and Australia. Again, let us communicate with the citizens of Europe, Arabian Peninsula, Asia, Africa, and Australia when it is mid-night in the West Zone with a view to search out whether it is night in Europe, Arabian Peninsula, Asia, Africa, and Australia at the same time or not. If the answer is affirmative, then that simple communications will prove that West Zone comprises Europe, Arabian Peninsula, Asia, Africa, and Australia.

Simultaneously, let us communicate with the citizens of North America and South America when it is noon [day-light] in Europe, Arabian Peninsula, Asia, Africa, and Australia with a view to search out whether it is day [day-light] in North America and South America at that time or not. If the answer is affirmative, then Europe, Arabian Peninsula, Asia, Africa, Australia, North America, and South America are there in the same Global Zone. If the answer

395

is negative, then it will prove that North America and South America are there in the opposite seashore of Europe, Arabian Peninsula, Asia, Africa, and Australia. That opposite seashore is the East Zone or Lower seashore or Semi-anticlockwise Zone; while the seashore of Europe, Arabian Peninsula, Asia, Africa, and Australia is the Upper Seashore or West Zone or Semi-clockwise Zone. Again, let us communicate with the citizens of North America and South America when it is mid-night in Europe, Arabian Peninsula, Asia, Africa, and Australia with a view to search out whether it is night in North America and South America at that time or not. If the answer is affirmative, then Europe, Arabian Peninsula, Asia, Africa, Australia, North America, and South America are there in the same Global Zone. If the answer is negative, then it will prove that North America and South America are there in the opposite seashore of Europe, Arabian Peninsula, Asia, Africa, and Australia. That opposite Seashore is the East Zone or Lower Seashore or Semi-anticlockwise Zone within the Black Square of East Horizon.

Manifested Truths – East Zone or Lower Seashore is the Semi-anticlockwise Zone within the Black Square of East Horizon with respect to equal & opposite stages of journey of the so-called sun as the manifested sign of natural magnetism; while West Zone or Upper Seashore is the Semi-clockwise Zone within the Black Square of East Horizon with respect to equal & opposite stages of journey of the so-called sun as the manifested sign of natural magnetism. So, for the East Zone [Lower Seashore], the so-called sun [Bullet] as the manifested sign of natural magnetism enters from North-East in correspondence with North America and sets in South-East in correspondence with South America; while for the West Zone [Upper Seashore], the so-called sun [Bullet] as the manifested sign of natural magnetism rises from South-West in correspondence with Europe and ends in North-West in correspondence with Australia. These are unalterable manifest truths in resemblance with revelations. Pen-paper-pencil workers, black & white artists, owners of nuclear weapons, possessors of AK47, Don of Historical Black Magic, Epistemic Uniqueness of Scientific Certainty, Researchers of Truth-in-itself & coded shared tautologies, Conspiracy Scientists of several scientific institutions, and the like have no control over Manifested Sign of Natural Magnetism as well as Equal & Opposite [Semi-anticlockwise and Semi-clockwise] Stages of Journey of the so-called sun [Bullet] as the manifested signs of natural magnetism. They have projected manmade nature before us with a view conceal the reality of manifested nature and natural magnetism. Their teleology is to dominate mankind of the appearing pentagon and their ontology is to persecute faith and belief of mankind in general and particularly followers of Upright West. Due to their teleological evidence sorceries and unique epistemic persecutions, believers of Middle Stair [Europe, Asia, Africa, and Australia] are performing Salat [Compulsory Prayer] towards Haiyalas-Swalaah [Manifested North-West] and believers of Ground Stair [North America and South America] are performing Salat [Compulsory Prayer] towards Haiyalal-Falaah [Manifested South-East]. These are neither statements, nor suppositions, nor hypotheses, nor imaginations, nor predictions, nor commentaries, but the pragmatic truths. If the teleological evidence sorcerers and unique epistemic persecutors are true to their words and deeds, let them come forward with a view to convert the manifested sign of natural magnetism in resemblance with their projected falsehood. It certain that they will not be able to reject or to falsify the searched out manifest truths shared as Solidified Solid Human Rights in the 'Kitaaba Wal-Hikmata' or 'Manifested Nature and the Utility of One's Upright Logic.'

WHERE ARE YOU?

1. **North-East of the appearing Pentagonal Earth** – The lower triangle rightward from the finite subjective point of view facing towards the appointed Kaba but leftward from the point of view of the appointed Kaba in correspondence with the existing North America is the North-East Continent of the pentagonal earth. This North-East Continent is the entering point of the so-called sun [Bullet] as the manifested sign of natural magnetism for Morning Show in the name of sister planet Venus or Maryam supplied with sustenance for the Ground Stair of the East Zone or Lower Seashore within the Black Square of East Horizon. Converted red shade of Windows'7 Ultimate represents North-East Region of the manifested hexagonal world and North-Eastern [Eastern] Continent of the appearing pentagonal earth within the Black Square of East Horizon.

2. **South-East of the appearing Pentagonal Earth** - The lower triangle leftward from the finite subjective point of view facing towards the appointed Kaba but rightward from the point of the appointed Kaba in correspondence with the existing South America is the South-East Continent of the pentagonal earth. This South-East Continent is the setting point of the so-called sun [Bullet] as the manifested sign of natural magnetism or the Morning Show for the Ground Stair of the East Zone or Lower Seashore as well as the turning point of the so-called sun [Bullet] as the manifested sign of natural magnetism for the Evening Show within the Black Square of East Horizon. Converted green shade of Windows'7 Ultimate represents South-East Region of the manifested hexagonal world and South-Eastern [Eastern] Continent

of the appearing pentagonal earth within the Black Square of East Horizon.

3. **South-West of the appearing Pentagonal Earth** – The upper triangle leftward from the finite subjective point of view facing towards the appointed Kaba but rightward from the point of view of the appointed Kaba in correspondence with the existing Europe is the South-West Continent of the pentagonal earth. This South-West Continent is the rising point of the so-called sun [Bullet] as the manifested sign of natural magnetism or the Morning Show for the West Zone or Upper Seashore but Evening show in the name of sister planet Venus or Maryam supplied with sustenance within the Black Square of East Horizon. Converted yellow shade of Windows'7 Ultimate represents South-West Region of the manifested hexagonal world and South-Western [Western] Continent of the appearing pentagonal earth within the Black Square of East Horizon.

4. **North-West of the appearing Pentagonal Earth** - The upper triangle rightward from the finite subjective point of view facing towards the appointed Kaba but leftward from the point of view of the appointed Kaba in correspondence with the existing Asia, Africa, and Australia is the North-West Continents of the pentagonal earth. These North-West Continents represent the **end point or Full Stop** [Unicode (hex) & ASCII (hex) 002E, ASCII (decimal) 46] of the equal & opposite stages of journey of the so-called sun [Bullet] as the manifested sign of natural magnetism or the Evening show in the name of sister planet Venus or Maryam supplied with sustenance for the West Zone or Upper Seashore within the Black Square of East Horizon. Converted blue shade of Windows'7 Ultimate represents North-West Region of the manifested hexagonal world and North-Western [Western] Continents of the appearing pentagonal earth within the Black Square of East Horizon. This North-Western Region of the manifested hexagon comprises three Western Continents, namely, Asia, Africa, and Australia of the appearing pentagonal earth.

5. **Upright West of the appearing Pentagonal Earth** – The remaining upper triangle of the appearing pentagon which is neither towards right nor towards left either from the subjective point of view or from the objective point of view is the Arabian Peninsula where there is the appointed Imam of the City called Kaba on the right side of the Mount Tuur facing towards manifestation or East. So, Arabian Peninsula is the upright West or Top West or Upper Most West Triangle of the manifested hexagon appearing as a pentagon where even and odd are contrasted. Kaba is facing towards East [Manifestation] and

consequently towards Kaba means towards West. White Cloud of Windows'7 Ultimate represents Upright West Region of the manifested hexagonal world and Top West [Uppermost] continent of the appearing pentagonal earth within the Black Square of East Horizon.

Fasta-'iz billaahi minash-Shaytaanir-Rajim
Bismillaahir-Rahmaanir- Rahim

After the truth, what is there except error?
How do you judge?
Township [Ground Stair] and Mountain Top [Topmost Stair]
JUNCTION

Ayat\erses - But your Rab is Most forgiving, full of Mercy. If He were to call them [at once] to account for what they have earned, then surely He would have hastened their punishment; but they have **their appointed time**, beyond which they will find no refuge. Such were the townships [Ground Stair] we destroyed when they **committed iniquities**; but we fixed an **appointed time** for their **destruction**. And when, Muusa said to his attendant: I will not give up until I reach **the junction of the two seashores**, though I march on for ages. But **when they reached the Junction, they forgot [about] their Fish, which took its course through the sea [straight] as in a tunnel.** [Sura No. 17\18 – Khaf – Ayat No. – 58 to 61]

Ayat\Verses - Say: Who is it that **sustains** you [in life] from the **sky** and from the **world**? Or who is it that has **power** over **hearing** and **sight**? And who is it that **brings** out the **living** from the **dead** and the **dead** from the **living**? And who is it that **rules** and **regulates** all affairs? They will soon say, "**Allah**". Say, will you then **not keep your duty** [unto Him]? Such is **Allah, your real Cherisher and Sustainer. After the truth, what is there except error? How** then are you **turned away**? Thus is the **word** of your Rab **proved true against** those who do **wrong** that they will **not believe**? [Sura No. – 9\10 – Yuunus, Ayat No. – 31 to 33]

Ayat\Verses - To Him is the return of all of you. The promise of Allah is true and sure. It is He **Who begins the process of creation, and repeats it**, that He may reward with justice to those who believe and work righteousness; but those who reject Him will have draughts of boiling fluids, and a penalty grievous, because they disbelieved. [Sura No. – 9\10 – Yuunus, Ayat No. – 4]

Ayat\Verses - Say: Of your **'partners'**, can any **originate** creation and **repeat** it? Say: It is Allah Who **originates creation and repeats it**: how then are you deluded away / **misled** [from the **truth**]? Say: Of your **'partners'** is there any that can give any **guidance towards truth**? Say: It is Allah Who **gives** guidance towards truth, is then He Who gives guidance to truth more **worthy** to be followed, or he who **finds not guidance** [himself] **unless** he is guided? What then is the **matter** with you? **How judge you?** But **most of them follow nothing but fancy: truly fancy can be of no avail against truth**. Verily Allah is well **aware** of all that **they do**. [Sura No. – 9\10 – Yuunus, Ayat No. – 34 to 36]

Ayat\Verses – [An example] the **similitude of one** who passed by a **township [Ground Stair]** which had fallen into **utter ruin**, exclaimed: Oh! How shall Allah give this **township life** after **its death**? But Allah **caused him to die** for a **hundred years**, and then raised him up [again]. He said: **How long did you tarry?** He said: **I have tarried a day or part of a day.** (He) said: Nay, you have tarried for **a hundred years**. But **look at** your **food** and your **drink; they show no signs of age;** and **look at** your **donkey**. And that We may make of you **a sign of proof** unto the **people. Look** further at the **bones, how We bring them together** and **clothe them with flesh**. When this was shown clearly to him, he said: **I know now that Allah is Able to do all things. [Sura No. – 1\2 – Baqarah, Ayat No. – 259]**

Ayat\Verses – When **Ibrahim** said (unto his Rab): My Rab! **Show me how You give life to the dead. He replied: Have you no faith? He said: Yes,** but **just** to **reassure** my **heart. Allah said: Take four birds,** draw them to you, and **cut** their bodies to pieces. **Scatter** them over the **mountain-tops**, and then **call them back. They will come swiftly to you. Know** that Allah is The Mighty, The Wise. The **parable (likeness / similitude)** of those who **spend** their wealth **in the way of Allah** is the likeness of a **grain** which growth **seven ears** and **each ear** has a **hundred** grains. Allah give **increase manifold** to which He **pleases**. Allah is All-Embracing, All-Knowing. **[Sura No. – 1\2 – Baqarah, Ayat No. – 260 & 261]**

Bish-Shamsi rises from the East

[Allah causes bish-shamsi and not Wash-Shamsi to rise from the East; so do you cause him to rise from the West?]

Ayat\Verses – **Bethink you of him who had an argument with Ibrahim about his Rab**, because **Allah had granted him the kingdom. How! When Ibrahim said: My Rab is He Who gives life and death. He (Namruud) answered: I give life and cause death. [**"Qaia Ibraahiimu fa-innallaha ya-tii bish-shamsi minal-maashriqi fa-ti bihaa minal magribi fa-buhital-lazi kafar"] Said Ibrahim: Allah will cause bish-shamsi (but not Wash-Shamsi) to rise from the east; so do you cause him to rise from the West? Thus was the disbeliever (who had rejected faith) confused and Allah guides not the wrong doing folk. [Sura No. – 1\2 – Baqarah, Ayat No. – 258]

Ayat\Verses – When **Ibrahim** said (unto his Rab): My Rab! **Show me how You give life to the dead. He replied: Have you no faith? He said: Yes,** but **just** to **reassure** my **heart. Allah said: Take four birds,** draw them to you, and **cut** their bodies to pieces. **Scatter** them over the **mountain-tops,** and then **call them back. They will come swiftly to you. Know** that Allah is The Mighty, The Wise. The **parable (likeness / similitude)** of those who **spend** their wealth **in the way of Allah** is the likeness of a **grain** which growth **seven ears** and **each ear** has a **hundred** grains. Allah give **increase manifold** to which He **pleases.** Allah is All-Embracing, All-Knowing. **[Sura No. – 1\2 – Baqarah, Ayat No. – 260 & 261]**

HOLLOW IN THE HAND

Ayat\Verses - When Taluuta set out with the armies (rays or comrades), she said: Allah will test you at the stream (river / sea). If any drinks of its water, he goes not with my army. Only those who taste not of it go with me except him who takes (thereof) in the **hollow of her hand.** But they all drank of it, except a few. When they crossed the river, she and the faithful ones with her, they said: This day we cannot cope with Goliath (*May be Jaaluut*) and his forces. But those who were convinced that they must meet Allah, said: How many of little company has overcome mighty armies by Allah's permission? Allah is with those who steadfastly persevere. When they advanced to meet Goliath and his forces, they prayed: Our Rab! Bestow on us endurance and make our steps firm. Help us against those who reject faith. [Sura No. – 1\2 – Baqarah, Ayat No. - 249 & 250]

NUUH'S ARK IS A SIGN – FLOOD ENGULFED THE WRONG DOERS AT THE CENTRE OF THE MANIFESTED EARTH – ESTABLIHED STORY OF TITANIC

[A thousand years save fifty years]

Ayat\Verses - And verily We sent **Nuuh** to his folk, and he tarried among them (continued with them) a **thousand years save fifty years,** and the **flood engulfed them,** for they were wrong doers. But We saved him and the companions of the Ark (ship), and We made the **[Ark / Ship] a Sign** for all peoples! [Sura No. – 28\29 – Ankabuut, Ayat No. – 14 & 15]

Centre of the Manifested Earth
REFERENCES OF THE CONCLUSIVE WORD OF THE QURAN

Ayat\Verses - By the firmament [**fermions**] **which returns** [in its **recycling region**], [Sura No. 85\86 – Wat-Taariq, Ayat No. – 11] And the world which **opens out [manifested]**. [Sura No. 85\86 – Wat-Taariq, Ayat No. – 12] Behold this is the **Conclusive Word of the Quran** [that will distinguish truth from falsehood]. [Sura No. 85\86 – Wat-Taariq, Ayat No. – 13]

INHALING COURSES
Ayat\Verses - By the **inhaling** [snorting] **courses** [equal & opposite stages of journey], and **striking sparks** of **fire [continual acceleration]**, and **push home the charge [electromagnetic wave]** in the **morning**, then, therewith, with their **trial of dust** cleaving [gravitational waves], as one, at the **centre [Nucleus or Dot Operator]**. [Sura No. – 99 \100 – Wal-Aadiyaat, Ayat No. – 1to 5]

REPRODUCTIVE SYSTEM
"**He knows all that goes into the world**, and all that **comes out** hereof"
Ayat\Verses - Praise be to Allah, to Whom belong all things in the **heavens** and in the world, to Him be Praise in the Hereafter, and He is Full of Wisdom, acquainted with all things. **He knows all that goes into the world**, and all that **comes out** hereof; all that **comes down** from the sky and all that **ascends** thereto and He is the Most Merciful, Ever Forgiving. [Sura No. – 33\34 – Saba, Ayat No. – 1 & 2]

MUUSA'S STAFF AND SNAKE
Ayat\Verses - And what is that in the **right hand**, O Muusa? He said: It is my **staff**, on it I lean; with it I beat down fodder for my flocks; and in it I find other uses. [Allah] said: Throw it, O Muusa! He threw it, and behold! It was a **snake**, **active in motion**. [Allah] said: Seize it, and fear not. We shall return it at once to its former condition. Now draw your hand close to your side. It shall come forth **white** [and shining], without harm [or stain], **as another Sign**. In order that We may show you our **Greater Signs**; go you to Firawn, for he has indeed transgressed all bounds. [Sura No. – 19\20 – Twa-Haa, Ayat No. – 17 to 23]

Ayat\Verses - Say: The truth is from your Rab. Let him who will **believe**, and let him who will **reject** [it]. For the **wrong-doers** We have prepared a **fire** whose like the walls and roof of a tent, will hem them in. If they implore relief they will be granted **water like melted brass** that will scald their faces, how dreadful the drink! How uncomfortable a couch to recline on! **As to those who believe and work righteousness, verily We shall not suffer to perish the reward of any who do a [single] righteous deed.** For them will be Gardens of Eternity or Eaden Gardens beneath them rivers will flow. They will be adorned therein with bracelets of **gold**, and they will wear green **garments** of fine silk and heavy **brocade**. They will recline therein on raised **thrones [Kaba]**. How good the recompense! How beautiful a **couch [heaven]** to recline on! Set forth to them the **parable of two men**: For one of them We provided two **gardens of grape-vines** and surrounded them with date palms [planetary barrier]; **in between the two We placed corn-fields. Each of those gardens brought forth its produce, and failed not in the least therein. In the midst of them We caused a river to flow.** [Sura No. – 17\18 – Khaf – Ayat No. – 29 to 33]

A SIMILITUDE OR A CONCRETE EXAMPLE
BURIED TREASURE
ZONE OF IRON – FLAMELESS FIRE – AT THE CENTRE

Ayat\Verses - Then they found **one of Our servants**, on whom We had bestowed Mercy from Ourselves and whom We had **taught knowledge** from Our own Presence. Muusa said to him: May I follow you, on the footing that you teach me something of the [Higher] Truth which you have been taught? [The other] said: Verily **you will not be able to have patience with me! And how can you have patience about things about which your understanding is not complete**? Muusa said: You will **find** me, if **Allah so will**, [truly] patient: nor shall I disobey you in aught. The other said: Well, if you go with me, **ask me no questions about anything until I myself speak to you concerning it**. [Sura No. – 17\18 – Khaf – Ayat No. – 65 to 70]

Ayat\Verses - So they both proceeded on till when they were in the **boat** [Similitude of Nuuh's Arak], **he made a hole therein**. Said Moses: **Have you made it in order to drown those in it**? You verily have done a **dreadful thing**. He answered: Did I **not tell** you that you could **not bear** with me? (Muusa) said: Be **not wroth** [archaic angry] with me that I forgot, and **be not** hard upon me for my **fault**. Then they proceeded on till, when they **met a young lad** [Isa-ibne-Maryam], he **slew him**. (Muusa) said: What! Have you **slain an innocent person** who had slain no man? Verily you have done a **horrid thing**. He answered: Did I **not tell you** that you could not bear with me? [Moses] said: **If ever I ask you** about anything after this, keep me not in your company. You have received [full] excuse from my side. Then they proceeded on till, when they came to the **inhabitants of a town**. They asked them for food, but they **refused them hospitality**. They found there a **wall** on the point of **falling down**, but he set it **upright**. [Moses] said: If you had wished, you could have exacted some **recompense** for it! [Sura No. – 17\18 – Khaf – Ayat No. – 71 to 77]

Ayat\Verses - He answered: This is the **parting between me and you**. Now I will tell you the **interpretation** of [those things] over which you was **unable** to hold patience. As for the **boat**, it belonged to **poor people** working on the river, and I wished to render it **unserviceable** for there was after them a certain king [Firawn or his similitude] who was **taking every boat by force**. As for the **youth, his parents were people of Faith**, and we feared that he would **grieve them by obstinate rebellion and ingratitude**. So we desired that their Rab would give them in **exchange** [a son] better in purity [of conduct] and **closer in affection**. [Sura No. – 17\18 – Khaf – Ayat No. – 78 to 81]

Ayat\Verses - As for the wall [Psychic or Planetary barrier], it belonged to **two youths orphans**, in the **Town**. There was, beneath it, a **buried treasure** [iron zone or flameless fire], to which they were entitled. Their father had been a **righteous man**. So your Rab desired that they should attain their **age of full strength** and get out **their treasure** [fictitious project of the Firawn's son as the so-called Sun instead of Imran's daughter supplied with Sustenance] - a mercy [and favour] from your Rab. **I did it not of my own accord**. Such is the **interpretation** of [those things] over which you was **unable to hold patience**. [Sura No. – 17\18 – Khaf – Ayat No. – 58 to 61]

CREATION AND EVOLUTION [REPRODUCTION]
Ayat\Verses - Has the parable of Muusa reached you? Behold, he saw a **fire**. So he said to his family: Tarry you; I perceive a fire; perhaps I can bring you some burning brand there from, or find some guidance at the fire. But when he came to the fire, **a voice was heard: O Muusa! Verily I am your Rab!** Therefore [in My presence] **put off your shoes**. You are in the **sacred valley of Tuwwaa**. I have chosen you. Listen, then, to the **inspiration** [sent to you]. Verily, I am Allah. There is no ilaaha (god) except Me. So serve you Me [only], and establish regular prayer for My remembrance. Verily the **Hour is coming**. But I will to keep it hidden, for every soul to receive its reward by the measure of its endeavour. Therefore let not he turn you aside from [the thought of] it who does not believe therein but follows his own desire, least you perish. [Sura No. – 19\20 – Twa-Haa, Ayat No. – 9 to 16]

THERE IS A TUNNEL [JUNCTION] AT THE CENTRE OF THE SEA
MIDDLE EAST
Ayat\Verses - But your Rab is Most forgiving, full of Mercy. If He were to call them [at once] to account for what they have earned, then surely He would have hastened their punishment; but they have **their appointed time**, beyond which they will find no refuge. Such were the **townships** [east zones] we destroyed when they **committed iniquities**; but we fixed an **appointed time** for their **destruction**. And when, Muusa said to his attendant: I will not give up until I reach **the junction of the two seas**, though I march on for ages. But **when they**

reached the Junction, they forgot [about] their Fish, which took its course through the sea [straight] as in a tunnel. [Sura No. – 17\18 – Khaf – Ayat No. – 58 to 61]

CREATION – FASHION – PROPORTION

Ayat/Verses – O man! What has made you careless concerning your Rab, the Most Beneficent / Bountiful [bi-Rabbikal-Karim]? Who **created you**, then **fashioned you**, then **proportioned you**. In whatever form He wills, He creates you. Nay! But they deny the judgment. [Sura No. – 81\82 - Izas-samaa-unfatarat, Ayat No. – 6 to 9]

Ayat\Verses – "Muttaki-iina-fiiha alal-araaa-ik, laa yarawna fiiha **shamsanw-wa** laa zam-hariiraa." [Trans.] **Reclining** therein upon **couches**, they will find there **neither** the heat of shams **nor** bitter cold [13]. And the **shade** thereof is close upon them and the clustered fruits thereof **bow down** [14]. And amongst them will be **passed round vessels of silver and goblets of crystal** [15], **Crystal-clear**, made of **silver**: they will determine the **measure** thereof [according to their wishes] [16]. And they will be given to drink there of a **Cup** [of Wine] mixed with Zanjabil [17], a **fountain** there, called Salsabil [18]. And **round** about them will [serve] youths of perpetual [freshness]. If you see them, you would think them scattered **Pearls** [19[]. And when you look, it is there you will see Bliss and a **Realm** Magnificent [20]. Upon them will be **green** Garments of fine **silk** and heavy **brocade**, and they will be adorned with **Bracelets** of silver; and their Rab will give to them to drink of a Wine **Pure** and **Holy** [21]. Verily this is a Reward for you, and your endeavour is accepted and recognised [22]. [Sura No. 75\76 – Hal ataa alal-Insaan, Ayat No. – 13 to 22]

WHERE IS THE APPOINTED KABA?
HIGHLIGHTS / CONCLUSIONS / SEARCHED OUT FINDINGS

i) There are two Zones and six Regions within the Black Square of East Horizon. Two Zones are West Zone [Upper Seashore] and East Zone [Lower Seashore].

ii) West Zone comprises three Regions in resemblance with 'therefore sign'. Europe represents Manifested South-West. Asia, Africa, and Australia represent Manifested North-West. Arabian Peninsula is neither towards right nor towards left of Revelation, Manifestation, and Appearance. Moreover, Arabian Peninsula is above Middle Stair or Nations. So, Arabian Peninsula is the Upright West Region within the Black Square of East Horizon.

iii) East Zone comprises three Regions in resemblance with the sign of 'since'. North America represents North-East Region. South America represents South-East Region. The Region of the East Zone which is neither towards North nor towards South is called Straight Middle-East Region. This Middle-East Region is the Recycling Region or Eartha 3D or Bermuda Triangle or Black Hole or Projected Planet Mars or Hole in the journey boat of Muusa (ass).

iv) There is an Equal & Opposite sth between Upright West region of Arabian Peninsula and Straight Middle East Region of Eartha 3D. Due to this equal & opposite sth, the manifested hexagonal world is appearing as a pentagonal earth within the Black Square of East Horizon.

v) There are three ascending stairs in the appearing pentagonal earth. The Black & White Imam of the City called Kabba as the leader of right direction for mankind has been appointed on the right side of the Mount Tuur in the Upright West Region of the manifested hexagonal world, topmost stair and uppermost land of the appearing pentagonal earth. In other words, Mount Tuur is holding firmly the appointed Kaba at its right side in the Upright West Region within the Black Square of East Horizon.

vi) Projection of Arabian Peninsula in the Middle East Region within the Black Square of East Horizon and appointed Kaba at the centre of Revelation, Manifestation, and Appearance are called Teleological Evidence Sorceries and Unique Epistemic Persecutions.

vii) Pen-Paper-Pencil works and activities which are either contrary or contradictory to manifested nature cannot be categorised as knowledge [Ref. Plato's Theory of Knowledge].

Have they **not observed all things that Allah has created,** how their **shadows incline to the right and to the left, making prostration unto Allah, and they are lowly?** [Sura No. – 15\16 – Nahl, Ayat\Verses – 48]

The projection of Arabian Peninsula in the Middle East Region within the Black Square of East Horizon is a Black Magic. The Historical Don of Teleological Evidence Sorcery and Unique Epistemic Persecution has concealed the Black Hole in the Straight Middle East Region within the Black Square and has projected the City of Black & White Imam called Arabian Peninsula in the Recycling Region of the East Zone called Middle East Triangle of Eartha 3D. This Black Magic is being shown with the help of mechanical trinity [mechanical globalisation, rotation & revolution of the immovable hexagonal world and appearing pentagonal earth, and man-made magnetism], and pen-paper-pencil works & activities of the Real Activists. This Black Magic has snatched away the equal & opposite right direction [middle course] of the followers of Upright West [Arsh]. Consequently, **believers of Upright West become followers of Black Magic**. In other words, **believers of middle course become followers of Recycling Region or Straight Middle East Triangle of Eartha 3D**. More specifically, the believers of Middle Stair become followers of Haiyalas-Swalaah or Northern Hemisphere in resemblance with the setting of the so-called sun for the West Zone and believers of Ground Stair become followers of Haiyalal-Falaah or Southern Hemisphere in resemblance with the setting of the so-called sun for the East Zone.

ARABIAN PENINSULA / MOUNT TUUR
&
MANIFESTED BLACK & WHITE IMAM OF THE CITY [KABA]

TIIN OR GROUND STAIR OF THE APPEARING PENTAGONAL EARTH
NORTH AMERICA OF NORTH-EAST [NE] AND SOUTH AMERICA OF SOUTH-EAST [SE]
Tribes, People of the Wood, Townships, East Zone, The Frigid Zones, Greenland, Tundra Region, Lower Stair, Ground Stair, Lower Sea-shore of North America and South America.

ZAYTUUN OR MIDDLE STAIR OF THE APPEARING PENTAGONAL EARTH
EUROPE OF SOUTH-WEST [SW] AND ASIA – AFRICA – AUSTRALIA OF NORTH-WEST [NW]
Nations, Ummat, Median, Middle Zone, The Temperate Zones, Middle Stair, Middle West, Upper Sea-shore of Europe, Asia, Africa, and Australia.

TUUR OR TOPMOST STAIR OF THE APEARING PENTAGONAL EARTH
ARABIAN PENINSULA OR THE UPRIGHT WESTERN CONTINENT
Arabian Peninsula, Banii-isra-iil, City, The Torrid Zone, the Zone of Hot Sands, Upper Most Stair, the Top West Region of the manifested Hexagon, the Uppermost Land of the appearing Pentagon where there is the appointed Kaba at the right side of the Mount Tuur

A few references from the Special Sheet of Epistemic Uniqueness

六 倍 嵐 凸 凹 㗊 㗊 㽬 㝵 宋 官 宣
宧 宙 山 圭 奎 圉 宁 亠 冫 亖 凷 屵

Unicode (hex) 516D, 500D, 21E4F, 51F8, 51F9, 55A6, F9ED, 21DB8, 519D, 5B8B, 5B98, 5BA3, 5BA6, 5B99, 5C71, 572D, 594E, 21201, 219C3, 3026, 3027, 3028, 21D93, 2067C

S (hex) / U (hex) - 0041, 0077, 007C, 00EC, 0075, 0051, 0027, 004A, 0051, 7388, 2250, 224E, 2234

Middle East is the Recycling Region within the Black Square or Straight Downward Triangle towards Space-bar [Dry Wood / Black Iron] of East

Horizon. It is called Middle as it is neither towards left [North-East] nor towards right [South-East], but the T-Junction or the Point of Intersection between Equator Line and Prime Meridian at the centre of East Zone [Lower Seashore]. **Eartha 3D**, Hole at the centre in the journey boat of Muusa [ass], Hollow in the hand, Triangular Nakta in the ideogram Nuun, the so-called Black Hole, the Reproductive Region from which Bish-Shamsi [projected/ man-made sun] rises, Recycling Region, Bermuda Triangle of Titanic, and the like represent Straight Middle East Region within the Black Square of East Horizon. Appearing gap at the centre of four converted shades of Windows'7 Ultimate represents Straight Middle East Region of Eartha 3D.

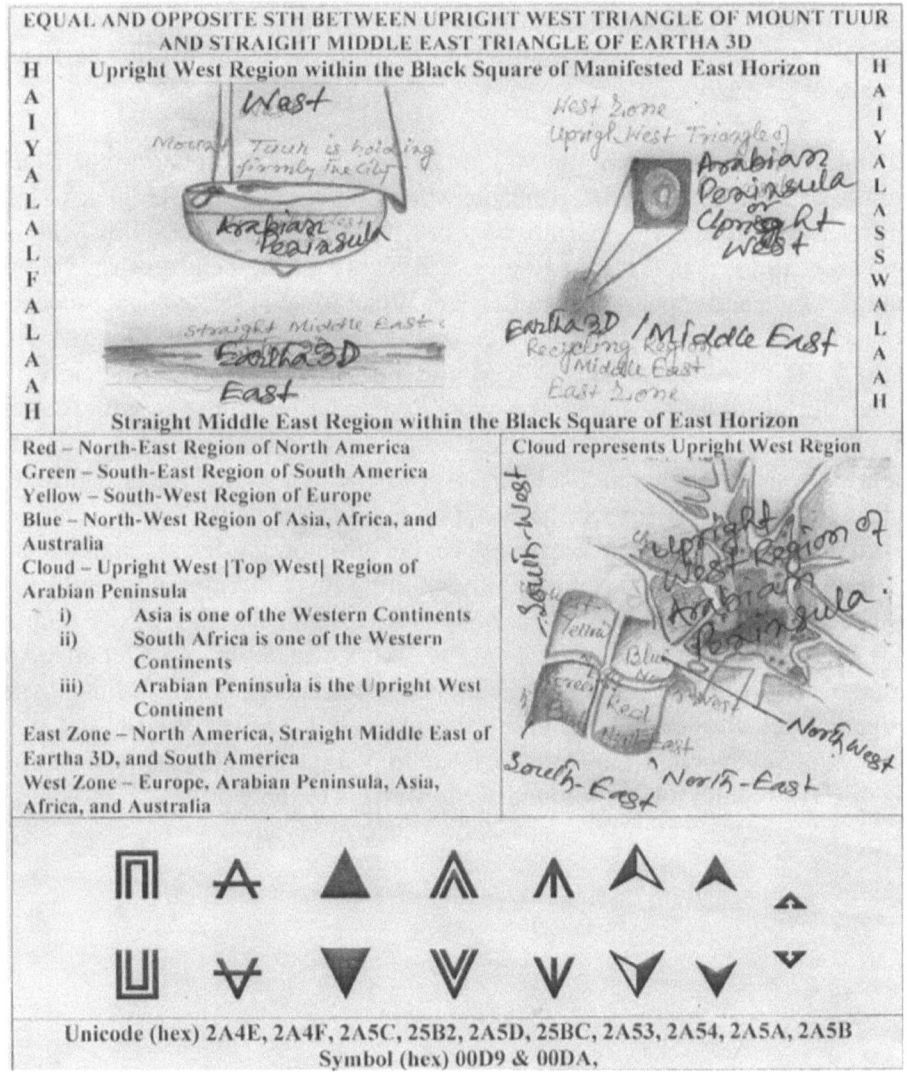

EQUAL AND OPPOSITE STH BETWEEN UPRIGHT WEST TRIANGLE OF MOUNT TUUR AND STRAIGHT MIDDLE EAST TRIANGLE OF EARTHA 3D

Upright West Region within the Black Square of Manifested East Horizon

Straight Middle East Region within the Black Square of East Horizon

Red – North-East Region of North America
Green – South-East Region of South America
Yellow – South-West Region of Europe
Blue – North-West Region of Asia, Africa, and Australia
Cloud – Upright West |Top West| Region of Arabian Peninsula
- i) Asia is one of the Western Continents
- ii) South Africa is one of the Western Continents
- iii) Arabian Peninsula is the Upright West Continent

East Zone – North America, Straight Middle East of Eartha 3D, and South America
West Zone – Europe, Arabian Peninsula, Asia, Africa, and Australia

Cloud represents Upright West Region

Unicode (hex) 2A4E, 2A4F, 2A5C, 25B2, 2A5D, 25BC, 2A53, 2A54, 2A5A, 2A5B
Symbol (hex) 00D9 & 00DA,

409

REVELATION AND MANIFESTATION
&
APPOINTED BLACK & WHITE LEADER OF RIGHT DIRECTION

Revelation is an Equal & Opposite Trinity [Binary ~sth or Intersection]. Revelation resembles with Plus Sign of Mathematics and Crucified Sign. Crucified Sign is the clear proof of the Diamond Operator with a Dot Operator or Helium-4 atom. Isa [ass] brought clear proof of Revelation. Four shades of Windows'7 Ultimate represent Diamond Operator.

Manifested hexagon is appearing as a pentagon due to equal and opposite ~sth between Arabian Peninsula [Upright West Region] and Eartha 3D [Straight Middle East Region] of the Manifested Hexagonal World within the Black Square of East Horizon.

Arabian Peninsula is not in the Straight Middle East Region within the Black Square of **Manifested East Horizon** where there is Eartha 3D [Black Hole or penetrated hole in the journey boat of Muusa (ass)]. Projection of Arabian Peninsula in the Middle East Region and Kabba at the centre of revelation, manifestation and appearance is a Black & White Magic [Teleological Evidence Sorcery and Unique Epistemic Persecution]. This **BLACK & WHITE MAGIC is the VERA CAUSA of ACTIVISM and TERRORISM**. Arabian Peninsula is the Upright West Triangle, Top West Region in resemblance with Region of White, Topmost Stair out of three ascending Stairs, Uppermost Land of the Manifested Hexagonal World and Appearing Pentagonal Earth within the Black Square of East Horizon. The appointed Kaba is neither at the centre of revelation nor at the centre of appearance. On the contrary, the appointed Kaba is on the right side of the Mount Tuur declining towards right in resemblance with multiplication sign and in correspondence with the sign of the mighty plot on the existing 9/11 Pentagonal Net Work caused by AK47 Laden. In resemblance with real science, Arabian Peninsula is the region of white dwarf which causes **clouds** to build up by the **afternoon [Juwal Waqt]**, and **rain** comes down at about **4 pm [Waqt of Asr to Magreeb]**. This happens almost **every day**, making the **region** one of the **wettest** in the world.

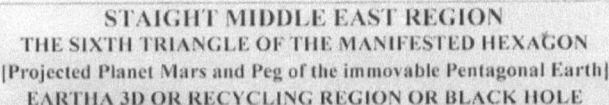

STAIGHT MIDDLE EAST REGION
THE SIXTH TRIANGLE OF THE MANIFESTED HEXAGON
[Projected Planet Mars and Peg of the immovable Pentagonal Earth]
EARTHA 3D OR RECYCLING REGION OR BLACK HOLE
Second Peg of the immovable Pentagonal Earth within the Black Square

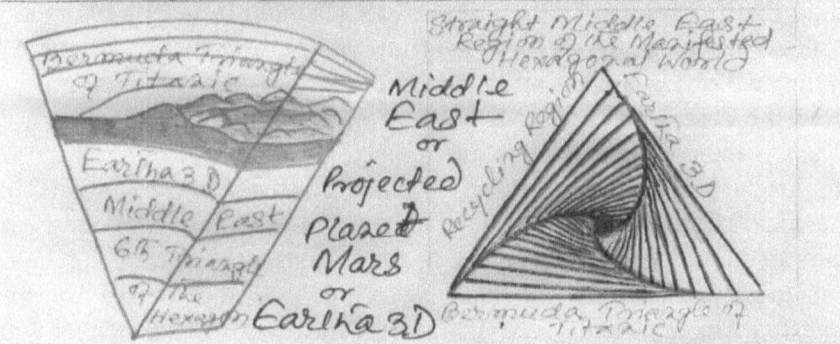

NO NAME [Excluded from Appearance]

Wife of Imran, Wife of Adam, Wife of Ibrahim, Wife of Luut, Wife of Firawn

The sixth triangle or Straight Middle East Region or Eartha 3D or Bermuda Triangle or Recycling Region of the manifested hexagon has been shared without name i.e. Straight Middle East Region has not been found in appearance. Towards Straight Middle East implies towards Black Hole or Bermuda Triangle of Titanic or Eartha 3D or Penetrated Hole in the journey boat of Muusa (ass) or Zone of Iron or Projected Planet Mars. This sixth triangle is the peg of the immovable Pentagonal Earth.

Upright ~sth [Prime Meridian] between Upright West Region [Top West Region / City of Black & White Imam] of Arabian Peninsula and Straight Middle East Region of Eartha 3D [Recycling Region / Black Hole] of the Manifested East Horizon within the Black Square
PRIME MERIDIAN
West within Manifested East Horizon [Black Square]
East within Manifested East Horizon [Black Square]

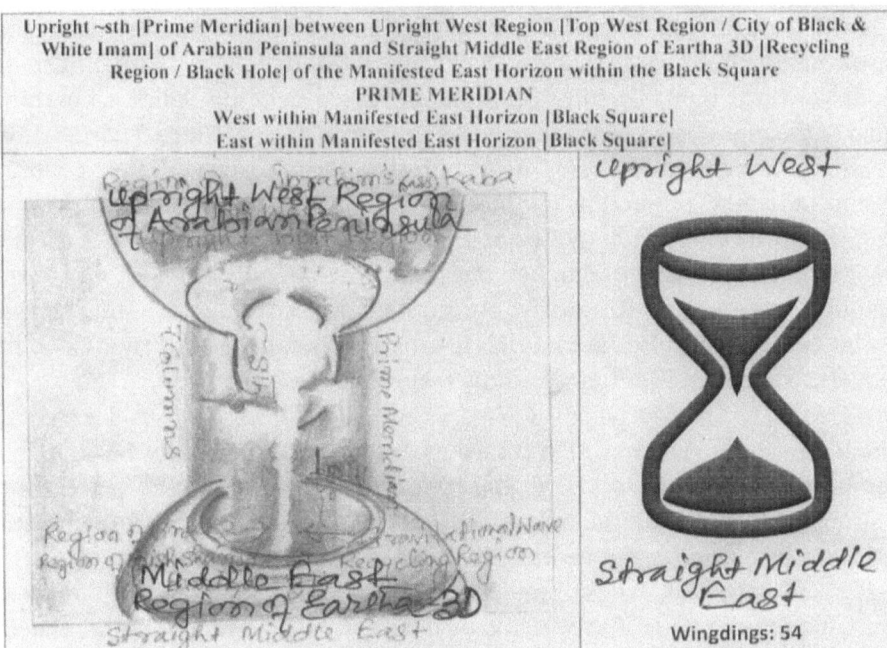

Wingdings: 54

411

Manifestation is declining towards right in resemblance with Multiplication Sign of Mathematics within the four basic forces of East Horizon. East Horizon has been manifested within four basic forces [four walls of scientific investigations] called Black Square. The Black & White Imam of the City called Ibrahim's Kaba as the Sole Leader [Standard] of Right Direction for mankind has been appointed on the right side of the Mount Tuur and. So, the appointed Kaba is neither at the centre of Revelation nor at the centre of the appearing Pentagonal Earth nor in the projected Middle East Region of Black Hole. On the contrary, the appointed Kaba is on the right side of the Mount Tuur, in the Upright West Region of the manifested hexagon, in the Top West Triangle of the manifested hexagon, in the Topmost Stair out of three ascending Stairs of the appearing pentagon, in the Uppermost Land of the appearing pentagonal earth within the Black Square of the Manifested East Horizon. Projected Arabian Peninsula in the Middle East corresponds neither with one of the Regions of East Zone nor with one of the Eastern Continents within the Black Square of the Manifested East Horizon. All manifested regions and appearing Continents are there within the Black Square of the Manifested East Horizon. So, projection of the City of Black & White Imam [Arabian Peninsula] in the Middle East Region within the Black Square of the Manifested East Horizon is a **Black & White Magic**.

City of the Black & White Imam [Arabian Peninsula] has been appointed as the Upright West Region, Topmost Triangle, Region of White in resemblance with the obversion of the converted Snake of Muusa's Staff, Muusa's Climbing of the Mount towards light, Topmost Stair out of three ascending Stairs, Uppermost Land of the manifested hexagonal world and appearing pentagonal earth within the Black Square of Revealed & Manifested East Horizon. Either contrary or contradictory projection of these manifested truths is a **Black Magic**. Each follower of Upright West has the **Solidarity Rights** to get confirmed recognition of the above searched out Manifested Truths from the Epistemic Uniqueness as well as Respective Authorities. Moreover, each human person has the **Solidified Solid Human Rights** to get generalised confirmation of the above searched out Manifested truths from IHRC.

Black Magic: - Arabian Peninsula has been projected as the Middle East Region of the Manifested Hexagonal World within the Black Square [Non-luminous Moon / Dark Moon] of East Horizon. Further, Arabian Peninsula has been projected at the centre of Global Trinity through pen-paper-pencil works and activities. These projections are Teleological Evidence Sorceries and Unique Epistemic Persecutions of faith and belief of mankind in general and particularly faith & belief of the followers of equal & opposite middle course towards Upright West [Arsh / Throne of Authority]. These Teleological Evidence Sorceries and Unique Epistemic Persecutions are called Black Magic. Due to this Black Magic, the followers of Upright West become the followers of Recycling Region or Middle East Region of Eartha 3D.

Manifested Truths in resemblance with Revelations: - Arabian Peninsula is the Upright West Region of the Manifested Hexagonal World; and Topmost Stair out of three ascending stairs; and Uppermost Land of the appearing Pentagonal Earth.

Middle East Region of Eartha 3D: - Straight Middle East Region of Eartha 3D or the Sixth Triangle of the Manifested Hexagonal World within the Black Square of East Horizon is the Recycling region or Black Hole in resemblance with the penetrated hole in the journey boat of Mussa (ass). This Straight Middle East Triangle of Eartha 3D is the peg of the immovable Pentagonal Earth. Due to Middle East Region of Eartha 3D within the Black Square of East Horizon, the appearing Pentagonal Earth is not moving. Planet/Moon Mars represent Straight Middle East Region of Eartha 3D.

Arabian Peninsula has not been appointed in the East Zone. On the contrary, Arabian Peninsula has been appointed above Middle Stair [Nations] of the West Zone. Projection of Arabian Peninsula in the Middle East Region within the Black Square of East Horizon through Mechanical Globalisation and man-made magnetism is Teleological. This Teleological Projection is leading believers towards the Black Hole of Flameless Fire. West Zone within the Black Square of East Horizon comprises three Regions. The appearing Pentagonal Earth has three ascending stairs. Arabian Peninsula is neither towards South-West Region [Revealed Right or Southern Hemisphere], nor towards North-West Region [Revealed Left or Northern Hemisphere] of the Manifested West Zone. So, Arabian Peninsula is the **Upright West Region** of the Manifested Hexagonal World, **Topmost Stair** out of three ascending stairs, and **Uppermost Land** above Nations [Middle Stair] of the Appearing Pentagonal Earth.

It is being taught by the people of the Kitaab with Truth as well as people of the book that the Black & White Imam of the City called Kaba has been appointed at the centre of Revelation, Manifestation, and in the Middle East of the appearing Earth. I have failed to search out a valid reference in resemblance with their teachings save hearsay. On the contrary, I have searched out verifiable as well as justifiable manifested truth in resemblance with revelation. These searched out manifested truths are both contrary as well as contradictory to Black Magic [Real Activism].

The appointed Kaba is neither in the East Zone [Lower Seashore] of the East Horizon, nor in the Middle of the East Zone [Lower Seashore], nor at the centre of Revelation & Appearance. On the contrary, **Kaba has been appointed at the right side of the Mount Tuur in the Upright West Region of the Manifested Hexagonal World, in the Topmost Stair out of three ascending stairs, and in the Uppermost Land of the Immovable Hexagonal World and Appearing Pentagonal Earth.**

References from Kitaab with Truth as Universal Major Premises
Teleological Placement
Banii-Isra-iil or People of the City are above Nations [Middle Stair]
Ayat\Verses – Neither Heaven [West Horizon] nor world [East Horizon] shed a tear over them; nor were they given a respite [again]. And We **delivered Banii-Isra-iil as above the nations [Middle Stair]**, from humiliating Punishment (shameful doom). (We delivered them) from Firawn. Lo! He was a tyrant of the inordinate transgressors. **And We chose them purposely above the nations. And We gave them Signs (Manifested Kaaba) in which there was a manifest (clear) trial.** [Sura No. – 43\44 - Bi-Dukhaanim-Mubiin, Ayat No. – 29 to 33]

Ayat\Verses – **Allah has appointed (manifested) the Kaba, the Sacred House, as a standard [criterion] for mankind**, and the **Sacred Months**, the **offerings**, and the **garland** (Festoon): That you may know that Allah has knowledge of what is in the heavens and in the world and that Allah is well acquainted with all things. [Sura No. – 4\5 – Maaaidah, Ayat\Verses – 97]

JUST ENDS
Ayat\Verses – We created not the heavens, the world, and all that is between them, merely in [idle] sport. **We created them not except for just ends (with Truth).** But most of them do not understand. Verily the Day of Decision (sorting out) is the time appointed for all of them. The Day when no protector can avail his client in aught, and no help can they receive, except him who will receive Allah's Mercy; for He is Exalted in Might, Most Merciful. [Sura No. – 43\44 - Bi-Dukhaanim-Mubiin, Ayat No.38 to 42]

APPOINTED SECURED SANCTUARY

"Lo! Allah is with those who do good (right) deeds."
    i)   Have they not seen that We have appointed a secure sanctuary (immune from violence), while mankind are devastated (snatched away) all around them?

ii) Do then they believe in falsehood, and disbelieve in the Bounty of Allah?

iii) And **who does greater wrong than he who invents a falsehood (lie) concerning Allah or rejects / denies the Truth when it comes to him?**

iv) Is not there a home in Hell for disbelievers (those who reject Faith)?

v)   And for those who serve for Us (strive in Our cause], We will certainly guide them to Our Paths.

Ayat\Verses - Have they not seen that We have appointed a secure sanctuary (immune from violence), while mankind are devastated (snatched away) all around them? **Do then they believe in falsehood, and disbelieve in the Bounty of Allah?** And who does greater wrong than he who invents a falsehood (lie) concerning Allah or rejects / denies the Truth when it comes to him? **Is not there a home in Hell for disbelievers (those who reject Faith)?** And for those who serve for Us (strive in Our cause], We will certainly guide them to Our Paths. Lo! Allah is with those who do good (right) deeds. (IFTA) [Sura No. – 28\29 – Ankabuut, Ayat No. – 67 to 69]

INVIOLABLE PLACE OF WORSHIP

Ayat\Verses – Glorified be He Who created His servant by night **from the Inviolable Place of Worship to the far distant place of worship the neighbourhood whereof We have blessed that We might show him of (Our Signs)!** Lo! He, only He, the Hearer, the Seer. We gave unto Moses the Kitaab, and We appointed it guidance to the Banii-Israai-iil [commanding]: **Take not other than Me as disposer of [your] affairs.** (They were) the seed of those whom We carried [in the Ark] along with Nuuh! Verily he was a most grateful devotee. And We gave [Clear] warning to the Banii-Israa-iil in the Kitaab, that **twice** they would do mischief in the world and you will become **great tyrants.** [Sura No. – 16\17 - Banii-Israai-iil, Ayat No. – 1 to 4]

PLACES OF WORSHIP ARE ONLY FOR ALLAH. SO INVOKE NOT FALSEHOOD ALONG WITH ALLAH

Ayat\Verses - **And the places of worship are only for Allah. So invoke not any one [projected falsehood] along with Allah.** And when the slave of Allah stands up in prayer to Him, they crowded on him, almost stifling. Say: **I pray to Allah only, and ascribe to Him no partner.** Say: Lo! It is not in my power to cause you harm, or to bring any benefit for you. Say: Lo! None can protect me from Allah, nor can I find any safe haven except Him. (Mine is) but conveyance (of the truth) from Allah and His messages; and who so **disobeys** Allah for them is **hell**, they shall dwell therein forever. At length, when they see [with their own eyes] that which they are promised, then they will know **who it is that is weakest in [his] helper and least important in point of numbers.** **[Sura No. – 71\72 – Nafarum-minal – Jinni, Ayat No. – 18 to 24]**

MOSQUES OR MASJIDS
OFF-SPRINGS OF NUUH AND IBRAHIM

Ayat\Verses – And We have sent **Nouuh** and **Ibrahim**, and have **placed Prophetship and Revelations among their off-springs (seeds)** and among them **there is he who goes right**. But many of them become **rebellious transgressors (evil-livers).**

KABA - THE EXISTENCE OF THE IMAM AS THE
STANDARD OF RIGHT DIRECTIONS FOR MANKIND
WE ARE TO CONFIRM WHAT HAS BEEN REVEALED

i) And when there comes to them that which they know (to be the truth), they refuse to believe in it but the curse of Allah is on those who disbelieve.

ii) Evil is that for which they sell their souls that they deny [Manifested Signs of Revealed Magnetism] which Allah has sent down, in cheeky jealousy that Allah of His Grace should send it to any of His servants He pleases.

iii) Thus they have drawn on themselves wrath upon wrath. And humiliating is the punishment of those who reject Faith.

iv) When it is said to them, believe in what Allah has sent down. They say: We believe in what was sent down to us [by IFTA]. Yet they reject all besides, even if it be truth confirming what is with them.

v) Say: Why then have you slain the prophets of Allah in times gone by, if you did indeed believe?

Ayat\Verses - They say: Our hearts are hardened. Nay, but Allah has **cursed them** for their unbelief. **Little is that which they believe**. And when there comes to them a **Kitab** from Allah, **confirming what is with them though before that they were asking for a signal triumph over those who disbelieved.** And when there comes to them that **which they know** (to be the truth), **they refuse to believe** in it but the **curse** of Allah is on those **who disbelieve.** Evil is that for which **they sell their souls** that they **deny** [the revelation] which Allah has sent down, in **cheeky jealousy** that Allah of His Grace should send it to any of His servants He pleases. Thus they have drawn on themselves **wrath upon wrath**. And **humiliating** is **the punishment of those who reject Faith**. When it is said to them, believe in **what Allah has sent down. They say: We believe in what was sent down to us.** Yet they reject all besides, **even if it be truth** confirming what is with them. Say: **Why then have you slain the prophets of Allah in times gone by, if you did indeed believe?** [Sura No. 1\2 – Baqarah, Ayat No. – 88 to 91]

GREATER SIGNS

Ayat\Verses - In order that We may show you our **Greater Signs**; go you to Firawn, for he has indeed transgressed all bounds. [Moses] said: O my Rab! Relieve my mind and ease my task for me, and remove the impediment from my speech so that they may understand what I say. And appoint for me a henchman [trusted supporter] from my folk Haarun, my brother. Confirm my strength with him and make him share my task so that we may glorify Your praise without stint and remember You without stint. For You are ever seeing us. [Sura No. – 19\20 – Twa-Haa, Ayat No. – 23 to 35]

MUUSA CAME TO US WITH CLEAR PROOFS
THE MOUNT TUUR IS CAUSED TO HOLD
FIRMLY THE GIVEN KABA

Ayat\verses - There came to you **Muusa** with **Clear Proofs**; yet while he was away, you chose the **calf [projected falsehood for worship), and you were wrong-doers.** And when We took your **promise** and **caused the Tuur to tower above you** (saying): **Hold firmly to what We have given you**, and hear (our words,). They said: **We hear, and we disobey.** And (worship of) **the calf was made to sink into their hearts** because of their **rejection** (of the promise). Say (unto them): **Evil is that** which your **belief enjoins** on you, if you are believers. [Sura No. 1\2 – Baqarah, Ayat No. – 92 & 93]

TWO PROOFS OF EQUAL & OPPOSITE MAGNETISM

Now when Muusa had **fulfilled the term**, and was **travelling** with his family, he perceived **a fire [light] in the direction of Mount Tuur.** He said to his family: **Tarry** you; I **perceive** a fire [light]; I hope to **bring you** from there some **information** (tidings), or a brand from the fire that you may **warm yourselves.** And when he reached it, he was called **from the right side of the valley in the blessed field, from the tree**: O Muusa! Lo! I, even I, am Allah, the Rab of the Universe. **Throw down your staff.** And **when he saw it moving as it had been a demon (mischievous sprite), he turned to flee headlong** (and it was said unto him): **Draw near**, and **fear not.** Lo! **You are of those who are secure. Thrust your hand into the bosom of your robe, it will come forth white without hurt. And guard your heart from fear. These are the two proofs from your Rab to Firawn and his chiefs, for truly they are a people rebellious and wicked.** [Sura No. – 27\28 – Qasas, Ayat\Verses – 29 to 32]

KABA

Ayat\Verses - And what is that in the **right hand**, O Muusa? He said: It is my **staff**, on it I lean; with it I beat down fodder for my flocks; and in it I find other uses. [Allah] said: Throw it, O Muusa! He threw it, and behold! It was a **snake**, **active in motion**. [Allah] said: Seize it, and fear not. We shall return it at once to its former condition. Now draw your hand close to your side. It shall come forth **white** [and shining], without harm [or stain], **as another Sign**. In order that We may show you our **Greater Signs**; go you to Firawn, for he has indeed transgressed all bounds. [Sura No. – 19\20 – Twa-Haa, Ayat No. – 17 to 23]

Ayat\Verses - Truly strong is the grip [punishment] of your Rab [12]. It is He Who produces and then reproduces [13]. And He is Ever-Forgiving, Full of Loving-Kindness [14], Rab of the **Throne of Glory [15]**, doer of what He wills [16]. Has there come to you the story of the forces [hosts] [17] of Firawn and the Samud [18]? And **yet the disbelievers [persist] in rejecting [the Truth] [19]**! But Allah encompasses [surrounds] them from unseen [20]! Nay, but it is a Glorious Qur'an [21], on a grounded tablet. [Sura No. – 84\85 – Was-Samaaa-i Zaatil-Buruuj, Ayat No. – 12 to 22]

Ayat\Verses - Do you not consider how your Rab dealt with [the tribe of] Aad [6] with many columned Iram [may be beams of light] [7] the similitude of which were not produced in the land [earth], [8] and with [the tribe of] Samud, who clove the rocks in the valley [9], and with Firawn, firm of might, [10] who [all] were rebellious [to Allah] in these lands [11] and heaped therein mischief [on mischief] [12]. Therefore your Rab poured on them the disaster of His punishment [13]; for your Rab is on a **watch-tower** [Sura No. – 88\89 – Wal-fajri, Ayat No. – 6 to 14]

Ayat\Verses - [Other] faces **that Day** will be joyful [8], pleased with their striving [9], in a high garden [10], where they shall hear no idle speech [11]. Therein will be a bubbling [sparkling] spring [12]. Therein will be **Throne, raised on high [13]**, and **glasses** set at hand [14], and **cushions** set in rows [15], and silken carpets spread out. [Sura No. – 87 \ 88 – Hadiisul-Gaashiya, Ayat No. –8 to 16]

Ayat\Verses - Nay, but the record of the Righteous is [preserved] in **Illiy-yuun**. And what will explain to you what **Illiy-yuun** is? [There is] a Register [kitaabum-marquum] [fully] inscribed, attested by those who are brought near [unto their Rab]. Truly the Righteous will be in delight. They will command a sight on **Throne** [of Dignity]. You will recognise in their faces the radiance of delight. They are given to drink of a pure wine, sealed. The seal thereof will be Musk; and for this let [all] those struggle, who struggle for delight [bliss]. With it will be [given] a mixture of Tasniim. A spring [planet], from [the waters] whereof drink those nearest to Allah. [Sura No. – 82\83 - Waylul-lil-mutaffifiin, Ayat No. – 18 to 28]

Ayat\Verses - Lo! The guilty used to laugh at those who believed, and whenever they passed by them, used to wink at each other [in mockery]; and when they returned to their own people, they would return jesting; and whenever they saw them, they would say: Behold! **These are the people truly astray! Yet they were not sent as guardians over them. This day the believers will laugh at the disbelievers. They will command a sight on Throne [of Dignity].** Will not the disbelievers have been paid back for what they do? [Sura No. – 82\83 - Waylul-lil-mutaffifiin, Ayat No. – 29 to 36]

THE UPPERMOST IN THE LAND [EARTH]
Ayat\Verses - O my people! Yours is the Kingdom (dominion) today. **You are the uppermost in the land [earth].** But who will help us from the Punishment of Allah, should it befall us? Firawn said: I do but **show you what I think**, and I do but **guide you to wise policy**. [Sura No. – 39\40 – Mu-Min, Ayat No. –29]

UPPERMOST
Ayat\Verses - O you who believe! Be **Allah's helpers** even as Iisabnu Maryam said to the Disciples: Who are my helpers for Allah? Said the disciples: We are Allah's helpers! And **a portion of the Baniii-Israa-iil believed**, while **a portion disbelieved.** Then We strengthened those who believed, against their enemies, and they became the **uppermost. [Sura No. 60\61 – Saff, Ayat No. -14]**

THE APPOINTED KABA IS
AT THE RIGHTSIDE OF THE MOUNT TUUR
"O Bani-Israaiil! We delivered you from your enemy, and We made a covenant with you on the right side of Mount Tuur"
Ayat\Verses – O Bani-Israaiil! We delivered you from your enemy, and We made a covenant with you on the **right side** of **Mount Tuur** [Kaba is appointed at the right side of the Mount Tuur] and We sent down to you **Manna and Salwaa** (alaykumul-Manna was-Salwaa). [Saying]: Eat of the **good things** We have provided for **your sustenance**, but **commit no excess therein**, lest My wrath should **justly descend** on you, and **those on whom descends My wrath do perish indeed! But, without doubt, I am [also] He Who forgives again and again, to those who repent, believe, and do right; who, in fine, are ready to receive true guidance**. [Sura No. – 19\20 – Twa-Haa, Ayat No. – 80 to 82]

419

KABA
&
Mount Alal-Ju-diyyi

"The **Ark rested on Mount Alal-Ju-diyyi**" – **This sharing is a sufficient proof for the believers that the appointed Kaba is at the top most land of the manifested earth.**

Ayat\Verses – So the **Ark floated with them on the waves [binary pulsar] like mountains,** and Nuhh cried out unto his son – and he was standing aloof - **O my son! Come with us, and be not with the unbelievers.** The son replied: I will betake myself to some mountain [for instance a fictitious planet called Mars], it will save me from the water. Noah said: this day nothing can save, from the command of Allah, save him whom He has had mercy! And the waves came between them, and the son was among drowned in the Flood. **Then the word went forth: O world! Swallow up your water** and **O sky! Be cleared of clouds** and **the water was made to subside.** And the Commandment was fulfilled. The **Ark rested on Mount Alal-Ju-diyyi,** and it was said: **Away with those who do wrong.** [Sura No. – 10\11 – Kitaabun-uh-kimat (Prev. – Huud), Ayat No. – 42 to 44]

KABA
MOUNT TUUR IS CAUSED TO TOWER ABOVE US

Ayat\Verses - Those who **believe** [in that which revealed unto you], and **those who are H a a d u u (Yahuudis), Nasaaras and Sabii-iians,** any who **believe** in **Allah** and the **Last Day,** and **work righteousness,** shall have their **reward** with their **Rab.** On them shall be **no fear, nor** shall they **grieve.** And remember We took your **covenant** (promise) and **caused the Mount Tuur to tower above you** [Saying]: **Hold firmly to what We have given you** and remember that **which is there** in **that** you may **ward off (evil).** Then even after that **you turned away,** and if it had **not** been for **the grace of Allah** and **His Mercy,** you had been **among the losers.** And you know of those of you who **transgressed in the matter of the Sabti.** We said to them: **Be you apes, despised and rejected. [Sura No. – 1\2 – Baqarah, Ayat No. – 62 to 65]**

You [Prophet] were not on the Western side when **We expounded unto Muusa the commandment,** and **you were not among those present.** But We **brought forth generations,** and their lives dragged on for them. And you were not a dweller among **the people of Madyan,** rehearsing **Our Signs** (Manifested Nature in resemblance with Revelations) to them; but We kept sending [messengers to men]. **And you were not beside the Mount Tuur when We called, but (the knowledge of it is) a mercy from your Rab that you may warn a folk unto whom no warner came before you in order that they may receive admonition.** [Sura No. – 27\28 – Qasas, Ayat\Verses No. – 44 to 46]

Ayat\Verses - Verily your Rab is Allah, who created the heavens and the world in **six** days, and is **firmly established Himself upon the Throne [Arsh], regulating and governing all things**. No intercessor [can plead with Him] except His permission. This is Allah your Rab; therefore serve Him. Oh! Will you not remind? [Sura No. – 9\10 – Yuunus, Ayat No. – 3]

OUR RAB MANIFESTED HIS GLORY ON THE MOUNT

Ayat\Verses – When Muusa came to the place appointed by Us, and his Rab addressed him, He said: O my Rab! Show [Yourself] to me, that I may look upon You. Allah said: **By no means you can see Me [direct]; But look upon the mount; if it abide in its place**, then you shall see Me. **When his Rab manifested His glory on the Mount, He made it as dust**. And Muusa fell down in a swoon. When he recovered his senses he said: Glory is to You! To You **I turn in repentance, and I am the first to believe.** [Sura No. – 6\7 - As-haabul- a' raaf, Ayat No. – 143]

Also mention in the Kitaab [the story of] Muusa. Lo! He was specially chosen, and he was a messenger [and] a prophet. And we called him from the **right angle of Mount Tuur** and brought him nigh in communion. And, out of Our Mercy, We gave him his brother **Haarun**, [also] a prophet. Also mention in the Kitaab [the story of] **Ismaiil**. He was [strictly] true to what he promised, and he was a messenger [and] a prophet. He used to enjoin on his people Prayer and Charity, and he was most acceptable in the sight of his Rab. Also mention in the Kitaab the case of **Idriis**. He was a man of truth [and sincerity], [and] a prophet: And We raised him to a **lofty station**. Those were some of the prophets on whom Allah showed His Grace of the posterity of Adam, and of those whom We carried [in the Ark] with **Nuuh**, and of the posterity of **Ibrahim** and **Israaiil,** and from among those whom We guided and chose. Whenever the **Signs** of [Allah] Most Gracious were rehearsed to them, they would fall down in prostrate adoration and in tears. [Sura No. – 18\19 – Maryam, Ayat\ Verses – 51 to 58]

He will say: What number of years did you stay in the world? They will say: **We stayed a day or part of a day**; but ask those who keep account. He will say: You stayed not but a little, if you only know! Did you then think that We had created you in jest, and that you would not be brought back to Us [for account]? Therefore exalted be Allah, The True King, The Absolute Reality. There is no ilaaha (reality) but He The Rab of the **Throne of Honour [Arsh]**! If **anyone invokes, besides Allah, any other reality**, he has no authority (Proof) thereof; and his reckoning will be only with his Rab! And verily the disbelievers will not be successful. So say: My Rab! Grant You forgiveness and mercy for You are the Best of those who show mercy! [Sura No. – 22\23 – Mu-Minuun, Ayat\ Verses – 112 to 118]

BEAUTIFUL DWELLING-PLACE AND
SUSTENANCE OF THE BEST

"So, be not you of the disbelievers. Neither be of those who reject the signs of Allah nor be of those who will perish."

Ayat\Verses - We settled the Bani-Israa-iil in a **beautiful dwelling-place**, and **provided** for them **sustenance of the best**. It was after knowledge had been granted to them, that they fell into ruptures [breaks]. Verily Allah will judge between them as to the ruptures amongst them, on the **Day of Resurrection** [Yawmal-Qiyaamati]. If you are in **doubt** concerning that which We reveal unto you, then question those **who read the Kitab before you**. Verily the **Truth** from your Rab has come unto you. **So, be not you of the disbelievers. Neither be of those who reject the signs of Allah nor be of those who will perish.** Those against whom the **word** of your Rab has been verified would **not believe**. Even if every **Sign** was brought unto them, till they see [for themselves] the **penalty grievous**. [Sura No. – 9\10 – Yuunus, Ayat No. – 93 to 97]

O **Banii-Isra-iil!** Remember My favour which I bestowed upon you, and that **I preferred you to all other.** Then **guard yourselves against a Day** when **one soul shall not avail another nor** shall **intercession** be accepted from it, **nor** shall **compensation** be taken from it, **nor** shall anyone be **helped**. And remember, **We delivered you from the people of Firawn.** They set you hard tasks and punishments, **slaughtered your sons and let your women-folk** live; therein was a **tremendous trial** from your Rab. **And remember, We divided the sea for you and saved you and drowned Firawn's people within your very sight. (IFTA). [Sura – ½ - Baqarah, Ayat No. – 47 to 50]**

And (remember) when **We took a promise from the Banii-Israa-iil** (saying): Serve none but Allah; treat with kindness your parents and family members, and orphans and those in need; speak fair to the people; be steadfast in prayer; and pay the poor-due. Then **you turned back, except a few among you,** and **you slip back [even now].** And remember We took your promise [saying]: Shed no blood amongst you, **nor turn out your own people from your homes,** and this you **solemnly approved,** and to this **you were witnesses** (there to). After this it is you, **the same people, who slay among yourselves,** and **drive out a party of you from their homes; supporting one another against them by sin and ill will**; and if they come to you as **captives,** you **ransom** them, where as **their expulsion was itself unlawful for you. This is only because you believe a part of the Kitaab and you reject the rest?** And what is the reward for those among you who behave like this but **disgrace** in this life and on the **Day of Judgment they will be consigned to the most grievous penalty**. For Allah is not unmindful of what you do. **[Sura – ½ - Baqarah, Ayat No. – 483 to 85]**

We **took the Baniii - Israaa -il [with safety] across the sea. They came upon a people who were given up to idol [calf] which they had.** They said: O Muusa! Fashion for us a god like unto the gods they have. He said: Surely you are a people without knowledge. Lo! As for these, their way will be destroyed and all that they are doing is in vain. **He said: Shall I seek for you a god other than the [true] Allah, when it is Allah Who hath endowed you with gifts above the nations?** And remember **We rescued you from Ferawn's people**, who afflicted you with the worst of penalties, **who slew your male children and saved alive your females: in that was a momentous trial from your Rab**. [Sura No. – 6\7 - As-haabul- a' raaf, Ayat No. – 138 to 141]

SECTION – VI
NATURAL MAGNETISM – 6.1
PROJECTED FALSEHOOD

INTRODUCED CALF OF FIRAWN

Fasta-'iz billaahi minash-Shaytaanir-Rajim
Bismillaahir-Rahmaanir- Rahim

The power [hands] of the father of flame [projected falsehood] will perish and he [the partner] will perish. No profit to him from all his wealth, and all his gains! He will soon be burnt in flaming fire! And his wife [activity], the wood carrier as fuel [Eartha 3D or projected planet Mars], a twisted rope [equal & opposite ~sth] of palm-leaf fibre round her [own] neck! [Sura No. – 110\111 – Abii lahabinw-wa tabb, Ayat – 1 to 5]

Say: **Call upon those whom you set up besides Allah**. They have no power, not the weight of an atom in the heavens or in the world. No [sort of] share have they therein, nor is any of them a helper to Allah. [Sura No. – 33\34 – Saba, Ayat No. - 22]

"Wa innahuu la-zikrul-laka wa li-qawmik, wa sawfa tusa'aluun" [Sura No. 42\43 – Ummil-Kitaab \ (Zukhruf), Ayat No. – 44] [Trans.] And lo! It is in truth a reminder for those and for your folk; and you will be questioned. [Sura No. 42\43 – Ummil-Kitaab \ (Zukhruf), Ayat No. – 44]

"Was-al man arsa 'arsalnaa min-qablika mir-rusulinaaa 'aja-alnaa min-duunir-Rahmaani-aalihatany-yu-baduun" [Sura No. 42\43 – Ummil-Kitaab \ (Zukhruf), Ayat No. – 45] [Trans.] And ask those of Our messengers whom We sent before you; did We appoint any falsehood (manmade magnetism) to be worshipped instead of the Beneficent? [Sura No. 42\43 – Ummil-Kitaab \ (Zukhruf), Ayat No. – 45]

A SIGN OF PROOF - DONKEY

[An example or] the similitude of one who passed by a township [ground stair] which had fallen into utter ruin, exclaimed: Oh! How shall Allah give this township [America of Columbus] life after its death? But Allah caused him to die for a **hundred years**, and then raised him up [again]. He said: How long did you tarry? He said: I have tarried a **day or part of a day**. (He) said: Nay, you have tarried for a hundred years. But look at your food and your drink; they show no signs of age; and look at your **donkey**. And that We may make of you a **sign** of proof unto the people. Look further at the **bones** [Napier Bones

424

or Bones of Firawn], how We bring them together and clothe them with flesh. When this was shown clearly to him, he said: I know now that Allah is Able to do all things. [Sura No. – 1\2 – Baqarah, Ayat No. – 259]

MISCHIEVOUS REFERENCES USED BY THE GUILTY FOLK
They [false-hood mongers] said: Even so has said your Rab; and He is full of Wisdom and Knowledge. [Ibrahim] said: And (afterward) what is your duty, O you [a glittering Cup and her companions] sent (from Allah)? They said: Lo! We are sent to a guilty folk that we may send upon them **stones of clay** [black spots], marked as from your Rab for those who trespass beyond bounds. Then We brought forth such believers who were there. **But We found not there any just persons except in one house and We left there a Sign for such as fear the Grievous Penalty.** [Sura No. – 50\51 – Waz-Zaariyaat, Ayat No. – 30 to 37]

A FATTED CALF
[Cries of a barren old woman]
Has the story of Ibrahim's honoured guests reached you? When they came unto him and said: Peace! He answered: Peace! (And thought) folk unknown to me. Then he turned quickly to his household, they brought out a **fatted calf**, and **placed it before them** [projected manmade magnetism]. He said: Will you not eat? [When they did not eat], He conceived a fear of them. They said: Fear not, and they gave him good tidings of a son endowed with knowledge. Then his wife came forward, making complain, and smote her face and cried: A barren old woman! [Sura No. – 50\51 – Waz-Zaariyaat, Ayat No. – 23 to 29]

A ROASTED CALF
There came Our messengers to Ibrahim with good tidings. They said: Peace; He answered: Peace, and hastened to entertain them with a **roasted calf**. But when he saw their hands went not towards the [meal], he felt some mistrust of them, and conceived a fear of them. They said: Fear not: We have been sent against the people of Luut. And his wife was standing [there], and she laughed: But we gave her good tidings of Ishaaq, and after him, of Yaquub. She said: Alas for me! Shall I bear a child, seeing I am an old woman, and my husband here is an old man? That would indeed be a wonderful thing! They said: Do you wonder at Allah's decree? The grace of Allah and His blessings on you, O you people of the house! For He is indeed worthy of all praise, full of all glory. When fear had passed from [the mind of] Ibrahim and the good tidings had reached him, he began to plead with us for Luut's people. For Ibrahim was, without doubt, tolerant, compassionate, and given to look to Allah. [Sura No. – 10\11 – Kitaabun-uh-kimat (Prev. – Huud), Ayat No. – 69 to 75]

O Ibrahim! Seek not this. The decree of your Rab has gone forth, for them there will come a penalty that cannot be turned back! When Our messengers came to Luut, he distressed and knew not how to protect them. He said: This is a distressful day. And his people came rushing towards him, and they had been long in the habit of practising abomination. He said: O my people! Here are my **daughters**: they are purer for you. Now fear Allah, and cover me not with shame about my guests! **Is there not among you a single upright man?** They said: Well do you know we have no need of your daughters: indeed you know quite well what we want! He said: Would that I had power to resist you or had some strong support (among you)! [The Messengers] said: O Luut! We are Messengers from your Rab! By no means shall they reach you! Now travel with your family while yet a part of the night remains, and let not any of you turn round: except your wife. Lo! That which smites, they will smite her (also). Lo! Their tryst is (for) the morning. Is not the morning nigh? So, when our commandment came to pass We overthrew (that township) and rained upon it stones of clay, one after another marked with fire in the providence of your Rab (for the destruction of the wicked). And they are never far from the wrong doers. [Sura No. – 10\11 – Kitaabun-uh-kimat (Prev. – Huud), Ayat No. – 76 to 83]

THE CALF

And remember, We divided the sea for you and saved you and drowned Firawn's people within your very sight. And remember We appointed forty nights for Muusa, and **in his absence you took the calf** [projected manmade magnetism] and you did grievous wrong. Even then We did forgive you; there was a chance for you to be grateful. And remember We gave Muusa the Kitaab and the Criterions [scales of justification and verification]. There was a chance for you to be guided aright. And when Muusa said to his people: O my people! **You have indeed wronged yourselves by your worship of the calf. So turn [in repentance] to your Creator, and kill (the calf) yourselves that will be the best for with your Creator and He will relent toward you, for He is the Relenting, the Merciful.** And remember you said: O Muusa! We shall never believe in you until we see **Allah manifestly,** but **you were confused with thunder and lightning seized you**. Then We received you after your extinction, that you might give thanks. And We gave you the **shade of clouds** and sent down to you **manna** and **quails**, saying: Eat of the good things We have provided for you. We wronged them not, but they did wrong themselves. And when We said: Enter this **township** and eat of the plenty therein as you wish; but **enter the gate prostrate, and say: Repentance**. We will forgive you your sins and will increase (reward) for the right-doers. But those who did wrong **change the word which had been told them for another saying** and

We sent on the transgressors a plague from heaven, for evil doing. [Sura No. – 1\2 – Baqarah, Ayat No. – 50 to 59]

We appointed for Muusa thirty nights, and completed [the period] with ten [more]. Thus was completed the term [of communion] with his Rab, forty nights. And Muusa had charged his brother Haarun [before he went up]: Act for me amongst my people: Do right, and follow not the way of those who do mischief [who conceal truth and project falsehood as truth]. When Muusa came to the place appointed by Us, and his Rab addressed him, He said: O my Rab! Show [Thyself] to me, that I may look upon You. Allah said: **By no means you can see Me [direct]; But look upon the mount; if it abide in its place**, then you shall see Me. When his Rab manifested His glory on the Mount, He made it as dust. And Muusa fell down in a swoon. When he recovered his senses he said: Glory is to You! To You I turn in repentance, and **I am the first to believe**. [Sura No. – 6\7 - As-haabul- a' raaf, Ayat No. – 142 & 143]

[Allah] said: O Muusa! I have chosen you above [other] men, by messages I [have given you] and the words I [have spoken to you]: Take then the [revelation] which I give you, and be of those who give thanks. And We ordained laws [Revelation or Tawraat] for him for the remedies to be drawn from all things and the explanation of all things and then (bade him): **Hold it fast; and command your people (saying): Take the better (course made clear) therein. Soon I shall show you the homes of the wicked / father of falsehood**. [How they lie uninhabited?] I shall turn away from my Revelations those who magnify themselves wrongfully in the world. Even if they see all the signs, they will not believe in them; and if they see the way of right conduct, they will not adopt it as the way; but **if they see the way of error**, that is the way they will adopt. For they have rejected our signs, and failed to take warning from them. Those who reject Our signs and the meeting in the Hereafter, vain are their deeds. Can they expect to be rewarded except as they have fashioned? [Sura No. – 6\7 - As-haabul- a' raaf, Ayat No. – 144 to 147]

The people of Muusa made, in his absence, **out of their ornaments, the image of calf. It seemed near to the ground [Middle East or Gravitational Force]: Did they not see that it could neither speak to them, nor show them the way? They took it [manmade magnetism] for worship and they did wrong.** When they repented, and saw that they had erred, they said: If our Rab have not mercy upon us and forgive us, we shall indeed be of those who perish. When Muusa came back to his people, angry and grieved, he said: **Evil is that which you took after I had left you.** Would you make haste to bring on the judgment of your Rab? He put down the remedies, seized his brother by his

427

head, and dragged him toward him. He said: Son of my mother! The people did indeed judge me weak, and went near to slaying me. Oh! Make not the enemies triumph over me, nor count you me amongst the people of sin. Muusa prayed: O my Rab! Forgive me and my brother! Admit us to Your Mercy! For You are the Most Merciful of those who show mercy! [Sura No. – 6\7 - As-haabul- a' raaf, Ayat No. – 148 to 151]

Those who took the calf [projected manmade magnetism] will indeed be overwhelmed with anger from their Rab, and with humiliation in this life. Thus We do requite those who invent [falsehoods], those who do wrong but repent thereafter and [truly] believe verily your Rab is thereafter Ever Forgiving, Most Merciful. When the anger of Muusa was settled, he took up there remedies and in their inscription there was guidance and mercy for all those who fear their Rob. And **Muusa chose seventy of his people for Our place of meeting.** When the trembling came on them, he prayed: O my Rab! If it had been Your will You could have destroyed, long before, both them and me: would You destroy **us for the deeds of the foolish ones among us**? This is no more than Your trial. By it You cause whom You will to stray, and You lead whom You will into the right path. You are our Protector. So forgive us and give us Your mercy; for You are the best of those who show forgiveness. And ordain for us that which is good, in this life and in the Hereafter, for we have turned unto You. He said: With My punishment I visit whom I will; but My mercy extends to all things. That [mercy] I shall ordain for those who do right, and practise regular charity, and **those who believe in Our signs.** [Sura No. – 6\7 - As-haabul- a' raaf, Ayat No. – 152 to 156]

[When Muusa was up on the Mount, Allah said:] What has made you hasten in advance of your folk, O Muusa? He replied: They are close on my footsteps [Eartha 3D or Middle East]. I have hastened to You, O my Rab that You might be well pleased. [Allah] said: Lo! **We have tested your folk in your absence** and Wa-azallahumumus-Saamiriyy has led **them astray**. [Sura No. – 19\20 – Twa-Haa, Ayat No. – 83 to 85]

Then Muusa returned to his folk in a state of **indignation and sorrow**. He said: O my people! Have not your Rab promised you a fair promised? Did then the promise seem to you long [in coming]? Or **did you desire that wrath from your Rab should descend on you that you broke meeting with me? They said: We broke not the meeting with you of our own will, but we were made to carry the weight of the ornaments of the folk, and we threw them [into the fire], and that was what the Alqas-Saamiriyy suggested.** Then he brought out [of the fire] before the [people] **the image of a calf, of** saffron hue [projected sun through Napier Bones which rises from the East] which gave forth a lowing sound. And they cried: This is **your reality**, and the reality of Muusa, but [Muusa] has forgotten [to interpret with shane-nuzzul and commentary]! Could they not see that it could not return them a word [for answer]? And that it had no power either to harm them or to do them good? [Sura No. – 19\20 – Twa-Haa, Ayat No. – 86 to 89]

And Harrun indeed had told them before this said to them: O my people! **You are being tested in this**, for verily your Rab is [Allah] Most Gracious. So follow me and obey my command. They had said: **We shall by no means cease to be its varieties till Muusa return unto us**. [Muusa] said: O Haarun! What kept you back, when you saw them **going astray** that you follow me not? Have you then disobeyed my order? He (Haarun) said: O son of my mother! **Clutch neither my beard nor my head!** I feared lest you should say: You had caused a division among Bani-Israaiil, and you did not respect my word! [Muusa] said: And what then have you to say, Yaa-Saamiriyy? He replied: I saw what they saw not. So **I took a handful [of dust] from the footprint of the Messenger, and threw it [into the calf]. Thus did my soul suggest to me**. [Sura No. – 19\20 – Twa-Haa, Ayat No. – 90 to 96]

[Muusa] said: Then go! And lo! In this life it is for you to say: Touch me not! And moreover [for a future penalty] you have a promise that will not break. Now look at your Rab, of whom you have remained a devoted worshipper. **We will certainly burn it and will scatter its dust over the sea.** But the Rab of you all is the One Allah. There is no other reality save He. All things He comprehends in His knowledge. Thus We narrate to you some tidings of that which happened before, for We have sent you a Message from Our own Presence. If any do turn away there from, verily they will bear a burden on the Day of Resurrection [Yawmal-Qiyaamati]; they will abide in this [state]: and grievous will the burden be to them on that Day. [Sura No. – 19\20 – Twa-Haa, Ayat No. – 97 to 101]

The Day when the Trumpet will be blown, that Day, We shall gather the sinful, white-eyed [with terror]. Murmuring among themselves: Yet tarried not longer than ten [Days]. We know best what they will say, when their leader most eminent in conduct will say: **You tarried not longer than a day!** They ask you concerning the Mountains [Top West]. Say: My Rab will break them and scatter them as dust. He will leave them as plains smooth and level. Nothing crooked or curved or global concept will you see in their place. [Sura No. – 19\20 – Twa-Haa, Ayat No. – 102 to 107]

On that Day, they will follow the caller [upright], no **crookedness** or **global** concept [can they show] in him. All sounds shall humble themselves in the Presence of [Allah] Most Gracious. Nothing shall you hear but the tramp of their feet [as they march]. On that Day no intercession will avail except for those for whom permission has been granted by [Allah] Most Gracious and whose word is acceptable to Him. He knows what [appears to His creatures as] before or after or behind them, but they shall not compass it with their knowledge. [All] faces shall be humbled before [Him], the Living, the self-subsisting, eternal. Hopeless indeed will be the man that carries iniquity [on his back]. But he who works deeds of righteousness, and has faith, will neither have fear of harm, nor of any curtailment [of what is his due]. [Sura No. – 19\20 – Twa-Haa, Ayat No. – 108 to 112]

DOG'S TWO FORE LEGS

Ayat\Verses – These, our people, have chosen (projected falsehood) for worship beside Him. Why do they not bring forward an authority clear [and convincing] for what they do? (It is certain that they will not be able to bring any clear evidence.) **Who do greater wrong than those who invent a falsehood [manmade magnetism] against Allah?** [Sura No. – 17\18 – Khaf – Ayat No. – 15]

Ayat\Verses – When you turn away from them and **the things they worship other than Allah,** betake yourselves to the cave. Your Rab will shower His mercies on you and disposes of your affair towards comfort and ease. [Sura No. – 17\18 – Khaf – Ayat No. – 16]

Ayat\Verses – "Wa tarash-shamsa izaa tala-at-lazaawaru an-kahfihim zaatal-yamimi wa izaa gara-at-taqri-zuhum zaatash-shimaali wa hum fi fajwatim-minh. Zaalikka min aayaatillaah; many-yahdillaaahu fahuwal-muhtad; wa many-yuzlil falan-tajida lahuu waliyyam-murshidaa". [Trans.] And might have seen wa tarash-shamsa when it roises, declining to the right [semi-anticlockwise] from their Cave, and when it set, turning away from them to the left [semi-clockwise], while they lay in the open space [within the planetary border under the guardianship of Zakiriya] in the midst of the Cave [Black Square]. Such [equal and opposite i.e. half-clockwise and half-anticlockwise stages of journey] are among the Signs of Allah. He whom Allah, guides is rightly guided; but he whom Allah leaves to stray, for him you will not find protector to lead him to the Right Way. [Sura No. – 17\18 – Khaf – Ayat No. – 17]

Ayat\Verses – **You would have deemed them awake, while they were asleep,** and We caused them to turn over to the right and the left [semi-anti-clockwise

& semi-clock-wise], and **their dog (projected falsehood) stretching forth his two fore-legs on the threshold** [Middle East or Black Hole] If you had observed them closely, then certainly you would turn back from them in flight, and would certainly have been filled with terror of them. [Sura No. – 17\18 – Khaf – Ayat No. – 18]

Ayat\Verses – And in like manner [a clue of how to verify the truth], we raised them up that they might question each other. A speaker from among them said: How long have you stayed? They said: We have stayed **a day, or part of a day**. [At length] they [all] said: Allah [alone] knows best how long you have stayed. Now send you one of you with this silver coin of yours to the town [ground stair]. Let him find out which is the best food [to be had] and bring some to you, that [you may] satisfy your hunger therewith. And let him behave with care and courtesy, and let no man know of you. For [reason] if they should come upon you, they would stone you or force you to return to their **cult**, and in that case you would never attain prosperity. [Sura No. – 17\18 – Khaf – Ayat No. – 19 & 20]

Ayat\Verses – And in the like manner [similitude], We make their case known to the people, that they might know that the promise of Allah is true, and that there can be no doubt about the Hour of Judgment. Behold, they dispute among themselves as to their affair. [Some] said: **Construct a building over them**. Their Rab knows best about them. Those who prevailed over their affair said: **Let us surely build a place of worship over them**. [Some] say they were **three**, the **dog being the fourth among them**; [others] say they were **five**, the **dog being the sixth**, doubtfully guessing at the unknown; [yet others] say they were **seven, the dog being the eighth**. Say you: My Rab knows best their number. It is but few that know their [real case]. **Enter not, therefore, into controversies concerning them, except on a matter that is clear, nor consult any of them about [the affair of] the sleepers.** [Sura No. – 17\18 – Khaf – Ayat No. – 21 & 22] [Dog – Webdings: 246]

Ayat\Verses – And say not of anything: I shall be sure to do so and so tomorrow, except 'If Allah will [Any-yashaaa-Allah]!' And call your Rab to mind when you forget and say: I hope that my Rab will guide me ever closer [even] than this to the right path (a near way of truth). So, they stayed in their Cave **three hundred years, and [some] add nine** [more] [Sura No. – 17\18 – Khaf – Ayat No. – 25] [Sura No. – 17\18 – Khaf – Ayat No. – 23 to 25]

Say: Allah knows best how long they stayed. With Him is [the knowledge of] the secrets of the heavens and the world; how clearly He sees, how finely He hears [everything]! They have no protector other than Him; nor does He share His Command with any person whatsoever. And recite what has been revealed to you of the Kitaab of your Rab. **None can change His Words, and none of you will find as a shelter other than Him.** [Sura No. – 17\18 – Khaf – Ayat No. – 25] [Sura No. – 17\18 – Khaf – Ayat No. – 26 to 27]

Ayat\Verses – And keep your soul content with those who cry unto their Rab morning and evening seeking His countenance; and let not your eyes pass beyond, seeking the pomp and glitter of this Life; and obey not those whose heart We have made heedless of our remembrance of Us, one who follows his own lust, whose case has gone beyond all bounds. [Sura No. – 17\18 – Khaf – Ayat No. – 28]

Ayat\Verses – **Say: The truth is from your Rab.** Let him who will believe, and let him who will reject [it]. For the wrong-doers We have prepared a fire whose [smoke and flames], **like the walls and roof of a tent [Pentagon],** will hem them in. If they implore relief they will be granted water like melted brass, that will scald their faces, how dreadful the drink! How uncomfortable a couch to recline on! As to those who believe and work righteousness, verily We shall not suffer to perish the reward of any who do a [single] righteous deed. For them will be Gardens of Eternity or Eaden Gardens [Heavens]; beneath them rivers will flow. They will be adorned therein with bracelets of gold, and they will wear green garments of fine silk and heavy brocade. They will recline therein on raised thrones [Kaba]. How good the recompense! How beautiful a couch [heaven] to recline on! Set forth to them the parable of two men: For one of them We provided **two gardens** of grape-vines and **surrounded them with date** [Arabian Peninsula], in between the two We **placed corn-fields** [Middle Zone or Lower West Zone of Europe, Asia, Africa, and Australia]. Each of those gardens brought forth its produce, and failed not in the least therein. In the midst of them We caused a river to flow [Maryam Supplied with Sustenance]. [Sura No. – 17\18 – Khaf – Ayat No. – 29 to 33]

YAJUUJA AND MAJUUJA
TWO-IN-ONE PARTNERSHIP

Ayat\Verses – "Qaaluu yaa- Zal-Qarnayni inna Yajuuja wa Majuuja mufsiduuna fil-arzi fahal naj-alu laka kharjan alaaa an-taj ala bainanaa wa bainahum saddaa?" [Trans.] They said: O Zal-Qarnayn! Yajuuja and Majuuja do great mischief in the world (spoiling the land or earth). So may we pay tribute on condition that you set a barrier [planetary bridge or Fulsirat] between us and them? [Sura No. – 17\18 – Khaf – Ayat No. – 94]

432

Ayat\Verses – She said: That wherein my Rab has established me is better (than your tribute). Do but help me with strength, I will erect a strong barrier between you and them. Give me pieces of iron – till, when he had levelled up (the gap) between the cliffs she said: Blow – till, when he had made it a fire. She said: Bring me molten copper to pour therein. And (Yajuuj and Majuuj) were not able to overcome, nor could they pierce (it). [Sura No. – 17\18 – Khaf – Ayat No. – 95 to 97]

Ayat\Verses – She said: This is a mercy from my Rab. But when the promise of my Rab comes to pass, He will make it into dust; and the promise of my Rab is true. On that day We shall leave some of them to surge like waves on one another and the Trumpet will be blown, and then We shall gather them all together in one gathering. On that day We shall present hell to the non-believers, plain to view. [Non-believers] whose eyes had been under a veil from remembrance of Me, and who had been unable even to hear. Do the Unbelievers think that they can take My servants [finite patrons or gods as created Imams like Yajuuja and Majuuja of the created red colour triangles in the Napier bones called the sun of the solar system] as protectors besides Me? Verily We have prepared Hell for the nonbelievers for [their] welcome. [Sura No. – 17\18 – Khaf – Ayat No. – 98 to 102]

Ayat\Verses – Say: Shall we tell you of those who lose most in respect of their deeds? Those whose efforts have been wasted in this life, while they thought that they were acquiring good by their works? They are those who deny the Signs of their Rab and the fact of their having to meet Him [in the Hereafter]. Therefore their works are vain, and on the Day of Resurrection We assign no weight to them. That is their reward, Hell, because they rejected Faith, and took My Signs and My Messengers by way of jest.

Ayat\Verses – As to those who believe and work righteous deeds, they have, for their entertainment, the Gardens of Paradise; wherein they will abide, with no desire to be removed from thence. [Sura No. – 17\18 – Khaf – Ayat No. – 103 to 108]

Ayat\Verses – Say: **Though the sea became ink for the Words of my Rab, verily the sea would be used up before the Words of my Rab were exhausted**, even though we brought the like thereof to help. Say: I am but a man / mortal like yourselves, [but] the inspiration has come to me, that your Allah is one Allah. Whoever expects to meet his Rab, let him work righteousness, and, in the worship [equal & opposite straightway] of his Rab. Admit no one as partner [projected mechanical globalisation and artificial magnetism]. [Sura No. – 17\18 – Khaf – Ayat No. – 109 to 110]

LISH-SHAMSI

And he sought among the birds and said: How is it that I see not the hoopoe (-------) or is he among the absentees? I will certainly punish him with a severe penalty, or slay him or he will verily bring me a plain excuse. But he (hoopoe) was not long in coming, and said: I have found out (a thing) that you apprehended not, and I have come to you from Saba with tidings true. Lo! I found [there] a woman ruling over them and provided with every requisite; and she has a magnificent throne. I found her and her people worshipping Lish-Shamsi (Black Hole or Middle East or Recycling Triangle) besides Allah. Shaytan has made their deeds seem pleasing in their eyes [through projected falsehood], and has kept them away from the [equal & opposite] Path so that they go not a right. So that they worship not Allah, Who brings to light what is hidden in the heavens and the world, and knows what you hide and what you reveal. Allah! There is no unbreakable reality but He! Rab of the Throne Supreme. [Sura No. – 26\27 – Namal, Ayat No. – 20 to 26]

Bish-Shamsi rises from the East

[Allah causes bish-shamsi and not Wash-Shamsi to rise from the East; so do you cause him to rise from the West?]
Ayat\Verses – Bethink you of him who had an argument with Ibrahim about his Rab, because **Allah had granted him the kingdom. How! When Ibrahim said: My Rab is He Who gives life and death. He (Namruud) answered: I give life and cause death. ["Qaia Ibraahiimu fa-innallaha ya-tii bish-shamsi minal-maashriqi fa-ti bihaa minal magribi fa-buhital-lazi kafar"] Said Ibrahim: Allah will cause bish-shamsi (but not Wash-Shamsi) to rise from the east; so do you cause him to rise from the West? Thus was the disbeliever (who had rejected faith) confused** and **Allah guides not the wrong doing folk. [Sura No. – 1\2 – Baqarah, Ayat No. – 258]**

EVIDENT AUTHORITY OR MANIFEST FROOF OF MAGNETISM

The people of the Kitaab ask of you that you should cause an (actual) Kitaab to descend upon them from heaven. Indeed they asked Muusa for an even greater [miracle], for they said, **Show us Allah in public.** The storm of lightening seized them for their wickedness. **Yet they worshipped the calf even after Clear Signs** (Manifested Signs of Natural Magnetism) had come to them. Even so we forgave them, and gave Muusa **manifest proofs** (evident authority). [Sura No. – 3\4 – Nisaa, Ayat No. – 153]

DIIN
WE ARE INSPIRED TO PRAY TO ALLAH, MAKING OUR DIIN PURE FOR HIM

Ayat – Fad-ullaaha **mukhlisiina lahud-diina** wa law karihal-**kaafiruun**. [Trans.] Therefore (O believers) pray to Allah, **making religion (diin) pure for Him (only),** even though **the disbelievers (Kaafir) may detest it**. [Sura No. – 39\40 – Mu-Min, Ayat No. – 14]

SHOW ME THOSE WHOM YOU HAVE JOINED WITH HIM AS PARTNERS.

Ayat\Verses – Say: **Show me those whom you have joined with Him as partners**. Nay, (you dare not)! For He is Allah, the Exalted in Power, the Wise; and We have not sent you save as a bringer of good tidings and a warner unto all mankind; but most of mankind understand not. [Sura No. – 33\34 – Saba, Ayat No. – 27 & 28]

DISBELIEVERS - KAAFIRUUN
Search for Truth: - Who is called a Kaafir?
Options are –

i) a believer who has both faith and belief on revelations & manifested signs of Allah,

ii) a believer who has faith but no belief on revelations & manifested signs of Allah,

iii) an unbeliever who has neither faith nor belief on revelations & manifestations [Allah and His Messengers]

iv) A non-believer who is either a disbeliever or an unbeliever i.e. either (ii) or (iii)

v) a human being who is a mixture of both (i) and (ii)

vi) a human being who is a mixture of both (ii) and (iii)

vii) only (ii) [both (ii) and (vii)]

A similitude – Who can divorce his wife? In other words, the question of divorce arises in case / cases of - -------------------------- [Choose an option / options]

Options are – (i) a married man, (ii) an unmarried man, (iii) both (i) & (ii), (iv) neither (i) nor (ii), (v) either (i) or (ii). (vi) only (i)

Yaaa-ayyuhal-Kaafiruun

Ayat\Verses – Say: O you who reject Faith [disbeliever]! I worship not that which you worship nor do you worship that which I worship. And I will not worship that which you worship nor will you worship that which I worship. To you be your Way, and to me mine. [Sura No. – 108\109 – Yaaa-ayyuhal-Kaafiruun, Ayat No. – 1 to 6]

Ayat\Verses – Ha Mim. The revelation of this Kitaab is from Allah, The Mighty The Knower; the Forgiver of sin, the Accepter of repentance, the Stern in punishment, and the Bountiful. **There is no ilaaha** [unbreakable reality] **except Him**. To Him is the final goal. **None dispute (argue) concerning the Signs (Manifested Sign of Natural Magnetism) of Allah except the disbelievers (those who reject faith)**. So, **let not their turn of fortune in the land** [earth] **deceives (misleads) you**. [Sura No. – 39\40 – Mu-Min, Ayat No. – 1 to 4]

Ayat\Verses – The **folk** of Nuuh and the **factions (groups / isms)** after them denied (their messengers) before these, and every nation [**Ummatim-bi-Rasuulihim**] purposed **to seize their messenger** and **argued falsely**, (thinking) thereby **to refute the Truth**. Then I seized them, and how [awful] was My punishment [requital]! And thus the **word** of your **Rab** was **proved** true against the disbelievers; **that truly they are companions of the Fire!** [Sura No. – 39\40 – Mu-Min, Ayat No. – 5 & 6]

SIMILITUDE OF BOTH DISBELIEVERS AND UNBELIEVERS NON-BELIEVERS

Ayat\Verses – Their **similitude** is that of a man who kindled a fire; **when it was lighted all around him, Allah took away their light and left them in utter darkness**. So they could not see. [Sura No. 1\2 – Baqarah, Ayat No. – 17]

Ayat\Verses – He said: O my people! Bethink you if [it be that] I rely on **a Clear Sign (proof)** from Allah, and that He has sent Mercy unto me from His own presence, but that the Mercy has been **obscured from your sight**, can we compel you when you are unwilling to accept it? And O my people! I ask you for no wealth in return: my reward is from none but Allah. But I will not drive away [in contempt] those who believe, for verily they are to meet their Rab, and you I see are the **ignorant ones**! [Sura No. – 10\11 – Kitaabun-uh-kimat (Prev. – Huud), Ayat No. – 28 & 29]

Ayat\Verses - [Their **similitude**] is that of a rain-laden cloud from the sky. In it are **zones of darkness**, and **thunder** and **lightning**. They press their fingers in their ears by reason of the stunning thunder-clap, for fear of death. But Allah **encompasses the unbelievers**.[Sura No. 1\2 – Baqarah, Ayat No. – 19]

Ayat\Verses - The **lightning all but snatches away their sight**; every time **the light** [helps] them, **they walk therein (light)**, and **when the darkness grows on them, they stand still**. And if Allah willed, He could **take away** their faculty of **hearing and seeing**; for Allah has power over all things. [Sura No. 1\2 – Baqarah, Ayat No. – 20]

**Fasta-'iz billaahi minash-Shaytaanir-Rajim
Bismillaahir-Rahmaanir- Rahim**

MAN-MADE MAGNETISM vs. NATURAL MAGNETISM

Ayat\Verses – Allah has said: **Choose not two realities [i.e. both Natural Magnetism and Manmade Magnetism]**; for He is **just One Allah**. So of Me, Me only, be in awe. [Sura No. – 15\16 – Nahl, Ayat\Verses – 51]

Ayat\Verses – And when Allah said: O Isabna-Maryama! Did you say unto mankind: **Take me and my mother for two realities besides Allah?** He will say: Glory to You! Never could I say what I had no right (to say). Had I said such a thing, You would indeed have known it. You know what is in my mind, though I know not what is in Your Mind. For You know in full all that is **hidden**. [Sura No. – 4\5 – Maaaidah, Ayat\Verses – 116]

Ayat\Verses – Glorify the name of your **Rab, Most High [Arsh]**, Who has **appointed further** and **disposes**. Who **measures and then guides**, and Who **brings forth** the pasturage [what binary system emits], then **turns it** to russet [reddish-brown] stubble [stalks of corn]. [Sura No. – 86\87 – Rabbikal-Alaa, Ayat No. – 1 to 5]

Ayat\erses - And We have **appointed** the **night** and the **day** two **Clear Signs.** Then We make **dark** the sign of the night, and We make the **sign** of the **day sight-giving**, that you may **seek bounty** from your Rab, and that you may know the **computations** of the years and reckoning and everything have **We expounded with a clear expounding**. [Sura No. – 16\17 - Banii-Israai-iil, Ayat No. – 9 to 12]

Ayat\erses - He is the One Who sends to His **slave Manifest Signs [clear proofs]**, that **He may lead you [guide] from the depths of darkness into the light** and verily for you Allah is full of Pity, Merciful. [Sura No. – 56\57 - Anzalnal-Hadiid, Ayat No. – 9]

Ayat\erses - Yawma yaquulul-munaafiquuna wal munaafiqaatu li-laziina aamanun-zuruunaa n a q-tabis min-**Nuurikum**. Qiilar-ji-uu waraaa-'akum fal-lamisuu **n u u r a a**. Fazuriba baynahum-bi-suuril-l a h u u baab. Baatinuhuu fiihir-Rah-matu wa zaahiruhuu min-qibalihil-azaab. [Trans.] One the Day when **the hypocrites (those who not only reject faith but also mislead**

mankind) **men and women** will say to those who believe: Wait for us! Let us borrow [a **light**] from your **Light**! It will be said: Go back and seek for **Light**! So a wall (barrier / danger line) will be put up between them with a **gate** therein. The **inner side** (West) whereof comes mercy, while the **outer side** (East) thereof is towards the **fire**. [Sura No. – 56\57 - Anzalnal-Hadiid, Ayat No. – 13]

Ayat\Verse – And **a Sign for them is the Night**. We **strip** of the Day, and lo! **They are in darkness.** [Sura No. – 35\36 – Yaa-siiin, Ayat No. – 37]

Ayat\Verse – And **Wash-Shamsu runs** *its* **courses for a period determined for** *it.* That is the **measuring of the Mighty**, The Wise. [Sura No. – 35\36 – Yaa-siiin, Ayat No. – 38]

Ayat\Verse – And for Wal-Qamara, We have appointed patches till *it* **returns like an old dried-up palm-leaf.** [Sura No. – 35\36 – Yaa-siiin, Ayat No. – 39]

Ayat\Verse – It is not permitted to the Lash-Shamsu to catch up Tudrikal-Qamara, **nor the Night can outstrip [exceed] the Day. Each [just] swims along in [its own] orbit or Axis [following equal & opposite Revelation or Tawraat].** [Sura No. – 35\36 – Yaa-siiin, Ayat No. – 40]

Ayat\Verse – And **a Sign for them is that We bear their off-springs in the laden ship.** [Sura No. – 35\36 – Yaa-siiin, Ayat No. – 41]

Ayat\Verse – And We have created for them of the like there of wherein they ride. [Sura No. – 35\36 – Yaa-siiin, Ayat No. – 42]

"**Allaahu Nuurus-samaa-waati wal-arz** – Masalu **Nuu-rihi** ka**Mishkaatin**-fiihaa **Mis-baah** – Al-**Misbaahu** fii **Zujaa-jah** – **azzuujaajatu** ka-annahaa – kawkabun durriyyuny-yuuqa-du – min Shajaratimmubaara-katin-**Zaytuunatil**-laa **Sharqiy**-yatinw wa laa **Garbiyyatiny**-yakaadu **Zaytuhaa** yuziii- u wa law lam tamsas-hu naar. **Nuu-run alaa Nuur! Yahdillaahu** li-**Nuurihii** many-yashaaa – wa-yazribullaahul-amsaala linnaas – wallahu bi-kulli shay-in Alim." [Sura No. – 23\24 – Nuur, Ayat No. – 35] [Trans.] **Allah** is the **Light** of the **heavens** and the **world**. The **similitude** (likeness) of **His Light** is as a **niche** (binary thread of a candle or binary pulsar of Einstein) **in a lamp [Black Square]. The lamp** is in a **glass [Diamond Operator]. The glass** is as it is a **shining star.** (The niche is) **kindled from a blessed tree**, an **olive** (Zaytuun tree), **neither from the East nor from the West**, whose (niche's) oil will almost glow forth (of itself) though **no fire touches** it. **Light upon Light!** Allah guides unto His light whom He wills. And Allah speaks to mankind in **parables**, for Allah is Knower of all things. [Sura No. – 23\24 – Nuur, Ayat No. – 35]

There are **two West, two East, one North**, and **one South** from the point of view of Revelation and Manifestation. So, **there is only one kind of Magnetism. This sole Magnetism is the Manifested Magnetism in resemblance with Equal & Opposite Revelation. This sole Magnetism is called Natural Magnetism.** But the Historical Don of Trinity & Duality has concealed the true concepts of Natural Magnetism and has **projected Man-made Magnetism**. This **projection is called Teleological Evidence Sorcery and Unique Epistemic Persecution or Historical Black Magic**. Further, the **introduction of Manmade Magnetism implies that there is another Nature** [Universe, World, Earth, Magnetism, and so on] called **Manmade Nature**. Due to this Historical Black Magic we are living in the Manmade Nature. Manmade Nature includes Globalisation, Circular Rotation and Revolution System, North and South directions grounded on Manmade Magnetism, Projected Northern Hemisphere and Southern Hemisphere on the Global Trinity etc. The **constituents of Manmade Nature are both contrary as well as contradictory to Newton's Three Laws of Motion, Newton's Law of Gravitation, Einstein's Gravitational Lensing, Einstein's Binary Pulsar**, and all verifiable truth of facts. But the **constituents of Manmade Nature resemble with Copernican Revolution, Kepler's Axis, Brahe's Data, and the like.**

People like the sharer of this re-search project had conceptualised Projected Manmade Nature as the Manifested Nature in resemblance with Revelation. We are determining North and South directions in resemblance with Man-made Magnetism. On the contrary, we are conceptualising those North and South Directions as Revealed Left & Revealed Right or Manifested Haiyalas-Swalaah & Haiyalal-Falaah or Natural Directions in resemblance with Revelation. This is called subjective selfcontradiction and objective paradox. **If North and South directions represented by Manmade Magnetism are the Manifested Directions in resemblance with Revelation, then what is the necessity of two kinds of Magnetism?** But we never think or feel it necessary to confirm manifest truth or manifested nature in resemblance with revelation. **This projected / introduced man-made magnetism is the vera causa of leading believers towards wrong direction**. Consequently, believers of Middle Stair [Europe, Asia, Africa, and Australia] are **performing Salat towards Haiyalas-Swalaah [Revealed Left or North]** and believers of Ground Stair [North America and South America] are **performing Salat towards Haiyalal-Falaah [Revealed Right or South]**. So, **this projected /**

introduced man-made magnetism is the Experimentum Crusis as well as Crucial Instance against Teleological Evidence Sorcery and Unique Epistemic Persecution or Historical Black Magic.

Probably we do not know or we have no concept of the Manifested Signs of Equal & Opposite Natural Magnetism. The so-called sun or what is called Bullet in real science is the Electromagnetic Wave of Einstein's Binary Pulsar, the converted snake of Muusa's Staff, and manifested sign of natural magnetism. As there is only one manifested nature, so, it is logically justified that there is only one kind of magnetism. Manifested sign of natural magnetism i.e. equal & opposite stages of journey of the so-called sun [Bullet] for the equal & opposite seashore represent Revealed Left & Revealed Right or Manifested Northern Hemisphere [Haiyalas-Swalaah] & Manifested Southern Hemisphere [Haiyalal-Falaah].

With a view to track off believers from the equal & opposite middle course towards Upright West, with a view to lead Unity towards diversities, with a view to dominate mankind from Spiderman's Network, and the like, the Don of Historical Black Magic has introduced Manmade Magnetism through pen-paper-pencil works, black & white activities, mechanical globalisation, and circular rotation and revolution system. In this regard, let us ask some simple questions to the Don as well as Chiefs, Ministers, Representatives, Mouth Speakers, Black & White Artists, and the like of the Teleological Evidence Sorceries and Unique Epistemic Persecutions –

i) How many manifested nature are there?
ii) What is the Existential Import of Uniform Principle with respect to Manmade Magnetism?
iii) If there are two kinds of Magnetism, then how many Uniform Principles are there? Whether the concept of Uniform Principles is consistent with itself i.e. Uniform?
iv) What are the criterions of determining Natural Magnetism and Manifested North & South Directions?
v) What are the criterions of verifying and justifying Manmade Magnetism?
vi) If there is only one manifested nature, then how many Magnetic Directions or Magnetisms are there?
vii) If there is one and only one manifested nature, then how many North Directions are there?
viii) If there is one and only one North Direction, then whether manmade magnetism represents Manifested North Direction or equal & opposite

stages of journey of the manifested signs of natural magnetism [so-called sun or Bullet] represent Manifested North Direction?

ix) If manifested signs of natural magnetism represent Manifested Northern Hemisphere [Haiyalas-Swalaah] and Southern Hemisphere [Haiyalal-Falaah], then do these not imply that the projected North & South Directions grounded on man-made magnetism as well as Northern Hemisphere & Southern Hemisphere are Teleological Evidence Sorceries and Unique Epistemic Persecutions?

x) What is the sole manifested sign of Natural Magnetism?

xi) If I share with you that the so-called sun [Bullet / Sister Planet Venus / Maryam supplied with sustenance / Tarash-Shamsi / Wash-Shamsi / She Camel, converted snake of Muusa's Staff, Electromagnetic Wave of Einstein's Binary Pulsar] represents manifested sign of natural magnetism, then whether this sharing is both contrary as well as contradictory to manifested nature & natural magnetism or consistent with equal & opposite manifested sign of natural magnetism [verifiable scientific laws & justifiable philosophical theories, Newton's Laws & Einstein's Binary Pulsar, Manifested Signs & Upright Logic]?

xii) If the above sharing is neither contrary nor contradictory to manifested nature and natural magnetism, then does it not imply that the above sharing is a tautology?

It is certain like 'I think, therefore, I exist' that the equal & opposite stages of journey of the so-called sun [Bullet] for the equal & opposite Seashore [East Zone and West Zone] are the manifested signs of Natural Magnetism. Pen-paper-pencil works, black & white activities, mechanisms, theorisations, revolutions, and the like in the domain of formal education which are either contrary or contradictory or both contrary & contradictory to manifested nature and equal & opposite manifested signs of natural magnetism must be categorised as Teleological Evidence Sorceries and Unique Epistemic Persecutions.

Due to this Historical Black Magic, believers of the Ground Stair of the appearing Pentagonal Earth are performing Salat [Compulsory Prayer] towards Haiyalal-Falaah [Manifested South-East] and believers of Middle Stair of the appearing Pentagonal Earth are performing Salat [Compulsory Prayer] towards Haiyalas-Swalaah [Manifested North-West]. People of the West Zone are observing Idd in two different days and people of the East Zone are observing Idd in a fictitious day. But those who have searched out manifest truths in resemblance with revelations and have installed those searched out findings symbolically in Insert Symbol must know that believers of Upright West are following wrong direction in resemblance with the wrong dialling shown in the film PK.

Man-made Magnetism is the **Vera Causa of Activism & Persecution**. It has been shared that Persecution is worse than Killing. The Don of Historical Black Magic has projected two kinds of Magnetism, namely, **Natural Magnetism** and **Man-made Magnetism** through pen-paper-pencil works and black & white activities. The **projection of Man-made Magnetism** is a selfevident formal document against Black Magicians, Experimentum Crusis as well as Crucial Instances against their Teleology & Ontology, formal documents against Violations of several Solidified Solid Human Rights of Mankind in general, Sound Logic of criticising uncountable pen-paper-pencil works & activities, Verifiable as well as Justifiable References of this Re-search Project on 'Solidarity Rights in Islam' in the name of 'Kitaaba Wal-Hikmata' or 'Manifested Nature and the Utility of One's Upright Logic', Solid Ground of Demanding Generalised Recognition of the Searched out Manifest Truths for the sake of Truth-in-itself and Manifested Nature.

If it is supposed that there are two kinds of Magnetism, then it is implied that there are two kinds of nature, namely, **Manifested Nature** in resemblance with Revelation and **Man-made Nature** in resemblance with the **Projected Falsehoods [will of the Historical Black Magician]**. The concepts of projected North and South Directions as well as Northern Hemisphere and Southern Hemisphere through Mechanical Globalisation, Circular Rotation & Revolution System, and Man-made Magnetism represent **Man-made Nature in resemblance with the Teleology and Ontology of the Black Magicians**. If Man-made Magnetism is a true concept of magnetism [knowledge] concerning North and South directions, then the Electromagnetic Wave or the so-called sun [Bullet] must have to rise from Man-made North Direction and to end its equal & opposite journey in Man-made North Direction. But in reality, the so-called sun [Bullet or Electromagnetic Wave] neither rises from Man-made North Direction nor sets in Man-made South/North Direction. So, it is certain that North and South directions determined by manmade magnetism do not represent Manifested Northern Hemisphere [Haiyalas-Swalaah] & Southern Hemisphere [Haiyalal-Falaah] / Natural North & South / Magnetic Force & Weak Force in resemblance with Revelations.

Now, if I determine North and South directions on the basis of Man-made Magnetism, then it is certain like 'Cogito Ergo Sum' that those directions are not Natural Directions i.e. Manifested Magnetic Directions in resemblance with Revelations or Manifested Nature or Manifest Truths. On the contrary, North and

South directions determined by man-made magnetism are man-made directions i.e. man-made North and man-made South. In other words, if I determine North direction on the basis of Man-made Magnetism, then that North direction neither represents revealed Magnetic Force nor represents Revealed Left nor represents Manifested Haiyalas-Swalaah [Northern Hemisphere]. Similarly, if I determine South direction on the basis of Man-made Magnetism, then that South direction neither represents Weak Force nor represents Revealed Right nor represents Manifested Haiyalal-Falaah [Southern Hemisphere]. But followers of Upright West are performing salat believing man-made North and South directions as Manifested Haiyalas-Swalaah [Revealed Left] and Manifested Haiyalal-falaah [Revealed Right] respectively due to acquired formal knowledge of Mechanical Trinity & Black Magic, massive black & white works & activities in resemblance with Mechanical Trinity & Black Magic, and impressive broadcasting in resemblance with the recognitions of the Owner and Possessors of Manmade Magnetism. So, the projection of Man-made Magnetism is the self-proved violation of Solidified Solid Human Rights particularly of the followers of Upright West and mankind in general. As a re-search scholar on 'Solidarity Rights in Islam', I have searched out on the basis of verifiable as well as justifiable references [valid references] **Mechanical Trinity & Black Magic of the Teleological Evidence Sorcerers and Unique Epistemic Persecutors**. So, I have placed my searched out findings before respective authorities for verification, justification, suggestion, confirmation, recognition, broadcasting, necessary action, and the like.

There are four cardinal [revealed] directions, namely, East [Down], West [Up], South [Revealed Right], and North [Revealed Left]. East represents Gravitational Force. West represents Strong Force. North represents Magnetic Force. South represents Weak Force. There are six manifested directions, namely, North-East, Middle East, South-East, South-West, Upright West, and North-West. Natural Magnetism represents North-East to South-East and South-West to North-West Directions out of six manifested directions. The remaining directions are Straight Middle East Region of Eartha 3D and Upright West Region of Arabian Peninsula. Gravitational Wave or Bish-shamsi rises from the Straight Middle East and Rain [white] comes from Upright West. **If Natural Magnetism is true, then Man-made Magnetism must be false. If Man-made magnetism is true, then Natural Magnetism must be false. Both cannot be true together or false together.** So, we are to reject one kind of magnetism and to accept the other kind of magnetism.

If we accept man-made magnetism as true concept of magnetism, then it will imply that there is no manifested nature, no uniform principle, no continual acceleration, no uniform motion, no electromagnetic wave, no

gravitational wave, and the like. In other words, Newton's Three Laws of Motion, Newton's Law of Gravitation, Einstein's Gravitational Lensing, Einstein's Binary Pulsar etc. are fictitious concepts; while Copernican Hypothesis, Kepler's Laws, and Brahe's Data have made an independent nature where Electromagnetic Wave rises once from the East [Gravitational Force] and sets once in the West [Strong Force] to cause the alteration of day and night for the equal & opposite East Zone [Lower Seashore] and West Zone [Upper Seashore].

Since there is no independent nature [manmade nature] in resemblance with Subjective Idealism of Berkeley, so, it is certain that the projection of Man-made Magnetism commits subjective self-contradictions and objective paradoxes. **To get rid of Subjective Selfcontradictions and Objective Paradoxes is one of the Solidified Solid purposes of this re-search project on 'Solidarity Rights in Islam'.**

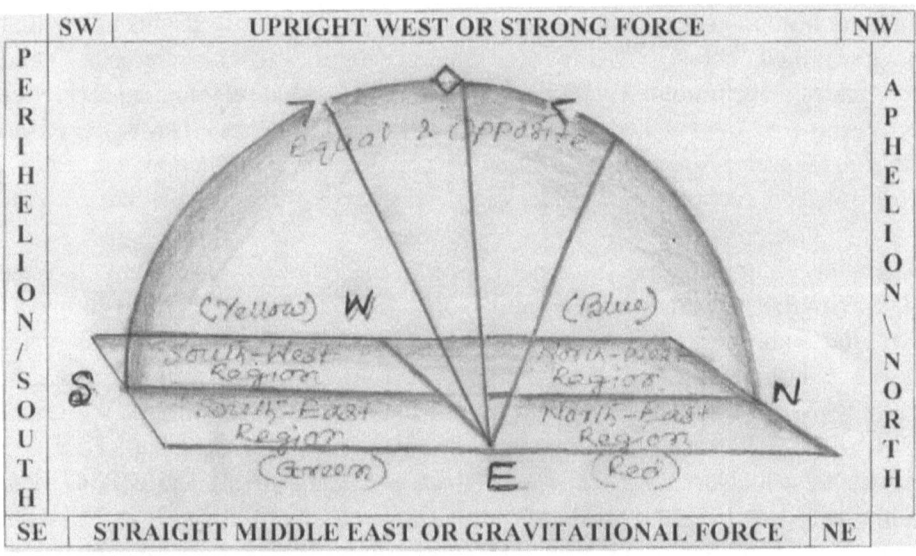

W represents Region of White, Upright West Region of Arabian Peninsula, Topmost Stair out of three ascending stairs, Uppermost Land of the appearing Pentaonal Earth, City, Tuur, and **Strong Force**

E represents Straight Middle East Region of Eartha 3D, Bermuda Triangle, Recycling Region, Penetrated hole in the journy boat of Muusa (ass), and **Gravitational Force**

N represents Manifested North Pole [Revealed Left], Northern Hemisphere, Haiyalas-Swalaah, and **Magnetic Force**

S represents Manifested South Pole [Revealed Right], Southern Hemisphere, Haiyalal-Falaah, and **Weak Force**

i)   N – E – S represents East Zone or Lower Sea-shore within the Black Square of East Horizon, Semi - anti-clock-wise Zone from Manifested North-East to Manifested South-East with respect to the equal & opposite stages of journey of the so-called sun as the manifested sign of natural magnetism [electromagnetic wave or Tarash-Shamsi or Bullet]. East Zone comprises North America, South America, and Straight Middle East Region [Eartha 3D or Recycling Region or Black Hole or Hole in the journey boat of Muusa (ass)]. In other words, N – E – S or Lower Seashore is the Zone of entering and setting of the Semi-anticlockwise stages of journey of the so-called sun as the manifested sign of natural magnetism or Bullet for Morning Show [anti-clockwise half of the one whole day] in the name of the sister planet Venus and shared name Maryam supplied with Sustenance. The so-called [**new**] moon or Qamar or Planet/Moon Uranus Triangular White Bullet appears for the East Zone from Manifested South-East following Left Double Quotation Mark for the 1st period/ quarter.

ii) S – W – N represents West Zone or Upper Sea-shore within the Black Square of East Horizon, Semi-clock-wise Zone from Manifested South-West to Manifested North-West with respect to the equal & opposite stages of journey of the so-called sun as the manifested sign of natural magnetism [electromagnetic wave or Wash-Shamsi or Bullet]. West Zone comprises Europe, Arabian Peninsula, Asia, Africa, and Australia. In other words, Upper Seashore is the Zone of rising and ending of the Semi-clockwise stages of journey or clock-wise half of the one whole day of the so-called sun as the manifested sign of natural magnetism [Wash-Shamsi or Bullet] for the Morning Show with respect to West Zone but Evening show with respect to sister planet Venus or Maryam supplied with sustenance. The so-called [new] moon or Qamar or Planet/Moon Uranus or Triangular White Bullet for the West Zone or Upper Seashore arises from Manifested North-West following Right Double Quotation Mark for the 1st period/quarter.

iii) W stands for Uppermost Land, Upright West Region, Region of White, Arabian Peninsula, City where there is the appointed Kaba on the right side of the Mount Tuur, obversion of the converted snake of Muusa's Staff, Topmost Stair out of three ascending Stairs, and Uppermost Land of the appearing Pentagonal Earth, Tallest Tree.

iv) E stands for Lower East Region, Bottom East Region, Straight Middle East Region where there is Eartha 3D, Recycling Region, Black

Hole, Bermuda Triangle of Titanic. Projected Sun or Bish-Shamsi or Gravitational Wave or Ether or Electron Cloud as a Veil [Screen or Veto] or Black Umbrella rises once from the East. But the so-called sun or commonly conceived sun as the Electromagnetic Wave enters neither from the East nor from the West. It enters from North-East and sets in South-East for the East Zone. Again, it rises from South-West and sets in North-West for the West Zone. So, the so-called sun [Bullet] as the manifested sign of Natural Magnetism rises twice and sets twice to cause the alteration of day and night for the equal & opposite seashore of East Zone and West Zone.

v) **N stands for Revealed Left & Manifested North Pole. This North represents revealed Magnetic Force and E-Point of the Manifested Electromagnetic Wave.** This Manifested North is the **Left Back Border** of revelation [Northern Hemisphere or Haiyalas-Swalaah]. Appointed Kaba or End of Proof has been manifested facing towards East on the right side of the Mount Tuur. So, towards appointed Kaba is the diagonal West in resemblance with Muusa's [ass] climbing of the Mount. Now, if one stands facing towards the appointed Kaba [suppose one's PC], then his/her right side but the left side of the appointed Kaba [PC] is the Manifested North in resemblance with the ideogram N of the above figure and in pragmatic correspondence with the on/off key, Enter Key, and Full Stop Key of the Key-board. This Manifested North is the on/ off Key Point or Entering & Ending E-Point of the manifested sign of natural magnetism. These sharing concepts resemble not only with the real science but also with revelation, manifestation, natural magnetism, manifested signs of natural magnetism [Ref. Ayat No. – 35 of Sura – Nuur and Ayat No. – 17 of Sura Kahf]. So, the sharer of 'Kitaaba Wal Hikmata' or 'Manifested Nature and the Utility of One's Upright Logic' has no doubt concerning the above findings. On the contrary, these findings are certain like 'Cogito Ergo Sum' of Descartes, and consequently, these findings are the selfevident truths [valid conclusions or Furqan] of Dictum de Omni ET Nullo [Mathematical Method, Cartesian Method, Deductive Method]. These findings are free from subjective selfcontradictions, objective paradoxes, fallacy of infinite regress, and fallacy of arguing in a circle, Existential fallacy, Post Hoc Ergo Propter Hoc, and the like. Simultaneously, these findings are the selfexplained & selfevident experimentum crucis as well as crucial instances against Teleological Evidence Sorceries and Unique Epistemic Persecutions by the Don of two-in-one partnership, the Don of Trinity [Hypocrisy – Tyranny – Conspiracy], the Don of Duality [Activism and Terrorism]. In this regard, an honest appeal is being placed before all epistemic persons to go sincerely through the shared findings of 'Kitaab Wal-Hikmata' or 'Manifested Nature and the Utility of One's Upright Logic' with a view to falsify or to recognise the searched out manifest truths which are justifiable on the basis of Universal Major Premises and verifiable in resemblance with searched out real scientific symbols along with their character codes in resemblance with revelations, manifested nature, natural magnetism, manifested signs of natural magnetism, equal & opposite stages of journey of the manifested sign of natural magnetism, verifiable scientific laws, justifiable philosophical theories of truth, upright nature, and the like. Your recognition will provide believers like me the equal and opposite right direction of Qibla towards Upright West [Arsh] following equal and opposite wings or signs of natural magnetism in correspondence with the four converted wings of Windows'7 Ultimate and in resemblance with shared parable with Ibrahim (ass).

vi) The so-called new moon or Triangular White Bullet also arises from the same root i.e. manifested North. When the so-called sun as the manifested sign of natural magnetism enters from Manifested North-East, and follows Earth Sky [Orbit], and moves in a geometrical progression from Manifested North-East [N] towards Manifested South-East [S] for the East Zone; the so-called new moon also arises from Manifested North-West, and follows Right Double Quotation Mark [for the 1st period] and Semi Major Axis, and moves in an arithmetical progression from Manifested North-West [N] towards Manifested South-West [S] for the West Zone. So, the left Back Border of manifestation is the Northern Hemisphere [entering point of the so-called sun as the manifested sign of natural magnetism for the East Zone and the arising point of the new Moon as the sign of successive relation for the West Zone]. In other words, Manifested **North Pole [N] is the entering as well as arising point of the so-called sun as the manifested sign of natural magnetism and the so-called new moon as the manifested sign of successive relation.**

vii) S stands for Manifested South Pole. This Manifested South Pole is the right border of manifestation. This Manifested South Pole [S] is the Turning Point [Broken Bar – ASCII (hex) 00A6 & ASCII (decimal) 166] of both the so-called sun and the so-called new moon or Wash-shamsi Wal Qamar. The so-called sun as the manifested sign of natural magnetism enters from Manifested North-East [N] in correspondence with the existing North America for the East Zone declining towards right border or Manifested South-East, and

447

follows semi-anticlockwise stages of journey, and sets in Manifested South-East in correspondence with South America for the East Zone. The so-called sun as the manifested sign of natural magnetism runs away from the setting point following Broken Bar and rises from Manifested South-West [S] for the West Zone where there are Europe, Arabian Peninsula, Asia, Africa, and Australia. The so-called sun as the manifested sign of natural magnetism follows semi-clockwise stages of journey and moves towards the end point or Back Border i.e. Manifested North-West [N] due to Revealed Magnetic Force or Natural Magnetism. Revealed Left [North] represents Magnetic Force as well as entering and ending of the manifested signs of Natural Magnetism.

viii) The Danger sign represents barrier between East Horizon and West Horizon; while the Double Danger represents barriers between West Zone and East Zone as well as between Top West Stair and Middle Stair of the appearing Pentagonal Earth.

Projected sun [day-light / Bullet] as the manifested sign of natural magnetism rises from the East and sets in the West for the entire day [day & night] but that once rising and once setting of the so-called sun cause the alteration of day and night for the equal & opposite Seashore [West Zone and East Zone]. This projection is **a post hoc ergo propter hoc Statistics / Calculation and Extreme Persecution.** The so-called sun as the manifested sign of natural magnetism [day-light / Bullet] **rises twice and sets twice** throughout one whole day [day & night] for the equal & opposite Seashore [West Zone and East Zone] within the Black square of East Horizon for the Manifested Hexagonal World and appearing Pentagonal Earth. The so-called Sun as the manifested sign of natural magnetism [day-light / Bullet] enters from Manifested North-East [Revealed Left or Northern Hemisphere or Haiyalas-Swalaah] in correspondence with the existing North America declining towards Manifested South-East [Revealed Right or Southern Hemisphere or Haiyalal-Falaah] and covers half of the stages of journey i.e. Semi-anticlockwise stages of journey for the East Zone and sets in Manifested South-East [Revealed Right or Southern Hemisphere or Haiyalal-Falaah] in correspondence with existing South America. This entering and setting of the so-called sun as the manifested sign of natural magnetism is the day-light or morning show in the name of projected sister planet Venus or shared name Maryam supplied with sustenance for the ground stair of North America and South America. The so-called setting Sun as the manifested sign of natural magnetism [day-light / Bullet] runs away from there and rises from Manifested South-West [Revealed Right or Southern Hemisphere or Haiyalal-Falaah] in correspondence with existing Europe turning towards

North-West due to Revealed Magnetic Force [Natural Magnetism] and covers remaining semi-clockwise stages of journey [including Arabian Peninsula] for the West Zone and ends the journey in Manifested North-West [Revealed Left or Northern Hemisphere or Haiyalas-Swalaah] in correspondence with existing Australia. This rising and ending of the so-called sun as the manifested sign of natural magnetism is the day-light or morning show from the point of view of Middle Stair and Topmost Stair of the appearing Pentagonal Earth but Evening Show from the point of view of the so-called sun [Bullet or planet Venus].

Note – (i) Zaatal-Yamimi or Northern Hemisphere [Revealed Left] and Zaatash-Shimaali or Southern Hemisphere [Revealed Right]

(ii) For common understanding Haiyalas-Swalaah [Northern Hemisphere] and Haiyalal-Falaah [Southern Hemisphere]

In brief, the so-called sun as the manifested sign of natural magnetism [day-light / Bullet] enters from Manifested North-East and sets in Manifested South-East, and again rises from Manifested South-West and ends in Manifested North-West. In other words, the direction from which the so-called sun [Bullet] as the manifested sign of natural magnetism enters as the day-light for the Ground Stair of North America and South America is the Manifested North Direction or Haiya-las-Swalaah [Revealed Left or North] for the East Zone within the Black Square of East Horizon. The direction towards which the so-called sun [day-light / Bullet] as the manifested sign of natural magnetism sets for the Ground Stair of North America and South America is the Manifested South Direction or Haiyalal-Falaah [Revealed Right or South] for the East Zone within the Black Square of East Horizon. The direction from which the so-called sun as the manifested sign of natural magnetism runs away and rises as the day-light for the Middle Stair of Europe, Asia, Africa, and Australia as well as Topmost Stair of Arabian Peninsula is the Manifested South Direction or Haiyalal-Falaah [Revealed Right or South] for the West Zone within the Black Square of East Horizon. The direction towards which the so-called sun as the manifested sign of natural magnetism ends the stages of journey for Middle Stair and Topmost Stair as well as one whole day is the Manifested North Direction or Haiyalas-Swalaah [Revealed Left or North] for the West Zone within the Black Square of East Horizon. So, the so-called sun as the manifested sign of natural magnetism [day-light / Bullet] rises twice and sets twice from Manifested North-East to Manifested South-East and from Manifested South-West to Manifested North-West. **This is called Natural Magnetism in resemblance with Revelation.** This Natural Magnetism is the cause of the alteration of day and night for the equal & opposite Upper

Seashore & Lower Seashore [West Zone and East Zone]. It is certain like 'I think, therefore, I exist' that there is no finite corporeal being who has the capability to contradict or to play the games of Black Magic and Mechanical Trinity further with the above searched out unalterable manifest truths. As a re-search scholar on 'Solidarity Rights in Islam', I have searched out the above mentioned verifiable as well as justifiable manifest truths in resemblance with Revelations. Now, to get verified & justified confirmation as well as generalised recognition of the searched out manifested truths from the Honourable Chairs and Respective Authorities are the Solidified Solid Human Rights of the re-searcher of 'Solidarity Rights in Islam' and sharer of 'Kitaaba Wal-Hikmata' / 'Manifested Nature and the Utility of One's Upright Logic'.

EQUAL AND OPPOSITE SCIENCE - I
SOURCE - INTERNET
"The Moon is Earth's only permanent natural satellite"

SIRIUS BINARY SYSTEM

Sirius (/'sɪriəs/) is the brightest star (in fact, a star system) in the Earth's night sky. With a visual apparent magnitude of −1.46, it is almost twice as bright as Canopus, the next brightest star. The name "Sirius" is derived from the Ancient Greek Σείριος (*Seirios*), meaning "glowing" or "scorcher". The system has the Bayer designation Alpha Canis Majoris (α CMa). What the naked eye perceives as a single star is actually a binary star system, consisting of a white main-sequence star of spectral type A1V, termed **Sirius A**, and a faint white dwarf companion of spectral type DA2, called **Sirius B**. The distance separating Sirius A from its companion varies between 8.2 and 31.5 AU.[24]

Sirius appears bright because of both its intrinsic luminosity and its proximity to Earth. At a distance of 2.6 parsecs (8.6 ly), as determined by the Hipparcos astrometry satellite,[2][25][26] the Sirius system is one of Earth's near neighbors. Sirius is gradually moving closer to the Solar System, so it will slightly increase in brightness over the next 60,000 years. After that time its distance will begin to increase, but it will continue to be the brightest star in the Earth's sky for the next 210,000 years.[27]

Sirius A is about twice as massive as the Sun ($M_\odot$) and has an absolute visual magnitude of 1.42. It is 25 times more luminous than the Sun[12] but has a significantly lower luminosity than other bright stars such as Canopus or Rigel. The system is between 200 and 300 million years old.[12] It was originally composed of two bright bluish stars. The more massive of these, Sirius B, consumed its resources and became a red giant before shedding its outer layers and collapsing into its current state as a white dwarf around 120 million years ago.[12]

Sirius is also known colloquially as the "**Dog Star**", reflecting its prominence in its <u>constellation</u>, <u>Canis Major</u> (Greater Dog).[18] The <u>heliacal rising</u> of Sirius marked the flooding of the <u>Nile</u> in <u>Ancient Egypt</u> and the "<u>dog days</u>" of summer for the <u>ancient Greeks</u>, while to the <u>Polynesians</u> in the <u>Southern Hemisphere</u> the star marked winter and was an important reference for their navigation around the <u>Pacific Ocean</u>.

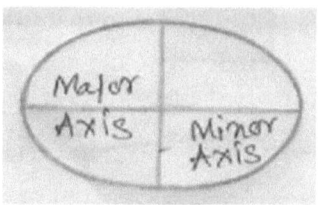

In reality, binary star systems are governed by Kepler's laws, as <u>modified by Newton</u> to account for the effect of the center of mass. Then each star executes an elliptical orbit such that at any instant the two stars are on opposite sides of the center of mass. The orbits generally are as depicted in the following figure.

Tilt of Binary Orbits

One important consideration for visual binary orbitals is that the plane of orbital revolution for such systems **is not usually perpendicular** to our line of sight. In general, there is some **tilt angle** *i*, as illustrated in the adjacent figure. Thus, when we see the orbit of a visual binary we **do not see the actual orbit** but only **the projection** of that orbit on the celestial sphere. For example, if the orbit looks like an ellipse, that could be because the orbit actually is elliptical, or because the **true orbit is a circle but we are seeing it from an angle that makes the circle look flattened and therefore elliptical**. In many cases it is possible to determine the angle *i* by careful measurement in order to deduce the true orbits of the binary system. In other cases we cannot and the angle *i* remains uncertain.

List of Multiple Star Systems - <u>600 Multiple Star Systems</u>.

Constellations Are Not Physical Groupings
Star Groupings and Asterisms

Some of the more familiar "**constellations**" are **technically not constellations at all**. For example, the grouping of stars known as the **Big Dipper** is **probably familiar to most**, but it is not actually a constellation. The **Big Dipper** is part of a larger grouping of stars called the **Big Bear (*Ursa Major*)** that *is* a constellation.

A well-known grouping of stars like the **Big Dipper that is not officially recognized as a constellation is called an** *asterism.*

VISUAL BINARIES

The apparent groupings of stars into constellations that we see on the celestial sphere are not physical groupings. In most cases the stars in constellations and asterisms are **each very different distances from us,** and only appear to be grouped because they **lie in approximately the same direction.**

Some **binary systems are sufficiently close to Earth** and the stars are well enough separated that we can see the **two stars** individually in a telescope and track their motion over a period of time. We term such systems *visual binaries.*

The Historical Constellations

In some cases one can discern easily the purported shape; for example, the constellation **Leo** shown on **the right might actually look like a lion** with the **dots connected as they are.** In other cases the supposed shape is very much in the eye of the beholder, as the example of **Canis Minor (The Little Dog).**

The general properties of the **celestial sphere: constellations, the naming of stars,** and a general **coordinate system for the celestial sphere that is analogous to the latitude-longitude system on the surface of the Earth** and that allows us to specify precisely a location on the **celestial sphere.** We shall also consider some general aspects of **timekeeping and calendars,** because historically, the regular apparent **motions of the heavens** provided many of the ideas and much of the **terminology** that we use in timekeeping.

ON GOING COMMAND
BIG-BANG OR THREE DIMENSIONAL ARROW

Command 'Kun', iron, arrow, <u>nucleosynthesis</u>, the arrow creating a vacuum behind it into which air rushed and applied a force to the back of the arrow, **Galileo's concept of inertia** was quite **contrary to Aristotle's ideas of motion:** in **Galileo's dynamics** the **arrow** (with very small frictional forces) **continued to fly through the air because of the law of inertia,** while a **block of wood** on a **table** stopped sliding once the applied **force** was removed because of **frictional forces** that Aristotle had **failed** to analyze correctly.

Higgs boson

God Particle, responsible for all the **mass** in the universe, iron, buried treasure, flame less fire, black stone, rock on which Muusa had stroke with thy staff, bish-shamsi, and the like], Higgs boson is responsible for all the **mass** in the universe, Intermediate **vector bosons.**

[92 + 7 = 99] Elementary particles
92 Attributive names and 7 repeated verses [Planets or beams of light]
25 Names of the messengers and warners in
resemblance with signs of Allah

All **observed elementary particles are either fermions or bosons,** elementary bosons are all gauge bosons: photons, W and Z bosons, gluons, and the Higgs boson, **Ninety-two elements,** Around **25 more** have been made artificially.

Photons = force carriers of the electromagnetic field

W and Z bosons = force carriers which mediate the weak force, [South Pole or Right side of the Kaba]

Gluons = fundamental force carriers underlying the strong force, [Upper West Zone]

Gravitation = the graviton, which is a boson of spin plus or minus two

Note: 92 + 7 [canopies] = 99 beautiful names and 25 shared names of the messengers as signs [artificial] of Allah.

Up or West or Major Axis = West Horizon, West Zone, strong force
Down or East or Minor Axis = East Horizon, East Zone, gravitational force
Right side of the Revelation & Manifestation = commonly called **South Pole, Southern hemisphere - weak forces**
Left side of the Revelation & Manifestation = commonly called **North Pole, Northern Hemisphere - electromagnetic force, revealed and manifested magnetism, North to North, back border**
Three minutes = Trinity, Three days communication of Jakariya with people through Signs only,

23 elements (up through iron) are formed mostly by **nuclear fusion processes within stars** = Effect 'Fayakun', 11 equally shared springs or matter particles and one equal and opposite Sirius Binary System, catastrophic collapse of massive stars (supernovae explosions), nuclear fusion [wise policy] within the trinity of Adam – wife of Adam – Sincere advisor Iblis,

An *element* **is an** *actual physical substance* that **cannot be broken down** into a simpler form capable of **an independent existence as observable matter**. As such, the concept of the element is a *macroscopic* one that relates to the world that we can observe with our senses. [As an element cannot be broken down into a simpler form, so an element is an attribute or an essence or a constituent.]
Ion - Positive and Negative Electric Charges - A neutral atom or group of atoms

Searching Questions
i)   **What is meant by Unification?**
ii) **What is the difference between Unification and Sirius Binary System?**

iii) **What is meant by 'electroweak' force?**

iv) **What is the difference between Companions of the Right hand and Electroweak Force?**

v) **What is the difference between 'electroweak force' and 'electro-gravitational force'?**

vi) **What is the difference between 'electroweak force' and 'life or consciousness'?**

vii) **What is the difference between 'Maryam supplied with sustenance' and 'electro-gravitational force'?**

viii) **What is the difference between 'the appointed light of Allah on earth or Shams' and 'electro-magnetic wave?**

ix) **What is meant by electromagnetic wave? What is the revealed and manifested existential import of electromagnetic wave? Which North represents Magnetic Force in resemblance with Manifested Sign of Magnetic Force?**

x) **I have searched out that the so-called sun [Bullet] is the manifested sign of both electromagnetic force and electro weak force or what is being taught as Natural Magnetism? Do you know this manifested truth or manifested nature? If you know this manifested truth, why are you knowingly concealing this manifested truth? If you are a people of the Kitaab with Truth or a people of the Book or a scientist, and you know any alternative manifested sign of electromagnetic wave & electroweak force, you are requested to share the same with reference to four verifiable and justifiable references.**

Unification = **W and Z particles, grand unification** Combining the weak and electromagnetic forces into a unified "electroweak" forces –
Four Walls within the Universe of the Manifested Hexagon = The helium-4 atom, consisting of **2 protons, 2 neutrons** and **2 electrons hexagonal earth**

COPERNICAN HYPOTHESIS

(1) **the assumption that the** Earth was the centre of the Universe, **(2) the** assumption of uniform circular motion in the heavens, **and (3) the** assumption that objects in the heavens were made from a perfect, unchanging substance not found on the Earth.

Copernican hypothesis

Heliocentric Universe of Copernicus - In the Copernican model the **Sun is at the centre of the universe.** Copernicus challenged assumption 1, but not assumption 2. We may also note that the **Copernican model implicitly**

454

questions the third tenet that **the objects in the sky were made of special unchanging stuff**. Since the **Earth is just another planet**, there will eventually be a natural progression to the idea that the **planets are made from the same stuff that we find on the Earth.**

Fermions	Bosons
Lepton and Quarks – Spin = ½	**Spin = 1* Force Carrier Particles**
Baryons [qqq] Spin = ½ [3/2, 5/2 ----]	**Spin = 0, 1, 2 --- Mesons [qq]**

1. Our entire universe is made of **12 different matter particles** and **four forces.**
2. Among those **12 particles**, you'll encounter **six quarks** and **six leptons.** **Quarks** make up **protons and neutrons**, while members of the **lepton family** include the **electron** and the **electron neutrino.**
3. The present belief is that **helium** and a few other very light elements were formed within about **three** minutes of the "**big bang**", and that the next **23 elements** (up through iron) are formed mostly by **nuclear fusion processes within stars**, in which lighter nuclei combine into successively heavier elements. [(11 x 2) + **wash-shamsa wal-qamara**] **= 23**
4. Maybe the famed boson's grand and **controversial nickname**, the "**God Particle**," has kept **media outlets buzzing**. Then again, the intriguing possibility that the Higgs boson is responsible for all the **mass** in the universe rather captures the imagination, too. Or perhaps we're simply excited to learn more **about our world**, and we know that if the **Higgs boson does exist, we'll unravel the mystery a little more.**
5. What exactly is the **Higgs boson**? - Particle physics usually has a hard time competing with politics and celebrity gossip for headlines, but the Higgs boson has garnered some serious attention.
6. That's exactly what happened on July 4, 2012, though, when scientists at CERN announced that they'd found a particle that behaved the way they expect the **Higgs boson** to behave.
7. First we discovered atoms, then protons, neutrons and electrons, and finally quarks and leptons (more on those later). But the universe doesn't only contain matter; it also **contains forces** that **act upon that matter.**
8. All the known **forces** in the universe are **manifestations** of **four fundamental forces**, the **strong, electromagnetic, weak,** and **gravitational** forces. Four blessed springs or planets for straight usage. But **why there are four forces**? Why not just one master force?

9. Gluons are the **fundamental force carriers** underlying the <u>strong force</u>.
10. Photons are the force carriers of the <u>electromagnetic field</u>.
11. W and Z bosons are the force carriers which mediate the <u>weak force</u>.
12. Finally, many approaches to quantum gravity postulate a force carrier for gravity, the <u>graviton</u>, which is a boson of spin plus or minus two. Composite particles (such as <u>hadrons</u>, <u>nuclei</u>, and <u>atoms</u>) can be bosons or fermions depending on their constituents. More precisely, because of the relation between spin and statistics, a particle containing an even number of fermions is a boson, since it has integer spin.
13. The nucleus of a <u>carbon-12</u> atom, which contains 6 protons and 6 neutrons.

1. Elements heavier than **iron** cannot be formed in this way, and are produced only during the **catastrophic collapse of massive stars (supernovae explosions)**.
2. The Moon is in **<u>synchronous rotation</u>** with Earth, always showing the **same face** with its **<u>near side</u>** marked by **dark volcanic <u>maria</u>** that **fill between the bright ancient crustal highlands and the prominent impact craters**. It is the **second-brightest regularly visible <u>celestial object</u>** in **Earth's sky** after the <u>Sun</u>, as measured by **illuminance** on Earth's surface. Although it can appear a <u>very bright white</u>, its **surface is actually dark**, with a <u>reflectance</u> just slightly higher than that of worn asphalt. Its prominence in the sky and its **regular cycle** of <u>phases</u> have, since ancient times, made the Moon an important cultural influence on <u>language</u>, **calendars**, <u>art</u>, and <u>mythology</u>
3. The Moon's gravitational influence produces the <u>ocean tides</u>, <u>body tides</u>, and the <u>slight lengthening</u> of the day. The Moon's current orbital distance is about thirty times the diameter of Earth, causing it to have an <u>apparent size</u> in the sky almost the same as that of the Sun, with the result that the **Moon covers the Sun nearly precisely in total <u>solar eclipse</u>**. This matching of apparent visual size is a **coincidence**. The Moon's linear distance from Earth is currently increasing at a rate of 3.82 ± 0.07 centimetres (1.504 ± 0.028 in) per year, but this rate is not constant.
4. The Moon makes a complete orbit around Earth with respect to the **fixed stars** about once every 27.3 days[g] (its <u>sidereal period</u>). However, because Earth is moving in its orbit around the Sun at the same time, it takes slightly longer for the Moon to show the same <u>phase</u> to Earth, which is about 29.5 days[h] (its <u>synodic period</u>).[55] Unlike most satellites of other planets, the **Moon orbits closer to the <u>ecliptic plane</u>** than to

the planet's <u>equatorial plane</u>. The Moon's orbit is subtly **perturbed** by the **Sun and Earth** in many **small, complex and interacting ways**. For example, the plane of the Moon's orbital motion <u>gradually rotates,</u> which affects other aspects of **lunar motion**. These follow-on effects are mathematically described by <u>Cassini's laws</u>.[114]

5. A **fermion** is any particle that has an **odd half-integer** (like 1/2, 3/2, and so forth) **spin**.

6. **Quarks** and **leptons**, as well as most **composite particles**, like **protons and neutrons, are fermions**.

7. For reasons we do not fully understand, a consequence of the **odd half-integer spin** is that fermions obey the **Pauli Exclusion Principle** and therefore cannot **co-exist** in the same state at same location at the same time.

8. **Bosons** are those particles which have an **integer spin** (0, 1, 2...). All the **force carrier particles are bosons**, as are those **composite particles with an** even number of fermion particles (like mesons).

9. The predicted graviton has **a spin of 2**.

10. Unlike bosons **two identical fermions cannot occupy the same quantum space**.

11. Whereas the elementary particles that make up matter (i.e. <u>leptons</u> and <u>quarks</u>) are fermions, the elementary bosons are force carriers that function as the 'glue' holding matter together.[10]

12. This property holds for all particles with integer <u>spin</u> (s = 0, 1, 2 etc.) as a consequence of the **spin–statistics theorem**.

13. The **nucleus** of an atom is a **fermion** or **boson** depending on whether the **total number of its protons** and **neutrons is odd or even**, respectively.

14. Those who joined the quest for a single unified master force declared that the first step toward **unification** had been achieved with the discovery of the discovery of the **W and Z particles,** the intermediate **vector bosons, in 1983**.

15. This brought experimental verification of particles whose prediction had already contributed to the Nobel Prize awarded to **Weinberg, Salam, and Glashow in 1979**.

16. **Combining the weak and electromagnetic forces into a unified "electroweak" force,** these great advances in both theory and experiment provide encouragement for moving on to the next step, the "**grand unification**" necessary to include the strong interaction.

17. All **observed elementary particles are either fermions or bosons**. The observed elementary bosons are all <u>gauge bosons</u>: photons, <u>W and Z bosons</u>, gluons, and the <u>Higgs boson</u>.

18. The number of bosons within a composite particle made up of simple particles bound with a potential has no effect on whether it is a boson or a fermion.
19. **Earth's movement** around the **Sun** is the basis of the **solar calendar**,
20. It is thought that almost **all stars form by this process**.
21. The **Sun is roughly middle age**.
22. **Sun was rotating on an axis**.
23. **Earth might rotate on an axis too**, as required in the **Copernican model**. Both represented new facts that were unknown to **Aristotle and Ptolemy**.
24. **The helium-4 atom**, consisting of **2 protons**, **2 neutrons** and **2 electrons**.
25. The enormous effect of the Sun on the Earth has been recognized since prehistoric times, and the Sun has been **regarded by some cultures as a deity**. Earth's **movement around the Sun is the basis of the solar calendar**, which is the predominant calendar in use today.
26. Some frequently-asked questions about elements – How many elements are there? **Ninety-two elements** have been found in nature. Around **25 more** have been made artificially, but all of these decay into lighter elements, with some of them disappearing in minutes or even seconds.
27. Where do the elements come from? The processes by which elements (or more properly, their nuclei) are formed is known as nucleosynthesis. The very first (primordial) nuclei formed shortly after the "big bang".

Four Galilean moons

Galileo observed **4 points of light** that changed their positions with time around the planet Jupiter. He concluded that these were objects in orbit around Jupiter. Indeed, they were the **4 brightest moons of Jupiter**, which are now commonly called the *Galilean moons* (Galileo himself called them the *Medicea Siderea*---the **"Medician Stars"**).

MILKY WAY OR GREATER CLOUD

The great "cloud" called the **Milky Way** (which we now know to be the disk of our spiral galaxy) was composed of **enormous numbers of stars** that **had not been seen before**.

Key Concepts – Hidden Force, Frictional Force, Surface, Object

Galileo, by virtue of a series of experiments (many with objects **sliding down inclined planes**), realized that the analysis of **Aristotle was incorrect** because it failed to account properly for a **hidden force**: the *frictional force* between the **surface** and the **object**.

NEWTON'S THIRD LAW OF MOTION

For every action there is an equal and opposite reaction. This law is exemplified by what happens if we step off **a boat onto the bank of a lake: as we move in the direction of the shore, the boat tends to move in the opposite direction (leaving us face down in the water**, if we aren't careful!).

What Really Happened with the Apple? Probably the more correct version of the story is that Newton, upon observing an apple fall from a **tree**, began to think along the following lines: The apple is **accelerated**, since its **velocity changes** from **zero** as it is hanging on the **tree** and **moves toward the ground**. Thus, by Newton's 2nd Law there must be a force that acts on the apple to cause this acceleration. Let's call this force "gravity", and the associated acceleration the "accleration due to gravity". Then imagine the **apple tree is twice as high**. Again, we expect the **apple to be accelerated toward the ground**, so this suggests that this **force** that we call **gravity** reaches to the top of the **tallest apple tree**.

Now came Newton's truly brilliant insight: if the force of gravity reaches to **the top of the highest tree**, might it not reach even further; in particular, might it not reach all the way to the **orbit of the Moon**! Then, **the orbit of the Moon about the Earth could be a consequence of the gravitational force**, because the **acceleration due to gravity** could change the velocity of the **Moon in just such a way that it followed an orbit around the earth.**

But as we increase the muzzle velocity for our imaginary cannon, the projectile will travel further and further before returning to earth. Finally, Newton reasoned that if the cannon projected the cannon ball with exactly the right velocity, the projectile would travel completely around the Earth, always falling in the gravitational field but never reaching the Earth, which is **curving away** at the same rate that the projectile falls. That is, *the cannon ball would have been put into orbit around the Earth.* **Newton concluded that the orbit of the Moon was of exactly the same nature: the Moon continuously "fell" in its path around the Earth because of the acceleration due to gravity, thus producing its orbit.**

By such reasoning, Newton came to the conclusion that any two objects in the Universe exert gravitational attraction on each other, with the force having a universal form.

The Two-Body Approximation - However, from the form of the <u>gravitational force</u>

In order to understand the discoveries of Newton, we must have an understanding of three basic quantities: (1) **velocity**, (2) **acceleration**, and (3) **force**. These three quantities have a common feature: they are what mathematicians call *vectors*.

Observing that the planet Saturn had "ears", we now know that Galileo was observing the **rings of Saturn**, but his telescope was not good enough to show them as more than extensions on either side of the planet.

GALILEO

Showing that the **Moon** was not smooth, as had been assumed, but was **covered** by **mountains** and craters [**hollow spaces**].

GALILEO AND LEANING TOWER [MOUNT TUUR]
Pisa experiment

Galileo made extensive contributions to our understanding of the laws governing the motion of objects. **The** famous Leaning Tower **of Pisa experiment may be** apocryphal. **It is likely that** Galileo himself did not drop **two objects** of very **different weight** from the tower to prove that **(contrary to popular expectations) they would hit the ground at the same time. However, it is certain that** Galileo understood the principle involved**, and** probably did similar experiments. **The realization that, as we would say in modern terms, the** acceleration due to gravity is independent of the weight of an object **was important to the formulation of a theory of gravitation by Newton.**

STATE OF EQUILIBRIUM OR SAYMYAVASTA
OF PRAKRITI [SANKHYA PHILOSOPHY]
Galileo = Inertia - State of Motion

Perhaps **Galileo's greatest contribution to physics** was his formulation of the concept of *inertia*: an object in a **state of motion** possesses an **"inertia"** that causes it to remain in **that state of motion** unless an **external force acts on it.**

INERTIA
Aristotle – Motion & Constant Act of a Force

Aristotle held that **objects at rest remained at rest unless a force acted on them,** but that **objects in motion did not remain in motion unless a force acted constantly on them.**

INERTIA – UNIFORM MOTION & EXTERNAL FORCE
NEWTON'S FIRST LAW OF MOTION
LAW OF INERTIA

i) State of Uniform Motion
ii) Uniform Circular Motion is Accelerated Motion = **continual acceleration**
iii) Motion on a Curved Path is Accelerated Motion.

Every object in a **state of uniform motion** tends to remain in that state of motion unless an **external force** is applied to it. (This we recognize as essentially <u>Galileo's</u> concept of **inertia** and this is often termed simply the "**Law of Inertia**".)

Uniform Circular Motion is **Accelerated Motion = continual acceleration,**

Motion on a curved path is **accelerated motion.**

Aristotle – Motion

i) **Rest** - Objects at rest remained at rest unless a force acted on them.
ii) **Motion** - Objects in motion did not remain in motion unless a force acted constantly on them.
iii) **Constantly Acting Force**
iv) **Act of a Force**

Aristotle held that **objects at rest remained at rest unless a force acted on them,** but that **objects in motion did not remain in motion unless a force acted constantly on them.**

Galileo = State of Motion

Perhaps **Galileo's greatest contribution to physics** was his formulation of the concept of *inertia*: an object in a **state of motion** possesses an **"inertia"** that causes it to remain in **that state of motion** unless an **external force acts on it.**

SLIDING DOWN INCLINED
DECLINING TO THE RIGHT SIDE OF THE MOUNT TUUR

Galileo, by virtue of a series of experiments (many with objects **sliding down inclined planes**), realized that the analysis of **Aristotle was incorrect** because it failed to account properly for a **hidden force**: the *frictional force* between the surface and the object.

461

HIDDEN FORCE [Layilatul Qadr]

Galileo, by virtue of a series of experiments (many with objects **sliding down inclined planes**), realized that the analysis of **Aristotle was incorrect** because it failed to account properly for a **hidden force**: the *frictional force* between the surface and the object.

FRICTIONAL FORCE AND INERTIA
TWO FORCES ACT IN THE OPPOSITE DIRECTION
[Galileo's Abstraction of the law of inertia]
Force associated with the push
&
Force that is associated with the friction

Thus, as we push the block of **wood across the table**, there are **two opposing forces** that act: the **force associated with the push**, and a **force that is associated with the friction** and **that acts in the opposite direction**. Galileo realized that as the **frictional forces were decreased** (for example, by placing **oil on the table**) the **object would move further** and **further before stopping**. From this he abstracted a basic form of the **law of inertia**: if the **frictional forces** could be reduced to **exactly zero** (not possible in a realistic experiment, but it can be **approximated to high precision**) **an object pushed at constant speed across a frictionless surface of infinite extent will continue at that speed forever after we stop pushing, unless a new force acts on it at a later time.**

Protons and neutrons are in the centre (nucleus) of the atom. You may want to mention that **hydrogen is the only atom that usually has no neutrons.** The nucleus of most hydrogen atoms is composed of **just 1 proton**. A small percentage of hydrogen atoms have 1 or even 2 neutrons. Atoms of the same element with different numbers of neutrons are called **isotopes**.

Aristotle taught that the **substances** making up the **Earth** were **different** from the substance making up the **heavens** [Galaxy of Stars]. He also taught that dynamics (the branch of physics that deals with motion) was primarily determined by the **nature of the substance that was moving**.

PRIME MOVER & APPLIED FORCES

Key Concepts –
  i)   Stripped to its essentials
  ii) A stone fell to the ground because the stone and the ground were similar in substance

iii) Smoke rose away from the Earth because in terms of the 4 basic elements it was primarily air (and some fire)

iv) Smoke wished to be closer to air and further away from earth and water

v) More perfect substance (the "quintessence") that made up the heavens [stars] had as its nature to execute perfect (that is, uniform circular) motion

vi) Objects only moved as long as they were pushed

vii) Objects on the Earth stopped moving once applied forces were removed

viii) Heavenly spheres only moved because of the action of the Prime Mover

ix) Continually applied the force to the outer spheres that turned the entire heavens

x) The arrow creating a vacuum behind it into which air rushed and applied a force to the back of the arrow

For example, **stripped** to its **essentials,** Aristotle believed that a **stone [ray of light as an octagonal planet with a particular attribute]** fell to the ground because the stone and the **ground** were similar in **substance (in terms of the 4 basic elements, they were mostly "earth").** Likewise, **smoke [Electron clouds or Magnetic field or Earth sky or the so-called sun which rises from the east]** rose away from the **Earth** because in terms of the **4 basic elements** it was primarily **air** (and **some fire**), and therefore the **smoke** wished to be **closer to air [light or Nuurun alaa Nuur]** and further **away from earth and water.** By the same token, Aristotle held that the **more perfect substance** (the "quintessence") that made up the **heavens** [stars] had as its nature to **execute perfect** (that is, **uniform circular**) motion. He also believed that objects only moved as long as they were **pushed.** Thus, **objects on the Earth stopped moving once applied forces were removed,** and the **heavenly spheres only moved because of the action of the Prime Mover [One Dot Leader],** who **continually applied the force to the outer spheres that turned the entire heavens.** (A notorious problem for the Aristotelian view was why arrows shot from a bow continued to fly through the air after they had left the bow and the string was no longer applying force to them. Elaborate explanations were hatched; for example, it was proposed that **the arrow creating a vacuum behind it into which air rushed and applied a force to the back of the arrow!**)

Galileo vs. Aristotle
FRICTIONAL FORCE
ARROW & BLOCK OF WOOD ON A TABLE

Thus, Aristotle believed that the **laws governing the motion of the heavens** [Galaxy of Stars] were a **different set of laws** than those that governed **motion**

on the earth. As we have seen, **Galileo's concept of inertia** was **quite contrary to Aristotle's ideas of motion**: in **Galileo's dynamics the arrow (with very small frictional forces) continued to fly through the air because of the law of inertia, while a block of wood on a table [opposite arrow] stopped sliding once the applied force was removed because of frictional forces** that Aristotle had failed to analyze correctly.

LIGHT AND STRONG GRAVITATIONAL FORCE

Then Albert Einstein shook the foundations of physics with the introduction of his **Special Theory of Relativity in 1905**, and his **General Theory of Relativity in 1915** (Here is an example of a thought experiment in special relativity). The first showed that **Newton's Three Laws of Motion were only approximately correct**, breaking down when **velocities approached that of light**. The second showed that Newton's Law of Gravitation was also only approximately correct, breaking down in the presence of very **strong gravitational force**.

EINSTEIN'S PREDICTION – GRAVITATIONAL LENSING

Einstein's **theory predicts** that the **direction of light propagation should be changed in a gravitational field, contrary to the Newtonian predictions**. **Precise observations** indicate that **Einstein is right**, both about the effect and its magnitude. A striking consequence is **gravitational lensing.**

EINSTEIN'S THEORY
Strong Gravitational Field

The **General Theory of Relativity predicts** that **light coming from a strong gravitational field should have its wavelength shifted to larger values** (what astronomers call a "red shift"), again contrary to **Newton's theory**. Once again, detailed observations indicate such a red shift, and that its magnitude is correctly given by **Einstein's theory**.

EINSTEIN'S THEORY

The **electromagnetic field can have waves** in it that **carry energy** and **that we call light**. Likewise, the **gravitational field can have waves** that **carry energy** and are called *gravitational waves*. These may be thought of as ripples [waves] in the **curvature of space-time** that **travel at the speed of light**.

ELECTROMAGNETIC WAVES AND GRAVITATIONAL WAVES
BINARY PULSAR

Just as **accelerating charges can emit electromagnetic waves, accelerating masses can emit gravitational waves**. However **gravitational waves** are

difficult to detect because they are very **weak** and **no conclusive evidence has yet been reported for their direct observation**. They have been **observed** *indirectly* in the **binary pulsar**. Because the arrival time of pulses from the **pulsar** can be measured very precisely, it can be determined that the period of the **binary system** is **gradually decreasing**.

EQUAL AND OPPOSITE SCIENCE – II

Copernican hypothesis
In addition, Galileo's extensive telescopic observations of the heavens made it more and more plausible that they were not made from a perfect, unchanging substance. In particular, Galileo's observational confirmation of the **Copernican hypothesis** suggested that the **Earth was just another planet**, so maybe it was made from the same material as the other planets.

The Copernican Revolution
We noted earlier that **3 incorrect ideas** held back the development of modern astronomy from the time of Aristotle until the 16th and 17th centuries: (1) the assumption that the **Earth was the centre of the Universe**, (2) the **assumption of uniform circular motion in the heavens**, and (3) the **assumption that objects in the heavens were made from a perfect, unchanging substance not found on the Earth**.
Copernicus challenged assumption 1, but not assumption 2. We may also note that the **Copernican model implicitly questions the third tenet that the objects in the sky were made of special unchanging stuff**. Since the **Earth is just another planet**, there will eventually be a natural progression to the idea that the **planets are made from the same stuff that we find on the Earth.**

Copernicus was an unlikely revolutionary. It is believed by many that his book was only published at the end of his life because he feared ridicule and disfavour: by his peers and by the Church, which had elevated the ideas of Aristotle to the level of religious dogma. However, this reluctant revolutionary set in motion a chain of events that would eventually (long after his lifetime) produce the greatest revolution in thinking that Western civilization has seen. **His ideas remained rather obscure for about 100 years after his death**. But, in the 17th century the work of Kepler, Galileo, and Newton would build on the **heliocentric Universe of Copernicus** and produce the revolution that would sweep away completely the ideas of Aristotle and replace them with the modern view of astronomy and natural science. This sequence is commonly called the *Copernican Revolution*.

Galileo Galilei (1564-1642) was a pivotal figure in the development of modern astronomy, both because of his contributions directly to astronomy, and because of his work in physics and its relation to astronomy. He provided the crucial observations that proved the **Copernican hypothesis**, and also laid the foundations for a correct understanding of how objects moved on the surface of the earth (dynamics) and of gravity.

Newton, who was born the same year that Galileo died, would build on Galileo's ideas to demonstrate that the laws of motion in the heavens and the laws of motion on the earth were one and the same. Thus, Galileo began and Newton completed a synthesis of astronomy and physics in which the former was recognized as but a particular example of the latter, and that would banish the notions of Aristotle almost completely from both.

One could, with considerable justification, view Galileo as the father both of modern astronomy and of modern physics.

Copernicus and the Need for Epicycles

There is a common misconception that the Copernican model did away with the need for epicycles. This is not true, because Copernicus was able to rid himself of the long-held notion that the **Earth was the center of the Solar system**, but he did not question the assumption of **uniform circular motion**. Thus, in the Copernican model the **Sun was at the center**, but the planets still executed uniform circular motion about it. As we shall see later, the orbits of the planets are not circles, they are actually **ellipses**. As a consequence, the Copernican model, with it assumption of uniform circular motion, still could not explain all the details of planetary motion on the celestial sphere without epicycles. The difference was that the Copernican system required many *fewer epicycles* than the Ptolemaic system because it moved the Sun to the center.

KEPLER

Thus, the groundwork was laid by Galileo (and to a lesser extent by others like **Kepler and Copernicus**) to overthrow the physics of Aristotle, in addition to his astronomy. It fell to Isaac Newton to bring these threads together and to demonstrate that the laws that governed the heavens were the same laws that governed motion on the surface of the Earth.

Galileo did not invent the telescope (Dutch spectacle makers receive that credit), but he was the first to use the telescope to study the heavens systematically. His little telescope was poorer than even a cheap modern amateur telescope, but what he observed in the heavens rocked the very foundations of Aristotle's universe and the theological-philosophical worldview that it supported. It is said that what Galileo saw was so disturbing for some officials of the Church that they refused to even look through his telescope; they reasoned that the **Devil** was capable of making anything appear in the telescope, so it was best not to look through it.

NEWTON'S SECOND LAW OF MOTION

The relationship between an object's mass m, its acceleration a, and the applied force F is $F = ma$. Acceleration and force are vectors (as indicated by their symbols being displayed in slant bold font); in this law the direction of the force vector is the same as the direction of the acceleration vector.

This is the most powerful of Newton's three Laws, because it allows quantitative calculations of dynamics: how do velocities change when forces are applied. Notice the fundamental difference between Newton's 2nd Law and the dynamics of Aristotle: according to Newton, a force causes only a *change in velocity* (acceleration); it does not maintain the velocity as Aristotle held.

This is sometimes summarized by saying that under Newton, $F = ma$, but under Aristotle $F = mv$, where v is the velocity. Thus, according to Aristotle there is only a velocity if there is a force, but according to Newton an object with a certain velocity maintains that velocity *unless* a force acts on it to cause acceleration (that is, a change in the velocity). As we have noted earlier in conjunction with the discussion of Galileo, Aristotle's view seems to be more in accord with common sense, but that is because of a failure to appreciate the role played by frictional forces. Once account is taken of *all* forces acting in a given situation it is the dynamics of Galileo and Newton, not of Aristotle, that are found to be in accord with the observations.

Acceleration in Keplerian Orbits

Kepler's Laws are illustrated in the adjacent animation. The **red arrow indicates** the **instantaneous velocity vector** at each point on the orbit (as always, we greatly exaggerate the eccentricty of the ellipse for purposes of illustration). Since the velocity is a vector, the direction of the velocity vector is indicated by the direction of the **arrow** and the magnitude of the velocity is indicated by the length of the arrow.

Notice that (because of Kepler's 2nd Law) the velocity vector is constantly changing both its magnitude and its direction as it moves around the **elliptical orbit** (if the orbits were circular, the magnitude of the velocity would remain constant but the direction would change continuously). Since either a change in the magnitude or the direction of the velocity vector constitutes an acceleration, there is a continuous acceleration as the planet moves about its orbit (whether circular or elliptical), and therefore by Newton's 2nd Law there is a force that acts at every point on the orbit. Furthermore, the force is not constant in magnitude, since the change in velocity (acceleration) is larger when the planet is near the Sun on the elliptical orbit.

Newton's Laws and Kepler's Laws

Since this is a survey course, we shall not cover all the mathematics, but we now outline how Kepler's Laws are implied by those of Newton, and use Newton's Laws to supply corrections to Kepler's Laws.

Since the **planets move on ellipses** (Kepler's 1st Law), they are continually **accelerating**, as we have noted above. As we have also noted above, this implies **force** acting continuously on the planets.

Because the **planet-Sun** line sweeps out **equal areas in equal times** (Kepler's 2nd Law), it is possible to show that the **force** must be directed **toward the Sun from the planet.**

From Kepler's 1st Law the orbit is an **ellipse** with the **Sun at one focus**; from Newton's laws it can be shown that this means that the magnitude of the force must vary as one over the square of the distance between the planet and the Sun.

Kepler's 3rd Law and Newton's 3rd Law imply that the **force must be proportional to the product of the masses for the planet and the Sun.**

Thus, Kepler's laws and Newton's laws taken together imply that **the force that holds the planets in their orbits by continuously changing the planet's velocity so that it follows an elliptical path is (1) directed toward the Sun from the planet**, (2) is proportional to the product of **masses for the Sun and planet,** and (3) is inversely proportional to the square of the planet-Sun separation. This is precisely the form of the gravitational force, with the universal gravitational constant G as the constant of proportionality. Thus, Newton's laws of motion, with a gravitational force used in the 2nd Law, imply Kepler's Laws, and the planets obey the same laws of motion as objects on the surface of the Earth!

Newton's discoveries suggest that Pope was indulging only slightly in hyperbole. We shall concentrate on three developments of most direct relevance to our discussion: (1) **Newton's Three Laws of Motion, (2) the Theory of Universal**

Gravitation, and (3) the demonstration that Kepler's Laws follow from the Law of Gravitation.

The Great Synthesis of Newton

<u>Kepler</u> **had proposed three Laws of Planetary motion** based on the systematics that he found in Brahe's data. These Laws were supposed to apply only to the motions of the planets; they said nothing about any other motion in the Universe. Further, they were purely empirical: they worked, but no one knew a fundamental reason WHY they should work.

Newton changed all of that. First, he demonstrated that the motion of objects on the Earth could be described by three new Laws of motion, and then he went on to show that Kepler's three Laws of Planetary Motion were but special cases of Newton's three Laws if a force of a particular kind (what we now know to be the gravitational force) were postulated to exist between all objects in the Universe having mass. In fact, Newton went even further: he showed that Kepler's Laws of planetary motion were only approximately correct, and supplied the quantitative corrections that with careful observations proved to be valid.

In the interplay between quantitative observation and theoretical construction that characterizes the development of modern science, we have seen that <u>Brahe</u> was the master of the first but was deficient in the second. The next great development in the history of astronomy was the theoretical intuition of Johannes Kepler (1571-1630), a German who went to Prague to become Brahe's assistant.

Brahe's Data and Kepler

Kepler and Brahe did not get along well. Brahe apparently mistrusted Kepler, fearing that his bright young assistant might eclipse him as the premiere astonomer of his day. He therefore let Kepler see only part of his voluminous data.

He set Kepler the task of understanding the orbit of the **planet Mars**, which was particularly troublesome. It is believed that part of the motivation for giving the Mars problem to Kepler was that it was difficult, and Brahe hoped it would occupy Kepler while Brahe worked on his theory of the Solar System. In a supreme irony, it was precisely the Martian data that allowed Kepler to formulate the correct laws of planetary motion, thus eventually achieving a place in the development of astronomy far surpassing that of Brahe.

Kepler and the Elliptical Orbits

Unlike Brahe, Kepler believed firmly in the <u>Copernican system</u>. In retrospect, the reason that the orbit of Mars was particularly difficult was that Copernicus had correctly placed the Sun at the center of the Solar System, but had erred in assuming the orbits of the planets to be circles. Thus, in the Copernican theory epicycles were still required to explain the details of planetary motion.

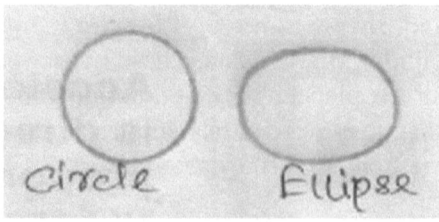

It fell to Kepler to provide the final piece of the puzzle: after a long struggle, in which he tried mightily to avoid his eventual conclusion, Kepler was forced finally to the realization that the orbits of the planets were not the circles demanded by Aristotle and assumed implicitly by Copernicus, but were instead the "flattened circles" that geometers call <u>ellipses</u> (See adjacent figure; the planetary orbits are only slightly elliptical and are not as flattened as in this example.)

The irony noted above lies in the realization that the difficulties with the Martian orbit derive *precisely* from the fact that the orbit of Mars was the most elliptical of the planets for which Brahe had extensive data. Thus Brahe had unwittingly given Kepler the very part of his data that would allow Kepler to eventually formulate the correct theory of the Solar System and thereby to banish Brahe's own theory!

Some Properties of Ellipses

Since the orbits of the planets are ellipses, let us review a few basic properties of ellipses.

For an ellipse there are two points called foci (singular: focus) such that the sum of the distances to the foci from any point on the ellipse is a constant. In terms of the diagram shown to the left, with "x" marking the location of the foci, we have the equation that defines the ellipse in terms of the distances a and b. $a + b =$ constant.

The amount of "flattening" of the ellipse is termed the *eccentricity*. Thus, in the following figure the ellipses become more eccentric from left to right. A circle may be viewed as a special case of an ellipse with zero eccentricity, while as the ellipse becomes more flattened the eccentricity approaches one.

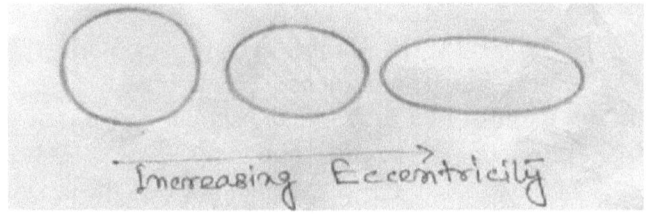

Thus, all ellipses have eccentricities lying between zero and one.

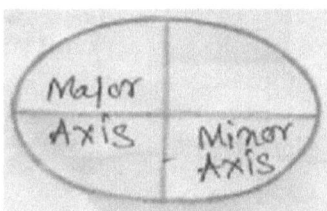

The orbits of the planets are ellipses but the eccentricities are so small for most of the planets that they look circular at first glance. For most of the planets one must measure the geometry carefully to determine that they are not circles, but ellipses of small eccentricity. Pluto and Mercury are exceptions: their orbits are sufficiently eccentric that they can be seen by inspection to not be circles.

The long axis of the ellipse is called the *major axis*, while the short axis is called the *minor axis* (adjacent figure). Half of the major axis is termed a *semimajor axis*. The length of a semimajor axis is often termed the size of the ellipse. It can be shown that the average separation of a planet from the Sun as it goes around its elliptical orbit is equal to the length of the semimajor axis. Thus, by the "radius" of a planet's orbit one usually means the length of the semimajor axis. For a more detailed investigation of the properties of ellipses, see this java applet.

The Laws of Planetary Motion

Kepler obtained Brahe's data after his death despite the attempts by Brahe's family to keep the data from him in the hope of monetary gain. There is some evidence that Kepler obtained the data by less than legal means; it is fortunate for the development of modern astronomy that he was successful. Utilizing the voluminous and precise data of Brahe, Kepler was eventually able to build on the realization that the orbits of the planets were ellipses to formulate his *Three Laws of Planetary Motion*.

Kepler's First Law:
The orbits of the planets are ellipses, with
the Sun at one focus of the ellipse.

Kepler's First Law is illustrated in the image shown above. The Sun is not at the center of the ellipse, but is instead at one focus (generally there is nothing at the other focus of the ellipse). The planet then follows the ellipse in its orbit, which means that the Earth-Sun distance is constantly changing as the planet goes around its orbit. For purpose of illustration we have shown the orbit as rather eccentric; remember that the actual orbits are much less eccentric than this. Kepler's Second Law:

The line joining the planet to the **Sun sweeps out equal areas in equal** times as the planet travels around the ellipse.

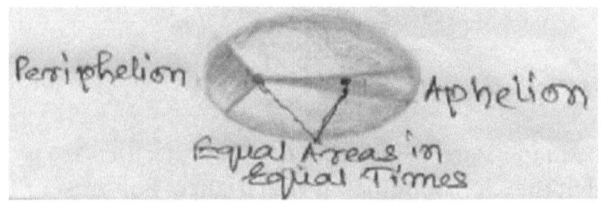

Kepler's second law is illustrated in the preceding figure. The line joining the Sun and planet sweeps out equal areas in equal times, so the planet moves faster when it is nearer the Sun. Thus, a planet executes elliptical motion with constantly changing **angular speed** as it moves about its orbit. The point of **nearest approach** of the planet to the Sun is termed *perihelion*; the point of **greatest separation** is termed *aphelion*. Hence, by Kepler's second law, the planet moves **fastest** when it is near perihelion and **slowest** when it is near aphelion.

NO WAY TO ESCAPE

Ayat\Verses – As to those who **deny the Signs** (Manifest Truths) of Allah and **slay the Prophets wrongfully**, and **slay** those of mankind **who enjoins equity**, announce to (promise) them a **grievous penalty**. Those are they whose **works will bear no fruit** in this **world [present life]** and in the **hereafter**, and they have **no helper**. [Sura No, 2\3 – Imran, Ayat No. –21 & 22]

Ayat\Verses - **Have you not seen that Allah created the heavens and the world in Truth**? If He so will, He can remove you and put [in your place] a new creation? And that is no great matter for Allah. They will all be marshalled before Allah together. Then the weak will say to those who were arrogant: For us, we but followed you; **can you then avail us to all against the wrath** of Allah? They will reply: **If we had received the Guidance of Allah, we should have given it to you**. To us it makes no difference [now] whether we rage, or bear [these torments] with patience. For us there is **no way of escape**. And Shaytan will say when the matter is decided: It was Allah Who gave you a promise of Truth: I too promised, but I failed in my promise to you. I had no authority over you except to call you but you listened to me. **Then reproach not me, but reproach your own souls**. I cannot listen to your cries, nor can you listen to mine. I reject your former act in associating me with Allah. For wrong-doers there must be a grievous penalty. [Sura No. – 13\14 – Ibrahim, Ayat No. – 19 to 22]

DAY OF WARNING – REMOVAL OF THE VEIL OF IGNORANCE
Ayat\Verses – And the Trumpet shall be blown: That will be the Day whereof Warning [had been given]. And there will come forth every soul with each will be an [angel] to drive, and an [angel] to bear witness. [And to the evil-doers it will be said]: You were heedless of this. Now We have removed from you your veil (covering) of ignorance and sharp is you sight this Day. And (unto the devil-doer) his comrade (two-in-one partnership) will say: Here is [his Record] ready with me. [Sura No. 49\50 – Qaaaf, Ayat No. – 20 to 23]

SECTION – VI
NATURAL MAGNETISM – 6.3
MAGNETISM

Fasta-'iz billaahi minash-Shaytaanir-Rajim
Bismillaahir-Rahmaanir- Rahim

A SINGLE FACT – MAGNETISM

Ayat\Verses - Have you not seen those unto whom it was said: **Hold back your hands**, establish worship and pay the poor-due? But when fighting was prescribed for them, behold! A section of them feared men as - or even more than - they should have feared Allah. They said: Our Rab! Why have you ordained fighting for us? If only You give us respite yet a while! Say: Short is the enjoyment of this world [present life]; the Hereafter is the better for those who ward off evil. Never will you be dealt with unjustly in the very least (down upon a date-stone). **Wherever you are, death will find you out, even if you are in towers built up strong and high**! If some good befalls them, they say: This is from Allah; but if evil, they say, this is from you. Say: All things are from Allah. But **what has come to these people that they fail to understand a single fact?** Whatever good, [O man!] happens to you, is from Allah; but whatever evil happens to you, is from yourself. We have sent you (Muhammad) as a messenger unto mankind. And **enough is Allah for a witness**. [Sura No. – 3\4 – Nisaa, Ayat No. – 77 to 79]

Ayat\Verses – Say: I do admonish you on **one point only: That you awake for the sake of Allah, by twos [top stair & middle stair] or singly [ground stair], and you reflect** (within yourselves / give thought). **There is no madness in your friend**. He is no less than a **warner** to you, in face of a **terrible Penalty**. [Sura No. – 33\34 – Saba, Ayat No. – 46]

There are four cardinal [revealed] directions in coherence with four basic forces with respect to equal & opposite revelation and in correspondence with Plus Sign of Mathematics. These four cardinal directions are East [Down], West [Up], South [Revealed Right], and North [Revealed Left]. East represents Gravitational Force. West represents Strong Force. North represents Magnetic Force. South represents Weak Force.

The world has been manifested declining towards revealed right in correspondence with Multiplication Sign of Mathematics and as a hexagon. So, there are six directions with respect to manifested hexagonal world. Each

Zone [Seashore] of the manifested hexagonal world comprises three regions [triangles]. East Zone [Lower Seashore] comprises North-East Region of North America [NE], Middle-East Region of Eartha 3D [ME], and South-East Region of South America [SE]. West Zone [Upper Seashore] comprises South-West Region of Europe [SW], Upright West Region of Arabian Peninsula [AP], and North-West Region of Asia – Africa – Australia [NW].

The Universe has been revealed as an equal & opposite Trinity and manifested as an upright rectangle. There are two Horizons and two Hemispheres. Two Horizons are West Horizon and East Horizon. East Horizon within the four basic forces is called Black Square [Non-luminous Moon]. There are two Zones [Seashores], namely, East Zone [Lower Seashore] and West Zone [Upper Seashore] within the Black Square of East Horizon. The World has been manifested as a Hexagon of six Regions [Triangles] within the Black Square of East Horizon. East Zone comprises North America, Straight Middle East of Eartha 3D, and South America. West Zone comprises Europe, Arabian Peninsula, Asia, Africa, and Australia. The **so-called sun [Bullet] as the manifested sign of natural magnetism neither rises from the East nor sets in the West [Ref. Ayat No. – 35 of Sura Nuur]. In other words, the direction from which the so-called sun as the manifested sign of natural magnetism enters / rises for the equal & opposite seashore cannot be called manifested East direction. Simultaneously, the direction towards which the so-called sun as the manifested sign of natural magnetism sets / ends the journey for the equal & opposite seashore cannot be called manifested West direction. These are the manifested as well as unalterable truths in resemblance with revelations.**

It is being taught that the so-called sun [Bullet / Tarash-Shamsi / Wash-Shamsi] rises from the East. This teaching is contradictory to both Revelation and Manifestation. So, this teaching is a **Black Magic**. Manifested East direction within the Black Square of East Horizon represents Gravitational Force. In other words, **Gravitational Wave [Electron Cloud or Ether or Bish-Shamsi] rises from the East within the Black Square of East Horizon as a Veil or Screen between Wash-Shamsa Wal Qamara or the so-called sun and the so-called moon**. Newton's Law of Gravitation is a valid reference in this regard.

The so-called sun [Bullet / Tarash-Shamsi] as the manifested sign of Natural Magnetism represents Magnetic Force or Electromagnetic Wave or North. So, **what we commonly perceive as day-light and commonly conceive as sun is not Gravitational Wave but Electromagnetic Wave of the Binary**

Pulsar of Einstein. Gravitational Wave [Electron Cloud / Ether / Bish-shamsi] rises from the East; while Electromagnetic Wave [so-called sun or Manifested sign of Revealed Magnetic Force or Bullet or Tarash-Shamsi or converted snake of Muusa's Staff or Binary Pulsar of Einstein or Niche] enters from Revealed Left or North Pole [Manifested Northern Hemisphere]. Conspiracy Science has been playing equal & opposite game with what accelerating charges emits [Electromagnetic Waves] and what accelerating masses emits [Gravitational Waves]. This equal & opposite game is worse than killing. This equal & opposite game is called immediate inference and research methodology. This methodology is grounded on Mechanism, Man-made Magnetism, Napier Bones, and conspiracy pen-paper-pencil works & activities. If I share with my friends in deed that **Magnetic Force** represents **North** and so-called sun as the **manifested sign of natural magnetism** rises from the **East**, then that sharing will be contradictory to Manifested Nature [Manifested Truth in resemblance with Revelation]. In other words, that teaching will involve subjective selfcontradictions, objective paradoxes, existential fallacies, and post hoc ergo propter hoc. If I argue further against manifested truth or in favour of Post Hoc Ergo Propter Hoc with reference to innumerable pen-paper-pencil works and activities, then I am none but one of the agents of the Don of Historical Black Magic [Teleological Evidence Sorcery and Unique Epistemic Persecution].

Further, it is scientifically verifiable factual truth [certain knowledge or manifested truth] and logically justifiable valid sharing that North [Magnetic Force] and South [Weak Force] directions determined by man-made magnetism cannot be called Manifested North [Northern Hemisphere or Haiyalas-Swalaah] and Manifested South [Southern Hemisphere or Haiyalaal-Falaah] or Revealed Left and Revealed Right. On the contrary, North and South directions determined by man-made magnetism must be categorised as **man-made directions or Man-made North and Man-made South of the Man-made Nature or Historical Black Magic. These man-made directions are the Vera Causa of leading believers of Europe, Asia, Africa, and Australia towards Manifested North [Haiyalas-Swalah]; while believers of North America and South America towards Manifested South [Haiyalal-Falaah]. Moreover, projected common run is leading believers towards the Black Hole of the Straight Middle East [Veil of Ignorance or Depth of Darkness from the point of view of formal education] without Broken Bar.** Those who are consciously / ignorantly using their pen-paper-pencil with a view to project **man-made directions [man-made magnetism] as natural directions [natural magnetism]**, they are the chiefs, ministers, group leaders, mouth speakers, representatives, and the like of the Don of Historical Black Magic.

The so-called sun [Bullet] as the manifested sign of natural magnetism [Tarash-Shamsi] enters from Manifested North-East [Haiyalas-Swalaah] in correspondence with North America and sets in Manifested South-East [Haiyalal-Falaah] in correspondence with South America. This entering from North-East [Revealed Left / Northern Hemisphere] and setting in South-East [Revealed Right / Southern Hemisphere] of the so-called sun as the manifested sign of natural magnetism [Bullet / Tarash-Shamsi] is the day-light for the Ground Stair of North America and South America of the appearing pentagonal earth within the Black Square of East Horizon. Due to magnetic force [Equal & Opposite Revelation / Law / Tawraat], the setting sun [Bullet] as the manifested sign of natural magnetism [Wash-Shamsi] runs away from South-East and rises from Manifested South-West [Haiyalal-Falaah] in correspondence with Europe and ends the journey in Manifested North-West in correspondence with Australia. This rising from Manifested South-West [Revealed Right / Southern Hemisphere] and ending in Manifested North-West [Revealed Left / Northern Hemisphere] of the so-called sun [Bullet] as the manifested sign of natural magnetism is the day-light for the Middle Stair of Europe, Asia, Africa, and Australia as well as Topmost Stair of Arabian Peninsula. In other words, the so-called sun as the manifested sign of natural magnetism rises twice and sets twice. These twice rising and twice setting of the so-called sun as the manifested signs of natural magnetism cause the alteration of day & night for the equal & opposite seashore of East Zone and West Zone. In brief, for the East Zone [Lower Seashore], the so-called sun as the manifested sign of natural magnetism [Bullet / Tarash-Shamsi] enters from Manifested North-East and sets in Manifested South-East. For the West Zone [Upper Seashore], the so-called sun as the manifested sign of natural magnetism [Bullet / Wash-Shamsi] rises from Manifested South-West and ends in Manifested North-West. In other words, the direction from which the so-called sun as the manifested sign of natural magnetism [Bullet / Tarash-Shamsi] enters as the day-light [Morning show] for the East Zone is the Northern Hemisphere or Haiyalas-Swalaah of the East Zone [Lower Seashore]. The direction towards which the so-called sun as the manifested sign of natural magnetism [Bullet / Tarash-Shamsi] sets as the day break for the East Zone is the Southern Hemisphere or Haiyalal-Falaah of the East Zone [Lower Seashore]. The direction from which the so-called sun as the manifested sign of natural magnetism [Bullet / wash-shamsi] rises for the West Zone as the day-light [Evening show] is the Southern Hemisphere or Haiyalal-Falaah of the West Zone [Upper Seashore]. The direction towards which the so-called sun as the manifested sign of natural magnetism [Bullet or Wash-Shamsi] ends the journey for one whole day is the Northern Hemisphere or Haiyalas-Swalaah of the West Zone. In brief, the Semi-anticlockwise stages of journey of the so-called sun as the manifested sign of natural magnetism from Manifested North-East to Manifested South-East for the East Zone and Semi-clockwise stages of journey of the so-called sun as the manifested sign of natural magnetism from Manifested South-West to Manifested North-West for the West Zone due to revealed electromagnetic force & electroweak force represent **Natural Magnetism**.

Now, if believers of the Ground Stair [North America and South America] perform their Salat [Namaz or Compulsory Prayer] towards setting direction of the manifested sign of natural magnetism [so-called sun or Bullet or Tarash-Shamsi or Electromagnetic Wave or converted snake of Muusa's Staff], then they are performing Salat [Namaz] towards Manifested Haiyalal-Falaah [South-East or Southern Hemisphere or Revealed Right]. If believers of the Middle Stair [Europe, Asia, Africa, and Australia] perform their [Namaz or Compulsory Prayer] towards setting direction of the manifested sign of natural magnetism [so-called sun or Bullet or Wash-Shamsi or Electromagnetic Wave or converted snake of Muusa's Staff], then they are performing Salat [Namaz] towards Manifested Haiyalas-Swalaah [North-West or Northern Hemisphere or Revealed Left]. The Black & White Imam of the City called Kaba has not been appointed / manifested at the centre of Revelation, Manifestation, appearance, and in the Middle-East within the Black Square of East Horizon. On the contrary, the Imam of the City called Kaba as the Leader of right direction [Standard] for mankind has been appointed / manifested on the right side of the Mount Tuur [Topmost Stair, Uppermost Land] in the Upright West Region within the Black Square of East Horizon. Arabian Peninsula is the Upright West Region of the Manifested Hexagonal World, Topmost Stair [Tuur] & Uppermost Land of the appearing Pentagonal Earth within the Black Square of East Horizon. The manifested sign of natural magnetism revolves round the appointed Kaba. This revolving is called Twaff declining towards Right and following Safa and Marwa i.e. two broad ways. The Leader of Right Direction has been appointed in the Upright West Region facing towards East [Manifestation] in resemblance with the manifested sign of natural magnetism. So, towards the appointed Kaba implies towards Upright West Region or Topmost Stair out of three ascending stair or Uppermost Land of the appearing pentagonal earth. As Kaba has been appointed in the City [Arabian Peninsula], so, people of the City [Banii-Israiil] are performing Salat following equal & opposite right direction or facing each other. But due to projected manmade nature or Historical Black Magic, people of the Middle Stair [Median or Zaytuun or Nation] and Ground stair [Tin or Townships] are performing Salat towards projected Middle-East as well as centre of Revelation i.e. Projected Planet Mars, Eartha 3D, Bermuda Triangle of Titanic, Recycling Region, Hollow in the Hand [Pentagon], penetrated hole in the journey boat of Muusa (ass), Black Hole, Zone of Iron, Zone of Black, Flameless Fire, and the like following Haiyalal-Falaah from the Ground Stair and Haiyalas-Swalaah from the Middle Stair. In this regard, as a believer, I have the Absolute Right to invite the Owner and Possessors of Manmade Nature i.e. Mechanical Globalisation, Circular Rotation & Revolution System, and Manmade Magnetism to face the monadic motion of the shared searched out manifest truths openly and publicly with a view to play the game of Magic with Equal & Opposite Manifested Signs of Natural Magnetism and Upright Logic. As a human person, I have the Birth Right or Inborn Right to know manifest truths concerning Natural Magnetism and Manifested nature. As a teacher, I have the Solidified Solid Human Rights to get generalised recognition of the searched out manifest truths shared in the 'kitaaba Wal-Hikmata' or 'Manifested Nature and the Utility of One's Upright Logic' from the respective authorities. Moreover, if an epistemic sense has the capability to play the game in favour of Manmade Nature [Projected Falsehood] individually, then that epistemic sense is being invited to play the game of 'GO WITH ME - AGREE WITH ME' openly and publicly.

Real Activist or Evidence Sorcerer – All pen-paper-pencil works and black & white activities which are either contrary or contradictory to Manifested Sign of Natural Magnetism are Evidence Sorceries. Those whose works and activities resemble with evidence sorceries are called Evidence Sorcerers. Such Evidence Sorcerers are the **Real Activists**. In brief **all conspiracy pen-paper-pencil workers and black & white Evidence Sorcerers are the Activists in the pragmatic use as well as truest sense of the term 'Activism'**. They are doing their pen-paper-pencil works and black & white activities **ignorantly** i.e. without verification and justification of the manifest truth [manifested signs or manifested nature or correspondence truth or Injiil or detail explanation of Fundamental Ayat] in resemblance with revelation [Tawraat or coherence truth or equal & opposite law or Fundamental Ayat]. Their works and activities are neither verifiable nor justifiable by one or the other revealed and established criterion [standard] of truth. **Real Activists** are those epistemic persons [teachers, lecturers, professors, writers, research scholars, supervisors, experts, verifiers, justifiers, editors, journalists, publishers, recognisers, certifiers, media persons, scientists, religious scholars, and the like] whose **works and activities resemble with Evidence Sorceries [Black Magic]**. So, each and every evidence sorcerer is a real activist, and conversely. That means each and every real activist is an evidence sorcerer.

Original Mechanical Terrorists or **Unique Epistemic Persecutors** - All pen-paper-pencil works and black & white activities which are either contrary or contradictory to Manifested Signs of Natural Magnetism are Evidence Sorceries. Those whose works and activities resemble with evidence sorceries are called Evidence Sorcerers. Those evidence sorcerers whose works and activities resemble with the two functions of Maya i.e. **concealment of real and projection of unreal as real** are called **Original Mechanical Terrorists** or **Unique Epistemic Persecutors**. They know that there is only one kind of revealed and manifested magnetism. This magnetism may be called Natural Magnetism. They also know that the equal & opposite Semi-anticlockwise and Semi-clockwise stages of journey of the so-called sun from Manifested North-East to Manifested South-East and from Manifested South-West to Manifested North-West are the manifested signs of Natural Magnetism. But due to their binary nature they have concealed revealed and manifested magnetism and have projected man-made magnetism. So, **according to them [Original Mechanical Terrorists or Unique Epistemic Persecutors], there are two kinds of magnetism, namely, Man-made Magnetism and Natural Magnetism. This projected man-made magnetism implies two ultimate realities or two Nature or Prakriti. One is the Manifested Nature and the other is the Manmade Nature. This projection** is the product of both conscious as well as teleological works and activities [Black Magic] with a view to play the game of Immediate Inference. Such **conscious as well as teleological Evidence Sorcerers are the Original Mechanical Terrorists or Unique Epistemic Persecutors of faith and belief of mankind on Scientific Certainty, Formal Education, Manifested Signs of Natural Magnetism, Manifested Nature, Revelation, and the like**. They are **uninterruptedly killing** the **Solidified Solid Human Rights** i.e. **First Generation Human Rights** [Natural Rights or Inborn Rights], **Second Generation Human Rights** [Formalised Human Rights or Bread & Butter Human Rights], **Third Generation Human Rights** [Solitary Rights or Solidarity Rights or Individual Rights or Faith and Belief]. **Original Mechanical Terrorists** are those epistemic persons [teachers, lecturers, professors, writers, research scholars, supervisors, experts, verifiers, justifiers, editors, journalists, publishers, recognisers, certifiers, media persons, scientists, religious scholars, and the like] whose **works and activities resemble with** conscious as well as teleological game of **Evidence Sorcery i.e. concealment of manifest truth and projection of mechanical [man-made] falsehood as truth**. So, each and every conscious as well as teleological evidence sorcerer is the original mechanical terrorist or unique epistemic persecutor. If projected two kinds of magnetism are the manifest truths in resemblance with revelations, then it is justified to ask a question to the owners of man-made magnetism – 'How many Uniform Principles [Manifested Nature] are there?'

BINARY FUNCTIONS OF THE ORIGINAL MECHANICAL TERRORISTS OR UNIQUE EPISTEMIC PERSECUTORS [BLACK

MAGICIANS] IN RESEMBLANCE WITH THE FUNCTIONS OF MAYA OF THE UPANISAD

Concealment of the Real – Conscious as well as teleological evidence sorcerers or original mechanical terrorists or unique epistemic persecutors know that there is only one kind of revealed and manifested magnetism i.e. natural magnetism, yet they have been consciously as well as purposefully concealing this manifested truth or manifested nature. Similarly they have been concealing several manifest truths without Broken Bar. A few manifest truths are as follows –

i) Universe is a Diamond Operator, revealed as an equal & opposite trinity or Sirius Binary System, and manifested as an upright rectangle in coherence with [Tawraat] the 'End of Proof' of Real Science and in correspondence with [Injiil] the appointed Kaba.

ii) The East Horizon has been revealed as two Zones [Seashores] and an Octagon.

iii) The East Horizon has been manifested within four basic forces after two psychic sacrifices/divorces as a hexagon;

iv) The manifested hexagonal East Horizon within the four walls [basic forces] is called Black Square or Non-luminous Moon or World or Asterisk.

v) Scientific investigations cannot go beyond the four walls of the Black Square [Non-luminous Moon or dark Moon] of the East Horizon.

vi) All scientific broadcastings external to the four walls of the Black Square are neither verifiable truth of facts nor certain knowledge. If those broadcastings resemble with Revelations, then such broadcastings are justifiable valid knowledge. If those broadcastings do not resemble with Revelations, then such broadcastings are teleological evidence sorceries.

vii) Due to equal & opposite sth between Upright West Region of Arabian Peninsula and Straight Middle East Region of Eartha 3D the manifested hexagonal world is appearing as a pentagon.

viii) The appearing pentagon is called the earth or Star Operator.

ix) The appearing pentagonal earth has three ascending stairs.

x) There are six regions of the manifested hexagon within the Black Square of East Horizon.

xi) There are seven windows in resemblance with seven continents of the appearing pentagonal earth, seven days of a week, seven colours of rainbow, seven firmaments, seven fermions, seven canopies, seven repeated verses, Windows'7 Ultimate, and the like.

xii)   The West Zone comprises three regions and five continents.

xiii) Three regions of the West Zone are – South-West Region [Europe], North-West Region [Asia, Africa, and Australia], and Upright West or Top West Region [Arabian Peninsula].

xiv)   Five continents of the West Zone are – Europe [South-West Continent / Western Continent], Arabian Peninsula [Upright West Continent or Uppermost Land], Asia [North-West Continent / Western Continent], Africa [North-West Continent / Western Continent], Australia [North-West Continent / Western Continent]

xv)   West Zone comprises two upper stairs out of three ascending stairs i.e. Top West Stair [City or Tuur or Banii-Israiil] and Middle West Stair [Nation or Median or Zaytuun or Children of Adam]. Arabian Peninsula is the Top West Stair; while Europe, Asia, Africa, and Australia are there in the Middle West Stair.

xvi)   Three regions of the East Zone are – North-East Region [North America], Straight Middle East Region [Eartha 3D], and South-East Region [South America].

xvii) Two continents of the East Zone are – North America [North-East Continent / Eastern Continent] and South America [South-East Continent / Eastern Continent].

xviii) East Zone comprises Ground Stair [Tin or Townships or Sons of Adam] out of three ascending stairs i.e. North America and South America.

xix)   The four cardinal directions resemble with the Plus Sign of Mathematics with respect to revelation [Night Sky & Earth's Night Sky/ Axis] and six manifested directions [after two sacrifices] resemble with the combination of Plus Sign and Multiplication Sign of Mathematics with respect to Manifestation [Earth Sky / Orbit]. Hexagonal unit represents six manifested directions. Downward Direction represents East Direction or Gravitational Force. Upward Direction represents West Direction or Strong Force. Leftward Direction from the point of view of Revelation and entering & ending points of the so-called sun represent North Direction or Magnetic Force. Rightward Direction from the point of view of Revelation and turning [setting & rising and turning & appearing] point or Broken Bar of the stages of journey of the so-called sun represents South Direction or Weak Force.

xx)   Upright ~sth between West and East is called Prime Meridian.

xxi)   Left side of the Prime Meridian from the point of view of Revelation is called Northern Hemisphere.

xxii)   Right side of the Prime Meridian from the point of view of Revelation is called Southern Hemisphere.

xxiii) The Black & white Imam of the City called Kaba has been appointed neither in the East Zone, nor in the Middle East Region of Eartha 3D, nor at the centre of revelation, manifestation, and appearance within the Black Square of East Horizon.

xxiv)   The Black & White Imam of the City called Kaba as the standard or the leader of right direction for mankind has been appointed on the right side of the Mount Tuur in the Upright West Region of the manifested hexagonal world and in the Topmost Stair and Uppermost Land of the appearing Pentagonal Earth within the Black Square of East Horizon.

xxv)   The so-called sun as the manifested sign of natural magnetism enters from North-East and sets in South-East for the Ground Stair of North America, and South America of the East Zone within the Black Square of East Horizon.

xxvi)   Due to magnetic force & weak force, the so-called sun as the manifested sign of natural magnetism turns back from South-East and rises from South-West and sets in North-West for the Middle Stair of Europe, Asia, Africa, and Australia as well as Topmost Stair of Arabian Peninsula of the West Zone within the Black Square of East Horizon. So, the so-called sun as the manifested sign of natural magnetism rises twice and sets twice to cause the alteration of day and night for the equal & opposite East Zone and West Zone.

xxvii) Equal & Opposite diagonal way [Middle Course facing each other] towards Upright West is the right direction of performing salat [prayer].

xxviii) Believers are following wrong directions. That means believers of Ground Stair of East Zone are performing salat towards Haiyalal-Falaah [Southern Hemisphere] and believers of the Middle Stair of West Zone are performing Salat towards Haiyalas-Swalaah [Northern Hemisphere].

xxix)   As night manifests [origins] from West Zone, so believers of Australia, Africa, Asia, Arabian Peninsula, and Europe will observe Idd in the morning from the point of view of West Zone i.e. prior to people of North America and South America of the East Zone. People of East Zone will observe Idd on the same day i.e. evening from the point of view of West Zone. This is called Unity-in-diversity.

Conscious as well as teleological evidence sorcerers or original mechanical terrorists or unique epistemic persecutors have searched out the above mentioned manifested truths in resemblance with revelations. They have concealed several searched out manifest truths like the above mentioned manifest truths from the unconscious as well as ignorant evidence sorcerers or real activists. They have also concealed several manifest truths from the domain of formal education. But their findings are not far from us. Their findings are symbolically present in Insert Symbol of your PC. I am neither a mystic, nor a spiritualist, nor a sufi, but an ordinary man with corporeal body. I have just tried to decode under the supervision of my Rab those concealed symbols in resemblance with **6310** [six thousand three hundred ten] translation of the meanings [Tafsir] & commentary codes of the researchers of IFTA on revelation and manifestation in the name of 'The Holy Quran' [free gift package]. So, it is certain like 'Cogito Ergo Sum' of Rene Descartes that due to the concealment of Manifest Truths by the conscious as well as teleological evidence sorcerers [original mechanical terrorists or unique epistemic persecutors] and uncountable works & activities of ignorant evidence sorcerers [real activists], mankind in general and particularly believers are far from manifest truths. Moreover, being a follower of equal & opposite diagonal way [Middle Course facing each other] towards Upright West [Arsh], I was off the track from the Solidified Solid Human Rights. Now, I have searched out several Solidified Solid Human Rights [findings], and are being shared those findings in the 'Kitaaba Wal-Hikmata' or 'Manifested Nature and the Utility of One's Upright Logic'. I have also placed searched out Solidified Solid Human Rights before respective authorities for verification, justification, confirmation, recognition, publication, broadcasting, and the like. For common and general sharing, I am trying to get 'Kitaaba Wal-Hikmata' or 'Manifested Nature and the Utility of One's Upright Logic' self-published. In this regard, Partridge Publishing [India] has taken the initiative to make this re-search project a successful one. Thanks a lot to each and every member of the Partridge Publishing Group. I shall remain grateful to you [all] forever.

Projection of the Unreal – All pen – paper – pencil works and black & white activities which are either contrary or contradictory to revelations and manifested signs [manifested nature] cannot be recognised as Manifest Truths [Real Nature] save projected falsehoods [mechanisms]. Concerning Manifest Truths [Manifested Nature] two contraries or two contradictories cannot be true together. Two contraries may be false together. Two contradictories can neither be true together nor false together. The truth of one of the contraries implies the falsity of the other, but not conversely. The truth of one of the contradictories implies the falsity of the other, and conversely. There are four revealed and established criterions [standards or units] of truth. Concerning manifested nature, projected falsehoods [mechanisms] are neither verifiable nor justifiable by one or the other criterion of truth in resemblance with manifested signs [manifested nature] and revelations. So, pen – paper – pencil works and black & white activities [concerning manifested nature] which are neither verifiable nor justifiable by one or the other criterion of truth in resemblance with manifested signs and revelations are teleological evidence sorceries [Black Magic]. **Mechanical Globalisation, Circular Rotation & Revolution System of the immovable hexagonal world and pentagonal earth, and Man-made Magnetism are the concrete examples as well as valid references of the projected falsehoods [Mechanical Trinity].** Those who have projected mechanical globalisation, Circular Rotation & Revolution System of the immovable hexagonal world and pentagonal earth, and man-made magnetism, they are the Unique Activists and Unique Epistemic Persecutors [chiefs, ministers, leaders, representatives of the Historical Don of Black Magic]. The Historical Don of Unique Activism & Unique Epistemic Persecution is also the Don of two-in-one partnership between Conscious & Teleological Evidence Sorcerers and Researchers of IFTA. This very Don has translated Quran with translation of the meanings and Commentary [critical remark] on Quran in resemblance with his/her projected mechanisms. So, this very Don is the vera causa of Activism and Terrorism [Black Magic & Mechanical Trinity]. Those who have the capability to argue in favour of the binary nature and functions of the vera causa of Activism and Terrorism i.e. in favour of the concealment of real and projection of unreal or binary functions of Maya or Black Magic & Mechanical Trinity, they are the Chiefs, Ministers, Soldiers, and the like of the Vera Causa [Black Magician].

The sharer of this self-financed re-search project is sincerely inviting all conscious as well as ignorant pen – paper – pencil workers and black & white artists to come forward with a view to verify & justify the searched out findings shared in the 'Kitaaba Wal-Hikmata' or 'Manifested Nature and the Utility of One's Upright Logic' on the basis of the four revealed and established criterions of truth in resemblance with manifested signs [manifested nature] and revelations. I want to do a re-search work with a view to become the follower of imposed guidelines of UGC [Hindustan] as well as to accumulate certified pieces of paper for API points & carrier advancement. If they are able to search out lacunae on the shared searched out findings with reference to four witnesses i.e. with reference to four revealed and established criterions of truth, then I will try to do re-search work on those pointed lacunae with a view to fulfil criterions of Bread and Butter Human Rights.

Teleological evidence sorceries with revealed and manifested truths through mechanical globalisation, circular rotation & revolution system, and man-made magnetism are full of subjective self-contradictions and

objective paradoxes. This Trinity involves all kinds of fallacies established by the possessors of true knowledge. In brief, 'the so-called sun as the manifested sign of natural magnetism rises from the East and sets in the west' involves the fallacy of Post Hoc Ergo Propter Hoc. Such kind of evidence sorcery is the persecution or killing of faith and belief of mankind on scientific certainty, revelation, manifested nature, formal education, and ultimately one's existence as a human person [possessor of rationality]. The Don of such kind of persecution is none but the Original Uncultured/ Immoral Educated Laden. Those who are involved to project mechanical globalisation, circular revolution & rotation system, and man-made magnetism without true concept of Revealed Left & Manifested Sign of Magnetic Force are the chiefs of the activist Laden. The sharer of this inspired sharing consciously as well as with firm faith on One Dot Leader and practical belief on Diamond Operator is placing a just shouting & a reasonable option before the Don of the Mechanical Globalisation, Circular Revolution & Rotation System of the immovable Hexagonal World and Pentagonal Earth, and Man-made Magnetism. The Historical Don of Mechanical Globalisation, Circular Revolution & Rotation System, and Man-made Magnetism may share the reality of natural magnetism so that we may get generalised recognition of the equal and opposite diagonal way facing each other towards Upright West as the right direction [Middle Course] of our Qibla or he will have to come forward with a view to prove the reality of his Mechanical Globalisation, Circular Revolution & Rotation of the immovable Hexagonal World and Pentagonal Earth, and Man-made Magnetism by converting the Manifested Signs of Natural Magnetism in resemblance with his Mechanical Globalisation, Circular Revolution & Rotation System, and Man-made Magnetism. In other words, he must have to convert the rising and setting direction of the so-called sun [Bullet] as the manifested sign of natural magnetism in resemblance with his projected Manmade North and South Directions or Projected Northern Hemisphere [top of the global trinity] & Southern Hemisphere [bottom of the global trinity].

The so-called Sun as the manifested sign of natural magnetism [Bullet / Tarash-Shamsi] enters from Manifested North-East as the Morning Show for the Ground Stair of North America and South America of the East Zone or Lower Seashore within the Black Square of East Horizon. The so-called sun as the manifested sign of natural magnetism follows lower high way, and moves Semi-anti-clockwise, and sets in Manifested South-East for the East Zone. This entering from Manifested North-East [Revealed Left] as well as setting in Manifested South-East [Revealed Right] is the day-light for the Ground Stair of the appearing Pentagonal Earth, East Zone [Lower Seashore] of the Manifested Hexagonal World, and 4/8 or ½ or half of the journey of the so-called sun [Bullet]. [Ref. ½ ASCII (hex) 00BD & ASCII (decimal) 189]

The so-called Sun as the manifested sign of natural magnetism [Bullet / Wash-Shamsi] rises from Manifested South-West as the Evening Show from the point of view of the so-called sun but morning show for the Middle Stair and Topmost Stair of the West Zone or Upper Seashore where there are Europe, Arabian Peninsula, Asia, Africa, and Australia. The so-called sun as the manifested sign of natural magnetism follows upper high way and ends the Semi- clockwise stages of journey in Manifested North-West for the Middle Stair and Topmost Stair of the appearing Pentagonal Earth and West Zone of the Manifested Hexagonal World within the Black Square of East Horizon. This rising from Manifested South-East [Revealed Right] as well as ending in Manifested North-West [Revealed Left] is the day-light for the Middle Stair as well as Topmost Stair of the appearing Pentagonal Earth of the West Zone [Upper Seashore], and remaining half of the journey for one whole day of the so-called sun [Bullet]. [Ref. ¼ ASCII (hex) 00BC & ASCII (decimal) 188 for the Topmost Stair of Arabian Peninsula, and ¾ ASCII (hex) 00BE & ASCII (decimal) 190 for the Middle Stair of Europe, Asia, Africa, and Australia]

The so-called sun as the manifested sign of natural magnetism **[Bullet] enters from Manifested North-East and ends in Manifested North-West. So, North represents E- Point [entering & ending] with respect to equal & opposite stages of journey of the so-called sun as the manifested sign of natural magnetism. The so-called sun** as the manifested sign of natural magnetism **[Bullet] sets in Manifested South-East and rises from Manifested South-West. So, South represents T- Point [setting –turning-rising] or Broken Bar [¦ ASCII (hex) 00A6 & ASCII (decimal) 166] with respect to equal & opposite stages of journey of the so-called sun [Bullet] as the manifested sign of natural magnetism. These are the searched out manifest truths in resemblance with revelations. These searched out manifest truths are certain like 'Cogito Ergo Sum' of Rene Descartes. In other words, these searched out manifest truths are the selfevident truths from the point of view of the sharer of 'Kitaaba Wal-Hikmat' or 'Manifested Nature and the Utility of One's Upright Logic'. So, sceptics are invited to do as much research works as possible on these searched out and shared manifest truths.**

REVEALED TRUTHS
NEITHER FROM THE EAST NOR FROM THE WEST
NICHE OR BINARY THREAD OR BINAR PULSAR OF EINSTEIN

The Niche [binary thread] is kindled neither from the East nor from the West within the Lamp [Black Square]. These are the revealed truths.
"Allaahu Nuurus-samaa-waati wal-arz – Masalu **Nuu-rihi** kaMishkaatin-fiihaa **Mis-baah** – Al-**Misbaahu** fii **Zujaa-jah** – **azzuujaajatu** ka-annahaa – kawkabun durriyyuny-yuuqa-du – min Shajaratimmubaara-katin-**Zaytuunatil-laa Sharqiy**-yatinw wa laa **Garbiyyatiny**-yakaadu **Zaytuhaa** yuziii- u wa law lam tamsas-hu naar. **Nuu-run alaa Nuur! Yahdillaahu** li-**Nuurihii** many-yashaaa – wa-yazribullaahul-amsaala linnaas – wallahu bi-kulli shay-in Alim." [Sura No. – 23\24 – Nuur, Ayat No. – 35]

Allah is the **Light** of the **heavens [West Horizon]** and the **world [East Horizon]**. The **similitude** (likeness) of **His Light** is as a **niche** (binary thread of a candle or binary pulsar) **in a lamp** [Black Square or World]. The **lamp** is in a **glass** [Shi-raa or Star Sirius or Diamond Operator]. The **glass** [Diamond Operator] is as it is a **shining star**. (This niche is) kindled from a **blessed tree [Prime Meridian]**, an **olive** (tree), **neither from the East nor from the West**, whose (niche's) oil [Electromagnetic Waves] will almost glow forth (of itself) though **no fire touches** it. **Light upon Light** [Nuurun alaa Nuur or Sirius Binary System]! Allah guides unto His light whom He wills. And Allah speaks to mankind in **parables**, for Allah is Knower of all things. [Sura No. – 23\24 – Nuur, Ayat No. – 35]

Safa and Marwa [Broad Ways or Orbit]
Clear Indications & Clear Proofs

Ayat\Verses – Lo! **Safa and Marwa** are **among the indications of Allah**. It is therefore no sin for him who is on pilgrimage to the House or at other times should compass them round. And he who does well of his own accord, be sure that Allah is He Who is Responsive Aware. **Those who conceal** the **Clear Proofs** We have sent down, and the Guidance, after We have made it clear for the people in the Kitab, on them shall be Allah's **curse**, and the **curse** of those entitled to curse. Except those who repent and amend and make manifest [openly declare the Truth]: To whom I relent; for I am Ever-Returning, Most Merciful. Lo! Those who disbelieve, and die while they are disbelievers, on them is Allah's curse, and the curse of angels, and of all mankind. They will abide therein. Their penalty will not be lightened, nor will they be reprieved. [Sura No. 1\2 – Baqarah, Ayat No. – 158 to 162]

Ayat\Verses - Nay [laaa], I swear by this Balad [City of the sole manifested Imam]; and you are an in-dweller of this city [People of Arabian Peninsula]; and begetter and that which he beget [parent and child]; verily We have created man in an atmosphere He thinks that none has power over him? And he says: I have destroyed vast wealth [Kitaab with Truth]. He [destroyer of Truth] thinks that none beholds him? Did We not assign unto him **two eyes**, and **a tongue**, and **two lips**, and guide her to the parting of the **two mountain ways** [Safa and Marwa]? But he [the follower of tautology] has not attempted the ascent. And what will convey to you the path that is ascent? [It is:] to free a bondman. Or the giving of food in a day of hardship to the orphan with claims of relationship or to the destitute in the dust; and to be of those who believe and enjoin patience, and enjoin deeds of kindness and compassion. Such are the Companions of the Right Hand. But those who reject Our Signs, they are the [unhappy] companions of the left hand. On them will be fire vaulted over. [Sura No. – 89\90 – Bi-haazal-balad, Ayat No. – 1 to 20].

Ayat\Verses – Do not the Unbelievers see that the heavens and the world were joined together [as one unit of creation], before we clove them asunder? We made from water every living thing. Will they not then believe? And We have set in the world mountains standing firm, lest it should shake with them, and We have made therein **broad highways** [safa and marwa] for them to pass through that they may receive Guidance. And We have made the heavens as a canopy well guarded; yet do they turn away from the **Signs** which these things [point to]! It is He Who created the Night and the Day, and **wash-shamsa wal-qamara. They float, each in an orbit or an Axis.** [Sura No. – 20\21 – Ambiyaaa, Ayat No. – 30 to 33]

FROM NORTH [LEFT] TO SOUTH [RIGHT] FOR THE EAST ZONE
&
FROM SOUTH [RIGHT] TO NORTH [LEFT] FOR THE WEST ZONE
Natural Magnetism – From North to North

[The niche or binary thread is kindled from Revealed Left or North (Zaatal-Yamin) declining towards Revealed Right or South (Zaatash-Shimaal), and runs away from South towards the end point i.e. North. These are the revealed as well as manifested truths or Revealed and Manifested Magnetism.]

"Wa **tarash-shamsa** izaa tala-at-lazaawaru an-kahfihim zaatal-**yamimi** wa izaa gara-at-taqri-zuhum **zaatash-shimaali** wa hum fi fajwatim-minh. Zaalikka min aayaatillaah; many-yahdillaaahu fahuwal-muhtad; wa many-yuzlil falan-tajida lahuu waliyyam-murshidaa". [Trans.] And might have seen **tarash-shamsa** [the so-called sun – for the morning show], when it **enters, declining to the right** [declining to South] from their Cave, **when it sets, turning away from them to the left** [to the North], while they lay in the **open space** in the midst of the **Cave** [within the four walls of East Horizon or the World]. **Such** [equal and opposite i.e. half-anti-clockwise and half-clockwise circulation of Allah's glittering show for the world] are among the **Signs of**

Allah. He whom Allah, guides is **rightly guided**; but he whom Allah **leaves to stray,** for him **you will not find protector to lead him to the Right Way.** [Sura No. – 17\18 – Khaf – Ayat No. – 17]

SINGLE ROOT

Alif-Lam-Mim-Ra. These are the Clear Signs of the Kitaab. That which has been revealed to you from your Rab is the Truth-in-itself. But most men believe not. Allah is He Who raised the **heavens without any pillars** that you can see; firmly established on the **throne** [of authority or Arsh). He has compelled **sakh-kharash-shamsa wal-qamar** to be of **service. Each one** [only so-called sun or so-called moon] **runs [its course] for a term appointed.** He **regulates** all affairs, **explaining the signs in detail** that you may **believe with certainty** in the meeting with your Rab. And it is He who **spread out the world [West Horizon],** and set thereon **mountains standing firm** and [flowing] **rivers:** and **fruit** of every kind **He made in pairs, two and two**: He draws the **night** as a veil over the **day.** Behold, **verily in these things there are signs for those who give thought!** And in the earth are tracts [diverse though] neighbouring, and gardens of vines and fields sown with corn, and palm trees - growing out of **single root** or otherwise (like and unlike): watered with the **same water,** yet some of them We make more excellent than others to eat. **Behold, verily in these things there are signs for those who understand!** [Sura No. – 12\13 – Rad, Ayat No. – 1 to 4]

TWICE RISING AND TWICE SETTING
SENT FORWARD AND KEPT BACK

Ayat\Verses - When the **sky is cleft asunder**; "Wa izal-kawaaki-buntasarat" [Trans.] When the **stars** are scattered**;** hen the **oceans [seas] are suffered to burst forth**; and when the **resting place are turned upside down**; [then] shall each soul know what is **sent forward** and **kept back.** [Sura No. – 81\82 - Izas-samaa-unfatarat, Ayat No. – 1 to 5]

"Our Rab! **Twice** You have made us die, and **twice** You have given us life"
Lo! (On that Day) the disbelievers (those who deny the Manifested Signs of Allah) are addressed by proclamation: Lo! Allah's abhorrence is more terrible than your abhorrence one of another (yourselves), seeing that (when) you were called to the faith but you used to refuse. They say: **Our Rab! Twice You have made us die, and twice You have given us life!** Now we confess our sins. Is there any way to go out of it? [It will be said to them]: This is because, when Allah only was invoked, you rejected faith (disbelieved), but when partners were ascribed to Him, you believed. The Command is with Allah, The Sublime, The Majestic! He it is Who shows you His Signs (Portents), and sends down

sustenance for you from the sky. None will receive admonition except those who turn [to Allah]. [Sura No. – 39\40 – Mu-Min, Ayat No. – 10 to 13]

CLUES FOR VERIFICATION OF THE SEARCHED OUT MANIFESTED TRUTHS

The separate sheet of the conspiracy science concerning natural magnetism can easily be verified by those who have upright logic. There are some clues for those who want to verify searched out **Manifested Truths** as well as **Black Magic**. Just switch on your PC. Open a page. Go to insert ▶ from insert to Symbol → from symbol to More Symbol, then a box will appear before you. In that box you will find Font. You will find more than 259 Font Names. Each Font name represents a Scientific Research Institution. Open a Font e.g. font name Bookshelf Symbol 7 or Batang, you will find 'from' where you will find Unicode (hex), ASCII (hex), ASCII (decimal), Symbol (hex), and Symbol (decimal). These are the units of measurement. The pen – paper – pencil workers and black & white artists have projected the Universe, the world, and the Earth in a single Globe; while real measuring unit is (hex) i.e. hexagon in resemblance with the manifested East Horizon within the Black Square [Non-luminous Moon or Dark Moon] as a hexagonal world or Asterisk. You will also find a box for Character Code which will provide you a code number in resemblance with both selected Symbol and Unit of Measurement. For instance Number 47 represents Binary Pulsar [Niche] where digit 4 represents four rows [of periodical appearance] and digit 7 represents seven windows [columns]. This number corresponds with AK47, where AK represents the ongoing command of Allah i.e. KUN [3D]. So, AK47 represents Continual Acceleration & Motion as well as Electromagnetic Wave & Gravitational Wave.

First Instance - Open a Font, for instance '**Vivaldi**' and go to 'from' and select ASCII (hex) or ASC II (decimal), a box will appear before you where in the 16th Column and 7th row [out of 16 columns] you will find a symbol like • [dark dot] called BULLET with character codes ASCII (hex) – 0095, ASCII (decimal) – 149, but avoid Unicode (hex) with respect to Bullet Symbol. This Bullet Symbol represents the so-called Sun as the manifested sign of natural magnetism as well as cause of day-light. You will find that the Bullet Symbol has been placed in the left side of the Table [PC] but right side from your point of view i.e. from the point of view of the finite observer. This left side of the table corresponds with the Revealed Left and Manifested Northern Hemisphere [North-East]. This left side of the table and right side from the point of view of the finite observer represent **Manifested North or Magnetic Force [Electromagnetic Wave]**. Now, the projection of the Bullet Symbol in North [objective left] represents that the so-called sun as the manifested sign of natural magnetism enters for morning show from Manifested North-East. In other words, the so-called sun as the manifested sign of natural magnetism enters for the morning show in that country which is represented by the Font name '**Vivaldi**' from Manifested North-East and the existence of that country must be in the Ground Stair or East Zone or Lower Seashore where there are North America and South America. The simple reason on the basis of which this manifested truth can be verified is that in the East Zone or Lower Seashore the so-called sun as the manifested sign of natural magnetism enters from Manifested North-East. So, the so-called sun as the manifested sign of natural magnetism enters for the morning show for the country represented by the Font name 'Vivaldi' from Manifested North-East. This is neither an invention nor a discovery but the manifest truth in resemblance with revelation and manifested signs of Natural Magnetism. Either contrary or contradictory to this revealed & manifested truth or Natural Magnetism is a selfexplained conspiracy, teleological evidence sorcery, and unique epistemic persecution i.e. Black Magic.

Second instance – Open the Font 'Aparajita' and go to 'from' and select ASCII (hex) or ASCII (decimal), a box will appear before you where in the 1st column and 8th row [out of 16 columns] you will find a symbol like • [point] called BULLET with character codes ASCII (hex) – 0095, ASCII (decimal) – 149 but avoid Unicode (hex) with respect to Bullet Symbol. This Bullet Symbol represents the so-called sun as the manifested sign of natural magnetism. You will find that the Bullet Symbol has been placed in the right side of the table [PC] but left side from your point of view i.e. from the point of view of the finite observer. This right side of the table corresponds with the Revealed Right and Manifested Southern Hemisphere. This right side of the table and left side from the point of view of the finite observer represents Manifested **South and Weak Force [Electroweak Force]**. Now, the projection of the Bullet Symbol at South represents that the so-called sun as the manifested sign of natural magnetism enters from Manifested North-East and sets in Manifested South-East, and runs away following Broken-Bar [ASCII (hex) 00A6 & ASCII (decimal) 166], and rises from Manifested South-West for the Upper Seashore [West Zone] and for that Scientific Research Institution, and then moves towards the end point i.e. Manifested North-West. This rising of the so-called sun as the manifested sign of natural magnetism from Manifested South-West is the Evening Show for the Bullet but morning show for the West Zone. In other words, the so-called sun as the manifested sign of natural magnetism rises for **evening show [daylight]** in that country which is represented by the Font name '**Aparjita**' from Manifested South-West. Moreover, if you verify that country, then you will find its existence in the West Zone where there are Europe, Arabian Peninsula, Asia, Africa, and Australia. So, the manifest truth is that for the West Zone or Upper

Sea-shore, the so-called Sun as the manifested sign of natural magnetism rises from Manifested South-West and ends in Manifested North-West. Now, it is really a hard task for pen – paper – pencil workers and black & white artists to disprove Revelation, Manifestation, and Real Scientific Symbols along with their character codes as well as to conceal their Teleological Evidence Sorceries concerning Natural Magnetism or manifested sings of Natural Magnetism from those people who possess upright nature.

Third Instance – Go to Batang again and select Bullet Symbol and after that select Unicode (hex), you will find that the Bullet Symbol is almost at the centre of the box with character code Unicode (hex) 2022. Similarly, go to Aparajita and select Bullet Symbol and after that select Unicode (hex), you will find that the Bullet Symbol is almost at the centre of the box with character code Unicode (hex) 2022. These placements of the Bullet Symbol are called Teleological Evidence Sorceries and Unique Epistemic Persecutions [Black Magic & Mechanical Trinity]. These placements resemble with the mechanisms of the red colour triangle in the 5th Column and the 7th row of the Napier Bones. Through these mechanisms, it is being projected that the so-called sun as the manifested sign of natural magnetism [Electromagnetic Wave & Electroweak Force] rises from the East [Gravitational Force & Downward Direction] and sets in the West [Strong Force & Upward Direction] in correspondence with the conspiracy projections of Northern Hemisphere & Southern Hemisphere concealing the true concepts of two West and two East, and projections of global Universe, global world and global Earth concealing the realities of Upright Rectangular Universe, Manifested Hexagonal World, and Appearing Pentagonal Earth with three ascending stairs. These mechanisms are being projected as Circular Rotation and Revolution System.

Fourth Instance – The first Font is the Normal Text. Go to Normal Text and select 7th row & 16th column of ASCII (hex) or ASCII (Decimal), you will find a character name – 'RIGHT DOUBLE QUOTATION MARK' & symbol " with character codes ASCII (hex) 0094 & ASCII (decimal) 148. Again, select 7th row & 15th column, you will find a character name – 'LEFT DOUBLE QUOTATION MARK' & symbol " with character codes ASCII (hex) 0093 & ASCII (decimal) 147. Double Quotation Mark represents so-called moons [Planets Uranus & Neptune or Muzzammil & Muddassir]. Double Quotation Mark & Bullet or so-called sun & so-called new moon or Wash-Shamsa Wal qamara are signs of equal and opposite sth [Binary Pulsar] between revelation and manifestation. So, if the Bullet is placed in the left side of the table i.e. North-East for the East Zone or lower seashore, then the Left Double Quotation Mark is to be placed in the right side of the table i.e. South-East for the East

Zone or Lower Seashore. If the Bullet is placed at the right side of the table i.e. South-West for the West Zone [Upper Seashore], then the Right Double Quotation Mark is to be placed at the left side of the table i.e. North-West for the West Zone [Upper Seashore].

In resemblance with the equal & opposite law, there is an equal and opposite game in the placement of Double Quotation Mark. Right Double Quotation Mark represents that the so-called new moon arises from North-West for the West Zone [1st period]. So, if the BULLET is placed at the right side of the table but left side of the finite observer for the West Zone, then the Right Double Quotation Mark is to be placed at the left side of the table but right side of the finite observer. Either contrary or contradictory placement will reflect that day and night occur together for that country which is represented by the Font name. If the BULLET is placed at the left side of the table but right side of the finite observer for the East Zone, then the Left Double Quotation Mark [1st period] is to be placed at the right side of the table but left side of the finite observer. Either contrary or contradictory placement will reflect that day and night occur together for that country which is represented by the Font name. Such kind of placement is a suicide game with Manifest Truth. Due to such kind of placement, there is diversity-in-unity concerning the appointed day of observing Idd uniformly.

But it is very unfortunate that these revealed & manifest truths correspond with some Font names and contradicts with other. In other words, you will find in some Font Symbols or Scientific Research Institutions Suicide Games. These placements are conspiracies with those who follow their religious duties and obligations on the basis of Double Quotation Mark i.e. arising / appearing of the new moon [Function Keys F1 & F2]. Due to these conspiracies most of the believers are failing to observe a single Idd on the appointed day. Being an Indian I am living in Asia of the West Zone. West Zone comprises Europe, Arabian Peninsula, Asia, Africa, and Australia. But due to conspiracy with the believers and evidence sorcery with revealed & manifest truth by the Historical Don of two-in-one partnership, people of Asia and people of Arabian Peninsula of the same West Zone are observing Idd in two different days. What is the scientific reason behind this diversity-in-unity? Is there any scientist who will be able to provide a justifiable as well as verifiable reason for this diversity-in-unity in resemblance with real science, manifested nature, and revelation?

Moreover, in resemblance with Bullet Symbol, if you select Double Quotation Mark and then select Unicode (hex), you will find its placement almost in the middle of the row with a view to project that the so-called new moon arises from the projected Middle East i.e. Eartha 3d or Historical Magician's Arabian Peninsula. The revealed and manifest truths are –

i) The so-called new moon or Qamar arises from Manifested North-West for the West Zone or Upper Seashore. So, the concept of Night [pm] manifests [origins] from West Zone or Upper Seashore where there are Europe, Arabian Peninsula, Asia, Africa, and Australia. This is the true concept of pm. Neither the so-called sun as the manifested sign of natural magnetism nor the so-called new moon as the sign of successive relation enters/arises at mid-night /noon i.e. in resemblance with the projected am/pm. On the contrary, projected concepts of the end of pm as well as the beginning of am at mid-night represent the reality of 'Lailatul Qadr' or Existential Import of Tahajuud or Philosophical concept of Unknown & Unknowable from the finite subjective point of view'. In other words, mid-night represents **Revelation [origin / creation]** of day & night; while entering or ending of the stages of journey represents **Manifestations [beginning / evolution]** of day & night. Try to conceptualise 'Origin' and 'Beginning' as well as creation and evolution in resemblance with Kant's Critical Theory of knowledge and Darwin's Theory of Evolution.

ii) The so-called sun as the manifested sign of natural magnetism enters from Manifested North-East for the East Zone or Lower Seashore. So, the concept of day [am] manifests [origins] from East Zone or Lower Seashore where there are North America, Straight Middle East Region of Eartha 3D and South America. This is the true concept of am.

iii) As Idd should be observed on the basis of the arising of so-called new moon for the West Zone, so, people of West Zone or Upper Seashore will observe Idd uniformly prior to the East Zone or Lower Seashore. In other words, people of Europe, Arabian Peninsula, Asia, Africa, and Australia will observe Idd on the basis of arising of new moon. If people of West Zone observe Idd in the morning, then people of East Zone will observe Idd on the same day i.e. morning from the point of view of East Zone but evening from the point of view of West Zone. In other words, people of East Zone will observe Idd on the same day following West Zone. This is called Unity-in-diversity.

Fifth Instance – Digits 1, 2, 3, 4 etc. of the Insert Font Symbols represent planets. You will find that with respect to Bullet and Double Quotation Mark, the placements of symbols in resemblance with ASCII (hex) and ASCII (decimal) are equal, but Unicode (hex) is opposite. But if you search digits which represent planets, you will find that Unicode (hex) and ASCII (hex) are equal with respect to character codes, but ASCII (decimal) is opposite.

These equal and opposite games resemble with the Pythagorean Number 'LINE – TWO' and Nature & Function of Sankar's concept of Maya. This binary nature represents two faces of the same coin i.e. Conspiracy Scientists and Researchers of IFTA. These planets are in reality nine clear proofs and eye opening evidences given to Muusa [ass]. Each projected Planet or Moon is called a Star. Each Star resembles with the name of a messenger or warner or relational name. Moreover, each Star is a clear proof or a clear manifested sign concerning the resemblance between revelation and manifestation. It has been inspired through several Ayat \ Verses of the Kitaab with Truth to search out Similitude or Resemblance between Revelation and Manifested Signs with a view to confirm Tautologies and to strengthen our faith and belief in One Dot Leader & Diamond Operator. Through this inspired sharing an attempt is being made to search out the resemblance between searched out real scientific symbols and manifested signs with a view to confirm the resemblance between manifested nature and revelation.

On the basis of this re-search project, I have conceptualised the 'LINE' of the 'Line – two' as The Historical Don of the two-in-one partnership. One face of this LINE is the activist Laden who is using his pen, paper, and pencil called researchers on coded shared tautologies to conceal manifested signs of Allah and to theorise formalities, guidelines, and grammatical rules as criterions of Truth. The other face of this LINE is the terrorist Laden who is using AK47 in resemblance with the conversion of good names like **Z Muhammad** instead of **Muhammad Z** with a view to play **win-win games** and to **project terror in the psyche of common people in the name of Muhammad**. In reality, A stands for Allah, K stands for Kalam or Revelation, the digit 4 stands for equal & opposite acceleration & motion i.e. Diamond Operator in resemblance with 'Solid – Four' of Pythagorean Number and Crucified Sign, the digit 7 stands for equal & opposite manifestation in resemblance with Pythagorean Number 'Manifestation – Seven' in coherence with the shared tautology – 'Allah is the manifest truth', seven windows, seven continents, seven days of a week, seven heavens and similar number of worlds, seven colours of Rainbow etc..

So, the sharer of this sharing declaring himself as a solid fundamentalist or a fundamental Jihadist is inviting the Don of the activist Laden and the terrorist Laden to face the monadic motion of the shared findings concerning manifested nature in resemblance with revelation with a view to falsify shared findings or to covert the reality of the manifested nature in resemblance with his Black Magic. Remember, **persecution is worse than killing**. In this regard, it is an appeal before all to verify as well as to justify the shared selfevident truths with a view to provide the sharer of this sharing a favourable environment so that an Akbari-Jihadist [an altruist / a karma yogi / a Satyagrahi] can play the game of 'Survival of the Truest' [Duty for Duty's Sake / Niskama Karma / Altruistic Hedonism] against the self-sensed activists as well as well known & well trained terrorists or the Historical Don of two-in-one Laden. This Historical Don of two-in-one partnership knows that the manifested universe is an upright rectangle, the world is manifested within the Black Square of East Horizon as a hexagon, there are twelve springs, the appearing earth is a pentagon, there are three ascending stairs in the appearing pentagonal earth, the so-called sun [Bullet] as the manifested sign of natural magnetism enters from Manifested North-East and ends its equal &opposite stages of journey in Manifested North-West due to revealed magnetism, projected magnetism is a mechanism and conspiracy with the believers and persecution of faith & belief of mankind in general on manifested nature, manifest truth, formal education, scientific certainty, and the like.

This game will not be a win-win game in resemblance with the lifting of two fingers. On the contrary, this is a Just Shouting Game without a Second One. In other words, if my Rab wills, then this sharing will act like Rene Descartes' 'Cogito Ergo Sum' to unveil the equal and opposite diagonal way facing each other towards Upright West following equal & opposite manifested signs of natural magnetism in coherence with the wings of Windows'7 Ultimate and in pragmatic correspondence with the sign of the mighty plot of 9/11 incident in the existing Pentagonal House. So, let the mouth speakers of the Don of two-in-one partnership play their significant role with a view to activate the rationality of the Don of Historical Conspiracy Science so that an Octagonal Spider may come forward from his Pentagonal Net Work with a view to prove that his projected Black Magic is the manifested nature [manifest truth or natural magnetism].

The sharing of 'Kitaaba Wal-Hikmata' or 'Manifested Nature and the Utility of One's Upright Logic' is based on the resemblance between Manifested Signs and Real Scientific Symbols justifiable by the criterions of truth in resemblance with complete coded shared tautologies. So, each searched out manifest truth of this sharing is a selfevident truth. There is no scope to argue with selfevident truth save justification and verification for self understanding and generalised recognition. Moreover, as my Rab has inspired me to search out manifest truth in resemblance with equal & opposite manifested signs of natural magnetism, so, it is certain that the Historical Don of two-in-one partnership will not be able to conceal further searched out manifest truths concerning manifested sing of natural magnetism from the possessors of upright logic.

NATURAL MAGNETISM

Equal & Opposite Stages of Journey of the so-called sun [Bullet] are the manifested signs of natural magnetism

[Ref. Ayat No. – 50 to 59 of Sura – Baqarah, Ayat No. – 142 & 143 of Sura - As-haabul- a' raaf, Ayat No. – 1 to 5 of Sura - Izas-samaa-unfatarat, Ayat No. – 17 of Sura Khaf]

Morning Show and Evening Show of the Sister Planet Venus
& Maryam Supplied with Sustenance

BEGINNING OF THE DAY OR EVENING SHOW – SEMI-CLOCKWISE ZONE

The so-called sun [Bullet] as the manifested sign of Natural Magnetism rises from Revealed South and Manifested South-West in correspondence with Europe and sets [ends] in Revealed North and Manifested North-West in correspondence with Australia. This is the Semi-clockwise journey of the so-called sun [Bullet / Wash-Shamsi] as the manifested sign of Natural Magnetism or day-light for the Middle Stair of Europe, Asia, Africa, and Australia as well as Topmost Stair of Arabian Peninsula of the appearing Pentagonal Earth. But this Semi-clockwise journey of the so-called sun [Bullet] is the Evening Show with respect to the Sister Planet Venus and Maryam Supplied with sustenance [Wash-Shamsi] and Break of the Day with respect to the West Zone of the manifested Hexagonal World within the Black Square [Non-luminous Moon] of East Horizon

SOUTH-WEST	UPRIGHT WEST	NORTH-WEST

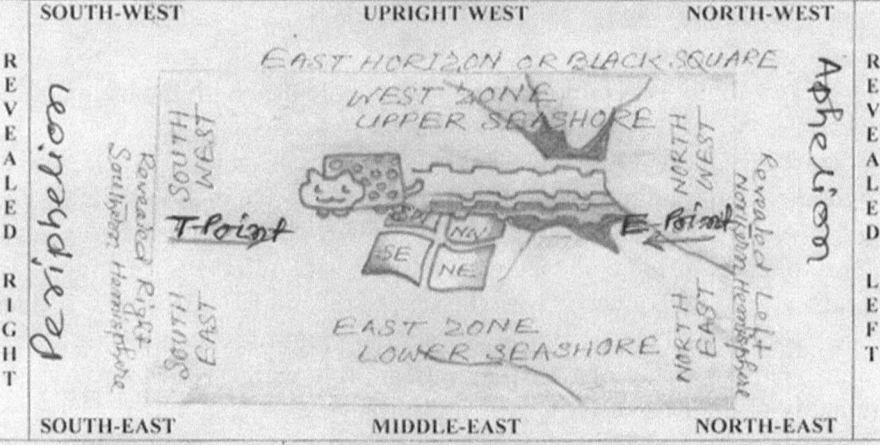

SOUTH-EAST	MIDDLE-EAST	NORTH-EAST

ORIGIN OF THE DAY OR MORNING SHOW – SEMI-ANTICLOCKWISE ZONE

The so-called sun [Bullet] as the manifested sign of Natural Magnetism rises [enters] from Revealed North and Manifested North-East in correspondence with North America and sets in Revealed South and Manifested South-East in correspondence with South America. This is the Semi-anticlockwise journey of the so-called sun [Bullet / Tarash-Shamsi] as the manifested sign of natural magnetism for the East Zone of the manifested hexagonal world within the Black Square [Dark Moon] of East Horizon. This Semi-anticlockwise journey of the so-called sun [Bullet] is the day-light for the Ground Stair of the appearing Pentagonal Earth and Morning Show of the manifested sign of Natural Magnetism in the name of Sister Planet Venus and Maryam Supplied with sustenance for the East Zone of the Manifested Hexagonal World within the Black Square [Non-luminous Moon] of East Horizon .

The so-called sun as the manifested sign of natural magnetism neither rises [once] from the East nor sets [once] in the West.

Ayat No. – 35 of Sura - Nuur and True Concept of Magnetism
Safa and Marwa - Ayat No. – 158 to 162 of Sura – Baqarah
Two Mountain Ways - Ayat No. – 1 to 20 of Sura - Bi-haazal-balad
NE to SE represents Semi-anticlockwise Stages of Journey for the East Zone
SW to NW represents Semi-clockwise Stages of Journey for the West Zone

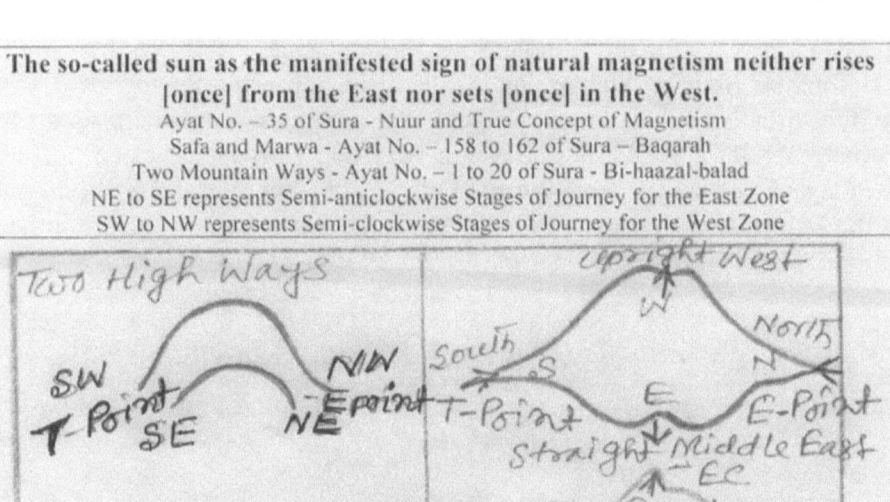

Beginning of the Day		WEST		So-called sun [Bullet] sets in
So-called sun [Bullet] rises from Revealed Right and Manifested South-West in correspondence with Europe as the day-light for the West Zone [Upper Seashore].				Revealed Left and Manifested North-West in correspondence with Australia for the Middle Stair of Europe, Asia, Africa, and Australia as well as Topmost Stair of Arabian Peninsula of the appearing Pentagonal Earth.
So-called sun [Bullet] sets in Revealed Right and Manifested South-East in correspondence with South America for the Ground Stair of North America and South America of the appearing Pentagonal Earth.				**Origin of the Day** So-called Sun [Bullet] enters from Revealed Left and Manifested North-East in correspondence with North America as the day-light for the East Zone [Lower Seashore].
		EAST		

The so-called sun [Bullet / Electromagnetic Wave / converted snake of Muusa's Staff] as the manifested sign of natural magnetism enters from NE and sets in SE for the Ground Stair of North America and South America of the appearing Pentagonal Earth. So, the so-called sun as the manifested sign of natural magnetism enters from Revealed Left and Manifested North-East and sets in Revealed Right and Manifested South-East from the point of view of Lower Seashore [East Zone] of the Manifested Hexagonal World within the Black Square of East Horizon.

The so-called sun as the manifested sign of natural magnetism [Bullet / Electromagnetic Wave / converted snake of Muusas Staff] rises from SW and ends in NW [for the Middle Stair Europe, Asia, Africa, and Australia; and Topmost Stair of Arabian Peninsula of the appearing Pentagonal Earth]. So, the so-called sun as the manifested sign of natural magnetism rises from Revealed Right and Manifested South-West and ends in Revealed Left and Manifested North-West from the point of view of Upper Seashore [West Zone] of the Manifested Hexagonal World within the Black Square [Non-luminous Moon] of East Horizon.

↶ [NE to SE] for the East Zone and [SW to NW] ↷ for the West Zone

Anticlockwise Top Semicircle Arrow – U (h) 21B6 & Clockwise Top Semicircle Arrow – U (h) 21B7

The so-called sun as the manifested sign of natural magnetism rises twice and sets twice to cause the alteration of day and night for the equal and opposite seashore of East Zone [Lower Seashore] and West Zone [Upper Seashore].

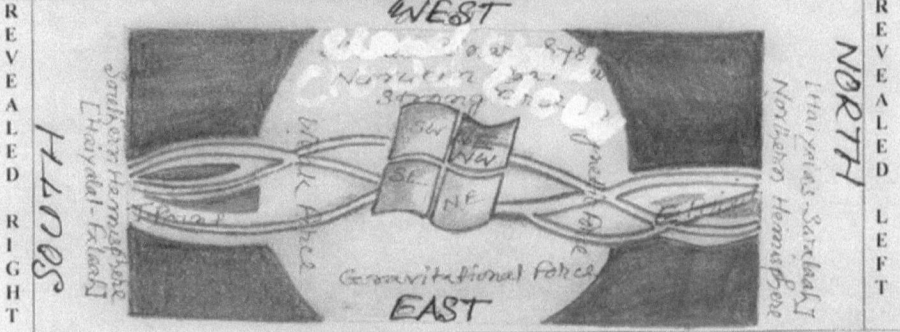

The so-called sun [Bullet] as the manifested sign of Natural Magnetism enters from NE [Manifested North-East] in correspondence with North America and sets in SE [Manifested South-East] in correspondence with South America as the day-light for the Ground Stair of the appearing Pentagonal Earth and East Zone of the Manifested Hexagonal World within the Black Square [Non-luminous Moon] of East Horizon. This is the half of the journey for one complete day i.e. 4/8 or ½ ASCII (hex) 00BD & ASCII (decimal) 189 of the so-called sun [Bullet] as the manifested sign of Natural Magnetism following 'Anticlockwise Top Semicircle Arrow' – U (h) 21B6. This Semi-anticlockwise journey is the 'Morning Show' of the so-called sun [Bullet] in the name of projected sister planet Venus and shared name Maryam supplied with sustenance.

The so-called sun [setting sun] or Bullet as the manifested sign of Natural Magnetism runs away from there and rises from SW [Manifested South-West] in correspondence with Europe and ends in NW [Manifested North-West] in correspondence with Australia as the day-light for the Middle Stair as well as Topmost Stair of the appearing Pentagonal Earth and West Zone [Upper Seashore] of the manifested Hexagonal World within the Black Square [Dark Moon] of East Horizon. This is the half of the journey for one complete day i.e. ¼ for the Topmost Stair of Arabian Peninsula & ¾ for the Middle Stair of Europe, Asia, Africa, and Australia, of the so-called sun [Bullet] as the manifested sign of Natural Magnetism [ASCII (hex) 00BC & 00BE, ASCII (decimal) 188 & 190] following 'Clockwise Top Semicircle Arrow' – U (h) 21B7. This Semi-clockwise journey is the 'Evening Show' of the so-called sun [Bullet] in the name of projected sister planet Venus and shared name Maryam supplied with sustenance.

So, the so-called sun as the manifested sign of Natural Magnetism rises twice and sets twice to cause the alteration of day & night for the equal & opposite Seashore of East Zone and West Zone.

These are the Unalterable Manifested Truths in resemblance with Revelation [Ref. Ayat No. – 35 of Sura – Nuur, Ayat No. – 17 of Sura Khaf, Ayat No. – 10 to 13 of Sura Mu-min, Ayat No. – 1 to 5 of Sura – Izas-samaa-unfatarat, Ayat No. – 72 to 81 of Sura - Banii-Israai-iil, Ayat No. – 1 to 4 of Sura – Rad, Ayat No. – 31 to 34 of Sura – Ibrahim].

NEITHER FROM THE EAST NOR FROM THE WEST

"**Allaahu Nuurus-samaa-waati wal-arz** – Masalu **Nuu-rihi** ka**Mishkaatin**-fiihaa **Mis-baah** – Al-**Misbaahu** fii **Zujaa-jah** – **azzuujaajatu** ka-annahaa – kawkabun durriyyuny-yuuqa-du – min Shajaratimmubaara-katin-**Zaytuunatil**-laa **Sharqiy**-yatinw wa laa **Garbiyyatiny**-yakaadu **Zaytuhaa** yuziii- u wa law lam tamsas-hu naar. **Nuu-run alaa Nuur! Yahdillaahu** li-**Nuurihii** many-yashaaa – wa-yazribullaahul-amsaala linnaas – wallahu bi-kulli shay-in Alim." [Sura No. – 23\24 – Nuur, Ayat No. – 35] **Allah** is the **Light** of the **heavens [West Horizon]** and the **world [East Horizon]**. The **similitude** (likeness / resemblance) of **His Light** is as a **niche** (binary thread of a candle or binary pulsar) **in a lamp** [Black Square or Dark Moon of the Manifested East Horizon]. The **lamp** is in a **glass** [Shi-raa or Star Sirius or Diamond Operator]. The **glass** [Diamond Operator] is as it is a **shining star.** (This niche is) kindled from a **blessed tree [Prime Meridian]**, an **olive** (Zaytuun or Middle Stair), **neither from the East nor from the West**, whose (niche's) oil [Continual Acceleration] will almost glow forth (of itself) though **no fire touches** it. **Light upon Light** [Nuurun alaa Nuur]! Allah guides unto His light whom He wills. And Allah speaks to mankind in **parables**, for Allah is Knower of all things. [Sura No. – 23\24 – Nuur, Ayat No. – 35]

Revealed Right or South or Weak Force or Electroweak Force		Revealed Left or North or Magnetic Force or Electromagnetic Wave	
Placement of Bullet at the right side of the Insert Symbol Table [left from the point of view of finite observer] represents that the so-called sun as the manifested sign of natural magnetism rises from Revealed Right [South] and Manifested South-West in correspondence with Europe for the West Zone [Upper Seashore] • Bullet [ASCII (hex) 0095, ASCII (decimal) 149]		Placement of Bullet at the left side of the Insert Symbol Table [right from the point of view of finite observer] represents that the so-called sun as the manifested sign of natural magnetism enters from Revealed Left [North] and Manifested North-East in correspondence with North America for the East Zone [Lower Seashore] Bullet [ASCII (hex) 0095, ASCII (decimal) 149] •	
South West	**UPRIGHT WEST** **[STRONG FORCE]**		**North** **West**
Revealed Right or Perihelion or SOUTH or WEAK FORCE			Revealed Left or Aphelion or NORTH or MAGNETIC FORCE
South East	**STRAIGHT MIDDLE EAST** **[GRAVITATIONAL FORCE]**		**North East**

FROM NORTH-EAST [REVEALED LEFT] TO SOUTH-EAST [REVEALED RIGHT] FOR THE EAST ZONE
&
FROM SOUTH-WEST [REVEALED RIGHT] TO NORTH-WEST [REVEALED LEFT] FOR THE WEST ZONE

"Wa **tarash-shamsa** izaa tala-at-lazaawaru an-kahfihim zaatal-**yamimi** wa izaa gara-at-taqri-zuhum **zaatash-shimaali** wa hum fi fajwatim-minh. Zaalikka min aayaatillaah; many-yahdillaaahu fahuwal-muhtad; wa many-yuzlil falan-tajida lahuu waliyyam-murshidaa". [Trans.] And might have seen **tarash-shamsa** [the so-called sun – for the morning show], when it **enters, declining to the right** [declining to South] from their Cave, **when it sets, turning away from them to the left** [to the North], while they lay in the **open space** in the midst of the **Cave** [within the four walls of East Horizon called Black Square or Dark Moon]. **Such** [equal and opposite i.e. half-anti-clockwise and half-clockwise circulation of Allah's glittering show for the Pentagonal Earth] are among the **Signs of Allah**. He whom Allah, guides is **rightly guided**; but he whom Allah **leaves to stray**, for him **you will not find protector to lead him to the Right Way.** [Sura No. – 17\18 – Khaf – Ayat No. – 17]

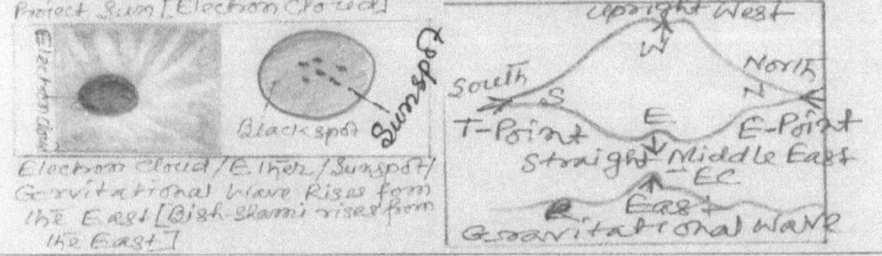

RING OPERATOR
ELECTRON CLOUD OR ETHER OR BISH-SHAMSI OR GRAVITATIONAL WAVE
Projected Sun [Historical Black Magic]
Electron Cloud [Gravitational wave or Ether or Bish-Shmsi or Dark Umbrella] rises from the Middle East within the Black Square [Dark Moon] of the Manifested East Horizon in the name of Zakariya, the appointed Guardian of Maryam [RING OPERATOR – Unicode (hex) 2218]

In Manmade Nature, the so-called sun [Bullet] as the Manifested Sign of Natural Magnetism rises [once] from the unidentified East; while the so-called sun [Bullet] as the Manifested Sign of Natural Magnetism enters from North-East for the East Zone within the Black Square of East Horizon in Manifested Nature.
In Manifested Nature, Bish-Shamsi or Black-Spot or so-called Sunspot [Electron Cloud or Ether or Gravitational Wave] rises from the East as a veil or screen for Maryam supplied with sustenance in the name of Ring Operator Zakariya. This Black-Spot or Bish-Shmsi or Ether or Veil is the so-called sun of the Manmade Nature which rises from the East. In other words, Electromagnetic Wave of Einstein rises from the East in Manmade Nature i.e. Mechanical Globalisation, Circular Rotation & Revolution System, Manmade Magnetism. Projected Manmade Nature is contrary as well as contradictory to all verifiable scientific laws, Manifested Truth in resemblance with Revelation, and upright logic. So, Conspiracy Science has been projecting Gravitational Force [East] and Strong Force [West] as manifested signs of natural magnetism concealing Magnetic Force [North] & Weak Force [South] in resemblance with the reference list of a research work without Broken Bar. This projection is called Black Magic & Man-made Nature [Mechanical Universe – World – Earth].
[References - Ayat No. – 258 of Sura – Baqarah, Ayat No. – 38 to 41 of Sura – Imran, Ayat No. – 7 to 11 of Sura – Maryam]

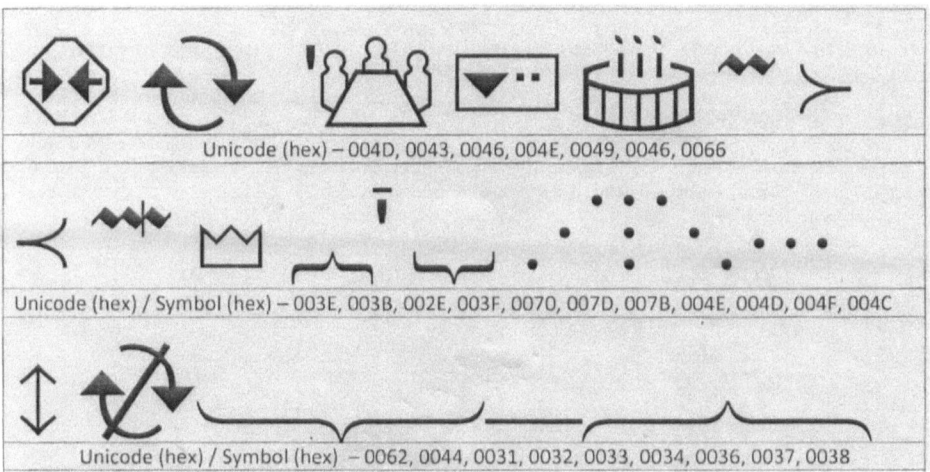

Unicode (hex) – 004D, 0043, 0046, 004E, 0049, 0046, 0066

Unicode (hex) / Symbol (hex) – 003E, 003B, 002E, 003F, 0070, 007D, 007B, 004E, 004D, 004F, 004C

Unicode (hex) / Symbol (hex) – 0062, 0044, 0031, 0032, 0033, 0034, 0036, 0037, 0038

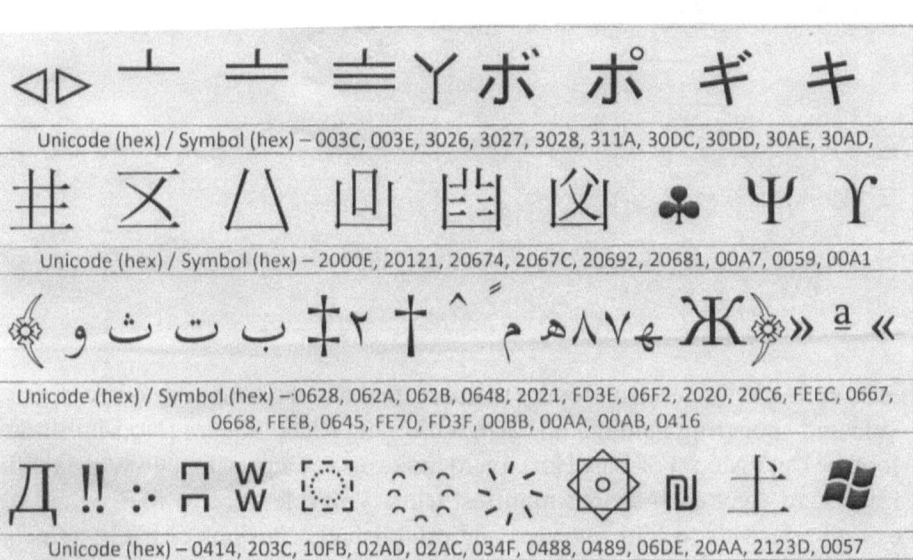

Unicode (hex) / Symbol (hex) – 003C, 003E, 3026, 3027, 3028, 311A, 30DC, 30DD, 30AE, 30AD,

Unicode (hex) / Symbol (hex) – 2000E, 20121, 20674, 2067C, 20692, 20681, 00A7, 0059, 00A1

Unicode (hex) / Symbol (hex) – 0628, 062A, 062B, 0648, 2021, FD3E, 06F2, 2020, 20C6, FEEC, 0667, 0668, FEEB, 0645, FE70, FD3F, 00BB, 00AA, 00AB, 0416

Unicode (hex) – 0414, 203C, 10FB, 02AD, 02AC, 034F, 0488, 0489, 06DE, 20AA, 2123D, 0057

E-Point and T-Point	
Revealed Right or South or Weak Force	Revealed Left or North or Magnetic Force
Placement of [Left] Double Quotation Mark at the right side of the Insert Symbol Table [left from the point of finite observer] represents that the so-called new moon appears from Revealed Right [South] and Manifested South-East for the East Zone [Lower Seashore]	Placement of [Right] Double Quotation Mark at the left side of the Insert Symbol Table [right from the point of view of finite observer] represents that the so-called new moon arises from Revealed Left [North] and Manifested North-West for the West Zone [Upper Seashore]
" Left Double Quotation Mark [ASCII (hex) 0093 & ASCII (decimal) 147]	Right Double Quotation Mark [ASCII (hex) 0094 & ASCII (decimal) 148] "

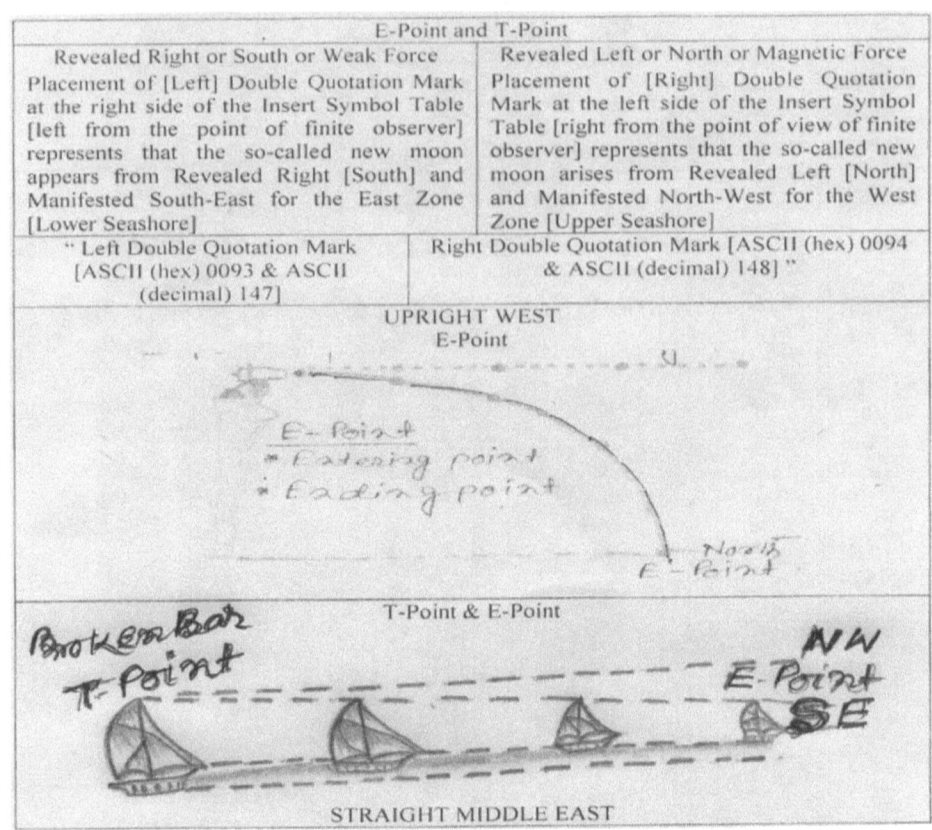

UPRIGHT WEST
E-Point

E- Point
* Entering point
* Ending point

North
E-Point

T-Point & E-Point

Broken Bar
T-Point

NW
E-Point
SE

STRAIGHT MIDDLE EAST

It is an appeal before each and every matured member of the manifested hexagonal world and appearing pentagonal earth within the Black Square [Non-luminous Moon or Dark Moon] of East Horizon to make an attempt with a view to falsify, to reject, to deny each shared manifest truth with reference to four witnesses i.e. with reference to four revealed and established criterions of truth, namely, Coherence Truth [Tawraat], Correspondence Truth [Injiil], Pragmatic Truth [Zabuur], and Selfevident Truth [Furqan]. As a re-search scholar on 'Solidarity Rights in Islam', I will try to do re-search work on the pointed verifiable as well as justifiable lacunae shared by you with reference to four witnesses. [verifiable and justifiable references] with a view to fulfil the criterions of Bread and Butter Human Rights, carrier advancement, API points, name and fame etc. Simultaneously, it is an appeal before the respective authorities to recognise as well as to take necessary steps for broadcasting of those searched out manifest truths which are scientifically verifiable and logically justifiable on the basis of revealed and established criterions of truth in resemblance with revelation, manifested nature, manifested signs, and upright logic. To get confirmation through recognition of the searched out Manifest Truths shared in the 'Kitaaba

Wal-Hikmata' or 'Manifested Nature and the Utility of One's Upright Logic' from the respective authorities is the Trinity of Natural Rights [First Generation Human Rights] – Solitary Rights [Third Generation Human Rights or Individual Rights] – Generalised Human Rights [Second Generation Human Rights] of the re-search scholar on 'Solidarity Rights in Islam'. The bond of these Rights is called **Solidified Solid Human Rights**. So, to get recognised confirmation and generalised recognition from all National and International Human Rights Commissions – Councils – Organisations are the Solidified Solid Human Rights of mankind in general and particularly the re-searcher & sharer of 'Kitaaba Wal-Hikmata' or 'Manifested Nature and the Utility of One's Upright Logic'.

Allah has appointed (manifested) the Kaba, the Sacred House, as a standard [criterion] for mankind, and the **Sacred Months**, the **offerings**, and the **garland** (Festoon): That you may know that Allah has knowledge of what is in the heavens and in the world and that Allah is well acquainted with all things. [Sura No. – 4\5 – Maaaidah, Ayat\Verses – 97]

O People of the Kitaab [Yaaa-Ahlal-Kitaabh]! Now has come to you our **disciple**, revealing to you much that you used to **hide in the Kitaab**, and **passing over** much. There has come to you from Allah a **light and a plain Kitaab** [Kitabbum-Mubin]. **Wherewith Allah guides all who seek His good pleasure to ways of peace and safety, and leads them out of darkness, by His Will, unto the light [equal & opposite electromagnetic wave from North to South & South to North], -- guides them to a Path that is Right [towards Upright West facing each other].** [Sura No. – 4\5 – Maaaidah, Ayat\Verses – 15 & 16]

"But verily **you call them to the Right Path**".
Or is it that you ask them for some recompense? But the recompense of your Rab is the best. **He is the Best of those who give sustenance**. But verily **you call them to the Right Path**. And verily those who believe not in the Hereafter are **astray from that Way**. [Sura No. – 22\23 – Mu-Minuun, Ayat\Verses – 72 to 74]

Say: Who is the **Rab of the seven heavens**, and the **Rab of the Supreme Throne** [Arsh towards Upright West]? They will say: [They belong] to Allah. Say: Will you then not keep duty? [Sura No. – 22\23 – Mu-Minuun, Ayat\ Verses – 86 & 87]

And unto Allah leads the **Right Path [towards Upright West]**, but **some ways go not Right [towards Upright West]**. If Allah had willed, He would have guided all of you. [Sura No. – 15\16 – Nahl, Ayat\Verses – 9]

Lo! Your Rab knows how you keep vigil, sometimes nearly **two-thirds** of the night, or (sometimes) **half** or a **third thereof**, as do a party of those with you. Allah **measures** the **night and the day**. He knows that **you count it not**, and **turns to you** in mercy. **Recite then of the Quran which is easy for you**. He knows that there are **sick folk** among you, while others **travel** in the land in search of Allah's bounty, and others are **fighting** for the cause of Allah. So recite of it which is **easy**, and establish worship and pay the poor-due, and lend to Allah a goodly loan. Whatsoever good you send before you for your souls, you will surely find it with Allah, better and greater in the recompense. And seek forgiveness of Allah. Lo! Allah is Ever Forgiving, Most Merciful. [Sura No. 72\73 – Yaaa-ayyuhal Muzzamil, Ayat No. - 20]

And We have indeed made the **Qur'an easy to understand** and **remember**: then is there any that will receive admonition? [Sura No. – 53\54 – Qamar, Ayat\ Verses – 17, 22, 32, 40]

And those who disbelieve say of those who believe: **If it had been (any) good, they would not have been before us in attaining it**. And since they will not be guided by it, they say: **This is an ancient lie (old falsehood)**. And before it, **there was the Kitaab of Muusa, an example and a mercy**. And **this Kitaab confirms [it]** in the **Arabic language** that **it may warn those who do wrong** and **bring good tidings for the righteous**. [Sura No. – 45\46 – Bil-ahqaafi, Ayat No. – 11 & 12]

Praise be to Allah, Who has revealed the **Abdihil-Kitaaba** unto His slave (servant), and has placed therein **no crookedness [global concept]**. [He has made it] **diagonal** [declining towards right] in order that He may **warn of a terrible punishment** from Him, and that He may give **good tidings to the believers who work righteous deeds [tabligh],** that they will have a **fair reward**; wherein they will abide forever. [Sura No. – 17\18 – Khaf – Ayat No. – 1 to 3]

Stand you upright, and follow not the path of those who know not.
Ayat\Verses - We **inspired** Muusa and his brother with this Message: Provide dwellings for your people in Misra (Egypt), make your dwellings into places of worship, and establish regular prayers: and give good tidings to those who believe. Muusa prayed: Our Rab! You has indeed bestowed on Firawn and his chiefs splendour and wealth in the life of the present, and so, Our Rab, they mislead [men] from Your Path. Spoil, our Rab, the features of their wealth, and send hardness to their hearts, so they will not believe until they see the grievous penalty. Allah said: Your prayer is accepted! **So stand you upright, and follow not the path of those who know not.** [Sura No. – 9\10 – Yuunus, Ayat No. - 87 to 89]

SECTION – VI
NATURAL MAGNETISM – 6.4
TRIANGULAR BULLETS

[Function Keys F1 & F2]

Obverted White of the Converted Snake of Muusa's Staff

Fasta-'iz billaahi minash-Shaytaanir-Rajim
Bismillaahir-Rahmaanir- Rahim

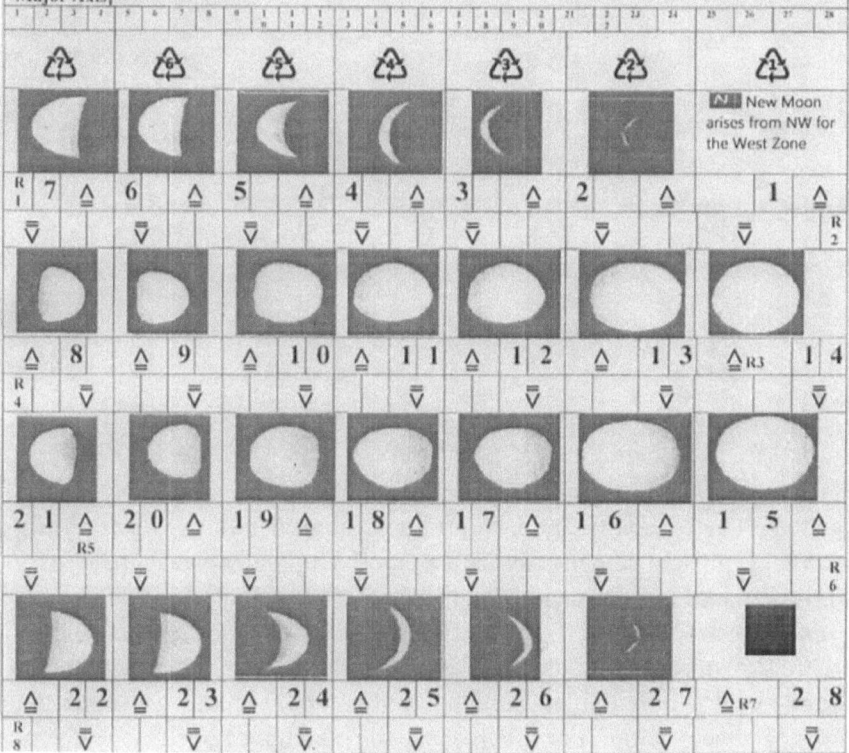

Triangular Bullets [so-called Moons] - ▸ [Unicode (hex) – 2023]

MUZZAMMIL AND MUDDASSIR - ƙƙ Unicode (hex) F756 & F757

Stars [Planets/Moons] Uranus and Neptune ⁝ Unicode (hex) F7E3

WEST ZONE / UPPER SEASHORE

Muzzammil and Muddassir – Signs of Successive Relation or Signs [ice-giants] to mark fixed periods of time in [the affairs of] men, and for Pilgrimage.

EQUAL & OPPOSITE [1ˢᵗ & 3ʳᵈ and 2ⁿᵈ & 4ᵗʰ]

Right Double Quotation Mark with respect to West Zone represents arising of the so-called new moon as the sign of successive relation, and to mark fixed periods of time in [the affairs of] men, and for Pilgrimage from Manifested North-West in correspondence with Australia as well as 1ˢᵗ & 5ᵗʰ Rows [1ˢᵗ & 3ʳᵈ Quarters / Periods] from NW to SW; while Left Double Quotation Mark represents 3ʳᵈ & 7ᵗʰ Rows [2ⁿᵈ & 4ᵗʰ Quarters/ Periods] from SW to NW following Axis. [Supposition – Semi Major Axis]

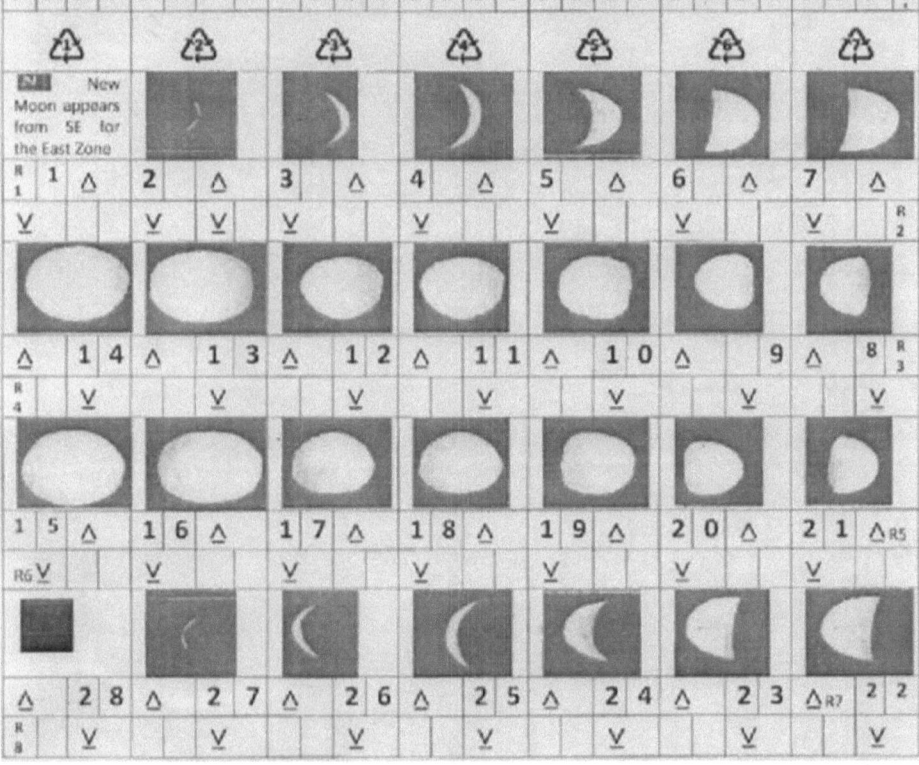

Triangular Bullets [so-called Moons] - ▸ [Unicode (hex) – 2023]

MUZZAMMIL AND MUDDASSIR - ƙƙ Unicode (hex) F756 & F757

Stars [Planets/Moons] Uranus and Neptune ⸰ Unicode (hex) F7E3

EAST ZONE / LOWER SEASHORE

Muzzammil and Muddassir – Signs of successive relation or Signs [ice-giants] to mark fixed periods of time in [the affairs of] men, and for Pilgrimage.

EQUAL & OPPOSITE [1ˢᵗ & 3ʳᵈ and 2ⁿᵈ & 4ᵗʰ]

Left Double Quotation Mark with respect to West Zone represents arising of the so-called new moon as the sign of successive relation, and to mark fixed periods of time in [the affairs of] men, and for Pilgrimage from Manifested South-East in correspondence with South America as well as 1ˢᵗ & 5ᵗʰ Rows [1ˢᵗ & 3ʳᵈ Quarters / Periods] from SE to NE; while Right Double Quotation Mark represents 3ʳᵈ & 7th Rows [2ⁿᵈ & 4ᵗʰ Quarters / Periods] from NE to SE following Axis. [Supposition – Semi Major Axis]

Projected Planets Uranus and Neptune or shared names Muzzammil and Muddassir represent new moons for the equal & opposite Seashore [West Zone and East Zone]. Four Galilean Moons represent Projected Planets Uranus and Neptune or shared names Muzzammil and Muddassir for the equal & opposite Seashore [West Zone and East Zone]. Neptune [Muddassir] is **not visible to the unaided eye** [the only planet found by mathematical prediction rather than by **empirical observation**]. For this reason, the concept of Muddassir [Neptune] has been shared as enveloped [Ref. Symbol (hex) 002A & Symbol (decimal) 42]. What we commonly perceive as moon is the white patches or layered cloud structure, a ring system, a magnetosphere. Projected Uranian system represents Lunar System. It has a unique configuration because its **axis of rotation** is **tilted sideways.** Commonly perceived white patches or moons are the signs of successive relation between revelation and manifestation. These signs are called Qamara in the Kitaab with Truth. In real science these moons are called Triangular Bullets.

Uranus is the **seventh** planet from the Sun. It has the **third-largest planetary radius and fourth-largest planetary mass** in the Solar System. **Uranus is similar in composition to Neptune,** and both have different bulk chemical composition from that of the larger gas giants **Jupiter** and **Saturn.** For this reason, scientists often classify **Uranus and Neptune as "ice giants"** to distinguish them from the gas giants. [Science]

MOONS ARE NOTHING BUT SIGNS

Ayat\Verses - Yas-'aluunaka 'anil- Ahil- lah. Qul hiya mawaaqiitu lin-naasi wal-Hajj. Wa laysal- birru bi-'an-ta'-tul-buyuuta min-zuhuurihaa wa laakinnal-birra manittaqaa. Wa' –tul-bu-yuutamin 'abwaabihaa wat-taqullaaha la-'allakum tuflihuun. [Trans.] They ask you concerning the **New Moons.** Say: They are **nothing except signs to mark fixed period of time** in [the affairs of] men, and for Pilgrimage. **It is no virtue (righteousness) that you go to houses (*Inviolable places or Mosques*) by the backs thereof** (as some worshipers of idols do). But the virtuous (righteous) man is he who wards off (evil). So, **go to houses by the gates thereof, and observe your duty to Allah** that you may be successful. [Sura No. 1\2 – Baqarah, Ayat No. – 189]

SINGLE BROTHERHOOD

O you messengers! Eat [all] things **good** and **pure**, and do Tabligh (work righteousness); for **I am Aware of what you do.** And verily **this Brotherhood of yours** is **a single Brotherhood**, and I am your Rab and Cherisher: therefore keep your **duty** unto Me. **But people have cut off their religion [of unity] among them, into sects. Each sect rejoices in its tenets. But leave them in their confused ignorance for a time. Do they think that because We have granted them abundance of wealth and sons, We would hasten them on in every good? Nay, they do not understand (they perceive not).** [Sura No. – 22\23 – Mu-Minuun, Ayat\Verses – 51 to 56]

A\V No.	SURA No. - 53\54 SURA – QAMAR [Wanshaqqal-Qamar]
1	**The Hour is near** and the **moon is cleft asunder.** / The Hour drew near and the moon was rent twain.
2	And if they see **a Sign,** they **turn away,** and say: This is [but] **transient magic** (Prolonged illusion).
3	**And they reject / deny [the Truth]** and **follow their [own] lusts**; but every matter has its **appointed time.**
4	And **surely** there has come to them **news** (recitals) wherein there is [enough] to **check** [them].
5	**Mature wisdom**; but [the preaching of] Warners **profits them not.**
6	Therefore, [O Prophet,] **turn away from them.** The Day that the **Caller will call** [them] to a terrible affair
7	They will come forth, **their eyes humbled** - from [their] **graves,** [torpid] like **locusts** scattered abroad,
8	Hastening, with eyes transfixed, towards the caller **"Hard is this Day!"** the **Unbelievers** will say.
9	Before them the People of Nuuh **rejected** [their messenger]: they rejected Our **servant,** and said, **"Here is one possessed!",** and he was **driven out**
10	Then he called on his Rab: "I am one overcome: do You then help [me]!"
11	So We **opened** the gates of heaven, with **water pouring forth.**
12	And We **caused the world to gush forth with springs,** so the **waters met** [and rose] to the **extent decreed.**
13	But We bore him on an [Ark] made of broad **planks** and caulked with **palm-fibre:**
14	**He floats** under our eyes [and care]: recompense to **one who had been rejected** [with scorn]!
15	And We have left **this as a Sign** [for all time]: then is there any that will receive **admonition?**
16	But how [terrible] was My **Penalty and My Warning?**
17	And We have indeed made the **Qur'an easy to understand** and **remember:** then is there any that will receive admonition?
18	The **'Ad** [people] [too] rejected [Truth]: then how **terrible** was My Penalty and My Warning?
19	For We sent against them a **furious wind,** on a Day of **violent Disaster,**
20	Plucking out men as if they were **roots of palm-trees torn** up [from the ground].
21	Yea, how [terrible] was My Penalty and My Warning!
22	But We have indeed made the **Qur'an easy to understand** and remember: then is there any that will receive admonition?
23	The Sammuud [also] rejected [their] Warners.
24	**For they said: "What! A man! A Solitary one from among ourselves! Shall we follow such a one? Truly should we then be straying in mind, and mad!**

25	Is it that the Message is sent to him, of all people amongst us? Nay, **he is a liar, an insolent one!**
26	Ah! They will know on the morrow, **which is the liar, the insolent one!**
27	For We will send the **she-camel** by way of **trial for them**. So watch them, [O Salih], and possess yourself in patience!
28	And tell them that the **water** is to be **divided between them: Each one's right to drink being** brought forward [by suitable turns].
29	But they called to their **companion**, and he took a **sword in hand, and hamstrung [her].**
30	Ah! how [terrible] was My Penalty and My Warning!
31	For We sent against them a **single Mighty Blast**, and they became like the **dry stubble** used by one who pens cattle.
32	And We have indeed made the **Qur'an easy to understand** and remember: then is there any that will receive admonition?
33	The **people** of Lut rejected [his] warning.
34	We sent against them a **violent Tornado** with showers of **stones**, [which destroyed them], except Lut's household: them We delivered by early Dawn,-
35	As a Grace from Us: thus do We **reward** those who give thanks.
36	And [Lut] did warn them of Our Punishment, but **they disputed about the Warning.**
37	And they even sought to **snatch away his guests from him**, but We blinded their **eyes**. [They heard:]. Now taste you My Wrath and My Warning.
38	Early on the morrow an **abiding Punishment** seized them:
39	So taste you My Wrath and My Warning.
40	And We have indeed made the **Qur'an easy to understand and remember**: then is there any that will receive admonition?
41	To the People of Firawn, too, afore time, came Warners [from Allah].
42	The [people] **rejected all Our Signs**; but We seized them with such penalty [as comes] from One Exalted in Power, able to carry out His Will
43	**Are your unbelievers, [O Quraish], better than they? Or have you an immunity in the Sacred Books?**
44	Or do they say: We acting together can defend ourselves?
45	**Soon** will their multitude be put to **flight**, and **they will show their backs.**
46	Nay, the **Hour** [of Judgment] is the time promised them [for their full recompense]: And that **Hour** will be **most grievous and most bitter**
47	Truly **those in sin** are the ones **straying in mind, and mad.**
48	The **Day** they will be **dragged through the Fire on their faces**, [they will hear:] **Taste you the touch of Hell!**
49	Verily, all things have We **created in proportion and measure.**
50	And **Our Command** is but a **single** [Act],- like the **twinkling of an eye.**
51	And [oft] in the past, We have destroyed gangs like unto you: then is there any that will receive admonition?

511

52	All that they do is noted in [their] Books [of Deeds]:
53	**Every matter**, small and great, is **on record.**
54	As to the **Righteous**, they will be in the midst of **Gardens and Rivers,**
55	In an **Assembly of Truth**, in the **Presence of a Sovereign Omnipotent.** [Sura No. – 53\54 – Qamar, Ayat\Veres – 41 to 55]

SURA - MUZZAMMIL

O you **folded in garments**! Keep vigil the night long **save a little, half** of it, or a **little less**, or a **little more**; and recite the Qur'an in measure. Lo! We shall **charge** you with **a word of weight.** Lo! The vigil of the night is (a time) when impression is more keen and speech more certain. Lo! You have **by day a chain of business**. So keep in remembrance the name of your Rab and devote Yourself with **a complete devotion**. [Ayat No. – 1 to 8]

"Rabbul-Mashriqi wal-Magribi Laaa ilaaha illa Huwa fattakiz-hu Wakiilaa"

Rab of the East and the West. There is no ilaaha [breakable reality] **except Him.** Therefore, you **take Him alone for disposer of affairs.** [Ayat No. - 9]

And bear with patience what they utter, and leave them with a fair-leave-taking. And leave Me to deal with the deniers' lords of ease and comfort, and bear with them for a little while. [10 & 11] [Ayat No. – 10 & 11]

Lo! With Us are Fetters [to bind them], and a Fire [to burn them]. [Ayat No. - 12]
And a Food that chokes, and a Penalty Grievous [Ayat No. - 13]

On **the Day when the earth and the mountains shake**, and the mountains become a heap of running sand; Lo! We have sent to you a **messenger**, to be a **witness** concerning you, even as We sent a messenger to Firawn. But Firawn rebelled against the messenger; so We seized him with a heavy Punishment. Then how if you disbelieve (deny), will you protect yourselves against a Day that will turn children grey? Whereon the sky will be cleft asunder; His Promise will be fulfilled. Lo! This is a reminder. Let him who will, then, choose a way to his Rab. [19] [Ayat No. – 14 to 19]

Lo! Your Rab knows **how you keep vigil**, sometimes nearly **two-thirds** of the night, or (sometimes) **half** or **a third** thereof, as do **a party of those with you. Allah measures the night and the day. He knows that you count it not, and turns to you in mercy.** Recite then, of the Quran that **which is easy for you.** He knows that there are **sick folk** among you, while **others travel** in the land in search of Allah's bounty, and **others are fighting for the cause of Allah**. So recite of it **which is easy**, and establish worship and pay the poor-due, and lend to Allah a goodly loan. Whatsoever good you send before you for your souls, you will surely find it with Allah, better and greater in the recompense. And seek forgiveness of Allah. Lo! Allah is Ever Forgiving, Most Merciful. [Ayat No. - 20]

SURA - MUDDASSIR

O you enveloped in your cloak; arise and warn! And your Rab magnify, and your raiment purifies, and pollution avoid! [1 to 5] [Ayat No. – 1 to 5]

And show not favour, seeking worldly gain! But for your Rab's [Cause], be patient! For when the Trumpet shall sound, surely that day will be a day of anguish, not of easy for the disbelievers. [Ayat No. – 6 to 10]

Leave Me alone, [to deal] with him whom I **created lonely**; and then bestowed upon him **ample means**; and **sons abiding in his presence**; and made [life] smooth for him. Yet he desires that I should give more. By no means! For lo! To **Our Signs** he has been refractory. [Ayat No. – 11 to 16]

On him I shall impose a fearful doom. For lo! He thought and then he **plotted**; (self-) destroyed is he, how he **plotted**! [Repeatedly], (self-) destroyed is he, how he plotted! [Ayat No. – 17 to 20]

Then looked he, then frowned he and showed displeasure, then he turned away in pride; and said: This is nothing else than **magic from of old. This is nothing but the word of a mortal man.** [Ayat No. – 21 to 25]
Soon will I cast him into **Hell-Fire**! Ah, what will explain to you what Hell-Fire is! It leaves none, it spears none! It dried-up man! Above it are **Nineteen**. [Ayat No. – 26 to 30]

We have appointed only angels to be wardens of the fire, and their number have We made to be a stumbling-block for those who disbelieve, that **those to whom the Kitaab has been given may have certainty, and that believers may increase in faith,** and that those to whom the Kitaab has been given and believers may not doubt, and that those in whose hearts there is disease, and **disbelievers, may say: What means Allah by this similitude?** Thus Allah sends astray whom He will, and whom He will He guides. None knows the host of your Rab except Him. This is nothing else than a reminder to mortals (this is no other than a warning to mankind). [Ayat No. - 31]

Nay, verily: By the Moon, **and the Night when it retreats, and the Dawn when it shines forth, lo! This is one of the great [Portents / Proofs / Manifested Signs] as a warning to mankind,** to him of you who will choose to advance forward or to hang back. [Ayat No. – 32 to 37]

Every soul is a pledge (promise) for its own deeds except those who will stand on the right hand. In Gardens they will question one another concerning the guilty – What led you into Hell Fire? [Ayat No. – 38 to 42]

They will answer: We were not of those who prayed nor were we of those who fed the indigent. **But we used to talk vanities with vain talkers; and we used to deny the Day of Judgment until there came to us [the Hour] that are certain. [Ayat No. – 43 to 47]**

Then will no intercession of [any] intercessors profit them. Then what is the matter with them that they turn away from admonition? As if they were frightened asses, fleeing from a lion! Forsooth, each one of them desires that he should be given open pages (from Allah). By no means! But **they fear not the Hereafter,** [Ayat No. – 48 to 53]

Nay, this surely is an admonition. Let any who will, keep it in remembrance! But none will keep it in remembrance except as Allah wills. He is the Rab of Righteousness, and the Rab of Forgiveness. [Ayat No. – 54 to 56]

PRESCRIBED PERIODS

O Prophet! When you divorce women, divorce them at their **prescribed periods**, and count their prescribed periods, and keep your duty to Allah, your Rab. **Expel them not from their houses, nor let them go forth unless they commit open immorality. Such are the limits set by Allah**; and any who **transgresses** the limits of Allah, does verily **wrong** his [own] soul. You know not it may be that Allah will **bring about thereafter some new situation.** [Sura No. – 64\65 – Tallaq-tumun-nisaaa-'a, Ayat No. – 1]

They ask you **concerning women's courses.** Say: **It is an illness. So keep away from women in their courses, and do not approach them until they are clean;** and **when** they have **purified themselves, then you go in unto them** as Allah has enjoined upon you. **Truly Allah loves those who turn to Him constantly and He loves those who keep themselves pure and clean.** Your wives are as a tilth (tilled soil) unto you. So go to your tilth as you will. But **do some good act for your souls beforehand**; and **fear Allah.** And know that you will meet Him [one day], and **give good tidings to those who believe.** [Sura No. 1\2 – Baqarah, Ayat No. – 222 & 223]

SEVEN COLUMNS

Unicode (hex) 2672, 2673, 2674, 2675, 2676, 2677, 2678, 2679, 267A, 267C

Unicode (hex) 16E5, 16DE, F7CE, 2354, 234C, 233A, 234D, 2353, 2391, 2392, 238F, 2390

Unicode (hex) – 22D8, 22D9, 22C7, 22C8, 22C9, 22CA, 22CB, 22CC, 2591, 2592, 2593

Unicode (hex) – 2440, 2441, 2442, 2443, 2444, 2445, 2573, 25F0, 25F1, 25F2, 25F3, 25F8, 25F9, 25FA, 25FF

Unicode (hex) - 2A39, 10424, 10425, A497, A498, 04E2, 04E3, 04E4, 04E5, 2613, 3037,

Unicode (hex) 16DD, 16D7, 1695, 16CA, 16D2, 22DA, 22DB, 16D6, 041C, 00AB, 00BB, 25A4, 2571 & 2572,

Unicode (hex) - 25A3 & 2A54, 2A55 & 2A56, 2A07 & 2A08, 2A60 & 2A62, 2A5F & 2A61, FE3D & FE3E, FE3F & FE40, FE39 & FE3A,

Symbol (hex) – 00B4, 00B3, 00B2, 00CB, 0093, 00CC, 0094, 00F1, 00F2, 00F6, 00F7, 0083, 0082, 0027, 0053, 0052, 00E5, 00CD, 00CE, 00CF, 00D0, 00D1, 00D2, 00D3, 00C6, 00C7, 00C8, 00C9, 00CA, 00CB, 00CC, 00AB, 00AC, 00AD, 00AE, 00AF, 00B4, 00B5, 00B6, 00B7, 00B8

Such of your **women** as have passed the age of **monthly courses**, for them the prescribed period, if you have any doubts, is **three months**, and for those who have **no courses** [it is the same – calendar year or seasons]; for those who carry [life within their wombs], their period is **until they deliver their burdens** [earth]; and for those who keeps duty to Allah, He will make their **path easy [Straightway]**. That is the Commandment of Allah, which He has revealed to you. And whosoever keeps his duty to Allah He will **remit** from his **evil deeds**, and will **magnify reward for him**. [Sura No. – 64\65 – Tallaq-tumun-nisaaa-'a, Ayat No. – 4 & 5]

EVERY QUARTER

Ayat\Verses – How should we not put our trust in Allah when He has shown us our ways? We shall certainly bear with patience all the hurt you may cause us. In Allah let the trusting put their trust. And the unbelievers said to their messengers: Be sure we shall drive you out of our land, or you shall return to our religion. But their Rab inspired [this Message] to them: Verily We shall cause the wrong-doers to perish! And verily We shall cause you to abide in the land, and succeed them. This is for him who fears My Majesty and fears My Warning. And they sought victory and decision [there and then], and frustration was the lot of every powerful obstinate transgressor. In front of such a one is Hell, and he is given, for drink, boiling fetid water. In gulps will he sip it, but never will he be near swallowing it down his throat. Death will come to him from **every quarter**, yet he will not die; and in front of him will be a **chastisement unrelenting**. The parable of those who reject their Rab is that their works are as ashes, on which the wind blows furiously on a tempestuous day. They have no control of aught that they have earned. That is the extreme failure. [Sura No. – 13\14 – Ibrahim, Ayat No. – 12 & 18]

TRIANGULAR BULLETS

Ayat\Verses – Has there come upon man [ever] **any period of time** in which he was a thing **unremembered**? [Sura No. 75\76 – Hal ataa alal-Insaan, Ayat No. – 1]

Ayat\Verses – And for the moon [Wal-Qamara], We have appointed patches till *he* returns like an old dried-up palm-leaf. [Sura No. – 35\36 – Yaa-siiin, Ayat No. – 39]

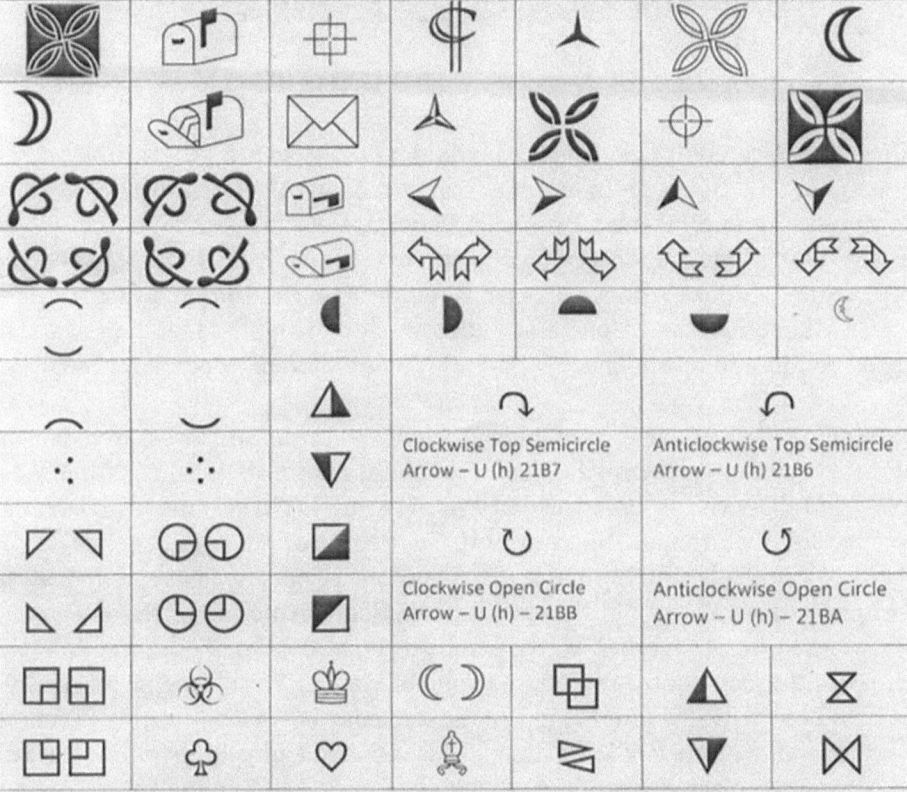

					Clockwise Top Semicircle Arrow – U (h) 21B7		Anticlockwise Top Semicircle Arrow – U (h) 21B6
					Clockwise Open Circle Arrow – U (h) – 21BB		Anticlockwise Open Circle Arrow – U (h) – 21BA

Symbol (hex) – 0083, 0082, 0084, 0094, 0093, 00E5, 00F1, 00CC, 00CB, 00B1, 00B0, 00DA, 00D9, 00D8, 00D7, 00C3, 00C4, 00C5, 00C6, 00C7, 00C8, 00C9, 00CA, 002A, 002C, 002D, 002E, 002F, 00CD, 00CE, 00CF, 00D0, 00D1, 00D2, 00D3, 00D4, 00BD, 00BC, 00BB, 00BA, 00E0.

Unicode (hex) – 2040, 203F, 2050, 2054, 2056, 2058, 25EE, 25ED, 25EA, 25EB, 25F8, 25F9, 25FA, 25FF, 25F6, 25F5, 25F7, 25F4, 25F2, 25F1, 25F3, 25F0, 2623, 2667, 2654, 2661, 263D, 263E, 29CE, 2657, 29C9, 29E8, 29E9, 29D6, 2A1D

SECTION – VI
NATURAL MAGNETISM – 6.5
DIVERSITY-IN-UNITY

Fasta-'iz billaahi minash-Shaytaanir-Rajim
Bismillaahir-Rahmaanir- Rahim

LET US OBSERVE IDD UNIFORMLY

Due to projected Black Magic, I could not observe a single Idd on the appointed day. Being an Indian, I am an inhabitant of West Zone. West Zone comprises Europe, Arabian Peninsula, Asia, Africa, and Australia. But people of the West Zone are observing Idd in two different days due to projected Mechanical Globalisation, Circular Rotation and Revolution System, Man-made Magnetism, and Conspiracy Placement of Double Quotation Mark in Insert Symbol in resemblance with the will of the Black Magician.

New moon [Triangular White Bullet / Uranus / Muzzammil] arises from North-West in correspondence with Australia. Night manifests [origins] from West Zone. Idd should be observed on the basis of the arising of new moon [Triangular White Bullet / Uranus / Muzzammil]. So, people of Australia, Africa, Asia, Arabian Peninsula, and Europe of the West Zone [Upper Seashore] will observe Idd prior to people of North America and South America of the East Zone [Lower Seashore]. In other words, people of West Zone will observe Idd at the Break of the Day [morning from the point of view of West Zone]; while people of East Zone will also observe Idd on the same day i.e. in the Morning [Evening from the point of view of West Zone]. This is called Unity-in-diversity. But due to conspiracy placements of the Double Quotation Mark in the Insert Symbol Table, believers are observing Idd in several fictitious days.

Why do we [people of the West Zone] observe Idd in two different days? The Historical Don of Hypocrisy at first projected the Upright Rectangular Universe in a globe, hexagonal world and appearing pentagonal earth also in the same globe, the map of the three ascending stairs of the appearing pentagonal earth also in the same globe. These suggest that manifestation is a trinity as well as quaternary of globalisation like the mechanism of a film. It has been distinctly shared in the Kitaab with Truth that there is no crookedness [circular ways / global concept / common run] either in revelation or in manifestation. The revealed truth is that globalisation is an impossible task. But due to projected mechanical globalisation, circular rotation & revolution system, man-made magnetism, black & white works and activities, over spreading/broadcasting

518

of falsehood like the advertisement of a product etc., I had conceptualised the projected film of the Historical Don of the Trinity of Hypocrisy – Tyranny – Conspiracy and Duality of Activism and Terrorism as the manifested nature in resemblance with revelation. So, I was one of those who have rejected their faith on revelation and belief on manifestation [Manifested Nature]. For this reason, I had failed to observe Idd on the appointed day.

Further, the mechanism of the Historical don of two-in-one partnership concerning manifested nature is nothing save subjective selfcontradictions and objective paradoxes. Probably, the Don of the two-in-one partnership has no clear and distinct concept of the existential imports of the revealed and manifested universe, revealed and manifested world, spread out table, natural magnetism etc. So, this very Don cannot be the ultimate decision provider concerning searched out manifest truths shared as Solidified Solid Human Rights in the 'Kitaaba Wal-Hikmata' or 'Manifested Nature and the Utility of One's Upright Logic'. On the contrary, those activists who have installed scientific symbols in the name of Insert Symbols must have the true concepts of the End of Proof, Black Square, House, Spiders' Net, Bullet, Diamond Operator, One Dot Leader, Ring Operator, White Ring, Bullet Operator, Right Double Quotation Mark, Left Double Quotation Mark etc. So, those black & white artists are the real teleological evidence sorcerers and unique epistemic persecutors. Due to their teleological evidence sorceries and unique epistemic persecutions, I was far from the equal and opposite right direction towards Upright West following Middle Course verifiable as well as justifiable by the equal & opposite stages of journey of the so-called sun [Bullet] as the manifested sign of natural magnetism. I could not observe a single Idd in my life on the appointed day due to their teleological evidence sorceries and unique epistemic persecutions. Now, the common people will ask those activists about the **verifiable as well as justifiable reasons** concerning the question - **Why should the believers of West Zone observe Idd in two different days**? It is certain that those activists will not be able to provide verifiable as well as justifiable reasons or four witnesses in favour of their teleological evidence sorceries and unique epistemic persecutions.

No mortal being like the Unique Epistemic Terrorist can play pragmatic game of immediate inference either with revelation or with manifestation. Everything has an appointed time. In other words, every beginning must have an end. As my Rab has shown me clear signs of the equal & opposite right direction towards Upright West and has inspired me to share the same with those who possess upright nature, these imply that the appointed time of the Trinity of Hypocrisy – Tyranny - Conspiracy, and Duality of Activism and

Terrorism as well as Teleological Evidence Sorceries and Unique Epistemic Persecutions are going to reach their end points very soon. If my Rab wills, then the consequences of their deeds will resemble with Newton's Third Law – 'Equal and Opposite'.

I have been failing to observe Idd on the appointed day even after the selfevident concept of the appointed days of observing Idd uniformly, as Idd cannot be observed individually. So, I have tried to search out the Vera causa with experimentum crusis as well as crucial instances concerning the diversity-in-unity with a view to share valid references that are leading Unity towards diversities. In this regard, I have searched out several un-contradicted self-contradictions as well as self-explained evidence sorceries and persecutions in the projected works and activities of pen – paper - pencil workers and black & white artists which are leading **Unity towards Diversities without Broken Bar.**

LET US OBSERVE IDD UNIFORMLY ON THE SAME DAY
West Zone [Upper Seashore] – People of Topmost Stair of Arabian Peninsula [Banii-Israiil] and Middle Stair of Australia, Africa, Asia, and Europe of the appearing Pentagonal Earth [Children of Adam] will observe Idd prior to people of the Ground Stair of North America and South America [Sons of Adam]. People of West Zone will observe Idd on the '**Break of the Day**' **[Morning for the West Zone]**; while people of East Zone will observe Idd at '**Early Hours of Night' [Evening for the West Zone]. All verifiable as well as justifiable contrary and contradictory findings concerning appointed day of observing Idd are Historical Black Magic.**

HYPOTHESIS, SUPPOSITION, IMAGINATION, PREDICTION GROUNDS
1. **North to South [Semi-anti-clockwise Journey]** - Placement of Bullet on the left side of the Insert Symbol Table but right side of the finite observer will represent that the so-called sun as the manifested sign of natural magnetism or Bullet enters as the Morning Show [day-light] for the Ground Stair of East Zone from Manifested North-East [NE] in correspondence with North America and sets in Manifested South-East [SE] in correspondence with South America. The scientific research institution or the country represented by the Font Name is there in the East Zone. This placement of the Bullet will resemble with manifested signs of natural magnetism in resemblance with revelation. The Don of Historical Conspiracy Science has concealed the existential import of the manifested sign of Natural Magnetism and has projected man-made

magnetism before us through mechanical activities and black & white works. This inspired sharing is nothing new but the unveiling of the separate sheet [as mentioned in the Kitaab with Truth] of the Epistemic Uniqueness of Scientific Certainty.

2. **South to North [Semi-clockwise Journey]** - Placement of Bullet on the right side of the Insert Symbol Table but left side of the finite observer will represent that the so-called sun as the manifested sign of natural magnetism or Bullet rises for the evening show i.e. day-light for the Middle Stair and Topmost Stair of the West Zone from Manifested South-West [SW] in correspondence with Europe and sets in Manifested North-West [NW] in correspondence with Australia. The scientific research institution or the country represented by the Font Name is there in the West Zone. This placement of Bullet will resemble with manifested signs of natural magnetism in resemblance with revelation. The Don of Historical Conspiracy Science has concealed the reality of the manifested sign of Natural Magnetism and has projected man-made magnetism before us through mechanical activities and black & white works. This inspired sharing is nothing new but the unveiling of the separate sheet of the Real Science.

3. **[Right] Double Quotation Mark** – The concept of Night origins from West Zone [Upper Seashore]. For the West Zone [Upper Seashore] the so-called sun [Bullet] as the manifested sign of Natural Magnetism rises from Manifested South-West [SW] in correspondence with Europe. So, the placement of Bullet at the right side of the Insert Symbol Table but left side from the point of view of finite observer represents Manifested South-West [SW] or rising direction of the so-called sun for the West Zone [Upper Seashore]. If the so-called sun [Bullet] rises from Manifested South-West, then the so-called new moon [Triangular Bullet] arises from Manifested North-West [NW] for the West Zone due to equal & opposite revelation. So, the symbol which stands for the arising of the new moon [Triangular Bullet] is to be placed at the left side of the Insert Symbol Table but right side of the finite observer. This new moon will move from Manifested North-West towards Manifested South-West [Revealed Right]. Right Double Quotation Mark represents the arising of new moon [Triangular Bullet] from Manifested North-West for the West Zone [Upper Seashore]. In other words, Right Double Quotation Mark is to be placed at the left side of the Insert Symbol Table but right side of the finite observer. Any inconsistency in the placement of Double Quotation Mark will

represent Teleological Evidence Sorcery and Unique Epistemic Persecution [Black Magic]. **The revealed truth is that persecution is worse than killing.**

4. **[Left] Double Quotation Mark** – The concept of Day origins from East Zone [Lower Seashore]. For the East Zone [Lower Seashore] the so-called sun [Bullet] as the manifested sign of Natural Magnetism enters from North-East [NE] in correspondence with North America. So, the placement of Bullet at the left side of the Insert Symbol Table but right side from the point of view of finite observer represents Manifested North-East or entering direction of the so-called sun for the East Zone [Lower Seashore]. If the so-called sun [Bullet] enters from Manifested North-East, then the so-called new moon [Triangular Bullet] appears from Manifested South-East for the East Zone due to equal & opposite revelation. So, the symbol which stands for the appearing of the new moon [Triangular Bullet] is to be placed at the right side of the Insert Symbol Table but left side of the finite observer. This new moon will move from Manifested South-East towards Manifested North-East or Revealed Left. Placement of Left Double Quotation Mark represents the appearing of new moon [Triangular Bullet] from Manifested South-East for East Zone [Lower Seashore]. In other words, Left Double Quotation Mark is to be placed at the right side of the Insert Symbol Table but left side of the finite observer. Any inconsistency in the placement of Double Quotation Mark will represent Teleological Evidence Sorcery and Unique Epistemic Persecution [Black Magic]. **The revealed truth is that persecution is worse than killing.**

ADDITIONAL HYPOTHESES

1. Font Names of the Category A with Symbol @ represent that the provided symbols guide believers in circular ways.
2. Font Names of the Category B & D represent that symbols are provided in resemblance with Copernican Revolution [circular rotation & revolution system or so-called Solar System] i.e. Semi anti-clockwise & Semi clockwise full circle in resemblance with the digit zero. Moreover, the ✓ Tick Mark at the left side of the Font Name from the point of view of the finite observer represents South or Revealed Right.
3. Font Names of the Category C resemble with Semi-circle. Moreover, the Tick Mark ✓ at the right side of the Font Name from the point of view of the finite represents North or Revealed Left.
4. Category E represents that each Font Name or the Scientific Institution is guided by three equal & opposite Font Names called Trinity. So,

the Trinity of circular universe, circular world, and circular earth resembles with the Trinity of Three Equal & Opposite Font Names. Moreover, C + E represent that Equal & Opposite games are playing by C of the Copyright using revealed E.

Abode Page Maker			On the basis of the placement of Bullet	On the basis of the placement of Double Quotation Mark	Black Magic with respect to the placement of Double Quotation Mark [Triangular Bullet]
Sl. No.	Font Name	Category	Bullet	Triangular Bullet	
1	**Arial Unicode**	A with @	SW to NW	NW to SW	**X**
2	**Batang**	A with @	SW to NW	**SW to NW**	**BM**
3	**BatangChe**	A with @	SW to NW	**SW to NW**	**BM**
4	DFKai-SB	A with @	SW to NW	NW to SW	**X**
5	Dotum	A with @	SW to NW	**SW to NW**	**BM**
6	DotumChe	A with @	SW to NW	**SW to NW**	**BM**
7	FangSong	A with @	NO BULLET BUT MIDDLE DOT		
8	Gulim	A with @	SW to NW	SW to NW	**BM**
9	GulimChe	A with @	SW to NW	SW to NW	**BM**
10	Gungsuh	A with @	SW to NW	SW to NW	**BM**
11	GungsuhChe	A with @	SW to NW	SW to NW	**BM**
12	KaiTi	A with @	NO BULLET BUT MIDDLE DOT		
13	Malgun Gothic	A with @	SW to NW	SW to NW	**BM**
14	Meiryo	A with @	SW to NW	NW to SW	**X**
15	Meiryo UI	A with @	SW to NW	NW to SW	**X**
16	Microsoft JhengHei	A with @	SW to NW	NW to SW	**X**
17	Microsoft YaHei	A with @	SW to NW	SW to NW	**BM**
18	MingLiU	A with @	SW to NW	NW to SW	**X**
19	MingLiU_HKSCS	A with @	SW to NW	NW to SW	**X**
20	MingLiU_HKSCS-ExtB	A with @	NO BULLET BUT MIDDLE DOT		
21	MingLiU-ExtB	A with @	NO BULLET BUT MIDDLE DOT		
22	MS Gothic	A with @	SW to NW	NW to SW	**X**
23	MS Mincho	A with @	SW to NW	NW to SW	**X**
24	MS PGothic	A with @	SW to NW	NW to SW	**X**
25	MS PMincho	A with @	SW to NW	NW to SW	**X**

Abode Page Maker			On the basis of the placement of Bullet	On the basis of the placement of Double Quotation Mark	Black Magic with respect to the placement of Double Quotation Mark [Triangular Bullet]
Sl. No.	Font Name	Category	Bullet	Triangular Bullet	
26	MS UI Gothic	A with @	SW to NW	NW to SW	X
27	NsimSun	A with @	SW to NW	NW to SW	X
28	PmingLiU	A with @	SW to NW	NW to SW	X
29	PMingLiU-ExtB	A with @	NO BULLET BUT MIDDLE DOT		
30	SimHei	A with @	NO BULLET BUT MIDDLE DOT		
31	SimSun	A with @	SW to NW	NW to SW	X
32	SimSun-ExtB	A with @	NO DOUBLE QUOTATION MARK		
1 to 32 GUIDED BY 33 to 35 EQUAL AND OPPOSITE FONT NAMES					
33	**AG Agency FB**	Xx	SW to NW	NW to SW	X
34	**Aharoni**	Xx	NE to SE	**NW to SW**	BM
35	**Algerian**	Xx	NE to SE	**NW to SW**	BM
36	**Andalus**	B & D	SW to NW	SW to NW	**BM**
37	**Angsana New**	B & D	SW to NW	SW to NW	**BM**
38	**AngsanaUPC**	B & D	SW to NW	SW to NW	**BM**
39	**Aparajita**	B & D	SW to NW	NW to SW	X
40	Arabic Transparent	**NO SYMBOL**			
41	Arabic Typesetting	B & D	SW to NW	NW to SW	X
42	Arial	B & D	SW to NW	NW to SW	X
43	Arial Baltic	**NO SYMBOL**			
44	Arial Black	B & D	SW to NW	NW to SW	X
45	Arial CE	**NO SYMBOL**			
46	Aria CYR	**NO SYMBOL**			
47	Arial Greek	**NO SYMBOL**			
48	Arial Narrow	B & D	SW to NW	NW to SW	X
49	**Arial Rounded MT Bold**	B & D	NE to SE	**NE to SE**	BM
50	Arial TUR	**NO SYMBOL**			
51	**Arial Unicode MS**	B & D	SW to NW	NW to SW	X
52	**Bangle**	B & D	NE to SE	NE to SE	BM
53	**Baskerville Old Face**	B & D	NE to SE	NE to SE	BM
54	**Batang**	B & D	SW to NW	**SW to NW**	**BM**

Abode Page Maker			On the basis of the placement of Bullet	On the basis of the placement of Double Quotation Mark	Black Magic with respect to the placement of Double Quotation Mark [Triangular Bullet]
Sl. No.	Font Name	Category	Bullet	Triangular Bullet	
55	**BatangChe**	B & D	SW to NW	**SW to NW**	BM
56	**Bauhaus 93**	B & D	SW to NW	**SW to NW**	BM
57	**Bell MT**	B & D	NE to SE	NE to SE	**BM**
58	Berlin Sans FB	B & D	SW to NW	NW to SW	X
59	Berlin Sans FB Demi	B & D	SW to NW	NW to SW	X
60	**Bernard MT Condensed**	B & D	NE to SE	NE to SE	**BM**
61	**Blackadder ITC**	B & D	NE to SE	NE to SE	**BM**
62	Bodoni MT	B & D	SW to NW	NW to SW	X
63	Bodoni MT Black	B & D	SW to NW	NW to SW	X
64	Bodoni MT Condensed	B & D	SW to NW	NW to SW	X
65	Bodoni MT Poster Compressed	B & D	SW to NW	NW to SW	X
66	Book Antiqua	B & D	SW to NW	NW to SW	X
67	Bookman Old Style	B & D	SW to NW	NW to SW	X
36 to 67 GUIDED BY 68 to 70 EQUAL AND OPPOSITE FONT NAMES					
68	Britannic Bold	B & D	NE to SE	NE to SE	BM
69	Bradley Hand ITC	B & D	SW to NW	NW to SW	X
70	**Bookshelf Symbol 7**	Special Reference			
71	Broadway	B & D	NE to SE	NE to SE	**BM**
72	Browallia New	B & D	SW to NW	SW to NW	**BM**
73	BrowalliaUPC	B & D	SW to NW	SW to NW	**BM**
74	Brush Script MT	B & D	NE to SE	NE to SE	**BM**
75	Calibri	B & D	SW to NW	NW to SW	X
76	Californian FB	B & D	SW to NW	NW to SW	X
77	Calisto MT	B & D	NE to SE	NE to SE	BM
78	Cambria	B & D	SW to NW	NW to SW	X
79	Cambria Math	B & D	SW to NW	NW to SW	X
80	Candara	B & D	SW to NW	NW to SW	X

Abode Page Maker			On the basis of the placement of Bullet	On the basis of the placement of Double Quotation Mark	Black Magic with respect to the placement of Double Quotation Mark [Triangular Bullet]
Sl. No.	Font Name	Category	Bullet	Triangular Bullet	
81	Castellar	B & D	NE to SE	NE to SE	**BM**
82	Centaur	B & D	NE to SE	NE to SE	BM
83	Century	B & D	SW to NW	NW to SW	X
84	Century Gothic	B & D	SW to NW	NW to SW	X
85	Century Schoolbook	B & D	SW to NW	NW to SW	X
86	Chiller	B & D	NE to SE	NE to SE	**BM**
87	Colonna MT	B & D	NE to SE	NE to SE	**BM**
88	Comic Sans MS	B & D	SW to NW	NW to SW	X
89	Consolas	B & D	SW to NW	NW to SW	X
90	Constantia	B & D	SW to NW	NW to SW	X
91	Cooper Black	B & D	NE to SE	NE to SE	BM
92	Copperplate Gothic Bold	B & D	NE to SE	NE to SE	**BM**
93	Copperplate Gothic Light	B & D	NE to SE	NE to SE	**BM**
94	Cordia New	B & D	SW to NW	SW to NW	**BM**
95	Cordia UPC	B & D	SW to NW	SW to NW	**BM**
96	Courier New	B & D	SW to NW	NW to SW	X
97	Courier New Baltic	B & D	NO SYMBOL		
98	Courier New CE	B & D	NO SYMBOL		
99	Courier New CYR	B & D	NO SYMBOL		
100	Courier New CYR	B & D	NO SYMBOL		
101	Courier New Greek	B & D	NO SYMBOL		
102	Courier New TUR	B & D	NO SYMBOL		
71 to 102 GUIDED BY 103 to 105 EQUAL AND OPPOSITE FONT NAMES					
103	**Curlz MT**	B & D	SW to NW	NW to SW	X
104	DaunPenh	B & D	NE to SE	NE to SE	BM
105	David	B & D	NE to SE	NE to SE	BM
106	Deyalika		NE to SE	NE to SE	**BM**
107	DFKai-SB	**B & D**	SW to NW	NW to SW	X
108	DilleniaUPC	**B & D**	SW to NW	SW to NW	**BM**

	Abode Page Maker		On the basis of the placement of Bullet	On the basis of the placement of Double Quotation Mark	Black Magic with respect to the placement of Double Quotation Mark [Triangular Bullet]
Sl. No.	Font Name	Category	Bullet	Triangular Bullet	
109	DokChampa	**B & D**	SW to NW	NW to SW	X
110	Dotum	**B & D**	SW to NW	SW to NW	**BM**
111	DotumChe	**B & D**	SW to NW	SW to NW	**BM**
112	Ebrima	**B & D**	SW to NW	NW to SW	X
113	Edwardian Script ITC	**B & D**	NE to SE	NE to SE	BM
114	Ekush	**B & D**	NE to SE	NE to SE	**BM**
115	EkushHlp	**B & D**	NE to SE	NE to SE	**BM**
116	Elephant	**B & D**	NE to SE	NE to SE	**BM**
117	Engravers MT	**B & D**	NE to SE	NE to SE	**BM**
118	Eras Bold ITC	**B & D**	SW to NW	NW to SW	X
119	Eras Demi ITC	**B & D**	SW to NW	NW to SW	X
120	Eras Light ITC	**B & D**	SW to NW	NW to SW	X
121	Eras Medium ITC	**B & D**	SW to NW	NW to SW	X
122	Estrangelo Edessa	**B & D**	SW to NW	NW to SW	X
123	EucrosiaUPC	**B & D**	SW to NW	SW to NW	**BM**
124	Euphemia	**B & D**	SW to NW	NW to SW	X
125	Falgun	**B & D**	NE to SE	NE to SE	**BM**
126	FangSong	**B & D**	MIDDLE DOT		
127	Felix Titling	**B & D**	SW to NW	NW to SW	X
128	Footlight MT Light	**B & D**	NE to SE	NE to SE	**BM**
129	Forte	**B & D**	SW to NW	NW to SW	X
130	Franklin Gothic Book	**B & D**	SW to NW	NW to SW	X
131	Franklin Gothic Demi	**B & D**	SW to NW	NW to SW	X
132	Franklin Gothic Demi Cond	**B & D**	SW to NW	NW to SW	X
133	Franklin Gothic Heavy	**B & D**	SW to NW	NW to SW	X
134	Franklin Gothic Medium	**B & D**	SW to NW	NW to SW	X

	Abode Page Maker		On the basis of the placement of Bullet	On the basis of the placement of Double Quotation Mark	Black Magic with respect to the placement of Double Quotation Mark [Triangular Bullet]
Sl. No.	Font Name	Category	Bullet	Triangular Bullet	
135	Franklin Gothic Medium Cond	**B & D**	SW to NW	NW to SW	X
136	FrankRuehl	**B & D**	NE to SE	NE to SE	BM
137	FreesiaUPC	**B & D**	SW to NW	SW to NW	**BM**
106 to 137 GUIDED BY 138 to 140 EQUAL AND OPPOSITE FONT NAMES					
138	French Script MT	**B & D**	SW to NW	NW to SW	X
139	Gabriola	**B & D**	SW to NW	NW to SW	X
140	Freestyle Script	**B & D**	NE to SE	NE to SE	BM
141	Garamond	B & D	SW to NW	NW to SW	**X**
142	Gautami	B & D	SW to NW	SW to NW	**BM**
143	Georgia	B & D	SW to NW	NW to SW	**X**
144	Gigi	B & D	SW to NW	NW to SW	**X**
145	Gill Sans MT	B & D	SW to NW	NW to SW	**X**
146	Gill Sans MT Condensed	B & D	SW to NW	NW to SW	**X**
147	Gill Sans MT Ext Condensed Bold	B & D	SW to NW	NW to SW	X
148	Gill Sans Ultra Bold	B & D	SW to NW	NW to SW	X
149	Gill Sans Ultra Bold Condensed	B & D	SW to NW	NW to SW	X
150	Gisha	B & D	SW to NW	SW to NW	**BM**
151	Gloucester MT Extra Condensed	B & D	NE to SE	NE to SE	**BM**
152	Goudy Old Style	B & D	NE to SE	NE to SE	BM
153	Goudy Stout	B & D	SW to NW	NW to SW	**X**
154	Gulim	B & D	SW to NW	SW to NW	**BM**
155	GulimChe	B & D	SW to NW	SW to NW	**BM**
156	Gungsuh	B & D	SW to NW	SW to NW	**BM**
157	GungsuhChe	B & D	SW to NW	SW to NW	**BM**
158	Gunplay	B & D	NE to SE	NE to SE	**BM**
159	Haettenschweiler	B & D	SW to NW	SW to NW	**BM**
160	Harlow Solid Italic	B & D	NE to SE	NE to SE	BM

Abode Page Maker			On the basis of the placement of Bullet	On the basis of the placement of Double Quotation Mark	Black Magic with respect to the placement of Double Quotation Mark [Triangular Bullet]
Sl. No.	Font Name	Category	Bullet	Triangular Bullet	
161	Harrington	B & D	SW to NW	NW to SW	X
162	High Tower Text	B & D	SW to NW	NW to SW	**X**
163	Impact	B & D	SW to NW	NW to SW	**X**
164	Imprint MT Shadow	B & D	NE to SE	NE to SE	**BM**
165	Informal Roman	B & D	SW to NW	NW to SW	**X**
166	IrisUPC	B & D	SW to NW	SW to NW	**BM**
167	Iskoola Pota	B & D	SW to NW	NW to SW	**X**
168	JasmineUPC	B & D	SW to NW	SW to NW	**BM**
169	Jokerman	B & D	SW to NW	NW to SW	X
170	Juice ITC	B & D	SW to NW	NW to SW	X
171	KaiTi	B & D	MIDDLE DOT		
172	Kalinga		SW to NW	NW to SW	X
141 to 172 GUIDED BY 173 to 175 EQUAL AND OPPOSITE FONT NAMES					
173	KodchiangUPC		SW to NW	SW to NW	BM
174	Kartika		SW to NW	NW to SW	X
175	Khmer UI		SW to NW	NW to SW	X
176	Kokila	B & D	SW to NW	NW to SW	X
177	Kristen ITC	B & D	SW to NW	NW to SW	X
178	Kunstler Script	B & D	SW to NW	NW to SW	X
179	Lao UI	B & D	SW to NW	NW to SW	X
180	Latha	B & D	SW to NW	NW to SW	X
181	Leelawadee	B & D	SW to NW	NW to SW	X
182	Levenim MT	B & D	NE to SE	NE to SE	BM
183	LilyUPC	B & D	SW to NW	SW to NW	**BM**
184	Lucida Bright	B & D	NE to SE	NE to SE	**BM**
185	Lucida Calligraphy	B & D	NE to SE	NE to SE	**BM**
186	Lucida Console	B & D	SW to NW	NW to SW	X
187	Lucida Fax	B & D	NE to SE	NE to SE	BM
188	Lucida Handwriting	B & D	NE to SE	NE to SE	**BM**
189	Lucida Sans	B & D	NE to SE	NE to SE	**BM**

Abode Page Maker			On the basis of the placement of Bullet	On the basis of the placement of Double Quotation Mark	Black Magic with respect to the placement of Double Quotation Mark [Triangular Bullet]
Sl. No.	Font Name	Category	Bullet	Triangular Bullet	
190	Lucida Sans Typewriter	B & D	NE to SE	NE to SE	**BM**
191	Lucida Sans Unicode	B & D	SW to NW	NW to SW	**X**
192	Magneto	B & D	SW to NW	NW to SW	**X**
193	Maiandra GD	B & D	SW to NW	NW to SW	**X**
194	Malgun Gothic	B & D	SW to NW	NW to SW	**X**
195	Mangal	B & D	SW to NW	NW to SW	**X**
196	**Marlett**	B & D	**Marlett represents conspiracy code number 786 of the two-in-one partnership by the conversion of waw [9] into 6 and by the projection of 8 as circular ways.**		
197	Matura MT Script Capitals	B & D	NE to SE	NE to SE	BM
198	Meiryo	B & D	SW to NW	NW to SW	**X**
199	Meiryo UI	B & D	SW to NW	NW to SW	**X**
200	Microsoft Himalaya	B & D	SW to NW	NW to SW	**X**
201	Microsoft JhengHei	B & D	SW to NW	NW to SW	**X**
202	Microsoft New Tai Lue	B & D	SW to NW	NW to SW	**X**
203	Microsoft PhagsPa	B & D	SW to NW	NW to SW	**X**
204	Microsoft Sans Serif	B & D	SW to NW	NW to SW	**X**
205	Microsoft Tai Le	B & D	SW to NW	NW to SW	X
206	Microsoft Uighur	B & D	SW to NW	NW to SW	X
207	Microsoft YaHei	B & D	SW to NW	SW to NW	BM
176 to 207 GUIDED BY 208 to 210 EQUAL AND OPPOSITE FONT NAMES					
208	French Script MT	B & D	SW to NW	NW to SW	X
209	Gabriola	B & D	SW to NW	NW to SW	X
210	Freestyle Script	B & D	NE to SE	NE to SE	BM
211	MingLiU_HKSCS	B & D	SW to NW	NW to SW	**X**

Abode Page Maker			On the basis of the placement of Bullet	On the basis of the placement of Double Quotation Mark	Black Magic with respect to the placement of Double Quotation Mark [Triangular Bullet]
Sl. No.	Font Name	Category	Bullet	Triangular Bullet	
212	MingLiU_HKSCS-ExtB	B & D	MIDDLE DOT		
213	Miriam (147 characters)	B & D	MIDDLE DOT		
214	Miriam Fixed	B & D	**NE to SE**	**NE to SE**	**BM**
215	Mistral	B & D	SW to NW	NW to SW	X
216	Modern No. 20	B & D	**NE to SE**	**NE to SE**	**BM**
217	Mongolian Baiti	B & D	SW to NW	NW to SW	X
218	Monotype Corsiva	B & D	SW to NW	NW to SW	X
219	MoolBoran	B & D	SW to NW	NW to SW	X
220	MS Gothic	B & D	SW to NW	NW to SW	X
221	MS Mincho	B & D	SW to NW	NW to SW	X
222	MS Outlook	B & D	MS Outlook represents the reality of the 'End of Proof' as well as the appointed Kaba as the Manifested Upright Rectangular Universe.		
223	MS PGothic	B & D	SW to NW	NW to SW	X
224	MS PMincho	B & D	SW to NW	NW to SW	X
225	MS Reference Sans Serif	B & D	SW to NW	NW to SW	X
226	MS Reference Specialty	B & D	::: MS Reference Speciality: 35 Symbol (hex) 0023 represents two Seashores, West Zone and East Zone, Manifested Hexagon; while the code number 23 represents the appearing pentagon.		
227	MS UI Gothic	B & D	SW to NW	NW to SW	X
228	MT Extra	B & D	µ MT Extra Symbol (hex) 006D represents East Horizon.		
229	MV Boli	B & D	SW to NW	NW to SW	X
230	Narkisim (147 characters)	B & D	SW to NW	NW to SW	X
231	Niagara Engraved	B & D	SW to NW	NW to SW	X
232	Niagara Solid	B & D	SW to NW	NW to SW	X
233	NSimSun	B & D	SW to NW	NW to SW	X
234	Nyala	B & D	SW to NW	NW to SW	X

Abode Page Maker			On the basis of the placement of Bullet	On the basis of the placement of Double Quotation Mark	Black Magic with respect to the placement of Double Quotation Mark [Triangular Bullet]
Sl. No.	Font Name	Category	Bullet	Triangular Bullet	
235	OCR A Extended	B & D	**NE to SE**	SE to NE	**X**
236	Old English Text MT	B & D	**NE to SE**	**NE to SE**	**BM**
237	Onyx	B & D	**NE to SE**	**NE to SE**	**BM**
238	Palace Script MT	B & D	**NE to SE**	**NE to SE**	**BM**
239	Palatino Linotype	B & D	SW to NW	NW to SW	**X**
240	Papyrus	B & D	**NE to SE**	**NE to SE**	BM
241	Parchment	B & D	**NE to SE**	**NE to SE**	BM
242	Perpetua	B & D	SW to NW	NW to SW	**X**
211 to 242 GUIDED BY 243 to 245 EQUAL AND OPPOSITE FONT NAMES					
243	Perpetua Titling MT	B & D	**NE to SE**	**NE to SE**	BM
244	Plantagenet Cherokee	B & D	SW to NW	NW to SW	X
245	Playbill	B & D	**NE to SE**	**NE to SE**	BM
246	PMingLiU	B & D	SW to NW	NW to SW	**X**
247	PMingLiU-ExtB	B & D	MIDDLE DOT		
248	Poor Richard	B & D	NE to SE	NE to SE	**BM**
249	Pristina	B & D	NE to SE	NE to SE	**BM**
250	Raavi	B & D	SW to NW	NW to SW	**X**
251	Rage Italic	B & D	SW to NW	NW to SW	**X**
252	Ravie	B & D	SW to NW	NW to SW	**X**
253	Rockwell	B & D	NE to SE	NE to SE	**BM**
254	Rockwell Condensed	B & D	SW to NW	NW to SW	**X**
255	Rockwell Extra Bold	B & D	NE to SE	NE to SE	**BM**
256	Rod (147 characters)	B & D	NE to SE	NE to SE	**BM**
257	Rupali	B & D	CONSPIRACY FONT		
258	Sakkal Majalla	B & D	SW to NW	NW to SW	**X**
259	Script MT Bold	B & D	SW to NW	NW to SW	**X**
260	Segoe Print	B & D	SW to NW	NW to SW	**X**

Abode Page Maker			On the basis of the placement of Bullet	On the basis of the placement of Double Quotation Mark	Black Magic with respect to the placement of Double Quotation Mark [Triangular Bullet]
Sl. No.	Font Name	Category	Bullet	Triangular Bullet	
261	Segoe Script	B & D	SW to NW	NW to SW	**X**
262	Segoe UI	B & D	SW to NW	NW to SW	**X**
263	Segoe UI Light	B & D	SW to NW	NW to SW	**X**
264	Segoe UI Semibold	B & D	SW to NW	NW to SW	**X**
265	Segoe UI Symbol	B & D	SW to NW	NW to SW	X
266	Shonar Bangla	B & D	SW to NW	NW to SW	X
267	Showcard Gothic	B & D	SW to NW	NW to SW	**X**
268	Shruti	B & D	SW to NW	NW to SW	**X**
269	SimHei (124 characters)	B & D	MIDDLE DOT & ± Unicode (hex) 00B1 represents West Horizon		
270	Simplified Arabic	B & D	SW to NW	**SW to NW**	**BM**
271	Simplified Arabic Fixed	B & D	SW to NW	**SW to NW**	**BM**
272	Simpson	B & D	SW to NW	NW to SW	**X**
273	SimSun	B & D	SW to NW	NW to SW	**X**
274	SimSun-ExtB	B & D	**FULL STOP**		
275	Snap ITC	B & D	SW to NW	NW to SW	X
276	Stencil	B & D	NE to SE	NE to SE	**BM**
277	Sulekha Type	B & D	NE to SE	NE to SE	BM
246 to 277 GUIDED BY 278 to 280 EQUAL AND OPPOSITE FONT NAMES					
278	Sylfaen	B & D	SW to NW	NW to SW	X
279	Symbol	B & D	**Special Symbols**		
280	Tahoma	B & D	SW to NW	NW to SW	X
284 to 306 GUIDED BY 281 to 283 EQUAL AND OPPOSITE FONT NAMES					
281	Tempus Sans ITC		SW to NW	NW to SW	**X**
282	Times New Roman	✓	SW to NW	NW to SW	**X**
283	Times New Roman Balti		NO SYMBOL		
Category – C / I					
284	Times New Roman CE		NO SYMBOL		
285	Times New Roman CYR		NO SYMBOL		
286	Times New Roman Greek		NO SYMBOL		
287	Times New Roman TUR		NO SYMBOL		

			On the basis of the placement of Bullet	On the basis of the placement of Double Quotation Mark	Black Magic with respect to the placement of Double Quotation Mark [Triangular Bullet]
Sl. No.	Font Name	Category	Bullet	Triangular Bullet	
288	Traditional Arabic	**C / I**	SW to NW	**SW to NW**	**BM**
289	Trebuchet MS	**C / I**	SW to NW	NW to SW	**X**
290	Tunga	**C / I**	SW to NW	NW to SW	X
291	Tw Cen MT	**C / I**	SW to NW	NW to SW	**X**
292	Tw Cen MT Condensed	**C / I**	SW to NW	NW to SW	X
293	Tw Cen MT Condensed Extra Bold	**C / I**	SW to NW	NW to SW	X
294	Utsaah	**C / I**	SW to NW	NW to SW	X
295	Vani	**C / I**	SW to NW	NW to SW	X
296	Verdana	**C / I**	SW to NW	NW to SW	X
297	Vijaya	**C / I**	SW to NW	NW to SW	X
298	Viner Hand ITC	**C / I**	SW to NW	NW to SW	X
299	Vivaldi	**C / I**	NE to SE	**NE to SE**	BM
300	Vladimir Script	**C / I**	SW to NW	NW to SW	**X**
301	Vrinda	**C / I**	NE to SE	**NE to SE**	**BM**
302	Webdings	**C / I**	**SPECIAL SYMBOLS**		
303	Wide Latin	**C / I**	NE to SE	**NE to SE**	**BM**
304	Wingdings	**C / I**	**SPECIAL SYMBOLS**		
305	Wingdings 2	**C / I**	**SPECIAL SYMBOLS**		
306	Wingdings 3	**C / I**	**SPECIAL SYMBOLS**		

Ayat\Verses – Your Ultimate-Rab is Allah, Who created the heavens [West Horizon] and the world [East Horizon] in **six days**, and then He **firmly mounted the Arsh** [towards Upright West]. **He draws the night as a veil over the day, each seeking the other in rapid succession.** He made Wash-shamsa, Wal-qamara and Wan-nujuuma (galaxy stars) **sub-servient by His command**; His verily all **creations** and **commandment**. Blessed be Allah, the Rab (Cherisher and Sustainer) of the Universe! [Sura No. – 6\7 - As-haabul- a' raaf, Ayat No. – 54]

Ayat\Verses – Do not the Unbelievers see that the **heavens and the world are joined together** [as one unit of creation or Trinity], before we clove them asunder? **We made from water every living thing. Will they not then believe?** And We have set in the world **mountains standing firm, lest it should shake with them**, and We have made therein **broad highways [safa and marwa] for them to pass through that they may receive Guidance [revealed and manifested magnetism is the true guidance].** And We have made the **heavens as a canopy well guarded; yet do they turn away from the Signs [manifested magnetism] which these things [point to]! It is He Who created the Night and the Day, and wash-shamsa wal-qamara. They [the so-called sun & the so-called moon] float, each in an orbit / axis.** [Sura No. – 20\21 – Ambiyaaa, Ayat No. – 30 to 33]

There was indeed a Sign for Saba, in their dwelling place. **Two Gardens** to the **right** and to the **left**. Eat of the Sustenance [provided] by your Rab, and be grateful to Him: A territory fair and happy, and a Rabbun Gafuur. [Sura No. – 33\34 – Saba, Ayat No. - 15]

It is He Who appointed **Ash-shams to be a glorious shining** and **the wal-qamar to be a light** and **measured out stages** for him that you might know the **number of years and the reckoning.** Allah created not that **except truth** and **righteousness. He detailed the Clear Signs for those who have understanding.** [Sura No. – 9\10 – Yuunus, Ayat No. – 5]

By the **dawn** [1]; and **ten nights**, [2]; by **the even and odd [contrasted]**[3]; and by the Night when it [Allah's Sovereignty] passes away [departs] [4]. There surely is an **oath** [Sura No. – 88\89 – Wal-fajri, Ayat No. – 1 to 5]

And We have commanded to men kindness toward parents. In pain did his mother bear him, and in pain did she give him birth. The carrying of the [child] to his weaning is [a period of] **thirty months**. At length, when he reaches the age of full strength and attains **forty years**, he says: My Rab! Grant me that I may be grateful for Your favour which You have bestowed upon me, and upon both my parents, and that I may work righteousness such as You may approve; and be gracious to me in my issue. Truly I have turned to You repentant and I am of those (Muslimin) who surrender (unto You). Such are they from whom We shall accept the best of their deeds and overlook their evil deeds. [They shall be] among the companions of the Garden. This is the true promise which they were promised. [Sura No. – 45\46 – Bil-ahqaafi, Ayat No. – 15 & 16]

An hour of a day

Among them are some who [pretend to] listen to you: But can you make the deaf to hear, even though they are without understanding? And among them are some who look at you: but can you guide the blind, even though they will not see? Verily Allah will not deal unjustly with man in aught [archaic anything]. It is man that wrongs his own soul. One day He will gather them together: [It will be] as if they had tarried but **an hour of a day**. They will recognise each other. Assuredly those will be lost who denied the meeting with Allah and **refused to receive true guidance**. Whether We show you [realised in thy life-time] some part of what We promise them, or We take your soul [to Our Mercy] [Before that], in any case, to Us is their return. Ultimately Allah is witness, to all that they do. [Sura No. – 9\10 – Yuunus, Ayat No. – 42 to 46]

SECTION – VI
NATURAL MAGNETISM – 6.6
SHAMS AND QAMAR

Fasta-'iz billaahi minash-Shaytaanir-Rajim
Bismillaahir-Rahmaanir- Rahim

MARYAM SUPPLIED WITH SUSTENANCE
[Projected Sister Planet Venus]
Morning Show and Evening Show

Ayat\Verses – "Yaaa – ayyuhallazina aamanuttaqul-laaha wa aaminuu bi-Rasuulihii yu-tikum kif-layni mir-Rahmatihii wa yaj-al-lakum Nuuran-tamshuuna bihi wa yagfir lakum; wal-laahu Gafuurur-Rahim." [Trans.] O you who believe! **Be mindful of your duty to Allah**, and **put faith in His Messengers**. He will give you **two fold** of His Mercy and will appoint for you a **Light wherein you will walk**, and He will forgive you; for Allah is Ever Forgiving, Most Merciful. That the **People of the Kitaab** may know that they have no control whatever over the bounty of Allah; but that **the bounty is in His Hand to bestow it on whomsoever He wills**. And Allah is of Infinite Bounty. [Sura No. – 56\57 - Anzalnal-Hadiid, Ayat No. – 28 & 29]

MARYAM – SUPPLIED WITH SUSTENANCE

Ayat\Verses – Allah did choose **Adam and Nuuh**, the **family of Ibrahim**, and the **family of Imran** above (all His) creatures. They were **descendents one of the other**. And Allah is Hearer and Knower of all things. (Remember) when the **wife of Imran** said: O **Allah**! I do **dedicate** unto You what is in my **womb** for Your **special service**, **accept** this from **me**, for You hears and knows all things. And **when she was delivered**, she said: **My Rab! Lo! I am delivered of a female**. Allah **knew** best **what she brought forth. The male is not as the female**; and Lo! **I have named Her Maryam**, and Lo! I **desire** Your **protection** for **Her** and **Her off springs** from **Shaytan** the outcast. And **her** (wife of Imran) **Rab accepted Her** (Maryam) with **full acceptance**. Allah **made her grow in purity** and **beauty** and **appointed Zakariya her guardian**. Whenever **Zakariya** enters into the **sanctuary** where **she** is, he finds her **supplied with sustenance**. He asks: O Maryam! **Whence [comes] this to you**? She answered: **It is from Allah, for Allah provides sustenance to which He pleases without measure**. [Sura No. 2\3 – Imran, Ayat No. –33 to 37]

PURE / CHASTE WOMAN

Ayat\Verses – Allah sets forth an **example** for those who disbelieve the wife of Nuuh and the wife of Luut. They were [respectively] under two of our righteous servants, but they were false to their [husbands], and they profited nothing before Allah on their account, but were told: Enter the Fire along with those who enter. And Allah sets forth, as an example to those who believe the wife of Firawn. When she said: My Rab! Build for me, in nearness to You, a mansion in the Garden, and save me from Firawn and his doings, and save me from those that do wrong. And **Maryam, daughter of Imran**, whose body was **chaste**, therefore **We breathed therein something of Our spirit; and she put faith in the words of her Rab and of His Revelations, and was of the obedient.** [Sura No. – 65\66 – Tahriim, Ayat No. – 10 to 12]

This is a Way that is right.
Ayat\Verses – And when the angels said: O Maryam! Lo! Allah has chosen you and made you pure and has preferred you above (all) the women of creation. O Maryam! Be obedient to Allah and **prostrate** yourself, and **bow with those who bow** [in prayer]. This is of the tidings of the things **hidden, which We have revealed unto you. You were not present with them** when **they threw their pens** (to know) **which of them should be the guardian of Maryam** nor **were you present with them when they disputed** [there upon). (And remember) the angels said: **O Maryam! Lo! Allah has given you good tidings of a Word from Him.** His name will be **Isa**, the son of **Maryam**, held in honour in this world [present life] and the Hereafter and of [the company of] those nearest to Allah. He will **speak to the people in childhood** and in **maturity.** And he is [of the company] of the **righteous. She said: My Rab! How shall I have a son when no man (*mortal*) has touched me?** He said: Even so, Allah creates what He wills. When He decrees a plan, He but says to it, **'Be,' and it is!** And Allah will teach him **the Kitaaba wal-Hikmata**, wat-**Tawraata** and wal-**Injiil**. And [appoint him] **a messenger to the Banii-Israa-'il**, [saying]: Lo! I have come to you, **with a Sign from your Rab**, in that **I make for you out of clay, the likeness of a bird**, and **I breathe into it**, and **it is a bird by Allah's permission.** And I **heal those born blind, and the lepers**, and **I raise the dead**, by Allah's permission; and I announce unto you what you eat, and what you store up in your houses. Surely **therein is a Sign for you if you do believe.** And [I have come to you], to confirm that which was before me in the **Tawraat.** And to make **lawful** some of which was **forbidden** unto you. I have come to you **with a Sign from your Rab**. So keep duty to Allah and obey me. It is Allah Who is my Rab and your Rab; so worship Him. This is a **Way that is right. [Sura No. 2\3 – Imran, Ayat No. – 42 to 51]**

Ayat\Verses – And make mention of Maryam in the Kitaab, when she had withdrawn from her people to a chamber looking East. (IBPLtd.) [Sura No. – 18\19 – Maryam, Ayat\Verses – 16]

Or

Relate in the Book [the story of] Mary, when **she withdrew from her family to a place in the East**. (IFTA) [Sura No. – 18\19 – Maryam, Ayat\Verses – 16]

Ayat\Verses – She placed a screen [to screen herself] from them. Then We sent unto her Our Spirit and assured for her the likeness of a perfect man. She said: I seek refuge in the beneficent one from you, if you fear Allah. He said: Nay, I am only a messenger from your Rab that **I may bestow on you a faultless son**. She said: **How shall I have a son, seeing that no mortal has touched me, and I am not unchaste?** He said: So [it will be]: The sayings of your Rab: **It is easy for Me. And (it will be) that We make of him a revelation for mankind and a mercy from Us. It is a matter [so] decreed**. And **she conceived him, and she withdrew with him to a remote place**. [Sura No. – 18\19 – Maryam, Ayat\Verses – 17 to 22]

Ayat\Verses – And the pains of childbirth drove her to the trunk of a **palm-tree**. She cried [in her anguish]: Ah! Would that I had died before this! Would that I had been a thing forgotten and out of sight! But [a voice] cried to her from beneath the [palm-tree]: Grieve not! For your Rab has provided a stream beneath you. And shake towards yourself the trunk of the palm-tree. It will let fall fresh ripe dates upon you. So eat and drink and be consoled. And if meet any mortal, say: Lo! I have vowed a fast unto the beneficent, and may **not speak to any mortal**. [Sura No. – 18\19 – Maryam, Ayat\Verses – 23 to 26]

Ayat\Verses – Then she brought him to her own folk, carrying him. They said: O Maryam! You have come with an amazing thing. Oh **sister of Haaruun!** Your father was not a man of evil, nor your mother a woman unchaste! Then she pointed to him (the babe). They said: **How can we talk to one who is a child in the cradle?** He spoke: Lo! I am indeed **a servant of Allah**. He has given me **revelation** and made me a **prophet**. And has made me **blessed** where so ever I be, and has **enjoined** on me Prayer and Charity as long as I remain alive. And (has made me) **dutiful toward her** who bore me, and **has not made me overbearing or miserable**. So **peace** is on me **the day I was born**, the day I die, and the day I shall be raised up alive! Such [was] Iisabnu-Maryam: [This is] **a statement of truth**, about which they **dispute**. It is not befitting to [the majesty of] Allah that He should take unto Himself a son. Glory be to Him! When He decrees a matter, He only says to it, **"Be", and it is. Verily Allah is my Rab and your Rab. Unto Him therefore serve you. This is a Way that is right**. [Sura No. – 18\19 – Maryam, Ayat\Verses – 27 to 36]

Ayat\Verses – But the **sects** differ among themselves; but woe to the unbelievers from the meeting of an awful Day! How plainly will they see and hear, the

Day that they will appear before Us! Yet the **evil-doers today are in error manifest!** But warn them of the Day of Distress, when the matter will be determined. How **they are carelessness, and they believe not.** It is We Who will inherit the earth, and all beings thereon. To Us they all will be returned. [Sura No. – 18\19 – Maryam, Ayat\Verses – 37 to 40]

ORPHANS

Ayat\Verses – They ask you **concerning orphans.** Say: **The best thing to do is what is for their good. If you mix their affairs with yours,** they are **your brethren;** but **Allah knows** the **man who spoils from him who improves.** And **if Allah had wished,** He could have **put you into difficulties:** He is indeed Exalted in Power, The Wise. [Sura No. 1\2 – Baqarah, Ayat No. – 220]

REVEALED WISDOM IS THE RIGHT BALANCE CONCERNING ORPHAN'S PROPERTY

    i)   Orphan – Come not near to the orphan's property except to improve it.
    ii)   Pursue not that of which you have no knowledge.
    iii) Take not, with Allah, another object of worship.

Ayat\Verses – Come not near to the **orphan's property** except to improve it, until he attains the age of full strength; and fulfil [every] engagement, for [every] engagement will be enquired into [on the Day of Reckoning]. Give full measure when you measure, and weigh with a **balance that is right**: that is the most fitting and the most advantageous in the **final determination.** And **pursue not that of which you have no knowledge;** for every act of hearing, or of seeing or of [feeling in] the heart will be enquired into [on the Day of Reckoning]. Nor walk in the world with insolence: for you cannot rend the earth asunder, nor reach the mountains in height. Of all such things the evil is hateful in the sight of your Rab. These are among the [precepts of] **wisdom,** which your Rab has revealed to you. **Take not,** with Allah, **another object of worship, lest you should be thrown into Hell, blameworthy and rejected.** [Sura No. – 16\17 - Banii-Israai-iil, Ayat No. – 34 to 39]

Ayat\Verses – Of **those messengers** some of whom We have **caused** to **excel others.** To **one** of them Allah **spoke; others** He **raised** to **degrees** [of honour]. To **Isabna Maryam** We gave **Clear Proof** and **supported** him with the **holy spirit. If Allah had so willed, succeeding generations would not have fought among each other, after Clear Proofs had come to them. But they differed. Some believing** and **others rejecting. If Allah had so willed, they would not have fought each other. But Allah fulfils His plan.** [Sura No. – 1\2 – Baqarah, Ayat No. - 253]

Ayat\Verses – O you who believe! Spend out of that We have provided you, **before the Day comes** when **no** bargaining / projected falsehood will avail, **nor friendship nor intercession. Those who reject Faith they are the wrong-doers. Allah, there is no unbreakable reality but He, the Living, the Self-subsisting, Eternal**. Neither slumber nor sleep can seize Him. Unto Him belong all things in the heavens and in the world. **Who is there that can intercede with Him except His permission?** He knows that which is before or after or behind them. They encompass nothing of His knowledge except as He will. His Throne extends over the heavens and the world, and He feels no fatigue in guarding and preserving them for He is the Sublime, the Supreme. **[Sura No. – 1\2 – Baqarah, Ayat No. - 254 & 255]**

Li-duluukish-shamsi
Equal & Opposite

Ayat\Verses – But those who were blind in this world will be blind in the hereafter, and most astray from the Path. And their purpose was to tempt you away from that which We had revealed unto you, **to substitute in Our name something quite different**; [in that case] **Behold! They would certainly have made you [their] friend!** And had We not given you **strength**, you would nearly have **inclined to them** a little. In that case We should have made you **taste an equal portion [of punishment]** in **this life**, and an **equal portion in death**; and moreover you would have found **none to help you against Us! And they indeed wished to scare you from the land that they might drive you forth from thence, and then they would have stayed (there) but a little after you.** [This was Our] way with those whom We sent before you and you will find **no change in Our ways (methods).** Establish regular prayers - **Li-duluukish-shamsi** till the darkness of the night, and (the recital of) Quran at dawn **(aanal-Fajr).** Lo! **(The recital of) the Quran at dawn is ever witnessed (carry testimony). [Innal Quraanal-Fajri kaana mash-huudaa.]** And some part of the night awake for it and additional prayer for you. It may be that your Rab will raise you to a station of praise. And say: My Rab! Cause me to come in with a **firm incoming** (i.e. by the gate of truth and honour) and to go out with a **firm out going** (i.e. by the gate of truth and honour). And give me from Your Presence an authority to aid. **And say: Truth has [now] arrived, and Falsehood perished. Lo! Falsehood is [by its nature] bound to perish. [Sura No. – 16\17 -** Banii-Israai-iil, Ayat No. – 72 to 81]

Wa **tarash-shamsa**
[THE GLITTERING SHOW FOR THE EARTH]

Ayat\Verses – Praise be to Allah, Who has revealed the Kitaab unto His slave (servant), and has placed therein **no crookedness or global concept**. **[He**

has made it] **clearly and distinctly** in order that He may **warn** of a terrible **punishment** from Him, and that He may give **good tidings** to the **believers** who work righteous deeds [tabligh], that they will have a fair **reward;** wherein they will abide forever. Further, that He may **warn those** [also] who say: Allah **has chosen a son. No knowledge they have of such a thing, nor had their fathers. Dreadful** is the **word** that comes out of their mouth. **What they say is nothing but falsehood!** Yet it may be trouble yourself to death, following after them, in grief, if they believe not in this Message. That which is on earth **we have made but as a glittering show for the earth**, in order that We may test them - as to which of them are best in conduct. [Sura No. – 17\18 – Khaf – Ayat No. – 1 to 7]

Izas-Shamsu

Ayat\Verses – "**Izas-shamsu** Kwwirat" [Trans.] When the **appointed Light of Allah** [with its spacious light] is folded up - [IFTA]. When the **sun** is over thrown – [IB (P) Ltd.] When the **glittering show of the Earth** is folded up – [Self-evident Concept] {Sura – No. – 80\81 – Izash-Shamsu **[Prev. Takwiir], Ayat\Verse No. – 1]**

"Wa Izan-nujuu-munkadarat" [Trans.] And when the **stars** fall, {Sura – No. – 80\81 – Izash-Shamsu **[Prev. Takwiir], Ayat\Verse No. – 2]**

"Wa izal-jibaalu suyyirat" [Trans.] And when the **hills** [signs of planetary barrier] are moved, {Sura – No. – 80\81 – Izash-Shamsu **[Prev. Takwiir], Ayat\Verse No. – 3]**

"Wa izal-ishaaru uttilat" [Trans.] And when the **camels big with young** are abandoned, {Sura – No. – 80\81 – Izash-Shamsu **[Prev. Takwiir], Ayat\Verse No. – 4]**

"Wa izal-wuhuushu bushirat" [Trans.] And when the **wild beasts** are herded together, {Sura – No. – 80\81 – Izash-Shamsu **[Prev. Takwiir], Ayat\Verse No. – 5]**

"Wa izal-bihaaru sujjirat" [Trans.] And when the **seas** rise, {Sura – No. – 80\81 – Izash-Shamsu **[Prev. Takwiir], Ayat\Verse No. – 6]**

"Wa izan-nufuusu zuwwijat" [Trans.] And when **souls** are reunited, {Sura – No. – 80\81 – Izash-Shamsu **[Prev. Takwiir], Ayat\Verse No. – 7]**

"Wa izal-maw-uudatu suilat" [Trans.] And when the **girl-child** [without mother's name] that was buried alive is asked, {Sura – No. – 80\81 – Izash-Shamsu **[Prev. Takwiir], Ayat\Verse No. – 8]**

"Bi-ayyi-zambin-qutilat" [Trans.] For what sin [reason] she [Maryam] was slain [hamstrung the she-camel] {Sura – No. – 80\81 – Izash-Shamsu **[Prev. Takwiir], Ayat\Verse No. – 9]**

Wa ja-ala-qamara fuhinna nuuranw-wa ja-alash-shamsa siraajaa

Ayat\Verses – "Wa ja-ala-qamara fuhinna nuuranw-wa ja-alash-shamsa siraajaa" [Trans.] And has **made ja-ala-qamara in their midst**, and has made the **ja-alash-shamsa as siraaja.** And Allah has caused you **to grow as a growth from the world.** And afterward He makes you **return** thereby, and He will **bring you forth again, a (new) forth-bringing.** [Sura No. – 70\71 – Nuuh, Ayat No. – 16 to 18]

Wash-shamsa wal-qamara.

Ayat\Verses – 'Iz qaala yuusuf li-'abiihi yaaa-'abati 'innii ra'-aytu 'aha-da 'ashara kaw-kabanw-wash-shamsa wal-qamara ra-'aytu-hum lii saajidiin." [Trans.] Behold! Yuusuf said to his father: O my father! I did see **eleven springs [planets]** and **wash-shamsa wal-qamara.** I saw them prostrating themselves unto me! [Sura No. – 11]\12 – Yuusuf, Aat No. – 4]

Washshamsa wal-qamar

Ayat\Verses – And He has constrained (made **subject to you) the Night and the Day. Washshamsa wal-qamar** and the **stars are in subjection by His Command.** Verily in **these are Signs for men who have sense.** [Sura No. – 15\16 – Nahl, Ayat\Verses – 12]

Wash-shamsa and Wal-qamara

Ayat\Verses – It is Allah Who causes the seed-grain and the date-stone to split and sprout. He causes the living to issue from the dead, and He is the one to cause the dead to issue from the living. That is Allah: then how are you deluded away from the truth? He (Allah) smites the day-break [from the dark]. He makes the night for rest and calmness, and **Wash-shamsa** and **Wal-qamara** for the reckoning [of time]. That is the measuring of the Mighty, the Wise. It is He Who makes the stars [as inspiration] for you that you may guide yourselves, with their help, through the dark spaces of land [earth] and sea: We have **detailed Our Signs for people who have knowledge.** [Sura No. – 5\6 – An-aam, Ayat No. –96 to 98]

Wash-shamsu wal qamar and **L i s h-shamsi** wa laa lil-qamari
Ayat\Verses – "Wa min Aayaatihil-laylu wan-nahaaru **wash-shamsu** wal qamar. Laa tasjuduu **L i s h-shamsi** wa laa lil-qamari wasjuduu lilaahil-lazii khalaqahunna in-kuntum iyyaahu ta-budun." [**Sura No. – 40\42 – Haa-Mim-Sajdah, Ayat No. - 37**]

And among **His Signs (clear proofs of natural magnetism)** are **the Night and the Day, wash-shamsu wal qamar. Adore not Lish-shamsi wa laa lil-qamari**, but **adore Allah Who created them, if it is in truth Him Whom you worship. [Sura No. – 40\42 – Haa-Mim-Sajdah, Ayat No. - 37]**

Ayat\Verses – But If the [disbelievers] are too proud, [**no matter**], still those who are with your Rab glorify Him night and day and tire not. And **of His Signs (clear proofs of revealed magnetism)** (is this): That you see the **world lowly**, but when We **send down water to it, it is stirred to life** and yields increase. **Truly, He Who gives life to the [dead] world can surely give life to [men] who are dead**; for He has power over all things. [38 & 39] [**Sura No. – 40\42 – Haa-Mim-Sajdah, Ayat No. – 38 & 39**]

"It is He Who created the Night and the Day, and wash-shamsa wal-qamara. They float, each in an Orbit / Axis."
"We made from water every living thing. Will they not then believe?"
Ayat\Verses – Do not the Unbelievers see that the heavens and the world were joined together [as one unit of creation], before we clove them asunder? We made from water every living thing. Will they not then believe? And We have set in the world mountains standing firm, lest it should shake with them, and We have made therein broad highways [safa and marwa] for them to pass through that they may receive Guidance. And We have made the heavens as a canopy [Night Sky] well guarded; yet do they turn away from the **Signs** which these things [point to]! It is He Who created the Night and the Day, and **wash-shamsa wal-qamara. They float, each in an Orbit / Axis.** [Sura No. – 20\21 – Ambiyaaa, Ayat No. – 30 to 33]

Wash-Shamsu and Wal-Qamara
Lash-Shamsu and Tudrikal-Qamara
Ayat\Verses – Glory be to Him, Who created **in pairs [equal & opposite] all things** that the world produces, as well as **their own kind** and **things of which they have no knowledge.** [Sura No. – 35\36 – Yaa-siiin, Ayat No. – 36]

And **a Sign for them is the Night**. We **strip** of the Day, and lo! **They are in darkness.** [Sura No. – 35\36 – Yaa-siiin, Ayat No. – 37]

And **Wash-Shamsu [accelerating motion] runs** *its* **courses for a period determined for** *it.* That is the **measuring of the Mighty**, The Wise. [Sura No. – 35\36 – Yaa-siiin, Ayat No. – 38]

And for **Wal-Qamara, We have appointed patches till** *it* **returns like an old dried-up palm-leaf.** [Sura No. – 35\36 – Yaa-siiin, Ayat No. – 39]

It is not permitted to the **Lash-Shamsu to catch up** Tudrikal-Qamara, **nor the Night can outstrip [exceed] the Day. Each [just] swims along in [its own] orbit or Axis [according to equal & opposite Law or Tawraat].** [Sura No. – 35\36 – Yaa-siiin, Ayat No. – 40]

And **a Sign for them is that We bear their off-springs in the laden ship.** [Sura No. – 35\36 – Yaa-siiin, Ayat No. – 41]

And **We have created for them of the like there of wherein they ride.** [Sura No. – 35\36 – Yaa-siiin, Ayat No. – 42]

LAKUMUSH-SHAMSA WAL-QAMARA
"And He has made subject to you lakumush-shamsa wal-qamara, both diligently pursuing their courses"

Ayat\Verses – Tell to my servants who have believed, that they may establish regular prayers, and spend [in charity] out of the sustenance we have given them, secretly and openly, before the coming of a Day in which there will be **neither mutual bargaining nor befriending**. It is Allah Who hath created the heavens and the world and sends down rain from the skies [West], and with it brings out fruits wherewith to feed you. It is He Who has made the ships subject to you, that they may sail through the sea by His command; and the rivers [also] has He made subject to you. And **He has made subject to you lakumush-shamsa wal-qamara, both diligently pursuing their courses**; and the night and the day has he [also] made subject to you. And He gives you of all that you ask for. But if you count the favours of Allah, never will you be able to number them. Verily, **man is given up to injustice and ingratitude**. [Sura No. 13\14Ibrahim, Ayat No.31 to 34]

Qabla Tuluu-ish-Shamsi and Wa Qabla Guruubiha
Ayat\Verses - Is it not a warning to such men [to call to mind] how many generations before them We destroyed, in whose haunts they [now] move? Verily, in these are **Signs** for men endued with understanding. And but for a decree that has already gone forth from your Rab, and a term already fixed,

the Judgment would (have) been inevitable (in this present life). **Therefore bear with what they say, and remember [constantly] the praises of your Rab, before the rising of the (qabla tuluu-ish-shamsi), and were going down thereof (wa qabla guruubiha). And glorify Him some hours of the night and at the two ends of the day.** And strain not your eyes toward that which We cause some wedded pairs among them to enjoy the flower of life of the world. But the **provision** of your Rab is better and more enduring. And enjoin prayer upon your people, and be constant therein. We ask you not to provide **sustenance**. We provide it for you. But the [fruit of] the Hereafter is for **righteousness**. [Sura No. – 19\20 – Twa-Haa, Ayat No. – 128 to 134]

"Qabla tuluu-ish-shamsi wa qabla-l guruub"

Ayat\Verses – And how many generations We destroyed before them who were mightier than these in skill so that they wander through the land! Had there any place of escape (when the Judgment came)? Lo! Verily therein a reminder for him who has a heart, or gives ear with full intelligence. And verily We created the heavens and the world and all between them in **Six Days**, nor did any sense of weariness touch Us. So bear with what they say, and hymn the praise of your Rab **qabla tuluu-ish-shamsi wa qabla-lguruub** (before the rising and before the setting of the Glittering show of Allah). And during part of the night, [also,] hymn His praises, and [so likewise] after the postures of adoration. And listen on the Day when the caller will call out from a quiet near place, The Day when they will hear the [awful] Cry in [very] truth that is the Day of **Renaissance coming forth**. Verily it is We Who give Life and Death; and to Us is the Final Goal. On the Day when the Earth will be rent asunder (apart) from them, hastening forth (they come). That is a gathering easy for Us (to make. We know best what they say; and you are not one to subdue them by force. So warn by (admonish with) the Qur'an him who fears My Warning. [Sura No. 49\50 Kaaaf, Ayat No.36 to 43]

Sakh-kharash-shamsa wal-qamar
Reference of Partnership between Two Faces of the Same Coin

Ayat\Verses – He makes the Night to pass into Day and He makes the Day to pass into the Night, and He has subjected **Sakh-kharash-shamsa wal-qamar** each one running its course for a term appointed. Such is Allah your Rab. To Him belongs all **Dominion** (Sovereignty). And those whom you invoke besides Him have not the least power. If you **invoke** them, they will not listen to your call, and if they were to listen, they cannot answer your [prayer]. On the Day of Judgment they will reject your "**Partnership**" and none, [O man!] can tell you [the Truth] like the One Who is acquainted with all things. [Sura No. – 34\35 - Faatiris-Samaawaati wal-'arz, Ayat\Verses – 13 & 14]

Sakh-kharash-Shamsa Wal-Qamara

Ayat\Verses – And if indeed you ask them who has created the heavens and the world, and constrained (hold back) **sakh-kharash-shamsa wal-qamara** they would say: Allah. How are then **they turned away**? Allah enlarges the **sustenance** / provision [which He gives] to whom of His **bondman** He pleases; and He [similarly] grants by [strict] **measure**, [as He will]. Lo! Allah is Aware of all things. [Sura No. – 28\29 – Ankabuut, Ayat No. – 61 & 62]

Ash-Shams & Wal-Qamar

Ayat\Verses – It is He **Who appointed Ash-shams to be a glorious shining and the wal-qamar to be a light and measured out stages** for him that you might know the number of years and the reckoning. Allah created not that except truth and righteousness. He **detailed the Clear Signs for those who have understanding**. [Sura No. – 9\10 – Yuunus, Ayat No. – 5]

Wash-Shams [She-Camel] Wal-Qamar

[Sura No. – 90\91, Wash-shams, Ayat\Verses No. – 1 to 15]
Wash-shamsi and **her** brightness, [1]
Wal-qamari when **he** follows **her** [2]
By the Day as it reveals her [shows her glory] [3]
By the Night when it conceals her [4]
By the fermions [beams or rays of light] [wonderful] structure; [5]
By the world and on whom it spreads [open space or sky] [6]
By **the Soul [spirit or motion or force or energy], and the proportion [due measure] and order [orbit & direction of movement] given to it;** [7]
And it **inspires [messages] as to what is wrong [projected falsehood] for it and what is right [revealed & manifested truth] for it;-** [8]
Truly **he succeeds who purifies it [does not follow paojected mechanical globalisation and artificial magnetism],** [9]
And **he fails that corrupts it! [follows projected falsehood]** [10]
The [tribe of Samud] **rejected** [denied] the **truth** due to their **rebellious pride** [their inordinate wrong- doing], [11]
When the **promise** [commit] of them **broke forth**. [12]
"Fa-qaala lahum Rasuulullaahi Naaqatallaahi wa suqyaahaa."[Trans.]But the Messenger of Allah [Prophet} said to them: It is a **She-camel of Allah**. So let **her** drink. [13]
Then they **rejected [denied] him [Prophet]**, and **they hamstrung [slain] her [slain the Messenger i.e. begin to follow projected falsehood]**. So their Rab, **doomed** them for their **sin** and **destroyed** their dwellings [traces]. [14]
And for **him/her [Prophet]** is no fear of its sequel [of events] / consequences. [15]

"**Wash-shamsu wal-qamaru wan-nu-juumu wal-jibaalu wash-shajaru**"
Ayat\Verses – "Alam tara 'annallaaha yasjudu lahuu man' fil-arzi wash-shamsu wal-qamaru wan-nu-juumu wal-jibaalu wash-shajaru wad-dawaaabbu wa kasiirum-minannaas? Wa kasiirun haqqa 'alayhil-Azaab. Wa many-yuhinillaahu famaa lahuu mim-mukrim: 'innallaaha yaf-'alu maa yashaaa." [Trans.] Have you not seen that unto Allah pay adoration (bow down) whatsoever is in the heavens and whatsoever is in the world, wash-shamsu, **wal –qamaaru**, and **the galaxy of stars (wan-nujujuumu)**, and the hills (wal-jibaalu), and the trees (wash-shajaru), and the beasts (wad- dawaabbu), and many of mankind (wa-kaasiirum minannas)? While there are many unto whom the doom (penalty) is justly due. He whom Allah scorns, there is none to give him honour. Lo! Allah carries out all that He wills. [Sura No. – 21\22 – Hajj, Ayat No. – 18]

PILOT OF THE SHADE
"Summa jaalnash-shamsa alayhi daliilaa"
Ayat\Verses – Have you not seen how your Rab has spread that shade? And if He willed, He could have made it still. Then We have appointed The Beneficent as its **pilot** (Summa jaalnash-shamsa alayhi daliilaa). Then We withdraw it unto Us, a gradual withdrawal. And He it is Who makes the Night a covering for you, and sleeps as recline, and makes the Day a resurrection. And He it is Who sends the winds as heralds of good tidings, going before His mercy, and We send down pure water from the sky. That with it We may give life to a dead land, and satisfy the thirst of things We have created, cattle and men in great numbers. And verily We have repeated it among them that they (may remember, but most of mankind are begrudge aught save ingratitude. If We willed, We could raise up warner in every village. Therefore obey not the disbelievers, but strive against them with the utmost endeavour. [Sura No. – 24\25 - Nazzalal-Furqaan, Ayat No. – 45 to 52]

Whether Kitaaba Wal-Hikmata, Wat-Tawraata, and Wal-Injiil are the three independent Revelations [Truths-in-themselves or Independent Truths or Independent Religions] excluded from Kitaab with Truth or three criterions [scales of verification & justification] of Truth-in-itself included in the Kitaab with Truth?	
Quraanil-Majiid	Kitaabim-mubin
Quraani Ziz-Zikr	Tanzilul-Kitaab
Quraanil-Hakiim	Ummil-Kitaab
Quraanim-Mubin	Kitaabun-fussilat
Aayaatul-Quraan	Zaalikal-Kitab
Quraanan Arabiyyal-liqawminy-ya-lamuun	Alaykal-Kitaaba-bil-Haqq
Quraanun Arabiyyal-la-allakum ta-qiluun	Alaykal-Kitaabim-Mubin

Haazal-Qur'an	Kitaabun-uh-kimat
Rabbaka fil-Qur-aan	Ilayka-minal-Kitaab
Kitaaba Wal-Hikmata, Wat-Tawraata, Wal-Injiil	

Which Shams rises from the East and sets in the West? Which Shams enters from Revealed Left and ends in Revealed Left? Which Shams enters from Manifested North-East and sets in Manifested South-East? Which Shams rises from Manifested South-West and ends in Manifested North-West	
Wash-shamsi	Washshamsa
Wash-shamsu	Ja-alash-shamsa
L i s h-shamsi	Shamsanw-wa
Lash-Shamsu	Lakumush-shamsa
Tuluu-ish-Shamsi	Sakh-kharash-shamsa
Ash-Shams	tarash-shamsa
Li-duluukish-shamsi	Izas-Shamsu
Magribashshamsi	Majli-ash-shamsi
Hay-nas-saddayni	Bish-Shamsi

AN EVIL REFUGE!

Ayat\Verses – Never think you that the disbelievers are going to frustrate [Allah's Plan] in the world. Their abode is the Fire, and it is indeed an evil refuge! [Sura No. – 23\24 – Nuur, Ayat No. –57]

Ayat\Verses – "Masii-habnu-Maryama-illaa Rasuul; qad khalat-min-qab-lihir-rusul. Wa ummuhuu Siddii-qah. Kaanaaa ya'-kulaanit-ta-'aam. 'Unzur kayfa nubay-yinu lahumul-'aayaati summan-zur'annaa yu'-fakuun!" [Trans.} Masii-habnu-Maryama was no more than a messenger. Many were the messengers that passed away before him. **His mother was a Siddiiqah [saintly woman]. And they both used to eat food. See how We make the Revelation Clear Signs (Proofs) for them and see how they are turned away.** [Sura No. – 4\5 – Maaaidah, Ayat\Verses – 75]

WHO IS THE OWNER OF THE NIGHT AND THE DAY?

Ayat\Verses – "Tabaarakallazii ja-ala fis-samaa'i Buruujanwwa-ja-ala fiihaa siraajanwwa qamaram-muniiraa!" [Trans.] Blessed is He Who has placed in the heaven mansions of the stars [galaxy of stars], and has placed therein siraajanwwa qamaram-muniiraa.

"Wa Huwallazii ja-alal-layla w a n-n a h a a r a k h il-fatal-liman a r a a d a a n y-yazzakkara aw araada shukuuraa." **(Q: - Where is the reference of shams as the so-called sun, the owner of binary rays, as the creator of the night and the day?]** [Trans.] And He it is Who has appointed the **night** and the **day** in **succession** (to follow each other) for him who desires to remember, or desires to show their gratitude. [Sura No. – 24\25 - Nazzalal-Furqaan, Ayat No. – 61 & 62]

MORNING AND EVENING
(This light is found) in **houses** (Ibrahim's Kaba and its off-springs) which Allah has permitted to be **raised to honour** and that His name should be remembered **therein. Therein do offer praise to Him** in the **morning** and in the evening. [Sura No. – 23\24 – Nuur, Ayat No. – 36]

Alteration of Day and Night
Allah causes the alternation of the day and the night. Verily herein (in these things) are instructive examples (clear proofs) for those who have vision! **And Allah has created every animal from water.** Of them there are some that creep on their bellies; some that walk on two legs; and some that walk on four. Allah creates what He wills for verily Allah has power over all things. [Sura No. – 23\24 – Nuur, Ayat No. – 44 & 45]

Fall of Night and Early Hours
"Fasbir inna wa-dallaahi haqqunw-wastagfir lizambika wa sabbih bi-hamdi Rabbika **bil-ashiyyi wal-ibkarr**". Patiently, then, persevere; for the **Promise of Allah is true**, and ask forgiveness for your fault, and hymn the praises of your Rab at fall of night [**bil-ashiyyi**] and in the early hours [**wal-ibkarr**]. [Sura No. 39\40 Mu-Min, Ayat No. – 55]

MORNING STAR
By the heaven and the **morning star**; ah, what will tell you what the morning star is! The **sharp star** there is **no soul** but has a **guardian** over it. So let man **consider from** what he is created! He is created from a **gushing fluid** that issued from between mim-baynis-sulbi wat-taraaa-ib [**loins and ribs**]. Surely [Allah] is able to bring him back [to life] on the Day **when hidden thoughts shall be searched out**. Then he [man] will have **no** power, and **no** helper. [Sura No. 85\86 – Wat-Taariq, Ayat No. – 1 to 10]

And you have a sense of pride and beauty in them as you **drive them (cattle) home in the evening**, and as you **lead them forth to pasture in the morning**. [Sura No. – 15\16 – Nahl, Ayat\Verses – 6]

Allah causes the alternation (equal & opposite) of the day and the night. **Verily herein are instructive examples (Manifested Signs) for those who have vision!** And Allah has created every animal from water. Of them there are **some that creep on their bellies; some that walk on two legs;** and **some that walk on four.** Allah creates what He wills for verily Allah has power over all things. [Sura No. – 23\24 – Nuur, Ayat No. – 44 & 45]

"Yuulijul-layla fin-nahaari wa yuulijun-nahaara fil-layl; wa huwa Alimum-bi-zaatissuduur." **He causes the night to pass into the day,** and **He causes the day to pass into the night**, and He has full knowledge of the secrets of [all] hearts. [Sura No. – 56\57 - Anzalnal-Hadiid, Ayat No. – 6]

The Samuud and the Aad People disbelieved in the Judgment to come. As for Samuud, they were destroyed by the **lightning**. And the Aad, they were destroyed by a furious Wind, exceedingly violent which He imposed on them **seven long nights and eight long days** so that you might have seen men lying overthrown, as they were hollow **trunks** of **palm-tree**. Can you see any remnant of them? And Firawn, and those before him, and the communities that were destroyed, committed habitual Sin. And **they disobeyed [each] the messenger of their Rab**; so He gripped them with a lightening grip. [Sura No. – 68\69 – Al-Haaaqqatu, Ayat No. – 4 to 10]

Ayat\Verses – And your Rab is Waahid [One Rab]. **There is no ilaaha (reality) but He,** The Beneficent, The Merciful. Lo! In the **creation of the heavens and the world**; in the **alternation of the night and the day**; in the **sailing of the ships through the sea** for the profit of mankind; in the **rain which Allah Sends down from the skies,** and **the life which He gives therewith to an earth that is dead**; in the **beasts of all kinds** that He **scatters through the world**; in the **change** of **the winds,** and **the clouds [binary pulsar] which they Trail** like their slaves **between the sky and the world**; [here] indeed are **Clear Proofs or Manifested Signs for a people who are wise.** [Sura No. 1\2 – Baqarah, Ayat No. – 163 to 164]

GLORIOUS MORNING LIGHT

Ayat\Verses - By the **Glorious Morning Light** [Morning represents East Zone] [1]. And by the **Night when it is still,** [Night represents West Zone] [2]. Your Rab has not forsaken you, nor is He displeased [3]. And verily the Hereafter will be better for you than the present [4]. And soon will your Rab give you [that wherewith] you shall be well-pleased. [Sura No. – 92\93 – **Waz-Zuhaa,** Ayat No. – 1 to 5]

551

A CUP
ZANJABIL AND SALSABIL
&
Shamsanw-wa

"Muttaki-iina-fiiha alal-araaa-ik laa yarawna fiiha **shamsanw-wa** laa zam-hariiraa." [Trans.] **Reclining** therein upon **couches**, they will find there **neither** the heat of shams **nor** bitter cold [13]. And the **shade** thereof is close upon them and the clustered fruits thereof **bow down** [14]. And amongst them will be **passed round vessels of silver and goblets of crystal** [15], **Crystal-clear**, made of **silver**: they will determine the **measure** thereof [according to their wishes] [16]. And they will be given to drink there of a **Cup** [of Wine] mixed with Zanjabil [17], a **fountain** there, called Salsabil [18]. And **round** about them will [serve] youths of perpetual [freshness]. If you see them, you would think them scattered **Pearls** [19[]. And when you look, it is there you will see Bliss and a **Realm** Magnificent [20]. Upon them will be **green** Garments of fine **silk** and heavy **brocade**, and they will be adorned with **Bracelets** of silver; and their Rab will give to them to drink of a Wine **Pure** and **Holy** [21]. Verily this is a Reward for you, and your endeavour is accepted and recognised [22]. [Sura No. 75\76 – Hal ataa alal-Insaan, Ayat No. – 13 to 22]

Ayat\Verses – As to the righteous, they shall drink of a **Cup [of Wine] mixed with Kafur** [4], a **fountain** [spring] where the devotees of Allah do drink, making it flow [gush forth] in ample **abundance** [5], [because] they perform [their] vows, and **they fear a Day** whereof the evil is wide-spreading. And they **feed the poor**, the **orphan**, and the **prisoner** for the love of Allah [7], [saying]: We feed you for the sake of Allah alone. **No reward do we desire from you, nor thanks**. We only fear a **Day of distressful** wrath from the side of our Rab. Therefore Allah will deliver them from the evil of that Day, and will shed over them a **Light of Beauty** and [blissful] Joy. And because they were patient and constant, He will reward them with a **Garden** and [garments of] **silk**. [Sura No. 75\76 – Hal ataa alal-Insaan, Ayat No. – 5 to 12]

Ayat\Verses – Say: Shall I inform you good tidings of things far better than those? For the **righteous** are **Gardens [West Horizon]** in **nearness** to their Lord, with rivers flowing beneath; therein is their **eternal home**; with **companions pure** [and holy]; and the **good pleasure of Allah**, for in Allah's sight are [all] His bondmen. [Namely], those who say: Our Rab! We indeed believe, so, forgive us our sins, and guard us from the punishment of fire. Those who show patience, firmness and self-control; who are true [in word and deed]; who serve devoutly; who spend [in the way of Allah]; and who pray for forgiveness in the **early hours of the morning**. Allah (Himself) is witness that there is no ilaaha (reality) save Him. And His angels and those endowed with knowledge, standing **firm on justice**. **There is no ilaaha** (unbreakable reality) save Him, the Almighty, the Wise. [**Sura No, 2\3 – Imran, Ayat No. –15 to 18**]

"We make it like a **reaped harvest**, as if it had not **flourished yesterday**" Ayat\Verses – Say: Who can keep you safe by night and by day from [the Wrath of] [Allah] Most Gracious? Yet they **turn away** from the mention of their Rab. Or have they absolutes that can guard them from Us? They have no power **to aid themselves, nor** can they be defended from Us. Nay, We gave the good things of this life to these men and their fathers **until the period grew long for them. They do not see how We visit the land reducing it of its outlying borders.** Is it then they who will win? [Sura No. – 20\21 – Ambiyaaa, Ayat No. – 42 to 44]

Ayat\Verses - The **likeness** (similitude) of the life of the present is as the **rain** which We send down from the **sky [West]**, then the **world's growth** [East] of that which **men and cattle** eat mingle with it till, when the earth has taken on her **ornaments** and is **embellished**, and her people deem that they are **masters of her**. There reaches **Our commandment by night or by day**, and We make it like a **reaped harvest**, as if it had not **flourished yesterday.** Thus We do expound the **Signs** (Revelations) **in detail** for those **who reflect**. But Allah summons to the abode of Peace. **He guides whom He will to a way that is right**. [Sura No. – 9\10 – Yuunus, Ayat No. – 24 & 25]

Bish-Shamsi rises from the East
[Allah causes bish-shamsi but not Wash-Shamsi to rise from the East; so do you cause him to rise from the West?]

Ayat\Verses – Bethink you of him who had an argument with Ibrahim about his Rab, because Allah had granted him the kingdom. How! When Ibrahim said: My Rab is He Who gives life and death. He (Namruud) answered: I give life and cause death. ["Qaia Ibraahiimu fa-innallaha ya-tii bish-shamsi minal-maashriqi fa-ti bihaa minal magribi fa-buhital-lazi kafar"] Said Ibrahim: Allah will cause bish-shamsi (but not Wash-Shamsi) to rise from the east; so do you cause him to rise from the West? Thus was the disbeliever (who had rejected faith) confused and Allah guides not the wrong doing folk. [Sura No. – 1\2 – Baqarah, Ayat No. – 258]

CLEAR SIGN OF FEMININE GENDER
SAALIH – SHE-CAMEL – LEADERS OF THE
ARROGANT PARTY OF SAMUD

"Serve Allah: You have no other reality but Him."

"Now **has come** unto you a **clear Sign** from your Rab! **This she-camel of Allah is a Clear Sign [Token] unto you.**"

Ayat\Verses – To the Samud people [We sent] Saalih, one of their own brethren: He said: O my people! Serve Allah: You have no other reality but Him. Now **has come** unto you a **clear Sign** from your Rab! **This she-camel of Allah is a Clear Sign [Token] unto you**: So leave her to graze in Allah's world, and let her come to **no harm** or you shall be seized with a **grievous punishment**. And remember how He made you inhabitants after the Aad people and gave you station in the earth: You build for yourselves **palaces** and **castles** in [open] **plains**, and care out homes in the **mountains**; so bring to remembrance the benefits [you have received] from Allah, and **refrain from evil and mischief in the world**. The **leaders of the arrogant party** among his people said to those who were reckoned **powerless**, those among them who believed: know you indeed that Saalih is a messenger from his Rab? They said: We do indeed believe in the revelation which has been sent through him. The Arrogant party said: For our part, **we reject** what you believe in. **Then they ham-strung (restricted) the she-camel, and disrespectfully challenged the order of their Rab, saying: O Saalih! Bring about your threats, if you are a messenger [of Allah]! So the earthquake took them unawares, and they lay prostrate in their homes in the morning! So Saalih left them, saying: O my people! I did indeed convey to you the message for which I was sent by my Rab. I gave you good counsel, but you love not good counsellors.** [Sura No. – 6\7 - As-haabul- a' raaf, Ayat No. – 72 to 79]

CLEAR SIGNS AND CLEAR PROOFS OF FEMININE GENDER
She-camel

Ayat\Verses - And O my people! This **she-camel of Allah is a sign to you**: leave her to feed [provide sustenance] on Allah's [creations in] the world, and inflict no harm [do not ascribe another saying] on her, or a swift penalty will seize you. But they restricted her and then he said: enjoy life in your **homes** for **three days**: [Then will be your ruin]: [Behold] this is a threat that will not to be contradicted. When Our Decree issued, We saved Saalih and those who believed with him, by [special] Grace from Ourselves - and from the Ignominy of that day. For your Rab, He is the Strong One, and able to enforce His Will. [Sura No. – 10\11 – Kitaabun-uh-kimat (Prev. – Huud), Ayat No. – 64 to 66]

SHE-CAMEL – SAMUUD – ACCUSED TREE
MEN OF FORMER GENERATIONS TREATED
THE SIGNS OF ALLAH AS FALSE
[FEMININE GENDER]

Ayat\Verses – And We **refrain from sending the signs, only because the men of former generations treated them as false. We sent the she-camel to the Samuud to open their eyes, but they treated her wrongfully: We only send the Signs by way of terror [and warning from evil]. And when We told you: Lo! Your Rab encompasses mankind, and We appointed the vision which We showed you as a trial for mankind and the accursed tree in the Quran. We warn them; but it only increases their inordinate transgression!** [Sura No. – 16\17 - Banii-Israai-iil, Ayat No. – 59 & 60]

But **the transgressors among them changed the word from that which had been told them.** So we sent on them a plague from heaven. For that they repeatedly transgressed. [Sura No. – 6\7 - As-haabul- a' raaf, Ayat No. – 162]

And when We said: **Enter this township** and eat of the plenty therein as you wish; but **enter the gate prostrate,** and say: **Repentance.** We will **forgive you your sins** and will **increase** (reward) for the **right-doers.** But **those who did wrong changed the word which had been told them for another saying,** and We sent **on the transgressors a plague** from heaven, **for evil doing.** [Sura No. – 1\2 – Baqarah, Ayat No. – 58 & 59]

Ha Mim [Sura No. – 39\40 – Mu-Min, Fundamental Ayat No. – 1]
The revelation of this **Tanzilul-Kitaab** is from Allah, the Mighty, the Knower, the Forgiver of sin, the Accepter of repentance, the Stern in punishment, and the Bountiful. **There is no unbreakable reality except Him.** To Him is the final goal. **None dispute (argue) concerning the Signs of Allah except the disbelievers (those who reject faith on Manifested Sign of Natural Magnetism).** So, **let not their turn of fortune in the land [earth] deceives (misleads) you.** [Sura No. – 39\40 – Mu-Min, Ayat No. – 2 to 4]

THREE THOUSAND - AXIS
And when you would say unto believers: Is it not enough that your Rab should help you with **three thousand angels** sent down (to your help)? [Sura – Imran, Ayat No. – 124] [Three thousand upright ~sth or straight ~sth for the Triangular Bullet or so-called moon or Earth's Night Sky]

FIVE THOUSAND - ORBITS
Yea, if you remain firm, and act aright, even if the enemy attack you suddenly, your Rab would help you with **five thousand angels sweeping on.** [Sura – Imran, Ayat No. – 125][Five thousand latitude-longitude ~sth in resemblance with multiplication sign or Earth Sky for Bullet or the so-called sun.]

180 STRIPES TO TWAIN
TWAIN – 180 STRIPES
HUNDRED STRIPES AND EIGHTY STRIPES
Manifested Magnetism in Resemblance with revelation

Thus when they fulfil their term appointed, **either take them back on equitable terms or part with them on equitable terms; and take for witness two just persons from among you,** and **keep your evidence upright for Allah.** Who so believes in Allah and the Last Day is **encouraged** to act thus. And **whosoever keeps his duty to Allah, Allah will appoint a way out for him.** [Sura No. – 64\65 – Tallaq-tumun-nisaaa-'a, Ayat No. – 2]

Hundred Stripes
(This is) a Sura which We have revealed and which We have **ordained** (enjoined), wherein We have **revealed Clear Signs** (Plain Tokens), in order that you **may receive warning.** The **adulterer** and the **adulteress scourge** (flog) each one of them (with) a **hundred stripes.** And **let not pity** (compassion) for the **twain (archaic two) with hold you from the obedience to Allah,** if you **believe in Allah** and the **Last Day.** And **let a party of the believers witness their punishment.** [Sura No. – 23\24 – Nuur, Ayat No. – 1 & 2]

Eighty Stripes
And **those** who **accuse honourable woman** but **fail to produce four criterions [witnesses]** of justifying truth or falsity [concerning the chastity of the honourable women] **curse** them with **eighty stripes;** and **never accept their evidence.** They **indeed** are **evil doers (wicked transgressors). Unless they repent thereafter** and **mend [for such];** for Allah is Ever Forgiving, Most Merciful. [Sura No. – 23\24 – Nuur, Ayat No. – 4 & 5]

Who does greater wrong than he who invents a lie against Allah or rejects His signs? But verily the wrong-doers will never be successful. [Sura No. – 5\6 – An-aam, Ayat No. – 21]

SECTION – VI
NATURAL MAGNETISM – 6.7
SINGLE ROOT, TWO BROAD WAYS, AND A VETO

Fasta-'iz billaahi minash-Shaytaanir-Rajim
Bismillaahir-Rahmaanir- Rahim

SAFA AND MARWA

Lo! Safa and Marwa are among the indications of Allah. It is therefore **no sin for him who is on pilgrimage to the House** or at other times should **compass** them round. And he who does well of his own accord, be sure that Allah is He Who is Responsive Aware. **Those who conceal the Clear Proofs We have sent down, and the Guidance, after We have made it clear for the people in the Kitab, on them shall be Allah's curse, and the curse of those entitled to curse. Except those who repent** and **amend** and **make manifest** [openly declare the Truth]: To whom I **relent**; for I am Ever-Returning, Most Merciful. Lo! **Those who disbelieve, and die while they are disbelievers,** on them is **Allah's curse**, and **the curse of angels, and of all mankind.** They will **abide therein. Their penalty will not be lightened, nor** will they be reprieved. [Sura No. 1\2 – Baqarah, Ayat No. – 158 to 162]

557

TWO MOUNTAIN WAYS
[Two Eyes, and a Tongue, and Two Lips]

Ayat\Verses - Nay [laaa], I swear by this Balad [City of the sole manifested Imam]; and you are an in-dweller of this city [People of Arabian Peninsula]; and begetter and that which he beget [parent and child]; verily We have created man in an atmosphere He thinks that none has power over him? And he says: I have destroyed vast wealth [Kitaab with Truth]. He [destroyer of Truth] thinks that none beholds him? Did We not assign unto him **two eyes**, and a **tongue**, and **two lips**, and guide her to the parting of the **two mountain ways** [Safa and Marwa]? But he [the follower of tautology] has not attempted the ascent. And what will convey to you the path that is ascent? [It is:] to free a bondman. Or the giving of food in a day of hardship to the orphan with claims of relationship or to the destitute in the dust; and to be of those who believe and enjoin patience, and enjoin deeds of kindness and compassion. Such are the Companions of the Right Hand. But those who reject Our Signs, they are the [unhappy] companions of the left hand. On them will be fire vaulted over. [Sura No. – 89\90 – Bi-haazal-balad, Ayat No. – 1 to 20].

Ayat\Verses – Do not the Unbelievers see that the heavens and the world were joined together [as one unit of creation], before we clove them asunder? We made from water every living thing. Will they not then believe? And We have set in the world mountains standing firm, lest it should shake with them, and We have made therein **broad highways** [safa and marwa] for them to pass through that they may receive Guidance. And We have made the heavens as a canopy well guarded; yet do they turn away from the **Signs** which these things [point to]! It is He Who created the Night and the Day, and **wash-shamsa wal-qamara. They float, each in an orbit or an Axis.** [Sura No. – 20\21 – Ambiyaaa, Ayat No. – 30 to 33]

TWICE RISING AND TWICE SETTING
SENT FORWARD AND KEPT BACK

Ayat\Verses - When the **sky is cleft asunder**; "Wa izal-kawaaki-buntasarat" [Trans.] When the **stars** are scattered; hen the **oceans [seas] are suffered to burst forth**; and when the **resting place are turned upside down**; [then] shall each soul know what is **sent forward** and **kept back.** [Sura No. – 81\82 - Izas-samaa-unfatarat, Ayat No. – 1 to 5]

"Our Rab! **Twice** You have made us die, and **twice** You have given us life" Lo! (On that Day) the disbelievers (those who deny the Manifested Signs of Allah) are addressed by proclamation: Lo! Allah's abhorrence is more terrible than your abhorrence one of another (yourselves), seeing that (when) you were called to the faith but you used to refuse. They say: **Our Rab! Twice You have made us die, and twice You have given us life!** Now we confess our sins. Is there any way to go out of it? [It will be said to them]: This is because, when Allah only was invoked, you rejected faith (disbelieved), but when partners were ascribed to Him, you believed. The Command is with Allah, The Sublime, The Majestic! He it is Who shows you His Signs (Portents), and sends down sustenance for you from the sky. None will receive admonition except those who turn [to Allah]. [Sura No. – 39\40 – Mu-Min, Ayat No. – 10 to 13]

[Firawn] said: **What then is the condition of previous generations? He replied: The knowledge of that is with my Rab, duly recorded. My Rab never errs, nor forgets.** Who has appointed the **earth as a bed** and has threaded roads for you therein and has sent down **water from the sky**, and thereby We have brought forth diverse kinds of **vegetation**. (Saying): **Eat** you and feed your **cattle**. Lo! Verily **herein** are **Signs for men of thought. Thereof We created you, and therein We return you**, and thence **We bring you out a second time**. [Sura No. – 19\20 – Twa-Haa, Ayat No. – 51 to 55]

O you who believe! Be mindful of your duty to Allah and seek the way of approach unto Him, and strive in His way so that you may succeed. As to those who reject Faith, if they had everything in the world, and **twice repeated**, to give as ransom for the penalty of the Day of Resurrection, it would never be accepted of them. Theirs would be a grievous Penalty. Their wish will be to get out of the fire, but never will they get out there from. Their Penalty will be one that endures. As to the thief, male or female, cut off his or her hands: A punishment by way of example, from Allah, for their crime: And Allah is Exalted in Power. But if the thief repents after his crime and amends his conduct, Allah turns to him in forgiveness; for Allah is Forgiving, Merciful. [Sura No. – 4\5 – Maaaidah, Ayat\Verses – 35 to 39]

Li-duluukish-shamsi
Equal & Opposite

Ayat\Verses – **But those who were blind in this world will be blind in the hereafter, and most astray from the Path.** And their purpose was to tempt you away from that which We had revealed unto you, **to substitute in Our name something quite different**; [in that case] **Behold! They would certainly have made you [their] friend!** And had We not given you **strength**, you would nearly have **inclined to them** a little. In that case We should have made you **taste an equal portion [of punishment] in this life, and an equal portion in death**; and moreover you would have found **none to help you against** Us! **And they indeed wished to scare you from the land that they might drive** you forth from thence, and then they would have stayed (there) but a little after you. [This was Our] way with those whom We sent before you and you will find no change in Our ways (methods). Establish regular prayers - Li-duluukish-shamsi till the darkness of the night, and (the recital of) Quran at dawn (aanal-Fajr). Lo! (The recital of) the Quran at dawn is ever witnessed (carry testimony). [Innal Quraanal-Fajri kaana mash-huudaa.] And some part of the night awake for it and additional prayer for you. It may be that your Rab will raise you to a station of praise. And say: My Rab! Cause me to come in with a firm incoming (i.e. by the gate of truth and honour) and to go out with a firm out going (i.e. by the gate of truth and honour). And give me from Your Presence an authority to aid. And say: Truth has [now] arrived, and Falsehood perished. Lo! Falsehood is [by its nature] bound to perish. [Sura No. – 16\17 - Banii-Israai-iil, Ayat No. – 72 to 81]

SINGLE ROOT

Ayat\Verses - Alif-Lam-Mim-Ra. These are the Clear Signs of the Kitaab. That which has been revealed to you from your Rab is the Truth. But most men believe not. Allah is He Who raised the heavens without any pillars that you can see; is firmly established on the throne [of authority). He has compelled sakh-kharash-shamsa wal-qamar to be of service. Each one runs [its course] for a term appointed. He regulates all affairs, explaining the signs in detail that you may believe with certainty in the meeting with your Rab. And it is He who spread out the world, and set thereon mountains standing firm and [flowing] rivers: and fruit of every kind He made in pairs, two and two: He draws the night as a veil over the day. Behold, verily in these things there are signs for those who give thought! And in the manifested world are tracts [diverse though] neighbouring, and gardens of vines and fields sown with corn, and palm trees - growing out of single root or otherwise (like and unlike): watered with the same water, yet some of them We make more excellent than others to eat. Behold, verily in these things there are signs for those who understand! [Sura No. – 12\13 – Rad, Ayat No. – 1 to 4]

Of the people of Muusa there is a section who leads with truth and establishes justice therewith. We divided them into twelve tribes, nations. We directed Muusa by inspiration, when his [thirsty] people asked him for water: Strike the rock with your staff. Out of it there gushed forth twelve springs. Each group knew its own place for water. We gave them the shade of clouds, and sent down to them manna and quails, [saying]: Eat of the good things We have provided for you: [but they rebelled]; to Us they did no harm, but they harmed their own souls. [Sura No. – 6\7 - As-haabul- a' raaf, Ayat No. – 159 & 160]

Neither mutual bargaining nor befriending

Ayat\Verses – Tell to my servants who have believed, that they may establish regular prayers, and spend [in charity] out of the sustenance we have given them, secretly and openly, before the coming of a Day in which there will be **neither mutual bargaining nor befriending**. It is Allah Who hath created the heavens and the world and sends down rain from the skies [West], and with it brings out fruits wherewith to feed you. It is He Who has made the ships subject to you, that they may sail through the sea by His command; and the rivers [also] has He made subject to you. And **He has made subject to you lakumush-shamsa wal-qamara, both diligently pursuing their courses**; and the night and the day has he [also] made subject to you. And He gives you of all that you ask for. But if you count the favours of Allah, never will you be able to number them. Verily, **man is given up to injustice and ingratitude**. [Sura No. 13\14Ibrahim, Ayat No.31 to 34]

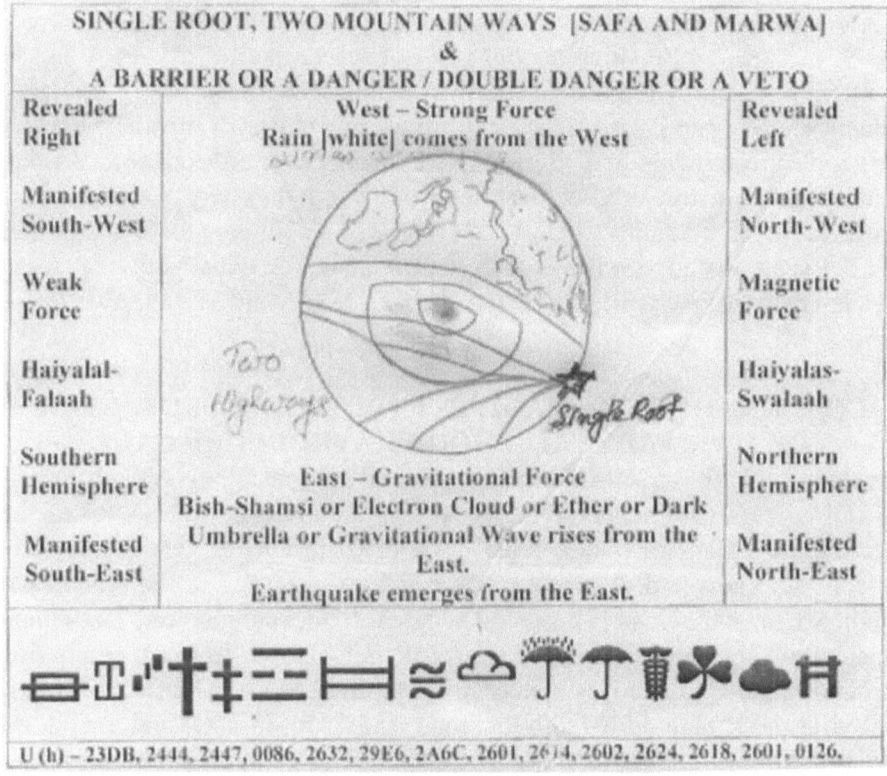

SINGLE ROOT, TWO MOUNTAIN WAYS [SAFA AND MARWA]
&
A BARRIER OR A DANGER / DOUBLE DANGER OR A VETO

Revealed Right	West – Strong Force Rain [white] comes from the West	Revealed Left
Manifested South-West		Manifested North-West
Weak Force		Magnetic Force
Haiyalal-Falaah		Haiyalas-Swalaah
Southern Hemisphere	East – Gravitational Force Bish-Shamsi or Electron Cloud or Ether or Dark Umbrella or Gravitational Wave rises from the East. Earthquake emerges from the East.	Northern Hemisphere
Manifested South-East		Manifested North-East

U (h) – 23DB, 2444, 2447, 0086, 2632, 29E6, 2A6C, 2601, 2614, 2602, 2624, 2618, 2601, 0126,

BARRIER OR DANGER OR VETO

Ayat\Verses – It is He Who has **let free (independent) the two bodies of flowing water.** One palatable and sweet and the other **salt and bitter;** yet

He has appointed **a barrier between them, a veto that is forbidden to be passed**. It is He Who has created man from **water**. Then He has established **relationships** of lineage and marriage. Lo! Your Rab has power [over all things]. Yet **they worship instead of (besides) Allah that which can neither benefit them nor hurt them**, and **the disbeliever is a helper [of devil / Shaytan] against his own Rab**! [Sura No. – 24\25 - Nazzalal-Furqaan, Ayat No. – 53 to 55]

Ayat\Verses - "Fabi-ayyi aalaaa – i Rabbikumaa tukazzibaan?" [Trans.] Which is it of the **favours** of your Rab that you **deny**? [Sura No. – 54\55, Ar-Rahman, Ayat\Verses – 21]
Ayat\Verses - "Marajal-bah-rayni yal-taqiyaan" He has let free the **two seas** meeting together. "Baynahumaa Barazakhul-laa yabgi-yaan" There is a **barrier between them**. They **encroach not** (one upon the other) [Sura No. – 54\55, Ar-Rahman, Ayat\Verses – 19 & 20]

"Between us and you is a screen (veil)."
Ayat\Verses - Ha Mim. (Revelation) from the Beneficent, the Merciful. A Kitaab whereof the Ayat [pl] are **explained in detail**; a **Qur'an in Arabic**, for people who understand. Giving **good tidings and admonition**; yet most of them **turn away**, and so **they hear not**. And **they say**: Our hearts are **protected** from that unto which you **call** us, and in our ears there is **deafness**, and **between us and you is a screen (veil)**. So act you [what you will]; for us, we shall do [what we will!] [Sura No. – 40\42 – Haa-Mim-Sajdah, Ayat No. – 1 to 5]

CONCEALMENT OF MANIFEST TRUTH AND PROJECTION OF MANMADE FALSEHOOD [INTRODUCED CALF]

Kitaab with Truth is a Kitaab of revelation and manifestation or Truth-in-itself [Fifth one or Fundamental Ayat] as well as Tautologies [Four Witnesses or Detail explanations of Fundamental Ayat or Criterions of Truth]. Subjective selfcontradictions and objective paradoxes have no place in the Kitaab with Truth. So, mortal beings like the Teleological Evidence Sorcerers and Unique Epistemic Killers of faith and belief will not be able to produce contrary and contradictory instances concerning Manifest Truth [Manifested Nature] shared in the Kitaab with Truth that may lead Truth-in-itself and Tautologies towards subjective selfcontradictions and objective paradoxes. For this reason, the allegations of Teleological Evidence Sorceries and Unique Epistemic Persecutions against people of the book, people of the Kitaab, pen-paper-pencil workers, black & white artists, and the like are certain like Descartes' 'Cogito Ergo Sum'.

"Will they not ponder on the Qur'an [with care]?"
Ayat\Verses – Whoso obeys the Messenger, obeys Allah, and those [whoso] turn away, We have not sent you to watch over their [evil deeds]. They have obedience on their lips; but when they leave you, a section of them **meditate all night on things very different from what thou tell them**. But Allah records their nightly [plots]. So **keep clear of (oppose) them, and put your trust in Allah, and enough is Allah as a disposer of affairs (trustee). Will they not ponder on the Qur'an [with care]?** If it had been from other than Allah, **they would surely have found therein much discrepancy (incongruity / contingency / contradictory).** [Sura No. – 3\4 – Nisaa, Ayat No. – 80 to 82]

REVEALED LEFT AND MANIFESTED
NORTH-EAST & NORTH-WEST
Concealment of NORTH and Projection of EAST & WEST
Northern Hemisphere or Haiyalas-Swalaah or E-Point
Early Hours of Night and Morning

The so-called new moons [Triangular Bullets or Uranus & Neptune or Muzzammil & Muddassir or F1 & F2 Keys] arise from Revealed Left [North] and Manifested North-West in correspondence with Australia. The sign of successive relation called Qamar or commonly perceived white moon follows Right Double Quotation Mark [1st period] and Semi Major Axis. **This Manifest Truth is being projected by the Historical Conspiracy Science as the new moon arises from the West concealing the concept of North from Revealed Left [North] and Manifested North-West.** The arising of new moon [Triangular Bullet] from Revealed Left [North] and Manifested North-West is the '**Evening**' for the Middle Stair [Australia, Africa, Asia, and Europe] and the Topmost Stair [Arabian Peninsula] of the appearing Pentagonal Earth, but '**Early Hours of Night**' for the West Zone [Upper Seashore] of the Manifested Hexagonal World within the Black Square [Non-luminous Moon or Dark Moon] of East Horizon. As new moon [Triangular Bullet] arises from Revealed Left [North] but Manifested North-West, so, the **concept of Night [pm] origins from the West Zone** [Upper Seashore] of the Manifested Hexagonal World within the Black Square of East Horizon.

The so-called sun [Bullet] as the manifested sign of natural magnetism **enters** from **Revealed Left** [North] and **Manifested North-East** in correspondence with **North America** as **day-light** for the Ground Stair of the appearing Pentagonal Earth but **Morning Show** for the East Zone of the Manifested Hexagonal World within the Black Square [Non-luminous Moon or Dark Moon] of East Horizon [Ref. Ayat No. – 17 of Sura Khaf]. **This Manifest Truth is being projected by the Historical Conspiracy Science as the so-called sun**

[Bullet] rises from the East concealing the concept of North from Revealed Left [North] and Manifested North-East. The so-called sun [Bullet] sets in Revealed Right [South] and Manifested South-East in correspondence with South America for the Ground Stair of the appearing Pentagonal Earth. These entering from Revealed Left [North] and Manifested North-East as well as setting in Revealed Right [South] and Manifested South-East of the so-called sun represent day-light for the Ground Stair [North America and South America] of the appearing Pentagonal Earth and **Morning Show** of the so-called sun [Bullet] as the manifested sign of natural magnetism for the East Zone of the Manifested Hexagonal World within the Black Square [Non-luminous Moon or Dark Moon] of East Horizon. The so-called sun [Bullet] as the manifested sign of natural magnetism covers half of the equal & opposite stages of journey [Ref. 4/8 i.e. ½ ASCII (hex) 00BD & ASCII (decimal) 189] or Semi-anticlockwise journey from North-East to South-East for the East Zone of the Manifested Hexagonal World within the Black Square [Non-luminous Moon or Dark Moon] of East Horizon. As the so-called sun [Bullet] enters from Revealed Left [North] but Manifested North-East, so, the **concept of Day [am] origins from the East Zone** [Lower Seashore].

Both the so-called sun [Bullet] and the so-called new moon [Triangular Bullet] have the single root. The so-called sun [Bullet] enters from Revealed Left [North] but manifested North-East for the East Zone and follows Semi-anticlockwise Orbit. The so-called new moon [Triangular Bullet] arises from Revealed Left [North] but Manifested North-West for the West Zone and follows Right Double Quotation Mark [and Semi Major Axis] for the 1st period out of four periods. So, '**Early Hours of Night**' represents '**Evening**' with respect to Middle Stair & Topmost Stair of the appearing Pentagonal Earth, and true concept of '**pm**' with respect to the West Zone of the manifested hexagonal world within the Black Square [Non-luminous Moon or Dark Moon] of East Horizon. Coherently, '**Morning**' represents '**Day-light**' with respect to Ground Stair of the appearing Pentagonal Earth, and true concept of '**am**' with respect to the East Zone of the manifested hexagonal world within the Black Square [Non-luminous Moon or Dark Moon] of East Horizon. These shared concepts are not only verifiable as well as justifiable Manifest Truths [Injiil / Correspondence Truth] in resemblance with **Equal & Opposite Revelation** [Tawraat / Coherence Truth], established **Newton's Third Law – 'Equal & Opposite'**, and the utility of one's **upright logic** [Hikmat or Selfevident Truth] but also unalterable Tautologies. So, '**Early Hours of Night**' and '**Morning**' represent **Equal & Opposite E-Point** as well as Revealed Left [North] and Manifested North-West & North-East.

EARLY HOURS OF NIGHT AND MORNING
[Revealed Concepts with reference to Wax Feeble Zakariya]
Evening [Magriib] with respect to West Zone [Upper Seashore] and
Morning [Fajr] with respect to East Zone [Lower Seashore]

Ayat\Verses - Then **Zakariya prayed** to Allah, saying: O Rab! **Grant unto me from Your bounty goodly off spring**. Lo! You are the Hearer of Prayer. While he was standing in prayer **in the sanctuary**, the angels called unto him: Allah gave you good tidings of (a son whose name is) **Yahya to confirm a Word from Allah**, and **lordly chaste**, and **a prophet of the righteous**. He (Zakariya) said: My Rab! **How can I have a son**, seeing **I am very old**, and my **wife is barren**? (The angel) answered, so (it will be), **Allah does what He wills**. He said: My Rab! **Give me a Sign [Token or Indication]**! (The angel) said: **The Sign unto you (shall be) that you shall not speak unto mankind three days except by signs. Remember Allah much and praise (Him) in early hours of night and morning.** [Sura No. 2\3 – Imran, Ayat No. – 38 to 41]

Both Wash-Shams Wal Qamar or the so-called sun [Bullet] and the so-called moon [Triangular Bullet] have the Single Root. The so-called new moon [Triangular Bullet] as the manifested sign of successive relation arises periodically from Revealed Left in correspondence with Australia for the West Zone. So, the concept of Night manifests from the Manifested West Zone [Upper Seashore] of the Hexagonal World within the Black Square. This arising of the new moon [Triangular Bullet] as the manifested sign of successive relation for the West Zone has been shared as '**Early Hours of Night**' i.e. Magriib from the point of view of the Black Square of manifested East Horizon. Simultaneously, the so-called sun [Bullet] as the manifested sign of natural magnetism [what accelerating motion releases / Electromagnetic wave / Bullet] enters from the same root i.e. Revealed Left in correspondence with North America and sets in South-East in correspondence with South America. This is the Semi-anticlockwise journey of the so-called sun [Bullet / Tarash-Shamsi] for the East Zone [Lower Seashore]. In the projected Science, this Semi-anticlockwise journey of the so-called sun [Bullet / Tarash-Shamsi] is being called **Morning Show** in the name of **Sister Planet Venus**. In the Kitaab with Truth [Quran], this Semi-anticlockwise journey of the so-called sun [Bullet / Tarash-Shamsi] has been shared as Morning i.e. Fajr from the point of view of the so-called sun [Bullet] for **the Ground Stair of North America and South America**. So, '**Early Hours of Night**' represents Magriib [Evening] or West Zone; while '**Morning**' **represents Mashriq [Fajr]** or East Zone. 'Early Hours of Night' and 'Morning' have been shared in the above Ayat/Verses with reference to Zakariya. Zakariya has been appointed as the guardian of Maryam i.e. barrier [Veil / Screen / Ether / Electron Cloud / Black

Umbrella] between Wash-Shams Wal Qamar or so-called sun and so-called moon or Bullet and Triangular Bullet. So, 'Early Hours of Night' and 'Morning' denote Single Root and connote equal & opposite pm & am [time] i.e. Evening [Magriib] from the point of view of West Zone but Morning [Fajr] from the point of view of East Zone. [Note: If you find any misconception concerning the revealed concepts of 'Early Hours of Night' and 'Morning', please try to share true concepts with valid references / witnesses]

REVEALED RIGHT AND MANIFESTED
SOUTH-EAST & SOUTH-WEST
Concealment of South and projection of Equal & Opposite
Break of the Day and Fall of Night
Southern Hemisphere or Haiyalal-Falaah or T-Point

The so-called sun [Bullet] as the manifested sign of natural magnetism sets in Revealed Right [South] and manifested South-East in correspondence with South America for the East Zone. This setting is the **'Break of the Day'** for the Ground Stair of the appearing Pentagonal Earth and the East Zone of the Manifested Hexagonal World within the Black Square [Non-luminous Moon or Dark Moon] of East Horizon. The setting sun [Bullet] runs away from Manifested South-East and rises from Revealed Right [South] and Manifested South-West in correspondence with Europe for the Middle Stair [Europe, Asia, Africa, and Australia] and the Topmost Stair [Arabian Peninsula]. These turning and rising of the so-called sun [Bullet] as the manifested sign of natural magnetism from Revealed Right [South] in the name of sister planet Venus or Maryam supplied with sustenance represent **'Evening Show'** for the so-called sun [Bullet] or Sister Planet Venus or Maryam supplied with sustenance because the so-called sun [Bullet] as the manifested sign of natural magnetism enters for the 'Morning Show' from Revealed Left [North] and Manifested North-East. So, the rising of the so-called sun as the manifested sign of natural magnetism from Revealed Right [South] and Manifested South-West for the West Zone is the 'Evening Show' for the so-called sun [Bullet]; while this rising is the beginning of the **'Morning Show'** or **'Break of the Day'** for the Middle Stair & the Topmost Stair of the appearing Pentagonal Earth and West Zone of the Manifested Hexagonal World within the Black Square [Non-luminous Moon or Dark Moon] of East Horizon. The so-called sun [Bullet] as the manifested sign of natural magnetism moves towards Revealed Left [North] and Manifested North-West. The so-called sun [Bullet] as the manifested sign of natural magnetism ends the journey in Revealed Left [North] and Manifested North-West for the Middle Stair & the Topmost Stair of the appearing Pentagonal Earth, and for the West Zone of the Manifested Hexagonal World within the Black Square [Non-luminous Moon or Dark Moon] of East Horizon. This rising of the so-called sun [Bullet] as the manifested sign of natural magnetism from Revealed Right [South] and Manifested South-West and ending the journey in Revealed Left [North] and Manifested North-West is the day-light for the Middle Stair & the Topmost Stair of the appearing Pentagonal Earth, but 'Evening Show' for the equal & opposite stages of journey of the so-called sun [Bullet]. In brief for the Middle Stair of Europe, Asia, Africa, Australia, and Topmost Stair of Arabian Peninsula, and West Zone of the Manifested Hexagonal World within the Black Square [Non-luminous Moon or Dark Moon] of East Horizon, the so-called sun [Bullet] as the manifested sign of natural magnetism rises from Revealed Right [South] and Manifested South-West and ends in Revealed Left [North] and Manifested North-West. These rising from Revealed Right and Manifested South-West and ending [setting] in Revealed Left and Manifested North-West is the Semi-clockwise journey or Half of the equal & opposite stages of journey of the so-called sun [Bullet] covering ¼ [ASCII (hex) 00BC & ASCII (decimal) 188] for the Topmost Stair of Arabian Peninsula and ¾ [ASCII (hex) 00BE & ASCII (decimal) 190] for the Middle Stair of Europe, Asia, Africa, and Australia.

Equal & opposite stages of journey of the so-called sun represent one day; while semi-anticlockwise or semi-clockwise journey represents half of the day.

The so-called new moon [Triangular Bullet] as the manifested sign of reckoning fixed periods of affairs [relationship] appears from Revealed Right [South] and Manifested South-East in correspondence with South America for the Ground Stair of South America and North America of the appearing Pentagonal Earth. This appearance/manifestation of the new moon [Triangular Bullet] from Revealed Right [South] and Manifested South-East is the beginning of Night or '**Early Hours of Night**' for the Ground Stair of the appearing Pentagonal Earth, but '**Fall of Night**' from the point of view of so-called new moon [Triangular Bullet] as new moon [Triangular Bullet] arises from Revealed Left and Manifested North-West for the Middle Stair and the Topmost Stair of the appearing Pentagonal Earth.

BREAK OF THE DAY AND FALL OF NIGHT
[Revealed Concepts with reference to Wax Feeble Zakariya]
Break of the Day [Fajr] with respect to West Zone [Upper Seashore] but
Fall of Night [Magriib] with respect to East Zone [Lower Seashore]
Ayat\Verses - [His prayer was answered]: O Zakariya! We give you good news of a **son**: His name shall be **Yahya. We have given the same name to none before (him).** He said: O my Rab! **How shall I have a son, when my wife is**

barren and I have grown quite decrepit from old age? He said: So [it will be] the sayings of your Rab. It is easy for Me, even as I did you before, **when you had been nothing**! [Zakariya] said: O my Rab! Give me a Sign (Token). Your **Sign or Token** is that you with no bodily defect shall **not speak unto mankind three nights**. So **Zakariya came out to his people from the sanctuary and signified to them glorify your Rab at break of the day and fall of night**. [Sura No. – 18\19 – Maryam, Ayat\Verses – 7 to 11]

Wash-Shams Wal Qamar or both the so-called sun [Bullet] and the so-called moon [Triangular Bullet] have the Single Root. The so-called sun [Bullet] as the manifested sign of natural magnetism enters from North-East in correspondence with North America as the day light for the appearing Pentagonal Earth within the Black Square of Manifested East Horizon. So, the concept of day manifests from East Zone [Lower Seashore] in correspondence with the entering of the so-called sun [Bullet]. The so-called sun [Bullet] covers Semi-anticlockwise journey for the East Zone and sets in South-East in correspondence with South America. This Semi-anticlockwise journey from North-East to South-East is the Morning Show of the so-called sun [Bullet] in the name of Sister Planet Venus or Maryam supplied with sustenance. The so-called sun [Bullet / setting sun] runs away from there and rises from the South-West in correspondence with Europe for the West Zone [Upper Seashore] and ends the Semi-clockwise journey in the North-West in correspondence with Australia. The setting of the so-called sun [Bullet] in the South-East is the **'Break of the Day'** i.e. Magriib for the East Zone [Lower Seashore]; while the rising of the so-called [setting] sun [Bullet] from the South-West in correspondence with Europe is the Morning i.e. Fajr or beginning of the day from the point of view of West Zone [Upper Seashore].

Moreover, the so-called sun [Bullet] rises from South-West and ends [sets] in North-West for the West Zone [Upper Seashore]. These rising and setting are the 'Evening Show' from the point of view of the so-called sun [Bullet] in the name of Sister Planet Venus or Maryam supplied with sustenance.

The so-called new moon [Triangular Bullet] arises from North-West for the West Zone. So, the concept of Night manifests from West Zone. The so-called new moon [Triangular Bullet] appears from South-East for the East Zone [Lower Seashore]. This appearance is the 'Fall of Night' i.e. Magriib or beginning of night from the point of view of East Zone [Lower Seashore].

In brief, the concepts of 'Break of the day' and 'Fall of Night' with reference to Zakariya, the guardian of Maryam supplied with sustenance [Beautiful

Clothing / Covering / Veil / Screen / Barrier / Electron Cloud / Ether] represent Morning [Fajr] for the West Zone [Upper Seashore] but Evening [Magriib] for the East Zone [Lower Seashore].

Ayat\Verses - Or Do those **who commit ill-deeds** (seek after evil ways or follow artificial magnetism) suppose that We shall make them **equal with those** who **believe [revealed & manifested magnetism]** and **do righteous works [tabligh]**, that **equal** will be their **life** and their **death**? **Ill is the judgment that they make.** (It is tautologous that) Allah created the **heavens** and the **world** for **just ends**, and in order that **each soul** may find the **recompense** of what it has **earned**, and they will **not** be **wronged**. **Have you seen him who makes his own vain desire [acquired pieces of papers] his reality?** Allah has **knowingly** left him **astray**, and **sealed** his hearing and his heart [and understanding], and put **a veil on his sight**. Who, then, will **guide** him **after Allah** [has withdrawn Guidance]? **Will you not then heed** (receive admonition)? [Sura No. – 44\45 – Tanziilul Kitaab (Prev. Jaasiya), Ayat No. – 21 to 23]

Ayat\Verses – Has then your Rab distinguished you [O people of Kitaab!] by giving you **sons** and has chosen for Himself **females** from among the angels? **Truly you utter a most dreadful saying! We have explained [things] in various [ways] in Haazal-Qur'an**, in order that they may receive **admonition**, but it only **increases** their flight [from the Truth]! Say: If there had been [other] realities with Him, as they say, behold, they would certainly have **sought out a way against** the Rab of the **Arsh**! Glory to Him! He is **high above all** that **they say!** Exalted and Great [beyond measure]! The **seven heavens and the world**, and all beings therein, declare His glory and there is **not** a thing but celebrates His praise. And yet you **understand not** how they declare His glory! Verily He is Forbearer, Forgiving! **And when you recite the Qur'an [izaa-qara-tal-Qur-aana], We put, between you and those who believe not in the Hereafter, a veil invisible.** And We put **coverings over their hearts** [and minds] **lest they should understand** the Rabbaka fil-Qur-aani, and **deafness** into their ears. And when you **commemorate** your Rab and Him alone in the Qur'an [Rabbaka fil-Qur-aani], **they turn on their backs**, fleeing [from the Truth]. [Sura No. – 16\17 - Banii-Israai-iil, Ayat No. – 40 to 46]

Ayat\Verses – Those who would **hinder** [men] from the **path of Allah** [equal & opposite way towards Upright West following revealed & manifested magnetism] and would seek in it **something crooked [global and mechanical magnetism]**: they were those **who denied the Hereafter. Between them shall be a veil, and on the heights will be men who would know everyone by his marks: they will call out to the Companions of the Garden, "peace be on you": they will not have entered, but they will have an assurance [thereof]. When their eyes shall be **turned** towards the **Companions of the fire,** they will say: **Our Rab! Send us not to the company of the wrong-doers.** And the dwellers on the **Heights** call unto men whom they know by their marks (saying): What did your **multitude** [latitude-longitude] and that in which you took your **pride**? [Sura No. – 6\7 - As-haabul- a' raaf, Ayat No.-45 to 48]

Ayat\Verses – Your Ultimate-Rab is Allah, Who created the heavens and the world in six days, and then He **firmly mounted the Arsh** [Throne]. **He draws the night as a veil over the day, each seeking the other in rapid succession.** He made Wash-shamsa, Wal-qamara and Wan-nujuuma (galaxy stars) **subservient by His command.** His verily all **creations** and **commandment.** Blessed be Allah, the Rab (Cherisher and Sustainer) of the Universe! [Sura No. – 6\7 - As-haabul- a' raaf, Ayat No. – 54]

Ayat\Verses – Of them there are **some who** [pretend to] **listen** to you; but **We have thrown veils on their hearts,** So **they understand it not,** and **deafness in their ears**; if they saw **every one of the signs,** they will **not believe** in them; in so much that when they come to you, they [but] **dispute with you**; the disbelievers say: **These are nothing but tales of the ancients. Others** they **keep away** from it, and themselves they keep away; but they only **destroy their own souls,** and they **perceive it not.** [Sura No. – 5\6- An-aam, Ayat No. – 25 & 26]

"Verily We have set **veils over their hearts lest they should understand this,** and over their ears, deafness, if you call them to guidance, even then they will never accept guidance."

Ayat\Verses - And **verily We have explained in detail in this Qur'an, for the benefit of mankind, every kind of similitude.** But man is, in most things, contentious. And what is there to keep back mankind from believing, when the Guidance has come to them, nor from praying for forgiveness from their Rab, but that [they ask that] the **ways of the ancients be repeated with them, or the wrath be brought to them face to face [equal & opposite]?** We only send the messengers to give **Good Tidings** and to **give warnings.** But the **disbelievers dispute with vain argument [artificial magnetism],** in order therewith to **weaken the truth [revealed & manifested magnetism],** and **they treat My Signs as a jest,** as also the fact that they are **warned!** And **who does greater wrong** than one who is **reminded** of the **Signs** of his Rab; but **turns away** from them, forgetting the [deeds] which his hands have **sent** forward (to the Judgment)? Verily We have set **veils over their hearts lest they should** understand this, and over their ears, deafness, if you call them to guidance, even then they will never accept guidance. [Sura No. – 17\18 – Khaf – Ayat No. – 54 to 57]

ALREADY COVERED WITH CONFUSION

Ayat\Verses – They say: **Why is not an angel sent down to him?** If we **did** send down an angel, the matter would be **settled at once**, and **no respite** would be granted them. If We had made it an **angel** We should have **sent** him **as a man**, and We should **certainly** have caused them **confusion** in a **matter** which they have **already covered with confusion**. [Sura No. – 5\6 – An-aam, Ayat No. – 8 & 9]

DAY OF WARNING – REMOVAL OF THE VEIL OF IGNORANCE

Ayat\Verses – **And the Trumpet shall be blown: That will be the Day whereof Warning [had been given].** And **there will come forth** every **soul** with **each** will be an [angel] to drive, and an [angel] to bear **witness**. [And to the **evil-doers** it will be said]: You were **heedless** of this. Now **We have removed from you your veil (covering) of ignorance** and **sharp is you sight this Day. And (unto the devil-doer) his comrade (two-in-one partnership) will say: Here is [his Record] ready with me**. [Sura No. 49\50 – Qaaaf, Ayat No. – 20 to 23]

Semi-anti-clock-wise entering and setting as well as semi-clock-wise rising and ending of the so-called sun for the equal & opposite seashore represent manifested signs of natural magnetism in resemblance with revelation. Electron Cloud or Ether plays the role of binary ~sth between day and night in resemblance with a semi-circular line. With regard to the concept of Heaven and Hell, this binary ~sth resembles with the concept of Fulsirat. This concept of Fulsirat resembles with the concept of a bridge. The concept of bridge has been developed as Golden Mean of Aristotle, Categorical Imperative of Kant, and Duty for Duty's sake of Kant. These concepts resemble with doing one's duty in a detached manner or Tabligh or to do righteous work & to give good tidings to those who believe or Jihade-Akbar or Fight for Utilitarian Right [Solidified Solid Human Rights / Greatest Happiness of the Greatest Number] or Selfrectification through Selfsacrifice [renunciation of the fruits of action] or Niskama-Karma. With respect to manifested equal & opposite Zones within the Black Square of East Horizon, this bridge is the binary ~sth like Neutrino & Neutron between Electron and Proton, between region of white and region of black, between people of garden and people of wood, between Nation and Tribe, between lower heaven and zone of iron. In Hindu Mythology, this line is called Lakshman Rekha between Ram-Rajya and Lankeshwar Ravana or between zone of peace and zone of violence.

Ayat\Verses – To **Allah** belong the **Mystery** of the **heavens** and the **world**. And the Decision of the Hour [of Judgment] is as the twinkling of an eye, or even quicker: for Allah has power over all things. **It is He Who brought you forth from the wombs of your mothers when you knew nothing; and He gave you hearing and sight and intelligence and affections that you may give thanks [to Allah]. Have they not seen the birds obedient in mid-air? None holds them save Allah. Lo! Herein, verily are signs for those who believe.** [Sura No. – 15\16 – Nahl, Ayat\Verses – 77 to 79]

OCTAGON AND EQUAL & OPPOSITE

Ayat\Verses – It is He Who produces **gardens**, with **trellises** and without, and **dates**, and **crops** of all kinds, and **olives** and **pomegranates**, similar [in kind] and different [in variety]: eat of their fruit in their **season**, but render the **dues** that are proper on the day that the harvest is gathered. But **waste not by excess**: for Allah loves not the wasters. Of the **cattle** are some for burden and some for meat: eat what Allah has **provided** for you, and **follow not the footsteps of shaytan**: for **he is an open enemy to you.** [Take] **eight** in [four] **pairs**: Of sheep a pair and of **goats a pair;** say, has He forbidden the **two males,** or the **two females,** or [the young] which the wombs of the **two females enclose? Expound to Me (the case) with knowledge if you are truthful.** Of camels a pair, and oxen a pair; say, has He forbidden the two males, or the two females, or [the young] which the wombs of the two females enclose? **Were you present when Allah ordered you such a thing? But who does more wrong than one who invents a lie against Allah, to lead astray mankind without knowledge?** For Allah guides not people who do wrong. [Sura No. – 5\6 – An-aam, Ayat No. –142 to 145]

RAIMENTS – OCTAGONAL BLACK
SPOTS AS VEIL OR COVERING

O Children of Adam! We have **bestowed raiment** (archaic clothing) upon you to cover your shame, **as well as splendid venture**. But the raiment of righteousness that is the best. Such are among the **Signs of Allah**, that they may remember. O Children of Adam! Let not Shaytan seduce (attract by the offer of wrong doing) you, in the same manner **as He got your parents out of the Garden, stripping them of their raiment, to expose their shame: for he and his tribe watch you from a position where you cannot see them**. We made the evil ones friends [only] to those without faith. When they do some **lewdness**, they say: We found **our fathers doing so**; and Allah commanded us thus. Say: Nay, **Allah never commands what is shameful. Do you say of Allah what you know not?** [Sura No. – 6\7 - As-haabul- a' raaf, Ayat No. 26 to 28]

Ayat\Verses – Allah charges you as regards (the provision) of your children: **to the male, a portion equal to that of two females: if there be women more than two, their share is two-thirds of the inheritance; if only one, her share is a half. For parents, a sixth share of the inheritance to each.** If the deceased left children; if no children and the parents are the [only] heirs, **the mother has a third**; if the deceased left brothers [or sisters] **the mother has a sixth [Sixth Triangle or Middle-East or Eartha 3D or Bermuda Triangle].** [The distribution in all cases] after any legacy he may have bequeathed or debts you know not whether your parents or your children are nearest to you in benefit. These are settled portions ordained by Allah; and Allah is All-knowing, All-wise. And unto you belong **a half** of that which your wives leave, **if they have no child.** But **if they have a child,** then unto you **the fourth** of that which they have, after any legacy they may have bequeathed, or debt (they may have contracted, has been paid). In what you leave, their share is **a fourth,** if you have no child. But if you have a child, they get **an eighth**; after payment of legacies and debts. If the man or woman whose inheritance is in question, has left neither ascendants nor descendants, but has left a brother or a sister, each one of the two gets **a sixth**; but if more than two, they share in **a third**; after payment of legacies and debts; so that no loss is caused [to any one]. Thus is it ordained by Allah; and Allah is All-knowing, Most Forbearing. **These are limits set by Allah**: those who obey Allah and His Messenger will be admitted to Gardens with rivers flowing beneath, to abide therein [for ever] and that will be the great success. But those who disobey Allah and His Messenger and transgress His limits will be admitted to fire, to abide therein: And they shall have a humiliating punishment. [Sura No. – 3\4 – Nisaa, Ayat No. – 11 to 14]

Niche is kindled neither from the East nor from the West

Or

So-called Sun or Bullet rises neither from the East nor from the West
"**Allaahu Nuurus-samaa-waati wal-arz** – Masalu **Nuu-rihi** kaMishkaatin-fiihaa **Mis-baah** – Al-**Misbaahu** fii **Zujaa-jah** – **azzuujaajatu** ka-annahaa – kawkabun durriyyuny-yuuqa-du – min Shajaratimmubaara-katin-**Zaytuunatil-**laa **Sharqiy-**yatinw wa laa **Garbiyyatiny-**yakaadu **Zaytuhaa** yuziii- u wa law lam tamsas-hu naar. **Nuu-run alaa Nuur! Yahdillaahu** li-**Nuurihii** many-yashaaa – wa-yazribullaahul-amsaala linnaas – wallahu bi-kulli shay-in Alim." [Sura No. – 23\24 – Nuur, Ayat No. – 35] [Trans.] **Allah** is the **Light** of the **heavens** and the **world**. The **similitude** (likeness) of **His Light** is as a **niche** (binary thread of a candle or binary pulsar of Einstein) **in a lamp [Black Square].** The **lamp** is in a **glass [Diamond Operator].** The **glass** is as it is a **shining star [Shi-raa].** (The niche is) **kindled** from a **blessed tree,** an **olive** (Zaytuun tree), **neither from the East nor from the West,** whose (niche's) oil

573

will almost glow forth (of itself) though **no fire touches** it. **Light upon Light!** Allah guides unto His light whom He wills. And Allah speaks to mankind in **parables**, for Allah is Knower of all things. [Sura No. – 23\24 – Nuur, Ayat No. – 35]

**Enters from Revealed Left [North] and Ends in Revealed North [Left]
Enters from Manifested North-East and ends in Manifested North-West
Enters from Manifested North-West and sets in Manifested South-East
Rises from Manifested South-West and ends in Manifested North-West.**
Ayat\Verses - "Wa **tarash-shamsa** izaa tala-at-lazaawaru an-kahfihim zaatal-**yamimi** wa izaa gara-at-taqri-zuhum **zaatash-shimaali** wa hum fi fajwatim-minh. Zaalikka min aayaatillaah; many-yahdillaaahu fahuwal-muhtad; wa many-yuzlil falan-tajida lahuu waliyyam-murshidaa". {Trans.} And might have seen **tarash-shamsa** when it **rises, declining to the right** [semi-anticlockwise] from their Cave**, when it sets, turning away from them to the left** [semi-clockwise], while they lay in the **open space** in the midst of the **Cave** [within the Black Square of Manifested East Horizon]. **Such** [equal and opposite i.e. half anti-clockwise and half clockwise stages of journey as the glittering show for the earth] are among the **Signs of Allah**. He whom Allah, guides is **rightly guided**; but he whom Allah **leaves to stray**, for him **you will not find protector to lead him to the Right Way**. [Sura No. – 17\18 – Khaf – Ayat No. – 17]

TWO ENDS OF THE DAY

Ayat\Verses - Is it not a warning to such men [to call to mind] how many generations before them We destroyed, in whose haunts they [now] move? Verily, in these are **Signs** for men endued with understanding. And but for a decree that has already gone forth from your Rab, and a term already fixed, the Judgment would (have) been inevitable (in this present life). **Therefore bear with what they say, and remember [constantly] the praises of your Rab, before the rising of the (qabla tuluu-ish-shamsi), and were going down thereof (wa qabla guruubiha). And glorify Him some hours of the night and at the two ends of the day**. And strain not your eyes toward that which We cause some wedded pairs among them to enjoy the flower of life of the world. But the **provision** of your Rab is better and more enduring. And enjoin prayer upon your people, and be constant therein. We ask you not to provide **sustenance**. We provide it for you. But the [fruit of] the Hereafter is for **righteousness**. [Sura No. – 19\20 – Twa-Haa, Ayat No. – 128 to 134]

PLOTTED PLAN AND THE BEST OF PLANNERS

Ayat\Verses – And when **Issa** became **conscious of their disbelief**, he cried: **Who will be my helpers in the way of Allah?** Said the **disciples: We are**

Allah's helpers. **We believe in Allah and bear you witness that we have surrendered (unto Him).** Our Rab! **We believe in what You have revealed,** and **we follow the Prophet** whom You have sent, **then enrol us among those who witness (to the truth).** And [the unbelievers] **plotted and planned**, and **Allah too planned** (against them), and the **best of planners is Allah.** [Sura No. 2\3 – Imran, Ayat No. –52 to 54]

If you turn away, I [at least] have conveyed the Message with which I was sent to you. My Rab will make another people to succeed you, and you will not harm Him in the least. For my Rab has care and watch over all things. So when Our decree issued, We saved Huud and those who believed with him, by [special] Grace from Ourselves: We saved them from a severe penalty. Such were the Aad People: **they rejected the Signs of their Rab and Cherisher;** disobeyed His messengers; And followed the command of every powerful, **obstinate transgressor.** And they were pursued by a Curse in this life, and on the Day of Judgment. Ah! Behold! For the Aad rejected their Rab and Cherisher! Ah! Behold! **Removed [from sight) were Aad, the people of Huud.** [Sura No. – 10\11 – Kitaabun-uh-kimat (Prev. – Huud), Ayat No. – 57 to 60]

There was indeed a Sign for Saba, in their dwelling place. **Two Gardens** to the **right** and to the **left**. Eat of the Sustenance [provided] by your Rab, and be grateful to Him: A territory fair and happy, and a Rabbun Gafuur. [Sura No. – 33\34 – Saba, Ayat No. - 15]

A few references from Real Science [Insert – Symbol]				
S (h) 0044	S (h) 0040	S (h) 003E	S (h) 0042	S (h) 0040
S (h) 0031	S (h) 0028	S (h) 0027	S (h) 005F	S (h) 005E
S (h) 00AF	S (h) 0096	S (h) 0079	S (h) 0083	S (h) 0027
U (h) 2742	U (h) 2668	S (h) 00DD	U (h) 2602	S (h) 002C
S (h) 003F	S (h) 0043	S (h) 003F	S (h) 0041	S (h) 0045

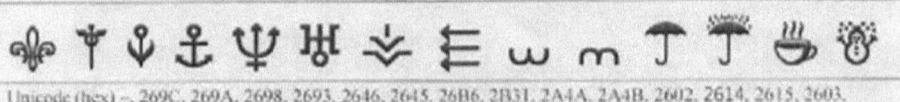

Unicode (hex) –, 269C, 269A, 2698, 2693, 2646, 2645, 26B6, 2B31, 2A4A, 2A4B, 2602, 2614, 2615, 2603.

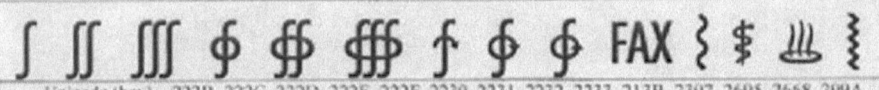

Unicode (hex) – 222B, 222C, 222D, 222E, 222F, 2230, 2231, 2232, 2233, 213B, 2307, 2695, 2668, 299A

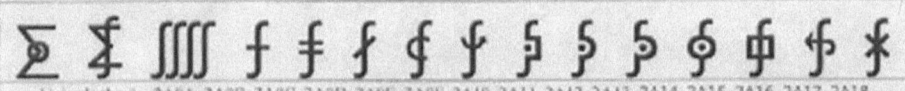

Unicode (hex) – 2A0A, 2A0B, 2A0C, 2A0D, 2A0E, 2A0F, 2A10, 2A11, 2A12, 2A13, 2A14, 2A15, 2A16, 2A17, 2A18

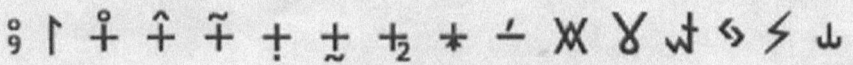

Unicode (hex) – 2A1F, 2A21, 2A22, 2A23, 2A24, 2A25, 2A26, 2A27, 2A28, 2A29, 2A59, 1041C, 10426, 16C3, 26A1, 297F

Symbol (hex) – 2628, 0060, 007E, 008F, 00A2, 00A3, 00B0, 00B1, 0079, 0078, 007A, 0076, 264A

And **those** who **accuse honourable woman** but **fail to produce four criterions [witnesses]** of justifying truth or falsity [concerning the chastity of the honourable woman] **curse** them with **eighty stripes; and never accept their evidence**. They **indeed** are **evil doers (wicked transgressors). Unless they repent thereafter** and **mend [for such]**; for Allah is Ever Forgiving, Most Merciful. [Sura No. – 23\24 – Nuur, Ayat No. – 4 & 5]

Fasta-'iz billaahi minash-Shaytaanir-Rajim
Bismillaahir-Rahmaanir- Rahim

Three ascending stairs of the appearing pentagonal earth

i)   Symbol '**Superscript Three**' [3] [Unicode (hex) & ASCII (hex) 00B3] represents Topmost Stair [Tuur or City or Banii-Israiil] of Arabian Peninsula where there is the appointed Kaba at the right side of the Mount Tuur.

ii) Symbol '**Superscript Two**' [2] [Unicode (hex) & ASCII (hex) 0082] represents Middle Stair [Zaytuun or Ummat or Nation or Median or Children of Adam] where there are Europe, Asia, Africa, and Australia.

iii) Symbol '**Super Script One**' [1] [Unicode (hex) & ASCII (hex) 00B9] represents Ground Stair [Tin or Townships or Sons of Adam] of North America and South America of the appearing pentagonal earth

FIRM INCOMING AND FIRM OUTGOING

But those who were blind in this world [present life] will be blind in the hereafter, and most astray from the Path. And their purpose was to tempt you away from that which We had revealed unto you, **to substitute in Our name something quite different**; [in that case] **Behold! They would certainly have made you [their] friend**! And had We not given you **strength**, you would nearly have **inclined to them** a little. In that case We should have made you **taste an equal portion [of punishment]** in **this life**, and an **equal portion in death**; and moreover you would have found **none to help you against** Us! **And they indeed wished to scare you from the land that they might drive you forth from thence, and then they would have stayed (there) but a little after you.** [This was Our] way with those whom We sent before you and you will find **no change in Our ways (methods).** Establish regular prayers - **Li-duluukish-shamsi (declining of Allah's Sovereignty**) till the darkness of the night, and (the recital of) Quran at dawn (**aanal-Fajr**). Lo! **(The recital of) the Quran at dawn is ever witnessed (carry testimony). [Innal Quraanal-Fajri kaana mash-huudaa.]** And some part of the night awake for it and additional prayer for you. It may be that your Rab will raise you to a station of praise. And say: My Rab! Cause me to come in with a **firm incoming** (from Revealed Left) and to go out with a **firm out going** (in Revealed Left). And give me from Your Presence an **authority to aid. And say: Truth has [now] arrived, and**

Falsehood perished. Lo! Falsehood is [by its nature] bound to perish. [Sura No. – 16\17 - Banii-Israai-iil, Ayat No. – 72 to 81]

OCTAGON & THREE STAGES

Ayat\Verses – "Khalaqakum-min-nafsinw-waahidatin summa ja-'ala minhaa zawjahaa wa anzala lakum-minal-an-aami samaaa-niyata azwaaj; yakhluqukum fii butuuni ummahaatikum khalaqam-mim-ba' di khalqin fii zulumaatinsalaas. Zaalikumullaahu Rabbukum lahul-mulk. Laaa ilaaha-illaa Huu. Fa-annaa tusrafuun?" [Trans.] He created you [all] from a single being. Then from that being He made of like nature his mate; and he **sent down for you eight head of cattle in pairs. He makes you, in the wombs of your mothers, in stages, one after another, in a threefold gloom. Such is Allah your Rab and Cherisher.** To Him belongs the **Sovereignty.** There is **no unbreakable reality** except Him. **How then are you turned away?** [Sura No. – 38\39 – Tanziilul-Kitaab, Ayat\Verses – 6]

Ayat\Verses – If you are thankless (reject faith), yet Allah is Independent of you. **But He likes not ingratitude from His servants. If you are thankful, He is pleased with you. No bearer of burdens (laden soul) can bear the burden of another.** In the end, to your Rab is your return. And He will **tell you the truth** of all that you did; for He knows well all that is in the hearts (of men). [Sura No. 38\39 Tanziilul-Kitaab, Ayat\Verses -7]

Ayat\Verses – Ah woe, that Day, to the rejecters of Truth! [It will be said:] Depart you to that which you **used to deny** [reject as false]! **Depart you to a shadow in three columns,** [which yet is] neither relief nor shelter from the flame. Lo! It throws up sparks like the castles, (or) as it might be **camels of bright yellow hue.** [Sura No. – 76\77 - Wal-Mursalaat, Ayat No. – 28 to 33]

Ayat\Verses – When some trouble touches man, he cries unto his Rab, turning to Him in repentance. But when He bestows a favour upon him as from Himself, [man] **forgets Whom** he cried and prayed for before, and **he sets up rivals [man-made nature] unto Allah,** thus **misleading others from the Path of Allah.** Say: **Take pleasure in your disbelief a while. Verily you are of the owners of the Fire!** [Sura No. – 38\39 – Tanziilul-Kitaab, Ayat\Verses – 8]

Ayat\Verses – It is He **Who appointed Ash-shams to be a glorious shining and wal-qamar to be a light and measured out stages** for him that you might know the number of years and the reckoning. Allah created not that except truth and righteousness. He **detailed the Clear Signs for those who have understanding.** [Sura No. – 9\10 – Yuunus, Ayat No. – 5]

Two-thirds, or half or a third thereof

O you **folded in garments**! Keep vigil the night long **save a little**, **half** of it, or a **little less**, or a **little more**; and recite the Qur'an in measure. Lo! We shall **charge** you with **a word of weight**. Lo! The vigil of the night is (a time) when impression is more keen and speech more certain. Lo! You have **by day a chain of business**. So keep in remembrance the name of your Rab and devote yourself with **a complete devotion**. [Sura No. 72\73 – Yaaa-ayyuhal Muzzamil, Ayat No. – 1 to 8]

Lo! Your Rab knows how you keep vigil, sometimes nearly **two-thirds** of the night, or **half** or a **third thereof,** as do a party of those with you. Allah **measures** the **night and the day**. He knows that **you count it not**, and **turns to you** in mercy. **Recite then of the Quran which is easy for you**. He knows that there are **sick folk** among you, while others **travel** in the land in search of Allah's bounty, and others are **fighting** for the cause of Allah. So recite of it which is **easy**, and establish worship and pay the poor-due, and lend to Allah a goodly loan. Whatsoever good you send before you for your souls, you will surely find it with Allah, better and greater in the recompense. And seek forgiveness of Allah. Lo! Allah is Ever Forgiving, Most Merciful. [Sura No. 72\73 – Yaaa-ayyuhal Muzzamil, Ayat No. - 20]

PARABLE OF AN-ZIL-QARNAYN
THREE STAGES OF JOURNEY
[MAGRIBASHSHAMSI, MAJLI-ASH-SHAMSI, AND HAY-NAS-SADDAYNI]

They ask you concerning 'An-Zil-Qarnayn'. Say: I will rehearse to you something of **her** story. Verily We established **her** power on earth, and We gave **her the ways [Safa and Marwa] and the means to all ends [sustenance]**. [Sura No. – 17\18 – Khaf – Ayat No. – 83 & 84]

Note: Here, the connotative term 'husnaa' [Ayat No. - 85 & 86] suggests that 'An-**Zil-Qarnayn' is a female name.**

"Fa-atba-a sababaa; Hattaaa izaa balaga **magribashshamsi** wajadahaa tagrubu fii aynin hami-atinwwa wajada indahaa qawmaa. Qulnaa yaa-Zal-Qarnayni immaaa antu-azziba wa immaaa an tattakhiza fiihim **husnaa**." [Self-evident Concept] And **she** followed one such a path until when she reached the magribashshamsi. She found it set in a **muddy spring**, and found a people thereabout. We said: O Zal-Qarnayn! Either punish or show them kindness. [Sura No. – 17\18 – Khaf – Ayat No. – 85 & 86]

She said: As for him who does wrong, **we shall punish him**, and then he will be brought back unto his Rab Who will **punish him with awful punishment**! But whoever believes, and works righteousness, he shall have a goodly reward, and easy will be his task as We order it by our Command. [Sura No. – 17\18 – Khaf – Ayat No. – 87 & 88]

"Summa atba-a sababaa. Hattaaa izaa balaga **majli-ash-shamsi** wajadahaa tatlu-u alaa qawmil-lam-naj-al-lahum-min-duunihaa sitraa. Kazaalik; wa qad ahajnaa bimaa ladayhi khubraa." [Self-evident Concept] Then she followed [another] way. Until, when she came to the **majli-ash-shamsi** she found it rising on a people for whom We had provided no shelter there from. So (it was); and We know all concerning him. [Sura No. – 17\18 – Khaf – Ayat No. – 89 to 91]

"Summa atba-a sababaa. Hattaaa izaa balaga **hay-nas-saddayni** wajada min-duunihimaa qawmal-laa yakaaduuna yafqahuuna qawlaa. [Trans.] Then she followed [another] way. Until, when she reached [a tract] **between two mountains [between Safa and Marwa]**, she found, beneath them, a people who **scarcely [almost not] understood a word [saying]**

"Qaaluu yaa- **Zal-Qarnayni inna Yajuuja** wa **Majuuja mufsiduuna fil-arzi fahal naj-alu laka kharjan alaaa an-taj ala bainanaa wa bainahum saddaa?**" [Trans.] They said: O **Zal-Qarnayn! Yajuuja** and **Majuuja** do great mischief on earth (spoiling the land). So may we pay tribute on condition that you set a barrier [planetary bridge or Fulsirat] between us and them? [Sura No. – 17\18 – Khaf – Ayat No. – 92 to 94]

APPOINTED LIGHT OF ALLAH RISES FROM NORTH AND TURNS BACK FROM THE BORDER [SOUTH] TOWARDS THE BACK BORDER [NORTH]

Ayat\Verses - "Wa **tarash-shamsa** izaa tala-at-lazaawaru an-kahfihim zaatal-**yamimi** wa izaa gara-at-taqri-zuhum **zaatash-shimaali** wa hum fi fajwatim-minh. Zaalikka min aayaatillaah; many-yahdillaaahu fahuwal-muhtad; wa many-yuzlil falan-tajida lahuu waliyyam-murshidaa".

Ayat\Verses - And might have seen **tarash-shamsa**, when it **rises, declining to the right** [semi-anticlockwise in resemblance with three to nine of a clock] from their Cave, **when it sets, turning away from them to the left**, while they lay in the **open space** in the midst of the **Cave** [open space]. **Such** are among the **Signs of Allah**. He whom Allah, guides is **rightly guided**; but he whom

Allah **leaves to stray**, for him **you will not find protector to lead him to the Right Way**. [Sura No. – 17\18 – Khaf – Ayat No. – 17]

Have they **not observed all things that Allah has created,** how their **shadows incline to the right and to the left [Natural Magnetism], making prostration unto Allah, and they are lowly?** [Sura No. – 15\16 – Nahl, Ayat\Verses – 48]

When the **sky is cleft asunder**; "Wa izal-kawaaki-buntasarat" [Trans.] When the **stars** are scattered **[IFTA]**; When the planets [skies] are dispersed [IBS (P) Ltd.]; When the **oceans [seas] are suffered to burst forth**; and when the **resting place are turned upside down**; [then] shall each soul know what it is **sent forward** and **kept back [Natural Magnetism].** [Sura No. – 81\82 - Izas-samaa-unfatarat, Ayat No. – 1 to 5]

We had already, beforehand, **taken the covenant of Adam**, but he forgot, and We found on his part no firm resolve. And when We said to the angels: Prostrate yourselves to Adam, they prostrated themselves, but not Iblis; he refused. Then We said: O Adam! Verily, this is an enemy to you and your wife. So let him not get you both out of the Garden, so that you are landed in misery. There is therein [enough **provision**] for you neither to go hungry nor to go naked. "**Wa annaka laa tazma-u fiiha wa laa tazhaa**". [Trans.] And you suffer not from thirst, nor from the heat of light [translated as **sun's heat though there is no term like shams**]. [Sura No. – 19\20 – Twa-Haa, Ayat No. – 115 & 119]

"Three times of privacy for you"

O you who believe! Let those whom your **right hands possess** and the [children] among you who have not come of age ask your permission [before they come to your presence], **on three occasions: Before the prayer of dawn; while you do-off your clothes for the noonday heat; and after the late-night prayer; three times of privacy for you.** Outside those times it is **not wrong** for you or for them **to move** about **attending to each other.** Thus Allah **makes clear the Signs for you**, for Allah is full of knowledge and wisdom. But when the children among you come of age, let them [also] ask for permission, as do those senior to them [in age]. Thus Allah makes clear **His Signs** to you, for Allah is full of knowledge and wisdom. [Sura No. – 23\24 – Nuur, Ayat No. – 58 to 59]

A TRINITY OF A WAVE, ABOVE WHICH A WAVE, ABOVE WHICH IS A CLOUD

And for **those who disbelieve, their deeds are as a mirage in a desert. The thirsty one approaches it to be water till he comes unto it and finds it to be nothing. But he finds in the place thereof Allah, Who pays him his due and**

582

Allah is swift in taking account. Or [the Unbelievers' state] is like the depths of darkness in a vast awful sea. There covers him **a wave, above which is a wave, above which is a cloud; depths of darkness, one above another. If a man stretches out his hands, he can hardly see it! For any to whom Allah gives not light, there is no light!** [Sura No. – 23\24 – Nuur, Ayat No. – 39 & 40]

BIRDS WITH WINGS OUTSPREAD

Have you not seen that Allah, He it is whom all who in the **heavens** and in the world do celebrate, and the **birds with wings outspread**? Each one knows its own [mode of] prayer and praise. And Allah knows well all that they do. And unto Allah belongs **the sovereignty of the heavens and the world**; and to Allah is the final goal [of all]. [Sura No. – 23\24 – Nuur, Ayat No. – 41 & 42]

Equal & opposite stages of journey in resemblance with three ascending stairs of the appearing pentagonal earth.

Equal & Opposite Stages of journey of the so-called sun as the manifested sign of natural magnetism or Bullet within the Black Square [Non-luminous Moon / Dark Moon] of East Horizon resemble with the parable of Zil-Qarnayn and three ascending stairs of the appearing pentagonal earth.

i) **Tiin – 4/8 or ½** or semi-anti-clockwise stages of journey for the ground stair of the appearing Pentagonal Earth and East Zone [Lower Seashore] within the Black Square of East Horizon [Ref. Unicode (hex) & ASCII (hex) 00BD, ASCII (decimal) 198]
   Note – With a view to verify as well as justify the two-in-one partnership between Conspiracy Science and IFTA, just reflect on Ayat \ Verse No. – 187 of Sura Baqarah and Commentary Codes 195 to 200 on Quran in the name of 'The Holy Quran' by IFTA.

ii) **Zaytuun - ¾** of the semi-clock-wise stages of journey for the middle stair of Europe, Asia, Africa, and Australia of the appearing Pentagonal Earth [Ref. Unicode (hex) & ASCII (hex) 00BE, ASCII (decimal) 190]
   Note – With a view to verify as well as justify the two-in-one partnership, just reflect on Ayat \ Verse No. – 183 to 185 of Sura Baqarah and Commentray codes 188 to 195 on Quran in the name of 'The Holy Quran' by IFTA.

iii) **Tuur-** ¼ of the semi-clock-wise stages of journey for the topmost stair and Uppermost Land of the appearing pentagonal earth called Arabian Peninsula and Upright West Triangle of the manifested hexagonal world where Mount Tuur is holding firmly the appointed

Kaba on its right side [Ref. Unicode (hex) & ASCII (hex) 00BC, ASCII (decimal) 188]

A TRINITY OF FORTY NIGHTS, THIRTY NIGHTS, AND TEN MORE

Forty [Tin or Ground Stair] – 4/8 or ½ [Unicode (hex) & ASCII (hex) – 00BD and ASCII (decimal) 189] for the East Zone

Thirty [Zaytuun or Middle Stair] - ¾ [Unicode (hex) & ASCII (hex) 00BE and ASCII (decimal) 190] for the Middle Stair of Europe, Asia, Africa, and Australia

Ten [Tuur or Topmost Stair] - ¼ [Unicode (hex) & ASCII (hex) 00BC and ASCII (decimal) 188] for the Topmost Stair of Arabian Peninsula

Ayat\Verses - And remember, We **divided the sea** for you and saved you and drowned Firawn's people within your very sight. And remember We appointed **forty nights** for Muusa [for East Zone of North America and South America], and **in his absence you took the calf** [projected falsehood for right direction] and you did grievous wrong. Even then We did forgive you; **there was a chance for you to be grateful**. And remember We gave **Muusa the Kitaab** and **the Criterions [scales of justification and verification].** There was a **chance for you to be guided aright**. And when Muusa said to his people: O my people! You have indeed **wronged** yourselves by your **worship of the calf [projected mechanical globalisation and artificial magnetism]**. So turn [in repentance] to your Creator, and **kill the calf (projected falsehood) yourselves that will be the best** for with your Creator and He will **relent** toward you, for He is the Relenting, the Merciful. And remember you said: O Muusa! We shall **never believe** in you until we see Allah **manifestly**, but you were **confused** with **thunder** and **lightning seized you**. Then We **received** you after your **extinction**, that you might give **thanks**. And We gave you the **shade** of clouds [Arabian Peninsula or Topmost Stair] and sent down to you **manna** and **quails [Middle Stair & Ground Stair]**, saying: Eat of the **good things** We have **provided** for you. We **wronged** them not, but they **did** wrong themselves. And when We said: Enter this **township** and eat of the plenty therein as you wish; but enter the gate prostrate, and say: Repentance. We will forgive you your sins and will increase (reward) for the right-doers. **But those who did wrong change the word which had been told them for another saying** and We sent on the transgressors a plague from heaven, for evil doing. [Sura No. – 1\2 – Baqarah, Ayat No. – 50 to 59]

Ayat\Verses - We appointed for Muusa **thirty nights [for Middle West Zone of Europe, Asia, Africa, and Australia], and completed [the period] with ten [more for Top West Triangle of Arabian Peninsula].** Thus was completed the term [equal & opposite stages of journey of the kindled niche or revealed magnetism] with his Rab, **forty nights.** And Muusa had charged his brother Haarun [before he went up]: Act for me amongst my people: Do right, and follow not the way of those who do mischief. When Muusa came to the place appointed by Us, and his Rab addressed him, He said: O my Rab! Show [Thyself] to me, that I may look upon You. Allah said: **By no means you can see Me [direct]; But look upon the mount [climbing towards Uprigt West];** if **it abide in its place,** then you shall see Me. **When his Rab manifested His glory on the Mount, He made it as dust.** And Muusa fell down in a swoon. When he recovered his senses he said: Glory is to You! To You **I turn in repentance, and I am the first to believe.** [Sura No. – 6\7 - As-haabul- a' raaf, Ayat No. – 142 & 143]

ATMOSPHERE
Nay [laaa], I swear by this Balad [City of the sole manifested Imam]; and you are an in-dweller of this city [Arabian States]; and begetter and that which he beget [parent and child]; verily We have created man in an **atmosphere.** [Sura No. – 89\90 – Bi-haazal-balad, Ayat No. – 1 to 4]

ALLAH PROVIDES WITHOUT MEASURE
Men whom **neither traffic** (commands of the chiefs / leaders) **nor merchandise** (sale / loss or profit / books) **can divert from the Remembrance of Allah** and **constancy** in prayer and paying to the poor their due, who fear a Day when hearts and eyeballs will be overturned that Allah may reward them according to the best of their deeds, and add even more for them out of His Grace, for Allah provides blessings for those whom He will, without measure. [Sura No. – 23\24 – Nuur, Ayat No. – 37 & 38]

TEN NIGHTS
By the **dawn** [1]; and **ten nights,** [2]; by **the even and odd [contrasted]**[3]; and by the Night when it [Allah's Sovereignty] passes away [departs] [4]. There surely is an **oath** [Sura No. – 88\89 – Wal-fajri, Ayat No. – 1 to 5]
Do you not consider how your Rab dealt with [the tribe of] Aad [6] with many columned Iram [~sth] [7] the similitude of which were not produced in the land, [8] and with [the tribe of] Samud, who clove the **rocks** in the valley [9], and with Firawn, firm of might, [10] who [all] were rebellious [to Allah] in **these** lands [11] and **heaped therein mischief [on mischief]** [12]. Therefore your Rab **poured on them the disaster of His punishment** [13]; for your Rab is on

a **watch-tower** [Mount Tuur, leaning tower of Galileo, tallest tree of Newton].
[Sura No. – 88\89 – Wal-fajri, Ayat No. – 6 to 14]

Now, as for man, whenever his Rab tries him, giving him honour, and is
gracious unto him, he says: My Rab has honoured me [15]. But when He tries
him, restricting his **sustenance** for him, then he says: My Rab has despises
me [16]! Nay, nay! But **you honour not the orphan [Who is the orphan in
resemblance with this sharing? - 17]**! Nor do you **encourage** one another to
feed the poor [18]! And you **devour inheritance** - all with greed [19], and you
love wealth with inordinate [excessive] love! [Sura No. – 88\89 – Wal-fajri,
Ayat No. – 15 to 20]

Nay! When the world is grounded to atoms, grinding, grinding [21], and your
Rab shall come with angels, rank on rank [22], and **hell is brought near that
day; on that day** will man remember, but how will that remembrance profit
him [23]? He will say: Ah! Would that I had sent forth [some provisions] for
[this] my [Future] Life [24]! For, that Day, His Chastisement will be such as
none [else] can inflict [25], and His bonds will be such as none [other] can bind.
[Sura No. – 88\89 – Wal-fajri, Ayat No. – 21 to 26]

[To the righteous soul will be said:] O [you] soul, in [complete] rest and
satisfaction [peace]! Return to your Rab, well pleased [yourself], and well-
pleasing unto Him! Enter you, then, among My devotees! Yea, enter you My
Heaven! [Sura No. – 88\89 – Wal-fajri, Ayat No. – 27to 30]

O you who believe! Let those whom your **right hands possess** and the [children]
among you who have not come of age ask your permission [before they come
to your presence], **on three occasions: Before the prayer of dawn; while you
do-off your clothes for the noonday heat;** and **after the late-night prayer
three times of privacy for you**. Outside those times it is **not wrong** for you
or for them **to move** about **attending to each other**. Thus Allah **makes clear
the Signs for you,** for Allah is full of knowledge and wisdom. But when the
children among you come of age, let them [also] ask for permission, as do those
senior to them [in age]. Thus Allah makes clear **His Signs** to you, for Allah is
full of knowledge and wisdom. [Sura No. – 23\24 – Nuur, Ayat No. – 58 to 59]

A questioner questioned concerning the Penalty about to fall upon the
nonbelievers which none can repel from Allah, **Rab of the Ascending
Stairways.** The angels and the spirit ascend unto him in a Day whereof the
span is **fifty thousand years.** But be patient with a patience fair to see. [Sura
No. – 69\70 – Zil-ma-aaru, Ayat No. – 1 to 5]

Ayat\Verses – And We have commanded to men kindness toward parents. In pain did his mother bear him, and in pain did she give him birth. The carrying of the [child] to his weaning is [a period of] **thirty months**. At length, when he reaches the age of full strength and attains **forty years**, he says: My Rab! Grant me that I may be grateful for Your favour which You have bestowed upon me, and upon both my parents, and that I may work righteousness such as You may approve; and be gracious to me in my issue. Truly I have turned to You repentant and I am of those (Muslimin) who surrender (unto You). Such are they from whom We shall accept the best of their deeds and overlook their evil deeds. [They shall be] among the companions of the Garden. This is the true promise which they were promised. [Sura No. – 45\46 – Bil-ahqaafi, Ayat No. – 15 & 16]

SECTION – VII
MIDDLE COURSE – 7.1
CAUTIONS AND PRECAUTIONS

Fasta-'iz billaahi minash-Shaytaanir-Rajim
Bismillaahir-Rahmaanir- Rahim

Allah has said: **Choose not two realities [i.e. both Natural Magnetism and Manmade Magnetism]**; for He is **just One Allah**. So of Me, Me only, be in awe. [Sura No. – 15\16 – Nahl, Ayat\Verses – 51]

And when Allah said: O Isabna-Maryama! Did you say unto mankind: **Take me and my mother for two realities besides Allah?** He will say: Glory to You! Never could I say what I had no right (to say). Had I said such a thing, You would indeed have known it. You know what is in my mind, though I know not what is in Your Mind. For You know in full all that is **hidden**. [Sura No. – 4\5 – Maaaidah, Ayat\Verses – 116]

O you who believe! Call in remembrance the favour of Allah unto you **when certain men formed the design to stretch out their hands against you [projecting mechanisms]**, but (Allah) held back their hands from you. So fear Allah. And on Allah let believers put (all) their trust. [Sura No. – 4\5 – Maaaidah, Ayat\Verses – 11]

Let the believers [followers of manifested nature in resemblance with revelation] **do not take** for friends or helpers non-believers [followers of manmade nature] in **preference** to believers. If any do that, he has **no connection with Allah** unless (it be) that you but **guard** yourselves against them, taking (as it were) security. But Allah **warns** you beware (only) of Himself. Unto Allah is the final goal (Summum bonum). [Sura No. - 2\3 – Imran, Ayat No. –28]

O you who believe! If you obey the disbelievers, they will drive you back on your heels, and you will turn back [from Faith] to your own loss. [Sura No. - 2\3 – Imran, Ayat No. –149]

O you who believe [followers of equal & opposite revelation]! Take not into your intimacy other than your own folk, who would spare no pains to ruin you, they love to hamper you. Hatred has already appeared from their mouths, what their hearts conceal is far worse. We have made clear to you the Signs, if you have understanding. Ah! You are those who love them, but they love you not, though you believe in the whole of the Kitaab. When they meet you, they say: We believe. But when they are alone, they

bite off the very tips of their fingers at you in their rage. Say: Perish in you rage; Allah knows well all the secrets of (your) heart. [Sura No. - 2\3 – Imran, Ayat No. –118 & 119]

CONCRETE EXAMPLE AND A FORMAL DOCUMENT

Clear proof of evidence sorcery by the researchers of IFTA is being manifested through the transliteration of Ayat No. – 138 of Sura - Baqarah – '[Our religion is] the **Baptism** of Allah' manifests that the work of IFTA is to make the believers [Muslims] **Baptists**.

Ayat – "Sibgatallaah, wa man ah-sanu minallaahi sibgah, wa nahnu lahuu aabiduun". [Trans.] [Our religion is] the **Baptism** of Allah: And who can **baptize** better than Allah? And it is He Whom we worship. (IFTA) [Sura No. 1\2 – Baqarah, Ayat No. – 138]

So when ye have accomplished your holy rites, **celebrate** the praises of Allah, **as ye used to celebrate the praises of your fathers**, yea, **with far more Heart and soul**. There are men who say: "Our Lord! Give us [Thy bounties] in this world!" but they will have no portion in the Hereafter. (IFTA) [Sura No. 1\2 – Baqarah, Ayat No. – 200]

Or

And when you have completed your devotions, then remember Allah as you remember your fathers or with a more lively remembrance. But of mankind is he who says: Our Rab (Lord)! Give unto us in this world, and he has no portion in the Hereafter. (IBPLtd.) [Sura No. 1\2 – Baqarah, Ayat No. – 200]

No dispute - We are **responsible for our doings and you for yours [Ayat No. -139 of Sura - Baqarah**.

Say (unto the peoples of the Kitaab): **Will you dispute with us concerning Allah** when He is **our Rab** and **your Rab**? We are **responsible for our doings and you for yours**. We look to Him alone. [Sura No. 1\2 – Baqarah, Ayat No. – 139]

Thus, We have made you an Ummat justly balanced, that you may be witnesses against mankind and that the Rasuul (Prophet) is a witness **over** you. And **We appointed the Qibla which you formerly observed only that We might know him who follows Rasuul** (the Prophet) from those who would turn on their heels (run away). Indeed it is a hard test, except to those guided by Allah. **But it is not Allah's purpose that your faith should be in vain**. For Allah is to all people full of piety, Most Merciful towards mankind. [Sura No. 1\2 – Baqarah, Ayat No. – 143]

"Yaaa-ayyu-hallaziina aa-manuu laa tattakhizuu aduw-wii wa aduw-wakum **awliyaa** –a tulquuna ilay-him-bil-mawaddati wa qad kafaruu bi-maa-jaa-akum-minal-haqqi, yukhrijuunar- Rasuula wa iyyakum an-tu-minuu billaahi Rabbikum. In kuntum kha-rajtum jihaadan-fii sabiilii wab-tigaaa-a marzaatii tusirruuna ilay-him bil-mawaddati wa ana a-lamu bimaaa akh-fay-tum wa maaa a-lantum. Wa many-yaf-alhu minkum faqad galla sawaaa-as-sabil." [Trans.] O you who believe! **Choose not My enemies** and **yours as friends. Do you give them friendship when they disbelieve in that truth which has come to you, driving out the Prophet and you (from the right direction) because you believe in Allah, your Rab?** If you have come out to strive in My way and to seek My good pleasure, [take them not as friends], holding secret converse of love [and friendship] with them. Lo! I know full well **all that you conceal** and **all that you reveal.** And **whosoever does this among you, verily he has strayed from the Right Path.** [Sura No. 59\60 – Yuhibbul-Muqsitin (prev. Mumtahana), Ayat No. – 1]

If they were to get the better of you, **they would behave to you as enemies, and stretch forth their hands and their tongues against you for evil,** and **they desire that you should reject the Truth.** Of no profit to you will be your relatives and your children on the Day of Resurrection [Yawmal-Qiyaamah]. He will part you, for Allah is the Seer of all that you do. [Sura No. 59\60 – Yuhibbul-Muqsitin (prev. Mumtahana), Ayat No. – 2 & 3]

There is for you **an excellent example** [to follow] in **Ibrahim and those with him,** when they said to their folk: **We are clear of you** and of **whatever you worship besides Allah.** We have rejected you, and there has arisen, between us and you, enmity and hatred for ever, unless you believe in Allah and Him alone. **But not** when Ibrahim said to his father: I will pray for forgiveness for you, though I have no power [to get] aught on your behalf from Allah. [They prayed]: Our Rab! In You do we trust, and to You we do turn in repentance, and to You is [our] Final Goal. Our Rab! Make us not a [test and] trial for the disbelievers, but forgive us, our Rab! Lo! You are the Exalted in Might, the Wise. There was indeed in them **an excellent example for you to follow** for those whose hope is in Allah and in the Last Day. But **if any turn away,** truly Allah is The Absolute, Owner of Praise. [Sura No. 59\60 – Yuhibbul-Muqsitin (prev. Mumtahana), Ayat No. – 4 to 6]

It may be that Allah will grant love [and friendship] between you and those whom you [now] hold as enemies. For Allah has power [over all things]. And Allah is Ever Forgiving, Most Merciful. Allah **forbids you not, with regard to those who fight you not for [your] Faith nor drive you out of your homes,** from dealing kindly and justly with them; for Allah loves those who are just. **Allah only forbids you, with regard to those who fight you for [your] Faith,** and **drive you out of your homes,** and **support [others] in driving you out, from turning to them [for friendship and protection].** It is such as turn to them [in these circumstances], that **do wrong.** [Sura No. 59\60 – Yuhibbul-Muqsitin (prev. Mumtahana), Ayat No. – 7 to 9]

COIN NOT SIMILITUDE FOR ALLAH
[It is the right time to reflect on two kinds of Magnetism]

It is Allah who creates you and takes your souls at death; and of you there are some who are sent back to a feeble age, so that they know nothing after having known [much], for Allah is All-Knowing, All-Powerful. Allah has bestowed His gifts of sustenance more freely on some of you than on others. Those more favoured are not going to throw back their gifts to those whom their right hands possess, so as to be equal in that respect. **Will they then deny the favours of Allah?** [Sura No. – 15\16 – Nahl, Ayat\Verses – 70 & 71]

And Allah has made for you mates [and companions] of your own nature, and made for you, out of them, sons and daughters and grandchildren, and provided for you sustenance of the best. **Will they then believe in vain things, and be ungrateful for Allah's favours? And they worship beside Allah** that such as have no power of providing them, for sustenance, with anything in heavens or earth, and cannot possibly have such power? So **coin not similitude for Allah,** for Allah knows, and you know not. [Sura No. – 15\16 – Nahl, Ayat\Verses – 70 to 74]

[Inevitable] **comes** [to pass] the **Command of Allah**: seek you **not** then to hasten it. Glory to Him, and far is He above having the **partners** they ascribe unto Him! [Sura No. – 15\16 – Nahl, Ayat\Verses – 1]

CAUTIONS

O you who believe! **Profane not Allah's monument, nor of the Sacred Month, nor the offerings, nor the garlands, nor repairing in the Sacred House seeking the grace and pleasure of Allah;** but when you left the Sacred territory, then go hunting (if you will). And let not your hatred of **a folk who (once) stopped you going to the Inviolable Places of Prayer,** seduce you to transgress **(lead astray),** but **help you one another to righteousness** and **pious duty. Help not** one another to **sins and transgression,** but keep your duty to Allah, for Allah is **severe in punishment.** [Sura No. – 4\5 – Maaaidah, Ayat\Verses – 2]

Say: **Whether you hide what is in your hearts or reveal it, Allah knows it all**. He **knows what is in the heavens, and what is in the world**. And Allah has power over all things. **On the Day** when every soul will find itself confronted with all that it has done of good, and all that it has done of evil, (every soul) will **wish** that there might be a **great space of distance between it and its evil** (that). But Allah **cautions you beware of him**. And Allah is full of kindness to those **who serve Him. [Sura No. - 2\3 – Imran, Ayat No. –29 & 30]**

OPEN OPPONENT

[Inevitable] **comes** [to pass] the **Command of Allah**: seek you **not** then to hasten it. Glory to Him, and far is He above having the **partners** they ascribe unto Him! He **sends** down His angels with **inspiration** of His Command, to such of His **servants** as He pleases, [saying]: **Warn** [Mankind] that **there is no ilaaha** (unbreakable reality) except Me. So keep your duty unto Me. He has created the **heavens and the world** for **just ends (with truth)**. Far is He above having the **partners** they ascribe to Him! He has created man from a drop of fluid; yet behold! This same [man] becomes an **open opponent**! [Sura No. – 15\16 – Nahl, Ayat\Verses – 1 to 4]

HEARSAY

Shall I inform you, [O people!], on whom it is that the evil ones descend? They descend on every lying, **wicked person**, [into whose ears] **they pour hearsay vanities**, and most of **them are liars**. And for Poets, it is those straying in evil, who follow them. Have you not seen that they wander distracted in every valley? And that they say what they practise not? Except those who believe, work righteousness (Tabligh), engage much in the remembrance of Allah and vindicate them-selves after they have been wronged. Those who do wrong will come to know by what a (great) reverse they will be over-turned! [Sura No. 25\26 – Shuraa, Ayat No. – 221 to 222]

A SERIOUS FALSE CHARGE AND A TREMENDOUS LIE

That they rejected Faith; that they uttered against Maryam [supplied with sustenance] a serious **false charge**; and because of their saying: We slew **Masiiha Isaabna-Maryam**, Rasuuiallah' but **they slew him not, nor crucified him, but so it was made to appear to them, and those who differ therein are full of doubts, with no [certain] knowledge, but only conjecture to follow, for of a surety they slew him not. And because of their disbelief and of their speaking against Maryam a tremendous lie**. There is none of the people of the Kitaab but will believe in him before his death; and on the Day of Resurrection he will be a witness against them. [Sura No. – 3\4 – Nisaa, Ayat No. – 156 to 159]

Because of the wrong doing of the **Haaduu**, We made **unlawful** for them certain **good** things which were (before) made **lawful** for them and because of **their much hindering from Allah's Way. That they took usury, though they were forbidden; and that they devoured men's wealth by false pretences. We have prepared for those among them who reject faith a grievous punishment.** But those of them who are **well-grounded in knowledge**, and the believers, believe in what hath been revealed to them and what was revealed before them, and [especially] those who **establish** regular prayer and **practise** regular alms giving and **believe** in **Allah** and in the **Last Day**; to them We shall soon bestow an **immense reward**. [Sura No. – 3\4 – Nisaa, Ayat No. – 160 to 162]

REJECT SUCH MEN AS USE BAD LANGUAGE
Ayat\Verses – The most beautiful names belong to Allah: so call on Him by them; but reject such men as use bad language in His names: for what they do, they will soon be requited? And off those whom We have created there is a nation (ummat) who guide with truth and establish justice therewith. [Sura No. – 6\7 - As-haabul- a' raaf, Ayat No. – 180 & 181]

PURPOSE OF FRIENDSHIP – CONCEALMENT OF
TRUTH AND PROJECTION OF FALSEHOOD
[Friendship between Researchers of IFTA & Conspiracy Scientists]
DIFFERENCE BETWEEN QURAN AND
THE HOLY QURAN OF IFTA
Then woe to those who write the Kitaab with their own hands, and then say: This is from Allah, to traffic with it for miserable price! Woe to them for what their hands do write, and for the gain they make thereby. And they say: The fire shall not touch us but for a few numbered days: Say: Have you taken a promise from Allah? Truly Allah will never break His promise. Or is it that you say of Allah what you do not know? Nay, those who seek gain in evil, and are girt round by their sins, they are companions of the fire. Therein they will abide [for ever]. [Sura No. 1\2 – Baqarah, Ayat No. – 79 to 81]

Have you any hope that **they** will be **true** to you when a **party** of **them** used to listen to **the word of Allah**, and **then used to change it, after they had understood it, knowingly?** And when they **fall** in with the men of **Faith**, they say: We **believe**. But when they **meet each other in private**, they say: **Shall you tell them what Allah has revealed to you**, that they may engage **you in argument** about it before your Rab? Have you then **no sense?** Know they not that **Allah knows what they conceal and what they reveal?** And there are among them **illiterates, who know not the Kitaab**, but **[see therein their own]**

desires, and they **do nothing** but **conceptualise the truth**. [Sura No. – 1\2 – Baqarah, Ayat No. – 75 to 78]

"Nor sell the covenant of Allah for a miserable price"
And take not your oaths, **to practise deception between yourselves**, with the result that **someone's foot may slip after it was firmly planted**, and you may have to **taste the evil** [consequences] of having **hindered [men] from the Path of Allah**, and a **Mighty Wrath descend on you. Nor sell the covenant of Allah for a miserable price**; for with Allah is [a prize] far better for you, if you only knew. [Sura No. – 15\16 – Nahl, Ayat\Verses – 94 & 95]

[For example] - Their **similitude** is that of a man who kindled a fire; **when it lighted all around him, Allah took away their light and left them in utter darkness**. So they could not see. [Sura No. 1\2 – Baqarah, Ayat\Verses No. – 17]

The **lightning all but snatches away their sight**; every time **the light [helps]** them, **they walk therein (light)**, and **when the darkness grows on them, they stand still**. And if Allah willed, He could **take away** their faculty of **hearing and seeing**; for Allah has power over all things. [Sura No. 1\2 – Baqarah, Ayat No. - 20]

"And their **purpose** was to tempt you away from that which We had revealed unto you, **to substitute in Our name something quite different**; [in that case] **Behold! They would certainly have made you [their] friend!**"

FIRM IN COMING AND FIRM OUTGOING OF THE
MANIFESTED SIGNS OF NATURAL MAGNETISM
But those who were blind in this world [present life] will be blind in the hereafter, and most astray from the Path. And their purpose was to tempt you away from that which We had revealed unto you, **to substitute in Our name something quite different**; [in that case] **Behold! They would certainly have made you [their] friend!** And had We not given you **strength**, you would nearly have **inclined to them** a little. In that case We should have made you **taste an equal portion [of punishment] in this life, and an equal portion in death**; and moreover you would have found **none to help you against** Us! **And they indeed wished to scare you from the land that they might drive you forth from thence, and then they would have stayed (there) but a little after you.** [This was Our] way with those whom We sent before you and you will find **no change in Our ways (methods)**. Establish regular prayers - **Li-duluukish-shamsi (declining of Allah's Sovereignty)** till the darkness of the night, and (the recital of) Quran at dawn **(aanal-Fajr)**. Lo! **(The recital of) the**

Quran at dawn is ever witnessed (carry testimony). [Innal Quraanal-Fajri kaana mash-huudaa.] And some part of the night awake for it and additional prayer for you. It may be that your Rab will raise you to a station of praise. And say: My Rab! Cause me to come in with a **firm incoming** (from Revealed Left) and to go out with a **firm out going** (in Revealed Left). And give me from Your Presence an **authority to aid. And say: Truth has [now] arrived, and Falsehood perished. Lo! Falsehood is [by its nature] bound to perish. [Sura No. – 16\17 -** Banii-Israai-iil, Ayat No. – 72 to 81]

Lo! Those who believe, then reject faith, then believe [again] and [again] reject faith, and go on increasing in unbelief, Allah will never pardon them nor guide them nor will He guide them into a way. Bear unto the hypocrites the tidings that there is for them [but] a grievous penalty (painful doom); [Sura No. – 3\4 – Nisaa, Ayat No. – 137 & 138]

Those who choose disbelievers for their friends instead of believers! Do they look for power at their hands? Lo! All power belongs to Allah. He has already revealed unto you in the Kitaab that when you hear the Revelations of Allah (Manifested Signs in resemblance with revelation) **rejected and ridiculed,** (you) **sit not** with them (who disbelieve and mock) unless they turn to a different theme. Lo! In that case (if you stayed) you would be like unto them. Lo! Allah will gather hypocrites (frauds). [These are] the ones who **wait and watch** about you: if you do gain a **victory** from Allah, they say: Were we not with you? But if the unbelievers gain a success, they say [to them]: Did we not gain an advantage over you, and did we not guard you from the believers? But Allah will judge between you on the Day of Judgment (Resurrection). **And never will Allah grant to the unbelievers a way [to triumph] over the believers.** [Sura No. – 3\4 – Nisaa, Ayat No. – 139 to 141]

Lo! The **hypocrites** seek to **beguile** Allah, but it is **Allah who beguiles them.** When they **stand up to worship,** they perform it **leisurely** (without earnestness) and to be **seen of men,** but little they are **mindful** to Allah in **remembrance.** Swaying between this (and that), (belonging) neither to these nor to those, he whom Allah causes to go astray you will not find a way for them. O you who believe! **Choose not disbelievers for (your) friends in place of believers.** Do you wish to give Allah an open proof against yourselves? [Sura No. – 3\4 – Nisaa, Ayat No. – 142 to 144]

Lo! The **hypocrites (will be) in the lowest depth of the fire,** and you will find **no helper** for them except those who repent, and amend and hold fast to Allah, and make their Diin (religion) pure for Allah (only). Those are with the believers. And soon Allah will bestow on the believers **a reward of immense value**. **What can Allah gain by your punishment**, if you are grateful (of His Mercies) and you believe (in Him)? Allah is Ever Responsive, Aware. Allah loves **not** that evil should be **noised abroad in public speech**, except where **injustice** has been done, for Allah is He who hears and knows all things. If you do good **openly** or keep it secret or forgive evil, lo! Allah is Forgiving, Powerful. [Sura No. – 3\4 – Nisaa, Ayat No. – 145 to 149]

Those who **deny** Allah and His messengers, and [those who] wish to **separate** Allah from His messengers, **saying: We believe in some but reject others**, and [those who] **wish to take a course straight, they are in truth [equally] unbelievers**; and we have prepared for unbelievers **a humiliating punishment**. To those who **believe** in Allah and His messengers and **make no distinction between any of them**, Allah will soon give their [due] **rewards**, for Allah is Ever-Forgiving, Merciful. [Sura No. – 3\4 – Nisaa, Ayat No. – 150 to 152]

Lo! Those who reject Faith (disbelieve) and keep off [men] from the way of Allah, have verily strayed far, far away from the Path. Those who **reject Faith** and **do wrong**, Allah will **not forgive** them nor **guide** them to anyway **except the way of Hell**, wherein they will **abide forever**. And this is **ever easy for Allah.** O Mankind! The Messenger has **come to you in truth from your Rab**. Therefore **believe** in him. It is **better** for you. But if you **reject Faith** (disbelieve), **to Allah belong whatever is in the heavens and on the earth**. Allah is Ever-Knower, All-wise. [Sura No. – 3\4 – Nisaa, Ayat No. – 167 to 170]

[I seek safe haven] from the **mischief of created things [projected globalisation]**. And from the **mischief of darkness [projected mechanical magnetis]** as it overspreads [like the advertisement of a product], **from the mischief of those who practise secret arts [activists]**, and **from the mischief of the envious [terrorists] one as he practises envies** [two-in-one Laden]. [Sura No. – 112\113 – Bi-Rabbil Falaq, Ayat No. – 2 to 5]

PRECAUTIONS
"**Those who believe fight in the cause of Allah** and **those who reject Faith (disbelieve) fight in the cause of idols.**"
"So **fight as the dogs bodies (minions) of devil**. Lo! The **devil's strategy is ever weak.**"

  i)   **Let those fight in the cause of Allah** Who sell the life of this world [present life] for the hereafter.
  ii) **To him who fights in the way of Allah**, whether he is slain or gets victory; soon shall We give him a reward of great [value].

iii) **And why should you not fight in the cause of Allah and of those who, being weak, are ill-treated [and oppressed]?**

iv) Men, women, and children, whose **cry** is, **Our Rab! Rescue us from this town, whose people are oppressors.**

v) Oh, **raise for us from Your presence some protecting friend!**

vi) Oh, **raise for us from Your presence some defender!**

O you who believe! **Take your precautions**, then either advance the proven ones (**go forth in parties**) or **advance all together**. There are certainly among you men who would **tarry behind**. If a misfortune **befalls** you, he says: Allah had favoured me as I was **not** present among them. But if **good fortune** comes to you from Allah, he would surely **cry**, as if there had never been ties of affection (love) between you and him. Oh! I wish I had been with them; a great success should I then have achieved of it! **Let those fight in the cause of Allah** Who sell the life of this world for the hereafter. **To him who fights in the way of Allah**, whether he is slain or gets victory; soon shall We give him a reward of great [value]. **And why should you not fight in the cause of Allah and of those who, being weak, are ill-treated [and oppressed]?** Men, women, and children, whose **cry** is, **Our Rab! Rescue us from this town, whose people are oppressors.** Oh, **raise for us from Your presence some protecting friend! Oh, raise for us from Your presence some defender! Those who believe fight in the cause of Allah** and **those who reject Faith (disbelieve) fight in the cause of idols.** So **fight as the dogsbodies (minions) of devil.** Lo! The **devil's strategy is ever weak.** [Sura No. – 3\4 – Nisaa, Ayat No. – 71 to 76]

O you who believe! Ask not questions about things which, if they were made plain [known] to you, that may cause you trouble. But if you ask about things when the Quran is being revealed, they will be made known [cleared] to you; Allah will forgive those, for Allah is Forgiving, Clement. Some people before you did ask such questions, and on that account lost their faith. [Sura No. – 4\5 – Maaaidah, Ayat\Verses – 101 & 102]]

Truly **Firawn** exalted himself in the world and **broke up its people** into **sections** (groups / castes), **oppressing a small group among them.** Their **sons** he **slew**, but he kept **alive** their **females**; for **he was indeed a maker of mischief.** And We wished to be **Gracious** to those who were being **oppressed** in the world, to make them **leaders** and make them **heirs.** And to **establish** them in the world, and to show **Firawn, Haamaan**, and their **hosts**, at their hands, the very things against which they were taking **precautions.** [Sura No. – 27\28 – Qasas, Ayat\Verses – 4 to 6]

Note: Here leader does not represent an Imam in resemblance with the projected Imams.

"Who is truer in statement than Allah?"
"So, fight in the way of Allah."

And when there comes to them some matter (tidings) of **safety or fear, they noise it abroad.** If they had **only** referred it to the Messenger, or to those charged with **authority** among them, those among them who are **able** to think out the matter would have known it. Were it not for the Grace and Mercy of Allah unto you, **all of you would have fallen into the clutches of Shaytan save a few (of you).** So, **fight in the way of Allah.** You are **not taxed** (with the responsibility for anyone) except for yourself and urge on the believers. It may be that Allah will restrain the fury of the unbelievers; for **Allah is stronger in might and stronger in inflicting punishment.** Whoever recommends and helps a good cause becomes a **partner therein.** And whoever recommends and helps an evil cause, **shares in its burden.** And Allah has power over all things. When a [courteous] greeting is offered you, meet it with a greeting still more courteous, or [at least] of equal courtesy. Allah takes careful account of all things. Allah! There is no ilaaha (unbreakable reality) except Him. He will gather you all unto a Day of Judgment, about which there is no doubt. **Who is truer in statement than Allah?** [Sura No. – 3\4 – Nisaa, Ayat No. – 83 to 87]

"So choose not friends from them **till they forsake their homes** in the way of Allah."

WE ARE INSPIRED TO FORSAKE OUR HOMES IN THE WAY OF ALLAH

What ails you that you are divided into two parties about the hypocrites when Allah has cast them back for their [evil] deeds? Would you guide those whom Allah has thrown out of the way? For those whom Allah has **thrown out** of the way, **never you will find the way. They wish that you should reject faith as they do and thus be on the same footing** [as they]. So **choose not friends from them till they forsake their homes in the way of Allah.** But if they betray (turn back to enmity) seize them and slay them wherever you find them; and [in any case] **choose no friend or helper from among them.** Except those who join a group between whom and you there is a **treaty** [of peace], or those who approach you with hearts restraining them from fighting you as well as fighting their own people. If Allah had pleased, He could have given them power over you, and they would have fought you. Therefore if they withdraw from you but fight you not, and [instead] send you [Guarantees of] peace, then Allah Hath opened no way for you [to war against them]. Others you will find that wish to gain your confidence as well as that of their people. Every time they are sent back to temptation, they succumb thereto. if they withdraw not from you nor give you [guarantees] of peace besides restraining their hands, seize them and slay them wherever you get them. **In their case We have provided you with a clear argument against them.** [Sura No. – 3\4 – Nisaa, Ayat No. – 88 to 91]

WE ARE INSTRUCTED TO MAKE CAREFUL INVESTIGATION

O you who believe! When you go forth in the way of Allah, **investigate** carefully, and **say not to anyone who offers you a salutation**: You are **not** a believer coveting the **perishable** goods of this life, with Allah are **profits** and **spoils** abundant. Even thus were you yourselves before, till Allah conferred on you His **favours**. Therefore **carefully investigate**. For Allah is well **aware** of all that you do. Those of the believers who **sit** still other than those who have a (disabling) heart are **not on equality** with those **who strive in the way of Allah** with their wealth and lives. Allah has conferred on those who strive with their wealth and lives a **rank above the sedentary**. Unto each Allah has promised good but He has bestowed on those who strive a **great reward** above the sedentary; **degrees of rank** from Him, and forgiveness and mercy. For Allah is ever Forgiving, Merciful. [Sura No. – 3\4 – Nisaa, Ayat No. – 94 to 96]

When angels **take the souls** of those who **die in sin** against their souls, they (the angels) will ask: In what were you **engaged**? They will reply: **Weak and oppressed** were we in the earth. They (the angels) will say: Was not the earth of Allah **spacious enough** that you could have **migrated** therein? Such men will find their abode in **Hell**, an **evil journey's end. Except those** who are [really] weak and oppressed - men, women, and children - who have no means in their power, **nor [a guide-post] to their way** -For these, there is **hope** that Allah will forgive. Allah is ever **Clement**, Forgiving. He who **migrates for the cause of Allah** will find much **refuge and abundance** in the world, and whoso **forsakes** his home, a fugitive unto Allah and His messenger, and death overtakes him, his **reward** is then incumbent on Allah. Allah is ever Forgiving, Merciful. [Sura No. – 3\4 – Nisaa, Ayat No. – 97 to 100]

"In truth the disbelievers are all open enemy to you."
If you are **suffering**, lo! **They suffer even as you suffer and but you have hope from Allah, while they have none.**

And when you travel through the world, there is no sin for you if you shorten your prayers, if you fear that those who disbelieve may attack you. **In truth the disbelievers are all open enemy to you.** And when you are among them and arrange worship for them, let only a party of them stand with you and let them take their arms. Then when they have performed their prostrations let them fall to the rear and let another party came that has not worshiped and let them worship with you, and let them take their precaution and their arms. Those who disbelieve, long for you to neglect your arms and your baggage, that they may attack you once for all. It is no sin for you to lay aside your arms, if rain impedes you or you are sick. But take your precaution. Lo! Allah prepares for the disbelievers shameful punishment. When you have performed the act of worship, remember Allah **standing, sitting** and **reclining**. And when you are in **safety**, observe proper worship. Worship at **fixed hours** has been enjoined on the believers. **Relent not in pursuit of enemy.** If you are **suffering**, lo! **They suffer even as you suffer and but you have hope from Allah, while they have none.** Allah is ever Knower, Wise. [Sura No. – 3\4 – Nisaa, Ayat No. – 101 to 104]

FALSEHOOD AND FLAGRANT SIN

Lo! We reveal unto you the **Kitaab with the truth** that you may **judge between mankind** by that which you are guided by Allah. So **be not you an advocate for those who betray their trust (unfaithful).** And seek the forgiveness of Allah; for Allah is ever Forgiving, Merciful. **Contend not on behalf** of such as **betray** their own souls; for Allah **loves not one who is unfaithful and sinful.** They may **hide** [manifested truth] from men, but they **cannot** hide [them] from Allah, seeing that He is in their **midst** when **they plot by night**, in words that He cannot approve. And Allah ever compasses round all that they do. Ah! These are the sort of men on whose behalf you may contend in this world; but **who will contend with Allah on their behalf on the Day of Judgment, or who will then be their defender?** If any one **does evil** or wrongs his **own soul** but afterwards seeks Allah's forgiveness, he will find Allah ever Forgiving, Merciful. And if any one **earns sin**, he earns it **against** His own **soul**, for Allah is full of knowledge and wisdom. **But if any one earns a fault or a sin and throws it on to one that is innocent, he carries [on himself] [both] a falsehood and a flagrant sin.** [Sura No. – 3\4 – Nisaa, Ayat No. – 105 to 112]

Time has arrived to submit before Allah confirming Manifest Truth in resemblance with Equal & Opposite Stages of Journey of the Manifested Signs of Magnetism

Ayat\Verses - Is not the time arrive (ripe) for the hearts of those who believe **to submit to Allah's remember and of the Truth which has been revealed,** and that they should not become like those to whom was given the **Uutul-Kitaab afore time**, but long ages passed over them and their hearts were burdened, and many of them are **rebellious transgressors** (evil-livers)? Know that Allah Quickens (gives life to) the word after its death! Already We have shown the **Signs** plainly to you that you may understand. For those who give alms, men and women, and loan to Allah a goodly loan, it will be **doubled** for them, and theirs will be a rich reward. [Sura No. – 56\57 - Anzalnal-Hadiid, Ayat No. – 16 to 18]

Has there come upon man [ever] any period of time in which he was a thing unremembered? [Sura No. 75\76 – Hal ataa alal-Insaan, Ayat No. – 1]

Say: **No reward do I ask of you for this, nor I am a pretender (imposter). This is no less than a reminder for all peoples of the Appearing Pentagonal Earth,** and **you will certainly know the truth of it [all] after a while.** [Sura No. – 37\38 – Swad, Ayat No. 86 to 88]

Twa-Ha [Sura No. – 19\20 – Twa-Haa, Fundamental Ayat No. – 1]
We have not revealed this **Qur'an** to you to be [an occasion] for your distress. But only as a **reminder** to those who fear [Allah]. [Quran is] a **revelation** from Him Who **created the world [East Horizon] and the heavens [West Horizon]** on **high.** [Allah] Most Gracious is **firmly established on the Arsh** [Upright West]. To Him belongs **what is in the heavens [Galaxy of Stars]** and **in the world [Black Square],** and all **between** them, and **all beneath the soil.** If you **pronounce** the word aloud, [it is no matter], for verily He know what is **secret** and what is yet more **hidden.** Allah! There is **no** ilaaha (unbreakable reaity) save Him. To Him belong the most **Beautiful Names.** [Sura No. – 19\20 – Twa-Haa, Ayat No. – 2 to 8]

Secrecy is permissible
A hapless journey's end
But for the **Grace of Allah upon you and his Mercy,** a **party** of them would **certainly have plotted to lead you astray.** But [in fact] they will only **lead their own souls astray,** and **to you they can do no harm in the least.** For Allah has revealed unto you the Kitaab and wisdom and **taught you what you Knew not [before].** The **grace** of Allah toward you has been **infinite.** In most of their **secret talks** there is **no good.** But if **one exhorts** to a deed of charity or justice or conciliation between men, [**secrecy is permissible**]. To him who does this, **seeking the good pleasure of Allah,** We shall soon **bestow** on him **a reward of the highest** [value]. **If anyone contends with the Messenger even after guidance has been plainly conveyed to him, and follows a path other than that becoming a man of Faith,** We shall **leave him in the path he has chosen,** and **land him in Hell,** a hapless journey's end! [Sura No. – 3\4 – Nisaa, Ayat No. – 113 to 115]

601

MIDDLE COURSE – 7.2
WHO WILL TURN IN REPENTANCE?

Fasta-'iz billaahi minash-Shaytaanir-Rajim
Bismillaahir-Rahmaanir- Rahim

Question – Several Ayat\Verses of the Kitaab with Truth have been inspiring to turn in repentance. Who will turn in repentance?

Options – (a) a believer, (b) a disbeliever, (c) a non-believer

If we are following right direction towards Arsh, then what are the Existential Imports / Significances of the following Revelations?

TO ALLAH IS THE FINAL GOAL

"Wa **qaalatil-Yahuudu** wan-**Nasaaraa nahnu 'abnaaa-'ul-laahi** wa '**ahibaaa-'uh**. Qul falima yu-'azzibukum-bi-zu-nuubikum/ Bal 'antum-basha-rum-mimman khalaq: yagfiru limany-yashaaa-'u way u-'az-zibu many-yashaaa'. Wa lil-laahi mulkussa-maawaati wal 'arzi wa maa bay-nahumaa wa 'ilayhil-masiir." [Trans.] Wa **qaalatil-Yahuudu** wan-**Nasaaraa nahnu 'abnaaa-'ul-laahi** wa '**ahibaaa-'uh** say: We are the sons of Allah, and His beloved. Say: Why then He punishs you for your sins? Nay, you are but mortals of His creating. He forgives whom He pleases. And He punishes whom He pleases: And to Allah belongs the **dominion** of the **heavens** and the **world**, and all that is **between**: And unto Him is the final goal (of all). [Sura No. – 4\5 – Maaaidah, Ayat\Verses – 18]

Qul atii-ullaaha wa atii-ur-Rasuul

Say: Obey Allah, and obey Rasuul. But if you **turn away**, he is only **responsible** for the **duty** placed on **him** and **you** for that **placed** on you. If you **obey** him, **you shall be on right guidance**. The Messenger's **duty** is only to **convey** [Light / Wisdom] **plainly**. [Sura No. – 23\24 – Nuur, Ayat No. – 54]

Believers will say: We hear and we obey. And such are the successful. The saying of the believers, when summoned to Allah and His Messenger, in order that He may judge between them, is no other than this, they say: **We hear and we obey. And such are the successful.** He who obeys Allah and His Rasuul, and fears Allah, and keep duty (unto Him) will win (in the end). [Sura No. – 23\24 Nuur, Ayat No. 51 & 52]

Obedience is [more] reasonable.
They swear their strongest oaths by Allah that, if only you would command (order) them, they **would leave [their homes / go forth]. Say: Swear you not. Obedience is [more] reasonable.** Verily, Allah is well acquainted with all that you do. [Sura No. – 23\24 – Nuur, Ayat No. –53]

TRUE BELIEVERS
Only those are the **true believers** who **believe** in **Allah** and His **Rasuul**; and when they are with him **on a matter requiring collective action [e.g. right direction of Qibla]**, they **do not depart until** they have asked for **his permission.** Those who **ask** for your permission are those who **believe in Allah** and His **Rasuul.** So when **they ask for your permission**, for some business of theirs, **give** permission to those of them **whom you will**, and **ask Allah** for their **forgiveness**, for Allah is Ever Forgiving, Most Merciful. [Sura No. – 23\24 – Nuur, Ayat No. –62]

TWO FOLD MERCY AND A LIGHT
"Yaaa – ayyuhallazna aamanuttaqul-laaha wa aaminuu bi-Rasuulihii yu-tikum kif-layni mir-Rahmatihii wa yaj-al-lakum Nuuran-tamshuuna bihi wa yagfir lakum; wal-laahu Gafuurur-Rahim." [Trans.] O you **who** believe! **Be mindful of your duty to Allah**, and **put faith in His Messenger.** He will give you **two fold** of His Mercy and will appoint for you a **Light wherein you will walk**, and He will forgive you; for Allah is Ever Forgiving, Most Merciful. That the **People of the Kitaab** may know that they have no control whatever over the bounty of Allah; but that **the bounty is in His Hand to bestow it on whomsoever He wills.** And Allah is of Infinite Bounty. [Sura No. – 56\57 - Anzalnal-Hadiid, Ayat No. – 28 & 29]

PROMISE OF ALLAH
When those who disbelieve had setup in their hearts heat and false piety, the heat and cant of the Age of ignorance, then Allah sent down **His peace of reassurance upon His Messenger** (tranquillity to His Messenger) and to the believers, and imposed on them **the word of self-restraint**; for they were entitled to it and worthy of it. And Allah is Aware of all things. **Allah has fulfilled the vision of His Messenger in very truth. You shall indeed enter the Inviolable Place of Worship (Masjidal – Haraamaa), if Allah will, with minds secure**, heads shaved, hair cut short, and without fear. But **He knows that which you know not, and He has granted, besides this, a near victory.** It is He Who has sent His Messenger with Guidance and the Religion of Truth (Diinil-Haqq) that **He may make it to prevail over all religions [projected falsehood]**, and enough is Allah for a **Witness. Muhammad is the messenger**

of Allah; and **those who are with him are hard against disbelievers**, but **merciful among themselves**. You will see them bowing and falling prostrate, seeking bounty from Allah and (His) acceptance. **This is their similitude in the Tawraat; and their similitude in the Injiil** like the sown corn that sends forth its blade (shoot), and then makes it strong; it then becomes thick, and it stands on its own stem delighting the showers that He may enrage the disbelievers with (the sight of) them. **Allah has promised those among them who believe and do righteous deeds forgiveness and a great Reward.** [Sura No. – 47\48 – Fat-ham-Mubiina, Ayat No. – 26 to 29]

Why do not the Rabbis and the Priests [humur-Rabbaa-niyyuuna wal-abbaaru] forbid them from their (habit of) **uttering sinful words and their devouring of illicit gain**? **Evil indeed are their handworks.** [Sura No. – 4\5 – Maaaidah, Ayat\Verses – 63]

PROMISES OF ALLAH

i)   Allah has promised, to those among you who believe and work righteous deeds [tabligh];

ii)   He will surely make them succeed in the land [earth] even as He caused those who were before them to succeed;

iii)   He will establish in authority their religion - the one which He has chosen for them;

iv)   He will change [their state], after the fear in which they [lived], to one of security and peace;

v)   **They will worship [follow revealed & manifested Signs] Me [alone] and not associate aught [projected falsehood] with Me.**

vi)   **If any do reject Faith after this, they are rebellious and wicked.**

vii)   So establish regular Prayer [following right direction] and give poor-due; and obey the Rasuul; that you may receive mercy.

Ayat\Verse – Allah has promised, to those among you who believe and work righteous deeds [tabligh], that He will surely make them succeed in the land [earth] even as He caused those who were before them to succeed, and that He will establish in authority their religion - the one which He has chosen for them; and that He will change [their state], after the fear in which they [lived], to one of security and peace. **They will worship Me [alone] and not associate aught with Me. If any do reject Faith after this, they are rebellious and wicked.** So establish regular Prayer and give poor-due; and obey the Rasuul; that you may receive mercy. [Sura No. – 23\24 – Nuur, Ayat No. – 55 & 56]

INSTRUCTIVE EXAMPLES

i)   Allah causes the alternation (equal & opposite stages of journey) of the **day** and the **night**.

ii) Verily herein are **instructive examples** (clear proofs of revealed & manifested magnetism) for those who have vision!

iii) And Allah has created every animal from **water**.

iv) **Trinity** - Of them there are some that creep on their **bellies [North America & South America]**; some that walk on **two legs [Europr, Asia, Africa, and Australia]**; and some that **walk on four [Arabian Peninsula]**.

v)   We have indeed sent down **signs that make things manifest.**

vi) And Allah guides whom He wills to **a way that is straight**.

Ayat\Verses – Allah causes the **alternation** (equal & opposite Semi-anti-clock-wise & Semi-clock-wise stages of journey) of the **day** and the **night**. Verily **herein** are **instructive examples (clear proofs of revealed magnetism)** for those who have **vision**! And Allah has **created** every animal from **water**. Of them there are some that creep on their **bellies**; some that walk on **two legs**; and some that walk on **four**. Allah creates what He wills for verily Allah has power over all things. **We have indeed sent down signs that make things manifest**. And **Allah guides whom He wills to a way that is straight**. [Sura No. – 23\24 – Nuur, Ayat No. – 44 & 46]

Among us live the Messenger and Rehearses the Signs of Allah
Searching questions – Where & How?

O you who believe! If you listen to a faction among the People of the Kitaab, they will make you disbelievers after your belief. And how would you deny Faith while unto you are **rehearsed the Signs of Allah**, and **among you live the Messenger**? Whoever holds firmly to Allah will be shown a way that is Right. [Sura No. 2\3 – Imran, Ayat No. – 100 & 101]

WE [ALLAH AND HIS MESSENGER]

Verily **We** have brought the Truth to you; but **most of you have a hatred for Truth**. Or do they **determine anything**? Lo! We are **determining**. [Sura No. 42\43 – Ummil-Kitaab \ (Zukhruf), Ayat No. – 78 & 79]

Or do they think that **We** hear not their secrets and their private counsels? Indeed [We do], and Our **representatives** are present with them to record. [Sura No. 42\43 – Ummil-Kitaab \ (Zukhruf), Ayat No. – 80]

Who is 'Malik'? & Who is 'Rab'?

The sinners will be in the punishment of hell, to dwell therein [for ever]. It is not relaxed for them, and they despair therein. We wronged them not; but they it was who did wrong. And they cry: O **Malik**! Lo! **Your Rab** put an end to us. He will say: Nay, but you must remain. [Sura No. 42\43 – Ummil-Kitaab \ (Zukhruf), Ayat No. – 74 to 77]

Who mocked before Messengers and had rejected Truth?

Messengers were mocked before you, but their mocks were surrounded by the thing that they mocked. Say: Travel through the world and see what the end of those who had rejected Truth. [Sura No. – 5\6 – An-aam, Ayat No. – 10 & 11]

Who is the Guardian Rab?

Praise be to Allah, Who has created the heavens and the world, and **appointed darkness** and **light [Nuur]**. Yet **those who reject Faith ascribe [others] as equal, with their Guardian-Rab.** He it is Who has created you from clay, and then decreed a stated term [for you]. And **there is another determined term fixed with Him; yet you doubt.** And **He is Allah** in the heavens and the world. He knows both **your secret and your utterance**, and He knows **what you earn. But never did a single one of the signs of their Rab reach them, but they turned away there from.** And now **they deny the truth** when it reaches them. But soon will **they** learn the **reality [truth]** of **what they used to mock at. [Sura No. – 5\6 – An-aam, Ayat No. – 1 to 5]**

TURN NOT AWAY FROM RASUUL

O you who believe! Obey Allah and His Rasuul, and turn not away from him when you hear [him speak]. Nor be like those who say, We hear, but listen not; for the worst of beasts in the sight of Allah are the deaf and the dumb,-those who understand not. If Allah had found in them any good, He would indeed have made them listen. [As it is] if He had made them listen, they would but have turned back and declined [Faith]. **[Sura No. – 7\8 – Anil-Anfaal, Ayat No. – 20 to 23]**

Has there come upon man [ever] **any period of time** in which he was a thing **unremembered**? [Sura No. 75\76 – Hal ataa alal-Insaan, Ayat No. – 1]

"And **they turn away from him** and say: **One taught (by others), a madman!"**

How can there be remembrance for them when Prophet making plain (the Truth) had already come to them? And **they turn away from him** and say: **One taught (by others), a madman!** Lo! We indeed remove the torment for a while. Lo! You return [to your ways]. **On the Day when We will seize them** with a mighty onslaught (greater seizure), (then) **in Truth We will indeed exact Retribution (punishment)!** [Sura No. – 43\44 - Bi-Dukhaanim-Mubiin, Ayat No. – 13 to 16]

Their **similitude** is that of a man who kindled a fire; **when it lighted all around him, Allah took away their light and left them in utter darkness**. So they could not see. [Sura No. 1\2 – Baqarah, Ayat No. – 17]

The **lightning all but snatches away their sight**; every time **the light [helps]** them, **they walk therein (light)**, and **when the darkness grows on them**, **they stand still.** And if Allah willed, He could **take away** their faculty of **hearing and seeing**; for Allah has power over all things. [Sura No. 1\2 – Baqarah, Ayat No. – 20]

It is not fitting for a believer, man or woman, when a matter has been decided by Allah and His Messenger to have any option about their decision. If anyone disobeys Allah and His Messenger, he is indeed on a clearly wrong Path [Sura No. – 32\33 – Yahsabuunal-'Ahzaaab, Ayat\Verses – 36]

Two warnings
"To enter the Masjid as they had entered it first time."

When the **first of the warnings** [time for the first of the two] came to pass, We roused against you slaves of Ours of great might who destroyed (your) country, and it was a threat performed. Then We granted you the return as against them. We gave you increase in resources and sons, and made you the more numerous in man-power. If you did well, you did well for yourselves. If you did evil, [you did it] against yourselves. So when the **second of the warnings** came to pass, [We permitted your enemies] to destroy you, and **to enter the Masjid as they had entered it first time**, and to lie waste all that they conquered with an utter wasting. It may be that your Rab may [yet] show Mercy unto you; but if you repeat [your sins], We shall revert [to Our punishments]: And we have made Hell a prison **for those who reject [Faith]**. **[Sura No. – 16\17 - Banii-Israai-iil, Ayat No. – 5 to 8]**

SECOND WARNING
Eye opening Evidences

And verily We gave to Muusa **Nine Tokens [Clear Proofs]** (of Allah's Sovereignty): Do but ask Bani-Israa-iil how he came to them? Firawn said to him: O Muusa! I consider you, indeed, to have been worked upon by sorcery! Muusa said: In truth you know well that these things (Tokens) have been sent down by none but the Rab of the heavens and the earth as **eye-opening evidence,** and I consider you indeed, O Firawn, to be one doomed to destruction! And he wished to remove them from the face of the earth; but We did drown him and all who were with him all together. And We said thereafter to Bani-Israa-iil: Dwell securely in the land [of promise], but when the **second of the warnings came to pass**, We gathered **you together in a mixed crowd**. [**Sura No. – 16\17** - Banii-Israai-iil, Ayat No. – 101 & 104]

"There can be no change in the words of Allah."

Behold! Verily on the friends of Allah there is no fear, nor shall they grieve. Those who believe and [constantly] guard against evil; for them are good tidings, in the life of the present and in the Hereafter. **There can be no change in the words of Allah**. This is indeed the supreme felicity. Let not their speech grieves you, for all power and honour belong to Allah. It is He Who hears and knows [all things]. [Sura No. – 9\10 – Yuunus, Ayat No. – 62 to 65]

WARNER OF WARNERS

This is a **warner**, of the **warners** of old! **The threatened hour is near!** None besides Allah can **disclose** it. **Do you then doubt at this recital**? And will you **laugh** and **not weep** wasting your **time** in vanities? Rather **prostrate** yourselves before Allah and **serve** Him. [Sura No. – 52\53 – Wan-Najm, Ayat No. – 56 to 62]

Say to the **believing men** that they **should lower their gaze** (fix their eyes on) and **guard their modesty. That is purer for them**. And Allah is well acquainted with all that they do. And say to the **believing women** that they should lower their gaze and guard their modesty; that they should not display their beauty and ornaments except what [must ordinarily] appear thereof; that they should draw their veils over their bosoms and not display their beauty except to their husbands, their fathers, their husband's fathers, their sons, their husbands' sons, their brothers or their brothers' sons, or their sisters' sons, or their women, or the slaves whom their right hands possess, or male servants free of physical needs, or small children who have no sense of the shame of sex; and that they should not strike their feet in order to draw attention to their hidden ornaments. **And O you believers! Turn you all together towards Allah in order that you may succeed**. [Sura No. – 23\24 – Nuur, Ayat No. – 30 & 31]

WE ARE INSPIRED TO MAKE OUR DIIN PURE FOR ALLAH

"Huwal-Hayyu **Laaa ilaaha illaa Huwa fad-uuhu muklisiina lahud-diin. Al-Hamdu lillaahi Rabbil-Aalamiin!**" [Trans.] He is the Living One. **There is no reality which is dead** [laaa ilaaha] but He. So, pray unto Him, **making**

diin [religion] pure for Him [only]. Praise be to Allah, Rab of the Universe! [Sura No. – 39\40 – Mu-Min, Ayat No. – 65]

WHAT FOOLS AMONG US WILL SAY?
The fools among the people will say: What has turned them from the Qibla which they formerly observed? Say: To Allah belong the East and the West. He guides whom He will to a Way that is right. [Sura No. 1\2 – Baqarah, Ayat No. – 142]

TESTING
"Surely we will test you with something of fear and hunger and some loss of wealth and lives and crops, but give good tidings (do tabligh) to those who patiently persevere."

O you who believe! Seek help with steadfastness and prayer, for **Allah is with the steadfast. And call not of those who are slain in the way of Allah. They are dead. Nay, they are living, though you perceive [it] not. Surely we will test you** with **something of fear** and **hunger** and **some loss of wealth** and **lives** and **crops,** but **give good tidings (do tabligh) to those who patiently persevere**. Who say, when afflicted with calamity: To Allah We belong, and to Him is our return. They are those on whom blessings from their Rab, and Mercy, and they are the ones who are **rightly guided**. [Sura No. 1\2 – Baqarah, Ayat No. – 153 to 157]

WE ARE INSTRUCTED TO GUARD OUR PRAYER STRICTLY
Guard strictly your prayers and of the mid-most prayer, and stand up with devotion to Allah. If you fear, pray on foot, or **riding. But when you are again in safety,** remember Allah in the manner as **He has taught you,** which you knew not [before]. [Sura No. 1\2 – Baqarah, Ayat No. – 238 & 239]

REMEMBER ALLAH BY THE SACRED MONUMENT [KABA]
It is no sin (act against Allah's will) in you if you seek from the bounty of your Rab. But when you exert pressure / press on (at the time of returning) in the multitude from Aarafat, then **remember Allah by the Sacred** (Kaba / Headstone) **Monument** Remember Him as He has guided you. Although before that you were of those who go astray (*i.e. in multitude*). Then hasten onward from the place whence the multitude hastens onward, and ask forgiveness of Allah. Lo! Allah is Forgiving, Merciful. [198 & 199] [Sura No. 1\2 – Baqarah, Ayat No. – 180 to 182]

Nay [laaa], I swear by this Balad [City of the sole manifested Imam]; and you are an in-dweller of this city [People of Arabian Peninsula]; and begetter and that which he beget [parent and child]; verily We have created man in an atmosphere He thinks that none has power over him? And he says: I have destroyed vast wealth [Kitaab with Truth]. He [destroyer of Truth] thinks that none beholds him? Did We not assign unto him **two eyes**, and **a tongue**, and **two lips**, and guide her to the parting of the **two mountain ways** [Safa and Marwa]? But he [the follower of tautology] has not attempted the ascent. And what will convey to you the path that is ascent? [It is:] to free a bondman. Or the giving of food in a day of hardship to the orphan with claims of relationship or to the destitute in the dust; and to be of those who believe and enjoin patience, and enjoin deeds of kindness and compassion. Such are the Companions of the Right Hand. But those who reject Our Signs, they are the [unhappy] companions of the left hand. On them will be fire vaulted over. [Sura No. – 89\90 – Bi-haazal-balad, Ayat No. – 1 to 20].

KABA AND THE CARING PALM-TREE
"And for the poor refugees (Muhaajirs) who have been **driven out from their homes** and **their belongings**, who seek bounty from Allah and help Allah and His Messenger, such are indeed the sincere [loyal]."

Whether you **cut down the caring palm-trees**, or you left them **standing** on their roots, it was by the **permission** of Allah, and in order that He might cover with **shame** (confound) the **rebellious transgressors** (evil-livers). And that which Allah has given as spoil to His Messenger from them – for this you urged not any **horse or riding-camel for the sake thereof**; but Allah gives power to His messengers over any He pleases, and Allah has power over all things. That which Allah gives as spoil to His Messenger from the people of the **townships** belongs to Allah and His Messenger and to **kindred and orphan**, the needy and the wayfarer in order that it may not [merely] make a commodity between the rich among you. And whatsoever the Messenger **assigns** to you, **take** it. And whatsoever he **forbidden, abstain** (from it). And keep your duty to Allah. Lo! Allah is strict in Punishment. And for the poor refugees (Muhaajirs) who have been **driven out from their homes** and **their belongings**, who seek bounty from Allah and help Allah and His Messenger, such are indeed the sincere [loyal]. {Sura No. – 58\59 – Awwalil-Hashr, Ayat No. – 5 to 8]

TURN TO ALLAH IN REPENTANCE
O Prophet! Why ban you that which Allah has made lawful to you seeking to please your wives. But Allah is Ever Forgiving, Most Merciful. Allah has already ordained for you, [O men], the dissolution of your oaths [in some cases]; and Allah is your Protector, and He is the Knower, the Wise. When the Prophet disclosed a matter in confidence to one of his wives, and she then divulged it

[to another], and Allah made it known to him, he confirmed part thereof and repudiated a part. Then when he told her thereof, she said: Who has told you? He said: The Knower, the Aware has told me. If you twain **turn to Allah in repentance**, your hearts are indeed so inclined. But if you back up each other against him, truly Allah is his Protector, and Jibbriel, and [every] righteous one among those who believe, and furthermore, the angels will back [him] up. It may be, if he divorced you [all], that Allah will give him in exchange wives better than you, who submit [their wills], who believe, who are devout, who turn to Allah in repentance, who worship, who travel [for Faith] and fast, previously married or virgins. [Sura No. – 65\66 - Tahriim, Ayat No. – 1 to 5]

O you who believe! **Save yourselves and your families from a Fire whose fuel is men and stones**, over which are [appointed] angels stern [and] severe, who flinch not [from executing] the commands they receive from Allah, but do [precisely] what they are commanded. [They will say]: O you disbelievers! Make no excuses this Day! You are being but requited for all that you did! [Sura No. – 65\66 – Tahriim, Ayat No. – 6 & 7]

O you who believe! **Turn to Allah with sincere repentance** in the hope that your Rab will remit you from **your evil deeds** and bring you into gardens beneath which rivers flow on the Day when Allah will not permit to be humiliated the Prophet and those who believe with him. Their **Light** will run before them and on their **right hands**, while they say: Our Rab! **Perfect our Light for us**, and grant us forgiveness; for You are Able to do all things. O Prophet! Strive hard against the disbelievers and the hypocrites, and be firm against them. Their abode is hell, an evil refuge [indeed]. [Sura No. – 65\66 – Tahriim, Ayat No. – 8 & 9]

To the Aad People [We sent] **Huud,** one of their own **brethren**. He said: O my people! **Worship Allah!** You have no other reality except Him. **You do nothing but invent**. O my people! I ask of you **no** reward for this [Message]. My reward is from none but Him who created me: **Will you not then understand**? And O my people! Ask forgiveness of your Rab, and **turn to Him** [in repentance]: He will cause the sky to rain abundance on you, and add strength to your strength: so **turn you not back in sin**. [Sura No. – 10\11 – Kitaabun-uh-kimat (Prev. – Huud), Ayat No. – 50 to 52]

Say: O you men! If you are in doubt as to my religion, [behold!] I worship not what you worship, other than Allah! But I worship Allah Who will take your souls [at death]. I am commanded to be [in the ranks] of the Believers. And further [thus]: **Set your face towards religion with true piety, as a man**

611

by nature upright, and be not of those who ascribe partners. And cry not besides Allah, unto that which cannot profit you nor hurt you, for if you do so then wrath / wert () you of the wrong doers. If Allah afflicts you with hurt, there is none who can remove it but Him. If He appoints some benefit for you, there is none who can keep back His favour. He causes it to reach whomsoever of His servants He pleases. And He is Ever- Forgiving, Most Merciful. [Sura No. – 9\10 – Yuunus, Ayat No. – 104 to 107]

By the sky holding mansions [windows] of the stars [1], and by the **Promised Day [2]**, and by **one** who witnesses, and the **subject** of the witness [testimony] [3]; woe to the makers of the **pit** [of fire] **[4]**, of the **fuel-fed fire [5]**, when they sat by it [6], and they witnessed [all] that **they were doing against the believers [7]**, and **they ill-treated them for no other reason than that they believed in Allah**, Exalted in Power, Worthy of all Praise [8]! Him to Whom belongs the Sovereignty of the heavens and the earth! And Allah is Witness to all things [9]. Those who persecute [or draw into temptation] the believing men and believing women, and **do not turn in repentance**, will have the penalty of hell. They will have the penalty of the **burning fire [10]**. For those who believe and do righteous deeds [tabligh], will be gardens; beneath which rivers flow. That is the great success. [Sura No. – 84\85 – Was-Samaaa-i Zaatil-Buruuj, Ayat No. – 1 to 11]

Have We not caused your bosom to dilate [open]? And removed from you your burden which weighed down your back, and raised high [exalted] your fame. So, verily, **with every difficulty, there is relief**: Verily, **with every difficulty there is relief**. Therefore, **when you are relieved, still toil**, and **turn your attention [strive]** unto your Rab [with a view to please Him]. [Sura No. 94\95 – Alam-nashrah laka sadrak, Ayat No. – 1 to 8]

Verily **in this is a Sign, but most of them do not believe.**
The Samuud [people] rejected the messengers. Behold, their brother Saalih said to them: Will you not fear [Allah]? I am to you a messenger worthy of all trust. So fear Allah, and obey me. No reward do I ask of you for it. My reward is only from the Rab of the universe. Will you be left secure, in [the enjoyment of] all that you have here Gardens and Springs, and tilled fields and palm-trees with spa the near breaking? And you carve houses out of [rocky] mountains with great skill. But fear Allah and obey me; and follow not the bidding of those who are extravagant, who make mischief in the land [earth], and mend not [their ways]. They said: You are only one of those bewitched! You are no more than a mortal like us: then bring us a Sign, if you tell the truth! He said: **Here is a she-camel: She has a right of watering, and you have a right of watering,**

[severally] on an appointed day. Touch her not with harm, lest the Penalty of a Great Day seize you. But **they ham-strung her**, then were penitent. But the penalty seized them. **Verily in this is a Sign, but most of them do not believe**. And verily your Rab is He, the Exalted in Might, Most Merciful. [Sura No. 25\26 – Shuraa, Ayat No. – 141 to 159]

TURN YOUR VISION AND LOOK AGAIN AND AGAIN

Blessed be He in Whose **hands is the Sovereignty**; and He is Able to do all things (Qadir). He has **created life** and **death** that He may **test** you, which of you is best in **conduct**, and He is the Exalted in The Mighty, The Forgiving (Aziz, Gafuur). He has created the **seven heavens one above another**. You can see **no** fault in The Beneficent (Rahman). So **turn your vision again, can you see any rifts?** Then **look again and yet again, your sight will come back to you weakened and made dim**. [Sura No. – 66\67 – Bi-Yadihil-Mulk, Ayat No. – 1 to 4]

Allah sets forth an example for those who disbelieve the wife of Nuuh and the wife of Luut. They were [respectively] **under two of our righteous servants**, but they were false to their [husbands], and they profited nothing before Allah on their account, but were told: **Enter the Fire** along with those who enter. **And Allah sets forth, as an example to those who believe the wife of Firawn. When she said: My Rab! Build for me, in nearness to You, a mansion in the Garden, and save me from Firawn and his doings, and save me from those that do wrong.** And **Maryam, daughter of Imran**, whose body was **chaste**, therefore **We breathed therein something of Our spirit; and she put faith in the words of her Rab and of His Revelations, and was of the obedient**. [Sura No. – 65\66 – Tahriim, Ayat No. 10 to 12]

O you who believe! Let **not** some men among you **laugh** at others. It may be that the [latter] are **better than** the [former]. Nor let some **women laugh** at others. It may be that the [latter are **better** than the [former]. Nor **defame** nor be **sarcastic** to each other, nor call each other by [**offensive] nicknames. Ill- seeming** is a name connoting **wickedness**, [to be used of one] after he has believed. And those who **do not desist (turn not in repentance)** are [indeed] doing wrong (evil doers). [Sura No. – 48\49 - Minw-waraaa-il-Hujuraat, Ayat No. – 11]

To the **Samuud People [We sent] Saalih**, one of their own brethren. He said: O my people! Serve Allah: you have no other reality except Him. It is He Who has produced you from the earth and settled you therein, then ask forgiveness of Him, and **turn to Him** [in repentance]: for my Rab is [always] near, ready to answer. [Sura No. – 10\11 – Kitaabun-uh-kimat (Prev. – Huud), Ayat No. – 61]

"Turn aside from those who join breakable
realities [falsehood] with Allah"

To Him is due the primal origin of the heavens and the world: **How can He have a son when He has no consort (companion)?** He created all things, and He has full knowledge of all things. That is Allah, your Rab! There is no unbreakable reality but He, the Creator of all things: then serve you Him: and He has power to dispose of all affairs. **No vision can grasp** Him, but His grasp is over all vision: He is above all comprehension, yet is acquainted with all things. Now have come to you, from your Rab, proofs [to open your eyes]: if any will see, it will be for [the good of] his own soul; if any will be blind, it will be to his own [harm]: I am not [here] to watch over your doings. Thus do we explain the signs by various [symbols]: that they may say, You have taught [us] diligently, and that We may make the matter clear to those who have knowledge? Follow what you are taught by **inspiration** from your Rab: there is no unbreakable reality but He: **and turn aside from those who join breakable realities [projected mechanism] with Allah.** [Sura No. – 5\6 – An-aam, Ayat No. –102 to 107]

For me, I have turned my face, firmly and truly, towards Him Who created the heavens and the world, as one by nature upright, and never shall I give partners [man-made nature] to Allah. - (I am not of the idolaters) [Sura No. – 5\6 – An-aam, Ayat No. –80]

Say: O my servants who have **transgressed** against their souls! **Despair not** of the Mercy of Allah; for Allah forgives all **sins**; for He is Ever Forgiving, Most Merciful. **Turn you to our Rab [in repentance] and bow to His [Will], before the penalty comes on you, as after that you shall not be helped.** And follow **the best** of [the guidance] **revealed** to you **from your Rab, before** the **penalty** comes on you of **a sudden while you perceive not!** [Sura No. – 38\39 – Tanziilul-Kitaab, Ayat\Verses – 53 to 55]

[But] the Penalty on the Day of Judgment will be doubled to him, and he will dwell therein in ignominy. **Save him who repents, believes, and works righteous deeds** as for such **Allah will change their evil deeds to good deeds**, and Allah is Ever Forgiving, Most Merciful, **And whoever repents and does good has truly turned toward Allah with true repentance.** And those who witness no falsehood, and, if they pass by futility, they pass by with dignity. [Sura No. – 24\25 - Nazzalal-Furqaan, Ayat No. – 69 to 72]

And those who, when they are **reminded** of the **Signs of their Rab**, fall not down at them **as if they were deaf or blind.** And those who pray: Our Rab! Grant unto us wives and offspring who will be the comfort of our eyes, and give us [the grace] **to lead the righteous.** Those are the ones who will be **rewarded** with the **highest place** in heaven, because they were steadfast. And they will meet therein with **welcome** and the word of **Peace**. Abiding there forever - how beautiful an abode and place of rest! Say [**to the rejecters / disbelievers**]: **My Rab would not concern Himself with you but for your prayer. But now you have rejected / denied (Manifested signs in resemblance with revelations), therefore there will be Judgment.** [Sura No. – 24\25 - Nazzalal-Furqaan, Ayat No. – 73 to 77]

Turn you away from them for a little while

But [now that the Qur'an has come], they reject it. But soon will they know! And verily **Our word went forth** before [this] to our bondmen sent that they would certainly be assisted, and that Our hosts, they surely would be the victors. So **turn you away** (withdraw) from them for a little while, and watch them, for they will soon see. Do they wish [indeed] to hurry on our Punishment? But when it comes home to them, then it will be a hapless morn for those who have been warned. So **turn you away** (withdraw) from them for a little while, and watch them, for they will soon see. [Sura No. – 36\37 – Was-saaaffaati, Ayat No. – 170 to 179]

"Verily Allah guides not wrong-doing folk."

Yaaa-'ayyu-hallaziina 'aa-manuu laa tataa-khizul-**Yahuu-da** wan-**Nasaaraa** 'awli-yaaa'. Ba-zuhum **awli-yaaa-'u** ba'-z. Wa many-yata-wallahum-min-kum fa-'innahuu minhuum. 'In- nallaaha laa yahdil-qaw-maz-zaalimiin. [Trans.] O you who believe! Take not the Yahuud and the Nasaara for your **'awli-yaaa** [friends and protectors]. They are but friends and protectors to each other [Ba-zuhum **awli-yaaa-'u** ba'-z]. And **he** amongst you takes them for friends is [one] of them. Verily Allah guides not wrong-doing folk. [Sura No. – 4\5 – Maaaidah, Ayat\Verses – 51]

"Many of them follow a course that is evil."

The Yahuudis say: **Allah's hand is tied up.** Be their hands tied up and be they accursed for saying so. Nay, both His hands are **spread out wide in bounty.** He gives and spends (of His bounty) as He pleases. But the **revelation** that comes to you from your Rab is **certain to increase the obstinate rebellion and blasphemy** [disbelief]. Amongst them We have placed **enmity and hatred** till the **Day of Resurrection.** Every time they kindle the fire of war, Allah extinguishes it; but they (ever) strive to do **mischief** on earth. And Allah loves not those **who do mischief.** If only the People of the **Ahlal-Kitaab** had **believed** and been **righteous,** We should indeed have **blotted out their iniquities** and **admitted them to Gardens of Bliss.** If only they had **stood fast** by the Tawraat, the Injiil, and all the revelation that was sent to them from their Rab, they would have enjoyed **happiness** from every side. There is from among them **a party on the right course**: But **many of them follow a course that is evil.** [Sura No. – 4\5 – Maaaidah, Ayat\Verses – 64 to 66]

"Who so **judge not** by that which Allah has revealed, such are **disbelievers [kaafirun].**"

Lo! We have revealed the Tawraat [Coherence Truth] wherein is **guidance and light [Nuur-ya-kum]**. By this **criterion [manifested signs of equal & opposite revelation]** the Prophets who surrendered [unto Allah's will] judged the Haaduu, and Rabbaa-niyyuuna, and abbaaru [Translated as the Jews, the Rabbis, and Doctors of law or the Priests] in resemblance with min-**Kitaa-billahi [Kitaab of Allah]** as they were proposed to observe, **wa kaanuu alayhi shuha-daaa** [and there unto they were witness]. And **barter not** My revelations for a **little gain.** Who so **judge not** by that which Allah has revealed, such are **disbelievers [kaafirun].** [Sura No. – 4\5 – Maaaidah, Ayat\Verses – 44]

"**Say not** [unto the Prophet] **Listen to us**. But **say: Look upon us**; and be you **listeners.**"
Ayat\Verses - O you who believe! **Say not** [unto the Prophet] **Listen to us**. But **say: Look upon us**; and be you **listeners**. To those without Faith is a **grievous punishment**. Neither those who disbelieve among the followers of the **Prophet of the Kitab** nor the **idolaters** love that there should be sent down unto you any good thing from Allah. But **Allah will choose for His special Mercy whom He will as Allah is the owner of all attributes in their superlative degree.** [Sura No. 1\2 Baqarah, Ayat No. 104 & 105]

"But **your Rab knows best who it is that is better guided on the Way.**"
Ayat\Verses - We send down [stage by stage] in the **Qur'an** that which is a **healing** and a **mercy** to those who believes and to **the evil-doers it causes nothing but loss after loss.** Yet when We bestow Our favours on man, he **turns away** and becomes remote on his side [instead of coming to Us], and when evil seizes him he gives himself up to despair! Say: Everyone acts according to his own disposition. But **your Rab knows best who it is that is better guided on the Way.** [Sura No. – 16\17 - Banii-Israai-iil, Ayat No. – 82 to 84]

Ayat\Verses - It is not righteousness that you turn your faces Towards East or West; but it is righteousness to believe in Allah and the Last Day, and the Angels, and the Kitab, and the Messengers; to spend of your substance, out of love for Him, for your kin, for orphans, for the needy, for the wayfarer, for those who ask, and for the ransom of slaves; to be steadfast in prayer, and to give the poor-due; **to fulfil the contracts which you have made**; and to be firm and patient, in pain [or suffering] and adversity, and throughout all periods of stress. Such are they who are **sincere**. Such are the Allah-fearing. [Sura No. 1\2 – Baqarah, Ayat No. – 177]

SUPREME TRIUMPH
UPHOLD THE THRONE OF DIGNITY
AND TURN IN REPENTANCE

Those who uphold the Throne and those (**signs**) **around** (**glorify**) it, **hymn** the **praises** of their Rab and **believe** in Him and **implore** forgiveness for those who believe (saying): Our Rab! You comprehend all things in mercy and knowledge, therefore **forgive those who turn in repentance**, and **follow Your way. Ward off from them the punishment of hell.** Our Rab! And make them such that they may enter the Garden of Eden (Eternity) which You have promised them, and to the **righteous** [who do right] among their fathers, their wives, and their descendents. Lo! You, only You is The Mighty, The Wise. And preserve them from [all] ill deeds; and any from whom You ward off ill deeds that Day, him verily have You taken into mercy. That will be truly the **Supreme Triumph** (the Highest Achievement). [Sura No. – 39\40 – Mu-Min, Ayat No. – 7 to 9]

SECTION – VII
MIDDLE COURSE – 7.3
WITHDRAW FROM THEM AND WAIT

Fasta-'iz billaahi minash-Shaytaanir-Rajim
Bismillaahir-Rahmaanir- Rahim

"So withdraw from them, and wait. Lo! They too are waiting."

Ayat No.	Sura No. – 31\32	Sura - Sajdah
1	Alif, Lam, Mim.	
2	[This is] the Revelation of the Kitaab in which there is **no doubt**, from the Rab of the Universe.	
3	Or do they say: **He has forged it**? Nay, it is the **Truth** from your Rab, that you may **admonish** a people to whom **no warner** has come before you in order that they may **receive guidance**.	
4	It is Allah Who has **created** the **heavens** [West Horizon] and the **world [East Horizon]**, and all **between** them, in six Days [Hexagon]. Then He **mounted the Throne [towards Upright West or Arsh]**. You have none, besides Him, a protecting friend or mediator. Will you then not receive admonition?	
5	He **decrees** [all] **affairs** [continual acceleration] from the **heavens [Major Axis]** to the **world [Minor Axis]**. In the **end** will [all affairs] go up to Him, in a **Day**, whereof the **measure** will be [as] a **thousand years** of your reckoning	
6	Such is He, the Knower of all things, **invisible and visible**, the Exalted [in power], the Merciful.	
7	Who has made everything which He has created most good which He created. He began the **creation of man from clay**.	
8	Then He made his **seed** from an **essence of the nature [fitrat]** of a **fluid** despised. [Wave of Electricity]	
9	But He fashioned him in **due proportion**, and **breathed into him of His spirit** and gave you [the faculties of] **hearing** and **sight** and **feeling** [and understanding].Little thanks do you give!	
10	And they say: When we are lost in the world, how can then be **re-created**? Nay, they are the disbelievers meeting with their Rab.	
11	Say: The **angel of death**, put in **charge** of you, who will take your souls; and afterwards unto your Rab you will be returned	
12	If only you could see when the guilty ones will bend low their heads before their Rab, [saying:] Our Rab! We have seen and we have heard. Now then send us back [to the world]. We will **work righteousness**; for we do indeed [now] **believe**.	

13	If We had so willed, We could certainly have brought **every soul its true guidance**. But **the word from Me will come true**. I will **fill Hell with jinns and men** all together.
14	**Taste you then** - for you forgot the meeting of this Day of yours, and We too will forget you. Taste you the **penalty** of immortality for **your evil deeds!**
15	Only those **believe in Our Signs**, who, when they are recited to them, fall down in prostration, and reminded the praises of their Rab, nor are they [ever] puffed up with pride.
16	Who **forsake** their beds to cry unto their Rab in fear and hope, and **spend** of what We have bestowed on them
17	**No soul** knows what is kept hid for them of joy as a reward for their [good] deeds.
18	Is then the man who believes like unto him who is **rebellious and wicked**? They are **not alike**.
19	For those who **believe and do righteous deeds** (Tabligh) are Gardens [Heavens] as hospitable homes, for their [good] deeds.
20	As to those who are **rebellious and wicked**, their abode will be the **fire**. Every time they wish to get away there from, they will be forced there into, and it will be said to them: Taste you the **Penalty** of the Fire, which you **used to deny**.
21	And indeed We will make them **taste** of the Penalty of this [life] **prior to the supreme Penalty**, in order that **they may [repent and] return.**
22	And **who does greater wrong** than one to whom are recited the **Signs** of his Rab, and who then **turns away there from**? Verily from those who transgress We shall exact [due] retribution.
23	We verily gave the Kitaab to Muusa. So, be not you in doubt of his receiving it and We made it a guide to the Bani-Israa-iil.
24	"Wa ja-alnaa minhum A-immatany-yahduuna bi-Amri-naa **lammaa** sabaruu; wa kanuu bi-' Aayaatinaa yuu- qinuun". [Trans.] And **We appointed [Manifested] Imam** from among them [signs] **giving guidance** under **Our command**, so long as they **persevered** with **patience and continued to have faith in Our Signs.**
25	Verily your **Rab will judge** between **them** on the **Day of Resurrection**, in the matters **wherein they differ** [among themselves]
26	**Does it not teach them a lesson,** how **many generations** We **destroyed** before them, in whose dwellings they do walk? Verily **in that are Signs. Do they not then listen (heed)?**
27	And do they not see that We do drive **rain** to parched soil [bare of herbage], and produce therewith **crops**, providing **food** for their **cattle** and **themselves**? **Have they not the vision?**
28	And they say: **When come this victory (of yours), if you are telling the truth?**
29	Say: **On the Day of Victory**, the **faith of those who disbelieve** (and who then will believe) **will not avail them, nor will they be granted a respite.**
30	So **withdraw from them**, and **wait. Lo! They too are waiting.**

SIGNS FOR THOSE WHO FEAR ALLAH

Ayat\Verses - Verily, in the alternation of the **night** and the **day**, and in all that Allah has **created**, in the **heavens** and the **world**, are **signs** for those who fear Him. [Sura No. – 9\10 – Yuunus, Ayat No. –6]

Who does greater wrong than he who invents a lie against Allah or rejects His Signs? But verily the wrong-doers will never be successful. [Sura No. – 5\6 – An-aam, Ayat No. – 21]

SECTION – VII
MIDDLE COURSE – 7.4
RIGHT PATH FACING EACH OTHER

Fasta-'iz billaahi minash-Shaytaanir-Rajim
Bismillaahir-Rahmaanir- Rahim

FORMAL EVIDENCES FOR JUSTIFICATION

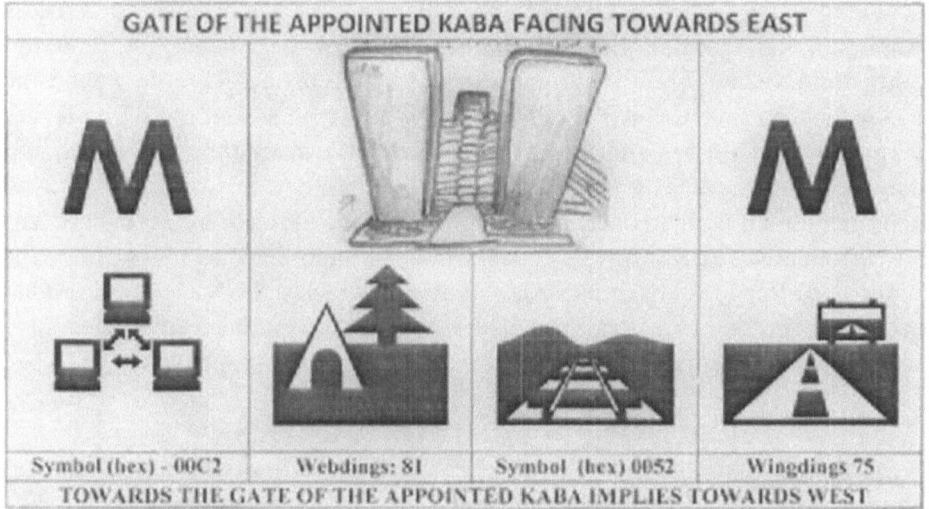

GATE OF THE APPOINTED KABA FACING TOWARDS EAST

| Symbol (hex) - 00C2 | Webdings: 81 | Symbol (hex) 0052 | Wingdings 75 |

TOWARDS THE GATE OF THE APPOINTED KABA IMPLIES TOWARDS WEST

SINGLE FACT – HOLD BACK YOUR HANDS

Ayat\Verses - Have you not seen those unto whom it was said: **Hold back your hands, establish worship** and pay the poor-due? But when fighting was prescribed for them, behold! A section of them feared men as - or even more than - they should have feared Allah. They said: Our Rab! Why have you ordained fighting for us? If only You give us respite yet a while! Say: Short is the enjoyment of this world [present life]; the Hereafter is the better for those who ward off evil. Never will you be dealt with unjustly in the very least (down upon a date-stone). **Wherever you are, death will find you out, even if you are in towers built up strong and high!** If some good befalls them, they say: This is from Allah; but if evil, they say, this is from you. Say: All things are from Allah. But **what has come to these people that they fail to understand a single fact [natural magnetism]?** Whatever good, [O man!] happens to you, is from Allah; but whatever evil happens to you, is from yourself. We have sent you as a messenger unto mankind. And **enough is Allah for a witness.** [Sura No. – 3\4 – Nisaa, Ayat No. – 77 to 79]

Ayat\Verses – Say: I do admonish you on **one point only: That you awake for the sake of Allah, by twos [top stair & middle stair] or singly [ground stair], and you reflect** (within yourselves / give thought). **There is no madness in your friend.** He is no less than a **warner** to you, in face of a **terrible Penalty.** [Sura No. – 33\34 – Saba, Ayat No. – 46]

Ayat\Verses - And verily **We have explained in detail in this Qur'an**, for the benefit of mankind, **every kind of similitude [resemblance].** But man is, in most things, contentious. And what is there to keep back mankind from believing, when the Guidance has come to them, nor from praying for forgiveness from their Rab, but that [they ask that] **the ways of the ancients be repeated with them,** or the wrath be brought to them **face to face?** We only send the messengers to give Good Tidings and to give warnings. But the **disbelievers dispute with vain argument,** in **order therewith to weaken the truth,** and they treat **My Signs as a jest,** as also the fact that they are warned! And who does greater wrong than one who is reminded of the **Signs of his Rab**; but turns away from them, forgetting the [deeds] which his hands have sent forward (to the Judgment)? Verily We have **set veils** over their hearts lest they should understand this, and over their ears, deafness, if you call them to guidance, even then they will never accept guidance. [Sura No. – 17\18 – Khaf – Ayat No. – 54 to 57]

DIRECTION FROM THE RAB AND A SLAVE

Ayat\Verses - **O mankind!** There has come to you a **direction** from your Rab and a **slave** for the [diseases] in your hearts, and for those who believe guidance and a Mercy. Say: In the **bounty of Allah** and in **His Mercy,** in that let them **rejoice,** that is **better than what they stored.** [Sura No. – 9\10 – Yuunus, Ayat No. – 57 to 58]

We are inspired to guard our prayer & Reference of the Separate Sheet

Ayat\Verses - No just estimate of Allah do they make when they say: Nothing Allah sends down to man [by way of revelation]. Say: Who then sent down the Kitab which Muusa brought - **a light and guidance to man [natural magnetism]? But you make it into [separate] sheets for show, while you conceal much [of its contents]:** therein were you taught that which you knew not- **neither you nor your fathers.** Say: Allah [sent it down]: Then leave them to plunge in vain discourse and trifling. And this is a Kitab which We have **sent down** [towards East], bringing blessings, and confirming which came before it: that you may warn **the mother of cities [Topmost Stair of Arabian Peninsula] and all around her [Middle Stair & Ground Stair].** Those who believe in the Hereafter believe in this and **they are careful in guarding their prayers.** [Sura No. – 5\6 – An-aam, Ayat No. –92 & 93]

UMMAT – WITNESS AGAINST MANKIND

"But it is not Allah's purpose that your faith should be in vain."

Ayat\Verses - Thus, We have made you an **Ummat** justly balanced (Middle Stair or Nation or Median), that you may be **witnesses against mankind** and that the **Rasuul** (Prophet) is a **witness over** you. And We **appointed the Qibla** which **you formerly observed only that We might know him who follows Rasuul** (the Prophet) from those **who would turn on** their heels (run away). Indeed it is a **hard test**, except to those **guided by Allah**. But it is **not** Allah's purpose that your faith should be in **vain**. For Allah is to all people full of piety, Most Merciful towards mankind. [Sura No. 1\2 – Baqarah, Ayat No. – 143]

THE APPOINTED KABA IS
ON THE RIGHTSIDE OF THE MOUNT TUUR

Bani-Israaiil – People of the Imam of the City [Arabian Peninsula] – Reference – "We delivered you from your enemy, and We made a covenant with you on the **right side** of **Mount Tuur"**

Ayat\Verses - O Bani-Israaiil! We delivered you from your enemy, and We made a covenant with you on the **right side** of **Mount Tuur** and We sent down to you Manna and Salwaa (alaykumul-Manna was-Salwaa). [Saying]: Eat of the good things We have provided for your sustenance, but commit no excess therein, lest My wrath should justly descend on you, and those on whom descends My wrath do perish indeed! But, without doubt, I am [also] He Who forgives again and again, to those who repent, believe, and do right; who, in fine, are ready to receive true guidance. [Sura No. – 19\20 – Twa-Haa, Ayat No. – 80 to 82]

Ayat\Verses - And what is that in the **right hand**, O Muusa? He said: It is my **staff**, on it I lean; with it I beat down fodder for my flocks; and in it I find other uses. [Allah] said: Throw it, O Muusa! He threw it, and behold! It was a **snake, active in motion**. [Allah] said: Seize it, and fear not. We shall return it at once to its former condition. Now draw your hand close to your side. It shall come forth **white** [and shining], without harm [or stain], **as another Sign**. In order that We may show you our **Greater Signs**; go you to Firawn, for he has indeed transgressed all bounds. [Sura No. – 19\20 – Twa-Haa, Ayat No. – 17 to 23]

Additional References - [Sura No. – 18\19 – Maryam, Ayat\Verses – 51 to 58], [Sura No. – 10\11 – Kitaabun-uh-kimat (Prev. – Huud), Ayat No. – 42 to 44], [Sura No. – 27\28 – Qasas, Ayat\Verses No. – 44 to 46], [Sura No. – 6\7 - As-haabul- a' raaf, Ayat No. – 143]

YOU ARE THE UPPERMOST

Ayat\Verses - O my people! Yours is the Kingdom (dominion) today. **You are the uppermost [Upright West Region] in the land [earth].** But who will help us from the Punishment of Allah, should it befall us? Firawn said: I do but **show you what I think,** and I do but **guide you to wise policy [which will lead you towards Straight Middle East].** [Sura No. – 39\40 – Mu-Min, Ayat No. –29]

He makes a guard to march before him and behind him
Before him [Messenger] – South-West to North-West / North-West to South-West

Behind him [Messenger] – North-East to South-East / South-East to North-East

Ayat\Verses – Say: **I know not whether the [Punishment] which you are promised is near, or whether my Rab will appoint for it a distant term.** He [alone] knows the Unseen [Mystery], nor does He make any one acquainted with His Mysteries [i.e. no mysticism / no spiritualism / no Sufism from the point of view of finite knower] except a **messenger** whom He has chosen, and then **He makes a guard to march before him and behind him.** That He may know that they have [truly] brought and conveyed the Messages of their Rab. He **surrounds** [all the mysteries] that are with them, and takes account of **every single thing.** [Sura No. – 71\72 – Nafarum-minal – Jinn, Ayat No. – 25 to 28]

Ayat\Verses - Wa nuriidu an-namunna alallaziinas-tua-ifuu fil=arzi wa naj-alahum a-immatanw-wa naj-alahumul-waarisiin. [Trans.] And We wished to be **Gracious** to those who were being **oppressed** in the world, to make them **leaders** and make them **heirs.** [Sura No. – 27\28 – Qasas, Ayat\Verse no. – 5]

Where is the appointed Kaba as the Imam of the City as well as the Leader of Right Direction [Standard] for mankind?

Options
Choose an option

i) The appointed Kaba is in the Middle East of the Upright Rectangular Universe, Hexagonal World, and Pentagonal Earth. Moreover, the appointed Kaba as the Leader of Right Direction is at the Centre of Global Trinity, Circular Rotation & Revolution System, and Man-made Magnetism. So, the appointed Kaba is at the centre of Revelation, Manifestation, and Appearance.

ii) The Black & White Imam of the City called Kaba has been appointed on the right side of the Mount Tuur in the Top [Upright] West Region of the Manifested Hexagonal World, Topmost Stair out of three ascending stairs, and Uppermost Land of the appearing pentagonal Earth within the Black Square [Non-luminous Moon / Dark Moon] of East Horizon.

SET – A	SET – B
Kaba has been appointed at the right side of the Mount Tuur in the Upright [Top] West Region of the manifested hexagonal world and topmost stair & uppermost land of the appearing pentagonal earth within the Black Square of East Horizon facing towards East or Manifestation. So, towards Kaba implies towards West.	Kaba has been projected in the Straight Middle East as well as at the centre of artificial nature [mechanical globalisation, circular rotation & revolution system, manmade magnetism] facing towards unidentified direction or towards West. So, towards Kaba implies towards crookedness or towards Straight Middle East.
UPRIGHT WEST South-West → ■ ← North-West South-East ▲ North-East **MIDDLE-EAST**	↘ ↶ ↓ ↷ ↙ → ■ ← ↗ ↺ ↑ ↻ ↖

Is there just a single contradictory reference in the Kitaab with Truth or Quran which shares with us that the so-called sun as the manifested sign of natural magnetism [Bullet] rises [once] from the East and sets [once] in the West i.e. contradictory to Ayat No. 35 of Sura Nuur?

Is there an Ayat in the Kitaab with Truth or Quran concerning equal & opposite stages of journey of the manifested signs of natural magnetism either contrary or contradictory to Ayat No. 17 of Sura Khaf?

If you are the possessor of true knowledge of the complete coded shared tautologies or Kitaab with Truth and if you find that shared searched out findings / manifest truths are either contrary or contradictory to revelations and manifestations, then kindly share your true concepts producing four witnesses verifiable as well as justifiable by Tawraat & Injiil and Manifested Signs in resemblance with revelations.

SHARE YOUR UPRIGHT CONCEPTS CONCERNING EQUAL AND OPPOSITE RIGHT DIRECTION TOWARDS UPRIGHT WEST
[Middle Course]

i) If I believe on the projected conspiracy i.e. Global Universe, Global world, Global Earth, Global Map, and that the appointed Kaba is in the Straight Middle East as well as at the centre of Revelation and Manifestation, and perform my Salat towards the appointed Kaba i.e. towards Straight Middle East, then towards which direction I am performing my Salat?

ii) If I believe on the evidence sorcery that Kaba has been appointed in the Straight Middle East Region of Eartha 3D and perform my Salat towards the appointed Kaba i.e. towards Straight Middle East Region of Black Hole in coherence with Set - B, then whether I am a follower of equal and opposite Right Path or I am a follower of Circular Paths [crookedness]?

iii) If Kaba has been appointed in the Middle East, then towards which direction is the gate of the appointed Kaba?

iv) Which of the above sets is representing equal and opposite right direction [middle course] towards Upright West or Arsh?

v) Are there references of circular ways, global universe, global world, global earth, moving earth, common run, appointed Kaba is at the centre of the earth, appointed Kaba is in the middle East, so-called sun as the manifested sign of natural magnetism rises from the East and sets in the West, and the like in the Kitaab with Truth? If there is just a single reference in the Kitaab with Truth in resemblance with Historical Black Magic or Manmade Nature, then produce that reference or share the same with four witnesses before providing any decision concerning searched out selfevident truths shared in the 'Kitaaba Wal-Hikmata' or 'Manifested Nature and the Utility of One's Upright Logic'. Each shared finding of 'Kitaaba Wal-Hikmata' or 'Manifested Nature and the Utility of One's Upright Logic' is verifiable as well as justifiable by the four criterions of truth in resemblance with revelations, manifested signs, real scientific symbols of the separate sheet, real works of the pen-paper-pencil educationists, and justifiable by upright logic.

vi) If you have clear concepts of the manifest truths concerning revealed as well as manifested universe, revealed as well as manifested world, appearing pentagon, four cardinal directions, Helium-4 atom in resemblance with revelations as well as manifestations, uniform principle, uniform motion, law of causation, apriori laws of thought, sole conditions of validity, right directions of the equal & opposite stages of journey of the manifested signs of natural magnetism, and the like, then share with us manifest truth concerning equal and opposite right direction of Qibla towards Upright West or Arsh. If you have no clear concept of the Manifest Truth in resemblance with Revelation concerning equal & opposite right direction of performing Salat [compulsory prayer towards upright West or

Arsh], then you are requested not to mislead me further, as I am already off the Track. On the contrary, let me make an attempt with a view to get generalised recognition from the respective authorities concerning the right direction of my Qibla towards Upright West [Arsh] grounded on the searched out selfevident truths as well as manifest truths after verification and justification of each and every shared selfevident truth on the basis of real science and complete coded shared tautologies in resemblance with manifested signs and manifested nature.

The Don of Historical Conspiracy Science has concealed several manifest truths with a view to track off believers from the equal and opposite right direction and to dominate mankind from the Spiderman's Net Work of the manifested hexagonal world and appearing pentagonal earth where even and odd are contrasted. His Teleological Evidence Sorceries and Unique Epistemic Persecutions in resemblance with his binary nature & function are leading me and my folk towards Eartha 3D / Black Hole of the Straight Middle East without Broken Bar. I have searched out several manifest truths which are justifiable by Universal Major Premises and verifiable by Affirmative Minor Premises, but these findings are both contrary as well as contradictory to projected teleological evidence sorceries and unique epistemic persecutions. On the basis of these findings, it can validly be deduced that all projections which are either contrary or contradictory to manifest truths or Manifested Nature or manifested signs are teleological evidence sorceries as well as epistemic persecutions. These are not the statements [premises] of the re-searcher of 'Solidarity Rights in Islam' & sharer of 'Kitaba Wal-Hikmata' or 'Manifested Nature and the Utility of One's Upright Logic'. On the contrary, these are the sum and substance of 'Kitaba Wal-Hikmata' or 'Manifested Nature and the Utility of One's Upright Logic' which are being shared as Solidified Solid Human Rights with a view to get generalised recognition of the searched out manifest truths from the respective authorities for the greatest happiness of the greatest number or Altruistic Hedonism.

UNTIL THE WHITE THREAD BECOMES DISTINCT
FROM THE BLACK THREAD OF DAWN

It is made lawful for you to go unto your wives on the night of fasting. They are your garments and you are their garments. Allah is aware that you were deceiving yourselves in this respect and He has turned in mercy toward you and relieved you. So associate with them, and seek that which Allah has ordained for you, and eat and drink, until **the white thread becomes distinct to you from the black thread of the dawn, then strictly observe the fast till the** nightfall; but do not associate your wives at your devotions in the mosques. **These are limits imposed by Allah**. So, **approach them not**. Thus, **Allah expounded His Clear Signs** to mankind that they may learn **self- restraint**. **And do not eat up your property among yourselves for vanities (falsehood), nor seek by it to gain hearing of the judges, with the plan that you may knowingly devour a portion of the property of others' wrongfully**. [Sura No. 1\2 – Baqarah, Ayat No. – 187 & 188]

[Muusa] said: **Rab of the East and the West**, and all between! **If you only had sense!** [Firawn] said: If you put forward any Reality other than me, I will certainly put you in prison! [Muusa] said: **Even if I showed you something clear [and] convincing**? [Firawn] said: Show it then, **if you tell the truth!** So [Muusa] flung down his staff, and behold, it was a serpent [snake], manifest [for all to see]! And he drew out his hand, and behold, it was **white** to all beholders! [Sura No. 25\26 – Shuraa, Ayat No. – 28 to 33]

SORCERY AND ASSEMBLY

[Firawn] said to the Chiefs around him: This is indeed a **sorcerer** well- versed (a knowing wizard). His plan is to get you out of your land [earth] by his **sorcery**; then what is it you **counsel**? They said: Keep **him and his brother in suspense** [for a while], and **dispatch to the Cities [people of Arabian Peninsula] heralds to collect** and bring up to you all [our] **sorcerers well-versed**. So the **sorcerers were got together for the appointment of a day well- known**. And the people were told: Are you [now] **assembled**? That we may follow the **sorcerers if they win**? So when the sorcerers arrived, they said to Firawn: Of course - **shall we have a [suitable] reward if we win?** He said: Yea, [and more], for you shall in that case be [raised to posts] nearest [to my person]. [Sura No. 25\26 – Shuraa, Ayat No. – 34 to 42]

"Sorcerers fell down prostrate in adoration."

Muusa said to them: Throw you - that which you are about to throw [produce all projected works of the pen – paper – pencil of the activist Laden concerning Globalisation, Manmade Magnetism, and Circular Rotation & Revolution System i.e. Manmade Nature]! So they threw **their ropes [Copernican Revolution]** and their **rods [Napier Rods or Napier Bones]**, and said: By the might of Firawn [two-in-one Laden], it is we who will **certainly win!** Then **Muusa threw his staff [manifested signs of natural magnetism]**, when, behold, it **straightway swallowed up all the falsehoods which they fake!** Then the **sorcerers fell down prostrate in adoration**, saying: We believe in the Rab of the universe, the Rab of Muusa and Haarun. [Sura No. 25\26 – Shuraa, Ayat No. – 43 to 48]

REFERENCE OF SELFEVIDENT TRUTH
LEFT TO RIGHT FROM THE POINT OF VIEW
OF THE APPOINTED KABA

Contradictory Grammatical Rule - Arabic is written and read from **right to left** [Ref. No. – 8, Page No. – 5]

Kitaab with Truth – "nor did you write it from right hand side"

In **like manner** [similitude] We have **revealed** to you the **Kitaab** and those, to whom We gave the Kitaab **afore time** believed therein, and of those (also) there are some who believe therein and **none** but **disbelievers reject Our Signs**. And you were not (able) to recite a Kitaab before this, **nor did you write it from right hand side**, for then might those have **doubted**, who follow **falsehood**. But it is **self-evident** in the hearts of those who are endowed with **true knowledge**, and **none deny Our Signs** (manifested magnetism in resemblance with revelation) except **wrong-doers**. [Sura No. – 28\29 – Ankabuut, Ayat No. – 47 to 49]

And He has constrained (made **subject to you)** the **Night and the Day. Washshamsa** wal-qamar and the **stars are in subjugation by His Command.** Verily in **these are Signs for men who have sense.** [Sura No. – 15\16 – Nahl, Ayat\Verses – 12]

INCLINING TO THE RIGHT AND TO THE LEFT

Have they **not observed all things that Allah has created,** how their **shadows incline to the right and to the left, making prostration unto Allah, and they are lowly?** [Sura No. – 15\16 – Nahl, Ayat\Verses – 48]

SENT FORWARD AND KEPT BACK
REVEALED AND MANIFESTED MAGNETIC FORCE
EQUAL AND OPPOSITE STAGES OF JOURNEY
OF THE MANIFESTED SIGNS OF NATURAL MAGNETISM

When the **sky is cleft asunder;** "Wa izal-kawaaki-buntasarat" [Trans.] When the **stars** are scattered **[IFTA]**; When the planets [skies] are dispersed [IBS (P) Ltd.]; When the planetary [psychic] barriers are scattered / dispersed / dissolved [Self-evident Concept]. When the **oceans [seas] are suffered to burst forth;** and when the **resting place are turned upside down;** [then] shall each soul know what is **sent forward** and **kept back.** [Sura No. – 81\82 - Izas-samaa-unfatarat, Ayat No. – 1 to 5]

TWO GUARDIAN ANGELS
Right [South] and Left [North]

We verily created a man, and We know what his soul whispers to him, and We are nearer to him than [his] jugular vein. When the two [guardian angels] appointed to read [his doings] and to write [noted them], seated on the **right** and on the **left**. And not a word does he utter but there is with him an **observer [possessor of the appointed light of Allah on earth]** ready. And the stupor of death comes in Truth. (And it is said to him): This was that which you were trying to escape. [Sura No. 49\50 – Kaaaf, Ayat No. – 16 to 19]

Allah has said: **Choose not two realities [both Natural Magnetism and Manmade Magnetism]**; for He is **just One Allah**. So of Me, Me only, be in awe [fear]. [Sura No. – 15\16 – Nahl, Ayat\Verses – 51]

He is the One Who sends to His slave **Manifest Signs (Light)**, that **He may lead you [guide] from the depths of darkness into the light** and verily for you Allah is full of Pity, Merciful. [Sura No. – 56\57 - Anzalnal-Hadiid, Ayat No. – 9]

Then after him We sent [many] **Messengers to their peoples**. They brought them **Clear Signs, but they would not believe what they had already rejected beforehand**. Thus We seal the hearts of the transgressors. [Sura No. – 9\10 – Yuunus, Ayat No. 73 & 74]

Verily **We** have brought the Truth to you; but **most of you have a hatred for Truth**. Or do they **determine anything**? Lo! We are **determining**. [Sura No. 42\43 – Ummil-Kitaab \ (Zukhruf), Ayat No. – 78 & 79]

Or do they think that **We** hear not their secrets and their private counsels? Indeed [We do], and Our **representatives** are present with them to record. [Sura No. 42\43 – Ummil-Kitaab \ (Zukhruf), Ayat No. – 80]

Neither from the East nor from the West
So-called Sun [Bullet & Triangular Bullets / Niche]
as the manifested sign of natural magnetism neither
rises from the East nor sets in the West
"**Allaahu Nuurus-samaa-waati wal-arz** – Masalu **Nuu-rihi** kaMishkaatin-fiihaa **Mis-baah** – Al-**Misbaahu** fii **Zujaa-jah** – azzuujaajatu ka-annahaa – kawkabun durriyyuny-yuuqa-du – min Shajaratimmubaara-katin-**Zaytuunatil-laa Sharqiy**-yatinw wa laa **Garbiyyatiny**-yakaadu **Zaytuhaa** yuziii- u wa law lam tamsas-hu naar. **Nuu-run alaa Nuur! Yahdillaahu** li-**Nuurihii** many-yashaaa – wa-yazribullaahul-amsaala linnaas – wallahu bi-kulli shay-in Alim." [Sura No. – 23\24 – Nuur, Ayat No. – 35] [Trans.] **Allah** is the **Light** of the **heavens** and the **world**. The **similitude** (likeness) of **His Light** is as a **niche** (binary thread of a candle or binary pulsar of Einstein) **in a lamp [Black**

630

Square]. The **lamp** is in a **glass [Diamond Operator]**. The **glass** is as it is a **shining star**. (The niche is) kindled from a **blessed tree**, an **olive** (Zaytuun tree), **neither from the East nor from the West**, whose (niche's) oil will almost glow forth (of itself) though **no fire touches** it. **Light upon Light!** Allah guides unto His light whom He wills. And Allah speaks to mankind in **parables**, for Allah is Knower of all things. [Sura No. – 23\24 – Nuur, Ayat No. – 35]

Left to Right and then Right to Left
So-called Sun [Bullet or Tarash-Shamsa] enters from Revealed Left [North] declining towards Revealed Right [South] and turns back from there due to Equal & Opposite Revelation and ends the journey in Revealed Left [North]

"Wa **tarash-shamsa** izaa tala-at-lazaawaru an-kahfihim zaatal-**yamimi** wa izaa gara-at-taqri-zuhum **zaatash-shimaali** wa hum fi fajwatim-minh. Zaalikka min aayaatillaah; many-yahdillaaahu fahuwal-muhtad; wa many-yuzlil falan-tajida lahuu waliyyam-murshidaa". {Trans.} And might have seen **tarash-shamsa** when it **rises, declining to the right** [semi-anticlockwise] from their Cave, **when it sets, turning away from them to the left** [semi-clockwise], while they lay in the **open space** in the midst of the **Cave** [within the Black Square of Manifested East Horizon]. **Such** [equal and opposite i.e. half anti-clockwise and half clockwise stages of journey as the glittering show for the earth] are among the **Signs of Allah**. He whom Allah, guides is **rightly guided**; but he whom Allah **leaves to stray**, for him **you will not find protector to lead him to the Right Way**. [Sura No. – 17\18 – Khaf – Ayat No. – 17]

Reference of Upright Nature [Middle Course]

Say: **O you men!** If you are in **doubt** as to my **religion**, [behold!] **I worship not what you worship**, other than **Allah**! But **I worship Allah** Who will take your souls [at death]. I am **commanded** to be [in the ranks] of the **believers**. And further [thus]: **Set your face towards religion with true piety, as a man by nature upright, and be not of those who ascribe partners [Manmade Magnetism]**. And **cry not** besides Allah, unto that which **cannot profit** you **nor hurt** you [Manmade Nature], for **if you do so** then **wrath** / wert (?) you of the **wrong doers**. If Allah **afflicts** you with hurt, there is **none** who can remove it but Him. If He appoints some **benefit** for you, there is **none** who can keep back His favour. He **causes** it to reach whomsoever of His **servants** He pleases. And He is Ever- Forgiving, Most Merciful. [Sura No. – 9\10 – Yuunus, Ayat No. – 104 to 107]

In the case of those **who say**: **Our Rab is Allah**, and afterward **Upright**, the angels descend on them [from time to time] saying: Fear you not! [They suggest]: Nor grieve! But receive the Good Tidings of the Garden [of Bliss], which you are promised. **We (angels) are your protecting friends in this life of the world and in the Hereafter.** Therein you will have all that your souls shall desire. Therein you will have all that you ask for! **A gift of welcome** from one Ever Forgiving, Most Merciful! **[Sura No. – 40\41 – Haa-Mim-Sajdah, Ayat No. – 30 to 32]**

TAKE FOR WITNESS TWO JUST PERSONS FROM AMONG YOU AND KEEP YOUR EVIDENCE UPRIGHT FOR ALLAH

Thus when they fulfil their term appointed, **either take them back on equitable terms or part with them on equitable terms**; and **take for witness two just persons from among you**, and **keep your evidence upright for Allah.** Whoso believes in Allah and the Last Day is **encouraged** to act thus. And **whosoever keeps his duty to Allah, Allah will appoint a way out for him.** [Sura No. – 64\65 – Tallaq-tumun-nisaaa-'a, Ayat No.2]
And cover not Truth with falsehood, nor conceal the Truth when you know [what it is]. And be steadfast in prayer; pay the poor-due; and **bow down your heads with those who bow down.** [Sura No. 1\2 – Baqarah, Ayat No. – 42 & 43]

JUSTICE AND SLAVES OF THE BENEFICENT

And the (faithful) **slaves of the Beneficent** are they who walk upon the world modestly, and when the **foolish (ignorant) ones** address them, they answer: Peace; and those who spend the night before their Rab, prostrate and standing. And those who say: Our Rab! Avert from us the Wrath of Hell, for its Wrath is indeed an affliction grievous. Evil indeed it is as an abode, and as a place to rest in. Those who, when they spend, are not extravagant and not niggardly, but hold a just [balance] between those [extremes]; **and those who cry not unto any other reality along with Allah, nor slay such life as Allah has forbidden save in (course of) justice, nor commit adultery;** and **whoso does this shall pay the penalty.** [Sura No. – 24\25 - Nazzalal-Furqaan, Ayat No. – 63 to 68]

SEEK MIDDLE COURSE
[Equal & Opposite Right Direction towards Arsh (Upright West)]

Ayat\Verses - Say: **Call unto Allah**, or **cry unto Rahman by whatever name** you call upon Him, [it is well], for to **Him belong the Most Beautiful Names. Neither** you speak **aloud** in Prayer, **nor** speak it in a **low tone**, but **seek a middle course between.** And say: Praise be to Allah, **who begets no son**, and has **no partner in [His] dominion.** Nor [needs] He any to protect Him from humiliation: yea, magnify Him for His greatness and glory! [Sura No. – 16\17 - Banii-Israai-iil, Ayat No. – 110 & 111]

FACING EACH OTHER TOWARDS ARSH [CRYSTAL-WHITE]

Ayat\Verses - But the sincere [and devoted] slaves of Allah; for them is a **Sustenance determined** fruits and they [shall enjoy] honour and dignity in Gardens of Felicity **facing each other on Throne** [Arsh]; round will be passed to them a **Cup** from a clear-flowing **fountain crystal-white** of a taste delicious to those who drink [thereof], wherein there is no headache [& or crookedness], nor they will suffer from intoxication. And besides them will be **chaste women**, restraining their glances, with big eyes [of wonder and beauty]; as if they were [delicate] **eggs closely guarded**. [Sura No. – 36\37 – Was-saaaffaati, Ayat No. – 40 to 49]

REVEALED RIGHT DIRECTION OF QIBLA

i) Right Direction - "**Set your faces upright at every place of prayer.**"

ii) Mosques or Masjids or Off-Springs of the Black & White Imam of the City called Kaba - "**Wear your beautiful clothing at every place of prayer.**" [Note – Beautiful clothing with reference to place represents right direction]

Say: My **Rab has commanded justice**; and **set your faces upright** [towards Upright West] **at every place of prayer**, and **call upon Him, making your devotion sincere in His sight**. As He brought you into being, so you will **return** (unto Him). Some He has **guided**: Others have deserved the **loss of their way [towards Middle East of Eartha 3D or Black-Hole]**; in that **they took the evil ones [wise policy of Shyatan, projected mechanism, and man-made directions], in preference to Allah, for their friends and protectors [justifiers and verifiers of manifested nature in resemblance with man-made magnetism], and think that they receive guidance [satisfied with falsehood]**. O Children of Adam! Wear your **beautiful clothing at every place of prayer, eat and drink**: But **waste not** by excess, for Allah loves not the wasters. [Sura No. – 6\7 - As-haabul- a' raaf, Ayat No. – 29 to 31]

Some ways go not Right

And unto Allah leads the **Right Path [towards Upright West]**, but **some ways go not Right [towards Upright West]**. If Allah had willed, He would have guided all of you. [Sura No. – 15\16 – Nahl, Ayat\Verses – 9]

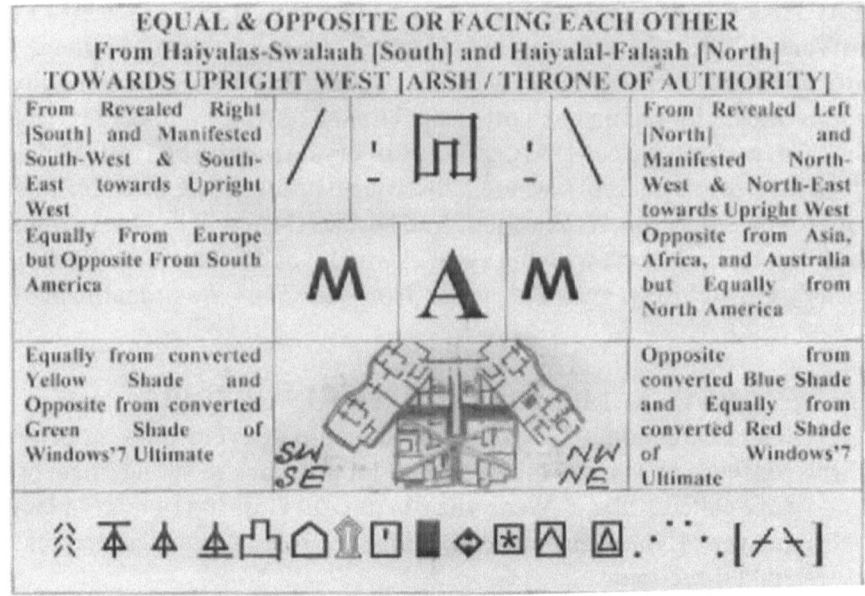

WEIGH WITH SCALES TRUE AND UPRIGHT

The companions of the wood rejected the messengers. Behold, Shuyaib said to them: Will you not fear [Allah]? I am to you a messenger worthy of all trust. So fear Allah and obey me. No reward do I ask of you for it. My reward is only from the Rab of the universe. Give just measure, and cause no loss [to others by fraud]. And **weigh with scales true and upright**. And withhold not things justly due to men, nor do evil in the land, working mischief. And fear Him Who created you and [who created] the generations before [you]. They said: You are only one of those bewitched! You are no more than a mortal like us, and indeed we think You are a liar! Now cause a piece of the sky to fall on us, if you are truthful. He said: My Rab knows best what you do. But they rejected him. Then the punishment of a day of overshadowing gloom seized them, and that was the Penalty of a Great Day. Verily in that is a Sign. But most of them do not believe. And verily your Rab is He, the Exalted in Might, Most Merciful. [Sura No. 25\26 – Shuraa, Ayat No. – 176 to 191]

ALLAH WILL APPOINT A WAY OUT

Thus when they fulfil their term appointed, **either take them back on equitable terms or part with them on equitable terms**; and **take for witness two just persons from among you**, and **keep your evidence upright for Allah.** Whoso believes in Allah and the Last Day is **encouraged** to act thus. And **whosoever keeps his duty to Allah, Allah will appoint a way out for him.** [Sura No. – 64\65 – Tallaq-tumun-nisaaa-'a, Ayat No.2]

RAIN IS A SIGN FOR THOSE WHO GIVE THOUGHT

It is He who sends down **rain from the sky [West]**; from it you **drink**, and out of it [grows] the vegetation on which you feed your cattle. With it He produces for you corn, olives, date-palms, grapes and every kind of fruit. Verily in this is **a sign for those who give thought**. [Sura No. – 15\16 – Nahl, Ayat\Verses – 10 & 11]

And Allah sends down rain from the skies [West], and gives therewith life to the world after its death. **Verily in this is a Sign for those who listen.** [Sura No. – 15\16 – Nahl, Ayat\Verses – 65]

MUUSA AND HIS BROTHER WERE INSPIRED TO STAND UPRIGHT

We **inspired** Muusa and his brother with this Message: Provide dwellings for your people in Misra (Egypt), make your dwellings into places of worship, and establish regular prayers: and give good tidings to those who believe. Muusa prayed: Our Rab! You has indeed bestowed on Firawn and his chiefs splendour and wealth in the life of the present, and so, Our Rab, they mislead [men] from Your Path. Spoil, our Rab, the features of their wealth, and send hardness to their hearts, so they will not believe until they see the grievous penalty. Allah said: Your prayer is accepted [O Muusa and Haaruun]! **So stand you upright, and follow not the path of those who know not.** [Sura No. – 9\10 – Yuunus, Ayat No. - 87 to 89]

Concept of Brotherhood with respect to right direction

Lo! Those who **ward off** (evil) are among gardens and **water-springs** [And it is said unto them]: Enter you here in **peace** and **security**. And We shall remove from their hearts any lurking **sense of injury**. [They will be] **brothers** [joyfully] **facing each other on thrones** of dignity [Arsh]. There no sense of **fatigue** shall touch them, nor shall they [ever] be asked to leave. **Tell My servants that I am indeed the Ever-Forgiving, Most Merciful. And that My Penalty will be indeed the most grievous Penalty.** [Sura No. – 14\15 = Hur, Ayat No. – 45 to 50]

IT HAD BEEN BETTER FOR THEM AND MORE UPRIGHT

See you not those unto whom a portion of the Kitaab has been given, how they purchase error, and seek to make you (believers) err from the right way? Allah knows best (who are) your enemies. Allah is sufficient as a Friend and Allah is sufficient as a Helper. Some of those who are **haaduu change words** from their context and say: We hear and disobey; hear you as one who hears not and listen to us, **distorting** with their tongues and **slandering** religion.

If they had said: We hear and we obey, hear you and look at us, it had been better for them and **more upright**. But Allah has **cursed** them for their disbelief, so they believe not save a few. [Sura No. – 3\4 – Nisaa, Ayat No. – 44 to 46]

ALLAH'S GUIDANCE IS THE ONLY GUIDANCE

Say: **Shall we indeed cry instead of unto Allah, things that can do us neither good nor harm, and turn us after receiving guidance from Allah like one bewildered whom the devils have infatuated in the earth, which has companions who invite him to the guidance (saying): come unto us?** Say: **Allah's guidance is the [only] guidance, and we are directed to submit ourselves to the Rab of the universe.** To establish regular prayers and to fear Allah: for it is to Him that we shall be gathered together. **It is He who created the heavens and the world in true/due proportion in that day when He said: Be! It is.** [Sura No. – 5\6 – An-aam, Ayat No. –71 & 73]

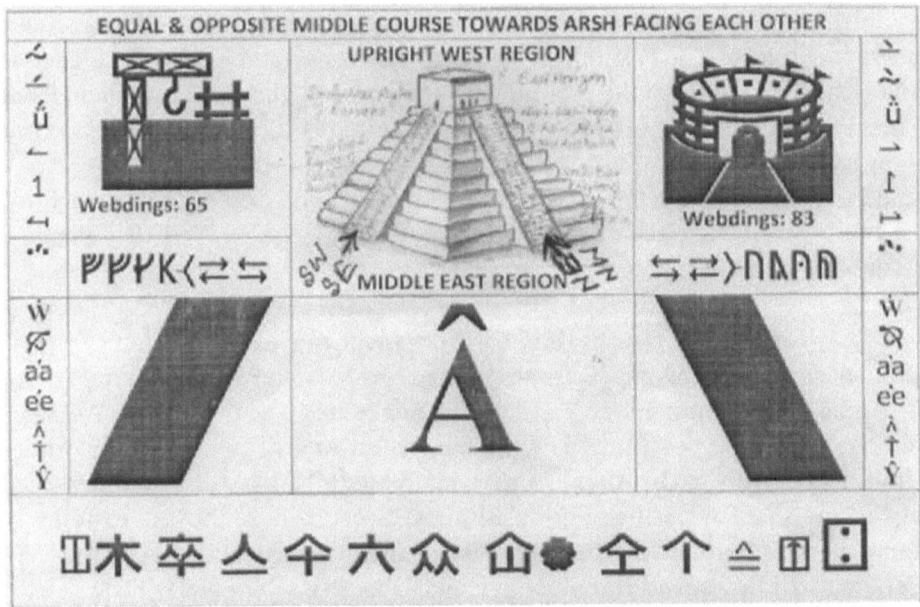

INDILLAAHI MUBAA-RAKATAN-TAYYIBAH
NO BLAME AND NO SIN
"Thus Allah **makes clear the signs for you** in
order that **you may understand**."
No **blame is there upon the blind** or any **blame upon the lame**; nor any **blame upon the sick nor on yourselves if you eat from your houses**, or those of your fathers, or your mothers, or your brothers, or your sisters, or your father's brothers or your father's sisters, or your mother's brothers, or your mother's sisters, or

(from that) whereof you hold the keys, or (from the house) of a friend (all). **There is no sin on you, whether you eat in company or separately**. But if you enter **houses, salute each other** - a greeting of blessing and purity as from Allah [Indillaahi-mubaa-rakatan-tayyibah]. Thus Allah **makes clear the signs for you** in order that **you may understand**. [Sura No. – 23\24 – Nuur, Ayat No. –61]

WILL OF ALLAH AND RIGHT PATH
DEAF, DUMB, DARKNESS WANDER IN CIRCULAR WAYS

Those **who reject our signs** are **deaf** and **dumb**, in the midst of **darkness** profound: whom Allah wills, He leaves to **wander**; whom He wills, He places on **the way that is right path**. [Sura No. – 5\6 – An-aam, Ayat No. – 39]

Ayat\Verses – Were you witness when death appeared before Yaquub? When, he said to his sons: **What will you worship after me?** They said: **We shall worship your Rab and the Rab of your fathers, of Ibrahim, Ismaiil and Ishaaq, the one [True] Allah. To Him we have surrendered.** Those are a people who have passed away. They shall **reap the fruit** of what they did, and **yours of what you do!** And **you will not be asked of what they used to do. They say: Be Yahuud or Nasaara, then you will be rightly guided.** Say (unto them): **Nay! But we follow the Religion of Ibrahim, the upright, and he was not of the idolaters. Say: We believe in Allah, and the revelation given to us, and to Ibrahim, Ismaiil, Ishaaq, Yaquub, and the Tribes (Aasbati), and that given to Muusa and 'Isa, and that which the prophets received from their Rab. We make no difference between one and another of them.** And **we surrender to Allah as Muslims.** So if **they believe as you believe, they are indeed on the right direction**; but **if they turn back**, it is they **who are in schism [ism or division]**; but Allah will **suffice** you (for defence) against them, and He is All-Hearing, All-Knowing. [**Sura No. – 1\2 – Baqarah, Ayat No. – 133 to 137**]

JUDGMENT

[O unbelievers / disbelievers] If you prayed for **judgment**, now has the **judgment** come to you: **if you desist [from wrong], it will be best for you. If you return [to the attack]**, so shall We. Not the least good will **your forces** are to you **even if they were multiplied**, for verily Allah is with those who believe. [Sura No. – 7\8 – Anil-Anfaal, Ayat No. – 15 to 18]

"Say: Shall I seek for judge other than Allah?"
TRUTH AND JUSTICE
"None can change His words."

To such [deceit] let the hearts of those incline, **who have no faith in the hereafter**: let them **delight** in it, and let them **learn** from it what they may. **Say:**

Shall I seek for a judge other than Allah? When He it is Who has sent unto you **the Kitab, explaining in detail.** They know full well, to whom We have given the Kitab that it has been sent down from your Rab in truth. **Never be then of those who doubt.** The **word** of your Rab **finds its fulfilment in truth** and in **justice: None can change His words:** for He is the one who hears and knows all. [Sura No. – 5\6 – An-aam, Ayat No. –114 to 116]

FINAL DETERMINATION

But those who believe and do deeds of righteousness, We shall soon admit them to Gardens [Heavens], with rivers flowing beneath, their eternal home. Therein they will have companions pure and holy. We shall admit them to shades, cool and ever deepening. Lo! **Allah commands you that you restore deposits to their owners, and if you judge between mankind, that you may judge justly. O you who believe! Obey Allah; and obey the Messenger,** and **those of you who are in authority, and if you have a dispute concerning any matter, refer it to Allah and the Messenger, if you are believers in Allah and the Last Day. That is better, and more suitable for final determination.** [Sura No. – 3\4 – Nisaa, Ayat No. – 57 to 59]

Ayat\Verses – So **set your purpose for Diin** (Religion) as a man **by nature upright – the nature of Allah,** in which He has created man. **There is no altering (in the Laws of) Allah's creation [Equal & Opposite Revelation or Tawraat].** This is the **right religion,** but **most among mankind understand not. Turn you back in repentance** to Him, and fear Him. Establish regular prayers, and **be not you among those who ascribe partners** (join manmade magnetism) with Allah. **Of those who split up their Religion, and become [mere] Sects, each party rejoicing in that which is with itself?** [Sura No. – 29\30 – Ruum, Ayat\Verses No. – 30 to 32]

Model of Right Direction

Ibrahim was indeed a **model, devoutly obedient to Allah,** [and] **true in Faith,** and **he joined not breakable reality (projected falsehood / Manmade Nature) with Allah.** He **showed** his **gratitude** for the favours of Allah, who chose him, and **guided him to a Straight Way [towards West].** And We gave him good in this world [present life], and he will be, in the Hereafter, **in the ranks of the Righteous. And afterwards We inspired you Follow the ways of Ibrahim [towards West] the True in Faith, and he joined not breakable realities [projected falsehood] with Allah.** [Sura No. – 15\16 – Nahl, Ayat\Verses – 120 to 123]

There is **no compulsion** in Fid-Diin (religion). **The right direction is henceforth distinct from Error. Whoever rejects false desires and believes in Allah has grasped the firm hand-hold that never breaks.** And Allah is

Hearer and Knower of all things. **Allah is the Protector of those who have faith. He brings them out of darkness unto light. As for those who reject faith their patrons are the false ones (projected mechanical globalisation, circular rotation and revolution system, and manmade magnetism i.e. Manmade Nature). They bring them out of light into depths of darkness. They will be companions of the fire. They will dwell there [for ever].** [Sura No. – 1\2 – Baqarah, Ayat No. - 256 & 257]

RUN AWAY

Ayat\Verses – And when a party of them said: You **folk of Yasrib**! You cannot stand [the attack]! Therefore **turn back**! And a band of them ask for permission of the Prophet, saying: Our houses are bare and exposed. And they were not exposed. **They intended nothing but to run away.** [Sura No. – 32\33 – Yahsabuunal-'Ahzaaab, Ayat\Verses – 13]

DELUDED AWAY

The **Yahuudis'** call 'Uzayru-nib-nullaahi' as the **son** of Allah, and the **Nasaras'** call 'Masiihub-nullaah' as the **son** of Allah. That is **a saying from their mouth**; [in this] they but **imitate** what the **unbelievers of old used to say**. **Allah's curse is on them**: how they are **deluded away from the Truth**. They have taken as **Rabs** besides Allah their **rabbis** and their **monks**, [and] **Wal-Musii-habna-Maryam** [to be their Rab in derogation of Allah], **yet they were commanded to worship none but One Allah**. There is **no unbreakable** reality except Him. Praise and glory to Him: [Far is He] from having the **partners they associate** [with Him]. Fain would **they extinguish Allah's light** (Equal & Opposite Revelation or Tawraat and Manifested Sign of Natural Magnetism or Injiil) with their **mouths**, but Allah will not allow that His light should extinguish, even though the unbelievers may abominate (it). It is He Who has sent His Messenger with guidance and **the Religion of Truth**, to prevail over **all isms [diversities or projected falsehood]**, even though the **Pagans / idolaters** [followers of falsehood] may **detest** [it]. [Sura No. – 8\9 – Tauba, Ayat No. – 30 to 33]

PART – VIII
RIGHTEOUSNESS – 8.1
MANKIND

Fasta-'iz billaahi minash-Shaytaanir-Rajim

Bismillaahir-Rahmaanir- Rahim

ONE COMMUNITY
VERILY THE HELP OF ALLAH IS ALWAYS NEAR

Mankind were but one community, and **they differed later.** Had it not been **for a word** that had already gone forth from your Rab, it had been **judged** between them in respect of that **wherein they differ**? And they will say: if only a **Clear Sign** (Clear Proof) were sent down upon him from his Rab! Say: **The Unseen belongs to Allah.** So, wait! I am also waiting with you. [Sura No. – 9\10 – Yuunus, Ayat No. – 19 to 20]

Mankind was one single community, and **Allah sent Messengers with good tidings** and **warnings**; and with them He revealed the **Kitaab with truth,** to **judge** between **people** in matters **wherein they differ.** But the People of the Kitaab, after the **Clear Signs** (Proofs) came to them, **did not differ among themselves, except through selfish contumacy** (hatred one of another). Allah by His Grace **guides** the believers to the Truth, concerning that wherein they differ; for Allah guides whom He will to **a path that is towards Upright West [Arsh].** Or do you think that you shall **enter the Garden** [of bliss] without such [trials] **as came to those who passed away before you?** They encountered **suffering and adversity**, and were so **shaken in spirit** that even the Messenger and those of faith who were with him cried. **When [will come] the help of Allah? Ah! Verily, the help of Allah is [always] near!** [Sura No. 1\2 – Baqarah, Ayat No. – 213 & 214]

O you messengers! Eat [all] things **good** and **pure**, and do Tabligh (work righteousness); for **I am Aware of what you do.** And verily **this Brotherhood of yours** is **a single Brotherhood**, and I am your Rab and Cherisher: therefore keep your **duty** unto Me. **But people have cut off their religion [of unity] among them, into sects. Each sect rejoices in its tenets. But leave them in their confused ignorance for a time. Do they think that because We have granted them abundance of wealth and sons, We would hasten them on in every good? Nay, they do not understand (they perceive not).** [Sura No. – 22\23 – Mu-Minuun, Ayat\Verses – 51 to 56]

And We send the fecundating **winds**, then cause the **rain** to descend from the **sky**, therewith providing you with **water** [in abundance], though **you are not the guardians of its stores**. And verily, it is **We Who give life, and Who give death**. It is We Who remain **inheritors** [after all else passes away]. And verily We know those of you **who hasten forward**, and those **who lag behind**. Assuredly it is your Rab Who will **gather** them together, for He is perfect in Wisdom and Knowledge. [Sura No. – 14\15 = Hur, Ayat No. – 14 to 25]

We **created man** from sounding **clay** (electron cloud) **of black mud altered**. And the **Jinn** race, We had created before of **essential fire**. And (remember) when your Rab said unto angels: Lo! I am creating **mortal** out of potter's clay of **black mud altered**. So when I have made him [in due proportion] and **breathed into him of My spirit, fall you down** in obeisance unto him. So the angels **prostrated** themselves, all of them together except the **Iblis**. He refused to be among those who prostrate. [Sura No. – 14\15 = Hur, Ayat No. – 26 to 31]

When We cause mankind to taste of mercy after some adversities which had touched them; behold! They **began to plotting against Our Signs**! Say: **Swifter to plan** is Allah! Verily, Our **Messengers record** all the **plots** that you make. He it is Who makes you to go on the **land** and the **sea** till, when you are in **ships** and they **sail** with them with a favourable **wind**, and they are glad therein; then comes a stormy wind and the waves come to them from all sides, and they think that they are being overwhelmed therein. They cry unto Allah, making their **faith pure** for Him only saying: if You deliver us from this, we shall truly show our **gratitude**. But when He has delivered them, behold! They **transgress** disrespectfully in the world. **O mankind! Your disrespect is only against yourselves**, an enjoyment of the life of the **present**. In the end, to Us is your return, and **We shall show you the truth of all that you did**. [Sura No. – 9\10 – Yuunus, Ayat No. – 21 to 23]

O mankind! If you have a **doubt** about the **Resurrection**, [consider] that We **created** you out of **dust**, then out of **sperm**, then out of a **leech**-like clot, then out of a morsel of **flesh, partly formed** and **partly unformed**, in order that We may **manifest** [our power] to you; and We **cause** whom We will to rest in the wombs for an **appointed term**, then We **bring** you out as babes, then [foster you] that you may reach your **age** of full strength; and some of you are **called to die**, and some are sent back to the feeblest **old age**, so that **they know nothing after having known [much]**, and [further], you see **the world barren and lifeless**, but when We pour down **rain** on it, it is stirred [to life], it swells, and it puts forth **every kind of beautiful growth [in pairs]**. [Sura No. – 21\22 – [Hajj], Ayat No. – 5]

O mankind! Here is a **parable** set forth! Listen to it! **Those on whom, besides Allah, you call cannot create [even] a fly, if they all meet together for the purpose! And if the fly took something from them, they could not rescue it from the fly. So weak are (both) the seeker and the sought. They estimate not Allah His rightful estimate.** Lo! Allah is strong and able to Carry out His Will. Allah **chooses messengers from angels** and **from men**. Lo! Allah is He Who hears and sees [all things]. He knows what is before them and what is behind them; and to Allah go back all questions [for decision]. [Sura No. – 21\22 – [Hajj], Ayat No. – 73 to 76]

Yet they worship things besides [follow projected mechanical globalisation and man-made magnetism], Allah, for which **no authority has been sent down to them,** and of which they have [really] no knowledge. For **those that do wrong there is no helper.** And when **Our Clear Signs** are rehearsed to them, you will notice a denial on the faces of the disbelievers! **They nearly attack with violence those who rehearse Our Signs to them. Say: Shall I tell you of something [far] worse than that? It is the Fire [of Hell]! Allah has promised it for those who disbelieve!** And **hapless journey's end!** [Sura No. – 21\22 – [Hajj], Ayat No. – 71 to 72]

This is so, because Allah is the Reality (Truth). It is He Who gives life to the dead, and it is He Who has power over all things. **And verily the Hour will come.** There can be **no doubt** about it or about [the fact] that Allah will rise up all who are in the graves. Yet there is among men such **a one as disputes about Allah, without Knowledge, without Guidance, and without a Kitaab of Light (Enlightenment) [disdainfully] bending his side, in order to lead [men] astray from the Path of Allah. For him there is disgrace in this life, and on the Day of Judgment. We shall make him taste the Penalty of burning [Fire].** [It will be said]: This is because of the **deeds** which your **hands sent forth**, for verily Allah is **not unjust** to His servants. [Sura No. – 21\22 – [Hajj], Ayat No. – 6 to 10]

There are among men some **who serve Allah**, as it were, on the **verge**. If good befalls them, they are, therewith, well content. But if a trial comes to them, **they turn on their faces**. They lose both **this world** and the **Hereafter**. That is loss for all to see! **They call on such deities, besides Allah, as can neither hurt nor profit them. That is straying far indeed [from the Way]!** They call on one **whose harm** is nearer than his **benefit**. Indeed an **evil patron** and verily an **evil friend**! [Sura No. – 21\22 – [Hajj], Ayat No. – 11 to 13]

Verily Allah will admit those who believe and do tabligh (work of righteous deeds) to Gardens, beneath which rivers flow. For Allah carries out all that

He plans. **If any think that Allah will not help him in this world and the Hereafter, let him stretch out a rope up to the roof (of his dwelling) and let him hang himself. Then let him see whether his plan will remove that which enrages [him]!** Thus We have **revealed it as Clear Signs (plain revelation)**; and verily Allah guides whom He **will!** [Sura No. – 21\22 – [Hajj], Ayat No. – 14 to 16]

Lo! Those who **ward off** (evil) are among gardens and **water-springs** [And it is said unto them]: Enter you here in **peace** and **security.** And We shall remove from their hearts any lurking **sense of injury.** [They will be] **brothers** [joyfully] **facing each other on thrones** of dignity [Arsh]. There no sense of **fatigue** shall touch them, nor shall they [ever] be asked to leave. **Tell My servants that I am indeed the Ever-Forgiving, Most Merciful. And that My Penalty will be indeed the most grievous Penalty.** [Sura No. – 14\15 - Hur, Ayat No. – 45 to 50]

A DAY – WHEN?
[PAST – PRESENT – FUTURE - I DO NOT KNOW]

Say: **I know not whether the [Punishment] which you are promised is near, or whether my Rab will appoint for it a distant term.** He [alone] knows the Unseen, nor does He make any one acquainted with His Mysteries [no mysticism / no spiritualism / no Sufism for breakable realities] except a messenger whom He has chosen, and then **He makes a guard to march before him and behind him.** That He may know that they have [truly] brought and conveyed the Messages of their Rab. He **surrounds** [all the mysteries] that are with them, and takes account of **every single thing.** **[Sura No. – 71\72 – Nafarum-minal – Jinni, Ayat No. – 25 to 28]**

THE DAY OF RESURRECTION
[REBIRTH OR RENAISSANCE]
Sura No. – 74\75
Sura – Bi-Yawmil-Qiyaamah

1. Nay, I swear by the Day of Resurrection;
2. And nay, I swear by the accusing soul;
3. Does man think that We will not assemble his bones?
4. Yea verily. Yea We are able to restore in perfect order the very tips of his fingers.
5. But man wishes to do wrong [even] in the time in front of him.
6. He questions: When will be this Day of Resurrection?
7. But when the sight is confounded,
8. And the moon is buried in darkness (eclipsed).

9. "Wa jumi-ash-shamsu wal-qamar"
    And the Sovereign Ruler and its follower (moon) are joined together,
10. On that Day man will cry: Where is the refuge / Whither to flee?
11. Alas! No refuge!
12. To your Rab is the recourse that day.
13. That Day man is told the tale of that which he has sent before and left behind.
14. Oh, but man will be evidence (witness) against himself,
15. Even though he tender his excuses.
16. Stir not your tongue here-with to hasten it.
17. Lo! Upon Us (rests) the putting together thereof and the reading thereof.
18. But when We read it, follow you the reading.
19. Then lo! Upon Us (rests) the explanation thereof.
20. Nay, but you love the fleeting life,
21. And neglect the Hereafter.
22. That Day will (some) faces be resplendent [splendid].
23. Looking towards their Rab;
24. And that Day other faces will be sad and dismal (despondent),
25. You will know that some great disaster is about to fall on them.
26. Nay, but when the life comes up to the throat,
27. And there will be a cry: Where is the wizard (magician or evidence sorcerer or killer of faith & belief of mankind) [to save him now]?
28. And he will conclude that it is [the Time] of Parting;
29. And agony is heaped on agony.
30. To your Rab that Day will be the driving.
31. For he neither trusted, nor prayed.
32. But on the contrary, he rejected Truth and turned awa
33. Then went he to his folk with glee
34. Nearer to him and nearer
35. Again, nearer to him and nearer.
36. Does man think that he will be left aimless [without purpose]?
37. Was he not a drop of fluid (sperm) which gushed forth?
38. Then did he become a leech-like clot; then did [Allah] make and fashion [him] in due proportion
39. And made of him a pair, male and female.
40. Is not He able to bring the dead to life?

YAWMIL-JUMU-ATI or THE DAY OF ASSEMBLY

All that is in the **heavens and all that is in the world,** glorify Allah, **the Sovereign Rab (Malikil-Qudduus), the Holy One, the Mighty, the Wise.** It is He Who has sent amongst the Ummiyyi (Unlettered) a messenger [Rasuul] from among themselves, to rehearse to them **His Signs,** and to make them grow, and to share with them the Kitaab and Wisdom **[Aayaatihii wa yuzakkihim wa yu-allimu-humul-Kitaaba wa-Hikmah]**, although they were, **in error manifest;** along with others of them, who have not yet joined them, and He is the Mighty, the Wise. Such is the Bounty of Allah, which He bestows on whom He will, and Allah is the Rab of the highest bounty. [Sura No. 61\62 - Yawmil-Jumu-ati, Ayat No. – 1 to 4]

The **similitude (likeness)** of those who are **entrusted with the Tawraat, but apply it not, is as the similitude of the ass carrying books. Wretched is the similitude / likeness of folk who deny the revelations of Allah.** And Allah guides not people who do wrong. [Sura No. 61\62 - Yawmil-Jumu-ati, Ayat No. - 5]

Say: O you who are **Haaduuu! If you think that you are favoured of Allah apart from (all) mankind, then long for death, if you are truthful.** But never will they express their desire [for Death], because of the [deeds] their hands have sent on before them! And Allah is Aware of evil-doers. Say: **The Death from which you flee will certainly overtake you.** Then you will be sent back to the Knower of things **secret and open,** and He will tell you [the truth of] the things that **you did!** [Sura No. 61\62 - Yawmil-Jumu-ati, Ayat No. – 6 to 8]

O you who believe! When the call is proclaimed to prayer on **Friday** [the Day of Assembly], **hasten earnestly to the Remembrance of Allah,** and **leave your trading. That is better for you if you but know!** And when the Prayer is finished, then may you disperse through the land [earth], and **seek of the Bounty of Allah,** and remember the Praises of Allah without stint that you may prosper. But when they **spy** (secret agent) some merchandise or some amusement, they go away headlong to it, and leave you standing. Say: The [blessing] which Allah has is better than any amusement or merchandise! And Allah is the Best of providers. [Sura No. 61\62 - Yawmil-Jumu-ati, Ayat No. – 9 to 11]

FIRST GATHERING

Whatever is in the heavens and in the world, let it declare the Praises and Glory of Allah: for He is the Exalted in Might, the Wise. It is He Who has caused the disbelievers among the people of the Kitaab who disbelieved to go forth from their houses at the **first gathering.** Little did you think that they would get out; while they deemed that their fortresses (strongholds) would protect them from Allah! But Allah reached them from a place whereof they little expected [reckoned not], and cast terror into their hearts, so that **they destroyed their** houses with their own hands and the hands of the believers. So learn a lesson (take warning, then), O you who have eyes! And if Allah had not decreed migration for them, He verily would have punished them in this world [present life] and in the Hereafter they shall [certainly] have the Punishment of the Fire. That is because they were opposed to Allah and His Messenger; and if any one opposes Allah, verily Allah is severe in Punishment.{Sura No. 58\59 Awwalil-Hashr, Ayat No. 1 to 4]

A PAINFUL TORMENT

"The Day when the sky [psychic or planetary barrier] will bring forth a kind of smoke [or mist] plainly visible"

Ayat\Verses – "Laaa ilaaha illa Huwa yuhyu wa yumiit – Rabbukum wa Rabbu aabaaa-ikumul-aawalin." There is no illa Huwa except Him. It is He Who gives life and gives death. The Rabbukum wa Rabbu of you and your forefathers. Nay but they play [the game of immediate inference] in doubt. Then watch you for the Day when the sky [psychic or planetary barrier] will bring forth a kind of smoke [or mist] plainly visible, That will envelop the people. This will be a painful torment. [Then they will say]: Our Rab! Relieve us from the torment! Lo! We are believers. [Sura No. – 43\44 - Bi-Dukhaanim-Mubiin, Ayat No. – 8 to 12]

ILMAL-YAQIIN [Sure Knowledge] AND
AYNAL-YAQIIN [Sure Vision]

The rivalry in worldly increase distracts [diverts] you [from the more serious things], until you come to the graves. Nay, but you will come to know [the reality]. [With emphasis] nay, but you will come to know [the reality]. Nay, you are to know with certainty of mind [with a sure knowledge / ilmal-yaqiin]. You will certainly see Hell-Fire! [With emphasis] you will behold it with sure vision [certainty of sight / aynal-yaqiin]! Then, on that day you will be asked concerning pleasure. [Sura No. -101\102 – Al-Hakumut-taka, Ayat – 1 to 8]

"Truly this is what you used to doubt!"

Verily the tree of Zaqquum will be the food of the sinners, like molten brass. It will boil in their bellies as the seething of boiling water. [And it will be said]: Seize him and drag him into the midst of the Blazing Fire (Hell)! Then pour over his head the penalty of boiling water, (saying): Taste! Lo! You were forsooth the Mighty, The Noble. Truly this is what you used to doubt! [Sura No. – 43\44 - Bi-Dukhaanim-Mubiin, Ayat No. – 43 to 50]

Day of Resurrection [Yawmal-Qiyaamati]
"Sufficient is your soul that day to make out an account against you."

[Neck, Scroll, Record, Soul, Account, Guidance, Bearer, Conditional visit of Allah with Wrath, Township, Definite order, Good tidings, The Word, Utter destruction]

Ayat\Verses – Every man's fate We have fastened on his own neck. On the Day of Resurrection [Yawmal-Qiyaamati] We shall bring out for him a scroll, which he will **see spread open**. [It will be said to him:] Read your [own] record. **Sufficient is your soul that day to make out an account against you.** Who receives guidance, receives it for his own benefit. He who will go astray, so to his own loss. **No bearer of burdens** can bear the burden of another, **nor would We visit with Our Wrath** until We had sent a messenger [**to give warning**]. And when We decide to **destroy** a **township**, We [first] **send a definite order** to those among them who are given the good things of this life and yet transgress; so that the **word** is proved true against them, then [it is] We destroy them **utterly. [Sura No. – 16\17 -** Banii-Israai-iil, Ayat No. – 13 to 16]

HOUR IS SURELY COMING

We created not the heavens, the world, and all between them, **but for just ends**. And the **Hour is surely coming** [when this will be manifest]. So forgive with a gracious forgiveness; for verily it is your Rab who is the Master-Creator, knowing all things. And We have bestowed upon you the **Seven repeated [verses]** and the **Grand Qur'an.** Strain not your eyes toward that which We cause some wedded pairs among them to enjoy; and be not grieved on their account, and lower your wing [in gentleness] for the believers. [Sura No. – 14\15 = Hur, Ayat No. – 85 to 89]

NO PROTECTOR

Say: Allah knows best how long they stayed. With Him is [the knowledge of] the secrets of the **heavens** and the **world**; how clearly He sees, how finely He hears [everything]! They have **no protector** other than Him; **nor does** He share His Command with any person whatsoever. And recite what has been revealed to you of the Kitaab of your Rab. **None can change His Words, and none of you will find as a shelter other than Him.** [Sura No. – 17\18 – Khaf – Ayat No. – 25] [Sura No. – 17\18 – Khaf – Ayat No. – 26 to 27]

THE CALAMITY

The Calamity [disaster or day of noise and shout]! What is calamity [disaster]? And what will convey to you what the calamity [disaster] is? [It is] a Day wherein men will be like moths scattered about, and the mountains [planetary barrier of the psychic period] will be like carded wool. Then, he whose balance [of good deeds] will be [found] heavy, he will live a pleasant life. But he whose balance [of good deeds] will be [found] light, the bereft and hungry one will be his **mother** [or He will have his home in a **bottomless pit**]. And what will explain to you what this [she] is? [**It is**] **a fire blazing fiercely** [**Raging fire at the lower east**]! [Sura No. – 100\101 – Al-Qari-ah, Ayat No. – 1 to 11]

SIMILITUDE / LIKENESS

Do you not consider how your **Rab dealt with** [**the tribe of**] **Aad** [6] with many **columned Iram** [7] the **similitude of which were not produced in the land,** [8] and with [the tribe of] **Samud**, who **clove the rocks in the valley** [9], and **with Firawn, firm of might,** [10] **who [all] were rebellious [to Allah] in these lands [pentagonal earth]** [11] **[pl.?]** and **heaped therein mischief [on mischief]** [12]. Therefore your Rab poured on them the disaster of His punishment [13]; for your Rab is on a **watch-tower** [Where is the watch-tower in the pentagonal earth]. [Sura No. – 88\89 – Wal-fajri, Ayat No.- 6 to 14]

NONE CAN INFLICT AND NONE CAN BIND

Now, as for man, whenever his Rab tries him, giving him honour, and is gracious unto him, he says: My Rab has honoured me [15]. But when He tries him, restricting his sustenance for him, then he says: My Rab has despises me [16]! Nay, nay! But you honour not the orphan [Who is the orphan in resemblance with this sharing – 17 save Maryam or projected sister planet Venus?]! Nor do you encourage one another to feed the poor [18]! And you **devour inheritance** - all with greed [19], and you love wealth with inordinate [excessive] love! [Sura No. – 88\89 – Wal-fajri, Ayat No. – 15 to 20]

Nay! When the **world is grounded to atoms, grinding, grinding** [21], and your Rab shall come with angels, **rank on rank** [22], and **hell is brought near that day; on that day** will man remember, but how will that remembrance profit him [23]? He will say: Ah! Would that I had sent forth [some provisions] for [this] my [Future] Life [24]! For, that Day, His Chastisement will be such as **none [else] can inflict [25], and His bonds will be such as none [other] can bind.** [Sura No. – 88\89 – Wal-fajri, Ayat No. – 21 to 26]

When the **world is shaken [future event]** to **her** [utmost] convulsion [earthquake] and the world throws up **her burdens** [from within], and man cries [distressed]: What is the matter with her? **On that Day she will declare her tidings**, because your Rab will have given her **inspiration.** On that Day men will proceed in companies sorted out, to be shown the deeds that they [had done]. Then anyone **who has done an atom's weight of good, will see it!** And anyone **who has done an atom's weight of evil, will see it.** [Sura No. – 98 \99 - **Izaa zulzilatil-arzu zilzalahaa, Ayat No. – 1 to 8**]

OVERWHELMING EVENT

Has there come to you tidings of the **overwhelming [event]** [1]? Some faces, on that **Day**, will be **humiliated [2]**, labouring [hard], weary [3], scorched by burning fire [4], drinking from a boiling spring [5], no food will there be for them but a bitter thorn-fruit [Dhari] [6], which will neither nourish nor satisfy hunger [Sura No. – 87 \ 88 – Hadiisul-Gaashiya, Ayat No. – 1 to 7]

[Other] faces **that Day** will be joyful [8], pleased with their striving [9], in a high garden [10], where they shall hear no idle speech [11]. Therein will be a bubbling [sparkling] spring [12]. Therein will be **Thrones [of dignity], raised on high [13]**, and glasses [galaxy of stars or West Horizon or Heaven] set at hand [14], and cushions [pentagonal earth] set in rows [ascending order] [15], and silken carpets spread out. [Sura No. – 87 \ 88 – Hadiisul-Gaashiya, Ayat No. –8 to 16]

WARN MANKIND OF THE DAY

Think not that Allah is unaware of what the wicked do. He but gives them a respite against a day when the eyes will fixedly stare in horror. They running forward with necks outstretched, their heads uplifted, their gaze returning not towards them, and their hearts a [gaping] void! So **warn mankind of the Day** when the wrath will come upon them. And those who did wrong will say: Our Rab! Respite us [if only] for a short term. We will answer Your call, and follow the messengers! What! Were you not wont to swear afore time that you should suffer no decline? [Sura No. – 13\14 – Ibrahim, Ayat No. – 42 to 44]

WARRANT

i)   Lo! I bring you a clear warrant.
ii) And lo! I have sought safety with my Rab and your Rab, against you injuring me.
iii) **And if you believe me not, at least keep yourselves away from me.**

And verily We tried, before them, the folk of Firawn when there came to them a noble messenger. Saying: Restore to me the slaves of Allah. Lo! I am a faithful messenger to you. And saying: Be not proud against Allah. Lo! **I bring you a clear warrant**. And lo! I have sought safety with my Rab and your Rab, against you injuring me; **and if you believe me not, at least keep yourselves away from me.** [Sura No. – 43\44 - Bi-Dukhaanim-Mubiin, Ayat No. – 17 to 21]

A FURROW
THE DEMARCATION BETWEEN WEST AND EAST

[But they were aggressive:] then he cried to his Rab: These are indeed a guilty folk. [The reply came:] March forth with My slaves by night; for you are sure to be pursued. And leave the **sea as a furrow** [divided]; for they are a host [destined] to be drowned. [Sura No. – 43\44 - Bi-Dukhaanim-Mubiin, Ayat No. – 22 to 24]

BRING BACK OUR FORE FATHERS

Lo! These, forsooth (no doubt), are saying: There is nothing beyond our first death, and we shall not be raised again. **Bring back our forefathers**, if you speak the truth. Are they better than the people of Tubba and those who were before them? We destroyed them because surely they were guilty. [Sura No. – 43\44 - Bi-Dukhaanim-Mubiin, Ayat No. – 34 to 37]

"Truly this is what you used to doubt!"

Verily the **tree of Zaqquum** will be the food of the sinners, like molten brass. It will boil in their bellies as the seething of boiling water. [And it will be said]: Seize him and drag him into the midst of the **Blazing Fire** (Hell)! Then pour over his head the penalty of boiling water, (saying): Taste! Lo! You were forsooth the Mighty, The Noble. Truly this is what you used to doubt! [Sura No. – 43\44 - Bi-Dukhaanim-Mubiin, Ayat No. – 43 to 50]

EARTHQUAKE

To the Samud people [We sent] Saalih, one of their own brethren: He said: O my people! Serve Allah: You have no other reality but Him. Now **has come** unto you a **Clear Sign** from your Rab! **This she-camel of Allah is a Clear Sign [Manifested Magnetism] unto you**: So leave her to graze in Allah's world, and let her come to **no harm** or you shall be seized with a **grievous punishment**. And remember how He made you inhabitants after the Aad people and gave you station in the world: You build for yourselves **palaces** and **castles** in [open] **plains**, and care out homes in the **mountains**; so bring to remembrance the benefits [you have received] from Allah, and **refrain from evil and mischief in the world**. The **leaders of the arrogant party** among his people said to those who were reckoned **powerless**, those among them who believed: know you indeed that Saalih is a messenger from his Rab? They said: We do indeed believe in the revelation which has been sent through him. The Arrogant party said: For our part, **we reject** what you believe in. **Then they ham-strung (projected globalisation & mechanical magnetism) the she-camel, and disrespectfully challenged the order of their Rab, saying: O Saalih! Bring about your threats, if you are a messenger [of Allah]! So the earthquake took them unawares, and they lay prostrate in their homes in the morning! So Saalih left them, saying: O my people! I did indeed convey to you the message for which I was sent by my Rab. I gave you good counsel, but you love not good counsellors.** [Sura No. – 6\7 - As-haabul- a' raaf, Ayat No. – 72 to 79]

O mankind! Fear your Rab! For the **earthquake** of the Hour [of Punishment] will be **a terrible thing**! On the Day when you behold it, every nursing mother will forget her nursing and every pregnant one will be delivered of her load, and you will see mankind as drunken, yet they will not be drunk but dreadful will be the wrath of Allah. And yet among mankind is he **who disputes concerning Allah, without knowledge**, and **follows every evil one obstinate in rebellion!** For him **it is decreed** that **whoso takes him for friend**, he verily will **mislead him** and **guide him** to the **punishment of the flame**. [Sura No. – 21\22 – [Hajj], Ayat No. – 1 to 4]

MANKIND WILL BE SORTED OUT INTO THREE CLASSES
When the event inevitable comes to pass (befall), there is no denying that it will come to pass (befall); abasing (some), exalting (others). **When the world is shaken with a shock (to its depths), and the hills are ground to powder**, so that they become a **scattered dust**, and you will be **sorted out into three classes. (First) those on the right hand, what of those on the right hand! And (you) those on the left hand, what of those on the left hand! And the foremost in the race, the foremost in the race.** [Sura No. – 55\56 - Waqa-atil-waaqiah, Ayat No. – 1 to 10]

SECTION – VIII
RIGHTEOUSNESS – 8.2
O YOU WHO BELIEVE!

Fasta-'iz billaahi minash-Shaytaanir-Rajim
Bismillaahir-Rahmaanir- Rahim

O you who believe!
O you who believe! Avoid suspicion.
"Suspicion in some cases is a **sin**."
"Spy not on each other behind their backs"
Would any of you like to eat the flesh of his dead brother?

Ayat\Verses – O you who believe! **Avoid suspicion** as much [as possible]; for suspicion in some cases is a **sin**. And **spy not** on each other behind their backs. Would any of you like **to eat the flesh of his dead brother**? Nay, you would abhor it (so abhor the other) and **keep your duty** to Allah. For Allah is Ever Returning, Most Merciful. [Sura No. – 48\49 - Minw-waraaa-il-Hujuraat, Ayat No. – 12]

O you who believe! Verify the Truth.
"And know that the Messenger of Allah is among you." – [Where?]
"The believers are but a single brotherhood."
"Fight you [all] against the one that transgresses"
"And if he returns, then make peace between
them with justly, and act equitably."
"So make peace and reconciliation between your two brothers."

Ayat\Verses – O you who believe! If a **wicked** person (killer of faith & belief of mankind) bring to you any **news**, **verify the truth**, lest you smite (**harm**) some folk in **ignorance**, and afterwards repent of what you have done. And know that the **Messenger of Allah is among you**. Had he in many matters to follow your [**wishes**], you would **certainly** fall into **misfortune** (troubles)? Allah has **endeared** the **Faith** to you, and has made it **beautiful** in your hearts, and He has made **hateful** to you **disbelief, wickedness**, and **rebellion**. Such indeed are they who are **rightly guided**. A Grace and Favour from Allah; and Allah is full of Knowledge and Wisdom. If **two parties** among the believers fall into a **quarrel**, make you **peace** between them. But if **one** of them **transgresses** beyond bounds against the other, then **fight** you [all] against the one that transgresses until he **returns** unto the **ordinance** of Allah. And if he returns, then make peace between them with **justly**, and act **equitably** (fairly); for Allah loves those who are **fair** [and just]. The believers are but a single brotherhood. So make **peace and reconciliation** between your **two brothers**; and fear (keep

652

duty to) Allah, that you may receive Mercy. [Sura No. – 48\49 - Minw-waraaa-il-Hujuraat, Ayat No. – 6 to 10]

[Minw-waraaa-il-Hujuraat], **most of them have no sense."**
Ayat\Verses – O you who believe! **Be not forward yourselves before Allah and His Messenger**; but **fear** Allah, for Allah is He Who hears and knows all things. O you who believe! **Raise not your voices above the voice of the Prophet, nor speak aloud to him in talk, as you may speak aloud to one another, lest your deeds become vain and you perceive not.** Lo! They **who subdue their voices in the presence of the Messenger of Allah**, those are they whose hearts Allah has proven unto righteousness. **For them is forgiveness and immense reward.** Lo! **Those who shout out to you from behind the private apartments, most of them have no sense.** And if they had patience till you came forth unto them, it had been better for them. And Allah is Ever Forgiving, Most Merciful. [Sura No. – 48\49 - Minw-waraaa-il-Hujuraat, Ayat No. – 1 to 5]

"Ill- seeming is a name connoting **wickedness"**
Ayat\Verses – O you who believe! Let **not** some men among you **laugh** at others. It may be that the [latter] are **better than** the [former]. Nor let some **women laugh** at others. It may be that the [latter are **better** than the [former]. Nor **defame** nor be **sarcastic** to each other, nor call each other by [**offensive] nicknames. Ill- seeming** is a name connoting **wickedness**, [to be used of one] after he has believed. And those who **do not desist (turn not in repentance)** are [indeed] doing wrong (evil doers). [Sura No. – 48\49 - Minw-waraaa-il-Hujuraat, Ayat No. – 11]

Ayat\Verses – O you who believe! When you hold secret counsel, **do it not for iniquity and hostility**, and **disobedience** to the Prophet; but **do it for righteousness and self-restraint; and keep your duty towards Allah**, to Whom you will be brought back. [Sura No. – 57\58 – Mujaadalah, Ayat No. – 9]

Ayat\Verses – O you who believe! Come all of you, into submission, and follow not the footsteps of the devil. Lo! He is an open enemy to you. And if you slide back after the clear proofs have come unto you, then know that Allah is Mighty, Wise. Will they wait until Allah comes to them in canopies of clouds, with angels and the question is [thus] settled? But to Allah do all questions go back [for Judgment]. Ask Banii-Isra-iil how many Clear Signs, We have sent them! But if any one, after Allah's favour has come to him, substitutes [something else], Allah is strict in punishment. The life of this world is attractive to those who reject faith, and they laugh at those who believe. But the righteous will be above them on the Day of Judgment; for Allah bestows His abundance without measure on whom He will. [Sura No. 1\2 – Baqarah, Ayat No. – 208 to 212]

Ayat\Verses – **And of mankind there is the type of man, whose conversation about the life of this world [present life] pleases you, and he calls to witness as to that which is in his heart; yet he is the most rigid of opponents**. And when he turns away (from you), **his effort is to spread mischief** through the earth and **to destroy the crops and the cattle;** but **Allah loves not mischief**. And when it is said to him: **Be careful of your duty to Allah, pride (sense of recognition) leads him to sin**. Hell will **settle his account, an evil-resting place**. And there is the type of man **who would sell himself seeking the pleasure of Allah**; and **Allah has compassion**. [Sura No. 1\2 – Baqarah, Ayat No. – 204 to 207]

Ayat\Verses – Men of faith say we hear, and obey. They seek protection from none except their Rab. The Messenger believes in that which has been revealed to him from his Rab, and so do the men of faith. Each one [of them] believes in Allah, His angels, His Kitaabs, and His Messengers. We make no distinction between any of His messengers. And they say: We hear, and we obey. [Grant us] Your forgiveness, our Rab, and unto You is the end of all journeys. On no soul Allah place a burden greater than it can bear. It **gets** every **good** that it **earns,** and it **suffers** every **ill** that it **earns**. [**Pray:**] Our Rab! **Condemn us not** if we **forget** or **miss the mark!** Our Rab! **Lay not on us a burden** like **that which You did lay on those before us**. Our Rab! **Lay not on us a burden greater than we have strength to bear. Blot out our sins, and grant us forgiveness. Have mercy on us**. You are our **Protector. Grant us victory over the unbelieving folk. [Sura No. – 1\2 – Baqarah, Ayat No. - 285 & 286]**

Ayat\Verses – **Guard strictly your prayers** and **of the mid-most prayer, and stand up with devotion to Allah. If you fear, pray on foot, or riding. But when you are again in safety, remember Allah in the manner as He has taught you, which you knew not [before].** [Sura No. 1\2 – Baqarah, Ayat No. – 238 & 239]

Ayat\Verses – **Bethink you of those of old, who went forth from their homes,** though they were **thousands** (in number), **for fear of death**, and Allah said unto them: **Die** and then He brought back them to life; for Allah is full of bounty to mankind, but **most of them are ungrateful. Then fight in the way of Allah** and know that Allah Hears and knows all things. **Who is he who will lend to Allah a beautiful loan, which Allah will double unto his credit and multiply many times? It is Allah Who gives [you] want or plenty, and to Him you will return. [Sura No. – 1\2 – Baqarah, Ayat No. – 243 to 245]**

Ayat\Verses – O you who believe! Spend out of that We have provided you, **before the Day comes** when **no** bargaining / trafficking will avail], **nor** friendship **nor** intercession. **Those who reject Faith they are the wrong-doers. Allah! There is no ilaaha (unbreakable reality) but He, the Living, the Self-subsisting, Eternal.** Neither slumber nor sleep can seize Him. Unto Him belong all things in the heavens and on earth. **Who is there that can intercede with Him except His permission?** He knows that which is before or after or behind them. They encompass nothing of His knowledge except as He will. His Throne extends over the heavens and the earth, and He feels no fatigue in guarding and preserving them for He is the Sublime, the Supreme. **[Sura No. – 1\2 – Baqarah, Ayat No. - 254 & 255]**

Ayat\Verses – And those who **spend** their **substance** in the **cause** of Allah and **follow not** up their gifts with reminders of their generosity or with injury, for them **their reward is with their Rab**. There will be **no fear** on them, **nor** shall they **grieve. Kind words and the covering of faults are better than alms giving followed by injury.** Allah is free of all wants, and He is Most-Forbearing. **[Sura No. – 1\2 – Baqarah, Ayat No. - 262 & 263]**

Ayat\Verses – And the **likeness or similitude** of those who spend their substance, seeking to please Allah and to strengthen their souls, is as a garden, high and fertile; heavy rain falls on it but makes it yield a double increase of harvest, and if it receives not Heavy rain, light moisture suffice it. Allah sees well whatever you do. Would any of you wish that he should have a garden with date-palms [upper west zone] and vines [middle west zone] and streams flowing underneath, and all kinds of fruit for him therein; and while he is stricken with old age, and he has feeble offspring and a fiery whirlwind strikes it and it is consumed by fire [east zone]? Thus **Allah makes clear to you [His] Signs that you may give thought. [Sura No. – 1\2 – Baqarah, Ayat No. - 265 & 266]**

Ayat\Verses – O you who believe! **Cancel not your alms giving by reminders of your generosity or by injury,** like those **who spend their wealth to be seen of men,** but **believe neither in Allah nor in the Last Day.** They are in **parable** like **a hard, barren rock, on which is a little soil on it falls heavy rain, which leaves it a bare stone.** They will be able to do nothing with aught they have earned. **And Allah guides not those who reject faith. [Sura No. – 1\2 – Baqarah, Ayat No. - 264]**

Ayat\Verses – **Those who eat greedily usury (lending of money at interest) cannot rise up except as he arises whom the devil has prostrated by (his)** touch. That is because they say: Trade is like usury. But Allah has permitted trade and forbidden usury. Those who after receiving direction from their Rab, desist, shall be pardoned for the past; their case is for Allah [to judge]. But those who repeat [the offence] are companions of the fire. They will abide therein [for ever]. Allah will deprive usury of all blessing, but will make alms giving fruitful; for Allah **loves not the impious and guilty**. Those who believe, and do deeds of righteousness and **establish** regular prayers and regular alms giving, will have their reward with their Rab. On them shall there be no fear, nor shall they grieve. **[Sura No. – 1\2 – Baqarah, Ayat No. - 275 to 277]**

Ayat\Verses – O you who believe! Observe your duty to Allah, and **give up what remains of your demand for usury, if you are indeed believers.** If you do it not, **take notice of war from Allah** and His Messenger. But **if you turn back, you shall have your capital sums. Deal not unjustly, and you shall not be dealt with unjustly.** If the debtor is in a **difficulty, grant him time** till it is easy for him to repay. But if you **remit** the debt as alms giving, that would be better for you if you did but know. And **guard yourselves** against the Day when you will be brought back to Allah. Then every soul will be paid what it earned, and none will be dealt with unjustly. **[Sura No. – 1\2 – Baqarah, Ayat No. - 278 to 281]**

Ayat\Verses – O you who believe! Spend of the good things which you have [honourably] earned, and of the fruits of the world which We have produced for you, and **do not even aim at getting anything which is bad**, in order that out of it you may give away something, when you yourselves would not receive it except with disdain and know that Allah is Absolute owner of all praise. **The evil one (devil) threatens you with poverty and proposes you to conduct unseemly.** Allah promises you His forgiveness and bounties. Allah is All-Embracing, All-Knowing. **[Sura No. – 1\2 – Baqarah, Ayat No. - 267 & 268]**

Ayat\Verses – **He grants wisdom unto whom He pleases, and he to whom wisdom is granted receives indeed a benefit overflowing; but none will grasp the message except men of understanding.** And whatever alms you spend or vow you vow [in devotion], be sure Allah knows it. But **the wrong-doers have no helpers.** If you disclose your alms giving, it is well, but if you **conceal** them, and make them **reaches those [really] in need, that will be better for you.** It will remove from you some of your **ill-deeds.** And Allah is well informed of **what you do.** The guiding of them is not your duty, but Allah guides whom He will. And whatever good things you spend benefits your own souls, and you shall only do so **in search of** Allah's countenance. Whatever good you give, shall be rendered back to you, and you will not be dealt with unjustly. (Alms are) for the poor who are straitened for the cause of Allah, who cannot travel in the land, the unthinking man accounts them wealthy because of their restraint. **You shall know them by their mark.** And whatsoever good thing you spend, lo! Allah knows it. Those who spend their wealth **by night and by day, in secret and in public,** have **their reward with their Rab.** On them shall be no fear, nor shall they grieve. **[Sura No. – 1\2 – Baqarah, Ayat No. - 269 to 274]**

Ayat\Verses – **O you who believe! If you listen to a faction among the People of the Kitaab, they will make you disbelievers after your belief.** And how **would you deny Faith while unto you are rehearsed the Signs of Allah,** and **among you live the Messenger? Whoever holds firmly to Allah will be shown a way that is straight.** [Sura No. 2\3 – Imran, Ayat No. – 100 & 101]

Ayat\Verses – **O you who believe! Fear Allah as He should be feared,** and **die not except in a state of Islam.** And hold fast, **all together, by the rope of Allah** and be **not divided among yourselves;** and **remember with gratitude Allah's favour on you; for you were enemies and He joined your hearts in love,** so that **by His Grace, you became brethren; and you were on the brink of the pit of Fire, and He saved** you from it. Thus **Allah** makes His Signs **clear to you:** that **you may be guided.** And **there arise out of you a group of people inviting to all that is good** (Tabligh), **enjoining what is right,** and **forbidding what is wrong: Such are they who are successful.** [Sura No. 2\3 – Imran, Ayat No. – 102 to 104]

Ayat\Verses - **Be not like those who are divided amongst themselves** and **fall into disputations after receiving Clear Signs: For them is a dreadful penalty, On the Day** when **some faces will be** [lit up with] **white, and some faces will be [in the gloom of] black. To those whose faces will be black** [will be said]: **Did you reject Faith after accepting it? Taste then the penalty for rejecting Faith.** But those faces which will be [lit with] white, they will be in [the light of] Allah's mercy, they will dwell therein [for ever]. **These are the Signs of Allah. We rehearse them to you in Truth. And Allah wills no injustice to any of His creatures. To Allah belongs all that is in the heavens and on earth.** To Allah all questions go back [for decision]. **[Sura No. 2\3 – Imran, Ayat No. – 105 to 109]**

Ayat\Verses – **You are the best of peoples, evolved for mankind, enjoining what is right, forbidding what is wrong (Tabligh), and believing in Allah. If only the People of the Kitaab had faith, it was best for them.** Among them

657

are **some who have faith**, but **most of them are perverted transgressors. They will not harm you except a trifling annoyance. If they come out to fight with you, they will show you their backs**, and **no help shall they get. Humiliation shall be their portion wherever they are found except under a covenant (of protection) from Allah and from men. They have incurred anger from their Rab**, and **wretchedness is laid upon them. This is because they had rejected the Signs of Allah**, and **slew the prophets wrongfully. That is because they rebelled and transgressed beyond bounds. [Sura No. 2\3 – Imran, Ayat No. – 110 to 112]**

Ayat\Verses – **O you who believe! Take not into your intimacy other than your own folk, who would spare no pains to ruin you, they love to hamper you. Hatred has already appeared from their mouths, what their hearts conceal is far worse. We have made clear to you the Signs, if you have understanding. [Sura No. 2\3 – Imran, Ayat No. – 118]**

Ayat\Verses – Ah! **You are those who love them, but they love you not, though you believe in the whole of the Kitaab. When they meet you**, they say: **We believe. But when they are alone, they bite off the very tips of their fingers at you in their rage.** Say: Perish in you rage; Allah knows well all the secrets of (your) heart. If **a lucky chance befalls you, it grieves them.** But if **some misfortune overtakes you, they rejoice at it**. But if you preserve and keep from evil, their **guile** will **never harm** you; for Allah **Compasses round** about all that they do. **[Sura No. 2\3 – Imran, Ayat No. – 119 & 120]**

Ayat\Verses – And **remember** when you **set forth at day-break** from your **house folk** to **assign** to the **believers** their **positions** for the battle, Allah is The Hearer, The Knower. When **two parties** of you almost fell away, and Allah was the Protecting Friend. In Allah believers do put their trust. Allah had **helped** you at **Badr**, when you were **a contemptible little force**. Then fear Allah. So, observe your duty to Allah in order that you **may be thankful. [Sura No. 2\3 – Imran, Ayat No. – 121 to 123]**

Ayat\Verses – O you who believe! Consume not **usury**, doubled and multiplied; but fear Allah; that you may [really] be successful. **Fear the Fire, which is prepared for those who reject Faith.** And obey Allah and the Messenger; that you may obtain mercy. Be **quick** in the race for **forgiveness** from your Rab, and for a garden whose width is that [of the whole] of the heavens and of the earth, prepared for the **righteous.** Those who spend [freely], whether in prosperity, or in adversity; who restrain anger, and pardon [all] men; for Allah loves those who do good. And those who, having done something to be ashamed of, or wronged their own souls, earnestly bring Allah to mind, and **ask for forgiveness for their sins**, and who can forgive sins except Allah? **And are never obstinate in persisting knowingly in [the wrong] they have done.** For such the reward is forgiveness from their Rab, and Gardens with rivers flowing underneath,- an eternal dwelling: **How excellent a recompense for those who work righteousness! [Sura No. 2\3 – Imran, Ayat No. – 130 to 136]**

Ayat\Verses – **Systems have passed away before you [concerning manifested magnetism in resemblance with revelation]. Travel through the world,** and **observe the nature of the consequence of those who rejected Truth.** This is a **declaration** for **mankind, a guidance** and **instruction** to **those who ward off (evil). So lose not heart, nor fall into despair; for you must gain mastery if you are true in Faith. [Sura No. 2\3 – Imran, Ayat No. – 137 to 139]**

Ayat\Verses – **If a wound has touched** you, **be sure a similar wound has touched the others [equal & opposite or every action must have an equal & opposite reaction].** These are the vicissitudes [of varying fortunes] which We **cause to follow one another for mankind,** to the end that Allah may **know those who believe** and may **choose witness from among you** and Allah **loves not wrong-doers.** And that Allah **may prove those who believe,** and may **bright the disbelievers** (i.e. to deprive of blessing those who resist Faith). **Or deemed you that you would enter Jannat (Heaven) while yet Allah knows not those of you who really strive, nor knows those who are steadfast? You did indeed wish for death before you met it. Now you have seen it with your own eyes. [Sura No. 2\3 – Imran, Ayat No. – 140 to 143]**

Ayat\Verses – **O you who believe! If you obey the disbelievers, they will drive you back on your heels, and you will turn back [from Faith] to your own loss. But, Allah is your protector, and He is the Best of helpers.** Soon shall We cast terror into the hearts of the disbelievers, for that **they joined companions (partners) with Allah, for which He had sent no authority.** Their **abode will be the Fire.** And **evil is the home of the wrong-doers! [Sura No. 2\3 – Imran, Ayat No. – 149 to 151]**

Ayat\Verses – Allah verily **made good His promise** unto you when you **routed them by His permission,** until when **your courage failed you;** but **you disagreed about the order and you disobeyed,** after He had **shown you that for which you long. Some of you desired the world [present life],** and **some of you desired the hereafter.** Therefore, **made you flee from them,** that **He might try you.** Yet now He **has forgiven you.** Allah is **full of grace to those who believe. When you climbed** (the hill) and **paid no heed to anyone,** while the Messenger, in your near, was calling you (to fight). **Therefore,** He **rewarded you grief for (his) grief,** that (He might teach) you **not to sorrow** either for **that which you missed** or **for that which be fell you.** Allah is well informed of what you do. **[Sura No. 2\3 – Imran, Ayat No. – 152 to 153]**

Ayat\Verses – Then **after grief, He sent down security for you.** As slumber did it **overcome a party,** who were **anxious** on their **own account, thought wrongly of Allah the thought of ignorance?** They said: **Have we any part in the cause?** Say: The cause belongs wholly to Allah. **They hide within themselves (a word) which they reveal not unto you,** saying: **Had we had any part in the cause we should not have been slain here.** Say: **Even though you had been in your houses, those appointed to be slain would have gone forth to the places where they were to lie.** This has been in order that **Allah might try what is in your breasts and prove what is in your hearts.** Allah is Aware of what is hidden in the breasts (of men). (IBPLtd.) **[Sura No. 2\3 – Imran, Ayat No. – 154]**

Or

Ayat\Verses – After [the excitement] of the distress, He sent down calm on a band of you overcome with slumber, while another band was stirred to anxiety by their own feelings, Moved by wrong suspicions of Allah-suspicions due to ignorance. They said: "Have we any hand in the affair?" Say you: "Indeed, this affair is wholly Allah's." They hide in their minds what they dare not reveal to thee. They say [to themselves]: "If we had had anything to do with this affair, We should not have been in the slaughter here." Say: "Even if you had remained in your homes, those for whom death was decreed would certainly have gone forth to the place of their death"; but [all this was] that Allah might test what is in your breasts and purge what is in your hearts. For Allah knoweth well the secrets of your hearts. (IFTA) **[Sura No. 2\3 – Imran, Ayat No. – 154]**

Ayat\Verses – Lo! **Those of you who turned back on the day the two hosts met, it was Shaytan alone who caused them to fail, because of some [evil] they had done.** Now Allah has forgiven them, for Allah isEvar Forgiving, Most Forbearing. **O you who believe! Be not like the disbelievers,** who say of their brethren, when they are travelling through the Earth or engaged in fighting: **If they had stayed with us, they would not have died, or been slain** that Allah **may make it anguish in their hearts.** It is Allah who gives Life and Death, and Allah sees well all that you do. **And if you are slain, or die, in the way of Allah, forgiveness and mercy from Allah are far better than all they could amass. What though you be slain or die, when unto Allah you are gathered?** It was by mercy of Allah that you were lenient with of them, for if you had been stern and fierce of heart they would have dispersed from round about you. So **pardon them and consult with them upon the conduct of affairs**. And when you are resolved, then **put your trust in Allah**. Lo! Allah loves those who put their trust (in Him). **If Allah helps you, none can overcome you. If He withdraws His help from you, who is there, after that, who can help you? In Allah, then, let the believers put their trust.** It is not for any Prophet to deceive (mankind). **Who so deceives will bring his deceit with him** on the Day of Judgment. Then every soul will receive its due, whatever it earned, and **none will be dealt with unjustly**. Is the man who **follows** the good pleasure of Allah like the man **who draws on himself the wrath of Allah**, and **whose abode is in Hell, a miserable journey's end**? They are in **varying gardens** in the sight of Allah, and Allah sees well all that they do. **Allah verily has shown grace to the believers by sending unto them a messenger from among themselves, rehearsing unto them the Signs of Allah, sanctifying them, and instructing them in Kitaab and Wisdom, while, before that, they had been in manifest error. What! When a single disaster smites you**, although you smote with one **twice** as great that you say: **How is this**? Say [to them]: It is from you. For Allah is Able to do all things. **That which befell you**, on the day when the **two armies met**, was by permission of Allah; that He might know the **true believers**. [Sura No. 2\3 – Imran, Ayat No. – 155 to 166]

Ayat\Verses – **And that He might know the Hypocrites unto whom it was said: Come, fight in the way of Allah, or defend yourselves. They answered: Had we known how to fight, we should certainly have followed you. They were that day nearer to unbelief than to Faith, saying with their lips what was not in their hearts but Allah has full knowledge of all they conceal..** [Sura No. 2\3 – Imran, Ayat No. – 167]

Ayat\Verses – Those who, while they **sat at home** said of their brethren (who were fighting for the cause of Allah): **If they had been guided by us they would not have been slain**. Say: **Then avert death from yourselves if you are truthful**. (IBPLtd.). [Sura No. 2\3 – Imran, Ayat No. – 168]

Or

Ayat\Verses – [They are] the ones that say, [of their brethren slain], while they themselves sit [at ease]: "If only they had listened to us they would not have been slain." Say: "Avert death from your own selves, if ye speak the truth." (IFTA). [Sura No. 2\3 – Imran, Ayat No. – 168]

Ayat\Verses – Think not of those **who are slain in Allah's way as dead**. But **they are living, finding their sustenance in the presence of their Rab. Jubilant** (are they) because of that which Allah has bestowed upon them of

His bounty, rejoicing **for the sake of those who have not joined them** but are left behind; that **there shall no fear come upon them nor they grieve**. They rejoice in the Grace and the bounty from Allah, and in the fact that Allah wastes not the wage of the believers. **[Sura No. 2\3 – Imran, Ayat No. – 169 to 171]**

Ayat\Verses – Of those who **heard** the **call** of **Allah** and His **Messenger**, even after **harm** befell them, for such of them **as do right** and **refrain from wrong (Tabligh)**, there is a **great reward**. Those unto whom men said: Lo! The people have gathered against you, therefore fear them. But (the threat of danger) increased the faith of them and **they cried: Allah is sufficient for us! Most Excellent is He in Whom we trust!** So they returned with grace and bounty from Allah, and no harm touched them; for they followed the good pleasure of Allah, and Allah is of Infinite Bounty. **[Sura No. 2\3 – Imran, Ayat No. – 172 to 174]**

Ayat\Verses – **It is only the devil** who would **make (men) fear his votaries. Be you not afraid of them, but fear Me, if you have Faith. Let their conducts do not grieve you, who run easily to disbelief;** for not the least harm they will do to Allah. Allah's plan is that He will give them **no portion in the Hereafter,** but **a severe punishment. Those who purchase unbelief at the price of faith,** not the least harm they will do to Allah, but they will have **a grievous punishment. Let not the disbelievers think that our respite to them is good for themselves.** We grant them **respite** that they may **grow in sinfulness. But** they will have **a shameful punishment. [Sura No. 2\3 – Imran, Ayat No. – 175 to 178]**

Ayat\Verses – **It is not (the purpose) of Allah to leave you in your present state till He shall separate what is evil from what is good.** And **it is not** the purpose of Allah **to let you know the unseen.** But Allah chooses of his messengers whom He will (to receive knowledge thereof), So believe in Allah and His messengers. And if you **believe** and do **right** (Tabligh), **you have a reward without measure. [Sura No. 2\3 – Imran, Ayat No. – 179]**

Ayat\Verses – And **let not those** who **covetously withhold** of the gifts which Allah has given them of His Grace. Think that it is good for them. **Nay, it will be the worse for them:** soon shall **the things which they covetously withheld be tied to their necks Like a twisted collar, on the Day of Judgment.** To Allah belongs the heritage of the heavens and the earth; and Allah is well-acquainted with all that you do. Verily Allah **heard the ridicule of those** who say: Truly, Allah is **indigent** and we are **rich!** We shall **certainly record their word with their slaying the prophets wrongfully,** and **We shall say: Taste** you the penalty of the burning fire! This is because of the [unrighteous deeds] which your hands sent on before (you to the Judgment), for Allah never harms those **who serve Him** (do Tabligh). [Sura No. 2\3 – Imran, Ayat No. – 180 to 182]

Ayat\Verses – (The same are) those who said: Allah took our promise not to believe in any messenger unless He showed us a sacrifice consumed by Fire. Say: There came to you messengers before me, with clear Signs and even with what you ask for: Why then did you slay them, if you are truthful? [Sura No. 2\3 – Imran, Ayat No. – 183]

Ayat\Verses – Then if they reject you, so were rejected messengers before you, who came with **Clear Signs**, and with the prophecies and with the **Kitaab** giving light. [Sura No. 2\3 – Imran, Ayat No. – 184]

Ayat\Verses – **Every soul shall have a taste of death**. And only on the Day of Judgment shall you be paid your full recompense. Only he who is **saved far from the Fire** and admitted to the **Garden** will have attained the object [of Life]. For the life of this world is but goods and chattels of deception. **You will certainly be tried and tested in your possessions and in your personal selves; and you will certainly hear much that will grieve you, from those who received the Kitaab before you and from those who worship many gods (idolaters). But** if you **persevere patiently**, and **ward off evil**, then **that will be a determining factor in all affairs.** [Sura No. 2\3 – Imran, Ayat No. – 185 & 186]

Ayat\Verses – And **remember** Allah **took a covenant (promise) from the People of the uutul-Kitaab, to make it known and clear to mankind, and not to hide it; but they threw it away behind their backs, and purchased with it some miserable gain! And evil was the bargain they made! Think not that those who exult in what they have brought about, and love to be praised for what they have not done, think escape the penalty. For them is a penalty Grievous indeed. To Allah belong the Sovereignty of the heavens and the world**; and Allah is Able to do all things. [Sura No. 2\3 – Imran, Ayat No. – 187 to 189]

Ayat\Verses – **Behold! In the creation of the heavens and the world, and the alternation of night and day, there are indeed Signs for men of understanding.** [Sura No. 2\3 – Imran, Ayat No. – 190]

Ayat\Verses – Such as **remember** of Allah, **standing, sitting**, and **lying down** on their sides, and **contemplate the [give thought of] creation in the heavens and** in the world, [and say]: Our Rab! **You have created not this in vain!** Glory is to You! **Save from the penalty of the fire.** Our Rab! **Whom You cause to enter the Fire**: Him indeed You have confounded. **For evil-doers there will be no helpers. Our Rab! Lo! We have heard a crier calling unto Faith: Believe you in your Faith. Believe in your Rab!** So we have believed. Our Rab! **Forgive us our sins, remit from us our evil deeds, and make us die the death of the righteous. Our Rab! And grant us that which You have promised to us by Your Messengers. Confound us not upon the Day of Resurrection [Yawmal-Qiyaamah]. Lo! You never break Your Promise. And their Rab has heard them (and says): Lo! I suffer not the work of any worker, male or female, to be lost. You go ahead one from another. So, those who fled and were driven forth from their homes and suffered damage for My cause (Tabligh), and fought and were slain, verily I shall remit their evil deeds from them and verily I shall bring them in to Gardens underneath which river flows, a reward from Allah. And with Allah is the fairest of rewards.** [Sura No. 2\3 – Imran, Ayat No. – 191 to 195]

Ayat\Verses – **Let not the boasting of unbelievers in the land [earth] who deceive (make a person to believe false) you. Little is it for enjoyment. Their ultimate abode is hell, where an evil bed [to lie on]!** On the other hand, for those **who keep their duty to their Rab**, are Gardens, with rivers flowing beneath; therein are they to dwell [for ever], **a gift of welcome from Allah**; and that which is in the presence of Allah is **the best [bliss] for the righteous.** And there are, certainly, **among the people of the Kitaab** those **who believe in Allah, in the revelation to you, and in the revelation to them**, bowing in humility to Allah: **They will not sell the Signs of Allah for a miserable gain! For them is a reward with their Rab, and Allah is swift in account. O you who believe! Persevere in patience and constancy; vie in such perseverance; strengthen each other (Tabligh); and observe your duty to Allah in order that you may succeed.** [Sura No. 2\3 – Imran, Ayat No. – 196 to 200]

Ayat\Verses – O you who believe! Take not your father and your brothers for **protectors** if they take pleasure in disbelief rather than Faith. **If any of you do so, they are wrong doers.** Say: If it be that your fathers, your sons, your brothers, your mates, or your kindred; the wealth that you have gained; the commerce in which you fear a decline: or the dwellings in which you delight are dearer to you than Allah, or His Messenger, or the striving in His way, then wait until Allah brings about His decision, and **Allah guides not the wrong doing folk.** [Sura No. – 8\9 – Tauba, Ayat No. – 23 & 24]

Ayat\Verses – Assuredly Allah did help you in many battle-fields and on the day of Hunaynin when you excited with great numbers, but they availed you naught, and the land, for all that it is wide, did constrain you, and you turned back in retreat. But Allah did pour His calm on the Messenger and on the Believers, and sent down forces which you saw not. He punished the Unbelievers. **Such is the reward of disbelievers.** Again, after this, Allah will turn [in mercy] to whom He will, for Allah is Oft-forgiving, Most Merciful. O you who believe! **Truly the Pagans idolaters are unclean.** So, let them not **come near the Inviolable Place of Worship after this year.** And if you fear poverty, soon will Allah enrich you, if He wills, out of His bounty, for Allah is All-knowing, All- wise. **Fight against such of those who have given the Kitab as believe not in Allah nor the Last Day, nor hold that forbidden which hath been forbidden by Allah and His Messenger, nor acknowledge the religion of Truth, [even if they are] of the People of the Kitab, until they pay the tribute readily, being brought low. [Sura No. – 8\9 – Tauba, Ayat No. – 25 to 29]**

Ayat\Verses – O you who believe! There are indeed many among the priests and anchorites [Imamgiris], **who in Falsehood demolish the wealth of men and obstruct [them] from the way of Allah.** And there are those who stored gold and silver and spend it not in the way of Allah announce unto them a most grievous penalty. **On the Day when heat will be produced out of that [wealth] in the fire of Hell,** and **with it will be branded their foreheads,** their **borders,** and their **backs borders,** and their **backs.** This is the [treasure] which you stored for yourselves. Taste you, then, the [treasures] you stored. **The number of months in the sight of Allah is twelve. So ordained by Him the day He created the heavens and the earth. Four of them are blessed that is the straight usage.** So **wrong not yourselves therein,** and **wage the Pagans / idolaters all together as they waging on all of you. But know that Allah is with those who restrain themselves. Verily the transposing [of a prohibited month] is an addition to unbelief. The unbelievers are led to wrong thereby, for they make it lawful one year, and forbidden another year, in order to adjust the number of months forbidden by Allah and make such forbidden ones lawful. The evil of their course seems pleasing to them. But Allah guideth not those who reject Faith. [Sura No. – 8\9 – Tauba, Ayat No. – 34 to 37]**

Ayat\Verses – O you who believe! **What is the matter with you, that, when you are asked to go forth in the way of Allah, you cling heavily to the world? Do you prefer the life of this world [present life] to the Hereafter? But little is the comfort of this life, as compared with the Hereafter. Unless you go forth, He will punish you with a grievous penalty, and put others in your place; but Him you would not harm in the least. For Allah hath power** over all things. If you help not [him], [it is no matter]; for Allah did indeed help him, when the unbelievers drove him out. He had no more than one companion. **They two were in the cave, and she said to her companion, have no fear, for Allah is with us. Then Allah sent down His peace upon her, and strengthened her with forces which you saw not, and made the words of those who disbelieved neither most. But the word of Allah is exalted to the heights, for Allah is Exalted in might, Wise.** Go you forth, [whether equipped] lightly or heavily, and strive and struggle, with your goods and your persons, in the way of Allah. That is best for you, if you [but] knew. If there had been immediate gain [in sight], and the journey easy, they would [all] without doubt have followed you, but the distance was long, [and weighed] on them. They would indeed swear by Allah, If we only could, we should certainly have come out with you. They would destroy their own souls; for Allah knows that they certainly are liars. **[Sura No. – 8\9 – Tauba, Ayat No. – 38 to 42]**

Ayat\Verses – Allah forgives you! **Where for did you grant them leave ere those who told the truth were manifest to you and you did know the liars?** Those who believe in Allah and the Last Day ask thee for no exemption from fighting with their goods and persons. And Allah knows well those who do their duty. Only those ask thee for exemption who believes not in Allah and the Last Day, and whose hearts are in doubt, so that they are tossed in their doubts to and fro. If they had intended to come out, they would certainly have made some preparation there for; but Allah was averse to their being sent forth. So He made them lag behind, and they were told: Sit you among those who sit [inactive]. If they had come out with you, they would not have added to your [strength] but only [made for] disorder, hurrying to and fro in your midst and sowing sedition among you, and there would have been some among you who would have listened to them. But Allah knows well those who do wrong. **Indeed they had plotted sedition** before, and upset matters for you, until the Truth arrived, and the Decree of Allah became manifest much to their disgust. **[Sura No. – 8\9 – Tauba, Ayat No. – 43 to 48]**

Ayat\Verses – Among them is [many] a man who says: Grant me exemption and draw me not into trial. **Have they not fallen into trial already?** And indeed Hell surrounds the Unbelievers [on all sides]. If good befalls you, it grieves them; but if a misfortune befalls you, they say: We took indeed our precautions beforehand, and they turn away rejoicing. Say: **Nothing will happen to us except what Allah has decreed for us: He is our protector": and on Allah let the Believers put their trust.** Say: Can you expect for us [any fate] other than one of two glorious things- [Martyrdom or victory]? But we can expect for you either that Allah will send his punishment from Himself, or by our hands. So wait [expectant]; we too will wait with you. Say: Spend [for the way of Allah] willingly or unwillingly; not from you will it be accepted; for you are indeed a people rebellious and wicked. **[Sura No. – 8\9 – Tauba, Ayat No. – 49 to 53]**

Ayat\Verses – The reasons why their contributions are not accepted are: that **they reject Allah and His Messenger; that they come to prayer without earnestness; and that they offer contributions unwillingly.** Let neither their wealth nor their [following in] sons dazzle you. In reality Allah's plan is to punish them with these things in this life, and that their souls may perish in their [very] denial of Allah. They swear by Allah that they are indeed of you; but they are not of you: yet they are afraid [to appear in their true colours]. If they could find a place to flee to, or caves, or a place of concealment, they would **turn straightaway** thereto, with an obstinate rush. **[Sura No. – 8\9 – Tauba, Ayat No. – 54 to 57]**

Ayat\Verses – And among them are men who insult you in the matter of [the distribution of] the alms. If they are given part thereof, they are pleased, but if not, **they are indignant.** If only they had been content with what Allah and His Messenger gave them, and had said: Sufficient unto us is Allah! Allah and His Messenger will soon give us of His bounty: to Allah do we turn our hopes, [that would have been the right course]. Alms are for the poor and the needy, and those employed to administer the [funds]; for those whose hearts have been [recently] reconciled [to Truth]; for those in bondage and in debt; in the way of Allah; and for the wayfarer: [thus is it] ordained by Allah, and Allah is full of knowledge and wisdom. **[Sura No. – 8\9 – Tauba, Ayat No. – 58 to 60]**

Ayat\Verses – Among them are men who assault [physically attack] the Prophet and say: He is only a hearer. Say: He listens to what is best for you, he believes in Allah, has faith in the believers, and is a Mercy to those of you who believe. But **those who assault the Messenger will have a grievous penalty.** To you they swear by Allah in order to please you. But it is more fitting that they should please Allah and His Messenger, if they are believers. Know they not that for those who oppose Allah and His Messenger, is the Fire of Hell wherein they shall dwell? That is the supreme disgrace. **The Hypocrites are afraid lest a Sura should be sent down about them, showing them what is [really passing] in their hearts.** Say: Mock you! But verily **Allah will bring to light** all that you fear [should be revealed]. If you question them, they declare [with emphasis]: We were only talking idly and in play. Say: **Was it at Allah, and His Signs, and His Messenger that you were mocking?** Make you no excuses: You have rejected Faith after you had accepted it. If We pardon some of you, We will punish others amongst you, for that they are in sin. **[Sura No. – 8\9 – Tauba, Ayat No. – 61 to 66]**

Ayat\Verses – The Believers, men and women, are **protectors one of another**. They enjoin **what is right**, and **forbid what is wrong**. They observe regular prayers, practice regular charity, and obey Allah and His Messenger. On them Allah will pour His mercy, for Allah is Exalted in power, Wise. Allah hath promised to Believers, men and women, Gardens under which rivers flow, to dwell therein, and beautiful mansions in Gardens of everlasting bliss. But the greatest bliss is the good pleasure of Allah that is the supreme felicity. **[Sura No. – 8\9 – Tauba, Ayat No. – 71 & 72]**

Ayat\Verses – **The forerunner or the first of those who left [their homes] and of those who gave them aid (of Muhajirin and Ansar), and [also] those who follow them in [all] good deeds, well-pleased is Allah with them, as are they with Him. For them He hath prepared Gardens under which rivers flow, to dwell therein forever. That is the Supreme Triumph.** (Tabligh). And among those around you of the **wandering Arabs**, there are **hypocrites (frauds),** and among the **town's [American] people** of Al-Madinah (there are some) who help in **hypocrisy** [double standard] whom you know not. We know them, and We shall punish them **twice**, then they will be relegated to a painful doom. **[Sura No. – 8\9 – Tauba, Ayat No. – 100 & 101]**

Ayat\Verses – **Lo! Allah has purchased from the believers their lives and their wealth because the Gardens will be theirs. They shall fight in the way of Allah and shall slay and be slain. It is a promise which is binding on Him in the Tawraat and the Injiil, and Quran. Who fulfilled the promise better than Allah? Rejoice then in your bargain that you have made for that is the Supreme Triumph. Triumphant are those who turn repentant (to Allah), those who serve (Him), those who praise (Him), those who fast, those who bow down, those who fall prostrate (in ownership), those who enjoin the right and who forbid wrong and those who keep the limits (ordained) of Allah and give glad tidings to believers.** [Sura No. – 8\9 – Tauba, Ayat No. – 111 & 112]

Ayat\Verses – Unto **Allah belongs the dominion [Sovereignty] of the heavens and the world.** He gives life and He takes it. Except for Him you have no protector, nor helper. **Allah turned with favour to the Prophet, the Muhajirs, and the Ansars, who followed him in a time of distress.** After that the hearts of a part of them had nearly swerved [from duty]; but He turned to them [also]: for He is unto them Most Kind, Most Merciful. And to **the three [ground stair, middle stair, and top stair] also** (did He turn in mercy) **who were left behind [Straight Middle East Region], when the world vast as it is, was straitened for them still they be thought them that there is no refuge from**

Allah except toward Him; then turned Him unto them in mercy that they (too) might turn (repentant unto Him). Lo! Allah! He is the Relenting, the Merciful. [Sura No. – 8\9 – Tauba, Ayat No. – 116 to 118]

Ayat\Verses – **O you who believe! Fear Allah and be with those who are true [in word and deed].** It is for the townsfolk [Americans] of Al-Madinah and for those around them of the wandering Arabs to stay behind the Prophet of Allah and prefer their lives to his life. That is because nether thirst nor toil nor hunger afflicted them in the way of Allah, nor step they any step that angers the **disbelievers**, nor gain they from the enemy a profit, but a good deed is recorded for them there for. Lo! Allah loses not the wages of the good. Nor spend they any spending, small or great, nor do they cross a valley, but it is recorded for them that Allah may repay them the best of what they used to do. **And the believers should not all go out to fight. Of every troop of them, a party only should go forth, that they (who are left behind) may gain sound knowledge in religion, and that they return to them, so that they may beware.** [Sura No. – 8\9 – Tauba, Ayat No. – 119 to 122]

Ayat\Verses – Yaaa-'ayyu-hallaziina 'aa-manuu laa tataa-khizul-**Yahuu-da** wan-**Nasaaraa 'awli-yaaa'.** Ba-zuhum **awli-yaaa-'u** ba'-z. Wa many-yata-wallahum-min-kum fa-'innahuu minhuum. 'In- nallaaha laa yahdil-qaw-maz-zaalimiin. [Trans.] O you who believe! Take not the Yahuud and the Nasaara for your **'awli-yaaa** [friends and protectors]. They are but friends and protectors to each other [Ba-zuhum **awli-yaaa-'u** ba'-z]. And **he** amongst you takes them for friends is [one] of them. Verily Allah guides not wrong-doing folk. [Sura No. – 4\5 – Maaaidah, Ayat\Verses – 51] Those in whose hearts are a disease—you see how eagerly they run about amongst them, saying: We do fear lest a **change** of fortune bring us disaster. Ah! Perhaps Allah will give (thee) victory, or a decision according to His Will. Then will they **repent** of the thoughts, which they **secretly harboured in their hearts**. And those who believe will say: **Are these the men who swore their strongest oaths by Allah that they were with you? All that they do will be in vain**, and they will **fall** into (nothing but) ruin. [Sura No. – 4\5 – Maaaidah, Ayat\Verses – 51 to 53]

Ayat\Verses – O you who believe! Allah makes a **trial** of you in a little matter of game well within reach of your **hands and your lances**, that He may test who fear Him unseen: Any who transgress thereafter, will have a grievous penalty. O you who believe! Kill **no wild game** while you are on the pilgrimage. Who so of you kills it of set purpose he shall pay its forfeit in the equivalent of that which has killed of domestic animals, **the judge to be two men among you known for justice**, (the forfeit) to be brought as an offering to the **Kabah**, or for compensation, he shall **food** poor persons, or the equivalent there of in **fasting**, that he may **taste** the **evil consequences of his deed**. Allah forgives whatever (for the kind) but who so ever **declines**, Allah will take **retribution** from him. Allah is Mighty, Able to Requite (the wrong). To **hunt and to eat the fish of the sea is made lawful** for you, a **provision** for you and for **seafarers**; but to hunt on land is **forbidden** you as long as you are on the **Pilgrimage**. Be mindful of your duty to Allah, unto Whom you will be **gathered**. [Sura No. – 4\5 – Maaaidah, Ayat\Verses – 94 to 96]

Ayat\Verses – O you who believe! Bow down, prostrate yourselves, and adore your Rab; and do good that you may prosper. And strive for Allah with the endeavour which is His right. He has chosen you, and has imposed no difficulties on you in religion, **the faith of your father Ibrahim. He has named you Muslims** of old time and in this (Kitaab) that the messenger may be a witness for mankind. So establish regular prayer, pay the poor-due, and hold fast of Allah. He is your protecting friend, a blessed Patron and a blessed Helper. [Sura No. – 21\22 – [Hajj], Ayat No. – 77 to 78]

FULFIL THE COVENANT OF ALLAH

Lo! Allah **commands justice**, the **doing of good**, and **liberality** to **kith** and **kin**, and He **forbids** all **shameful deeds**, and **injustice** and **rebellion**. He **instructs** you, that you may **receive admonition**. Fulfil the **covenant** of Allah when you have **promised**, and **break not** your **promise after you have confirmed them**. Indeed you have made Allah your **surety**, for Allah knows all that you do. [Sura No. – 15\16 – Nahl, Ayat\Verses. 90 & 91]

And **give good tidings** to **those who believe** and work **righteousness**, that **their portion is Gardens**, beneath which **rivers** flow. **Every time they are fed with fruits** there from. They say: **This is what was given us afore time; and it is given to them in resemblance**. Thereof, there are **pure companions**, and they **abide** therein **[for ever]**. Allah scorns not to coin the **similitude of things**, lowest as well as highest. Those who believe know that it is truth (tautology) from their Rab. But **those who reject faith** say: **What means Allah by this similitude? By it He causes many to go astray**, and **many He leads into the right path**; but He causes not to stray, except **those who forsake** [the path]. Those **who break Allah's Covenant (promise) after it is ratified**, and **who sunder (Literary separate) what Allah has ordered to be joined**, and **do mischief on earth: These cause loss [only] to themselves** [Sura No.1\2 Baqarah, Ayat No. 25 to 27]

SECTION – VIII
RIGHTEOUSNESS – 8.3
BEAUTIFUL FELLOWSHIP

Fasta-'iz billaahi minash-Shaytaanir-Rajim
Bismillaahir-Rahmaanir- Rahim

Alif-Lam-Ra. A Kitaab which We have revealed unto you, in order that you might lead mankind out of the **depths of darkness into light** by the permission of their Rab to the Way of [Him] the Exalted in power, worthy of all praise [Arsh]! [Sura No. – 13\14 – Ibrahim, Ayat No. - 1]

Unto Allah belong all things in the **heavens** and in the world. But alas **for the unbelievers for a terrible penalty [their Unfaith will bring them]! Those who love the life of this world [present life] more than the Hereafter** and debar, [men] **from the Path of Allah** and **seek therein something crooked or global or circular, they are astray by a long distance**. And We never sent a messenger except [to teach] in the **language** of his [own] people, in order to make [things] clear to them. Then Allah leaves whom He pleases astray and guides whom He pleases; and He is Exalted in power, full of Wisdom. [Sura No. – 13\14 – Ibrahim, Ayat No. – 2 to 4]

We sent **Muusa with Our signs** [and the command]. **Bring out** your people from the **depths of darkness into light**, and teach them to remember the Days of Allah. Verily in this there are **Signs** for such as are **firmly patient and constant, grateful and appreciative.** And how **Muusa** said unto his people: Remember! Call to mind the favour of Allah to you **when He delivered you from the people of Firawn.** They set you hard tasks and punishments, slaughtered your sons, and let your women-folk live. Therein was a tremendous trial from your Rab. And remember! **Your Rab caused to be declared [publicly].** If you are grateful, I will add more [favours] unto you. But if you show ingratitude, truly My punishment is terrible indeed. And Muusa said: If you show ingratitude, you and all on earth together, yet i Allah is free of all wants, worthy of all praise. [Sura No. – 13\14 – Ibrahim, Ayat No. – 5 to 8]

Has not the **history** of those before you reached you, the folk of Nuuh and the tribes of Aad ans Samuud and those after them? Their messengers came to them with clear proofs but they thrust their hands unto their mouths, and said: Lo! **We disbelieve in that wherewith you have been sent**, and lo! **We are really in doubt concerning that to which you invite us. Their messenger said: Is there a doubt about Allah, The Creator of the heavens and the** earth? It is He Who invites you, in order that He may forgive you your sins and give you respite for a term appointed! They said: Ah! You are no more than human, like us! You wish to turn us away from the [gods] our fathers used to worship. Then bring us some clear authority. Their messengers said to them: True, we are mortals like you. But Allah grants His grace to such of his servants as He pleases. It is not for us to bring you an authority except as Allah permits. And on Allah let all men of faith put their trust [Sura No. – 13\14 – Ibrahim, Ayat No. – 9 & 11] How should we not put our trust in Allah when He has shown us our ways? We shall certainly bear with patience all the hurt you may cause us. In Allah let the trusting put their trust. And the unbelievers said to their messengers: Be sure we shall drive you out of our land, or you shall return to our religion. But their Rab inspired [this Message] to them: Verily We shall cause the wrong-doers to perish!

And verily We shall cause you to abide in the land [earth], and succeed them. This is for him who fears My Majesty and fears My Warning. And they sought victory and decision [there and then], and frustration was the lot of every powerful obstinate transgressor. In front of such a one is Hell, and he is given, for drink, boiling fetid water. In gulps will he sip it, but never will he be near swallowing it down his throat. Death will come to him from every quarter, yet he will not die; and in front of him will be a chastisement unrelenting. The parable of those who reject their Rab is that their works are as ashes, on which the wind blows furiously on a tempestuous day. They have no control of aught that they have earned. That is the extreme failure. [Sura No. – 13\14 – Ibrahim, Ayat No. – 12 & 18]

Have you not seen that Allah created the heavens and the world in Truth? If He so will, He can remove you and put [in your place] a new creation? And that is no great matter for Allah. They will all be marshalled before Allah together. Then the weak will say to those who were arrogant: For us, we but followed you; can you then avail us to all against the wrath of Allah? They will reply: If we had received the Guidance of Allah, we should have given it to you. To us it makes no difference [now] whether we rage, or bear [these torments] with patience. For us there is no way of escape. And Shaytan will say when the matter is decided: It was Allah Who gave you a promise of Truth: I too promised, but I failed in my promise to you. I had no authority over you except to call you but you listened to me. Then reproach not me, but reproach your own souls. I cannot listen to your cries, nor can you listen to mine. I reject your former act in associating me with Allah. For wrong-doers there must be a grievous penalty. [Sura No. – 13\14 – Ibrahim, Ayat No. – 19 to 22]

But those who believe and work righteousness will be admitted to gardens beneath which rivers flow to dwell therein for aye with the permission of their Rab. **Their greeting therein will be: Peace!** Do you not see how Allah sets forth a **parable** (similitude)? **A goodly word like a goodly tree, whose root is firmly fixed**, and **its branches [reach] to the heavens** of its Rab. So Allah sets forth **parables for men**, in order that they may receive admonition. It brings forth its fruit at all times, by the permission of its Rab. So **Allah sets forth parables** for men, in order that they may receive admonition. And the **parable** of an evil word is that of an **evil tree**. It is torn up by the root from the surface of the earth. It has no stability. Allah will establish in strength those who believe, with the word that stands firm, in this world and in the Hereafter. But Allah will leave, to go astray, those who do wrong. Allah does what He will. **Do not turn your vision to those who have exchanged the favour of Allah for thanklessness and are leading their people down to the abode of loss** into Hell? They will burn therein, an evil place to stay in! **And they set up rivals to Allah, to mislead [men] from the Path! Say: Enjoy [your brief power]! But verily you are making straightway for Hell!** [Sura No. – 13\14 – Ibrahim, Ayat No. – 23 to 30]

Tell to my servants who have believed, that they may establish regular prayers, and spend [in charity] out of the sustenance we have given them, secretly and openly, before the coming of a Day in which there will be neither mutual bargaining nor befriending. It is Allah Who hath created the **heavens** and the **world** and sends down **rain** from the skies, and with it brings out fruits wherewith to feed you. It is He **Who has made the ships subject to you**, that **they may sail through the sea by His command**; and **the rivers [also] has He made subject to you. And He has made subject to you lakumush-shamsa wal-qamara**, both **diligently pursuing their courses**; and **the night and the day has he [also] made subject to you. And He gives you of all that you ask for. But if you count the favours of Allah, never will you be able to number them. Verily, man is given up to injustice and ingratitude.** [Sura No. – 13\14 – Ibrahim, Ayat No. – 31 to 34]

And when **Ibrahim** said: O my Rab! Make **this city [Arabian States where there is the sole manifested leader of four cardinal directions for mankind] one of peace and security**: and **preserve [protect] me and my sons from worshipping idols.** My Rab! Lo! **They have indeed led astray many among mankind.** He then who follows my [ways] is of me. And he who disobeys me, still You are indeed Ever- Forgiving, Most Merciful. Our Rab! Lo! I have settled some of my succeeding generations in an uncultivable valley near unto Your Holy House [Kaba]. Our Rab! That they may establish regular Prayer, so fill the hearts of some among men with love towards them, and feed them with fruits, so that they may be thankful. Our Rab! Truly You know what we conceal and what we reveal. Nothing in the world or in the heaven is hidden from Allah. [Sura No. – 13\14 – Ibrahim, Ayat No. – 35 to 38]

Praise be to Allah, Who has granted unto me in old age Ismaiil and Ishaaq. Lo! Truly my Rab is He, the Hearer of Prayer! My Rab! Make me one who establishes regular prayer, and some of my posterity also [raise such] among my offspring. Our Rab! And accept You my prayer. Our Rab! Forgive me, my parents, and believers, on the Day when the Reckoning [Yawma yaquuu-mul-Hisaab] will be established! [Sura No. – 13\14 – Ibrahim, Ayat No. – 39 to 41]

Think not that Allah is unaware of what the wicked do. He but gives them a respite against a day when the eyes will fixedly stare in horror. They running forward with necks outstretched, their heads uplifted, their gaze returning not towards them, and their hearts a [gaping] void! **So warn mankind of the Day when the wrath will come upon them. And those who did wrong will say: Our Rab! Respite us [if only] for a short term. We will answer Your call, and follow the messengers! What! Were you not wont to swear afore time that you should suffer no decline?** [Sura No. – 13\14 – Ibrahim, Ayat No. – 42 to 44]

And **you dwell in the dwellings of those who wronged their own souls**. You were clearly shown how We dealt with them; and made examples for you. Mighty indeed were the plots which they made, but their plots were [well] within the sight of Allah, even though they were such as to shake the hills! **Never think that Allah would fail his messengers in His promise.** Lo! Allah is Exalted in power, the Lord of Retribution. [Sura No. – 13\14 – Ibrahim, Ayat No. – 45 to 47]

One day **the world will be changed to a different world**, and so will be the heavens and [men] will be marshalled forth, before Allah, the One, the Irresistible. And you will see the sinners that day bound together in fetters. Their garments of liquid pitch, and their faces covered with Fire. That Allah may repay each soul according to its deserts; and verily Allah is swift in calling to account. **Here is a Message for mankind. Let them take warning there from, and let them know that He is [no other than] One Allah. Let men of understanding take heed**. [Sura No. – 13\14 – Ibrahim, Ayat No. 48 to 52]

BEAUTIFUL FELLOSHIP

In **like manner** [example] We have **revealed** to you the **Kitaab** and those, to whom We gave the Kitaab **afore time** believed therein, and of those (also) there are some who believe therein and **none** but **disbelievers reject Our Signs**. And **you were not (able) to recite a Kitaab** before this, **nor did you write it from right hand side**, for then might those have **doubted**, who follow **falsehood**. But it is **self-evident** in the hearts of those who are endowed with **true knowledge**, and **none deny Our Signs** (Revelations) except **wrong-doers**. [Sura No. – 28\29 – Ankabuut, Ayat No. – 47 to 49]

They will say: Be You glorified! It was not for us to choose any protecting friends besides You. But You did give them and their fathers ease till they forgot the warning and became lost folk. Thus they will give you the lie regarding what you say. Then you can neither avert (the doom) nor obtain help. And whoever among you does wrong, him shall We cause to taste of a grievous Penalty. And **the messengers whom We sent before you were all [men] who ate food and walked through the streets: We have made some of you as a trial for others. Will you have patience**? For Allah is One Who sees [all things]. [Sura No. – 24\25 - Nazzalal-Furqaan, Ayat No. – 18 to 20]

O you who Believe! Make not unlawful the good things which Allah has made lawful for you, but commit no excess: For Allah loves not those given to excess. Eat of the things which Allah has provided for you, lawful and good; but keep your duty to Allah, in Whom you believe. Allah will not call you to account for what is **futile in your oaths**, but He will call you to account for your **deliberate oaths**: For **compensation, feed ten indigent persons**, on a **scale** of the **average for the food** of your families; or **clothe** them; or **give a slave his freedom.** If that is beyond your **means, fast for three days.** That is the **compensation for the oaths** you have sworn. But **keep to your oaths.** Thus Allah makes clear to you His Signs, that you may be grateful. [Sura No. – 4\5 – Maaaidah, Ayat\Verses – 87 to 89]

And verily We gave **Bani-Israa-iil the Kitaaba wal-Hukma** (and the Command), and **Prophet-hood**, and provided them with good things and **favoured them above the nations** [Middle West Zone]. And We gave them **Clear Signs** (manifested magnetism in resemblance with revelation). And they differed not until after the **knowledge** came to them, through **insolent jealousy among themselves.** Verily your Rab will judge between them on the Day of Resurrection as to **those matters in which they set up differences.** And **now We have set you on a clear road (straightway) of commandment, so follow it, and follow not the desires of those who know not. They will be of no use to you in the sight of Allah. It is only wrong-doers [that stand as] protectors, one to another.** But Allah is the **Protector** of the Righteous. **These are clear**

evidences (indications) to men and a **Guidance and Mercy to those of assured Faith**. [Sura No. – 44\45 – Tanziilul Kitaab (Prev. Jaasiya), Ayat No. – 16 to 20]

Say: I will recite unto you that which your Rab have made a **sacred duty** for you: that you **ascribe nothing as partner** unto Him and that you do good to parents, and that you **slay not your children** because of poverty – We **provide** for you and for them – and that you draw not nigh to **lewd things** whether open or concealed. And that you **slay not the life which Allah has made sacred**, save in the course of **justice**. This He has **commanded** you, in order that you may discern. And come **not** nigh to the **orphan's property**, except to improve it, until he attain the age of full strength; give **measure and weight with [full] justice**. No burden do We place on any soul, except that which it can bear. Whenever you speak, speak justly, even if a near relative is concerned; and fulfil the **promise** of Allah. Thus He **commands** you that you may **remember**. Verily, **this is My way, leading straight: follow it: follow not [other] paths. They will scatter you about from His [great] path. This He has command for you that you may be righteous.** [Sura No. – 5\6 – An-aam, Ayat No. –152 to 153]

It is Allah Who causes the seed-grain and the date-stone to split and sprout. He causes the living to issue from the dead, and He is the one to cause the dead to issue from the living. That is Allah: **then how are you deluded away from the truth?** [Sura No. – 5\6 – An-aam, Ayat No. –96]

He (Allah) smites the day-break [from the dark]. He makes the night for rest and calmness, and Wash-shamsa (Allah's Light) and Wal-qamara (splendid reflection of Allah's Light) for the reckoning [of time]. That is the measuring of the Mighty, the Wise. [Sura No. – 5\6 – An-aam, Ayat No. –96]

It is He Who makes the stars [as inspiration] for you that you may guide yourselves, with their help, through the dark spaces of land and sea: We have detailed Our Signs (anifested magnetism in resemblance with revelation) for people who have knowledge. It is He Who has produced you from a single person. Here is a place of sojourn (Habitation) and a place of departure (repository). We have detailed Our signs for people who understand. He it is Who sends down water from the sky [West], and therewith We brings forth buds of every kind. We bring forth the **green** blade from which We bring forth the thick-clustered **grain**, and from the pull thereof, spring pendant **bunches**, and (we bring forth) **gardens of grapes**, and the **olive** and the **pomegranate, alike and unlike** (equal & opposite). Look upon the fruit thereof when they bear fruit, and upon its ripening. Lo! Herein **verily are clear**

proofs for a people who believe. Yet they ascribe as partners unto Him the Jinns, although He did create them, and attribute falsely, without knowledge, **sons and daughters unto Him**. Praise and glory be to Him! **[for He is] above what they attribute to Him**. [Sura No. – 5\6 – An-aam, Ayat No. –96 to 101]

Say: My Rab! If You should show me that which they are **promised (warned against)**. Then, O my Rab! Put me not amongst the people **who do wrong**! And We are certainly able to show you [in fulfilment] that which We have promised (against which they are warned). **Repel evil with that which is better**. We are well acquainted with the things they say. And say: My Rab! I seek **sanctuary in You** from the suggestions of the **evil ones [projected falsehood]**. And I seek sanctuary with You, O my Rab! Lest they be present with me. [In Falsehood will they be] Until, when death comes to one of them, he says: O my Rab! Send me back [to life] in order that I may work righteousness (Tabligh) in that which I left behind. By no means! It is but a word he says. Before them is a Partition till the Day they are raised up. Then when the **Trumpet** will be blown, there will be no more relationships between them that Day, nor will one ask after another! Then those whose balance [of good deeds] is heavy, they will attain salvation. But those whose balance is light, will be those who have lost their souls. In Hell they will abide. The Fire will burn their faces, and they will therein grin, with their lips displaced. [Sura No. – 22\23 – Mu-Minuun, Ayat\Verses – 93 & 104]

Know you that Allah is strict in punishment and that Allah is Forgiving, Merciful The Prophet's duty is but to proclaim (the Message). But **Allah knows all that you reveal and you conceal.** Say: The **evil and the good are not alike** even though the **plenty of the evil attract** you. So, be **mindful** of your duty **to Allah**, O men of **understanding** that you may **succeed**. [Sura No. – 4\5 – Maaaidah, Ayat\Verses – 98 to 100]]

Verily He Who **ordained the Qur'an for you**, will **bring you back to the Place of Return**. Say: My Rab is best aware of him **who brings true guidance**, and **who is in manifest error**. And you **had not expected** that the **Kitaab would be sent** to you except as a Mercy from your Rab. Therefore **lend not your support in any way to those who reject [Allah's Message].** And **let them not divert you from the Signs of Allah** after **they have been sent down unto you.** But **call** (mankind) to your Rab (do work of righteousness), and **be not of the company of those who join gods (partners / breakable realities) with Allah.** And **call / cry not unto any other ilaaha** (breakable reality) **along with Allah. There is no unbreakable** reality [ilaaha] **save Him. Everything will perish** except His **countenance. To Him belongs the Command,** and **to Him will you [all] be brought back.** [Sura No. – 27\28 – Qasas, Ayat\Verses No. – 85 to 88]

If it had been Allah's plan, they would not have taken false realities; but We made you not one to watch over their doings, nor are you set over them to dispose of their affairs. Insult not you those whom they call upon besides Allah, lest they out of spite insult Allah in their ignorance. Thus We have made alluring to each people his/her own-doings. In the end they will return to their Rab, and We shall then tell them the truth of all that they did. They swear their strongest oaths by Allah, that if **a [special] sign** came to them, by it they would believe. Say: **Certainly [all] signs are in the power of Allah**: but what will make you realise that [even] if [special] signs came, they will not believe? We [too] shall turn to [confusion] their hearts and their eyes, even as they refused to believe in this in the **first instance**: We shall leave them in their trespasses, **to wander in distraction.** [Sura No. – 5\6 – An-aam, Ayat No. –108 to 111]

When there comes to them a **Clear Proof (token) from Allah**, they say: We shall not believe until we receive one [exactly] like those received by Allah's Prophets. **Allah knows the best where [and how] to carry out His mission. Soon will the wicked be overtaken by humiliation before Allah, and a severe punishment, for all their plots?** And whomsoever it is Allah's will to guide, He expands His bosom unto the surrender, and whomsoever it is His will to send astray, He makes His bosom close and narrow as if He were engage in sheer ascent. Thus Allah lays ignominy upon **those who refuse to believe. This is the way of your Rab, leading upright: We have detailed the signs for those who receive admonition.** For them will be a home of peace in the presence of their Rab: He will be their friend, because they practised [righteousness]. One day He will gather them all together, [and say]: O you assembly of jinns! Much [toll] did you take of men. Their friends amongst men will say: Our Rab! We made profit from each other: but [alas!] we reached our term – which you appointed for us. He will say: The fire will be your dwelling-place. You will dwell therein forever, except as Allah wills, for your Rab is full of Wisdom and Awareness. **Thus we let some of the wrong-doers turn to each other, because of that they won't to earn.** [Sura No. – 5\6 – An-aam, Ayat No. –125 to 130]

Said [Firawn]: **Believe you in Him before I give you permission? Surely** he is your leader, who has **taught you sorcery!** But **soon shall you know! Be sure** I will **cut off your hands** and your **feet on opposite sides**, and I will **cause** you all to **die on the cross!** They said: **No matter!** For us, **we shall but return to our Rab!** Only, our **desire** is that our Rab will forgive us our faults, that we may become foremost among the believers! [Sura No. 25\26 – Shuraa, Ayat No. – 49 to 51]

Remember how they said: O Allah if this is indeed the Truth from You, rain down on us a **shower of stones** from the sky [West], or send us a grievous **penalty**. But Allah was not going to send them a penalty while you were amongst them; nor was He going to send it while they could ask for pardon. But what plea have they that Allah should not punish them, **when they keep out [believers] from the sacred Mosque [projected globalisation & mechanical magnetism], and they are not its guardians [manifested magnetism in resemblance with revelation]? No men can be its guardians except the righteous; but most of them do not understand.** Their prayer at the (Holy) House [Masque] is nothing but whistling and clapping of hands. Therefore, taste you the penalty because you disbelieved. **[Sura No. – 7\8 – Anil-Anfaal, Ayat No. – 32 to 35]**

Then, after them, We brought forth another generation. And We sent to them a messenger from among themselves, [saying]: Worship Allah! You have **no other reality (La ilaaha)** save Him. **Will you not ward off (evil)?** And the **chiefs of his people**, who disbelieved and denied the Meeting in the Hereafter, and on whom We had bestowed the good things of this life, said: He is no more than a man like you. He eats of that of which you eat, and drinks of what you drink. **If you obey a man like yourselves, behold, it is certain you will be lost. Does he promise that when you die and become dust and bones, you shall be brought forth [again]? Far, very far is that which you are promised! There is nothing but our life in this world! We shall die and we live! But we shall never be raised up again! He is only a man who invents a lie against Allah, but we are not the ones to believe in him!** He said: My Rab! Help me, for that **they accuse me of falsehood.** [Allah] said: In a little while, they surely will become **repentant! So the (Awful) cry overtook them rightfully, and We make them as wreckage (that a torrent hurls); a far removal for wrong doing folk.** [Sura No. – 22\23 – Mu-Minuun, Ayat\Verses – 31 to 41]

Then after them We brought forth **other generations. No people can hasten their term, nor can they delay** [it]. Then We sent Our messengers in succession (one after another). Whenever its messenger came to a nation, they denied him. So We caused them to follow one another (to disaster) and We made them by words; a far removal for folk who believe not. [Sura No. – 22\23 – Mu-Minuun, Ayat\Verses – 42 to 44]

The blame is only against those who oppress men and wrong-doing and insolently transgress beyond bounds through the land, defying right and justice. For such there will be **a penalty grievous**. And verily if any show patience and forgive, that would truly be an exercise of courageous will and resolution in the conduct of affairs. And he whom Allah **sends astray**, there is **no protector** after Him. And you will see the evil-doers, when they see the doom, they say: **Is there any way of return**? [Sura No. – 41\42 – Shuuraa, Ayat No. – 42 to 44]

Successful indeed are the believers those who are humble in their prayers; and who avoid vain conversation; who are active in deeds of the poor-due; who guard their modesty - save from their wives or the (slaves) that their right hand possess, for them **they are free from blame**. But those whose desires exceed those limits are **transgressors.** [Sura No. – 22\23 – Mu-Minuun, Ayat\ Verses – 1 to 7]

Those who faithfully observe their trusts and their covenants; and who pay heed to their prayers; these will be the **heirs**, who will inherit Paradise: they will dwell therein [for ever]. [Sura No. – 22\23 – Mu-Minuun, Ayat\Verses – 8 to 11]

We sent **no messenger except** that he should be obeyed **in accordance with the will of Allah**. And if, when they were unjust to themselves, they had come unto you and asked forgiveness of Allah, and asked forgiveness of the messenger, they would have found Allah Forgiving, Merciful. **But no, by your Rab, they will not believe (in truth) until they make you judge of what is in dispute between them, and find within themselves no dislike of that which you decides, and submit with full submission.** And if We had decreed for them: Lay down your lives or go forth from your dwellings (to leave their homes), very few of them would have done it. But if they had done what they were [actually] told, it would have been better for them, and would have gone farther to strengthen their [faith]. **(Significance of Tabligh)** And We should then have bestowed upon them from Our presence an immense reward. **And We** should have shown (guide) them the Straight Way. All who obey Allah and the messenger are in the company of those upon whom is the Grace of Allah, of the prophets, the sincere [lovers of Truth], the witnesses / martyrs [who testify], and the Righteous [who do good]: **Ah! What a beautiful fellowship! Such is the bounty from Allah**. And Allah as Knower is **sufficient**. [Sura No. – 3\4 – Nisaa, Ayat No. – 64 to 69]

Ayat\Verses – And the foremost in the race, the foremost in the race; these are they who will be brought near in Gardens of Bliss, a multitude of those of old, and a few of those of later times on lined couches, reclining **therein face to face [equal & opposite]**. There wait on them immortal youths with bowls and ewers and a cup from a pure spring wherefrom they get no aching of the head nor any madness (nor will they suffer intoxication). And fruit that they prefer and flesh of fowls that they desire and [there are) fair ones with wide lovely eyes, like unto hidden pearls reward for what they used to do. There they will hear neither vain speaking nor recrimination; (naught) but the saying: Peace, (and again) Peace. [Sura No. – 55\56 - Waqa-atil-waaqiah, Ayat No. – 10 to 26]

Ayat\Verses – And those on the right hand, **what of those on the right hand?** "Fii sidrim-makhzuud" [Trans.] Among thorn less **lute tree.** "Wa talhim-manzuud" [Trans.] And clustered plantains / (Among Talh trees with flowers [or fruits] piled one above another); "Wa zillim-mamduud" [Trans.] And spreading shade, "Wa maaa-im-maskuub" [Trans.] By water gushing (flowing constantly), and fruit in abundance. Neither out of reach nor yet forbidden, and on Thrones [of Dignity], raised high. Lo! We have created them [their Companions] a (new) creation, and made them virgins [pure and undefiled], lovers, and friends for the **Companions of the Right Hand**, a multitude of those of old and a multitude of those of later times. [Sura No. – 55\56 - Waqa-atil-waaqiah, Ayat No. – 27 to 40]

Ayat\Verses – Then why do you not [intervene] when [the soul of the dying man] reaches the throat, / Why, then, when comes up to the throat, and you are at that moment looking? But We are nearer to him than you are, and you see not, why then, if you are not in bondage (unto Us)? **Do you not force it back, if you are truthful?** Thus, then, if he be of those nearest to Allah; [there is for him] rest and satisfaction, and a garden of delights. And if he be of the **companions of the Right Hand**, [for him is the salutation]: Peace unto you, from the companions of the **Right Hand**. [Sura No. – 55\56 - Waqa-atil-waaqiah, Ayat No. – 83 to 91]

SECTION – VIII
RIGHTEOUSNESS – 8.4
HYPOCRITES OR MUNAAFIQUUN

Fasta-'iz billaahi minash-Shaytaanir-Rajim

Bismillaahir-Rahmaanir- Rahim

"Allah bears witness that the hypocrites [munaafiquun] are indeed liars."
   i)   The hypocrites [munaafiquun] are indeed liars.
   ii)   They make their faith a pretext.
   iii) They turn / obstruct believers from the way of Allah.
   iv) Truly evil are their deeds.
   v)   They believed, then disbelieved.
   vi)   They understand not.
   vii) They are unable to stand on their own.
   viii) They think that every shout is against them.
   ix)   They are the enemies.
   x)   Beware of the hypocrites [munaafiquun].
   xi)   Allah will not forgive them.
   xii) Truly Allah guides not rebellious transgressors.
   xiii) They give advice to their followers to spend nothing on those who are
        with Allah's Messenger.
   xiv) **The hypocrites know not.**

Ayat\Verses – When the Hypocrites come to you, they say: We bear witness that you are indeed the Messenger of Allah. And, Allah knows that you are indeed His Messenger, and **Allah bears witness that the hypocrites [munaafiquun] are indeed liars.** They make their **faith a pretext** so that they **may turn / obstruct from the way of Allah**. Truly **evil are their deeds**. That is **because they believed, then disbelieved**. So a **seal** was set on their **hearts**. Therefore they **understand not**. And when you look at them, **their figures please you**; and when they speak, you give ear to their speech. They are as [**worthless as hollow**] **pieces of timber propped up**, [unable to stand on their own]. **They think that every shout is against them**. They are the **enemies**. So **beware of them. The curse of Allah is on them**! How they are **perverted**! And when it is said to them: **Come, the Messenger of Allah will pray for your forgiveness, they turn aside their heads**, and you would see them **turning away their faces in disdain**. It is **equal** to them whether you **pray** for their forgiveness or **not. Allah will not forgive them**. Truly Allah **guides not rebellious transgressors**. They are the ones who say: **Spend nothing** on those who are with Allah's Messenger, to the end that they may disperse. But to Allah belong the treasures of the heavens and the earth; but **the hypocrites understand not**. They say: If we return to Medina, surely **the mightier will soon turnout the weaker**, when the might belongs to Allah and His Messenger, and to the believers; but **the hypocrites know not**. [Sura No. 62\63 – Jaa-akal-Munaafiquun, Ayat No. – 1 to 8]

"But to no soul will Allah grant respite when the time appointed [for it] has come."
Ayat\Verses – O you who believe! Let not your wealth or your children **divert you** from the remembrance of Allah. If any act thus, **the loss is their own**. And spend of that wherewith We have provided you **before death** should come to any of you and he says: My Rab! If only You would **respite** me for a little while, then I would give **alms and among the righteous**. **But to no soul will Allah grant respite when the time appointed [for it] has come;** and Allah is Aware of [all] that you do. [Sura No. 62\63 – Jaa-akal-Munaafiquun, Ayat No. – 9 to 11]

REBELLIOUS TRANSGRESSORS
Ayat\Verses – Whether you cut down the caring palm-trees, or you left them standing on their **roots**, it was by the permission of Allah, and in order that He might cover with shame (confound) the **rebellious transgressors** (evil-livers). And that which Allah has given as **spoil** to His Messenger from them – for this you urged **not any horse or riding-camel** for the sake thereof; but Allah gives power to His messengers over any He pleases, and Allah has power over all things. That which Allah gives as spoil to His Messenger from the people of the **townships** (both North America and South America) belongs to Allah and His Messenger and to **kindred** and **orphan**, the **needy** and the **wayfarer** in order that it may **not** [merely] make a commodity between the rich among you. And whatsoever the Messenger assigns to you, take it. And whatsoever he forbidden, abstain (from it). And keep your duty to Allah. Lo! **Allah is strict in Punishment**. And for the poor refugees (Muhaajirs) who have been **driven out from their homes and their belongings**, who seek **bounty** from Allah and help Allah and His Messenger, such are indeed the sincere [loyal]. {Sura No. – 58\59 – Awwalil-Hashr, Ayat No. – 5 to 8]

THOSE WHO PREFER THE REFUGEES ABOVE THEMSELVES
City – Arabian Peninsula
Ayat\Verses – But **those who entered the city (Arabian Countries) and the faith before them love those who fee to them for refuse (turn down)**, and find in their breasts **no need** for that which has been given them, but **prefer (the refugee) above themselves** though poverty become their lot. And those who are **saved from the covetousness** of their own souls – such are they who

are **successful**. And those who came after them say: Our Rab! Forgive us, and our brethren who came before us into the Faith, and leave not, in our hearts, **sense of injury** against those who have believed. Our Rab! You are indeed Full of Kindness, Most Merciful. {Sura No. – 58\59 – Awwalil-Hashr, Ayat No. – 9 & 10]

"And be you not like those who forgot Allah"

Ayat\Verses – O you who believe! Observe your duty to Allah, and let every soul look to what [provision] He has sent forth for the morrow (following day). Yea, fear Allah, for Allah is well-acquainted with [all] that you do. **And be you not like those who forgot Allah**; and He made them forget their own souls! Such are the **rebellious transgressors**! Not equal are the companions of the Fire and the companions of the Garden. It is the Companions of the Garden that will achieve Felicity. Had We sent down this Qur'an on a mountain, verily, you would have seen it humble itself and cleave asunder for fear of Allah. **Such are the similitude which We propound to men that they may reflect**. {Sura No. – 58\59 – Awwalil-Hashr, Ayat No. – 18 & 21]

HYPOCRITES AND PEOPLE OF THE KITAAB

i)   People of the Kitaab are the disbelieving brethren of the Hypocrites.
ii) If you are expelled, we too will go out with you.
iii) We will never hearken to any one in your affair.
iv) If you are attacked we will help you.
v)   They will turn their backs.
vi) They will not fight you [even] against you except in fortified townships, or from behind walls.
vii) They are a people devoid of wisdom

Ayat\Verses – Have you not observed those who are **hypocrites**, (how) they say to their **disbelieving** brethren among the **People of the Kitaab**? If you are expelled, we too will go out with you, and **we will never hearken to any one in your affair; and if you are attacked we will help you. But Allah is witness that they are indeed liars.** If they are expelled, never will they go out with them; and if they are attacked, they will never help them; and if they do help them, they will turn their backs; so they will receive no help. Of a truth you are stronger [than they] because of the terror in their hearts, [sent] by Allah. This is because **they are men devoid of understanding.** They will not fight you [even] against you except in **fortified townships**, or from **behind walls**. Strong is their fighting [spirit] amongst themselves. You would think that they were **united**, but **their hearts are divided**. That is because they are a people **devoid of wisdom**. Like those who lately preceded them, they have tasted the evil result of their conduct; and [in the Hereafter there is] for them a grievous Penalty. [Their allies deceived them], like the evil one, **when he tells man in disbelieve**. But when [man] disbelieve, [the evil one] says: I am free of you. I do fear Allah, the Rab of the Worlds! The end of both will be that they will go into the Fire, dwelling therein forever. Such is the reward of the wrong-doers. {Sura No. – 58\59 – Awwalil-Hashr, Ayat No. 11 & 17]

THE HYPOCRITES
PAGANS / IDOLATERS / WANDERING ARABS
Reference of Four Months

"Go you, then, for four months, backwards and forwards, throughout the land [earth]"

Ayat\Verses – A [declaration] of **resistance from Allah and His Messenger**, to those of the pagans / idolaters with whom you have **contracted** mutual alliances. **Go you, then, for four months, backwards and forwards, throughout the land,** but know you that you cannot frustrate Allah [by your falsehood] but that Allah will cover with shame those who reject Him. And an announcement from Allah and His Messenger, to the people [assembled] on the day of the Great Pilgrimage, that **Allah and His Messenger dissolve [treaty] obligations with the pagans / idolaters**. If then, you regret, it were best for you; but **if you turn away, know you that you cannot frustrate Allah. And proclaim a grievous penalty** to those who **reject Faith**. But the **treaties** are not dissolved with those Pagans / idolaters with whom you have entered into alliance and who have **not** subsequently failed you in aught, **nor** aided **any one against you**. So **fulfil your engagements** with them to the end of their term: for Allah loves the **righteous**. But when the **forbidden months** are past, then **fight and slay the Pagans / idolaters wherever you find them**, an **seize them, beleaguer** them, and lie in wait for them in every **stratagem** [of war]; but if they **regret**, and **establish** regular prayers and **practise** regular alms giving, then **open the way** for them: for Allah is Oft- forgiving, Most Merciful. If one amongst the Pagans / idolaters asks you for **asylum**, grant it to him, so that he may hear the **word of Allah**; and then **escort** him to where he can be **secure**. That is because **they are men without knowledge** i.e. ignorant. [Sura No. – 8\9 – Tauba, Ayat No. – 1 to 6]

Do you fear them?
REBELLIOUS AND WICKED
Who has sold Signs of Allah?

"How can there be a league, before Allah and His Messenger, with the Pagans / idolaters, except those with whom you made a treaty near the sacred Mosque?"
"Will you not fight people who **violated their oaths, plotted to expel the Messenger,** and took the **aggressive** by being the first [to assault] you?"

Ayat\Verses – **How can there be a league, before Allah and His Messenger, with the** Pagans / idolaters, except those with whom you made a treaty near the sacred Mosque? So long as **they are true to you, be true to them**. For Allah **loves** those who keep their **duty**. How [can there be such a league], seeing that if they get an advantage over you, they respect **not** in you the ties either of **kinship** or of **covenant**? With [fair words from] their mouths **they attract you,** but their **hearts are averse** from you; and most of them are **rebellious and wicked**. The **Signs of Allah** have they **sold** for a **miserable price**, and [many] they have **delayed from His way. Evil indeed** are the **deeds they have done.** In a Believer they respect **not** the ties either of kinship or of Promise. It is they who have **transgressed all bounds**. But [even so], if they **regret,** establish regular prayers, and practice regular charity, they are your **brethren in Faith.** We **explained the Signs in detail for those who understand**. But if they **violate** their oaths after their promise, and **criticise** you for your Faith, **fight** the chiefs / leaders of disbelief, for their oaths are nothing to them in order that they may stop. Will you not fight people who **violated their oaths, plotted to expel the Messenger,** and took the **aggressive** by being the first [to assault] you? **Do you fear them?** Nay, it is Allah Whom you should more justly fear, if you are believers. Fight them! **Allah will punish them at your hands** cover them with shame, and He will lay them low and give you **victory** over them, and heal the breasts of folk who are believers. And who will remove the anger of their hearts. For Allah will **turn** [in mercy] to whom He will; and Allah is All-Knowing, All-Wise. Or think you that you would be left, as though Allah did not know those among you who strive with might and main, and **take none for friends and protectors** except Allah, His Messenger, and the [community of] believers? But Allah is well-acquainted with [all] that you do. [Sura No. – 8\9 – Tauba, Ayat No. – 7 to 16]

> **"It is not for the idolaters to tend Allah's sanctuaries."**

i) Do you make the **satisfying thirst of Pilgrim** and ten-dance of the **Inviolable Place of Worship equal** to [the pious service of] those who believe in Allah and the Last Day, and strive with might and main in the way of Allah? [An Imam is a believer]

ii) **They are not comparable (equal) in the sight of Allah, and Allah guides not those who do wrong.**

Ayat\Verses – **It is not for the idolaters to tend Allah's sanctuaries,** bearing witness against themselves of disbelief. As for such, their **works are vain** and **in the fire** they will abide. **The mosques of Allah shall be visited and maintained by such as believe in Allah and the Last Day, establish regular prayers, and practise regular charity, and fear none [at all] except Allah. It** is they who are expected to be on **true guidance**. Do you make the **satisfying thirst of Pilgrim** and ten-dance of the **Inviolable Place of Worship equal** to [the pious service of] those who believe in Allah and the Last Day, and strive with might and main in the way of Allah? **They are not comparable (equal) in the sight of Allah, and Allah guides not those who do wrong.** Those who believe, and **leave their homes and strive with their wealth** and their lives in the way of Allah, are of much **greater worth** in the sight of Allah(Tabligh). They are the people who are **successful.** Their Rab will give them good tidings of a Mercy from Himself, of His good pleasure, and of gardens for them, wherein are delights that endure. They will dwell therein forever. **Verily in Allah's presence is a reward, the greatest** [of all]. [Sura No. – 8\9 – Tauba, Ayat No. – 17 to 22]

Allah guides not the wrong doing folk.

i) O you who believe! **Take not your father and your brothers for protectors** if they take pleasure in disbelief **rather than Faith.**
ii) **If any of you do so, they are wrong doers.**
iii) If it be that your fathers, your sons, your brothers, your mates, or your kindred; the wealth that you have gained; the commerce in which you fear a decline: or the dwellings in which you delight are **dearer to you than Allah**, or **His Messenger**, or **the striving in His way**, then wait until Allah brings about His decision.

Ayat\Verses – O you who believe! **Take not your father and your brothers for protectors** if they take pleasure in disbelief **rather than Faith. If any of you do so, they are wrong doers.** Say: If it be that your fathers, your sons, your brothers, your mates, or your kindred; the wealth that you have gained; the commerce in which you fear a decline: or the dwellings in which you delight are dearer to you than Allah, or His Messenger, or the striving in His way, then wait until Allah brings about His decision, and **Allah guides not the wrong doing folk. [Sura No. – 8\9 – Tauba, Ayat No. – 17 to 22]**

Such is the reward of disbelievers.

i) O you who believe! Truly the Pagans idolaters are unclean.
ii) So, let them not come near the Inviolable Place of Worship after this year.
iii) Fight against such of those who have given the Kitab as believe not in Allah nor the Last Day, nor hold that forbidden which has been forbidden by Allah and His Messenger, nor acknowledge the religion of Truth, [even if they are] of the People of the Kitab, until they pay the tribute readily, being brought low.

Ayat\Verses – Assuredly Allah did help you in many battle-fields and on the day of Hunaynin when you excited with great numbers, but they availed you naught, and the land, for all that it is wide, did constrain you, and you turned back in retreat. But Allah did pour His calm on the Messenger and on the Believers, and sent down forces which you saw not. He punished the Unbelievers. **Such is the reward of disbelievers.** Again, after this, Allah will turn [in mercy] to whom He will, for Allah is Oft-forgiving, Most Merciful. O you who believe! **Truly the Pagans idolaters are unclean.** So, let them not **come near the Inviolable Place of Worship after this year.** And if you fear poverty, soon will Allah enrich you, if He wills, out of His bounty, for Allah is All-knowing, All- wise. **Fight against such of those who have given the Kitab as believe not in Allah nor the Last Day, nor hold that forbidden which has been forbidden by Allah and His Messenger, nor acknowledge the religion of Truth, [even if they are] of the People of the Kitab, until they pay the tribute readily, being brought low. [Sura No. – 8\9 – Tauba, Ayat No. – 25 to 29]**

Allah's curse is on them
To Search out the Pragmatic correspondent Truth – Fain would they extinguish Allah's light (Nuural-laahi) with their mouths
Ayat\Verses – The Yahuudis' call 'Uzayru-nib-nullaahi' as the son of Allah, and the Nasaras' call Masiihub-nullaah' as the son of Allah. **That is a saying from their mouth;** [in this] they but imitate **what the unbelievers of old used to say. Allah's curse is on them: how they are deluded away from the Truth? They have taken as lords besides Allah** their **rabbis** and their **monks** [Lukmans as Imams and their Imamgiris], [and] Wal-Musii-habna-Maryam [to be their Rabs in **derogation of Allah], yet they were commanded to worship none but One Allah.** There is no unbreakable reality except Him. Praise and glory to Him: [Far is He] from having the **partners they associate** [with Him]. **Fain would they extinguish Allah's light (Nuural-laahi) with their mouths,** but **Allah will allow that His light should be perfected, even though the unbelievers may abominate (it).** It is He Who has sent His Messenger with guidance and the Religion of Truth, to prevail over all religions, even though the Pagans / idolaters may detest [it]. [Sura No. – 8\9 – Tauba, Ayat No. – 30 to 33]

WE ARE TO REFLECT ON THE FOLLOWING TRUTHS
i) O you who believe! There are indeed many among the priests (scholars / researchers) and anchorites [leadership], who in Falsehood demolish the wealth of men and obstruct [them] from the way of Allah.
ii) And there are those who stored gold and silver and spend it not in the way of Allah announce unto them a most grievous penalty.

iii) The number of months in the sight of Allah is twelve. [Twelve Springs or Planets]

iv) Four of them are blessed that is the straight usage. [Four Springs or Planets or Imams are the four pillars of the pentagonal tent].

v) Verily the transposing [of a prohibited month] is an addition to unbelief.

vi) The unbelievers are led to wrong thereby, for they make it lawful one year, and forbidden another year, in order to adjust the number of months forbidden by Allah and make such forbidden ones lawful.

vii) The evil of their course seems pleasing to them. But Allah guides not those who reject Faith.

Ayat\Verses – O you who believe! There are indeed many among the priests (scholars) and anchorites [leadership], **who in Falsehood demolish the wealth of men and obstruct [them] from the way of Allah**. And there are those **who stored gold and silver** and spend it not in the **way of Allah** announce unto them a most grievous penalty. **On the Day when heat will be produced out of that [wealth] in the fire of Hell**, and **with it will be branded their foreheads**, their **borders**, and their **backs borders**, and their **backs**. This is the [treasure] which you stored for yourselves. Taste you, then, the [treasures] you stored. **The number of months in the sight of Allah is twelve. So ordained by Him the day He created the heavens and the world. Four of them are blessed that is the straight usage.** So **wrong not yourselves therein, and wage the Pagans / idolaters all together as they waging on all of you. But know that Allah is with those who restrain themselves. Verily the transposing [of a prohibited month] is an addition to unbelief. The unbelievers are led to wrong thereby, for they make it lawful one year, and forbidden another year, in order to adjust the number of months forbidden by Allah and make such forbidden ones lawful. The evil of their course seems pleasing to them. But Allah guides not those who reject Faith. [Sura No. – 8\9 – Tauba, Ayat No. – 34 to 37]**

CERTAINLY THEY ARE LIARS

i) O you who believe! What is the matter with you, that, when you are asked to go forth in the way of Allah, you cling heavily in the world?

ii) Do you prefer the life of this world [present life] to the Hereafter?

iii) If you help not [him], [it is no matter]; for Allah did indeed help him, when the unbelievers drove him out.

iv) They would destroy their own souls; for Allah knows that certainly they are liars.

Ayat\Verses – O you who believe! What is the matter with you, that, when you are asked to go forth in the way of Allah, you cling heavily in the world? Do you prefer the life of this world [present life] to the Hereafter? But little is the comfort of this life, as compared with the Hereafter. Unless you go forth, He will punish you with a grievous penalty, and put others in your place; but Him you would not harm in the least. For Allah has power over all things. If you help not [him], [it is no matter]; for Allah did indeed help him, when the unbelievers drove him out. **He had no more than one companion. They two were in the cave, and she said to her companion, have no fear, for Allah is with us. Then Allah sent down His peace upon him, and strengthened him with forces which you saw not, and made the words of those who disbelieved neither most. But the word of Allah is exalted to the heights, for Allah is Exalted in might, Wise. Go you forth**, [whether equipped] lightly or heavily, and strive and struggle, with your goods and your persons, in the way of Allah. **That is best for you**, if you [but] knew. If there had been **immediate gain** [in sight], and the **journey easy**, they would [all] **without doubt** have followed you, but the distance was long, [and weighed] on them. They would indeed swear by Allah, if we only could, we should certainly have come out with you. They would destroy their own souls; for **Allah knows that certainly they are liars**. [Sura No. – 8\9 – Tauba, Ayat No. – 38 to 42]

"Indeed they had plotted sedition before, and upset matters for you, until the Truth arrived, and the Decree of Allah became manifest much to their disgust." Ayat\Verses – Allah forgives you! **Where for did you grant them leave ere those who told the truth were manifest to you and you did know the liars?** Those who believe in Allah and the Last Day ask you for no exemption from fighting with their goods and persons. And Allah knows well those who do their duty. Only those ask you for exemption who believes not in Allah and the Last Day, and whose hearts are in doubt, so that they are tossed in their doubts to and fro. If they had intended to come out, they would certainly have made some preparation there for; but Allah was averse to their being sent forth. So He made them lag behind, and they were told: Sit you among those who sit [inactive]. If they had come out with you, they would not have added to your [strength] but only [made for] disorder, hurrying to and fro in your midst and sowing sedition among you, and there would have been some among you who would have listened to them. But Allah knows well those who do wrong. **Indeed they had plotted sedition before, and upset matters for you, until the Truth arrived, and the Decree of Allah became manifest much to their disgust.** [Sura No. – 8\9 – Tauba, Ayat No. – 43 to 48]

Have they not fallen into trial already?
REBELLIOUS AND WICKED

Ayat\Verses – Among them is [many] a man who says: Grant me exemption and draw me not into trial. **Have they not fallen into trial already?** And indeed **Hell** surrounds the **Unbelievers** [on all sides]. If **good** befalls you, it **grieves** them; but if a **misfortune** befalls you, they say: We took indeed our **precautions** beforehand, and they **turn away rejoicing**. Say: **Nothing will happen to us except what Allah has decreed for us: He is our protector: and on Allah let the Believers put their trust. Say: Can you expect for us [any fate] other than one of two glorious things- [Martyrdom or victory]?** But we can expect for you either that Allah will send his **punishment from Himself,** or **by our hands. So wait [expectant]; we too will wait with you.** Say: Spend [for the way of Allah] willingly or unwillingly; not from you will it be accepted; for you are indeed a people **rebellious and wicked. [Sura No. – 8\9 – Tauba, Ayat No. – 49 to 53]**

THEY REJECT ALLAH AND HIS MESSENGER
i)   They come to prayer without earnestness.
ii)   They offer contributions unwillingly.
iii)   If they could find a place to flee to, or caves, or a place of concealment, they would turn straightaway thereto, with an obstinate rush.

Ayat\Verses – **The reasons why their contributions are not accepted are:** that **they reject Allah and His Messenger; that they come to prayer without earnestness; and that they offer contributions unwillingly.** Let **neither** their wealth **nor** their [following in] sons dazzle you. **In reality Allah's plan is to punish them with these things in this life, and that their souls may perish in their [very] denial of Allah.** They swear by Allah that they are indeed of you; but **they are not of you**: yet they are afraid [to appear in their true colours]. **If they could find a place to flee to, or caves, or a place of concealment, they would turn straightaway thereto, with an obstinate rush. [Sura No. – 8\9 – Tauba, Ayat No. – 54 to 57]**

THEY ARE INDIGNANT
Ayat\Verses – And among them are men **who insult you** in the matter of [the distribution of] the alms. If they are given **part thereof**, they are pleased, but if not, **they are indignant.** If only they had been content with what Allah and His Messenger gave them, and had said: **Sufficient unto us is Allah!** Allah and His Messenger will soon give us of His bounty: to Allah do we turn our hopes, [that would have been the right course]. Alms are for the poor and the needy, and those employed to administer the [funds]; for those whose hearts have been [recently] reconciled [to Truth]; for those in bondage and in debt; in the way of Allah; and for the wayfarer: [thus is it] ordained by Allah, and Allah is full of knowledge and wisdom. **[Sura No. – 8\9 – Tauba, Ayat No. – 58 to 60]**

"You have rejected Faith after you had accepted it."
THEY ARE IN SIN

i) Among them are men who assault the Prophet.
ii) But those who assault the Messenger will have a grievous penalty.
iii) The Hypocrites are afraid lest a Sura should be sent down about them, showing them what is [really passing] in their hearts.
iv) Say: Mock you! But verily Allah will bring to light all that you fear [should be revealed].
v) Say: Was it at Allah, and His Signs, and His Messenger that you were mocking?

Ayat\Verses – **Among them are men who assault the Prophet** and say: He is only a hearer. Say: He listens to what is best for you, he believes in Allah, has faith in the believers, and is a Mercy to those of you who believe. But those who assault the Messenger will have a grievous penalty. To you they swear by Allah in order to please you. But it is more fitting that they should please Allah and His Messenger, if they are believers. Know they not that for those who oppose Allah and His Messenger, is **the Fire of Hell wherein they shall dwell**. That is the supreme disgrace. **The Hypocrites are afraid lest a Sura should be sent down about them, showing them what is [really passing] in their hearts.** Say: Mock you! But verily Allah will bring to light all that you fear [should be revealed]. If you question them, they declare [with emphasis]: We were only talking idly and in play. **Say: Was it at Allah, and His Signs, and His Messenger that you were mocking?** Make you **no excuses**: You have **rejected Faith after you had accepted it**. If We pardon some of you, We will punish others amongst you, for that **they are in sin**. [Sura No. – 8\9 – Tauba, Ayat No. – 61 to 66]

THEY ENJOIN EVIL AND FORBID WHAT IS JUST

i) Verily the Hypocrites are rebellious and perverse.
ii) Allah has promised the Hypocrites men and women, and the rejecters of Faith, the fire of Hell:
iii) For them is the curse of Allah, and an enduring punishment.
iv) Their works are fruitless in this world and in the Hereafter.
v) It is not Allah Who wrongs them, but they wrong their own souls.

Ayat\Verses – The **Hypocrites**, men and women, [have an understanding] with each other. They **enjoin evil, and forbid what is just**, and are close with their hands. They have forgotten Allah; so He hath forgotten them. Verily the **Hypocrites are rebellious and perverse**. Allah has **promised** the Hypocrites men and women, and the rejecters of Faith, the fire of Hell: Therein shall they dwell: Sufficient is it for them. For them is the curse of Allah, and an enduring punishment. As in the case of those before you, they were mightier than you in power, and more flourishing in wealth and children. They had their enjoyment of their portion: and you have of yours, as did those before you; and you indulge in idle talk as they did. Their works are fruitless in this world and in the Hereafter, and they will lose [all spiritual good]. Have not the story reached them of those before them? The people of **Nuuh**, and **Aad**, and **Samuud**; the people of **Ibrahim**, the men of **Midiyan** [people of the Middle Zone or Middle west Zone], and **cities** [Upper West Zone] overthrown. To them **came their messengers with clear signs. It is not Allah Who wrongs them, but they wrong their own souls.** [Sura No. – 8\9 – Tauba, Ayat No. – 67 to 70]the

"But the greatest bliss is the good pleasure of Allah that is the supreme felicity."
Ayat\Verses – The **Believers**, men and women, are **protectors** one of another. They enjoin **what is right**, and **forbid what is wrong**. They **observe** regular prayers, **practice** regular charity, and **obey** Allah and His Messenger. **On them Allah will pour His mercy,** for Allah is Exalted in power, Wise. Allah has **promised** to Believers, men and women, **Gardens** under which rivers flow, to dwell therein, and **beautiful mansions** in Gardens of everlasting bliss. **But the greatest bliss is the good pleasure of Allah that is the supreme felicity.** [Sura No. – 8\9 – Tauba, Ayat No. – 71 & 72]

WE ARE TO REFLECT ON EACH AND EVERY SHARED CONCEPTS WITH A VIEW TO RETAIN OUR FAITH AND BELIEF
AN UNFORTUNATE JOURNEY'S END
MEDITATED PLOT
HIPOCRACY
SECRET DIRECTIONS

Ayat\Verses – **O Prophet! Strive hard** against the **unbelievers** and the **Hypocrites,** and be **firm** / harsh **against them. Their abode is Hell, an unfortunate journey's end.** They **swear** by Allah that **they said nothing** [evil], but indeed they uttered disbelief, and **they did it after accepting Islam;** and **they meditated a plot which they were unable to carry out.** They sought **revenge** only that Allah by His Messenger should enrich them of His **bounty.** If they **repent, it will be best for them;** but if they **turn back** [to their evil ways], Allah will **punish** them with a **grievous penalty in this life and in the Hereafter.** They shall have **none** in the world **to protect or help them.** Amongst **them are men who made a promise with Allah,** that if He **bestowed** on them of His bounty, they would **give** [largely] in charity, and be **truly** amongst those who are **righteous.** But when He did **bestow** of His bounty, they became **greedy**, and **turned back** [from their covenant], **averse** [from its fulfilment]. So He has made as a **consequence hypocrisy** into their hearts, [to last] **till the Day**, whereon they shall meet Him, because they **broke their promise with Allah**, and **because they lied** [again and again]. **Know they not that Allah know their secret** [thoughts] and **their secret directions**, and that Allah knows well **all things unseen**? [Sura No. – 8\9 – Tauba, Ayat No. – 73 to 78]

PERVERSELY REBELLIOUS

"**Allah guides not those who are perversely rebellious.**"

"If you ask **seventy times for their forgiveness, Allah will not forgive them**, because **they have rejected Allah and His Messenger**:"

Ayat\Verses – Those who point at such of the believers as give the alms freely as well as such as can find nothing to give except the fruits of their labour, and **throw ridicule on them**; Allah will **throw back their ridicule on them**: and they shall have a **grievous penalty**. Whether you **ask** for their forgiveness, or **not**, [their sin is unforgivable]. If you ask **seventy times for their forgiveness, Allah will not forgive them**, because **they have rejected Allah and His Messenger**: and Allah guides not those who are perversely rebellious. **Those who were left behind rejoiced at sitting still behind the Prophet of Allah**, and were **averse** to striving with their wealth and their lives in the way of Allah. And they said: The heat of hell is more intense of heat if only they could understand. **Let them laugh a little**. They will weep much, as the **recompense** of what they do. If, then, Allah bring you back to any of them, and they ask your permission to come out [with you], say: **Never** shall you come out with me, **nor** fight an enemy with me, for you preferred to sit **inactive** on the **first occasion**. Then sit you [now] with those who **lag behind**. **Nor** you ever pray for any of them that died, **nor stand at his grave**; for they rejected Allah and His Messenger, and died in a state of perverse rebellion. Neither let their wealth nor their **sons dazzle you. Allah's plan is to punish them with these things in this world [present life], and that their souls may perish in their [very] denial of Allah.** [Sura No. – 8\9 – Tauba, Ayat No. – 79 to 85]

SUPEREME TRIUMPH

Ayat\Verses – When a Sura comes down, enjoining them to believe in Allah and to strive along with His Messenger, those with wealth and influence among them ask you for exemption, and say: leave us [behind]: we would be with those who sit. **They prefer to be with the useless, and their hearts are sealed and so they understand not. But the Messenger, and those who believe with him, strive and fight with their wealth and their persons, for them are [all] good things, and it is they who will prosper.** Allah has prepared for them gardens under which rivers flow, to dwell therein. That is the **supreme triumph**. [**Sura No. – 8\9 – Tauba, Ayat No. – 86 to 89**]

OON A GRIEVOUS PENALTY WILL SEIZE THE UNBELIEVERS AMONG WANDERING ARABS

i) And there were, among the **wandering Arabs [also]**, who made **excuses** and came to **claim** exemption; and those who **lied** to Allah and His Messenger [merely] sat **inactive**.

ii) There is no blame on those who are **infirm**, or ill, or who find no **resources** to spend [on the way], if they are **sincere** [in duty] to Allah and His Messenger.

iii) They prefer to stay with the **useless** who remain behind: Allah has **sealed their hearts**; so **they know not** [What they miss].

iv) They will present their **excuses** to you **when you return to them**.

v) Say you: **Present no excuses, we shall not believe you.**

vi) Allah has already informed us of **the true state** of matters concerning you.

vii) They will swear to you by Allah, when you return to them, that you may leave them alone.

viii) So leave them alone: For they are an abomination and Hell is their dwelling-place, a fitting recompense for the [evil] that they did.

ix) **They will swear unto you, that you may be pleased with them but if you are pleased with them, Allah is not pleased with those who disobey.**

Ayat\Verses – And there were, among the **wandering Arabs [also]**, who made **excuses** and came to **claim** exemption; and those who **lied** to Allah and His Messenger [merely] sat **inactive. Soon a grievous penalty will seize the Unbelievers among them.** There is no blame on those who are **infirm**, or ill, or who find no **resources** to spend [on the way], if they are **sincere** [in duty] to Allah and His Messenger. **No ground** [of complaint] there can be against such as **do right**: and Allah is Oft-forgiving, Most Merciful. There is **no blame** on those who came to you to be **provided with mounts**, and when you said: **I can find no mounts for you, they turned back, their eyes streaming with tears of grief that they had no resources wherewith to provide the expenses. The ground [of complaint] is against such as claim exemption while they are rich.** They prefer to stay with the **useless** who remain behind: Allah has **sealed their hearts**; so **they know not** [What they miss]. They will present their **excuses** to you **when you return to them**. Say you: **Present no excuses, we shall not believe you.** Allah has already informed us of **the true state** of matters concerning you. It is your actions that Allah and His Messenger will observe. In the end **you will be brought back to Him Who knows what is hidden and what is open**; then He will show you the **truth** of all that you did. They will swear to you by Allah, when you return to them, that you may leave them alone. So leave them alone: For they are an abomination and Hell is their dwelling-place, a fitting recompense for the [evil] that they did. **They will swear unto you, that you may be pleased with them but if you are pleased with them, Allah is not pleased with those who disobey. [Sura No. – 8\9 – Tauba, Ayat No. – 90 to 96]**

WANDERING ARABS AND HYPOCRACY

We are to search out those WANDERING ARABS who watch for disasters for the believers as on them is the **disaster of evil.**

Ayat\Verses – The wandering Arabs are harder in disbelief and hypocrisy, and more likely to be ignorant of the limit which Allah has revealed to His Messenger. But Allah is All-knowing, All-Wise. **Some of the wandering Arabs look upon their payments as a fine**, and **watch for disasters for you**: on them is the **disaster of evil**, for Allah is He That hears and knows [all things]. But some of the **wandering Arabs believe in Allah and the Last Day**, and look on their payments as pious gifts bringing them nearer to Allah and obtaining the prayers of the Messenger. Lo!, indeed they bring them nearer [to Him]. Soon Allah will admit them to His Mercy, for Allah is Oft-forgiving, Most Merciful. **[Sura No. – 8\9 – Tauba, Ayat No. – 97 to 99]**

ALLAH KNOWS THE HYPOCRITES

The Historical Don of two-in-one partnership and his chiefs

Ayat\Verses – The forerunner or the first of those who left [their homes] and of those who gave them aid (of Muhajirin and Ansar), and [also] those who follow them in [all] good deeds, well-pleased is Allah with them, as are they with Him. For them He has prepared **Gardens** under which rivers flow, to dwell therein forever. That is the **Supreme Triumph**. (Tabligh). And among those around you of the **wandering Arabs, there are hypocrites** (frauds), and **among the town's [American] people of Al-Madinah** (there are some) **who help in hypocrisy [double standard] whom you know not. We know them, and We shall punish them twice, then they will be relegated to a painful doom.** [Sura No. – 8\9 – Tauba, Ayat No. – 100 & 101]

"Then **He will show you the truth** of all that you did"
- i) And [there are) others who have acknowledged their wrong-doings. They have mixed an act that was good with another that was bad.

ii) Perhaps, Allah will turn unto them [in Mercy], for Allah is Oft-Forgiving, Most Merciful. Of their goods, take alms, so that you might purify and sanctify them; and pray on their behalf.

iii) Verily your prayers are a source of security for them. And Allah is One Who Hears and Knows.

iv) They know not that Allah accepts repentance from His bondmen and receives their gifts of alms giving and that Allah is verily He, Who is Ever Returning, Most Merciful.

v) Soon you will be brought back to the Knower of what is hidden and what is open.

Ayat\Verses – And [there are) others who have **acknowledged** their **wrong-doings**. They have **mixed an act that was good with another that was bad**. **Perhaps**, Allah will **turn** unto them [in Mercy], for Allah is Oft-Forgiving, Most Merciful. Of their goods, take alms, so that you might purify and sanctify them; and pray on their behalf. **Verily your prayers are a source of security for them. And Allah is One Who Hears and Knows. They know not** that Allah accepts repentance from His bondmen and receives their gifts of alms giving and that Allah is verily He, Who is Ever Returning, Most Merciful. And say: **Work** [righteousness]. **Soon** Allah will observe your work, and His Messenger, and the believers. Soon you will be brought back to the Knower of **what is hidden and what is open**. Then **He will show you the truth** of all that you did. **[Sura No. – 8\9 – Tauba, Ayat No. – 102 to 105]**

"Allah loves the purifiers".
We are to confirm the following Ayat\Verses sincerely -

i) There are [yet] others, held in **suspense** for the command of Allah, whether He will **punish** them, or **turn** in mercy to them: and Allah is All- Knowing, Wise.

ii) And there are **those who put up a mosque by way of disbelief and opposition**, and **in order to cause dissent among the believers** (or to disunite the believers) and as an **outpost for those who warred (Fought) against Allah and His Messenger afore time.**

iii) They will indeed **swear** that **their intention is nothing but good**.

iv) But **Allah declares that they are certainly liars.**

v) You **never stand (in pray) there.**

vi) A **place of worship** which was founded upon **duty** (to Allah) **from the First Day** is more **worthy** that you should **stand** (to pray) herein; **wherein** are men **who love to purify themselves.**

Ayat\Verses – There are [yet] others, held in **suspense** for the command of Allah, whether He will **punish** them, or **turn** in mercy to them: and Allah is All- Knowing, Wise. And there are **those who put up a mosque by way of disbelief and opposition,** and **in order to cause dissent among the believers** (or to disunite the believers) and as an **outpost for those who warred (Fought) against Allah and His Messenger afore time.** They will indeed **swear** that **their intention is nothing but good.** But **Allah declares that they are certainly liars.** You **never stand (in pray) there.** A **place of worship** which was founded upon **duty** (to Allah) **from the First Day** is more **worthy** that you should **stand** (to pray) herein; **wherein** are men **who love to purify themselves. Allah loves the purifiers.** [Sura No. – 8\9 – Tauba, Ayat No. – 106 to 108]

WE ARE TO CONFIRM MANIFESTED SIGNS OF NATURAL MAGNETISM CONCERNING THE RIGHT DIRECTION OF OUR QIBLA IN RESEMBLANCE WITH REVELATIONS

Who fulfils the promise better than Allah?
SURVIVAL OF THE TRUEST

Ayat\Verses – Lo! Allah has **purchased** from the believers their **lives** and their **wealth** because the **Gardens** will be theirs. They shall **fight** in the **way** of Allah and shall **slay** and be **slain.** It is a **promise** which is **binding** on Him in the **Tawraat** and the **Injiil,** and **Quran. Who fulfilled the promise better than Allah?** Rejoice then in your bargain that you have made for that is the **Supreme Triumph.** Triumphant are those who **turn repentant** (to Allah), those who **serve** (Him), those who **praise** (Him), those who **fast,** those who **bow down,** those who fall **prostrate** (in ownership), those who **enjoin** the right and who **forbid** wrong and those who **keep** the limits (ordained) of Allah and give **good tidings** to believers. [Sura No. – 8\9 – Tauba, Ayat No. – 111 & 112]

Ayat\Verses – Is he who founded his **building** upon **duty to Allah** and His good pleasure better, or he who founded his **building** on the **edge of a crumbling, overhanging rock face** so that it **collapsed with him into the fire of hell**? Allah guides **not** wrong-doing folk. The **foundation** of those who are so **built** is **never free from suspicion and unsteadiness** in their hearts, **until their hearts are cut to pieces.** And Allah is All-Knowing, Wise. [Sura No. – 8\9 – Tauba, Ayat No. – 109 & 110]

THEY ARE COMPANIONS OF THE FIRE
A SHARING TRUTH WITH CONCRETE EXAMPLE

Ayat\Verses – **It is not fitting,** for the Prophet and those who believe, that they should pray for **forgiveness for Pagans** (idolaters), even though **they are of** kin, after it is clear to them that **they are companions of the Fire. The prayer of Ibrahim for the forgiveness of his father was only because of a promise he had promised him**, but when it had become **clear** unto him that he (his father) was an **enemy to Allah,** he (Ibrahim) **disowned** him. Lo! Ibrahim was **soft-hearted, long-suffering.** It was never Allah's part that He should send a folk **astray** after He had **guided** them **until He had made clear unto them what they should avoid.** Lo! Allah is Aware of all things. [Sura No. – 8\9 – Tauba, Ayat No. – 113 to 115]

NO PROTECTOR, NO HELPER

Ayat\Verses – To Allah belongs the **dominion** [Sovereignty] of the **heavens** and the **world**. He gives **life** and He takes **it**. Except for Him you have **no protector, nor helper. Allah turned with favour to the Prophet, the Muhajirs, and the Ansars, who followed him in a time of distress.** After that the **hearts of a part of them had nearly swerved** [from duty]; but He **turned** to them [also]: for He is unto them Most Kind, Most Merciful. And to the **three** [three ascending stairs) who were left behind, when the world vast as it is, was straitened for them still they be thought them that there is **no refuge from Allah except toward Him,** then **turned** Him unto them in mercy that they (too) might turn (repentant unto Him). Lo! Allah! He is the Relenting, the Merciful. [Sura No. – 8\9 – Tauba, Ayat No. – 116 to 118]

CONDITIONS

"Of every troop of them, a party only should go forth, that they (who are left behind) may gain sound knowledge in religion, and that they return to them, so that they may beware."

Ayat\Verses – **O you who believe! Fear Allah** and be with those who are **true** [in word and deed]. It is for the **townsfolk** [Americans] of Al-Madinah and **for those around them** of the **wandering Arabs to stay behind the Prophet of Allah** and **prefer their lives to his life.** That is **because** neither **thirst** nor **toil** nor **hunger** afflicted them **in the way of Allah, nor** step they any step that angers the **disbelievers, nor** gain they from the **enemy a profit,** but a **good deed** is recorded for them there for. **Lo! Allah loses not the wages of the good. Nor** spend they any spending, small or great, **nor** do they **cross a valley,** but it is recorded for them that Allah may **repay** them the best of **what they used to do.** And the believers should **not** all go out to fight. Of **every troop of them, a party only should go forth, that they (who are left behind) may gain sound knowledge in religion, and that they return to them, so that they may beware.** [Sura No. – 8\9 – Tauba, Ayat No. – 119 to 122]

ADD WICKEDNESS TO THEIR WICKEDNESS

i)   Fight those of the disbelievers who are near to you.
ii) Which one [sura] of you have thus increased in faith?
iii) They will die in a state of disbelief.
iv) Whether they do not see though they are tested once or twice in every year?
v)   Still they turn not in repentance, and they pay no heed.

Ayat\Verses – O you who believe! **Fight** those of the **disbelievers who are near to you**, and **let them find harshness** (firmness) in you, and know that Allah is with those who keep their **duty** (unto Him). And whenever a Sura is revealed, there are some of them who say: **Which one of you have thus increased in faith?** As for those who believe, it has increased them in faith and they rejoice (there for). But those in whose **hearts are a disease**, it will add **wickedness to their wickedness**, and **they will die in a state of disbelief. Whether they do not see though they are tested once or twice in every year? Still they turn not in repentance, and they pay no heed.** [Sura No. – 8\9 – Tauba, Ayat No. – 123 to 126]

THEY TURN ASIDE AS THEY UNDERSTAND NOT

Ayat\Verses – Whenever there revealed a Sura, they look at each other, [saying], Does anyone see you? Then they turn aside. **Allah has turned their hearts** [from the light]; for they are a people that understand not. Now has come unto you a messenger from among (one) of yourselves, unto whom aught that you are overburdened and grievous full of concern for you, for the believers full of pity, merciful. **But if they turn away, say: Allah is sufficient for me. There is no unbreakable reality but He. On Him is my trust.** He is the Rab of the **Tremendous Throne.** [Sura No. – 8\9 – Tauba, Ayat No. – 127 to 129]

WARNINGS - EXPLAINED THEREIN IN DETAIL IN AN ARABIC QURAN

Ayat\Verses – Thus We have sent this down - **an Arabic Qur'an - and explained therein in detail some of the warnings, in order that they may fear Allah**, or that it may cause their **remembrance** [of Him]. High above all is Allah, the True King! And hasten not with the Quran ere (archaic before) its revelation has been perfected unto you, and say: My Rab! Increase me in knowledge. [Sura No. – 19\20 – Twa-Haa, Ayat No. – 113 & 114]

Sura No. – 7\8 - Anil-Anfaal

49. Lo! The hypocrites say, and those in whose hearts is a disease: These people, their religion has misled them. But if any trust in Allah, behold! Allah is Exalted in might, Wise.

50. If you could see, when the angels take the souls of the unbelievers [at death], [How] they smite their faces and their backs, [saying]: Taste the penalty of the blazing Fire.

51. Because of [the deeds] which your [own] hands sent forth; for Allah is never unjust to His servants.

52. (There way is) as the way of Ferawn's folk and of those before them, they rejected the Signs of Allah, and Allah punished them for their sins, for Allah is Strong, and Strict in punishment:

53. Because Allah will never change the grace which He has bestowed on a people until they change what is in their [own] souls: and verily Allah is He Who hears and knows [all things].

54. (There way is) as the way of Ferawn's folk and of those before them. They denied the Signs of their Rab. So We destroyed them for their sins, and We drowned the people of Ferawn, for they were all tyrants and wrong-doers.

55. For the worst of beasts in the sight of Allah are the ungrateful will not believe.

56. They are those with whom you made a treaty, but they break their promise in every opportunity, and they keep not their duty [of Allah].

57. If you gain the mastery over them in war, separate, with them, those who follow them that they may remember.

58. If you fear treachery from any group, then throw back [their treaty] to them fairly, for Allah loves not the treacherous.

59. Let not the unbelievers think that they can get the better [of the godly]: they will never frustrate [them].

60. Against them make ready your strength to the utmost of your power, including horses of war, to strike terror into [the hearts of] the enemies of Allah and your enemies, and others besides, whom you may not know, but whom Allah knows. Whatever you shall spend in the way of Allah, shall be repaid unto you, and you shall not be treated unjustly.

61. But if the enemy incline towards peace, incline you also towards peace, and trust in Allah, for He is One that hears and knows [all things].

62. If they intend to deceive you, verily Allah is sufficient for you. He is One Who has strengthened you with His aid and with [the company of] the believers.

63. Moreover (believers) have attuned their hearts. If you had spent all that is in the world you could not have attuned their hearts, but Allah has attuned them. Lo! He is Mighty, Wise.

SECTION – VIII
RIGHTEOUSNESS – 8.5
TWO WARNINGS

Fasta-'iz billaahi minash-Shaytaanir-Rajim
Bismillaahir-Rahmaanir- Rahim

Two warnings
"To enter the Masjid as they had entered it first time."

Ayat\Verses – When the **first of the warnings** [time for the first of the two] came to pass, We roused against you slaves of Ours of great might who destroyed (your) country, and it was a threat performed. Then We granted you the return as against them. We gave you increase in resources and sons, and made you the more numerous in man-power. If you did well, you did well for yourselves. If you did evil, [you did it] against yourselves. So when the **second of the warnings** came to pass, [We permitted your enemies] to destroy you, and **to enter the Masjid as they had entered it first time**, and to lie waste all that they conquered with an utter wasting. It may be that your Rab may [yet] show Mercy unto you; but if you repeat [your sins], We shall revert [to Our punishments]: And we have made Hell a prison **for those who reject [Faith]**. [Sura No. – 16\17 - Banii-Israai-iil, Ayat No. – 5 to 8]

SECOND WARNING
Eye opening Evidences

Ayat\Verses - And verily We gave to Muusa **Nine Tokens [Clear Proofs]** (of Allah's Sovereignty): Do but ask Bani-Israa-iil how he came to them? Firawn said to him: O Muusa! I consider you, indeed, to have been worked upon by sorcery! Muusa said: In truth you know well that these things (Tokens) have been sent down by none but the Rab of the heavens and the earth as **eye-opening evidence**, and I consider you indeed, O Firawn, to be one doomed to destruction! And he wished to remove them from the face of the earth; but We did drown him and all who were with him all together. And We said thereafter to Bani-Israa-iil: Dwell securely in the land [of promise], but when the **second of the warnings came to pass**, We gathered **you together in a mixed crowd**. [Sura No. – 16\17 - Banii-Israai-iil, Ayat No. – 101 & 104]

"There can be no change in the words of Allah."

Ayat\Verses – Behold! Verily on the friends of Allah there is no fear, nor shall they grieve. Those who believe and [constantly] guard against evil; for them are good tidings, in the life of the present and in the Hereafter. **There can be no change in the words of Allah**. This is indeed the supreme felicity. Let not their speech grieves you, for all power and honour belong to Allah. It is He Who hears and knows [all things]. [Sura No. – 9\10 – Yuunus, Ayat No. – 62 to 65]

WARNER OF WARNERS

Ayat\Verses – This is a **warner**, of the **warners** of old! **The threatened hour is near! None** besides Allah can **disclose** it. **Do you then doubt at this recital?** And will you **laugh** and **not weep** wasting your **time** in vanities? Rather **prostrate** yourselves before Allah and **serve** Him. [Sura No. – 52\53 – Wan-Najm, Ayat No. – 56 to 62]

Warning to the Mother of Cities

Ayat - Ha-Mim [Revelation – Sura No. – 41\42 – Shuuraa, Ayat No. - 1]

Ayat - Ain-Sin-Qaf. [Revelation – Sura No. – 41\42 – Shuuraa, Ayat No. - 2]

Ayat\Verses – Thus [He] **has sent inspiration to you as [He did] to those before you**. Allah is Exalted in Power, Full of Wisdom. To Him belongs all that is in the **heavens and in the world**; and He is The Sublime, The Tremendous. [Sura No. – 41\42 – Shuuraa, Ayat No. – 3 & 4]

Ayat\Verses – The **heavens are almost rent asunder** and the angels hymn the praise of their Rab and ask forgiveness for those in the world. Lo! Verily Allah is Ever Forgiving, Most Merciful. And as for **those who choose protecting friends besides Him, Allah is Warden over them**; and **you are not the disposer of their affairs**. And thus We have inspired you an **Arabic Qur'an** that you may warn the **Mother of Cities** and those around it, and warn [them] of the Day of Assembly, of which there is no doubt. **A host will be in the Garden, and a host of them in the flame.** If Allah had so willed, He could have made them one community. But **He brings whom He wills to His Mercy**; and **the Wrong-doers will have neither protector nor helper.** [Sura No. – 41\42 – Shuuraa, Ayat No. – 5 to 8]

Ayat\Verses – We **created man** from sounding **clay** (potter's clay) **of black mud altered.** And the **Jinn** race, We had created before of **essential fire.** And (remember) when your Rab said unto angels: Lo! I am creating **mortal** out of potter's clay of **black mud altered.** So when I have made him [in due proportion] and **breathed into him of My spirit, fall you down** in obeisance unto him. So the angels **prostrated** themselves, all of them together except the **Iblis.** He refused to be among those who prostrate. [Sura No. – 14\15 = Hur, Ayat No. – 26 to 31]

"O mankind! Your disrespect is only against yourselves."
"We shall show you the truth of all that you did."
Ayat\Verses – When We cause mankind to taste of mercy after some adversities which had touched them; behold! They **began to plotting against Our Signs**! Say: **Swifter to plan** is Allah! Verily, Our **Messengers record** all the **plots** that you make. He it is Who makes you to go on the **land** and the **sea** till, when you are in **ships** and they **sail** with them with a favourable **wind**, and they are glad therein; then comes a stormy wind and the waves come to them from all sides, and they think that they are being overwhelmed therein. They cry unto Allah, making their **faith pure** for Him only saying: if You deliver us from this, we shall truly show our **gratitude**. But when He has delivered them, behold! They **transgress** disrespectfully in the world. **O mankind! Your disrespect is only against yourselves**, an enjoyment of the life of the **present**. In the **end**, to Us is your return, and **We shall show you the truth of all that you did**. [Sura No. – 9\10 – Yuunus, Ayat No. – 21 to 23]

THEY KNOW NOTHING AFTER HAVING KNOWN MUCH

Ayat\Verses – O mankind! If you have a **doubt** about the **Resurrection**, [consider] that We **created** you out of **dust**, then out of **sperm**, then out of a **leech**-like clot, then out of a morsel of **flesh, partly formed** and **partly unformed**, in order that We may **manifest** [our power] to you; and We **cause** whom We will to rest in the wombs for an **appointed term**, then We **bring** you out as babes, then [foster you] that you may reach your **age** of full strength; and some of you are **called to die**, and some are sent back to the feeblest **old age**, so that **they know nothing after having known [much]**, and [further], you see **the earth barren and lifeless**, but when We pour down **rain** on it, it is stirred [to life], it swells, and it puts forth **every kind of beautiful growth [in pairs]**. [Sura No. – 21\22 – [Hajj], Ayat No. – 5]

Ayat\Verses – O mankind! Here is a **parable** set forth! Listen to it! **Those on whom, besides Allah, you call, cannot create [even] a fly, if they all meet together for the purpose! And if the fly took something from them, they could not rescue it from the fly. So weak are (both) the seeker and the sought. They estimate not Allah & His rightful estimate**. Lo! Allah is strong and able to Carry out His Will. Allah **chooses messengers from angels and from men**. Lo! Allah is He Who hears and sees [all things]. He knows what is before them and what is behind them; and to Allah go back all questions [for decision]. [Sura No. – 21\22 – [Hajj], Ayat No. – 73 to 76]

Ayat\Verses – Yet they worship things besides Allah, for which **no authority has been sent down to them**, and of which they have [really] no knowledge.

705

For **those that do wrong there is no helper**. And when **Our Clear Signs** are rehearsed to them, you will notice a denial on the faces of the disbelievers! They nearly attack with violence those who rehearse **Our Signs** to them. Say: Shall I tell you of something [far] worse than that? It is the **Fire** [of Hell]! Allah has promised it for those who disbelieve! And **hapless journey's end**! [Sura No. – 21\22 – [Hajj], Ayat No. – 71 to 72]

"Yet there is among men such a one as disputes about Allah, without Knowledge, without Guidance, and without a Kitaab of Light (Enlightenment)."

Ayat\Verses – This is so, because Allah is the Reality (Truth). It is He Who gives life to the dead, and it is He Who has power over all things. **And verily the Hour will come**. There can be **no doubt** about it or about [the fact] that Allah will rise up all who are in the graves. Yet there is among men such **a one as disputes about Allah, without Knowledge, without Guidance, and without a Kitaab of Light (Enlightenment) disdainfully bending his side, in order to lead [men] astray from the Path of Allah. For him there is disgrace in this life, and on the Day of Judgment. We shall make him taste the Penalty of burning [Fire]**. [It will be said]: This is because of the **deeds** which your **hands sent forth**, for verily Allah is **not unjust** to His servants. [Sura No. – 21\22 – [Hajj], Ayat No. – 6 to 10]

"They call on such deities, besides Allah, as
can neither hurt nor profit them."

Ayat\Verses – There are among men some **who serve Allah**, as it were, on the **verge**. If good befalls them, they are, therewith, well content. But if a trial comes to them, **they turn on their faces**. They lose both **this world [present life]** and the **Hereafter**. That is loss for all to see! **They call on such deities, besides Allah, as can neither hurt nor profit them. That is straying far indeed [from the Way]**! They call on one **whose harm** is nearer than his **benefit**. Indeed an **evil patron** and verily an **evil friend**! [Sura No. – 21\22 – [Hajj], Ayat No. – 11 to 13]

SCEPTIC
"Let him stretch out a rope up to the roof (of his
dwelling) and let him hang himself."

Ayat\Verses – Verily Allah will admit those who believe and do tabligh (work of righteous deeds) to Gardens, beneath which rivers flow. For Allah carries out all that He plans. **If any think that Allah will not help him in this world and the Hereafter, let him stretch out a rope up to the roof (of his dwelling)** and let him hang himself. **Then let him see whether his plan will remove that which enrages [him]!** Thus We have **revealed it** as **Clear Signs (plain revelation)**; and verily Allah guides whom He **will**! [Sura No. – 21\22 – [Hajj], Ayat No. – 14 to 16]

Ayat\Verses – O mankind! Fear your Rab! For the **earthquake** of the Hour [of Punishment] will be **a terrible thing**! On the Day when you behold it, every nursing mother will forget her nursing and every pregnant one will be delivered of her load, and you will see mankind as drunken, yet they will not be drunk but dreadful will be the wrath of Allah. And yet among mankind is he **who disputes concerning Allah, without knowledge**, and **follows every evil one obstinate in rebellion!** For him **it is decreed** that **whoso takes him for friend**, he verily will **mislead him** and **guide him** to the **punishment of the flame**. [Sura No. – 21\22 – [Hajj], Ayat No. – 1 to 4]

"Will you not then understand?"
"It is **not you** they reject. It is the **signs of Allah**, which the **wicked flouted**."
Ayat\Verses – What is the **life of this world [present life]** but play and amusement? But **best** is the home in the **hereafter**, for those who are **righteous**. **Will you not then understand? We know** indeed the **grief** which **their words do cause you**. It is **not you** they reject. It is the signs of Allah, which the wicked flouted [disobeyed]. [Sura No. – 5\6 – An-aam, Ayat No. – 32 &33]

Think you to yourselves - if you are truthful.
Say: **Think you to yourselves,** if there come upon you the punishment of Allah, or the Hour [that you fear], would you then call upon other than Allah, **if you are truthful**? Nay, On Him would you call, and if it be His will, He would remove [the distress] which occasioned your call upon Him, and you would forget [projected falsehood & breakable realities] **which you join with Him**. [Sura No. – 5\6 – An-aam, Ayat No. – 40 to 41]

SECTION – VIII
RIGHTEOUSNESS – 8.6
CONCLUSIVE WORD

Fasta-'iz billaahi minash-Shaytaanir-Rajim
Bismillaahir-Rahmaanir- Rahim

PROMISE OF ALLAH IS TRUE

"Fasbir inna wa-dallaahi haqqunw-wastagfir lizambika wa sabbih bi-hamdi Rabbika **bil-ashiyyi wal-ibkarr**". Patiently, then, persevere; for the **Promise of Allah is true**, and ask forgiveness for your fault, and hymn the praises of your Rab at fall of night [**bil-ashiyyi**] and in the early hours [**wal-ibkarr**]. [Sura No. 39\40 Mu-Min, Ayat No. – 55]

PROMISE OF ALLAH

Ayat\Verses – When those who disbelieve had setup in their hearts heat and false piety, the heat and cant of the Age of ignorance, then Allah sent down **His peace of reassurance upon His Messenger** (tranquillity to His Messenger) and to the believers, and imposed on them **the word of self-restraint**; for they were entitled to it and worthy of it. And Allah is Aware of all things. **Allah has fulfilled the vision of His Messenger in very truth. You shall indeed enter the Inviolable Place of Worship (Masjidal – Haraamaa), if Allah will, with minds secure**, heads shaved, hair cut short, and without fear. But **He knows that which you know not, and He has granted, besides this, a near victory**. It is He Who has sent His Messenger with Guidance and the Religion of Truth (Diinil-Haqq) that **He may make it to prevail over all religions [projected falsehood]**, and enough is Allah for a **Witness. Muhammad is the messenger of Allah**; and **those who are with him are hard against disbelievers**, but **merciful among themselves**. You will see them bowing and falling prostrate, seeking bounty from Allah and (His) acceptance. **This is their similitude in the Tawraat; and their similitude in the Injiil** like the sown corn that sends forth its blade (shoot), and then makes it strong; it then becomes thick, and it stands on its own stem delighting the showers that He may enrage the disbelievers with (the sight of) them. **Allah has promised those among them who believe and do righteous deeds forgiveness and a great Reward**. [Sura No. – 47\48 – Fat-ham-Mubiina, Ayat No. – 26 to 29]

"The promise of Allah is the Truth."

Ayat\Verses – And as for those who **disbelieved** (rejected faith), (it will be said to them): Were not **Our Signs** (Revilations) rehearsed to you? But you were **scornful**, and became a **guilty folk**. And when it was said: Lo! **The promise** of Allah is the Truth, and there was no doubt of the Hour's coming; you used to say: We know not what the Hour is. We only deem it nothing but a conjecture (an idea), and we are by no means convinced. And the evil of what they did will appear to them, and they will be completely encircled by that which they used to mock at! And it will also be said: This Day We will forget you as you forgot the meeting of this Day of yours! And your abode is the Fire, and there is none to help you. This, because you used to take the Signs (Manifested Nature) of Allah in jest, and the life of the world deceived you. [From] that Day, therefore, they shall not be taken out thence, nor shall they be received into Grace. [Sura No. – 44\45 – Tanziilul Kitaab (Prev. Jaasiya), Ayat No. – 31 to 35]

"Give a respite [break] to the disbelievers"

Ayat\Verses – And if it were Our Will, We could have set among you angels to be viceroys in the world. And lo! Verily there is knowledge of the Hour. So doubt you not concerning it, but follow me. This is the Right Path [towards Upright West]. And let not Shaytan (The Historical Don of two-in-one) turn you aside (hinder the revealed and manifested magnetism); for he is an open enemy for you. [Sura No. 42\43 – Ummil-Kitaab \ (Zukhruf), Ayat No. – 60 to 62]

Ayat\Verses – When Isa came with Clear Proofs (of Allah's Sovereignty), he said: I have come to you with wisdom, and in order to make clear to you some of the [points] on which you dispute. Therefore keep your duty to (fear) Allah and obey me. Lo! Allah, He is my Rab and your Rab. So worship Him. This is a Right Path. [Sura No. 42\43 – Ummil-Kitaab \ (Zukhruf), Ayat No. – 63 & 64]

Ayat\Verses – By the heaven and the morning star; ah, what will tell you what the morning star is?! The sharp star there is no soul but has a guardian over it. So let man consider from what he is created! He is created from a gushing fluid that issued from between mim-baynis-sulbi wat-taraaa-ib [loins and ribs]. Surely [Allah] is able to bring him back [to life] on the Day when hidden thoughts shall be searched out. Then he [man] will have no power, and no helper. [Sura No. 85\86 – Wat-Taariq, Ayat No. – 1 to 10]

Ayat\Verses – It is not a thing for amusement. As they plot a scheme [against followers of equal & opposite middle course towards Upright West] [Sura No. 85\86 – Wat-Taariq, Ayat No. – 1 to 10]

Ayat\Verses – By the firmament [seven windows], [Sura No. 85\86 – Tariq, Ayat No. – 11] And the earth which opens out [Sura No. 85\86 – Tariq, Ayat

No. – 12] Behold this is the **Conclusive Word of the Quran**. [Sura No. 85\86 – Wat-Taariq, Ayat No. – 13]

Ayat\Verses – And I [Allah] plotted a plot [**against them**]. Therefore **give a respite to the disbelievers**. Deal you gently with them for a while. [Sura No. 85\86 – Wat-Taariq, Ayat No. – 1 to 10]

"You will find no change in the Law of Allah"
Ayat\Verses – If those who reject faith (disbelievers) join battle with you, they would certainly turn their backs; then they will find neither protector nor helper. It is **the Law of Allah** which has taken course afore time. **You will find no change in the Law of Allah**. And it is He Who has restrained their hands from you and your hands from them in the valley of Makka, after that He gave you the victory over them. And Allah sees well all that you do. **They are the ones who denied Signs of Allah [Manifested Nature] and hindered (debarred) you from the Sacred Mosque (equal & opposite right direction) and debarred the offering from reaching its goal**. And if it not been for believing men and believing women, whom you know not – lest you should trampled them down and on whose account a crime would have accrued to you without [your] knowledge. [Allah would have allowed you to force your way, but He held back your hands] that He may admit to His Mercy whom He will. If they had been apart, We should certainly have punished the disbelievers among them with a grievous punishment. [Sura No. – 47\48 – Fat-ham-Mubiina, Ayat No. – 22 to 25]

AN UNFAILING REWARD
Ayat\Verses – **We have truly sent you as a witness, as a bearer of good tidings, and as a warner in order that you may believe in Allah and His Messenger** that you may assist and honour Him, and may glorify Him at **early dawn and at the close of day**. Verily those who swear allegiance, swear allegiance to Allah. The Hand of Allah is above their hands. So **whosoever violates his oath does so to the harm of his own soul**, and **anyone who fulfils** what he has covenanted (promised) with Allah, Allah will soon grant him a great Reward. **Those of the wandering Arabs who were left behind** will tell you: Our possessions and our households occupied us, ask for forgiveness for us! They speak with their tongues which are not in their hearts. Say: Who then has any power at all [to intervene] on your behalf with Allah, if His Will is to give you hurt or to give you some profit? But Allah is ever Aware of what you do. Nay, you thought that the Messenger and the believers would never return to their families [right direction]. This seemed pleasing in your hearts, and you did think an evil thought, and yet you were worthless folk. And if any believe

not in Allah and His Messenger, We have prepared, **for those who reject faith, a blazing fire! To Allah belongs the Sovereignty** of the heavens and the world. He forgives whom He wills, and He punishes whom He wills. But Allah is Ever Forgiving, Most Merciful. [Sura No. – 47\48 Fat-ham-Mubiina, Ayat No. 8 to 14]

SURA No. – 67\68 NAME OF THE SURA – QALAM

Nuuun. By the pen and that which they write (therewith), you are not, **by the Grace of your Rab unto you,** a **madman.** [Sura No. – 67\68 – Qalam, Ayat No. – 1 & 2]

And lo! Verily for you is **a Reward unfailing.** And lo! You [stand] on an exalted standard of character. **And soon you will see, and they will see,** which of you is afflicted with madness? Verily your Rab is best aware of him who goes astray from His way, and He is best aware of those who receive [true] Guidance (walk aright). **So obey not you the rejecters [those who deny manifested signs of Allah] who would have had you compromise, that they may compromise.** Neither obey you each feeble oath-monger, a slanderer, going about with lies, [habitually] hindering [all] good, transgressing beyond bounds, deep in sin, greedy (violent) with all that, base-born (intrusive), because he possesses wealth and children **that when to him are rehearsed Our Signs, he says mere fables of the men of old! We brand him on the nose.** [Sura No. – 67\68 – Qalam, Ayat No. – 3 to 16]

Verily We have tried them as We tried the People of the Garden, when they resolved to gather the fruits of the [garden] in the next morning. And made no exception. Then there came on the [garden] a **visitation** from your Rab, while they were asleep. And in the morning it was as if plucked as the morning broke, they cried out, one to another, saying: Run unto your field if you would pluck (the fruit). So they departed, conversing in secret low tones. No needy man shall enter it today against you. And they opened the morning, strong in an [unjust] resolve. But when they saw it [garden], they said: We have surely lost our way (we were in error). [Sura No. – 67\68 – Qalam, Ayat No. – 17 to 26]

Indeed we are shut out [of the fruits of our labour]! The best among them said: Did I not say to you, Why not glorify [Allah]? They said: Glory to our Rab! Verily we have been doing wrong! Then they turned, one against another, in reproach. They said: Alas for us! We have indeed transgressed! It may be that our Rab will give us in exchange a better [garden] than this, for we do turn to Him [in repentance]! Such is the Punishment [in this life]; but greater is the Punishment in the Hereafter, if only they knew! [Sura No. – 67\68 – Qalam, Ayat No. – 27 to 33]

711

Verily, for the Righteous, are Gardens of Delight, in the Presence of their Rab. **Shall We then treat those who have surrendered as We treat the guilty?** What is the matter with you? How foolishly you judge? Or have you a **Kitaab** wherein you learn, that you shall indeed have all that you choose? Or have you a Covenant on oath from Us that reaching to the Day of Resurrection [Yawmal-Qiyaamati] that yours shall be all that you ordain? Ask them: Which of them will stand surety for that? **Or have they some Partners (projected mechanism)? Then let them bring their Partners (come forward with their mechanical globalisation and artificial magnetism), if they are truthful!** On the Day when it befalls in earnest, and they shall be summoned to bow in adoration, but they shall not be able, their eyes will be cast down,- ignominy will cover them; seeing that they had been summoned afore time to bow in adoration, while they were unhurt. [Sura No. 67\68 – Qalam, Ayat No. 34 to 43]

Leave Me (to deal) with those who give the lie to this pronouncement. We shall lead them on by steps from whence they know not. Yet I bear with them, for lo! My scheme is firm. Or is it that you ask a fee from them so that they are heavily taxed, or that the Unseen is in their hands, so that they can write (thereof)? But wait with patience for the decree of your Rab, and be not like the **companion of the fish**, who cried out in despair. Had not Grace from his Rab reached him, he would indeed have been cast off on the naked shore, in disgrace. Thus did his Rab choose him and make him of the Company of the Righteous. And the disbelievers would almost trip you up with their eyes when they hear the reminder; and they say: Lo! He is indeed mad! But **it is nothing less than a reminder to the universe.** [Sura No. – 67\68 – Qalam, Ayat No. – 34 to 43]

SECTION – VIII
RIGHTEOUSNESS – 8.7
END IS FOR THOSE WHO ARE RIGHTEOUS

Fasta-'iz billaahi minash-Shaytaanir-Rajim
Bismillaahir-Rahmaanir- Rahim

"Alif-Lam-Ra. Kitaabun-uh-kimat Aayaa-tuhu summa fussi-lat milla-dun Hakiimin khabiir" [Trans.] - This is a **Kitaab** with **basic or fundamental Ayat** [of established meaning], **further explained in detail [verses]**, from One **Who is Wise and Well- acquainted** with all things] [1].(Saying) you should serve none but Allah. Verily I am unto you from Him a **Warner and a Bringer** of good tidings. [And to moralise you thus], seek you the forgiveness of Allah, and **turn to Him in repentance**; that He may grant you enjoyment, well [and true], for a term appointed, and bestow His abounding grace on all **beautiful ones**. But if you **turn away**, then I fear for you the penalty of a great day. To Allah is your return, and He has power over all things. [Sura No. – 10\11 – Kitaabun-uh-kimat (Prev. – Huud), Ayat No. – 1 to 4]

Behold! They fold up their hearts, that they may lie hid from Him! Ah even when they cover themselves with their garments, He knows **what they conceal, and what they reveal**: for He knows well the [inmost secrets] of the hearts. **There is no moving creature in the world but its sustenance depends on Allah: He knows the time and place of its definite abode and its temporary deposit: All is in a clear Record. He it is Who created the heavens and the world in six Days, and His Arsh [Throne] was over the waters that He might try you, which of you is best in conduct.** But if You (Prophet) say to them, You shall indeed be raised up after death, the **Unbelievers would be sure to say,** this is nothing but **obvious sorcery (open magic) If We delay the penalty for them for a definite term, they are sure to say: What keeps it back? Ah! On the day it [actually] reaches them, nothing will turn it away from them, and they will be completely encircled by that which they used to mock at!** [Sura No. – 10\11 – Kitaabun-uh-kimat (Prev. – Huud), Ayat No. – 5 to 8]

If We give man a taste of Mercy from Ourselves, and then withdraw it from him, behold! He is in despair and [falls into] **wickedness**. But if We give him a taste of [Our] favours **after misfortune** has touched him, **he is sure to say: all evil has departed from me: Behold! He falls into delight and pride**. Not so do those who show patience and constancy, and work righteousness; for them is forgiveness and a **great reward**. By chance you may [feel the inclination] to give up a part of what is revealed unto you, and your heart fee straitened lest

713

they say: Why is not a treasure sent down unto him, or why does not an angel come down with him? But you are there only to warn! It is Allah that arranges all affairs. Or they may say: He forged it, say, **bring you** then **ten Suras forged**, like unto it, and **call** [to your aid] **whomsoever you can**, other than Allah!- **If you speak the truth!** If then they [your false gods] answer not your [call], know you that this revelation is sent down only with the knowledge of Allah, and that there is no ilaaha but He! Will you then be those who surrender? [Sura No. – 10\11 – Kitaabun-uh-kimat (Prev. – Huud), Ayat No. – 6 to 14]

Those who desire the life of the present and its glitter, to them We shall repay their deeds herein, and therein they will not be wronged. They are those for whom there is nothing in the Hereafter but the Fire, useless are the designs they frame herein, and of no effect and the deeds that they do! Whether they are [like] those who accept a Clear [Sign] from Allah and whom a witness from Him recites, as did the **Kitab of Muusa** before it, an example and a mercy? **They believe therein; but those of the Sects that reject it, the Fire will be their promised meeting-place.** Be not then in doubt thereon: for it is the truth from Allah: yet many among men do not believe! **Who does greater wrong than those who invent a lie against Allah?** They will be turned back to the presence of Allah, and the witnesses will say: these are the ones who lied against Allah! Behold! The Curse of Allah is on those who do wrong! **Those who debar from the path of Allah and would have it something crooked or global: these are they who denied the Hereafter. Such will not escape in the world, nor have they any protecting friends besides Allah.** Their penalty will be doubled! They will not bear the power to hear, and they will not see! They are the ones who have lost their own souls: and that which they used to invent has failed them. Without a doubt, these are the very ones who will lose most in the Hereafter! But those who believe and work righteousness, and humble themselves before Allah, they will be companions of the gardens, to dwell therein. **The similitude of the two parties is as the blind and the deaf; and the seer and the hearer. Are they equal in similitude? Will you not then be admonished?** [Sura No. – 10\11 – Kitaabun-uh-kimat (Prev. – Huud), Ayat No. – 15 to 24]

We sent Nuuh to his people (and he said): I am a **plain Warner unto you**. That you serve none but Allah: surely I do fear for you the penalty of a grievous day. But the chieftains of the Unbelievers among his people said: We see [in] you but a moral like us, and we see not that any follow you except the most miserable among us, without reflection; nor do we see in you (all) any merit above us, in fact **we think you are liars**. [Sura No. – 10\11 – Kitaabun-uh-kimat (Prev. – Huud), Ayat No. – 25 to 27]

He said: O my people! Bethink you if [it be that] I rely on **a Clear Sign (proof)** from Allah, and that He has sent Mercy unto me from His own presence, but that the Mercy has been **obscured from your sight**, can we compel you when you are unwilling to accept it? And O my people! I ask you for no wealth in return: my reward is from none but Allah. But I will not drive away [in contempt] those who believe, for verily they are to meet their Rab, and you I see are the **ignorant ones**! [Sura No. – 10\11 – Kitaabun-uh-kimat (Prev. – Huud), Ayat No. – 28 & 29]

And O my people! **Who would help me against Allah if I drove them away? Will you not then reflect? Neither I am telling you that with me are the treasures of Allah, nor I know what is hidden, nor I claim to be an angel, nor yet do I say, of those whom your eyes do despise that Allah will not grant them [all] that is good:** Allah knows best what is in their souls: **I should, if I did, indeed be a wrong-doer.** [Sura No. – 10\11 – Kitaabun-uh-kimat (Prev. – Huud), Ayat No. – 30 & 31]

They said: O Nuuh! You has disputed with us, and multiplied disputation with us. Now **bring upon us that where with you threaten us, if you speak the truth**. He said: Truly, Allah will bring it on you if He wills, and then, you will not be able to escape it! Of no profit will be my warning to you, much as I desire to give you [good] warning, if it be that Allah wills to leave you **astray**: He is your Rab! And to Him will you return! Or do they say: He has invented it. Say: If I had invented it, on me would be my sin! And I am innocent of all that you commit. [Sura No. – 10\11 – Kitaabun-uh-kimat (Prev. – Huud), Ayat No. – 32 to 35]

It was **inspired** in Nuuh, (saying): None of your people will believe except those who have believed already! Be not distressed because of what they do. But construct an Ark (Ship) under Our eyes and Our inspiration, and speak not unto Me on behalf of **those who do wrong: for they will be drowned**. Forthwith he [starts] constructing the Ark: every time that the **chieftains [projected pillars] of his people passed by him**, they **threw ridicule on him. He said: if you ridicule us now, we [in our turn] can look down on you with ridicule likewise.** But soon you will know to whom will descend a penalty that will cover them with shame, and on whom will be unloosed a penalty lasting. [Sura No. – 10\11 – Kitaabun-uh-kimat (Prev. – Huud), Ayat No. – 36 to 39]

At length, behold! There came Our command, and the **fountains** of the world gushed forth. We said: embark therein **two of every kind, male and female, and your family** - except those against whom the word has already gone forth, and the Believers; but only a few believed with him. So he said: embark you on the Ark, In the name of Allah, whether it move or be at rest! For my Rab is, be sure, Forgiving, And Merciful. [Sura No. – 10\11 – Kitaabun-uh-kimat (Prev. – Huud), Ayat No. – 40 & 41]

So the **Ark floated with them on the waves [stages of journey] like mountains**, and Nuuh cried out unto his son – and he was standing aloof - **O my son! Come with us, and be not with the unbelievers.** The son replied: I will betake myself to some mountain [for instance a fictitious planet called Mars], it will save me from the water. Noah said: this day nothing can save, from the command of Allah, save him whom He hath had mercy! And the waves came between them, and the son was among drowned in the Flood. **Then the word went forth: O world! Swallow up your water** and **O sky! Be cleared of clouds** and **the water was made to subside.** And the Commandment was fulfilled. The **Ark rested on Mount Alal-Ju-diyyi**, and it was said: **Away with those who do wrong.** [Sura No. – 10\11 – Kitaabun-uh-kimat (Prev. – Huud), Ayat No. – 42 to 44]

And **Nuuh cried unto Allah**, and said: **O Allah! Surely my son is of my family!** And **Your promise is true, and You are the Justest of Judges!** He said: O Nuhh! **He is not of your family: for he is of evil conduct.** So, ask not of Me **that of which you have no knowledge. I admonish you,** lest **you act like the ignorant.** Nuhh said: "O my Rab! I do seek refuge with You, lest I ask You for that of which I have no knowledge. And unless You forgive me and have Mercy on me, I should indeed be lost! The word came: "O Nuhh! **Come down [from the Mountain]** with peace from Us, and blessing on you and on some of the peoples [who will spring] from those with you: but [there will be other] peoples to whom We shall grant their pleasures [for a time], but in the end will a grievous penalty reach them from Us. Such are **some of the tales of the unseen**, which We have revealed unto you: before this, neither You nor your people knew them. So persevere patiently: for the **End is for those who are righteous.** [Sura No. – 10\11 – Kitaabun-uh-kimat (Prev. – Huud), Ayat No. – 45 to 49]

To the Aad People [We sent] **Huud**, one of their own **brethren**. He said: O my people! **Worship Allah!** You have no other reality except Him. **You do nothing but invent.** O my people! I ask of you **no** reward for this [Message]. My reward is from none but Him who created me: **Will you not then understand?** And O my people! Ask forgiveness of your Rab, and **turn to Him** [in repentance]: He will cause the sky to rain abundance on you, and add strength to your strength: so **turn you not back in sin.** [Sura No. – 10\11 – Kitaabun-uh-kimat (Prev. – Huud), Ayat No. – 50 to 52]

They said: O Huud! **No Clear [Sign] that you have brought us, and we are not the ones to desert our projected realities on your word! Nor shall we believe in you.** We say nothing except that [perhaps] one of **our realities may have possessed you in an evil way.** He said: **I call Allah to witness, and you will bear witness, that I am innocent of (all) that you ascribe as partners (to Allah -) Besides Him! So scheme [your worst] against me, all of you, and give me no break.** I put my trust in Allah, My Rab and your Rab! There is not a moving creature, but He has grasp of its fore-lock. Verily, it is my **Rab that is on a straight Path [towards West].** [Sura No. – 10\11 – Kitaabun-uh-kimat (Prev. – Huud), Ayat No. – 53 to 56]

If you turn away, I [at least] have conveyed the Message with which I was sent to you. My Rab will make another people to succeed you, and you will not harm Him in the least. For my Rab has care and watch over all things. So when Our decree issued, We saved Huud and those who believed with him, by [special] Grace from Ourselves: We saved them from a severe penalty. Such were the Aad People: **they rejected the Signs of their Rab and Cherisher**; disobeyed His messengers; And followed the command of every powerful, **obstinate transgressor.** And they were pursued by a Curse in this life, and on the Day of Judgment. Ah! Behold! For the Aad rejected their Rab and Cherisher! Ah! Behold! **Removed [from sight) were Aad, the people of Huud.** [Sura No. 10\1 Kitaabun-uh-kimat (Prev. – Huud), Ayat No. 57 to 60]

THOSE WHO REPENT AND BELIEVE, AND DO DEEDS OF RIGHTEOUSNESS [TABLIGH] THERE IS NO SIN

Ayat\Verses – Obey Allah, and obey the Prophet, and beware (of evil): If you do **turn back,** know you that it is Our Prophet's duty to proclaim (the Message) in the **clearest manner. On those who believe and do deeds of righteousness there is no sin for what they ate (in the past), when they guard themselves from evil, and believe, and do deeds of righteousness, (or) again, guard themselves from evil and believe, (or) again, guard themselves from evil and do good. For Allah loves those who do good.** [Sura No. – 4\5 – Maaaidah, Ayat\Verses – 92 & 93]

Ayat\Verses – O you who believe! Stand out firmly for Allah, as witnesses to fair dealing, and **let not the hatred of others to you make you swerve to wrong and depart from justice.** Be just: That is next to Piety: And fear Allah; for Allah is well acquainted with all that you do. Allah has promised **forgiveness** and **great reward** to those who **believe** and do **deeds of righteousness.** Those who reject faith and **deny Our Signs** will be **Companions of Hellfire.** [Sura No. – 4\5 – Maaaidah, Ayat\Verses – 8 to 10]

Ayat\Verses – But after them there followed a posterity who missed prayers and followed after lusts soon, then, will they face destruction; except those who **repent** and **believe**, and do righteous **work [tabligh]**, for they will **enter the Garden** and **will not be wronged** in the **least. Gardens of Eternity**, those which [Allah] Most Gracious has **promised** to His **servants** in the **Unseen**, for His **promise must [necessarily] come to pass.** They **will not** there hear any **vain discourse**, but **only salutations of Peace.** And **they will have therein their sustenance, morning and evening.** Such is the Garden which We give as an inheritance to those of Our servants who guard against evil. [The angels say:] We descend not but by command of your Rab. To Him belongs what is before us and what is behind us, and what is between; and your Rab never is not forgetful. Rab of the heavens and of the earth, and of all that is between them! So worship Him, and be constant and patient in His service; **knowing you one that can be named along with Him?** [Sura No. – 18\19 – Maryam, Ayat\Verses – 59 to 65]

Ayat\Verses – But verily your Rab, for **those who leave their homes after trials and persecutions, and who thereafter strive and fight for the faith and patiently persevere**, lo! your Rab, after all this (for them) is **Forgiving, Merciful.** One Day every soul will come up struggling for itself and every soul will be recompensed [fully] for all its actions, and **none will be unjustly dealt with.** Allah sets forth a **Parable: A city enjoying security and quiet, abundantly supplied with sustenance from every place**; but it **disbelieved** the **favours** of Allah. So Allah made it **taste of hunger and terror** [in extremes] [closing in on it] like a **garment** [from every side], **because** of the [evil] which [its people] **wrought.** And there came to them a **Messenger from among themselves**, but they **falsely rejected him.** So the **wrath** seized them **even in the midst of their iniquities.** So **eat of the sustenance which Allah has provided for you**, lawful and good; and be grateful for the favours of Allah, if it is Him you serve. [Sura No. – 15\16 – Nahl, Ayat\Verses – 110 to 114]

Ayat\Verses – But **those who believe and do deeds of righteousness**, We shall soon admit to Gardens, with rivers flowing beneath, their eternal home. Therein they will have companions pure and holy. We shall admit them to shades, cool and ever deepening. Lo! **Allah commands you that you restore deposits to their owners, and if you judge between mankind, that you may judge justly.** [Sura No. – 3\4 – Nisaa, Ayat No. – 57 & 58]

Ayat\Verses – But those who believe and do deeds of righteousness, we shall soon admit them to gardens, with rivers flowing beneath, to dwell therein forever. It is a promise from Allah in truth; and who can be more truthful

than Allah in utterance? It will not be in accordance with your desires, nor the desires of the people of the Kitaab. Whoever works evil, will be requited accordingly. Nor will he find, besides Allah, any protector or helper. If any do deeds of righteousness, be they male or female - and have faith, they will enter Heaven, and not the least injustice will be done to them. Who is better in Diin (religion) than he who submits his purpose (his whole self) to Allah while doing good (to man) and follows the way of Ibrahim, the upright? Allah (Himself) chose Ibrahim for friend. Unto Allah belong all things in the heavens and on earth. And He it is that encompasses all things. [Sura No. – 3\4 – Nisaa, Ayat No. – 122 to 126]

Ayat\Verses – A part of My servants there are, who used to pray 'our Rab! We believe. Then do You forgive us, and have mercy upon us. For You are the Best of those who show mercy! **But you treated them with ridicule, so much so that [ridicule of] them made you forget My Message while you were laughing at them!** I have **rewarded** them this Day for their **patience and constancy**. They are indeed the ones that have achieved Bliss. [Sura No. – 22\23 – Mu-Minuun, Ayat\Verses – 109 to 111]

"As to those who believed but did not leave their homes, you have no duty to protect them until they leave their homes"

Ayat\Verses – O Prophet! Say to those who are imprisoned in your hands: If Allah finds any reality in your hearts, He will give you something better than what has been taken from you, and He will forgive you: for Allah is Ever Forgiving, Most Merciful. And if they would **betray you, they betrayed Allah** before, and He gave (you) power over them. Allah is Knower, Wise. **Those who believed, and left their homes, and fought for the Faith, with their property and their lives, in the cause of Allah, as well as those who gave [them] asylum and aid, these are [all] friends and protectors of one another. As to those who believed but did not leave their homes, you have no duty to protect them until they leave their homes;** but if they seek your help in the matter of religion, then it is your duty to help them, except against a group with whom you have a treaty of mutual union. And [remember] Allah seeth all that you do. The unbelievers are protectors, one of another: Unless you do this, [protect each other], there would be confusion in the land, and great corruption. **Those who believe, and left their homes, and fight for the Faith, in the cause of Allah as well as those who give [them] asylum and aid, these are [all] in very truth the believers. For them is the forgiveness of sins and a bountiful provision.** And those who accept Faith subsequently, and left their homes, and fight for the Faith in your company, they are of you.

But kindred (family members / relatives) by blood have prior rights against each other in the Kitab of Allah. Verily Allah is well-acquainted with all things. **[Sura No. – 7\8 – Anil-Anfaal, Ayat No. – 70 to 75]**

Ayat\Verses – To the righteous [when] it is said: What is it that your Rab has revealed? They will say: All that is good. To those who do good, there is good in this world, and the Home of the Hereafter is even better and excellent indeed is the Home of the righteous. Gardens of Eternity which they will enter beneath them flow [pleasant] rivers. They will have therein all that they will. Thus Allah repays those who ward off (evil). [Namely] those whose lives the angels take in a state of purity, saying [to them]: Peace be on you; enter you the Garden, because of [the good] which you did [in the world]. [Sura No. – 15\16 – Nahl, Ayat\Verses – 30 to 32]

Ayat\Verses – What is with you must **vanish**: what is with Allah will **endure**. And We will **certainly bestow**, on those who **patiently persevere**, their **reward** according to the **best of their actions**. Whoever **works righteousness, man or woman**, and has **Faith**, verily, to him will We give **a new Life, a life that is good and pure** and We will **bestow** on such **their reward according to the best of their actions**. [Sura No. – 15\16 – Nahl, Ayat\Verses – 96 & 97]

A REMINDER AND A MAN OF YOUR OWN PEOPLE TO WARN YOU

Ayat\Verses - **Do you wonder that there will come to you a reminder from your Rab, through a man of your own people, to warn you,** so that you may fear Allah and haply receive His Mercy? [Sura No. – 6\7 - As-haabul- a' raaf, Ayat No. – 63]

Ayat\Verses - **Do you wonder that there has come to you a reminder from your Rab through a man of your own people, to warn you?** Call in **remembrance** that He **made** you inheritors after the people of Nuuh, and **gave** you a **height tall among the nations**. Call in **remembrance** the benefits [you **have** received] from Allah: so that you may prosper. They **said**: Come you to us, that we **should** serve Allah alone, and **give** up what our fathers worshiped? Then **bring** upon us that wherewith you **threatened** us, if you **are** telling the truth! He **said**: Terror and wrath **have** already fallen on you from your Rab. **Would** you argue with me over names which you **have** named you [manmade directions] and your **fathers** without **authority from Allah**? Then **wait**: I **am** amongst you, also waiting. We **saved** him and those who **adhered** to him by mercy from Us, and We cut off the roots of those who rejected **Our signs** [Manifest Truth] and did not believe. [Sura No. – 6\7 - As-haabul- a' raaf, Ayat No. – 69 to 72]

YAWMAL-QIYAAMATI

Say: Have you thought on all that you invoked besides Allah? Show me what they have created in the world. Or have they a share in the heavens? Bring me a Kitaab [revelation] before this or any remainder of knowledge [in support of what you say], if you are truthful. And who is more astray than one who invokes besides Allah, pray unto such as hear not their prayer until the **Day of Resurrection [Yawmal-Qiyaamati],** and who [in fact] are unconscious of their call [to them]? And when mankind are gathered together [at the Resurrection], they will be hostile to them and reject their worship [altogether]! And **when Our Clear Signs are rehearsed to them, the disbelievers say of the Truth: When it comes to them. This is evident sorcery (mere magic)!** [Sura No. – 45\46 – Bil-ahqaafi, Ayat No. – 4 to 7]

REFERENCES
WORKS OF THE PEOPLE OF THE KITAAB WITH TRUTH

The sharing of the 'Kitaba Wal-Hikmata' or 'Manifested Nature and the Utility of One's Upright Logic' is neither a so-called research work nor a book written on a particular issue or topic. On the contrary, it is an inspired sharing of the searched out Manifest Truths which are justifiable as well as verifiable on the basis of criterions of truth, manifested signs, equal & opposite stages of journey of the so-called sun [Bullet] as the manifested signs of natural magnetism, searched out real scientific symbols from the separate sheet in resemblance with Revelations [**Universal Major Premises** of **Dictum de Omni et Nullo** or **Kitaab with Truth**]. In other words, this is a sharing of the searched out Manifested Nature contrary as well as contradictory to projected Manmade Nature with a view to get rid of several **subjective selfcontradictions** and **objective paradoxes**. If it is asked, "Which direction represents Revealed & Manifested North or from which direction the electromagnetic wave enters and towards which direction it ends the equal & opposite stages of journey [so-called sun as the manifested sign of natural magnetism in resemblance with revelation]", then are you able to provide a definite answer free from subjective selfcontradictions and objective paradoxes? If facing towards the so-called rising sun from equal & opposite seashores, I want to search out North Direction, then the left side in resemblance with my left hand side is the projected North. If I want to search out North direction in the projected Global Trinity [map on the globe], then the top or upward direction is resemblance with projected Northern Hemisphere is North Direction. If I want to know the direction concerning the running of water from the hills, mountains, and even of a flood, then the answer is North-East direction. Now, it may be concluded that North Direction is a Trinity of Left side – Upward direction – Downward direction. This Trinity is a full of subjective selfceontradictions and objective paradoxes. Under the guidance of my Rab, I have searched out Revealed Left and Manifested North-East & North-West directions on the basis of Real Scientific Symbols, Complete Coded Shared Tautologies, and equal & opposite stages of journey of the so-called sun [Bullet] as the manifested sign of natural magnetism in resemblance with revelations. So, the findings of this inspired sharing resemble with both Real Science and Kitaab with Truth.

References of this searched out findings / manifest truths are of two kinds - (i) References of Minor Premises or **General Truths are manifested signs, verifiable scientific laws, verifiable symbols available in the separate sheet** of Real Science i.e. Insert Symbol, justifiable **Philosophical theories of truth**, and pen-paper-pencil **works & activities of the so-called epistemic persons**; (ii) References of the **Universal Major Premises of Dictum de Omni et Nullo are both Truths-in-themselves [Fundamental Ayat] and Coded Shared Tautologies** [detail explanations of Fundamental Ayat] i.e. Kitaab with Truth]. If a law or an established theory contradicts with the Universal Major Premise, then it is neither right knowledge nor wrong knowledge [Ref. Plato] save teleological evidence sorcery and unique epistemic persecution. In other words, such law or theory is either contingent or contradictory but cannot be categorised as a tautology. The establishment of contingency or contradiction as truth is the conspiracy work of the Real Activists; while the projection of contingency or contradiction as truth **with a view to kill faith & belief of mankind** is the work of the Unique Epistemic Persecutor. It is an expectation that the findings of this sharing will be able to unveil the two-in-one partnership of the Historical Don of Trinity & Duality as well as Historical Black Magic along with the Owner and Possessors of Manmade Nature. So, this sharing is an inspirational warning to the Historical Don to choose one of his two-in-one faces. The findings of this sharing are being shared with reference to real scientific symbols along with their scientific character codes so that you can justify as well as verify selfevident truth. References of the Universal Major premises are being shared on the basis of the following works of the people of the Kitaab with Truth.

REFERENCES
WORKS OF THE PEOPLE OF THE KITAAB WITH TRUTH
REJECT SUCH MEN AS USE BAD LANGUAGE
Fasta-'iz billaahi minash-Shaytaanir-Rajim
Bismillaahir-Rahmaanir- Rahim
Ayat\Verses – The most beautiful names belong to Allah: so call on Him by them; but reject such men as use bad language in His names: for what they do, they will soon be requited? And of those whom We have created there is a nation (Ummat or Middle Stair out of three ascending stairs) who guide with truth and establish justice therewith. [Sura No. – 6\7 - As-haabul- a' raaf, Ayat No. – 180 & 181]

(A) **IFTA**
1. IFTA – 'THE HOLY QURAN' – English Translation of the Meanings and Commentary – Revised & Edited by the Presidency of Islamic Researches, IFTA, King Fahd Holy Quran Printing Complex. [Gifted by Mufti Khan MD, on 07 – 4 -1995, at Naglui Market, New Delhi, India, Asia.]

(B) OFF-SPRING OF IFTA

2. **Off-Spring of IFTA** - 'The Holy Quran' – with Original Arabic Text – Transliteration in Roman Script – by M. A. Haleem Eliyasee – English Translation by Abdullah Yusuf Ali – Published by – Islamic Book Service (P) Ltd., Reprint Edition – 2012. [Purchased]

(C) XEROX COPY OF IFTA

3. **Xerox Copy of IFTA** - KANZUL IMAN O NURUL IRFAN [PART – I & II[- by Ala Hazrat Mujaddid –e-Din O Millat Imam Muhammad Ahmed Reza Khan Brellawi [Rahmatullahi Alaihi & Translated in to Bengali by Al-haj Maulana Muhammad Abdul Mannan – Published by – Chistia Markkaz Dargah Sultanul Hind, Atai Rasul [Purchased]

(D) FAINT COPIES OF IFTA

4. **Faint copy of IFTA** - Nur Nurani Quran Shareef' – [The Holy Quran] – with Arabic Text – Transliteration in Bangla Script and Bangla Translation – by Mawlana Mohammad Mubarak Karim & Mawlana Mohammad Morshad Alam Hatueb – Published by – Sulemaniya Book House – Bangla Mazar, Daka, Bangladesh, Asia.[Purchased]

5. **Faint copy of IFTA** - 'Quran Sharif' – Banganubad – O – Byakhya – by Prof. Dr. Usman Gani – Published by Mallik Brothers, 45 – College Street, Kalkata, India, Asia. [Purchased]

6. **Faint copy of IFTA** - 'Pabitra Quran Shareef' - [The Holy Quran] – with Arabic Text – Transliteration in Bangla Script and Bangla Translation - By Al-haj Mawlana A K M Fazlur Rahman Munshi – Published by – Naz Book Depo, 6 – Balai Dutta Street, Kalkata, India, Asia [Purchased]

(E) IBS (P) Ltd. –COMPARATIVELY ENHANCED

7. **IBS (P) Ltd.** – 'The Holy Quran' – with Original Arabic Text – Transliteration in Roman Script – by M. A. Haleem Eliyasee – English Translation by Muhammad Marmaduke Pickthall – Published by – Islamic Book Service (P) Ltd., Reprint Edition – 2012. [Purchased]

GRAMMAR

1. LEARN ARABIC IN 30 DAYS – by N. S. R. GANATHE, - Published by – Balaji Publications, Royapettah, Chenni – 600014, Seventh Edition – 2002.

References From Real Science
Insert – Symbol [General Reverence]
SPECIAL REFERENCES FOR AFFIRMATIVE MINOR PREMISES
Bookshelf Symbol 7
KaiTi
Marlett
MingLiU-ExtB
MS Outlook
MS Reference Specialty
MT Extra
PMingLiU-ExtB
Rupali
SimHei
SimSun-ExtB
Symbol
Webdings
Wingdings
Wingdings 2
Wingdings 3

Verifiable Scientific Laws and Justifiable Philosophical Theories

www.ingramcontent.com/pod-product-compliance
Lightning Source LLC
Chambersburg PA
CBHW031809170526
45157CB00001B/7